Dinosaur Tracks
and Traces

Dinosaur Tracks and Traces

Edited by

David D. Gillette

Division of State History
Salt Lake City, Utah

and

Martin G. Lockley

Geology Department
University of Colorado, Denver

The right of the
University of Cambridge
to print and sell
all manner of books
was granted by
Henry VIII in 1534.
The University has printed
and published continuously
since 1584.

Cambridge University Press

Cambridge
New York   Port Chester
Melbourne   Sydney

Published by the Press Syndicate of the University of Cambridge
The Pitt Building, Trumpington Street, Cambridge CB2 1RP
40 W. 20th Street, New York, NY 10011, USA
10 Stamford Road, Oakleigh, Melbourne 3166, Australia

First published 1989
First paperback edition 1991

Library of Congress Cataloging-in-Publication Data

Dinosaur tracks and traces.
Papers solicited and developed at the 1st International Symposium
on Dinosaur Tracks and Traces, held May 23–24, 1986 at the New
Mexico Museum of Natural History in Albuquerque.
Bibliography: p.
1. Dinosaurs – Congresses.   2. Footprints, Fossil – Congresses.
I. Gillette, David D.   II. Lockley, Martin G.   III. International
Symposium on Dinosaur Tracks and Traces (1st : 1986 : New
Mexico Museum of Natural History)
QE862.D5D496   1989   567.9'1   88-34049

British Library Cataloguing in Publication Data

Dinosaur tracks and traces.
1. Dinosaurs
I. Gillette, David D.
II. Lockley, Martin G.
567.9'1

ISBN 0-521-36354-3 hardback
ISBN 0-521-40788-5 paperback

Transferred to digital printing 2004

DEDICATION

to our children

Jennifer,

Peter and Katie,

and future generations of

dinosaur trackers.

Contents

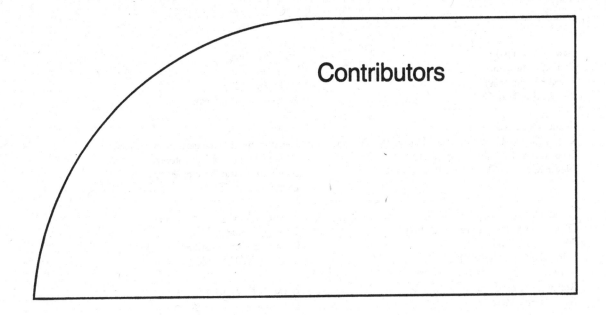

Contributors

Neville Agnew
The Queensland Museum
P.O. Box 300
South Brisbane, Queensland 4101, Australia
current address
Getty Conservation Institute
4503 Glencoe Avenue
Marina del Rey, CA 90292–6537

Ricardo N. Alonso
Universidad Nacional de Salta (and CONICET)
Buenos Aires 177
4400 Salta, Argentina

Youichi Azuma
Fukui Prefectural Museum
Fukui 910 Japan

John K. Balsley
P.O. Box 1008
Indian Hills, CO 80454

R. Dana Batory
402 E. Bucyrus Street
Crestline, OH 44827

Brooks Britt
Tyrrell Museum of Palaeontology
Box 7500
Drumheller, Alberta T0J 0Y0, Canada

Michel Brunet
Laboratoire de Paleontologie de Vertebres
Unite d'Enseignement et de Recherche
Unite Associete 12, CNRS, Universite de Poitiers
86022 Poitiers Cedex, France

Kelly Conrad
Department of Geology
University of Colorado at Denver
1200 Larimer Street
Denver, CO 80204

Phillip J. Currie
Tyrrell Museum of Palaeontology
Box 7500
Drumheller, Alberta T0J 0Y0, Canada

Jean Dejax
Laboratoire de Paleobotanique
Museum National d'Histoire Naturelle
8 Rue Buffon, 75005 Paris, France

Justin B. Delair
19 Cumnor Road
Wootton, Boars Hill
Oxford, England

Georges F. Demathieu
Centre des Sciences de la Terre
UA CNRS No. 157
Dijon 21100, France

James O. Farlow
Department of Earth and Space Sciences
Indiana–Purdue University
2101 Coliseum East Boulevard
Fort Wayne, IN 46805

Kathryn N. Flanagan
Department of Geology and Geophysics
University of Wyoming
P.O. Box 3006
Laramie, WY 82071

Lawrence J. Flynn
Peabody Museum
Harvard University
Cambridge, MA 02138

Toshihide Fujisaki
Osaka Electro–Communication University
Hatsumachi 18–8
Neyagawa City, Osaka 572, Japan

David D. Gillette
Division of State History/Antiquities
300 S. Rio Grande
Salt Lake City, Utah 84101

Robert G. Grantham
Nova Scotia Museum
1747 Summer Street
Halifax, Nova Scotia B3H 3A6, Canada

Heather Griffin
The Queensland Museum
P.O. Box 300
South Brisbane, Queensland 4101, Australia

Yoshikazu Hasegawa
Geological Institute
Yokohama National University
Yokohama 240, Japan

J. Michael Hawthorne
Department of Geology
Baylor University
Waco, TX 76706

Joseph Victor Hall
Institute de Recherche Geologique et Miniene
Boite Postale 333
Garoua, Cameroon

John S. Hines
3 Woodlands Way
Billingshurst
West Sussex RH14, 9TB, England

Karl F. Hirsch
University of Colorado Museum
Campus Box 218
University of Colorado, Boulder, CO 80309

Adrian P. Hunt
Department of Geology
University of New Mexico
Albuquerque, NM 87131

Shinobu Ishigaki
Arie 102
Wakayama–City
Wakayama, 640, Japan

Louis L. Jacobs
Department of Geological Sciences and
Shuler Museum of Paleontology
Southern Methodist University
Dallas, TX 75275

Sohan L. Jain
Geological Studies Unit
Indian Statistical Institute
Calcutta 700035, India

Kenneth K. Kietzke
Department of Geology
University of New Mexico
Albuquerque, NM 87131

Glen J. Kuban
14139 Pine Forest Drive #310
North Royalton, OH 44133

Sergei M. Kurzanov
Palaeontological Institute of USSR Academy of Sciences
Profsouznaya St. 123
Moscow, USSR 117321

Guiseppe Leonardi
SE/Sul Q. 801/B
Cx. P. 13–2067
70401 Brasilia, Brazil

Li Jianjun
Beijing Natural Museum
126 Tian Qiao Street
Beijing, Peoples Republic of China

Seong–Kyu Lim
Department of Earth Sciences
Kyungpook National University
Taegu 635, Korea

Martin G. Lockley
Department of Geology
University of Colorado at Denver
1200 Larimer Street
Denver, CO 80204

Spencer G. Lucas
Department of Geology
University of New Mexico
Albuquerque, NM 87131

Peggy J. Maceo
Texas Parks and Wildlife Department
Austin, TX 78744

Makoto Manabe
Department of Geology and Geophysics
Yale University
P.O. Box 6666
New Haven, CT 06511

Hiromi Maruo
Member of Pacific Artists Society
Tokyo 116, Japan

Niall J. Mateer
1467 N. 17th
Laramie, WY 82070

Masaki Matsukawa
Department of Earth Science
Faculty of Science, Ehime University
Matsuyama 790, Japan

James A. McAllister
Museum of Natural History
University of Kansas
Lawrence, KS 66045-2454

Konstantin I. Mikhailov
Palaeontological Institute of USSR
Academy of Sciences, Profsouznaya Street
123, Moscow, USSR 117321

Wade E. Miller
Department of Geology
Brigham Young University
Provo, UT 84602

Tsukumo Murakami
Member of Kyukobo Artists Society
Ibaraki 300-12, Japan

Ikowo Obata
Department of Paleontology
National Science Museum
Tokyo 160, Japan

Paul E. Olsen
Department of Geology
Lamont-Doherty Geological Observatory
of Columbia University
Palisades, NY 10964

Warren Oxnam
Queensland National Parks and Wildlife Service
Brisbane, Queensland, Australia

Kevin Padian
Department of Paleontology
University of California
Berkeley, CA 94720

Lee R. Parker
Department of Biological Sciences
California Polytechnic State University
San Luis Obispo, CA 93407

J. Michael Parrish
University of Colorado Museum
Campus Box 315
Boulder, CO 80309
current address
Department of Biological Science
Northern Illinois University
DeKalb, IL 60115

Jeffrey G. Pittman
Department of Geological Sciences and
Vertebrate Paleontology Lab
Building 6, Balcones Research Center
10100 Burnet Road
University of Texas, Austin, TX 78758

Nancy K. Prince
Department of Geology
University of Colorado at Denver
1200 Larimer Street
Denver, CO 80204

Rao Chenggang
Beijing Natural Museums
126 Tian Qiao Street
Beijing, Peoples Republic of China

Robert E. Reynolds
Department of Earth Sciences
San Bernardino County Museum
2024 Orange Tree Lane
Redlands, CA 92374

David H. Riskind
Texas Parks and Wildlife Department
Austin, Texas 78744

Robert L. Rowley Jr.
305 South 100 East
Price, UT 84501

Ashok Sahni
Centre of Advanced Study in Geology
Panjab University
Chandigarh - 160 014, India

William A. S. Sarjeant
Department of Geological Sciences
University of Saskatchewan
Saskatoon, Saskatchewan S7N 0W0, Canada

Kenneth L. Stadtman
Earth Science Museum
134 Page School
Brigham Young University
Provo, UT 84602

R. Ted Steinbock
715 Grand Avenue NE
Albuquerque, NM 87102

Toshinori Tanaka
Member of Kyukobo Artists Society
Ibaraki 300-12, Japan

Terence Tebble
The Queensland Museum
P.O. Box 300
South Brisbane, Queensland 4101, Australia

Hiroyuki Terakado
Member of Modern Artists Society
Tokyo 114, Japan

David A. Thomas
3205 Alcazer N.E.
Albuquerque, NM 87110

Richard A. Thulborn
Department of Zoology
University of Queensland
St. Lucia, Queensland 4067, Australia

David M. Unwin
Room 603, Department of Zoology
University of Reading
P.O. Box 228, Reading, Berks, RG6 2AY
United Kingdom

Mary Wade
The Queensland Museum
P.O. Box 300
South Brisbane, Queensland 4101, Australia

Ken E. Woodhams
60 Penwortham Road
Sanderstead, South Croyden
Surrey, C42 0QS, England

Seong-Young Yang
Department of Earth Science
Kyungpook National University
Taegu 635, Korea

Zhen Shounan
Beijing Natural Museum
126 Tian Qiao Street
Beijing, Peoples Republic of China

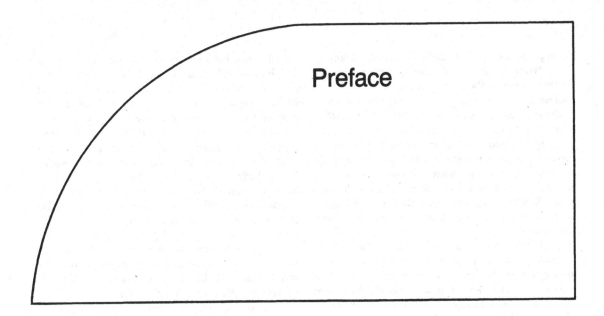

Preface

The science of vertebrate paleoichnology is emerging from the shadows of vertebrate paleontology after over a century of obscurity. It has long been a neglected field outside the mainstream of traditional paleontology and unappreciated as a legitimate branch of ichnology and sedimentology. The unprecedented resurgence of research in dinosaur ichnology, as a specialty within the broader discipline of vertebrate ichnology, has been stimulated by numerous new discoveries of dinosaur eggs and tracksites. These in turn have led to provocative interpretations related to dinosaur behavior and dinosaur ecology.

The sudden revival of interest in this dormant area has resulted in a paradoxical situation. On the one hand ichnologists are being overwhelmed by the sheer quantity of new data available, while on the other hand they lack well-established guidelines and precedents for data synthesis and interpretation. Largely because of this situation, we realized the need for a meeting of minds and convened the "First International Symposium on Dinosaur Tracks and Traces" held in Albuquerque, New Mexico, at the New Mexico Museum of Natural History on May 23–24, 1986. The goals were to bring together a majority of the world's dinosaur ichnologists for the first time, define the scope of the subdiscipline of "Dinosaur Tracks and Traces" and identify new trends in current research. Most participants also registered for an extended field trip through New Mexico, Arizona, Utah, Colorado, Oklahoma and Texas, observing new and classic localities, generally for the first time (Lockley 1986, Farlow 1987).

The symposium and field trip raised many new questions pertaining to track and trackway definitions, trackmaker affinity, and behavioral interpretation of trackways, as well as to the paleoecological, biostratigraphic and paleoenvironmental utility of tracks (Padian 1986, Unwin 1986, Alonso 1987, Lockley and Gillette 1987, Lockley 1987). Sixty abstracts were presented by seventy authors and with the present publication in mind two categories of paper were solicited: 1) longer thematic and synthetic papers of the type generally included in Sections III through VII herein, and 2) brief descriptive site reports dealing with specific new discoveries. The response was enthusiastic, and because of the fertile exchange of views at the symposium and the post-meeting manuscript submission date, the contributions to this volume are more than a faithful rendition of the symposium's proceedings. They include a number of topics not alluded to or addressed in depth during the symposium (cf. Gillette 1986) and they benefit from a degree of standardization not always evident when the participants met for the first time. Among the more significant of these new contributions are a major revision of the systematics of brontosaur tracks (Chapter 42), and several contributions that pertain to trackbearing and eggbearing strata that extend for tens or hundreds of kilometers (Chapters 11, 15, 18 and 50). This latter topic promises to be one of the most important areas of ichnology which demand integration with sedimentology and stratigraphy.

Other topics covered ranged from "History of Ichnology" through "Behavior", and "Eggs and Nests" to "Experimental Studies" and to a large number of "Site Reports" dealing almost exclusively with new discoveries (see Section VIII herein). Several participants addressed important questions regarding track recognition and trackmaker affinity. Both Langston and Seilacher demonstrated that tracks are not always what they seem; they may be undertracks, not true tracks (Chapter 50 this volume) and vary in morphology with different substrate conditions. Kuban (Chapter 7) clearly demonstrated the dinosaurian (theropod) affinity of the problematic so-called "man tracks" of Texas, thereby settling a long standing evolution-creation debate. Thulborn dismissed a hopping dinosaur trackway as that of a turtle (Chapter 5). Leonardi (Chapter 17) and Lockley and Conrad (Chapter 14) demonstrated the utility of ichnofaunas in the interpretation of paleoecological associations and paleobiogeographic distributions, while Demathieu (Chapter 20) demonstrated the importance of tracks in tracing the origin of dinosaurs.

When sixty or more dinosaur ichnologists convened for the symposium, the majority of the world's workers in that subdiscipline were present under one roof. It was then a challenge to find more than a handful of qualified manuscript reviewers without relying heavily on the group itself, and running into the types of criticism sometimes levelled at inhouse reviews. To circumvent some of the inevitable shortcomings inherent in such a system, we solicited at least three and sometimes four reviews for each paper. We acknowledge with thanks the labor and diligence of all the reviewers, especially those who handled several manuscripts.

Furthermore, while it is the editors' responsibility to demand consistency and appropriate standardization of terminology, much of the credit for progress in this direction must go to the participants, and especially to Giuseppe Leonardi and his colleagues, R. M. Casamiquela, G. R. Demathieu, H. Haubold, and W. A. S. Sarjeant, who produced preprints of the Glossary and Manual of Tetrapod Footprint Paleoichnology for use by symposium participants. This useful volume has since been published, incorporating extensive input from the symposium participants (Leonardi 1987) and we have followed the terminology used therein wherever possible.

We extend our thanks and acknowledgments to a large number of people. Initially, the symposium was conceived in 1983 by the two editors during a trip to one of the world's largest dinosaur tracksites, the Purgatoire Valley site in southeast Colorado. On this occasion we were accompanied by Jeff Pittman, who, within a week, identified the Briar site (Chapters 15 and 34), Nancy Prince (Chapter 16), Dave Thomas (Chapter 36), and Steve and Sylvia Czerkas. We thank them all for their input at the conceptual stage, during the symposium and during the compiling of this volume. Many of the other contributors, but especially Kevin Padian (University of California) and Michael Parrish (University of Colorado), offered invaluable advice in the early planning stages.

A major editorial responsibility was initially undertaken by Michael Morales, of the Museum of Northern Arizona. He received all submitted manuscripts, sent them out for review, circulated them to his two co–editors for additional reviews, and returned them to the authors for revision. His considerable efforts are much appreciated and we are indebted to him for his contribution. We also wish to thank a number of colleagues, friends, and institutions for support and sponsorship. Foremost among these is Lani Duke, Museum of Western Colorado, whose unstinting typesetting and proofreading work has gone far beyond the call of duty. We also thank Carolyn Howard for compiling the index with great efficiency and dispatch.

Initial support for the symposium came from the New Mexico Friends of Paleontology, a support group for paleontology that had its roots in the New Mexico Museum of Natural History. The Friends of Paleontology raised funds that supported foreign participants, meeting expenses, and partial costs of publication of the Abstracts with Program and this volume. Mike and Charlene Dietz, Lynett Gillette, David and Regina Hunter, Bob and Linda Strong, and Pauline Ungnade were all instrumental in generating the support that came from the Friends of Paleontology, and to them we extend our heartfelt thanks. Similarly initial support for the symposium field trip came from the University of Colorado at Denver, Dinosaur Trackers Research Group. Kelly Conrad and Nancy Prince were particularly active in supplying information incorporated in the Field Guide (Lockley 1986), and thanks are also due to Debbie Adelsperger and Mark Jones for assistance in the field. The CU Denver group also provided financial support to bring Dr. Giuseppe Leonardi to the Symposium. We are particularly endebted to him for compiling preprints of the above–mentioned Glossary and Manual of Tetrapod Paleoichnology for circulation at the meeting.

Preparation of this volume was supported by the New Mexico Museum of Natural History Foundation, and the institutional support of the University of Colorado at Denver, the Museum of Western Colorado, and the Utah Division of State History. We also thank editor Peter–John Leone (Cambridge University Press) for his unfailing patience and good will through the publication of this book. Thanks also go to the following individuals, not for specific contributions, but instead for inspiration they provided on the long path that has led to this publication: Donald Baird, Wilson and Peggy Bechtel, Edwin and Margaret Colbert, Walter Coombs, Linda Dale Jennings, Ed and Mary Gavin, Jack Horner, Chet Jennings, Kevin Padian, the late Ted Schreiber, Ed and Rosemary Springer, and David Thomas.

We thank Jack Horner (Museum of the Rockies) for hosting the Second International Symposium on Dinosaur Tracks and Traces as part of the 1988 Symposium on Vertebrate Behavior, and Walter Coombs (Amherst College) for proposing a venue for the third meeting in 1990.

Finally we thank all of the contributors to the symposium and this volume for their participation, cooperation and support. They have helped dinosaur ichnology emerge from the shadows and stand as a respectable and recognized science.

References

Alonso, R. N. 1987. First international symposium on dinosaur tracks and traces. Boletin Informativo de la Associación Paleontológica Argentina 16:18.

Farlow, J. O. 1987. A Guide to Lower Cretaceous Dinosaur Footprints and Tracksites of Paluxy River Valley, Somervell County, Texas. (Baylor University) 50 pp.

Gillette, D. D. (ed.). 1986. First International Symposium on Dinosaur Tracks and Traces. Abst. Prog. (Albuquerque: New Mexico Museum of Natural History) 31 pp.

Leonardi, G. (ed.). 1987. Glossary and Manual of Tetrapod Footprint Paleoichnology. Dept. Nacional de Produçåo Mineral, Brasil. 75 pp.

Lockley, M. G. 1986. Dinosaur Tracksites. Univ. Colorado Denver, Geol. Dept., Spec. Issue #1. 56 pp.

1987. Tracking the dinosaurs. *Geology Today* 3:195.

Lockley, M. G., and Gillette, D. D. 1987. Dinosaur tracks symposium signals a renaissance in vertebrate ichnology. *Paleobiology* 13:246-252.

Padian, K. 1986. On the track of the dinosaurs. *Palaios* 1:519-520.

Unwin, D. 1986. Tracking the dinosaurs. *Geology Today* 2:168.

David D. Gillette
Salt Lake City, Utah

Martin G. Lockley
Denver, Colorado and
Grand Junction, Colorado

August 1988

Introduction

There is no branch of detective science so important and so much neglected as the art of tracing footsteps.

A. Conan Doyle 1891
(Study in Scarlet)

Dinosaur Tracks and Traces: An Overview

MARTIN G. LOCKLEY AND
DAVID D. GILLETTE

The journey of a thousand miles begins with a single step.
Lao Tse

Dinosaurian trace fossils are an extremely important if hitherto neglected branch of vertebrate ichnology; they generally provide information that is very different from that furnished by body fossils — information which is often not directly available from fossilized skeletal material. Like dinosaur skeletal remains, dinosaur tracks and traces — including footprints, eggs, nests and coprolites — have a world–wide distribution. Footprints are currently known from all continents except the Antarctic. Although sites yielding egg and nest remains are less numerous, they are almost as widely distributed. On the other hand, the distribution of coprolites is poorly documented (Fig. 0.1).

Footprints, nests, undisturbed eggs and coprolites represent in situ evidence of the dynamic activity of dinosaurs during life. Footprints can tell us approximately how many animals were present in a particular area, the directions they moved in, their relative abundance, size, and speed. They can also, like eggs and nests, reveal information about specific ancient environments and ecologies.

Identification of the trackmaker is one of the main objectives of tracking and it is often possible to identify trackmakers at the family level, as with footprints attributed to hadrosaurs or ceratopsians. In some cases a generic level identification is implied, as with the frequent designation of *Iguanodon* tracks. However, it has not yet been demonstrated that particular species of dinosaur can be identified from footprints with unique characteristics.

It is important to exercise caution in this endeavor and recognize the clear distinction between taxonomy based on skeletal remains and ichnotaxonomy based on track morphology. The considerable progress in tracksite documentation in recent years has led to significant improvement in this area and many mistakes in trackmaker identification have been rectified. Furthermore, even though trackmakers may not be identified below the family level, tracks often provide information on the morphology of hands and feet (mani and pedes) that is not available from the skeletal record. Similarly, recently advances in our understanding of the relationship between nests, eggs and skeletal remains has led to radical progress in our ability to interpret the affinity of eggshells and the behavior of

dinosaurs at nest sites (Horner 1987, Horner and Weishampel 1987, Mohabey 1987). It is also clear that coprolites are very abundant in some deposits (Jain this volume) and could benefit from systematic study.

At the present time dinosaur ichnology is making a significant number of new contributions to our understanding of dinosaurs and their habitats. Paleobiologists and restorers rely increasingly on trackways for authentic information on locomotion, stance and gait (e.g., Wade this volume), leading to radical reinterpretation of outdated reconstructions and exhibits. Nest sites and multiple parallel trackways are providing new insights into dinosaur social behavior and gregariousness (Sections IV and V herein). Regional scale synthesis of dinosaur tracks and traces shows a remarkable potential for suggesting temporal and spatial patterns of stratigraphic paleoecological and paleobiogeographic distribution of dinosaur faunas (e.g., Demathieu, this volume, Pittman this volume, Leonardi this volume, Lockley and Conrad this volume). As information emerges to suggest answers to other paleobiological issues, new questions arise. How important are experimental studies with modern trackmakers (Section VII)? How well do we understand the sedimentological context and preservation of tracks and nestsites (Section V)? And where does dinosaur ichnology go from here, after the rapid resurgence of the field in the last decade (Chapter 50)?

As the first comprehensive book ever compiled on dinosaur tracks and traces, this volume suggests various approaches, interpretations and answers, both to these new questions and to longer standing paleontological questions previously addressed without recourse to the burgeoning ichnological record.

Historical Perspective and Current Status

Early in the nineteenth century when dinosaurian and other Mesozoic tracks were first discovered in Europe and eastern North America, they were considered remarkable and important. As shown by the classic work of Hitchcock (1858), they were accorded much serious scientific attention, and the leading paleontologists and geologists of the day, including Owen, Huxley, Murchison and Lyell,

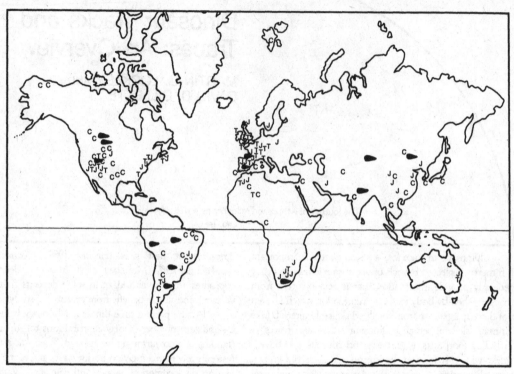

Figure 0.1. Global distribution of main dinosaur footprint sites and main dinosaur egg/nestsites.
T = Triassic, J = Jurassic and C = Cretaceous tracksites. Egg symbol represents dinosaur egg and nestsites.

all took a keen interest in their documentation and interpretation. Following Hitchcock's death in 1863, research in his field declined and, while our knowledge of dinosaurs began to increase dramatically, ichnology was generally neglected.

There is a perception that much of the early work was misleading or wrong; dinosaur tracks were identified as those of birds (*Ornithichnites*), a "mistake" which undermined the validity of ichnology, especially in the crucial area of trackmaker identification. Generally such a view is unfounded. Hitchcock and some of his contemporaries produced careful work which was exemplary in many respects, and, except in a minority of cases, much of the work, until comparatively recent times, was shoddy by comparison. The inference that three-toed dinosaur tracks were made by birds was *not* a result of thoughtless comparisons with modern tracks. Many early workers were well aware of the pitfalls of such a superficial approach, and only leaned towards suggesting avian affinity when large fossil birds like the "Moa" (*Dinornis*) and elephant bird (*Aepyornis*) were discovered. Knowing now what we do about dinosaur-bird relationships, the judgments of these early workers should not be called into question unduly, especially considering that dinosaur skeletal remains were then virtually unknown. If we accept that taxonomic categories like Ornithischia and Ornithopoda are firmly established and that the theropod-bird connection is widely accepted, Hitchcock and his contemporaries were essentially correct in their interpretation of tracks from theropod-dominated footprint assemblages.

Perhaps it was the dramatic increase in discoveries of dinosaur skeletal remains which directed workers away from ichnology through the end of the nineteenth century and through the middle of the twentieth century. In any event, apart from the celebrated Mongolian egg discoveries of the 1920's and a minor flurry of activity associated with the discovery of important footprints in North America in the 1930's (Sternberg 1932, Brown 1938, Bird 1939), until recently, there has been little sustained resurgence of interest in dinosaur ichnology.

Consequently the recent dramatic increase in research activity on dinosaur footprints and nestsites is entirely without precedent. It is in part a facet of the general renaissance in all aspects of dinosaur research and in part a result of advances in the study of ancient terrestrial lithosomes and ichnology in general.

A survey of known dinosaur tracksites indicates that many hundreds are known. In areas where ichnologists have compiled locality data in a systematic fashion, reliable statistics are readily available. In the New World for example, recently compiled statistics indicate that at least 400 tracksites are known. These are distributed as follows:

Region	Approximate Number of Sites	Reference
Western Canada	~ 30	Currie this volume
Colorado Plateau region	~ 200	Lockley and Conrad this volume
Texas	~ 40	Pittman this volume
Eastern seaboard	~ 100	miscellaneous
South America	~ 40	Leonardi this volume

In the Old World a high concentration of sites has been documented in northwest Europe. These include the many Triassic sites, some without actual dinosaur tracks, researched by Demathieu, Haubold, Sarjeant and other leading ichnologists. In Africa a number of classic footprint-rich Late Triassic-Early Jurassic sites were documented from the Stormberg Series in the southern part of the continent (Ellenberger 1972, 1974; Olsen and Galton 1984). In the northern part of the continent somewhat younger Jurassic sites have been well-documented in recent years (Montbaron et al. 1985, Ishigaki this volume). In Asia there are a disproportionately small number of sites for the large land area. However, though very few sites are known for the Soviet Union, at least one appears to be very large and extensive (Romashko 1986), and in China, where research in this field is on the increase, over twenty-five sites are now known (Zhen et al. this volume). In the last decade sites have also been discovered for the first time in Korea (Lim et al. this volume), Japan (two papers herein) and Thailand (Buffetaut et al. 1985). There is also a low density of tracksites in Australia although the intensively studied Winton site has become very well-known (Thulborn and Wade 1984, this volume).

A survey of the world's largest tracksites indicates that they have all been documented and in most cases discovered only recently. A "large" tracksite is arbitrarily defined as one in whicih at least 1000 tracks or about 100 trackways are mapped or reliably documented. In ascending stratigraphic order these large sites include:

Age	Site Name and Location	# Tracks–Trackways		Reference
Up Cret	Toro Toro, Bolivia	—	~ 100	Leonardi 1984
Up Cret	Lark Quarry, Australia	>4000	> 500	Thulborn and Wade 1984
Lr Cret	Peace River, Canada	~ 1200	~ 100	Currie 1983
Lr Cret	Jindong, South Korea	—	~ 250	Lim et al. herein
Lr Cret	Nei-Monggol, China	~ 1000	—	(Rong 1985)*
Lr Cret	Piau, Brazil	—	~ 150	Leonardi 1984
Up Jura	Purgatoire, USA	~ 1300	~ 100	Lockley et al. 1986
Up Jura	Mt Kugitang-Tau, USSR	~ 2700	—	(Romashko 1986)*
Mid Jura	Salt Valley, Utah	>1000	> 100	Lockley and Gillette herein
Mid Jura	Chaoyang, China	~ 4000	—	(Zhen et al. 1983)
Lr Jura	Rocky Hill, USA	~ 1000	—	Ostrom 1968
Up Trias	Peacock Canyon, USA	—	~ 100	Conrad et al. 1987

(* indicates brief and uncertain documentation)

The statistics listed above indicate that, in the last decade alone, tens of thousands of tracks, comprising hundreds of distinct trackways, have been mapped, counted or otherwise documented from only a handful of the larger known sites. There exist a bigger number of sites where several hundred tracks comprise trackways numbering in the dozens. When included with all known sites the available ichnological record of dinosaur footprints must approach hundreds of thousands of tracks representing evidence of the activity of tens of thousands of animals.

Such statistics, although impressive, are not particularly surprising. An individual animal was presumably capable of leaving hundreds or even thousands of footprints in environments where reasonable fossilization potential existed. It may also have laid eggs by the dozens. However, the same individual could never contribute more than one skeleton to the fossil record.

One outcome of the resurgence of dinosaur ichnology has been a significant accumulation of new data on a variety of Mesozoic tracks. This data can be compared and integrated with information derived from other field research for an improved understanding of dinosaur paleobiology. As suggested by Lockley (1986), ichnological data can contribute to an understanding of dinosaur paleobiology in many areas including behavior (locomotion, speed, social behavior), paleoecology (relative abundance, dinosaur community composition and diversity), paleobiogeography (latitudinal and paleoenvironmental range) and biostratigraphy (evolutionary longevity). As outlined below, ichnological data can also have interesting paleoenvironmental implications.

Before analyzing dinosaur behavior on the basis of trackway evidence, it is important for the ichnologist to be reasonably sure of the trackmaker's affinity. There were a number of important non-dinosaurian trackmakers which left footprints in the Mesozoic record. Some of these produced tracks which have proved controversial. At present about seven localities are known to yield true bird tracks. These are all Cretaceous in age and all represent discoveries which postdate 1930. Because the tracks all bear such a striking resemblance to modern birds, with high divarication angles of 110°–120° between digits II and IV, they have generally not been misidentified. However, in a reversal of the 19th century pattern, some have been incorrectly assigned to dinosaurs. Others remain problematic (Parker and Balsley this volume). Non-dinosaurian archosaurs and other reptiles have proved more problematical. Crocodilian tracks appear to have been interpreted as pterosaurian in origin on many occasions (Padian and Olsen 1984, Unwin this volume, Prince and Lockley this volume) but in other cases may be genuinely attributable to pterosaurs (Gillette and Thomas this volume). In one instance a turtle trackway has been interpreted as that of a hopping dinosaur (Thulborn this volume). It appears that other tracks, e.g., those of mammals (Sarjeant and Thulborn 1986), lepidosaurs and mammal-like reptiles are generally very rare and poorly preserved during most of the age of dinosaurs, particularly in the Jurassic and Cretaceous.

There are at least two important reasons that help explain the rarity and problematic nature of non-dinosaurian ichnites in the Mesozoic. Firstly non-dinosaurian trackmakers were generally small, at least after the Late Triassic. This means that there is a bias against the preservation of their tracks, whereas dinosaurs included many large species whose tracks were obvious and easily preserved (Lockley and Conrad this volume). The second point pertains to the different habits of terrestrial, aquatic and airborne creatures. We might expect that the tracks of swimmers and fliers are relatively rare in comparison with those of terrestrial walkers, although for fliers this may be a function of low diversity in the Mesozoic. The trackways of swimmers and fliers may also be much less complete, and, in the case of swimmers, modified by water action. While the size–related bias in favor of the preservation of large tracks is similar to the taphonomic biases affecting the skeletal record, ethological or behavioral biases represent specific group–related phenomena that are likely to be expressed differently in the ichnological and body fossil record. The difference in frequency between the traces and skeletal remains of aquatic and terrestrial animals is a good example. Terrestrial tracks may significantly outnumber body fossil remains, whereas the reverse appears to hold true for aquatic animals.

Although dinosaur egg material was discovered in Europe in the late nineteenth century, it was not until the 1920's that the Mongolian nest site discoveries confirmed that dinosaurs laid eggs in well ordered nests. As was the case in the history of footprint ichnology, the initial discoveries were followed by a lull. During this time, material came to light only sporadically. However, in the field of egg and nest research there has also been a dramatic resurgence of activity in recent years. Discoveries of large sites have been made in North America (Horner 1987), China (Mateer this volume) and on the Indian subcontinent (Hirsch this volume, Jain this volume, Sahni this volume), as well as at previously known localities in Europe. Collectively the sites reveal a variety of distinctive egg types based on microstructure and macrostructure, some of which can be assigned to specific dinosaurian egg layers on the basis of associated embryos. Others, however, are of unproven affinity and not necessarily dinosaurian in origin. The situation closely parallels the current state of knowledge in footprint ichnology, where considerable progress has been made. Many important questions have been answered but much still remains to be discovered.

While there remains considerable untapped material in the fields of track and eggsite ichnology, the field of coprolite research remains largely unexplored and the potential unknown. As indicated by Jain (this volume), coprolites are very abundant locally.

Footprint Paleobiology
(i) Behavior

Dealing first with strictly behavioral aspects of paleobiology we note that tracks are unequivocal evidence of the posture and locomotion of the trackmaker. The erectness of a particular group or its inclination towards digitigrade or plantigrade stance, or bipedal versus quadrupedal progression is clearly indicated by the trackways (e.g., Wade this volume, Kuban this volume). Incorrect postural

reconstructions can not be adhered to in the face of contrary trackway evidence.

In particular, dinosaur trackways provide evidence that dinosaurs stood and walked erect. Among bipedal trackways attributed to both theropods and ornithopods, feet were placed one in front of another leaving a narrow parasagittal trackway (Fig. 0.2). Pace angulation values are high and pes tracks invariably overlap the trackway midline to some degree. The main differences between trackways of bipedal ornithopods and theropods include: (i) foot shape, with toes generally more slender in theropods, (ii) degree of pedal rotation, with inward (positive) rotation more pronounced in ornithopods, and (iii) step and stride length generally shorter in ornithopods.

Generally bipeds were digitigrade. However, a number of examples exist of obligatory or facultative bipeds which left tracks and trackways exhibiting distinct metatarsus impressions (Fig. 0.3). This suggests that occasional plantigrade progression was employed by a variety of different theropods and ornithopods. The reasons for this are open to speculation but may have been a behavioral response to soft substrate conditions, or a function of changing from bipedal to quadrupedal progression (slowing, sitting, squatting or stalking). Such tracks provide useful information on metatarsus dimensions.

Trackways of large quadrupeds, including sauropods, ceratopsians, ankylosaurs and certain large ornithopods also indicate erect posture. However, the trackways are generally somewhat wider than those of bipeds of comparable size. In all cases the pes tracks are larger than the manus prints (heteropody), with the disparity being most pronounced in sauropods and ornithopods. Usually the pes tracks overlap the trackway midline, although this is not always the case. Invariably however, the manus tracks exhibit higher pace angulation values and are usually situated anterolaterally with respect to the pes. This evidence suggests that in general the front limbs hung vertically from the shoulder girdle or diverged slightly, while the hind limbs converged slightly beneath the body.

Tracks have also been used to derive speed estimates. The whole debate over dinosaur speeds has been quite lively and need not be reviewed here in detail (see Thulborn this volume). In essence, the ichnological record indicates that the vast majority of dinosaurs whose tracks are preserved were walking. A number of authors have pointed out that this is to be expected — animals do not expend energy unnecessarily. However a few examples of running dinosaurs do exist (Farlow 1981, Thulborn this volume), indicating that small bipeds were fleet of foot and that larger theropods attained speeds of about 30–40 km/hr. Trackways indicating rapid progression by large quadrupedal dinosaurs have yet to be documented. This should not be taken as an indication that these animals could not run. Rather it indicates that they ran infrequently, and that the probability of finding "maximum speed trackways" is very remote. The converse of this assumption is that data on walking dinosaurs is very abundant, allowing for the compilation of reliable statistics for this mode of progression (Thulborn this

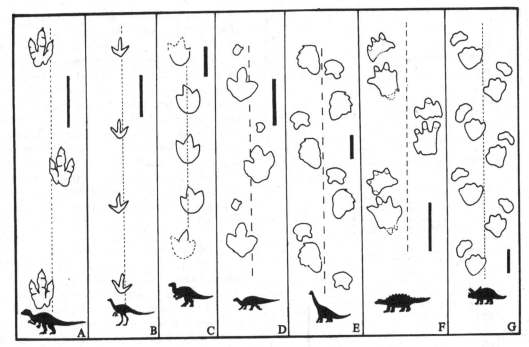

Figure 0.2. Representative dinosaur trackways. **A–C**, bipeds; **D–G**, quadrupeds. **A**, *Eubrontes* after Lull (1953). **B**, unnamed coelurosaur. **C**, unnamed ornithopod. **D**, *Caririchnium* (B-D) after Lockley (1987). **E**, *Brontopodus* after Bird (1944). **F**, *Tetrapodosaurus* after Sternberg (1932). **G**, unnamed ceratopsian trackway after Lockley (1988). Dotted lines indicate trackway midline. Scale bars = 50 cm.

volume).

(ii) Paleoecology

In the realm of paleoecology it is becoming increasingly clear that ichnofaunas can give insight into the composition and diversity of dinosaur communities in particular areas. As data accumulates to suggest that footprint assemblages, like other faunas, are facies related, it is clear that paleoecological census data derived from tracksites is intricately related to patterns of paleobiogeographic distribution (Leonardi this volume, Lockley and Conrad this volume).

The simplest examples of the application of ichnological data in paleoecological census studies are cases where distinctive tracks are known in otherwise unfossiliferous deposits. Here the ichnofauna is the only paleontological data available. However, many other examples exist in which the ichnofauna contributes a proportion of our knowledge on the composition of a fauna in a particular area or deposit.

The issue of the relative importance of trace and body fossils in such situations has led to two serious misconceptions: (i) that tracks do not occur in the same deposits as body fossils, and (ii) that tracks are only really important in situations where body fossils have not been found.

The first misconception is generally unfounded.

Although there are many trackbearing stratigraphic units that lack abundant body fossils, there are a large number of instances of substantial ichnofaunas originating from fossiliferous beds. For example, in the Late Triassic of North America and Europe, large ichnofaunas occur in association with diverse body fossil assemblages. The same is true for later Mesozoic deposits like the Morrison Formation of North America, the Wealden of Europe and a number of Late Cretaceous deposits in the Rocky Mountain region.

Having recognized that important trace and body fossil assemblages do co–occur, the question arises as to how well the different data sets correspond. If the tracks indicate a faunal assemblage which is similar to that derived from skeletal remains, the track data might be considered redundant, especially if it was discovered afterwards. Such a superficial approach should be discouraged because it ignores a number of important paleoecological and taphonomic considerations. An ichnofauna that is compatible with the skeletal record confirms that minimal taphonomic biases exist, and affords an opportunity to study the fauna in the dynamic context of trackmaking activity. It may also provide relative abundance data not available from the skeletal record.

In instances where the ichnofauna supplements the faunal data derived from the body fossil record, an obvious paleontological benefit accrues. However, the additional data do more than just enlarge the faunal list. They sug-

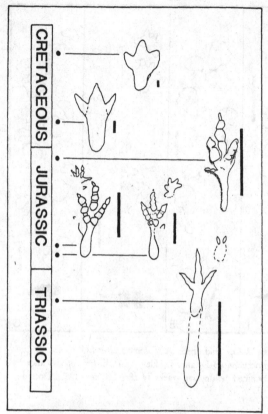

We do not claim that ichnofaunas can not be misleading. They may, for example, reflect the activity of a particular group in a particular setting rather than the overall composition of the fauna in a given region. Theropod abundance may appear over-represented because members of this group were more active than species in other niches. However, as data accumulate, consistent patterns do emerge. For example, the majority of Cretaceous ichnofaunas are dominated by large ornithopods (Lockley and Conrad this volume) and occur in lithosomes representing lowland, humid coastal plain settings. By contrast the theropod-sauropod ichnofaunas appear to be associated with semi-arid low latitude settings. Similar paleogeographic patterns are also noted elsewhere in the Mesozoic (Leonardi this volume).

(iii) Evolution

Finally in the area of biostratigraphy it is now possible to trace many of the major phases of dinosaur evolution through the ichnological record. Demathieu (this volume) has demonstrated that early dinosauroid forms emerged as early as the Middle Triassic, at which time they represented only a small proportion (about 3–5%) of the fauna. In the Late Triassic-Early Jurassic transition, clear ichnological changes are seen and can be used for biostratigraphic correlation on a global scale (Olsen and Galton 1984). Whereas Late Triassic ichnofaunas are characterized by chirothere tracks and various other distinct forms like *Atreipus* and the ubiquitous *Rhynchosauroides*, the Early Jurassic is characterized by an abundance of grallatorid tracks (*Grallator* and *Anchisauripus*) in association with distinctive forms like *Anomoepus* and *Batrachopus*. The rise of sauropod faunas in the Jurassic is also now discernible in the ichnological record (Monbaron et al. 1985), as is the distinctive transition to faunas dominated by large ornithopods in the Cretaceous (Lockley and Conrad this volume).

Figure 0.3. Dinosaur tracks exhibiting metatarsus impressions. *Agialopus* from the Late Triassic of Colorado. *Anomoepus* from the Lower Jurassic of New England (left) and southern Africa (right), *Jialingpus* from the Upper Jurassic of China, unnamed theropod from the Lower Cretaceous of Texas and "Dinosauropodes," a hadrosaurian ichnite from the Late Cretaceous of Colorado. Scale bars = 10 cm.

Paleoenvironmental Implications of Dinosaur Footprint Ichnology

In addition to being used for an improved understanding of dinosaur paleobiology, tracks may also be used to help interpret paleoenvironments. For example, the depth of tracks is an indication of the water content of sediments. Employing such a simple observation, we can use track depth to help determine the position of shorelines or paleogradients between firm dry ground and moist ground or bodies of water. In other words, they may be "paleo water-table" indicators.

Dinosaurs had considerable impact on the substrates over which they passed. While firm substrates resisted disturbance, soft yielding substrates were particularly susceptible to deformation. The resulting vertebrate bioturbation or "dinoturbation" is therefore a function of both paleobiological and paleoenvironmental factors. While the density and degree of activity of the dinosaur population represent contributing paleobiological factors, it is physical paleoenvironmental factors that ultimately determine the

gest the degree to which the skeletal record is incomplete and the probable nature of associated taphonomic biases. In the Popo Agie Formation of Colorado and Wyoming for example, an ichnofauna of small vertebrate and invertebrate traces exists in association with a skeletal record that only records the presence of a few large vertebrates. Apart from tripling the overall faunal list, the ichnofauna indicates a strong bias against preservation of the small vertebrate fauna in the skeletal record.

Because ichnofaunas are the in situ traces of animal activity, a strong case can be made in favor of their use in paleoecological census estimates. In the case of the Popo Agie fauna cited above, we can speak with some confidence of a fauna dominated by non-dinosaurian mesovertebrates and estimate the dinosaurian component as only a minimal proportion (about 5%, see Lockley and Conrad this volume). Whenever a large ichnofauna is known, the potential exists to estimate the composition of the fauna with some accuracy. Generally the data derived from this exercise conform with known patterns of faunal distribution, thus inspiring confidence in the ichnofaunal census method.

character of the impacted sediment. For example, extensive and heavily bioturbated or "dinoturbated" beds generally occur at lithological boundaries and imply long exposure to trampling rather than sudden influx of dense dinosaur populations.

Despite being widespread, "dinoturbation" has received little serious scientific attention and has often been overlooked or misinterpreted as soft sediment deformation of non-biogenic origin. As sedimentologists, stratigraphers and paleontologists look more closely at Mesozoic terrestrial successions, an increasing number of dinoturbated beds have been recognized, particularly in Late Jurassic through Late Cretaceous formations. Preliminary indications suggest that vertebrate bioturbation reached a Phanerozoic peak at this time due to the extensive impact of large gregarious herbivores, particularly sauropods, iguanodontids and hadrosaurs. Dinoturbated beds may be relatively localized in association with playa and larger perennial lake deposits; however, it may also be prevalent in coastal lowland lithosomes where extensive deposits are accumulating at base level. Because dinoturbation may radically alter sediments leading to their mixing and homogenization, the phenomenon should be of particular interest to soft rock geologists concerned with the interpretation of terrestrial deposits.

Paleobiological and Paleoenvironmental Aspects of Egg and Nestsite Ichnology

From a paleobiological perspective, egg and nestsites also provide important information about (i) behavior, (ii) paleoecology, and (iii) evolution. Interpretations of dinosaurian behavior based on egg and nestsite discoveries have proved quite dramatic. Dinosaurs apparently laid eggs in an orderly fashion, sometimes in neat spiral arrangements or linear rows. Clutch size was also quite specific for certain groups. Another aspect of nestsite behavior which has attracted much attention is the parental care scenario. This was first hinted at by the discovery of *Protoceratops*, the inferred egg layer, locked in a fatal struggle with *Velociraptor*, the presumed egg thief. Subsequent work by Horner (1987) and his colleagues has produced further evidence of parental care and protection at nestsites. Horner's work has also indicated that dinosaurs nested in colonies, sometimes sharing a site among several species, and that they returned to these chosen sites over extended periods of time. Similar "site fidelity" is also suggested by multiple nest bearing layers in China (Mateer this volume) and India (Jain this volume).

The choice of specific paleogeographic locations for nestsites may account, to some degree, for certain dinosaur distribution patterns. The rarity of small dinosaurs, whether real or perceived, in any given deposit or area, may relate to the location of nest colonies as much as to other factors such as preservation or population dynamics.

The evolution of the hard-shelled egg was a major evolutionary accomplishment, facilitating the diversification of reptilian faunas in Permo–Triassic terrestrial habitats. However, few diagnostic eggs are known before the Late Mesozoic. By this time they are morphologically distinct, like the dinosaurs that laid them.

Elongated ornithopod eggs appear to be quite distinct from the large subspherical eggs attributed to sauropods. Differences in external ornament and microstructure are also analogous to shell structure differences seen in different invertebrate classes. Because egg remains are less abundant than tracks, and associated only with reptiles and birds, it is not yet possible to discern with confidence major evolutionary events by analysis of existing egg fossils (Hirsch this volume). However, as information accumulates, there is no reason to doubt that eggshell studies will add to our knowledge of the evolution of dinosaurs, other reptiles and birds.

Summary

Dinosaur tracks and traces form a major and often spectacular part of the vertebrate ichnological record. Tracks outnumber egg and coprolite remains and provide important information on morphology gait, speed, dinosaur paleoecology and habitat zonation. Egg and nest remains provide specific information on hatchling morphology, colonial nesting strategies and parental care. Coprolites remain problematic and poorly documented despite their potential for analyzing dinosaurian diets.

Viewed in the broad context of vertebrate (tetrapod) ichnology, dinosaur tracks and traces are more abundant than those attributed to other classes, either during the "Age of Dinosaurs" (Middle Triassic to Late Cretaceous) or at other times. There are several explanations for this situation. In Late Paleozoic through Earliest Mesozoic time (Late Devonian to Middle Triassic), following the first appearance of land-based tetrapods, ichnites consist mainly of tracks, with few confirmed reports of eggs, nests or coprolites. Initially also trackmakers were small and confined to lowland humid paleoenvironments, thus minimizing their impact. With the Permo–Triassic radiation of reptiles, tracks become more diverse and abundant, but reports of other ichnites remain scarce.

By Jurassic and Cretaceous times, the evolution of dinosaurs resulted in a significant increase in vertebrate bioturbation. "Dinoturbation" reached a peak. Egg and nestsites also become more abundant at this time, presumably reflecting a better preservation potential for large and strategically located eggs than for the smaller and less robust eggs that were probably produced by dinosaur predecessors.

Following the demise of the dinosaurs there is a distinct shortage of vertebrate ichnites in the early part of the Age of Mammals. The reasons for this are fairly obvious and relate to the impoverishment of tetrapod faunas and the sharp reduction in egg laying groups. Despite the increased abundance of fossil ichnites in the latter part of the Age of Mammals, there is little doubt that dinosaur tracks and traces are an unusually important and abundant part of the overall ichnological record. They have also been studied with increasing intensity in recent years and are now understood in much greater depth.

References

Bird, R. T. 1939. Thunder in his footsteps. *Nat. Hist.* 43:254–261, 302.

Buffetaut, E. Ingavat, R., Sattayarak, N., and Suteetorn, V. 1985. Les Premières empreintes de pas de Dinosaures du Crétacé inferieur de Thailande. *C.R. Acad. Sci. Paris* t. 301, Ser. II (9) pp. 643–648.

Brown, B. 1938. The mystery dinosaur. *Nat. Hist.* 41:190–202, 235.

Ellenberger, P. 1972. Contribution a la classification des pistes de vertebres du Trias: Les types du Stormberg d'Afrique du Sud (I). *Paleovertebrata. Mem. Extraordinaire, Montpellier* 152 pp.

—— 1974. Contribution a la classification des pistes de vertebres du Trias: Les types du Stormberg d'Afrique du Sud (II): Le Stormberg Superieur. *Paleovertebrata. Mem. Extraordinaire Montpellier.*

Farlow, J. O. 1981. Estimates of dinosaur speeds from a new trackway site in Texas. *Nature* 294:747–748.

Hitchcock, E. 1858. A report on the sandstone of the Connecticut Valley, especially its fossil footmarks. *In Natural Sciences in America.* (Arno Press: 1974) 220 pp.

Horner, J. R. 1987. Ecological and behavioral implications derived from a dinosaur nesting site. *In* Czerkas, S. J., and Olsen, E. C. (eds.). *Dinosaurs Past and Present.* (Los Angeles County Museum) pp. 51–63.

Horner, J. R., and Weishampel, D. B. 1988. A comparative embryological study of two ornithischian dinosaurs. *Nature* 332:256–257.

Lockley, M. G. in press. Dinosaurs near Denver. *Field Guide, Geological Society of America Centennial Meeting.* (Colorado School of Mines)

Mohabey, D. M. 1987. Juvenile sauropod dinosaur from Upper Cretaceous Lameta Formation of Panchmahals District, Gujarat, India. *Jour. Geol. Soc. India* 30:210–216.

Monbarron, M., Dejax, J., and Demathieu, G. 1985. Longues pistes de dinosaures bipèdes à Adrar-n-ovglagal (Maroc) et repartition des faunes de grands Reptiles dans le domain Atlatique au cours du Mesozoique. *Bull. Mus. Nat. Hist. Paris* 4e Ser. 7 (A):229–242.

Olsen, P. E., and Galton, P. M. 1984. Review of the reptile and amphibian assemblages from the Stormberg of southern Africa, with special emphasis on the footprints and the age of the Stormberg. *Paleont. Africana* 25:87–110.

Romasko, A. 1986. Man — a contemporary of the dinosaurs? *Creation/Evolution* 6:28–29.

Sarjeant, W. A. S., and Thulborn, R. A. 1986. Probable marsupial footprints from the Cretaceous sediments of British Columbia. *Canadian Jour. Earth Sci.* 23:1223–1227.

Sternberg, C. M. 1932. Dinosaur tracks from the Peace River, British Columbia. *Ann. Report Nat. Hist. Museum Canada* (for 1930). pp. 59–74.

*References by authors contributing to this volume are only cited when they do *not* appear in the bibliographies of these authors' respective contributions.

II Historical Perspectives

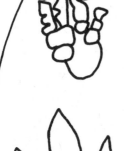

'"Not a track remains," says Dr Buckland, "or a single hoof, of all the countless millions of men and beasts whose progress spread desolation over the Earth. But the reptiles that crawled upon the half finished surface of our planet, have left memmorials of their passage enduring and indelible." And we may add, that the proudest monuments of human art will moulder down and disappear; but while there are eyes to behold them, the sandstone of the Connecticut valley will never cease to remind the observer of the gigantic races that passed over it while yet in an incipient state.'

Edward Hitchcock 1844 p. 321

The markings in question appear to have been observed by several persons at Hastings; but they have not been found consecutive....

They are of large size ... measuring sixteen inches in length; but there does not appear ... any decisive evidence as to their origin.

Tagart 1846

We may therefore be allowed provisionally to refer these tracks to the *Iguanodon*, who certainly wallowed in the Wealden waters and frequented their sand–bars and mud–banks — who had a great three–toed foot — and who ... may have ... planted his footprints uniserially, leaving as his spoor a single row of thick–toed trifid imprints....

Jones 1862

Dinosaur tracks were observed by prehistoric peoples who evidently regarded them with interest and sometimes carved their own symbols alongside tracks (Leonardi 1984). However the science of vertebrate ichnology dates back only a little over 150 years. We know that dinosaur tracks were observed by Pliny Moody as early as 1802, in the Connecticut Valley, and described by Hitchcock in 1836. Although Hitchcock is famous for assigning many to *Ornithichnites*, or stony–bird tracks, he and his contemporaries also recognized the affinities of many other trackmakers, as in *Sauroidichnites* (see quotation above). The accompanying figure of *Otozoum moodi* from the original 1802 locality represents a problematic trackmaker assigned by Hitchcock to the Amphibia and by subsequent workers to the Prosauropoda and the Pseudosuchia.

In Europe a different situation existed. By 1836 non–dinosaurian reptile tracks of Permo–Triassic age had already been described, and Cretaceous dinosaur remains had been described as fossil reptiles. By 1863, the year of Hitchcock's death, Cretaceous tracks from England had been discovered, scrutinized and assigned with some confidence to *Iguanodon* (see quotations by Tagart 1846 and Jones 1862). Although the class Dinosauria had been established in 1841 by Richard Owen, Hitchcock never lived to see a pre–Cretaceous dinosaur described or realize that skeletal remains would eventually be unearthed in the track–bearing Jurassic strata he knew so well.

In the latter part of the 19th century, when a wealth of dinosaur skeletal material was unearthed, the study of fossil footprints was all but abandoned. Even in the early part of the 20th century, when the rate of discovery of dinosaur skeletal remains slowed somewhat, the study of footprints was revived as a parttime activity by only a few workers, notably Richard Swann Lull, Charles Sternberg, Roland T. Bird and Barnum Brown. Consequently the late 20th century revival of interest in dinosaur tracks is unprecedented and long overdue. Although vertebrate ichnology is breaking new ground in this new age of dinosaur research, after a century of neglect, ichnologists still rely on the valuable contributions of many of the early contributors to the field.

References

Leonardi, G. 1984. Le impreinte fossili di dinosauri. *In* Bonaparte, J. F. et al. (eds.). *Sulle Orme de Dinosauri* (Venezia: Erizzo) 333 pp.

1 Sussex *Iguanodon* Footprints and the Writing of *The Lost World*

R. DANA BATORY AND
WILLIAM A. S. SARJEANT

Abstract

The discovery in 1909 of *Iguanodon* footprints in the Wealden Beds at Crowborough, Sussex, excited the attention of Sir Arthur Conan Doyle and served as a stimulus to his writing of *The Lost World*. One of the footprints was illustrated in a later edition of this work, done in the form of an expedition report — the only published report of the find.

"Look at this!" said [Lord John Roxton]. "By George, this must be the trail of the father of all birds!"

An enormous three-toed track was imprinted in the soft mud before us. The creature, whatever it was, had crossed the swamp and had passed on into the forest. We all stopped to examine that monstrous spoor. If it were indeed a bird — and what animal could leave such a mark? — its foot was so much larger than an ostrich's that its height upon the same scale must be enormous. Lord John

Figure 1.1. The cover of the deluxe edition of A. Conan Doyle's *The Lost World* [1912b], decorated by a line of stylized *Iguanodon* footprints.

looked eagerly round him and slipped two cartridges into his elephant–gun.

"I'll stake my good name as a shikaree," said he, "that the track is a fresh one. The creature has not passed ten minutes. Look how the water is still oozing into that deeper print! By Jove! See, here is the mark of a little one!"

Sure enough, smaller tracks of the same general form were running parallel to the large ones.

"But what do you make of this?" cried Professor Summerlee triumphantly, pointing to what looked like the huge print of a five-fingered human hand appearing among the three–toed marks.

"Wealden!" cried Challenger, in an ecstasy. "I've seen them in the Wealden clay. It is a creature walking erect upon three–toed feet, and occasionally putting one of its five-fingered fore–paws upon the ground. Not a bird, my dear Roxton — not a bird."

"A beast?"

"No; a reptile — a dinosaur. Nothing else could have left such a track. They puzzled a worthy Sussex doctor some ninety years ago; but who in the world could have hoped — hoped — to have seen a sight like that?"

This passage of intelligent ichnological deduction is to be found in the tenth chapter of Sir Arthur Conan Doyle's classic adventure story *The Lost World* (1912), a story in which four Englishmen climb to the summit of an isolated South American plateau in circa 1907 and discover that it is still populated by dinosaurs, pterodactyls, ape-men and cave–men. Perhaps more than any other book, this work has stimulated the imagination of young persons interested in fossils and has served as a dream wish-fulfillment even for professional paleontologists.

The passage is somewhat misleading, for the good Sussex doctor must have been Gideon Algernon Mantell (1790–1852), who described and named *Iguanodon* not from footprints, but from bones. The first discoverer of the foot-prints of that dinosaur was a Sussex clergyman, the Reverend Edward Tagart, who mistook them for those of giant birds (see Sarjeant 1974, p. 347–348, Fig. 31 and Woodhams and Hines this volume). Challenger was cor-rect, however, in interpreting the behavior of *Iguanodon*. While imprints of the hind foot (pes) of this dinosaur are common, those of its forefoot (manus) have not yet been reported with certainty (see discussion in Lockley 1985, p. 3–135 – 3–136). For the adventurers, Lord John's deduc-tion was soon to be shown wrong, and Professor Challenger's correct:

There were ... five of them, two being adults and three young ones. In size they were enormous. Even the babies were big as elephants, while the two large ones were far beyond all creatures I have ever seen. They had slate-colored skin, which was scaled like a lizard's and shim-mered where the sun shone upon it. All five were sitting up, balancing themselves upon their broad powerful tails and their huge three-toed hind feet, while with their small five-fingered front feet they pulled down the branches upon which they browsed. I do not know that I can bring their appearance home to you better than by saying that they looked like monstrous kangaroos, twenty feet in

Figure 1.2. Sir Arthur Conan Doyle in false beard, in the guise of George Edward Challenger (reproduced from A. Conan Doyle [1912b], *The Lost World*, plate between pp. 32 and 33).

length, and with skins like black crocodiles.

"Iguanodons," said Summerlee. "You'll find their footmarks all over the Hastings sands, in Kent, and in Sussex. The South of England was alive with them when there was plenty of good lush green–stuff to keep them going. Conditions have changed, and the beasts died. Here it seems that the conditions have not changed, and the beasts have lived."

What a pleasure, to have one's ichnological deduc-tions so immediately confirmed — though we might be sur-prised to find *slate-colored* reptiles dwelling in such lush green forests.

During Doyle's lifetime (1859–1930), the earlier age, when voyages of discovery were motivated by the desire for profit, was giving way to a newer one, when a develop-ing intellectual curiosity was causing a much more search-ing examination of the environments and inhabitants of the freshly discovered lands. Profit had ceased to be the sole motivation for investigation; knowledge was coming to be valued in its own right, as an enlargement of the horizons of humankind. Already a succession of scientists — Alexander von Humboldt, Alcide d'Orbigny, Thomas Huxley, Joseph Dalton Hooker and Charles Darwin were just a few of them — had, by making original observations in such lands, laid solid foundations for their scientific careers.

Yet scientific investigation in wild places involved considerable dangers. The archaeologists Frederick Cather-wood and John Lloyd Stephens, hacking their way through Central American jungles; the zoologists Henry Walter

Bates and Alfred Russel Wallace, paddling dugout canoes through Amazonian jungles in the quest for new insects; the scientific polymath Charles Darwin, carefully recording his observations in a Chilean earthquake and watching in fascination as an unfamiliar insect stung his finger (and, perhaps, implanted into him the germs of a disease from which he was never again to be free); and the geologist Frederick Vandiveer Hayden, making his observations under military escort and hastily retiring to a military fort when the Indians were rising — all of them encountered adventures enough, sometimes modest, sometimes hair-raising, to entertain Victorian readers only marginally interested in their scientific attainments.

Novelists began to perceive the potential of such adventures as the bases for saleable stories. Henry Rider Haggard took the brilliant first step into this new field when, in *King Solomon's Mines* (1885), he linked the story of a treasure hunt — familiar enough ground — with archaeological discoveries that evoked beguiling echoes of a lost civilization. In his later *She: A History of Adventure* (1887), this concept was developed into a dream-fantasy of a degenerate people living among the ruins of an ancient, much more advanced civilization.

"With this novel," wrote Douglas Menville (1974, p. ii), "he created an entire sub-genre of fantasy, now known as the 'lost world novel', and spawned many imitators who, over a span of more than 80 years, have produced numerous progeny of *She*...."

Conan Doyle is to be ranked among the successors to Haggard. Yet he was more innovator than imitator, for he expanded Haggard's idea in a new direction. The important element in *The Lost World* is, not the discovery of a surviving ancient human civilization, but of surviving ancient animals — the dinosaurs.

In the early part of this century, dinosaurs had not yet come to capture the imagination of children and adults at large. Only a handful of scientific books, and a smaller handful of popular books, had been published about these giant denizens of the world in the past. Yet it remained possible for ordinary citizens, if fortunately situated, to gaze with awe upon the giant skeletons mounted in a few museums; among these were the British Museum (Natural History), the Palaeontology Museum of the Jardin des Plantes in Paris, the Yale Peabody Museum and the American Museum of Natural History. Even more striking were those first reconstructions of dinosaurs, built for the 1851 Exhibition and, after its closure, displayed in the grounds of the Crystal Palace at Sydenham.

There can be little doubt that Doyle had visited the Crystal Palace and seen those fascinating, if inexact, recreations of vanished animals. However, in 1909 he became interested in paleontology by direct association. That year, shortly after moving to a house called "Windlesham" at Crowborough, Sussex, he noticed the fossil tracks of dinosaurs in a neighboring ironstone quarry into the Hastings Sands, a division of the Wealden Beds (Lower Cretaceous Hauterivian to Late Barremian stages). The discovery must have been in May, for on May 17th Doyle wrote to Arthur Smith Woodward (1864–1944) of the British Museum (Natural History) — Woodward, like Doyle himself, was later to be knighted — enclosing a sketch of one of the footprints and requesting that someone come down to Sussex to examine them.

Figure 1.3. "The Members of the Exploring Party" to Maple White Land. Left to right: Edward D. Malone, Professor Summerlee, Professor George E. Challenger and Lord John Roxton — Arthur Conan Doyle and friends in appropriate disguise! (Reproduced from A. Conan Doyle [1912b], *The Lost World*, plate between pp. 80 and 81.

The letter and sketch survive in the Museum Archives. A brief memo is attached, probably by Woodward's secretary, offering possible dates for a visit; and there is also a second, undated letter from Doyle to Woodward, expressing the fear that "the objects may be unworthy of your pains" but giving the train times to Groombridge Station and indicating that Woodward would be met by car. The exact date of the visit is uncertain, but it was probably in June 1909 and certainly on a Monday, for Doyle, writing to his mother, noted:

> "...I have another expert of the British Museum coming on Monday to advise me about the fossils we get from the quarry opposite. Huge lizard's tracks." (quoted in Nordon 1966, p. 329)

The tracks were, in fact, judged to be those of the herbivorous dinosaur *Iguanodon*. However, Woodward cannot have considered them of great interest, for similar tracks had already been reported widely from the Wealden sandstones of Sussex and the Isle of Wight and from the somewhat older Purbeck Beds (Late Jurassic [Portlandian] to Early Cretaceous [early Hauterivian]) of Dorset (see Sarjeant 1974, pp. 347–358). Consequently, no scientific account of the footprints was ever published.

Nevertheless, Doyle was interested enough to have casts made from these tracks and to display them in his home. He was also intrigued by the prehistoric relics to be found in the Sussex countryside. According to A. St. John Adcock (1912), it was the combined effect of these finds that set his imagination working and produced his classic novel. Describing a visit to Windlesham, Adcock wrote:

> ...on the floor of the billiard room stand two huge fossil feet [i.e. casts of footprints] of the prehistoric *Iguanodon*, and on the table above them is the flint head of an arrow

Figure 1.4. Footprint of *Iguanodon* from a quarry near Windlesham House, Crowborough, Sussex. (Reproduced from A. Conan Doyle [1912b], *The Lost World*, plate between pp. 168 and 169.).

possibility of opening up a new type of story of action. He contended that the possibilities had been exhausted and that with the pirate ship, the treasure hunt, and the other well-known forms of adventure books no new thrill was possible. The novelist, on the contrary, upheld the view that there was a large field which had not yet been worked, and that it should develop upon the lines of a combination of imagination and realism each pushed as far as the writer's capacity would carry him. The argument ended in a small bet and a promise by Sir Arthur that he would endeavour to vindicate his opinion by producing such a book. The result is *The Lost World*. It must be admitted that in his Sherlock Holmes tales Sir Arthur recast the stereotyped detective story of our childhood, and it will be interesting to see how far he succeeds in this new attempt at fresh methods of treatment. (Smith 1930, p. 395)

Doyle was extremely thorough in working up the backgrounds for his literary work; only when he felt he had properly mastered the necessary facts would he put pen to paper. He had become acquainted with Sir Edwin Ray Lankester (1847–1929), who had recently retired from the Directorship of the British Museum (Natural History); their acquaintance may well have been a consequence of Doyle's finding of the footprints. The text and illustrations of Lankester's book *Extinct Animals* (1905) served as further inspiration for Doyle.

He enjoyed very thoroughly the writing of the book:

Each evening through October–November 1911, Doyle would read aloud to his wife and any guests present what he had written during the day: laughing, gesturing — living the very part as he went along.

"I think it will make the very best serial (bar special S. Holmes values) that I have ever done, especially when it has its trimming of faked photos, maps, and plans," wrote Doyle to editor Smith. "My ambition is to do for the boys' book what Sherlock Holmes did for the detective tale. I don't suppose I could bring off two such coups. And yet I hope it may." (Carr 1949, p. 319)

The Lost World was serialized in *The Strand Magazine* from April through November 1912. It was quite lavishly illustrated, the drawings including several of dinosaurs that were based on Lankester's restorations. Doyle's accuracy in describing the details of that place of his imagination, and the plausibility of the reasons he advanced for the survival of such a prehistoric environment, alike received praise from scientists. A letter received in August 1912 from his "technical adviser" Dr. Lankester must have given him especial gratification:

You are perfectly splendid in your story of the 'lost world' mountaintop. I feel proud to have a certain small share in its inception as you indicate by quoting the book on extinct animals in the start. It is just sufficiently conceivable to make it 'go' smoothly. I notice that you rightly withhold any intelligence from the big dinosaurs, and also acute smell from the ape-men. (Carr 1949, p. 318)

that has survived from the Stone Age. It was the discovery of these relics on the downs that stretch for miles before his own door that set Sir Arthur's imagination at work on the period to which they belong and resulted in the creation of the astonishing Professor Challenger, the sending of him and his search party to that almost inaccessible plateau in the wilds of South America which they find still inhabited by men and animals of the prehistoric type, and, in a word, the writing of *The Lost World* which is at once one of the most realistic and one of the most romantic of his books — its wild imagination wearing an air of sheer reality from the Defoe-like matter-of-fact manner of their narration.

Yet the direct cause of Doyle's writing of the book was a wager. The story is recounted by H. Greenhough Smith, editor of *The Strand Magazine* (1930):

Anything like puffery was against the very nature of the man, and advertisements were apt to stir his ire. The following announcement he drew up himself, and it is of two-fold interest — it shows his own idea of how such a notice would be written, while it also tells the striking origin of another of his plots:

Sir Arthur Conan Doyle's serial story, "The Lost World," which begins in the April number of THE STRAND MAGAZINE, had its genesis in a curious way. A friend had been discussing with the author as to the

The Lost World was published in book form very shortly after its magazine appearance, by Hodder and Stoughton in London and simultaneously by other publishers in New York and Toronto. Though the earliest edition (1912) contained two maps, including one of the plateau, and many other illustrations, it was not in the "realistic" format of Doyle's hopes. These were attained instead in a magnificent edition, undated but published later in the same year, which was done in the fashion of a lavish expedition report. It was in generous format (7.0" X 9.5" X 2.3"), its cover decorated with a sequence of stylized iguanodont footprints, as if to stress the part that the dinosaur footprints had played in the genesis of Doyle's inspiration (Fig. 1.1). The illustrations included two photographs of the expedition's leader, George Edward Challenger, one in color and signed "Yours truly (to use the usual conventional lie!)" — and a black and white group photograph of the expedition's members (Figs. 1.2, 1.3). These were in fact Doyle himself, equipped with false beard as Challenger, and some of his friends suitably dressed up. Moreover, under the caption of "The First Footprint in Maple White Land" there is actually a photograph of one of Doyle's *Iguanodon* footprints from Crowborough — the only illustration of a part of this trackway ever to be published. It is here reillustrated (Fig. 1.4).

Doyle's book was to be made into a film by First National, in which the liberties with the text that are customary in the film industry were taken (Fig. 1.5); nevertheless, the film remains of great interest for its relatively early restorations of moving dinosaurs and pterodactyls. During the 1940's *The Lost World* was serialized on radio by the British Broadcasting Corporation (Fig. 1.6); this was a very faithful rendition of the story that kept the second author enthralled over many weeks. The book itself has gone through many editions and remains popular and readable today.

An unforeseen and unfortunate by-product of Doyle's story may have been to help implant in the popular mind the misleading idea that dinosaurs and early man were contemporaries. This is not implied in *The Lost World*; Doyle knew better, and quite evidently visualized the cavemen and apemen as having ascended at a much later

Figure 1.5. A still from the First Nation film version of *The Lost World*, showing the adventurers concealing themselves from an indignant brontosaur — an incident not to be found in the book! The scene shows the dinosaur's footprints and anticipates real demonstrations, from the study of fossil tracks, that saurpods were able to move freely on dry land! (Photo: courtesy of the museum of Modern Art Stills Archive, New York)

Figure 1.6. A herd of dinosaurs (*Iguanodon*) feeding in a glade of Maple White Land. (Illustration from *The Radio Times*, London, c. 1946, relating to a broadcast serialization of *The Lost World*).

date to the plateau. However, a careless reading of the book may perhaps have been one source for the numerous later stories in which men and dinosaurs are contemporaries and for the numerous cartoons depictng such impossible, but often amusing, situations as a caveman being pursued at high speed by a ravening sauropod dinosaur!

As to the footprints themselves, however, less can be said. Though an earlier find in a waterworks excavation at Crowborough had received brief published mention (Herries 1907; see also Sarjeant 1974, p. 350), the track which Doyle found has never been mentioned in any scientific publication. No specimens from Crowborough are to be found among the holdings of the British Museum (Natural History) or in any other major British collection (see Sarjeant 1983) and the fate of Doyle's own casts of the footprints is not known to us. However, even if they furnished no contribution to scientific knowledge, they have helped to enrich in unique fashion the field of literature. Moreover, they have contributed to the progress of science by serving to stimulate interest among the many budding paleontologists who, over the years, have enjoyed *The Lost World*.

Acknowledgment

The authors are indebted to Dr. Ron Cleevely for courteously checking Sir Arthur Conan Doyle's letters to the British Museum (Natural History) on our behalf.

References

Adcock, A. D. St. J. 1912. Sir Arthur Conan Doyle. *The Bookman (London)* November issue, pp. 95–110.

Carr, J. D. 1949. *The Life of Sir Arthur Conan Doyle.* (London: Murray) 361 pp.

Doyle, A. C. 1912a. *The Lost World.* (London: Hodder and Stoughton) 309 pp., frontis. + 6 illus., 1 map.

  n.d. [1912b]. *The Lost World.* New edition. (London: Henry Frowde, Hodder and Stoughton) vi–x + 11–319 pp., 2 color pls., 20 b & w pls., 2 maps, dec. endpapers.

Haggard, H. R. 1885. *King Solomon's Mines.* (London: Cassel) 320 pp., folding frontis. (map).

  1887. *She: A History of Adventure.* (London: Longmans, Green) 317 pp., 2 pls.

Herries, R. S. 1907. Excursion to Crowborough. *Proc. Geol. Assoc. London* vol. 20, pp. 163–166.

Lankester, E. R. 1905. *Extinct Animals.* (London: Constable) xxiii + 331 pp.

Lockley, M. G. 1985. Vanishing tracks along Alameda Parkway. Implications for Cretaceous dinosaurian paleobiology from the Dakota Group, Colorado. *In* Chamberlain, C. K. (ed.). *A Field Guide to the Environments of Deposition (and Trace Fossils) of Cretaceous Sandstones in the Western Interior.* (Golden, Colorado: Society of Economic Paleontists and Mineralogists) pp. 3-131 – 3-142.

Menville, A. D. 1974. *Introduction to H. Rider Haggard: Eric Brighteyes.* Forgotten Fantasy Library. (North Hollywood, Calif.: Newcastle Publishing Co.)

Nordon, P. 1966. *Conan Doyle.* (London: John Murray) x + 370 pp.

Sarjeant, W. A. S. 1974. A history and bibliogrpahy of the study of fossil vertebrate footprints in the British Isles. *Palaeogeogr., Palaeoclimat., Palaeoecol.* 18 (4): ii + 160 pp.

  1983. British fossil footprints in the collections of some principal British museums. *Geol. Curator* 3 (9):541–560.

Smith, H. G. 1930. Some letters of Sir Arthur Conan Doyle. *Strand Magazine*, October, pp. 390–395.

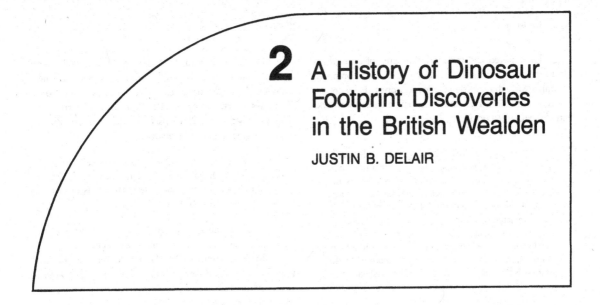

2 A History of Dinosaur Footprint Discoveries in the British Wealden

JUSTIN B. DELAIR

Abstract

This paper summarizes the known discoveries of British Wealden dinosaur footprints. The various finds are presented in chronological order and special attention is drawn to their geographical occurrence and the high percentage of inferred herbivore tracks. A general bibliography is also appended.

The Early Discoveries (1846–1860)

The earliest known discovery of dinosaur footprints in the British Wealden occurred in 1846, when the Rev. Edward Tagart presented a single imprint from the Hastings Sands to the Geological Society of London. The specimen was later transferred to the Geological Survey collections (GSM376). An abstract of a published letter informs us that not only had similar footprints been observed nearby, but that:

> The markings in question appear to have been observed by several persons at Hastings; but they have not been found consecutive, or having any distinct relation to one another. They are of large size, the one presented to the Society measuring sixteen inches in length; but there does not appear, either from this specimen, or from the account communicated by the author, any decisive evidence as to their origin. (Tagart 1846)

The title of this abstract reflected the tendency at that time to ascribe trifid markings of this kind to huge extinct birds. However, later, Alfred Tylor mentioned that Tagart's original letter contained the statement that a "Dr Harwood suspects them to be the footmarks of the Iguanodon" (Tylor 1862a). As is well known, *Iguanodon* was a large herbivorous dinosaur common during Wealden times and first described by Gideon Mantell in 1822 (Delair and Sarjeant 1975). Tagart's important historic specimen was figured for the first time by Sarjeant — in 1974.

Tagart's letter stimulated further searches for additional Wealdon specimens. In the four years to 1850 a large number came to light along the Sussex coast. These included "many natural casts and impressions of reptilian footprints" from the cliffs at Bexhill, and "fossil casts of large reptilian footprints on the undersides of a band of stone projecting in the clay cliff at Goldbury Point" near Fairlight (Dixon 1850). By 1850, therefore, these footprints were beginning to be regarded by some as being of reptilian rather than avian origin.

Among those who were then searching the Sussex coast for Wealden footprints was Samuel Beckles, who, in 1851, delivered to the Geological Society of London the first of a series of papers on discoveries made near his home town of St. Leonard's. His first paper stated:

> Certain trifid bodies, presenting a resemblance to the casts of the impressions of birds' feet, are rather numerous in the cliffs to the east and west of Hastings (from the latter locality, Mr Beckles has now obtained eight specimens, in a limestone containing *Cyrenae*, remains of *Lepidotus*, etc....).
>
> Several specimens, detached from the cliffs, have been taken from the beach; but at about four miles east of Hastings, where the cliffs are 200 feet high, the casts occur at about 40 feet above sea-level. They were found in a stratum of rock, overlying a bed of clay: which latter having been removed by rain and weather, the casts appeared in relief on the under-surface of the rock, just as if they were hanging from the ceiling of a room.... One detached block ... bears four of these trifid bodies in relief; they are arranged with the toes pointing in a uniform direction, so as to make out a nearly perfect square. A distance of 2 ft. 7 in. separate the two in front and 2 ft. 5 in. the hinder two; between the two on the right, from the toe of the hinder one to the heel of the foremost, there is a space of 2 ft. 3 in.; and between the other two, the distance is less by nearly two inches. The largest specimen has a length of 21 inches". (Beckles 1851)

Jones noted the tridactyl character of *Iguanodon*'s hind foot, and that its size in adult individuals compared favorably with that of the average footprint. He went on to become the first to definitely associate these footmarks with *Iguanodon* when he wrote:

> We may therefore be allowed provisionally to refer these tracks to the *Iguanodon*, who certainly wallowed in the Wealden waters and frequented their sand-bars and mud-banks — who had a great three-toed foot — and who, like some other quadrupeds (such as the Tapir, etc.), may have actually, if not always, planted his footprints uniserially, leaving as his spoor a single row of thick-toed trifid imprints, sometimes showing the marks both of toes and heels, sometimes of the toes only, according to the firmness of the mud or sand on which he walked. (Jones 1862a)

Figure 2.1 shows examples of the three kinds of footprint preservation that Jones evidently had in mind. They have also been discussed more recently by Edmonds (1979).

In 1862, Beckles delivered another paper which dealt with "natural casts ... nearly 3-1/2 feet long ... indicating not merely imprints of the toes, but also of the sloping metatarsals. The animal must have been of great size and weight, leaving deep imprints." The paper also reported others from the Isle of Wight, and, more importantly, from the foreshore at Swanage Bay (the first Wealden footprints from Dorset), which

> ... occur in two bands of sand-rock, usually about 1 foot thick, separated by about 20 feet of clay, and coming down to the seashore with the other beds. These casts are of the usual thick-toed trifidal shape, and of the usual size — about 15 inches long. (Beckles 1862)

Figure 2.1. Three common kinds of preservation of Wealden dinosaur footprints interpreted as: **a,** impression (or cast) produced by a sitting animal with most of its weight over the back part of the foot. **b,** impression (or cast) produced by a walking or standing animal. **c,** impression (or cast) produced by an animal running, the balance and weight being over the toes only.

So far as is known, none of these early Dorset specimens were ever collected. Beckles also recorded his discovery of smaller trifid casts "3 in. by 3 in. ... set about 15 inches apart." They apparently formed a portion of trackway. (Beckles 1862). Considering the question of their origin, Beckles cautiously observed:

> It is certain that other Dinosaurians besides the *Iguanodon* had the same modifications of structure; and we must not refer these pachydactylous trifids to that animal exclusively." (Beckles 1862)

It was to be several decades before other workers paid much heed to Beckles' cautionary remark.

In 1862 members of the Geologists' Association visited Hastings and viewed "Iguanodon footprints" (Deck 1862), while Jones again mentioned the occurrence of tridactyl footprints (Jones 1862b).

The association of these footprints with *Iguanodon* was rapidly gaining acceptance about this time:

> It represents what I believe to be the hind foot of an Iguanodon, resting upon a ripple-marked surface of sandy mud sufficiently hard to retain an exact impression. The pressure of the foot had raised the sand surrounding the impression about half an inch above the ripple-mark, at the same time turning over some shells of the genus *Cyrena*, which may be seen in the undisturbed mud. (Tylor 1862a)

Tylor also furnished a coastal section indicating the localities from which "Iguanodon" footprints had been reported, and mentioned discoveries in thin sandstones which he felt came from higher in the Wealden series, probably from the Wadhurst Clay (Tylor 1862a).

In a summary of this paper, Tylor noted that several footprints had been found at two horizons in the cliff (Tylor 1862b).

In 1865, an "Iguanodon" footprint specimen from Sussex was recorded as having been acquired for the Liverpool Museum. It had formerly been a part of the collection of Gideon Mantell (Moore 1865).

Twelve years elapsed before British Wealden dinosaur footprints were again mentioned in print (Harrison 1877). These were the footprints found by Ernest Westlake who was studying the geology of the Isle of Wight. His field notebooks (Delair 1982) included information on three large tridactyl footprints on low-tide sandstone ledges off Brook Point, which, even then, were being rapidly eroded by tidal scour. Westlake did not collect any of the specimens (Delair 1985).

Ten years later, Mansel-Playdell noted that footprints had been found in the Wealden at Worbarrow Bay, Dorset, along with dissociated *Iguanodon* remains (Mansel-Pleydell 1888).

The Geological Survey's memoir on the geology of the Isle of Wight (Bristow, Reid, and Strachan 1889), seems to have been the last nineteenth century publication to mention Wealden dinosaur footprints. By the close of the

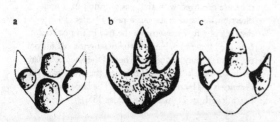

a   b   c

In the same paper, Beckles also referred to Gideon Mantell's then recent discovery of similar footprints in the Wealden of the Isle of Wight, Mantell himself commenting upon them in the 1854 edition of his book on the Island's geology thus:

> This specimen is a solid tripartite mass of fawn-coloured sandstone; the middle process is fifteen inches, and the two lateral projections are twelve inches in length; the greatest thickness is six inches; the processes are laterally compressed and rounded at the extremities, and united to one common base.... As the origin of these singular concretions is very problematical, every specimen should be preserved; and if several occur in the same bed, their relative position should be ascertained" (Mantell 1854)

Neither Beckles nor Mantell tells us the locality or precise horizon at which this early Isle of Wight discovery was made, although the latter was almost certainly Wealden.

In 1852, Beckles presented his second paper in which he described similar tridactyl specimens from several localities. He noted that large three-toed footprints occurred throughout the Wealden beds exposed along the Sussex coast for a distance of about 18 miles. He stated his belief that they had been made by large defunct birds, and referred to them as 'ornithoidichnites' (Beckles 1852).

In 1854 Beckles published his third paper, which dealt with four separate sets of footprints. It was the fourth set which afforded the greatest interest and information, for it comprised no less than 60 individual impressions scattered over a 400 square yard area of low–tide foreshore at Bexhill. These formed three distinct trackways of bipedal animals. Beckles removed a slab containing six of the footprints and exhibited it at one of the meetings of the Geological Society of London. He also recorded other isolated footprints and added that in the use of the term 'ornithoidichnites':

> ... I intend rather to convey an intimation that the trifid bodies are of organic origin, than to determine the affinities of the animals that produced them: I adopt the term, therefore, provisionally and most cautiously. Although the evidence seems to connect the footprints with the class *Aves*, yet I am not aware that it is such as positively to exclude animals of different organization. (Beckles 1854)

Beckles also referred to the scepticism voiced then in some quarters about these ichnites. He wrote:

> With the extensive accumulation of these natural casts in my collection, I felt much surprise that men of real science should still pronounce them mere *accidental concretions*. The cause, whatever it was, so uniformly produced the same effects, whether in clay-rock, sandstone, or shale, as to be inconsistent with our idea of an accident. To reject these trifid bodies as organic phaenomena, because they may not happen to come immediately within the types of existing organisation, would be a singular

disregard for all those researches which are daily revealing the wonders of former epochs". (Beckles 1854)

Beckles' pioneer studies were briefly referred in an editorial footnote to the seventh (1857) edition of Mantell's *Wonders of Geology*, although Tagart's find of 1846 was erroneously recorded as having occurred on "the shores of the Isle of Wight" (Jones 1857). Robert Damon also mentioned Beckles' discoveries (Damon 1860).

By the late 1850's, therefore, all the known Wealden footprint discoveries in Britain had been made in Sussex or the Isle of Wight, and had occurred exclusively in coastal exposures. Three differing views were also current as to their origin. These were that the footprints were accidental concretions, the footmarks of giant extinct birds, or those of enormous defunct reptiles (*Iguanodon* being the most favored trackmaker).

Later Nineteenth Century Discoveries (1861–1899)

Beginning in 1862, we encounter an upsurge of published opinion on British Wealden footprints, mostly occasioned, it would seem, by new evidence exposed in a cliff fall near Hastings early that year. Some of these publications also contain the first references to inland occurrences of these footmarks.

The earliest account of this cliff fall appeared on March 8th, 1862, and recorded how:

> The fall of the cliff near Hastings, last week, has brought to light an interesting slab of stone, bearing on its surface the clear impression of the foot of a gigantic bird. It has three toes, each of which is about 9 inches long in the tread, with a claw at the end, of perhaps two inches in length. The back of the foot, where the three toes meet as in a centre, does not appear: that part of the foot did not reach the ground. But still further back is the mark of the spur, or fourth toe. From the point of the middle claw to this mark of the spur it measures twenty-four inches, and in width twenty inches. The whole of the slab is covered with the lines of ripple made by the waves upon soft mud: and there are numerous other impressions more or less perfect of the same bird's claw on the other slabs of stone. The bird which has left us this footprint may be supposed to have been at least twelve feet high, and perhaps much more. (Anonymous 1862)

This report induced a reply from T. Rupert Jones on March 22nd; this summarized the history of the study of Wealden footprints up to that date and referred to other occurrences near Cuckfield and near Horsham. He emphasized that:

> There are other animals, however, belonging to the "Wealden" and far better known than the birds of that period, that may have had to do with the foot-tracks in question; namely, the gigantic reptiles, of which we see excellent models in the Crystal Palace Park — the *Iguanodon*, the *Hylaeosaurus*, and the *Megalosaurus*.

1800's, therefore, the older views that these footprints were accidental concretions or the footmarks of enormous extinct birds had been abandoned. But, despite Beckles' cautionary remarks, they were nearly always attributed to *Iguanodon*. Moreover, few if any proper measurements or plans had been made of the several distinct trackways which had come to light.

Early Twentieth Century Finds (1900–1949)

The earliest known twentieth century discoveries concerned unspecified "Iguanodon" footprints from Wealden strata near Haslemere, Surrey (Fowler 1904). Fowler's note constitutes the only record of dinosaurian footprints yet known from Surrey.

In 1907 R. S. Herries recorded an iguanodont footprint in Wealden beds at Crowborough (Herries 1907). In the same year Reginald Hooley discussed the nineteenth century footprint discoveries at Hastings and the Isle of Wight, referring to them as iguanodont rather than *Iguanodon* (Hooley 1907).

In 1918, Anthony Belt reviewed discoveries of ichnites made at Hastings, Cooden, Bulverhythe, Govers, Ecclesbourne, Fairlight, and Bexhill (Belt 1918). He revealed that footprints were being discovered relatively often along that stretch of the Sussex coast and that they apparently existed there in comparative abundance.

A Miss Rangel was associated with these trace fossils in 1920, when she exhibited footprint–bearing slabs from the Wealden of Galley Hill, Hastings (Anonymous 1922a). They were ascribed, quite wrongly, to *Labyrinthodon*, an amphibian genus long extinct by Wealden times. We may, however, confidently infer that Miss Rangel's specimens were iguanodont as only the next year footprints of that kind, from the *same* horizon and locality, were presented by H. W. Wilson to the Geological Survey Museum, London (GSM 37960 and GSM 37961). Natural casts of *Iguanodon* footprints from the Lower Wealden near Bexhill, and a series of photographs, showing the footprints *in situ* in the Bexhill cliffs, were exhibited in the Geologists' Association in 1922 (Anonymous 1922b).

In 1924, further discoveries included elongated or slender–toed forms (Thompson 1924). This was the first time that footprints recognizably different from those ascribed to *Iguanodon* had been met with in the Wealden, but no adequate description or illustration of them has ever appeared.

In 1925, another Geologists' Association field excursion to Hastings encountered additional dinosaur footprints.

> Beyond the point (Little Galley Hill) and the stack, large footprints of *Iguanodon* can be seen on the sandstone blocks on the foreshore. Some good specimens of these are preserved in Bexhill Museum. (Milner and Bull 1925)

The horizon at which these specimens occurred was in the Ashdown Sands division of the Hastings Sands.

A year later, Osborne–White recorded the occurrence of "saurian footprints" on the foreshore at Cooden

and Bexhill, and that specimens from various places along the East Sussex coast were in museum collections at Brighton and Bexhill (Osborne–White 1926).

The series of tridactyl footprints exposed the same year (1926) by a cliff fall at Bulverhythe was, however, of great importance. They were studied and mapped in detail by N. F. Ticehurst, who recorded fifty or more tracks (Sargent 1980). As no examples of this notable series were retrieved, and the whole set obliterated soon afterwards by marine erosion, Ticehurst's valuable report of the tracks (Ticehurst 1926) is reproduced here as Figure 2.2.

The Later Twentieth Century Discoveries (1950 to the present)

For almost forty years following Ticehurst's paper, there were no reports on British Wealden dinosaur ichnites. A brief reference to Frank Raw's acquisition for Birmingham University of an "*Iguanodon*" footprint from Swanage Bay showed that specimens were still obtainable (Anonymous 1962). It was not until 1965 that the next significant discovery occurred in the Middle Ashdown beds (Bazley and Bristow 1965). The specimens were collected and placed in the British Museum (Natural History), along with R.3619, bearing one complete and four incomplete tridactyl footprints from the Wealden at Hurtis Hill, also near Crowborough.

British Wealden footprints were reviewed by Hartmut Haubold (1971). Haubold's work facilitates comparison between British Wealden and other footprints.

William Sarjeant's extensive, but incomplete, survey of all British fossil vertebrate footprints (Sarjeant 1974), included many of the pre–1973 Wealden discoveries dealt with in this paper, and was particularly meritorious for citing numerous obscure references.

In 1977, William Blows found tridactyl impressions on a low–water sandstone ledge off Chilton Chine in the Isle of Wight. Subsequent mapping disclosed the presence of over thirty footprints comprising portions of ten separate trackways, thought at the time to be attributable to at least two kinds of large theropods — *Megalosaurus* and *Antrodemus* (Blows 1978). In 1982, however, some of the larger

Figure 2.2. The series of tridactyl footprints exposed on the foreshore near Bulverhythe, East Sussex, in 1926. After Ticehurst (1928).

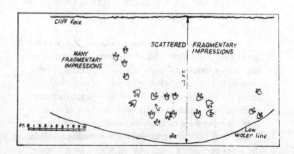

Table 2.1.

Date Found by	Sussex	Surrey	Isle of Wight	Dorset
1846	Hastings (E. of)	—	—	—
1850	Bexhill (cliffs)	—	—	—
1850	Goldbury Point	—	—	—
1851	Hastings (E. of)	—	—	—
1851	Hastings (W. of)	—	Unlocalized	—
1852	White Rock	—	—	—
1852	St. Leonards	—	—	—
1852	The Sluice (W. of St. Leonards)	—	—	—
1852	Tunnel between St. Leonards and Hastings	—	—	—
1854	Bexhill (W. of)	—	—	—
1854	Bexhill (foreshore)	—	—	—
1854	Near Galley Hill	—	—	—
1854	Near Bulverhythe	—	—	—
1854	Between Pevensey Sluice and Cooden	—	—	—
1862	Near Hastings	—	—	—
1862	Hastings (W. of)	—	Compton Bay	Swanage Bay
1862	Near Cuckfield	—	Sedmore Point	—
1862	Stammerham	—	Hanover Point	—
1862	Near Horsham	—	Between Brook and Brighstone	—
1862	Hastings (foreshore)	—	—	—
1862	Ecclesbourne Glen (East Cliff)	—	—	—
1862	Biggs' Farm, Cuckfield	—	—	—
1865	Unlocalized	—	—	—
1877	—	—	Brook Point (W. of)	—
1878	—	—	Brook Point (off)	—
1888	—	—	—	Worborrow Bay
1904	—	Harting Combe	—	—
1907	Crowborough (waterworks)	—	—	—
1918	Hastings	—	—	—
1918	Bulverhythe	—	—	—
1918	Cooden	—	—	—
1918	Govers	—	—	—
1918	Ecclesbourne	—	—	—
1918	Fairlight	—	—	—
1918	Bexhill	—	—	—
1920	Galley Hill	—	—	—
1922	Near Bexhill	—	—	—
1924	Galley Hill	—	—	—
1925	Near Little Galley Hill	—	—	—
1926	Cooden (foreshore)	—	—	—
1926	Bexhill (foreshore)	—	—	—
1926	Bulverhythe	—	—	—
1961	—	—	—	Swanage Bay
1965	Jarvis Brook	—	—	—
1977	—	—	Chilton Chine (off)	—
1978	Bexhill (foreshore)	—	—	—
1979	—	—	Yaverland (foreshore)	—
1980	Bexhill (foreshore)	—	—	—
1980	Cooden (foreshore)	—	—	—
1981	Glyne Gap, near Galley Hill	—	—	—
1981	Fairlight	—	—	—
1983	—	—	Brook (foreshore)	—
?	Hurtis Hill	—	—	—

Chronological list of British Wealden dinosaur footprint discoveries arranged by counties.

imprints were referred to *Iguanodon* (Insole 1982), and in 1985 all the footprints were ascribed to that dinosaur (S. Hutt in Delair 1985). Some of the footprints were later excavated and placed on permanent exhibit in Sandown Museum. A full account appears in Delair (1985), where Blows' scale map of the various in situ tracks is reproduced.

Blows' discovery marked the beginning of a new era in the study of British Wealden dinosaur ichnites, commencing in the Isle of Wight, and soon embracing coastal Sussex. Following a succession of gales in January 1978, numerous large trifid imprints became visible on a gray sandstone bed forming the foreshore near Bexhill. These rapidly disappeared through natural tidal scour and the entire site was soon buried under a thick shingle deposit (Sargent 1980), although not before a single specimen had been excavated and taken to Bexhill Museum, where it was identified as that of an *Iguanodon* (Anonymous 1980a; Delair and Sarjeant 1985; Woodhams and Hines, this volume).

During the spring of 1979, dinosaur footprints came to light on the foreshore near Yaverland, on the Isle of Wight (Delair 1985).

Large three-toed footprints discovered on the surface of soft mudstone at Bexhill proved to be part of an extensive series, from which one specimen was excavated (Anonymous 1980b,c; Chapman 1980; Delair and Sarjeant 1985). All the above footprints are now inaccessible to further study. It was therefore fortuitous that portions of two further trackways, composed of typical "*Iguanodon*" footprints, were found on a surface of Wealden sandstone on the foreshore at Cooden in 1980. This new evidence was photographed and carefully mapped by K. Woodhams and J. S. Hines, who also procured plaster-casts of the best preserved specimens (Delair and Sarjeant 1985; Woodhams and Hines, this volume).

In April 1981, twelve or more natural casts of iguanodont footprints were found at Glyne Gap, adjacent to well-known Wealden dinosaur footprint localities (Gibbons 1981, Delair and Sarjeant 1985). These ichnites are currently being studied by Woodhams and Hines.

Later the same year, better preserved tridactyl casts (width 30–50 cm) were studied by Hines and Woodhams in beds beneath the Lee Ness Sandstone at Fairlight. Only rarely do two or three of these occur as a trackway portion (Stewart 1981, Delair and Sarjeant 1985).

In 1983, Ian Barker figured and discussed one of the footprints from the Chilton Chine series removed to Sandown Museum in 1977 (Barker 1983). 1983 also saw the publication of Sarjeant's list of various fossil footprints in some British museums (Sarjeant 1983). Among those of relevance here were Tagart's original footprint (GSM 376), the Galley Hill specimens (GSM 37960–37961) and some Sussex "Iguanodon" footprints (no. 54.28) in the Castle Museum, Norwich.

Finally, in December 1983, the natural cast of a huge iguanodont footprint was discovered on the beach at Brook in the Isle of Wight (Anonymous 1984, Delair 1985).

Summary

Footprints of Wealden dinosaurs have been certainly known from Britain since 1846. They occur in the southern English counties of Sussex (45 finds), Surrey (1 find), the Isle of Wight (10 finds), and Dorset (3 finds). Of these 59 separate discoveries (see Table 2.1), comparatively few have been adequately studied or recorded and preserved for posterity.

There is a possibility that several of the Sussex and Isle of Wight finds, made at the same localities on different dates, represent different portions of the same trackways exposed on separate occasions. Many of the finds made at Hastings, Bexhill, Cooden, Galley Hill, and Brook may also fall into this category.

Nearly all the known specimens appear to represent the footmarks of herbivorous dinosaurs, such as *Iguanodon* could be expected to have made, although some specimens have been attributed to theropod dinosaurs (Woodhams and Hines, this volume).

The overwhelming majority of these ichnites have been preserved in sandstone and is as typical of inland exposures as of coastal outcrops. Beckles, however, mentions the ocurrence of some specimens in shale (Beckles 1854). Considerable research remains to be accomplished, however, on the correlation of these sandstones, and the several separate horizons at which the footprints are known to occur. Studies also need to be made of what changes, if any, occur between specimens throughout the Wealden stratigraphic succession.

Acknowledgments

Sincere thanks are extended to Mr. John Hines and Mr. Ken Woodhams for their assistance in furnishing information about the Sussex discoveries with which they have been so closely associated, and to Dr. Alan Insole and Mr. Stephen Hutt for similar service respecting several of the more recent Isle of Wight discoveries.

References

Anonymous. 1862. Notes of the week. *Literary Gazette (London)*, n.s., 8 (193), March 8:232.

— 1922a. Exhibits. *Report Proc. Hampstead Sci. Soc.* (1920–22):42.

— 1922b. (Fossils exhibited). *Proc. Geologists' Assoc.* 33:159–160.

— 1962. (Obituary of Frank Raw – died 2nd Sept. 1961). *Proc. Geologists' Assoc.* 73:158–159.

— 1980a. Giant footprints at bottom of a garden. *Bexhill Observer* November 29.

— 1980b. Major's success. *Bexhill Observer* December 6.

— 1980c. Alf's date with the dinosaurs. *Bexhill Evening Argus* December 5.

— 1984. Museum news. *Museum I.O.W. Geol. Newsletter* 5 (March):2.

Barker, I. 1983. Thoughts on a footprint. *Museum I.O.W. Geol. Newsletter* 3 (August):8–10.

Bazley, R. A., and Bristow, C. R. 1965. Field Meeting to the Wealden of East Sussex. *Proc. Geologists' Assoc.* 76:315–319.

Beckles, S. H. 1851. On the supposed casts of footprints in the Wealden. *Quart. Jour. Geol. Soc. London* 7:117 (abst.).

—— 1852. On the Ornithoidichnites of the Wealden. *Quart. Jour. Geol. Soc. London* 8:396–397.

—— 1854. On the Ornithoidichnites of the Wealden. *Quart. Jour. Geol. Soc. London* 10:456–464.

—— 1862. On some natural casts of reptilian footprints in the Wealden beds of the Isle of Wight and Swanage. *Quart. Jour. Geol. Soc. London* 18:443–447.

Belt, A. 1918. Prehistoric Hastings. *Hastings and East Sussex Naturalist* 3 (1):1–46.

Blows, W. T. 1978. *Reptiles on the Rocks.* (Newport: I.O.W.) 60 pp.

Bristow, H. W., Reid, C., and Strachan, A. 1889. The geology of the Isle of Wight. *Mem. Geol. Surv. Great Britain* 2nd edition.

Damon, R. 1860. *Handbook to the Geology of Weymouth and the Isle of Portland, with Notes on the Natural History of the Coast and Neighbourhood.* (London: Stanford) 200 pp.

Deck, A. 1862. Notes on the (Hastings) excursion. *Proc. Geologists' Assoc.* 1:248–251.

Delair, J. B. 1982. Ernest Westlake (1855–1922), geologist and prehistorian: with a synopsis of the contents of his field notebooks. *Geological Curator* 3 (3):133–152.

—— 1985. Cretaceous dinosaur footprints from the Isle of Wight: a brief history. *Proc. Isle of Wight Natural History and Archaeological Soc.* 7 (8), (1983):609–615.

Delair, J. B., and Sarjeant, W. A. S. 1975. The earliest discoveries of dinosaurs. *Isis* 66 (231):5–25.

—— 1985. History and bibliography of the study of fossil vertebrate footprints in the British Isles: Supplement 1973–1983. *Palaeogeog., Palaeoclimat., Palaeoecol.* 49:123–160.

Dixon, F. 1850. *The Geology and Fossils of the Tertiary and Cretaceous Formations of Sussex.* (London: F. Dixon) xxiv + 469 pp.

Edmonds, W. 1979. *The Iguanodon Mystery.* (Harmondsworth: Kestrel Books) 79 pp.

Fowler, J. B. 1904. A local geology for amateurs and beginners. *Haslemere Microscopical Nat. Hist. Soc., Science Paper* 3. 23 pp.

Gibons, W. 1981. *The Weald.* (London: Unwin Paperbacks) see p. 96.

Harrison, W. J. 1877. *A Sketch of the Geology of Hampshire.* (Sheffield) 12 pp.

Haubold, H. 1971. Ichnia Amphibiorum et Reptiliorum fossilim. *In* Kuhn, O. (ed.). *Handbuch der Palaeoherpetologie* 18. (Stuttgart: Fischer) 124 pp.

Herries, R. S. 1907. Excursion to Crowborough. *Proc. Geologists' Assoc.* 20:163–166.

Hooley, R. 1907. A brief sketch of the Wealden beds of the Isle of Wight, and the history they reveal. *Papers Proc. Hampshire Field Club.* 6 (1):90–105.

Hutt, S. 1985. *In* Delair, J. B. Cretaceous dinosaur footprints from the Isle of Wight: a brief history. *Proc. Isle of Wight Natural History and Archaeological Soc.* 7 (8), (1983):609–615.

Insole, A. N. 1982. The habitat of the Wealden dinosaurs. *Jour. Portsmouth and District Nat. Hist. Soc.* 3 (2):80–87.

Jones, T. R. 1857. Editorial footnotes to Mantell, G. A. *The Wonders of Geology.* 7th edition. (London: Bohm) 2 vol.

—— 1862a. Correspondence. *Literary Gazette, London* n.s., 8 (195) (22 March).

—— 1862b. Tracks, trails, and surface-markings. *The Geologist* 5:128–139.

Mantell, G. A. 1854. Geological Excursions around the Isle of Wight, and along the Adjacent Coast of Dorsetshire: Illustrative of the most interesting geological phenomena and organic remains. 3rd edition. (London: Bohn) xxiv + 356 pp.

—— 1857. *The Wonders of Geology.* 7th edition. (London: Bohn) 2 vol. (see vol. 1:383–384).

Milner, H. B., and Bull, A .J. 1952. The geology of the Eastbourne–Hastings coastline; with special reference to the localities visited by the Association in June, 1925. *Proc. Geologists' Assoc.* 36:291–320.

Moore, T. J. 1865. On a footprint probably of the *Iguanodon* in the Free Public Museum of Liverpool. *Proc. Liverpool Geol. Soc.* 6th session (1864–1865): 35–37.

Osborne-White, H. J. 1926. The geology of the country near Lewes. *Mem. Geolog. Surv. Great Britain.*

Sargent, H. J. 1980. *Leaflet no. 1.* (Bexhill Museum) 2 pp.

Sarjeant, W. A. S. 1974. A history and bibliography of the study of fossil vertebrate footprints in the British Isles. *Palaeogeog., Palaeoclimat., Palaeoecol.* 16 (4):265–379.

—— 1983. Some British fossil footprints in the collections of some principal British museums. *Geological Curator* 3 (9):541–560.

Stewart, D. J. 1981. A field guide to the Wealden Group of the Hastings area and Isle of Wight. *In* Elliot, T. (ed.). *Field Guide to Modern and Ancient Fluvial Systems in Britain and Spain.* Proceedings 2nd International Fluvial Conference, University of Keele. Sept. 1981.

Tagart, E. 1846. On markings in the Hastings Sands near Hastings, supposed to be the footprints of birds. *Quart. Jour. Geolog. Soc. London* 2:267.

Thompson, J.C. 1924. Local erosion of the coast. *Hastings and East Sussex Naturalist* 3 (3):154–155.

Ticehurst, N. F. 1928. Iguanodon footprints at Bulverhythe. *Hastings and East Sussex Naturalist* 4 (2):15–19.

Tylor, A. 1862a. On the footprints of an Iguanodon lately found at Hastings. *Quart. Jour. Geolog. Soc. London* 18:247–253.

—— 1862b. On the footprints of an Iguanodon lately found at Hastings. *The Geologist* 5:185.

3 Ichnology of the Connecticut Valley: A Vignette of American Science in the Early Nineteenth Century

R. TED STEINBOCK

Abstract

The major figure in the ichnology of the Connecticut Valley was Reverend Edward Hitchcock, a Congregational minister turned geologist and educator at Amherst College. His work on bird-like tracks in the Red Sandstone culminated in two elaborate monographs and numerous scientific reports. Hitchcock's grave concern for the atheism and pantheism expounded by a portion of the scientific community resulted in the classic *Religion of Geology and its Connected Sciences* — an elaborate attempt to reconcile and support revealed scripture with the latest geological theories and discoveries.

In these transitional years for the subdivision of natural history into narrower disciplines, broadly educated men still played an important role in scientific investigation. This is exemplified in the ichnologic work of James Deane, M.D., and John Collins Warren, M.D. The lives and times of these individuals constitute a most interesting chapter in the history of pre-Darwinian natural history.

Reverend Edward Hitchcock (1793–1864)

The discovery and interpretation of dinosaur footprints in the Connecticut Valley provides a fascinating vignette of American science in the early and mid-nineteenth century. The years 1825–1865 are an important transitional period in American science. It began as a preoccupation with William Paley's (1802) natural theology stressing the power, wisdom, and goodness of God as manifested in nature. An increasingly objective approach to natural history created a widening rift between scientific evidence and theological dogma. Out of this pre-Darwinian confrontation emerged a professionalism of science probing the processes of evolution and the genetic basis of life.

The major figure was Reverend Edward Hitchcock (1793–1864), a Congregational minister turned geologist and educator at Amherst College. Hitchcock's work with the bird-like tracks and other fossil impressions of the Lower Jurassic spanned a period of 30 years and resulted in 2 elaborate monographs and numerous scientific reports. Born in 1793 in Deerfield, Massachusetts, Hitchcock showed great aptitude for astronomy and military engineering fostered by his uncle, General Epaphras Hoyt. During the solar eclipse of 1811, Hitchcock determined the longitude of Deerfield, and his painstaking calculations uncovered 80 errors in Blunt's *Nautical Almanac* (Fig. 3.1).

Figure 3.1. Reverend Edward Hitchcock (1793–1864). Professor of Natural Theology and Geology at Amherst College.

Edward Hitchcock

Hitchcock was principal at Deerfield Academy for 4 years where he became interested in mineralogy and botany through the acquaintance and lectures of Amos Eaton (1776–1842). He then entered theological training at New Haven where he formed a life–long friendship with the renowned scientist and educator Benjamin Silliman (1779–1864)

Following a brief stint as Congregational minister, Hitchcock was appointed professor of chemistry and natural history at the fledgling Amherst College in 1825. He played a major role in the formative years at Amherst, 10 years as its president, then stepping down to fill a self–designed chair of Geology and Natural Theology (Hitchcock 1863b).

Hitchcock's professional life at Amherst spanned 38 years, and during that time he headed the first state geological surveys of Massachusetts and Vermont (Hitchcock 1835, 1858b). His text, *Elementary Geology*, passed through 30 editions. Hitchcock was the founding chairman of the Association of American Geologists and Naturalists, which 7 years later became the American Association for the Advancement of Science.

In 1835, Edward Hitchcock became the first American scientist to investigate the ornithichnites, or bird–like tracks, when several specimens were brought to his attention by Dr. James Deane of Greenfield, Massachusetts. These slabs of rock had come from Turner's Falls to be used for sidewalk paving. Actually, residents in nearby South Hadley had known of such curiosities since 1802 when young Pliny Moody turned up tracks of "Noah's Raven" while plowing his field (Figs. 3.2, 3.3).

In 1836, Hitchcock's first description of these tracks was published in Silliman's *American Journal of Science*, describing 7 species of what he termed giant birds. By 1863 his final listing included 17 pachydactylous birds, 17 lepto-dactylous birds, 21 ornithoid reptiles, 25 assorted reptiles and amphibians, 17 frogs or salamanders, 6 turtles, 2 fish, 1 marsupialoid, and 45 insects, crustaceans or larvae — all of these 151 species from about 38 localities (Hitchcock 1865).

Hitchcock's extensive ornithichnite collection was placed on exhibit in the specially designed Appleton Cabinet erected in 1855 (Fig. 3.4). He insisted on natural southern light to best illuminate the slabs with tracks, and by his own count over 8,000 tracks were on display. This important collection currently resides across the Amherst College yard in the basement of the Pratt Museum.

Edward Hitchcock defined 30 characters to be used in describing a set of tracks and placing the animal in the most closely related order or genus (Fig. 3.5). He realized the inadequacies of this method as stated in his classic 1858 monograph: "That it will be necessary to change the place of some of the species which I have described, I expect. If I could have had access to the large collections of comparative anatomy and zoology in Europe, I might have avoided some errors. Living in the midst of a region which has become classic ground for ichnology, I have done what I could in laying the foundations, and in gathering a storehouse of materials. Let others, with better light to guide them, carry up and complete the structure."

Hitchcock never realized that he was in a sense the first man to describe dinosaurs in North America. He actively collected and accurately described an immense number of dinosaur footprints but attributed the large three-toed tracks to pachydactylous or leptodactylous birds (Fig. 3.6). This concept was unfortunately reinforced by the 1861 discovery of the feathered *Archaeopteryx* in the limestone quarries of Solnhofen, Germany. Hitchcock noted a resemblance of the hind feet to his genus

Figure 3.2. View of Moody's Quarry in South Hadley.

Figure 3.3. First dinosaur track specimen discovered by young Pliny Moody in 1802. (Amherst specimen 16/2)

Figure 3.4. Interior and exterior views of Appleton Cabinet at Amherst College housing Hitchcock's extensive collection of dinosaur footprints and other organic imprints. The building was a model of neoclassical design.

Figure 3.5. Plate from Hitchcock's 1858 monograph illustrating physical aspects of forming a foot imprint as well as comparing fossil footprints and gaits with those of living animals.

Figure 3.6. Illustrations of *Brontozoum* and *Amblonyx* footprints. From Hitchcock's 1858 monograph.

Anomoepus. He nearly grasped the distinction when he commented on the affinities of the *Anomoepus* fore foot to a lizard or certain mammals (Colbert 1968).

Genesis and Geology

An interesting sidelight to Hitchcock's devotion to fossil footprints was his ability to discern in these strange tracks of marvelous prehistoric creatures the Hand of Almighty Providence. In Hitchcock's words: "His parental care shines forth illustriously in these anomalous forms of sandstone days, and awakens the delightful confidence that in like manner He will consult and provide for the wants of individuals" (Hitchcock 1858a, p. 190).

Hitchcock's grave concern for the atheism and pantheism expounded by a portion of the scientific community resulted in the classic *Religion of Geology and its Connected Sciences* published in 1851. As both Congregational minister and highly respected geologist, he felt a special obligation to articulate his twin faiths in science and religion. Hitchcock sought to reconcile and support revealed scripture with the latest geological theories and discoveries. Like fellow Protestant scientists and men of learning extending back to Isaac Newton and Thomas Browne, Hitchcock maintained both a faith in God's revealed word and a faith in the efficiency of God's physical laws to regulate His universe (Guralnick 1972).

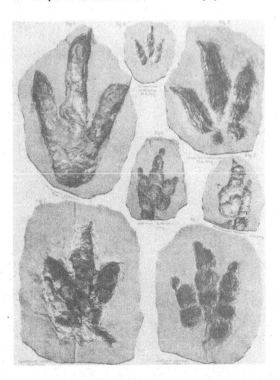

The six days of Genesis was interpreted symbolically by Hitchcock as a gradual work performed by successive exhibitions of Divine Power with long intervals of repose. Like nearly all scientists in this pre–Darwinian period, he perceived no connection between life forms in successive deposits saying that "advance has been by special creative acts, and not by infinitesimal development" (Hitchcock 1851).

In writing to the lay public and assimilating geologic discoveries with revealed scripture, Edward Hitchcock joined the ranks of such men as Hugh Miller (1847, 1857) and Reverend William Buckland (1823). Hugh Miller was a Scots stonemason turned poet and literary spokesman for the Evangelical Church of Scotland. He wrote *Footprints of the Creator* to counteract the religious heresy of *Vestiges of the Natural History of Creation*, published anonymously by Robert Chambers in 1844. Incidentally, Darwin considered Chamber's book of great help in preparing the public for his *Origin of Species* 14 years later.

Another European counterpart to Hitchcock was Reverend William Buckland, a distinguished cleric and first professor of geology at Oxford. His report on *Megalosaurus*, or giant fossil lizard, in 1824 provided one of the first descriptions of a dinosaur (a term coined by Richard Owen years later). Buckland's book *Geology and Mineralogy*, published in 1837, was part of the famous 8–part Bridgewater Treatise published to exhibit the "power, wisdom, and goodness of God, as manifested in the Creation" (Gillispie 1959).

Edward Hitchcock's final literary effort came shortly before his death and perhaps resembles St. George attempting to slay a final dragon — Darwin's law of natural selection. However, he disagreed with Darwin (1859) for lack of fossil evidence rather than on religious scruples. Hitchcock's integrity as a scientist is revealed in his comment on evolution: "the real question is, not whether these hypotheses accord with our religious views, but whether they are true" (Hitchcock 1863a, p. 524).

Dr. James Deane (1801–1858)

Returning to the ichnology of the Connecticut Valley, brief mention should be given to Dr. James Deane, the man who first brought the Turner's Falls slabs of turkey–like tracks to Edward Hitchcock's attention (Fig. 3.7). Deane obtained his medical degree through an apprenticeship in Deerfield and lectures in New York, and established a busy surgical practice in Greenfield, Massachusetts.

Following Hitchcock's initial description of the fossil footprints, Deane continued to collect specimens. He did not publish his findings until 8 years later, and at that time claimed credit as the first to scientifically recognize the significance of these fossil impressions (Deane 1844). This ignited a bitter Hitchcock–Deane controversy with a series of vituperative attacks and counterattacks so characteristc of that period (Hitchcock 1844).

To his credit, Deane collected many important specimens, providing a number to Edward Hitchcock and Dr. John Collins Warren, and sending a few to Gideon Mantell, the man who first described *Iguanodon* in 1825.

Figure 3.7. View of Turner's Falls where the earliest scientifically recognized dinosaur footprints were discovered.

Deane published a further report in 1849, and many of these tracks were reclassified by Hitchcock in 1858.

A posthumous work by Deane, meant for a Smithsonian report, was published in 1861 through the efforts of Thomas Bouvé and the Boston Society of Natural History. This included very early photographs of specimens tipped into each copy and also showed tracks made by a living alligator for comparison (Deane 1861).

Dr. John Collins Warren (1778–1856)

Finally, one other person deserves mention here, not so much for his important contribution to the ichnology of Connecticut Valley, but for the perfect example he sets of the important role of the broadly educated man in American science during these transitional years for the subdivision of natural history into narrower disciplines. John Collins Warren, M.D., was the Harvard Professor of Anatomy and Surgery for 32 years and very active in the Boston Society of Natural History (Fig. 3.8).

The son of Dr. John Warren, Founder of Harvard Medical School, John Collins Warren was instrumental in establishing Massachusetts General Hospital in 1821, and performed the first operation using ether anesthesia there in 1846. Warren trained in medicine at London, Edinburgh and Paris. He learned anatomy under Munro and the Bell brothers in Edinburgh and dissected extensively at Guy's Hospital with Sir Astley Cooper. He included comparative anatomy with Georges Cuvier as part of his training in Paris.

John Collins Warren published the first American work on comparative anatomy in 1822. He established the Warren Anatomical Museum at Harvard, including a large number of specially prepared specimens from Europe. In 1846, he obtained the most perfectly preserved mastodon skeleton yet discovered from the Hudson Valley near Newburgh, New York. He later acquired one of the Peale mastodons from the defunct Baltimore branch of their museum.

In 1852, Warren published the lavishly illustrated monograph on the skeleton of *Mastodon giganteus*, including comparative fossil and recent material from all over the world. Warren's mastodon is now proudly displayed at the American Musuem of Natural History.

Warren's 1854 *Remarks on Fossil Impressions from the Sandstone Rocks of the Connecticut River* is quite meager by comparison with the works of others. However, it includes the first use of a photographic illustration in an American scientific publication, and further emphasizes his wide-ranging interests. Like Hitchcock, John Collins Warren attributed the ostrich–like tracks to large birds. He was partially influenced by the recent discovery in Madagascar of remains of a giant bird reaching 13 feet in height and termed *Aepyornis maximus*.

The footprint slab in Warren's photograph is now at the American Museum of Natural History, presumably purchased along with the mastodon in 1907. The counterpart slab is specimen 26/10 at Amherst and figured in Hitchcock's 1858 monograph as Plate 40, number 1 (Figs. 3.9, 3.10).

Although Hitchcock, Deane, and Warren incorrectly attributed the dinosaur tracks to giant birds, recent workers

Figure 3.8. Dr. John Collins Warren (1778–1856). Harvard Professor of Anatomy and Surgery.

Figure 3.9. Early photograph of fossil footmarks providing the frontispiece for Warren's 1854 monograph. It includes several tridactylous "bird" tracks.

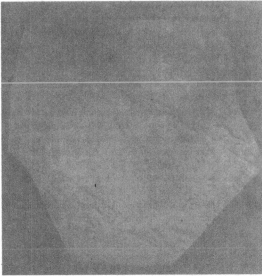

Figure 3.10. The counterpart slab to Warren's specimen as depicted in the upper left of Plate 40 (Hitchcock 1858a).

have now demonstrated a close dinosaur–bird relationship. More importantly, these paleontologic pioneers provided a firm foundation for subsequent studies through their taxonomic application to tracks, comparative anatomical approach to tracks and gaits, and mechanical analysis of footprint formation.

Acknowledgments

The author wishes to thank Dr. Donald Baird of Princeton University and Professor Farish A. Jenkins, Jr. of Harvard University for their advice during the initial phase of this historical research. Walter Coombs, Curator of the Amherst Ichnological Collection, also provided valuable assistance.

References

Buckland, W. 1823. *Reliquiae Diluvianae; or, Observations on the Organic Remains contained in Caves, Fissures, and Diluvian Gravel, and on Other Geological Phenomena, Attesting the Action of the Universal Deluge.* (London: John Murray) 303 pp., 27 pls.

1824. Notice on the *Megalosaurus* or great fossil lizard of Stonesfield. *Trans. Geol. Soc. London* 21:390–397.

1837. *Geology and Mineralogy, Considered with Reference to Natural Theology.* (London: Pickering) 2 vol., Bridgewater Treatise 6.

Chambers, R. 1844. *Vestiges of the Natural History of Creation.* (London: Churchill) 353 pp.

Colbert, E. H. 1968. *Men and Dinosaurs. The Search in Field and Laboratory.* (New York: E.P. Dutton) 283 pp.

Darwin, C. 1859. *On the Origin of Species by Means of Natural Selection; or, The Preservation of Favoured Races in the Struggle for Life.* (London: John Murray) 502 pp.

Deane, J. 1844. On the fossil footmarks of Turner's Falls, Mass. *Amer. Jour. Sci.* 46:73–77.

1849. Illustration of fossil footprints of the valley of the Connecticut. *Trans. Amer. Acad. Arts Sciences* 4:139–221.

1861. *Ichnographs from the Sandstone of the Connecticut River.* (Boston: Little–Brown) 61 pp., 46 pl.

Gillispie, C. 1959. *Genesis and Geology.* (New York: Harper) 306 pp.

Guralnick, S. 1972. Geology and religion before Darwin, the case of Edward Hitchcock, theologian and geologist. *Isis* 63:529–543.

Hitchcock, E. 1835. *Report on the Geology of Massachusetts.* (Amherst: J.S. Adams) 702 pp.

1836. Ornithichnology. Description of the footmarks of birds (ornithichnites) on New Red Sandstone in Massachusetts. *Amer. Jour. Sci.* 29:307–340.

1844. Rejoinder to the "Discovery of fossil footmarks" by J. Deane. *Amer. Jour. Sci.* 47:390–399.

1851. *Religion of Geology and its Connected Sciences.* (Boston: Crosby, Nichols, Lee and Company) 511 pp.

1858a. *Ichnology of New England. A Report on the Sandstone of the Connecticut Valley, especially its Fossil Footmarks.* (Boston: William White, Printer) 232 pp., 60 pls.

1858b. *Report on the geological survey of the state of Vermont.* (Burlington) 13 pp.

1863a. The law of nature's constancy subordinate to the higher law of change. *Bibliotheca Sacra* 20:489–561.

1863b. *Reminiscences of Amherst College, Historical, Scientific, Biographical, and Autobiographical.* (Northampton: Bridgman and Childs) 420 pp.

1865. *Supplement to the Ichnology of New England.* Hitchcock, C. H. (ed.). (Boston: William White, Printer) 106 pp. 20 pls.

Miller, Hugh. 1847. *Footprints of the Creator; or, The Asteroilepis of Stromness.* (Edinburg: Constable) 308 pp.

1857. *The Testimony of the Rocks; or, Geology in its Bearing on the Two Theologies, Natural and Revealed.* (Edinburgh: Constable) 420 pp.

Paley, W. 1802. *Natural Theology; or, Evidence of the Existence and Attributes of the Deity Collected from the Appearances of Nature.* (London: Baynes) 238 pp.

Warren, J. C. 1822. *A Comparative View of the Sensorial and Nervous Systems in Man and Animals.* (Boston, J.W. Ingraham Publisher) 159 pp.

1852. *The Mastodon giganteus of North America.* (Boston: John Wilson, Publisher) 219 pp., 26 pls.

1854. *Remarks on Some Fossil Impressions in the Sandstone Rocks of Connecticut River.* (Boston: Ticknor and Fields) 54 pp.

4 Roland T. Bird, Dinosaur Tracker: An Appreciation

JAMES O. FARLOW AND
MARTIN G. LOCKLEY

One of the most important collectors of dinosaur footprints of the early twentieth century was R. T. Bird, who worked as right–hand man to Barnum Brown of the American Museum of Natural History during the mid–late 1930's and early 1940's. What he lacked in formal training, Bird more than made up in enthusiasm and hard work (see Bird 1985 for his own account of his career).

Although Brown made his name as a collector of dinosaur bones, he had a fondness for the footprints of the great reptiles. "He [Brown] said he wanted me to think about making a collection of fossil dinosaur tracks and trackways to exhibit as flat slabs set vertically against an abundance of empty side wall space available along the concourse between the museum's two dinosaur halls .... He wanted to assemble sample trackways from all the available regions about the U.S. and Canada ... We already had on exhibition one fine trail from ... the Connecticut River Valley, and a huge slab of 18,000 pounds that I ... had taken from a coal mine in western Colorado." These and other quotations from Bird in this paper are from unpublished manuscript pages given to V. T. Schreiber after Bird's death. Schreiber was a close friend of Bird who traveled with him in the field and edited his posthumous autobiography (Bird 1985).

In 1986, Schreiber began to write a tribute to Bird for inclusion in this volume but was unable to complete and revise it due to his untimely death. Many of his notes and Bird's unpublished writings were passed on to Farlow at this time and we are pleased to be able to honor Schreiber's contribution to science by publishing some of the authentic information he had accumulated on Bird's unique career.

Bird's first formal involvement in dinosaur track work came in a Colorado coal mine near Cedaredge when Barnum Brown mounted a major excavation of a set of giant hadrosaur tracks for exhibit at the American Museum (Brown 1938). Brown was somewhat preoccupied with promoting spectacular dinosaur statistics regarding footprint size and step and stride dimensions, and as a result his interpretations have proved controversial. Bird's own account of the excavation and search for associated fossil plants (Bird 1985) is therefore a welcome addition to an otherwise oversensationalized chapter in footprint research (see Lockley and Jennings 1987, and Lockley et al. 1983 for details).

Through his employment as a dinosaur hunter, Bird kept his eyes open for new footprint discoveries that might yield specimens suitable for Brown's exhibit. Because Bird kept a photographic record of important sites, it is through examination of some of his pictures that we are able to determine the existence of various otherwise undocumented sites. A good example is the Dinosaur Canyon tracksite in the vicinity of Cameron, Arizona, which Bird visited in 1934. He documented the trip by taking a self portrait beside a spectacular dinosaur track–bearing bed exhibiting abundant theropod tracks. The picture was not published until much later (Colbert 1961, 1983), at which time it aroused the curiosity of paleontologists unfamiliar with such an extensive site. Following the publication of his posthumous autobiography (Bird 1985), Madsen (1986) was able to relocate the site and establish it as one of the first, and largest, ever recorded in the Moenave Formation.

In 1986 one of us (MGL) learned of the existence of a large site near Moab, Utah. After visiting the site and being informed by local residents that it had been discovered only recently, a photograph of the site was discovered (by JOF) among Bird's archives (Fig. 4.1). The site is very extensive and certainly represents the largest footprint assemblage known in the strata representing the Entrada–Morrison stratigraphical transition. In light of the above evidence it is possible that Bird's unpublished records may yet yield clues to other "long–lost" tracksites.

In November of 1938, at the end of a disappointing field season, Bird "... had hopes that Brown's final assignment might yet save the day. This was a report coming from Pueblo, Colorado, stating that some footprints as large

Figure 4.1. Photograph (left) taken by R. T. Bird in 1944 of theropod tracks in Upper Entrada Formation near Moab, Utah, with 1988 photo by M. Lockley (right) for comparison.

around as barrel hoops had been seen in the valley of the Purgatoire River east of the city. Sent in by a young man, John Stuart McClary, it was believed by all who knew of them that here were the tracks of some gigantic dinosaur, presumably a quadruped. Were they those of a brontosaur? Brought to the site by four friends of McClary ... my hopes were both raised to the skies and partly blasted in the same breath. The tracks were new all right, but they were almost round, and the gigantic trail was not that of a typical quadruped, in that impressions of all fours were not present." Bird had hoped to establish these as sauropod ichnites because brontosaur footprints had never previously been reported. In fact both Bird and MacClary published short accounts of the Purgatoire site (MacClary 1938, Bird 1939) before attention was diverted to the now famous Texas tracksites.

In view of the extensiveness of the Purgatoire site and the presence of brontosaur trackways (Lockey et al. 1986), Bird and Brown could have profited from a thorough study of the site. However, it is clear from Bird's photographs that the configuration of exposures at the site was different from that observed today. This suggests that Museum quality tracks were not readily available, thus encouraging Bird to hold out for something more spectacular. Bird also took accessibility into consideration and knew that the Arizona and Colorado sites were both at remote locations. Thus he had amassed considerable first-hand experience of Western tracksites before he made the footprint discovery for which he is most famous.

Although the Colorado tracksite had not proved interesting enough to salvage the disappointing 1938 field season, Bird had one more lead to follow. Through a curious chain of events (Bird 1985), he had learned of the possibility of obtaining theropod footprints from the bed of the Paluxy River, near Glen Rose, Texas. "Individual tracks would be comparatively large, three-toed, and spec-

tacular enough to warrant mention in possible future advertising campaigns put out by the Sinclair people — another point that Brown always kept in mind."

The tracks were there, all right. "I had just finished shoveling a crust of half-dried mud off a number of fossil footprints made by a large Cretaceous carnosaur .... To all outward appearances, the prints showed such a fine degree of clarity, it might seem they were made only yesterday .... I knew I had something good for the dinosaur track collection in the American museum that would fit in with Brown's plans."

But there was much more to come. "[While] eating lunch seated on a limestone bench, part of the overburden above [the] track zone, I noticed a large depression in the surface of the rock nearby [into] which I had thoughtlessly been shoveling the present mud of the Paluxy to get it out of the way. Why was this big depression ... on this same surface with the carnosaur trail? Was it a footprint of some sort? The very thought appalled me" — but not for long, as this turned out to be the first of several footprints of what Bird had missed in Colorado, the first clear sauropod trail known to science (Fig. 4.2).

Figure 4.2. R. T. Bird's photographic self-portrait with a sauropod left manus–pes set, Paluxy River, Texas. The pes footprint is the last track in the American Museum trackway slab, and the associated manus track is the first footprint in the Texas Memorial Museum slab. See Farlow et al. (this volume) for details.

"When I returned north and appeared at Brown's insistence before the annual meeting of the Society of Vertebrate Paleontology, bearing between us the plaster copy of ... [a manus track and] a sample print of a hind foot, over a yard in length ... the astonished assemblage wanted to know every detail of the discovery. Had the living sauropod left his assumed watery habitat to cross an open mudflat, perhaps? ... Had this gigantic creature crawled on its belly in the manner of many reptiles, or had it somehow managed to carry its great weight on perpendicular legs? How *wide* was the trackway? What about an inner claw on the forefoot? ... When at last the assemblage relaxed in a general discussion of the report, conceded to be the highlight of the year, the question arose: When can specimens be collected?"

Bird returned to Texas in the spring of 1940, supported by funds from the Sinclair oil company, the state of Texas, and the federal government, to collect a sauropod trail suitable for exhibition at the American Museum beneath a mounted skeleton of *Apatosaurus* (Bird 1985, Farlow 1987). A preliminary foray into south Texas yielded the Mayan and Davenport Ranch sauropod tracksites, with their wealth of insights into brontosaur behavior (Lockley 1987, Farlow et al. this volume), but no exhibition–quality trackways. So it was back to the Paluxy for Bird, and during the spring and summer he struggled against the weather and the occasional perversities of human nature to wrest from the river bed a spectacular trackway of a sauropod (Fig. 4.2) and a carnosaur that may have been following the big herbivore with dishonorable gastronmomic intent.

"To cut down on expense, the tonnage [of trackway pieces] going to the American Museum ... would be loaded on a ship in Houston. This nation was not yet in the great World War then going on, but ships were being torpedoed daily, and the Gulf Stream off Florida offered many fat targets for Germany's submarines. When at last, after many delays, the ship docked in New York, [I] could not refrain from going on board and down in the vast bowels of the big freighter, again renewing contact with at least one boxed portion of those two splendid trackways."

Barnum Brown's big wall display of dinosaur footprints from around North America never was erected, but no matter. The project had served as the stimulus that led to the biggest discovery of R. T. Bird's career, and when, as the final act of that career, Bird supervised the reassembly and installation of the Paluxy River trackway slab beneath Brown's *Apatosaurus* mount, he created an exhibit that was far more dramatic than any wall display could have been.

Bird's legacy goes beyond the trackway exhibit itself, however. Even more important are the ideas about the habitats and habits of sauropods that came from his discoveries in Texas. In Bird's day brontosaurs were generally seen as dull–witted brutes barely able to support their own weight outside the watery environments they were believed to have frequented. Bird's trackways showed otherwise, revealing the sauropods to have been perfectly capable of carrying their own bulk, and suggesting at least a degree of herding behavior in these perhaps–not–so–stupid–after–all reptiles.

Bird's interpretation of the Paluxy, Davenport and Mayan Ranch trackways may be open to question but there can be little doubt that he is the originator of the gregarious "brontosaur" hypothesis, and responsible for several perennial dinosaur behavior debates (Farlow et al., Ishigaki, Lockley and Conrad this volume) arising directly from the data he accumulated. There is no evidence that Bird intended to stimulate controversial debate. In fact he was by nature very unassuming and always conscientious about amassing reliable data. Bird's legacy of footprint maps, photographs and museum exhibits (Bird 1985, Lockley 1987, Farlow et al. this volume) have no parallel among workers of his generation. He was the most conscientious American vertebrate ichnologist since Edward Hitchcock, and with the exception of Charles Sternberg and Richard Swan Lull the only one to produce enduring work in the first half of the twentieth century.

We close this tribute with the emotional sentiments and words of Bird's friend, the late V. Theodore Schreiber. Bird set an example of personal and professional conduct that is worth emulating today. Paleontology has always had its share of prima donnas; R. T. Bird's unassuming professional modesty and generosity of spirit are in refreshing contrast to the self-serving antics of those who are unwilling to let their discoveries speak for themselves. But for all his gentleness, "he was ... one of the toughest little guys I ever met. I am the end product of four diverse lines of yeoman and peasant stock, and for much of my life I was exposed to many rugged forms of hard labor. Yet he used to embarrass me no end, the way he could gallop up a mountain goat trail without breathing hard at the top" (V. T. Schreiber written comm.). As a human being, R. T. Bird enriched the lives of all who knew him, just as his discoveries as a dinosaur hunter enriched the science of paleontology.

Acknowledgments
We thank Ted Schreiber for suggesting that this volume include a tribute to Roland T. Bird, and for providing so much valuable material to help us complete the task.

References
Bird, R. T. 1939. Letter, *Nat. Hist.* 43:245.

    1985. *Bones for Barnum Brown: Adventures of a Dinosaur Hunter* (Texas Christian University Press) 225 pp.

Brown, B. 1938. The mystery dinosaur. *Nat. Hist.* 41:190–202, 235.

Colbert, E. H. 1961. *Dinosaurs: Their Discovery and Their World.* (New York: E. P. Dutton).

    1983. *Dinosaurs, An Illustrated History* (Hammond Corp., A Dembner Book).

Farlow, J. O. 1987. *A Guide to Lower Cretaceous Dinosaur Footprints and Tracksites of the Paluxy River Valley, Somervell County, Texas.* Field trip guidebook, South-Central Section, Geological Society of America. Baylor University. 50 pp.

Farlow, J. O., Pittman, J. S., and Hawthorne, J. M. This volume. *Brontopodus birdi,* Lower Cretaceous sauropod footprints from the U.S. Gulf Coastal Plain.

Lockley, M. G. 1987. Dinosaur trackways. *In* Czerkas, S. J., and Olson, E. C. (eds.). *Dinosaurs Past and Present*, Vol. I. (Natural History Museum of Los Angeles County/University of Washington Press) pp. 81–95.

Lockley, M.G., Houck, K., and Prince, N. K. 1986. North America's largest dinosaur tracksite: implications for Morrison Formation paleoecology. *Geol. Soc. Amer. Bull.* 97:1163–1176.

Lockley, M. G., and Jennings, C. 1987. Dinosaur tracksites of western Colorado and eastern Utah. *In* Averett, W. (ed.). *Paleontology and Geology of the Dinosaur Triangle* (Grand Junction: Museum of Western Colorado Press) pp. 85–90.

Lockley, M. G., Young, B., and Carpenter, K. 1983. Hadrosaur locomotion and herding behavior: evidence from footprints in the Mesa Verde Formation, Grand Mesa Coalfield, Colorado. *Mountain Geol.* 20:5–13.

MacClary, J. S. 1938. Dinosaur trails of Purgatory. *Sci. Amer.* 158:72.

Madsen, S. 1986. The rediscovery of dinosaur tracks near Cameron, Arizona. *In* Gillette, D. D. *Abstracts with Program, First International Symposium on Dinosaur Tracks and Traces.* (Albuquerque: New Mexico Natural History Museum) p. 20.

III Locomotion and Behavior

"Leidy and Cope had no qualms about letting their "kangaroo" dinosaurs bound after one another unchecked, and the notion of an active on occasions fast moving dinosaur took root, and has been alluded to ever since."

Desmond 1975 p. 104–105

"All dinosaurs are characterized by a "fully improved" or "fully erect" position of the limbs (not unlike that found in higher mammals) in which the limbs support the body from beneath, holding it clear of the ground."

Charig 1979 p. 18.

Trackways are an important source of information on dinosaur locomotion. Despite contrary assertions and many incorrect museum reconstructions (Wade this volume), the vast majority of dinosaur trackways indicate fully erect, non-sprawling posture (Fig 0.2 this volume). As erect reconstructions have gained acceptance, there has been much speculation about dinosaur agility and speed. While it is reasonable to infer that dinosaurs were not slow and cumbersome, as often suggested, it is difficult if not impossible to make absolute speed estimates from trackways (Thulborn this volume). The vast majority of trackways indicate walking progression. Examples attributed to running dinosaurs are rare exceptions. The example (left) of a theropod with a 531 cm stride (after Farlow 1981 Fig. 1) was reported from a site where three trackways were used to estimate running speeds of about 30–40 km/hr. Such speeds indicate considerable cursorial ability, but not necessarily the maximum speeds attained by dinosaurs.

Trackways with district metatarsal impressions also shed new light on occasional plantigrade progression by dinosaurs (Kuban this volume; Introduction Fig. 0.3). Incomplete and irregular trackways raise controversial questions about the swimming ability and behavior of dinosaurs (Ishigaki this volume; McAllister this volume).

Parallel trackways permit speculations about gregariousness and herding, or social behavior among dinosaurs (Bird 1944, Farlow and Lockley this volume).

References

Charig, A. 1979. *A New Look at the Dinosaurs.* (New York: Facts on File Inc.) 160 pp.
Desmond, A. J. 1975. *The Hot Blooded Dinosaurs.* (New York: Warner Books) 352 pp.

5 The Gaits of Dinosaurs

RICHARD A. THULBORN

Abstract

The criteria used to define the gaits of living animals cannot be applied very easily to skeletons or trackways. Consequently the gaits of dinosaurian trackmakers are defined arbitrarily on a scale of relative stride length, λ/h, where λ represents length of stride and h represents the trackmaker's height at the hip. An accurate estimate of h is an essential prerequisite for ascertaining a trackmaker's gait. The most enlightening estimates of h are obtained by means of allometric equations. Analysis of trackway data confirms that bipedal dinosaurs favored a walking gait with λ/h about 1.3. Large bipedal dinosaurs were restricted to walking or trotting gaits, with λ/h less than 2.9, whereas smaller bipeds were capable of running with λ/h as high as 5.0. Theropods and ornithopods of similar size seem to have had roughly similar gaits. There is no convincing evidence that bipedal dinosaurs ever used a ricochetal (hopping) gait. Quadrupedal dinosaurs may have been restricted to a slow walking gait, with λ/h less than 1.0. Semibipedal dinosaurs (ornithopods, prosauropods and, perhaps, some theropods) adopted a somewhat sprawling and low-slung gait while progressing slowly on all fours, but reverted to a normal bipedal gait for faster progression.

Introduction

The study of dinosaur gaits is hampered by one major difficulty: the criteria used to define and describe the gaits of living animals cannot be applied very successfully to the skeletons or trackways of extinct animals. Fortunately this technical problem can be circumvented by classifying dinosaur gaits into three categories — walk, trot and run. Although this three-fold classification is somewhat arbitrary it does have some definite advantages: (i) it can be applied objectively to the evidence of trackways; (ii) it allows the gaits of dinosaurs to be compared to those of other animals; and (iii) it provides some revealing insights into the locomotor abilities of dinosaurs.

Criteria for Definition and Description of Gaits

The various gaits of living animals are distinguished by differences in the sequence and duration of footfalls. In other words each gait is defined by the sequence in which the limbs move and by the amount of time that each foot remains on the ground. These criteria have been used to describe numerous gaits, and variations of gaits, among living animals. For example, Muybridge (1899) recognized eight basic mammalian gaits whereas Brown and Yalden (1973) identified nine. In a more detailed analysis Hildebrand (1965) discerned no fewer than 39 variations in the *symmetrical* gaits of horses alone. Dinosaur gaits cannot be documented in similar detail because there is insufficient evidence about the sequence and duration of footfalls.

Sequence of Footfalls

The sequence of footfalls is obvious in the trackways of bipedal dinosaurs, where the regular alternation of left and right footprints reveals that these animals walked and ran like humans or ostriches (Fig. 5.1A). The notion that bipedal dinosaurs might have used a ricochetal (hopping) gait has not been substantiated by the discovery of appropriate trackway patterns (Fig. 5.1B,C).

It is difficult to ascertain the sequence of footfalls in the trackways of quadrupedal dinosaurs. However, it is noticeable that the left and right paces (whether for forefeet or for hindfeet) are fairly consistent in their length (Fig. 5.2A,C). Such consistency in pace length implies that the trackmaker used a regular or symmetrical gait — with a left foot half a stride out of phase with its right counterpart. Thus the two hindlimbs of a quadrupedal dinosaur would have moved like the legs of a walking human; the same would have been true for the forelimbs, so that a walking quadrupedal dinosaur might be likened to two humans walking in tandem (Fig. 5.3A). There remains the question of *when* the forelimbs were moved in relation to the hindlimbs. This question concerns relative phase, which may be defined as the time at which a foot is set down,

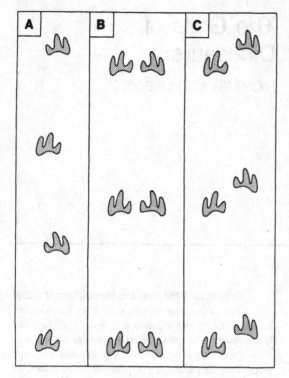

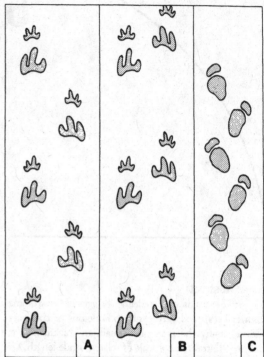

Figure 5.1. Diagrammatic trackway patterns characteristic of bipedal gaits. **A**, walking or running gait, with consistent pace lengths and regular alternation of left and right footprints. **B**, in–phase hopping, with synchronous movements of the hindfeet. **C**, out–of–phase hopping, with almost synchronous movements of the hindfeet; note the regular alternation of long and short paces. Tracks of bipedal dinosaurs invariably resemble pattern A.

Figure 5.2. Diagrammatic trackway patterns characteristic of quadrupedal gaits. **A**, symmetrical gait; pace length is consistent for hindfeet and for forefeet (but not necessarily the same for hindfeet and forefeet). **B**, asymmetrical gait, distinguished by regular alternation of long and short paces. **C**, part of sauropod trackway, with total length about 12 m (after Ishigaki 1985); this, like other tracks of quadrupedal dinosaurs, resembles pattern A rather than pattern B.

expressed as a fraction or percentage of the stride that has elapsed since the setting down of an arbitrarily–chosen reference foot (adapted from Alexander 1985, p. 16). In the case of the sauropod trackway shown in Figure 5.2C the relative phases of the four feet may be expressed as follows (as fractions of stride elapsed, and with the left forefoot selected as the reference foot):

left forefoot: 0   right forefoot: 0.5
left hindfoot: p right hindfoot: $p + 0.5$

Unfortunately the quantity p, which is some fraction between 0 and 1, cannot be ascertained from a trackway or from a dinosaur skeleton. At best this quantity can be estimated from restorations or scale–models of dinosaurs in what are assumed to be their most life–like postures. Alexander (1985) attempted to estimate relative phase in a sauropod trackmaker, but his results were inconclusive in that he obtained two very different values for p. The first of these estimates ($p = 0.73$) implied that the trackmaker used a standard walking gait (Fig. 5.3B); the second estimate ($p = 0.94$) implied that the ipsilateral limbs moved almost synchronously in the rolling and shambling gait ('walking pace' or 'amble') that is sometimes used by

elephants and by long–legged animals such as camels (Fig. 5.3C). At present there is insufficient evidence to allow a choice between these two possibilities.

Duration of Footfalls

The amount of time that a foot remains on the ground is termed its duty factor, which is usually expressed as a fraction or percentage of total stride duration. Duty factor decreases as animals accelerate (Brown and Yalden 1973) and is known to be roughly equal for the forefeet and hindfeet in quadrupedal mammals (Alexander and Jayes 1983). Similar generalizations doubtless applied to dinosaurs, although it is impossible to calculate duty factor from the evidence of their trackways.

Relative Stride Length

Dinosaur tracks provide limited information about the sequence of footfalls and no information at all about duty factor. Consequently the gaits of dinosaurs must be defined and described by some other criterion.

Most terrestrial vertebrates use relatively few gaits; often they have a slow gait (walking), an intermediate gait

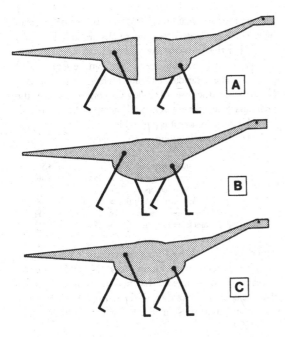

Figure 5.3. Diagrams to illustrate possible gaits of quadrupedal dinosaurs. **A,** forelimbs and hindlimbs separated and envisaged as bipeds with symmetrical gaits (as in Fig. 5.1A); note that forelimbs are shorter than hindlimbs. **B,** regular walking gait, where contralateral fore and hind limbs move almost synchronously. **C,** walking pace or 'amble', where ipsilateral limbs move almost synchronously.

(such as trotting) and a fast gait (such as galloping). Also it is well known that animals take relatively short strides while walking and that they take increasingly longer strides when they accelerate through their faster gaits. These generalizations apply to all land vertebrates, including dinosaurs, and they allow the definition of three gait categories — walk, trot and run — each characterized by successively longer strides. In practice the criterion of stride length must be modified to take into account the differing sizes of animals; this scaling is necessary because the strides of a small running animal (e.g., a mouse) may be far shorter, in absolute terms, than the strides of a big animal (e.g., an elephant) that is merely walking. Stride length scaled in accordance with the size of the animal is termed relative stride length (Alexander 1976).

Relative stride length is conveniently defined as λ/h, where λ is the length of the animal's stride and h is the animal's height at the hip. Stride length (λ) may be measured directly on a dinosaur's trackway, and h may be estimated from the dimensions or spacing of the trackmaker's footprints (by methods explained below).

Alexander (1976) found that living terrestrial vertebrates change from a walking gait to a trotting or running gait when λ/h reaches a value of about 2.0; and he suggested that the same was probably true for dinosaurs. In later studies of dinosaur locomotion (Thulborn 1982,

Thulborn and Wade 1984), Alexander's observations on the gaits of living vertebrates (1976, 1977) have been extended to define three dinosaurian gaits: *walk* (λ/h less than 2.0), *trot* (λ/h between 2.0 and 2.9) and *run* (λ/h greater than 2.9).

Relative stride length is not only an indicator of gait; it can be used to estimate the absolute speed of the trackmaker (Alexander 1976) and seems also to be one of the best available criteria for appraising and comparing the locomotor performances of dinosaurs (Thulborn and Wade 1984). Dinosaurs that shared the same value for λ/h may be regarded as having maintained equivalent locomotor performances (or as having shared a 'physiologically similar speed'), even though those dinosaurs may have been of different sizes and may have been moving at different absolute speeds. In short, estimates of λ/h can yield valuable insights into the locomotion, behavior and physiology of dinosaurs. The value of such estimates will depend, of course, on their accuracy. In practice there is no difficulty in measuring λ (stride length) on a dinosaur's trackway; inaccuracies are more likely to be introduced in attempting to estimate the trackmaker's height at the hip (h). Several methods are available to predict h, each depending on different assumptions.

Geometric Methods

In the simplest of these methods pace length for the hindfeet, measured directly on the trackway, is taken to represent the base of an isosceles triangle (X–Y in Fig. 5.4). The two equal sides of the triangle represent the dinosaur's hindlimbs, one extended forwards and the other extended backwards (H–X and H–Y in Fig. 5.4); these two sides enclose the angle of step (2θ, also known as the angle of gait). A vertical line bisecting this angle represents the height of the hip joint (h) above the ground. The height of the hip joint and the lengths of the estimated hindlimbs may be calculated by simple trigonometry ... *providing* that one knows the angle of step. The problem here, of course, is that the angle of step is unknown in dinosaurs. Exactly the same problem arises in more elaborate geometric methods that take into account lateral displacement of the hindfeet (e.g., Demathieu 1970) or that treat the hindlimb as a pendulum (e.g., Demathieu 1984).

In applying these geometric methods to dinosaur tracks one is obliged to make untestable assumptions about the angle of step in dinosaurs. For example, Demathieu (1984) assumed that this angle was about 40° in a variety of dinosaurs including *Iguanodon*, *Albertosaurus* and *Struthiomimus* whereas Avnimelech (1966) assumed it to be about 33° in an unknown dinosaur resembling *Struthiomimus*. The validity of these assumptions remains unknown and the results provided by these geometric methods should be treated with appropriate caution.

Morphometric Ratios

Alexander (1976) suggested that h could be calculated as approximately four times footprint length (abbreviation FL) in a variety of dinosaurs, both bipeds and quadrupeds,

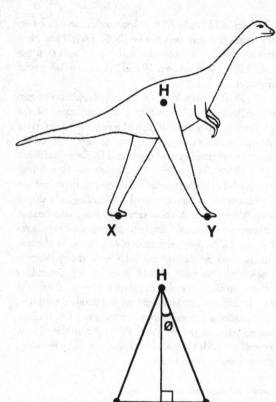

Figure 5.4. Trigonometric method to estimate dinosaurian trackmaker's height at hip. Distance of the hip joint (H) above the ground may be calculated by simple trigonometry on the basis of pace length (X–Y) and the angle of gait (2θ). Outline of dinosaur adapted from Avnimelech (1966).

allometric growth that prevails in land vertebrates; juveniles have relatively larger feet (and smaller ratio h/FL) than adults. For these reasons the indiscriminate application of a single ratio to all dinosaur tracks can be expected to generate some misleading estimates for h (see Figs. 5.9, 5.10). Nevertheless the use of ratios is convenient for the preliminary analysis of trackway data, especially in the field. In such circumstances it is preferable to use a separate ratio for each taxonomic group of trackmakers, as follows:

$$\text{small theropods (FL} < 25 \text{ cm): } h \simeq 4.5\text{FL}$$
$$\text{large theropods (FL} > 25 \text{ cm): } h \simeq 4.9\text{FL}$$
$$\text{small ornithopods (FL} < 25 \text{ cm): } h \simeq 4.8\text{FL}$$
$$\text{large ornithopods (FL} > 25 \text{ cm): } h \simeq 5.9\text{FL}$$
$$\text{small bipedal dinosaurs in general (FL} < 25 \text{ cm): } h \simeq 4.6\text{FL}$$
$$\text{large bipedal dinosaurs in general (FL} > 25 \text{ cm): } h \simeq 5.7\text{FL}$$

These ratios are derived from osteometric data and may need some revision in future work. No ratios are yet available for quadrupedal trackmakers, though I suspect that the commonest of these — the sauropods — had ratios rather similar to those in the bigger ornithopods (i.e., $h \simeq 5.9$FL).

Allometric Equations
      The use of allometric equations to predict h has been explained in detail elsewhere (Thulborn 1984, Thulborn and Wade 1984). Equations applicable to the trackways of bipedal dinosaurs are as follows:

$$\text{small theropods (FL} < 25 \text{ cm): } h \simeq 3.06\text{FL}^{1.14}$$
$$\text{large theropods (FL} > 25 \text{ cm): } h \simeq 8.6\text{FL}^{0.85}$$
$$\text{ornithomimid theropods ('ostrich dinosaurs'): } h \simeq 5.24\text{FL}^{1.02}$$
$$\text{theropods in general (excluding ornithomimids): } h \simeq 3.14\text{FL}^{1.14}$$
$$\text{small ornithopods (FL} < 25 \text{ cm): } h \simeq 3.97\text{FL}^{1.08}$$
$$\text{large ornithopods (FL} > 25 \text{ cm): } h \simeq 5.06\text{FL}^{1.07}$$
$$\text{ornithopods in general: } h \simeq 3.76\text{FL}^{1.16}$$

These equations are not always more accurate than ratios, but they do at least acknowledge the existence of allometry in dinosaurs. No equations are available for the semibipedal prosauropods or for the habitually quadrupedal dinosaurs (sauropods, stegosaurs, ankylosaurs, ceratopsians).

Gaits of Bipedal Dinosaurs
      Measurements gathered from several hundred trackways reveal that bipedal dinosaurs favored a walking gait ($\lambda/h < 2.0$) and that the tracks of trotting or running animals are uncommon (Fig. 5.5). This preference for a walking gait is scarcely surprising. It is evident in dinosaurs of all sizes, even though dinosaurs of any particular size may show considerable variation in λ/h (Fig. 5.6). Despite this variation it has been suggested that λ/h about 1.3 defines an 'average' walking gait for bipedal dinosaurs in general (Thulborn 1984).

Preferred Gaits
      Trackways made by fast-running bipedal dinosaurs seem to be rare; most of the world's examples were discovered at a single site in Queensland, Australia, where

and his suggestion has been rather widely adopted (e.g., Russell and Béland 1976, Tucker and Burchette 1977, Thulborn and Wade 1979, Farlow 1981, Kool 1981). However, Coombs (1978) expressed some reservations about this generalization and Alexander (1976) did mention that FL could represent anything from $0.23h$ to $0.28h$ in the bipedal dinosaurs that he examined. Lockley et al. (1983) noted that h was between five and six times the length of the foot in a skeleton of the hadrosaur *Anatosaurus*, and Sanz et al. (1985) found that the ratio h/FL ranged from about 3.4 to 5.2 in a variety of iguanodontid ornithopods. The observation that $h \simeq 4$FL is certainly easy to remember and easy to use, but it is best regarded as a rule of thumb rather than an infallible guide.
      The use of such a ratio assumes that the size of a dinosaur's foot (and hence of its footprint) bears a constant relationship to h. This assumption is likely to be wrong on two counts. First, the ratio h/FL probably varies in a systematic fashion between dinosaur taxa; so, for example, the averate ratio in coelurosaurs is unlikely to be identical to that in hadrosaurs. Second, the ratio h/FL is certain to have changed during ontogeny, on account of the

it seems that a large number of small ornithopods and coelurosaurs were startled into a stampede by the approach of a carnosaur (Thulborn and Wade 1979, 1984; Fig. 5.5E). Several other trackways of fast–moving bipedal dinosaurs were reported by Farlow (1981) from the Cretaceous of Texas. Overall it seems likely that bipedal dinosaurs normally used a walking gait and that they resorted to running only on rare occasions or in unusual circumstances.

It is noticeable that few bipedal dinosaurs used a trotting gait, with λ/h between 2.0 and 2.9 (Figs. 5.5, 5.6). The rarity of tracks made by trotting animals might be fortuitous or it might be an indication to the gait preferences of bipedal dinosaurs. If these animals did avoid the trotting gait they might have done so for any of three reasons — anatomical, behavioral or physiological. These possibilities were examined by Thulborn (1984) who concluded that the gait preferences of bipedal dinosaurs were most probably controlled by physiological factors: the speeds and gaits defined by λ/h 1.3 (walk) and 3.7 (run) might represent energetic optima whereas the trot (λ/h 2.0 to 2.9) might have been a transitional gait of high energetic cost. It is reasonable to suppose that dinosaurs, like other animals, should have selected energetically optimal gaits.

Hopping Dinosaurs?

The notion of hopping dinosaurs was largely dispelled in the latter half of the nineteenth century, when it was realized that 'ornithoidichnites' and similar markings were the tracks of dinosaurs and not of antediluvian birds. The tracks showed conclusively that bipedal dinosaurs moved their hindfeet alternately, like humans and ostriches. Nevertheless the notion of hopping dinosaurs is far from extinct (e.g., Raath 1972). It has been resurrected most recently by Bernier et al. (1984), who has described tracks of hopping bipedal dinosaurs from the Late Jurassic of southeastern France. Bernier (1984) has also published an appropriate restoration of such a hopping dinosaur (Fig. 5.8A), while

Figure 5.5. Frequency distributions for estimates of relative stride length (λ/h) in various samples of dinosaur trackways. Scale of percentage frequency (top right) applies to all samples. **A,** 85 trackways of bipedal dinosaurs, Lower Cretaceous of Canada (mean λ/h 1.05). **B,** 60 trackways of bipedal dinosaurs, Lower Cretaceous of Texas (mean λ/h 1.22). **C,** 120 trackways of bipedal dinosaurs, Upper Triassic to Lower Jurassic of southern Africa (mean λ/h 1.49). **D,** 175 trackways of bipedal dinosaurs, Upper Triassic through Lower Cretaceous, worldwide (mean λ/h 1.30). **E,** 92 trackways of bipedal dinosaurs, Mid–Cretaceous of Queensland, Australia (mean λ/h 3.73). **F,** 49 trackways of quadrupedal dinosaurs, Upper Triassic through Lower Cretaceous, worldwide (mean λ/h 0.95). **G,** Samples A, B, a and D combined (440 trackways of bipedal dinosaurs, mean λ/h 1.29). In sample F (quadrupeds) h was estimated as four times footprint length; in all other samples h was estimated by means of allometric equations listed in text. Based on data from Currie 1983 (A); J. O. Farlow pers. comm. (B); Ellenberger 1972, 1974 (C); Thulborn 1984 (D); Thulborn and Wade 1984 (E).

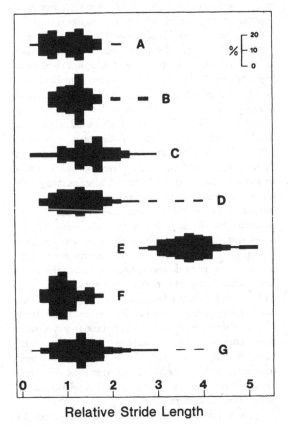

Relative Stride Length

Figure 5.6. Scatter diagram to illustrate relationship between estimated relative stride length and footprint length for bipedal dinosaurs. Based on data from 532 trackways of bipedal dinosaurs (samples A to E in Fig. 5.5). Horizontal lines at relative stride length 2.0 and 2.9 indicate walk–trot transition and trot–run transition respectively.

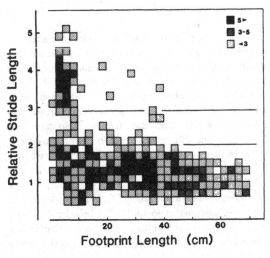

Footprint Length (cm)

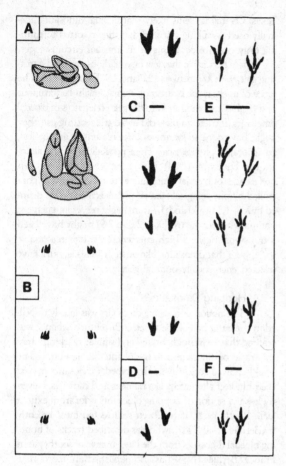

Figure 5.7. Tracks of *Saltosauropus latus* (**A,B**) compared to those of modern hopping animals (**C-F**). **A,** two footprints of *S. latus*, after Bernier et al. 1984; upper example is a foreshortened right footprint; lower example, a full left footprint. **B,** diagram to illustrate average spacing of footprints in trackways of *S. latus*. **C,D,** tracks of wallabies. **E,** track of hopping sparrow, in snow. **F,** track of unidentified hopping bird, in dry sand. Scale bar represents 5 cm in A,C,D; 2 cm in E and F.

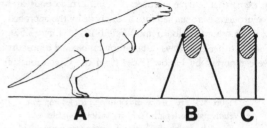

Figure 5.8. **A,** outline restoration of the *Saltosauropus* trackmaker envisaged by Bernier (1984). **B,** diagrammatic cross-section through the hip region of such a trackmaker, to reveal splayed-out attitude of the hindlimbs; proportions based on mean trackway data of Bernier et al. (1984). **C,** diagrammatic cross-section through the hip region of an idealized dinosaur, to show erect limb posture.

Demathieu (1984) has pondered the locomotor mechanics of such a creature and has even attempted to estimate the speed at which it hopped. The trackways, named *Saltosauropus latus*, were attributed to theropod dinosaurs of moderate size, either large coelurosaurs or small carnosaurs. The individual footprints are short and broad, with indications of three stubby digits terminating in stout claws (Fig. 5.7A), and their arrangement in left-right pairs certainly implies that they might have been produced by a hopping biped (Fig. 5.7B).

However, the *Saltosauropus* tracks differ from those of modern hopping animals (Fig. 5.C-F) in several respects. First, the *Saltosauropus* footprints are broader than long whereas those of modern hoppers are longer than broad. The relatively long foot is important for a hopper: it pro-

vides a stable base for take-off and especially for landing. Second, the *Saltosauropus* footprints show positive (inwards) rotation or have digital imprints that are distinctly curved (convex to the exterior). These features imply that the trackmaker's foot applied lateral (as well as downwards and backwards) forces to the substrate. By comparison the footprints of modern hoppers have rather straight digital imprints and show little or no rotation. A third and most important difference concerns the spacing of the footprints: modern animals hop with the left and right feet close together whereas the *Saltosauropus* trackmaker had its feet widely separated. In *Saltosauropus* the interpes distance (between left and right footprints) ranges between 40 and 100 percent of stride length; in the tracks of wallabies and hopping birds the interpes distance is less than 20 percent of stride length (Fig. 5.7C-F). Walking and running dinosaurs produced narrow trackways but the *Saltosauropus* trackmaker produced an astonishingly broad trackway. If the body proportions of the *Saltosauropus* trackmaker resembled those of any known dinosaur then the animal must have hopped with its legs splayed out sideways to a remarkable extent (Fig. 5.8B). Frankly it seems improbable that any animal could have hopped efficiently with its legs splayed out in this fashion. Here it is worth recalling that dinosaurs are characterized by their *erect* posture, with movements of the hindlimb largely restricted to a parasagittal plane (Fig. 5.8C). The same is true for modern hoppers, such as birds and kangaroos, where the major limb joints are adapted to resist any sideways flexures. Such comparisons lead me to doubt that *Saltosauropus* is the track of a hopping dinosaur. I suspect that it is actually the track of a broad-bodied animal, possibly a turtle that was swimming in shallow water and touching down almost synchronously with its front flippers. Tracks of turtles have been reported in the same sediments (Bernier et al. 1982) and some of them seem to bear a close resemblance to the *Saltosauropus* footprints.

It remains possible that some bipedal dinosaurs did

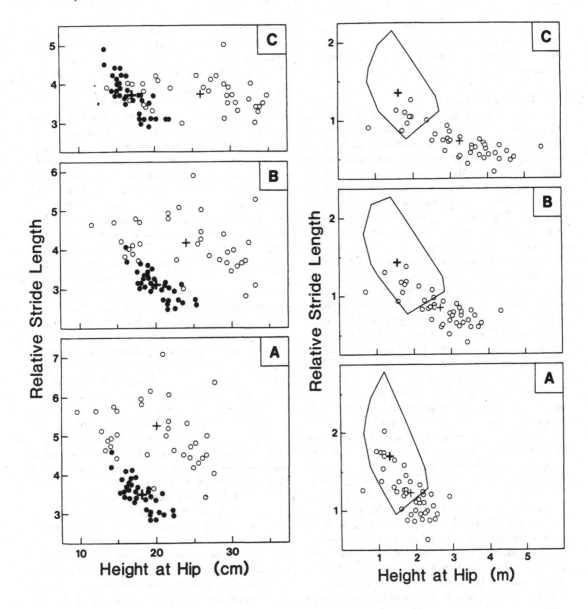

Figure 5.9. Scatter diagram to illustrate relationship of gait (relative stride length, λ/h) to body size (height at hip, h) for small bipedal dinosaurs at the Lark Quarry trackway site, Queensland. Solid circles: small theropods (N = 34, with means indicated by heavy cross). Open circles: small ornithopods (N = 35, with means indicated by light cross). **A,** with h estimated as 4 x FL in all cases. **B,** with h estimated as 4.5 x FL in theropods and as 4.8 x FL in ornithopods. **C,** with h estimated by means of allometric equations listed in text. Analysis A finds the two groups to be roughly similar in size (h) but very different in their gaits (λ/h); *analysis C finds the reverse. Based on data from Thulborn (1984).*

Figure 5.10. Scatter diagrams to illustrate relationship of gait (relative stride length, λ/h) to body size (height at hip, h) for bipedal dinosaurs at the Peace River trackway sites, Canada. Ornithopod distribution indicated by open circles (N = 41, with means indicated by light cross). Theropod distribution indicated by minimum convex polygon for sake of clarity (N = 44, with means indicated by heavy cross). **A,** with h estimated as 4 x FL in all cases. **B,** with h estimated as 4.9 x FL in theropods and as 5.9 x FL in ornithopods. **C,** with h estimated by means of allometric equations listed in text. Analysis C reveals a major difference in size between theropods and ornithopods. Based on data from Currie (1983).

use a hopping gait, but there is no very convincing evidence that they actually did so. Trackways such as *Saltosauropus* must be regarded as dubious evidence for hopping dinosaurs.

Running Dinosaurs

In their fastest gaits animals such as cheetahs, greyhounds and horses extend stride length by virtue of special anatomical and behavioral adaptations. Such cursorial adaptations are well–known in mammals and have also been described in dinosaurs (Coombs 1978, Thulborn 1982). One important stride–lengthening technique is the interpolation of an unsupported (or suspended) phase in each stride. This phase is an interval when the animal lifts all feet clear of the ground and 'floats' through the air under its own inertia. The ability to use a fast–running gait with unsupported phase is apparently related to the body mass of an animal. Full–grown shire horses (600–800 kg) appear to be close to the upper limit of body mass for galloping; heavier mammals such as rhinos (3–4 t) and elephants (4–6 t) do not attain a true galloping gait, with unsupported phase, but reach their maximum speeds in a fast trotting gait. Beyond the critical weight–limit an unsupported phase may place dangerously high stresses on joints and limb bones when an animal lands at the end of its stride. Moreover an unsupported phase requires such an input of vertical thrust (to overcome gravity) that it may be energetically uneconomical for very heavy animals.

Similar constraints probably applied to dinosaurs. Small and fast-moving dinosaurs clearly did exploit an unsupported phase, for their trackways may indicate relative stride lengths well over 3.0 and perhaps even higher than 5.0 (Thulborn and Wade 1984). Bipedal dinosaurs with body mass greater than 1–2 t were probably at, or above, the limit of use for an unsupported interval. Such heavyweight dinosaurs were probably incapable of running and are unlikely to have attained a relative stride length greater than about 2.9 (the trot–gallop transition). This conclusion seems to be borne out by analysis of dinosaur trackways (Fig. 5.6; none of the largest bipedal dinosaurs (with FL > 40 cm) used a gait faster than a walk, and all the trotting or running animals were of small or moderate size (FL < 40 cm). It may be concluded that large bipedal dinosaurs, including many carnosaurs, iguanodonts, and hadrosaurs, were restricted to walking or (at most) trotting gaits whereas small dinosaurs could, and did, make use of faster gaits. Overall there is likely to have been a negative correlation between λ/h and the size of a dinosaur (expressed by FL, h or body mass; see Figs. 5.6, 5.10C). The maximum limit of λ/h for small and fast-running dinosaurs may have been about 5.0 (Thulborn and Wade 1984).

Theropods versus Ornithopods

Some studies suggest that theropods and ornithopods may have shared similar gaits (e.g., Thulborn 1982) while others suggest that theropods may have moved faster than ornithopods (e.g., Currie 1983). By comparing the gaits of these two groups of dinosaurs it is possible to demonstrate certain points that are important for the study of dinosaur gaits in general.

Figure 5.9 shows three different analyses of data from the Lark Quarry site, Queensland, Australia (Thulborn and Wade 1979, 1984). The trackways selected for these comparisons were made by 34 small theropods and 35 small ornithopods, all with FL less than 7 cm. In the first analysis (Fig. 5.9A) the size (h) of all trackmakers was estimated by using a single ratio ($h \simeq 4FL$); consequently it appears that the theropods were roughly similar in size to the ornithopods, though they seem to have used a much slower gait (mean λ/h 3.5 as opposed to 5.3). [A similar distinction between theropods and ornithopods would emerge from the seemingly objective exercise of plotting stride length against footprint length.] This first analysis might well lead one to assume that small ornithopods moved faster than theropods of similar size. However, the use of a single ratio to estimate h in all trackmakers has ignored the fact that theropods differed from ornithopods in their body proportions. This difference is acknowledged in the second analysis (Fig. 5.9B) where separate ratios were used to estimate h in each group (4.5FL in theropods and 4.8FL in ornithopods). The ornithopods now seem to have been slightly larger (on average) than the theropods, and there is a less obvious difference in the gaits of the two groups (mean λ/h 3.1 in theropods as opposed to 4.2 in ornithopods). In the third analysis (Fig. 5.9C) h was estimated by means of allometric equations; it is now apparent that the theropods were considerably smaller than the majority of the ornithopods and that these two groups shared an almost identical gait (mean λ/h 3.7 in both cases).

Figure 5.10 shows similar analyses of trackway data from the Early Cretaceous of Canada (44 theropod trackways versus 41 ornithopod trackways). In all three cases it seems that the ornithopods used a slower gait than the theropods. However, the *reason* for this difference in gaits only becomes apparent when allometric equations are used to estimate the size (h) of the trackmakers: the ornithopods used a slower gait because they were, on the whole, much bigger than the theropods (Fig. 5.10C). This underlying difference in size is less obvious when h is estimated by means of ratios (Fig. 5.10A,B).

These comparisons (Figs. 5.9, 5.10) demonstrate some important guidelines for the study of dinosaur gaits in general. First, it is essential that the size of the trackmakers should be estimated as accurately as possible. Second, the most enlightening measure of a trackmaker's size is h rather than FL (or a multiple of FL). Third, and most important, the choice of method to predict h will affect the outcome in terms of λ/h. That is, the blanket application of a single predictive equation (such as $h \simeq 4FL$) may generate spurious differences in λ/h (e.g., Fig. 5.9A) simply because any set of trackmakers is unlikely to have been uniform in the ratio λ/FL. These spurious differences in λ/h can be suppressed to some extent by using a separate morphometric ratio to predict h in each taxonomic group of trackmakers (Figs. 5.9B, 5.10B). However, the effects of geometric and size dissimilarities among the trackmakers can best be mitigated by using allometric equations to predict *h* (e.g., Fig. 5.9C). Such equations do at least acknowledge the prevalence of allometry within and among dinosaur taxa.

Gaits of Quadrupedal Dinosaurs

The gaits of quadrupedal dinosaurs cannot be investigated in great detail because the trackways of these animals are not so well-known as those of bipedal dinosaurs. Nevertheless it is possible to reach some general and rather tentative conclusions.

Quadrupedal dinosaurs favored a walking gait, as did bipedal dinosaurs (Fig. 5.5F), but there is no evidence that any quadrupedal dinosaur used a trotting or running gait. This does not necessarily mean that these animals never ran, but they may have done so only on rare occasions.

Figure 5.5 shows that bipedal dinosaurs walked with relatively long strides (mean λ/h 1.29) while quadrupeds walked with relatively short strides (mean λ/h 0.95). This seems to be a valid distinction. [Here it might be objected that *h* was estimated by one method (allometric equations) in the bipeds and by a different method (ratio $h \simeq 4FL$) in quadrupeds. However, most of the data for Figure 5.5F were obtained from trackways of sauropods, where FL might represent as little as one–fifth or even one–sixth of *h*. Consequently λ/h may have been *over*estimated for these animals.] Evidently quadrupedal dinosaurs used a *slow* walking gait, a conclusion that is supported by some independent observations.

First, there is the observation that most quadrupedal dinosaurs were big animals. With body weights estimated at several tons, and often more, it is barely conceivable that the larger quadrupedal dinosaurs could have lengthened their stride by introducing an unsupported interval. These animals must have been restricted to walking or (at most) trotting gaits by virtue of their great body mass.

Second, the quadrupedal dinosaurs have forelimbs and hindlimbs of unequal length. Usually the forelimbs are shorter than the hindlimbs, and sometimes markedly so, as in *Stegosaurus*. In a walking quadrupedal dinosaur the forelimbs and hindlimbs, though of different sizes, would have taken strides that were equal in number and equal in length. So, at any given speed, relative stride length would have been greater for the (short) forelimb than for the (long) hindlimb. The magnitude of this difference in relative stride length would have depended on the ratio of shoulder height (*s*) to hip height (*h*) and on the absolute length of the stride (λ). These differences are plotted for two quadrupedal dinosaurs in Figure 5.11. In the case of the sauropod dinosaur *Diplodocus* the length of the forelimb is about 69 percent the length of the hindlimb; if the hindlimb reached a relative stride length of 2.0 (absolute stride length of about 6 m) the forelimb would have had a relative stride length of about 2.9. That is, the forelegs of *Diplodocus* would have been trotting while the animal's hindlegs were merely walking. The case of *Stegosaurus* is even more remarkable. In a skeleton of this dinosaur the forelimb is less than half the length of the hind limb; if *Stegosaurus* took strides 3 m long its hindlimbs would have been walk-

Figure 5.11. Predictions of relative stride length (λ/h) related to absolute stride length in two quadrupedal dinosaurs. Predictions for forelimb and hindlimb are shown separately in both cases. Left, the sauropod *Diplodocus*, with forelimb 69% the length of the hindlimb. Right, the stegosaur *Stegosaurus*, with forelimb 47% the length of the hindlimb. Horizontal lines at relative stride length 2.0 and 2.9 indicate walk–trot transition and trot–run transition respectively. Based on data from Thulborn (1982).

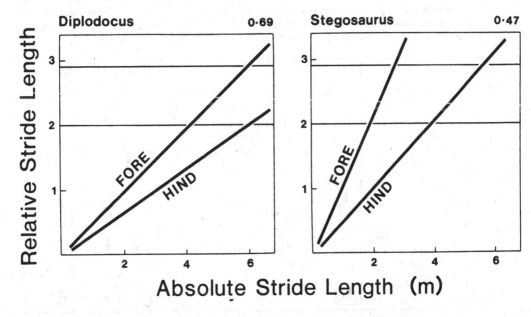

ing (λ/h 1.5) while its forelimbs would have been running (λ/s 3.3). Such differences between the gaits of forelimbs and hindlimbs would have become more pronounced as quadrupedal dinosaurs accelerated and took increasingly longer strides. At low speeds (with short strides) quadrupedal dinosaurs would have found it relatively easy to co-ordinate the striding of fore and hind limbs; but at higher speeds (with longer strides) it would have become increasingly more difficult to maintain such co-ordination. In summary, quadrupedal dinosaurs may have been restricted to short strides, and hence to slow gaits, for simple mechanical reasons: if they attempted to accelerate by taking longer strides they might have found it difficult to maintain co-ordination between the striding of their short forelegs and the striding of their long hindlegs.

Quadrupedal dinosaurs that were unable to run, or even to trot very fast, might have been easy prey for bipedal predators. Their only defense — aside from behavioral responses such as herding — might have lain in increasing body size, with or without the added protection of horns or bony spikes and plates. Such consequences might possibly explain the rarity of small quadrupedal dinosaurs in general and the almost total lack of quadrupedal predators among the dinosaurs.

Sauropods as Knuckle-Walkers?

The osteology of the sauropod manus is poorly known, though in all cases the digits were rather short and there was a large hook-like claw on the pollex (digit I). The posture of the manus was also unusual: the pes was semi-plantigrade, like the feet of elephants and mammoths (see Kubiak 1982), whereas the manus had an erect digitigrade (or perhaps even semi-unguligrade) posture. The most

curious fact is that prints of the forefeet show no obvious trace of the large pollex claw: instead the manus prints are horseshoe-shaped or semi-circular impressions that lack clear indications of separate digits (Fig. 5.2C). It was suggested by Beaumont and Demathieu (1980) that the pollex claw failed to leave an imprint because sauropods were knuckle-walkers ... that they supported themselves on the distal ends of the metacarpals, with the digits curled up behind. However, the sauropod pollex comprised only two phalanges (and, hence, two articulations) and may not have been sufficiently flexible to allow such knuckle-walking. Moreover the distal articular surface of the proximal phalanx faces dorso-medially, implying that the ungual phalanx could be retracted by hyperextension. Consequently it seems possible that the pollex claw was carried clear of the ground by means of dorso-medial hyperextension, such as it was in the related prosauropods (Galton 1971). Whatever its orientation this claw seems to have played no major role in sauropod locomotion (unlike the claws on the hindfeet). Presumably it has some other function; it is unlikely to have been an easily maneuvered weapon but might have served as a hook for pulling down vegetation.

Gaits of Semibipedal Dinosaurs

Semibipedal dinosaurs such as ornithopods and (perhaps) prosauropods were able to switch between bipedal and quadrupedal gaits according to circumstances. The shift from quadrupedal to bipedal gait entailed a number of correlated changes in the trackway pattern left by the hindfeet (Fig. 5.12D): an increase in stride length and pace angulation, and a concomitant decrease in trackway width. The increase in stride length may indicate that these

Figure 5.12. Tracks of quasi-bipedal dinosaurs.
A, *Moyenisauropus* (*?Anomoepus*) *natator*, after Ellenberger (1974); the trackway of an ornithopod walking on all fours. **B**, *Navahopus falcipollex*, after Baird (1980); the trackway of a prosauropod walking on all fours. **C,D**, *Moyenisauropus* (*?Anomoepus*) *natator*, after Ellenberger (1974); diagrams of two successive sections of an ornithopod trackway; in C the animal is fully quadrupedal, in D it switches from quadrupedal to bipedal gait; pes imprints indicated by solid circles and linked by tie-lines to demonstrate pace angulation; manus imprints indicated by ringed points.

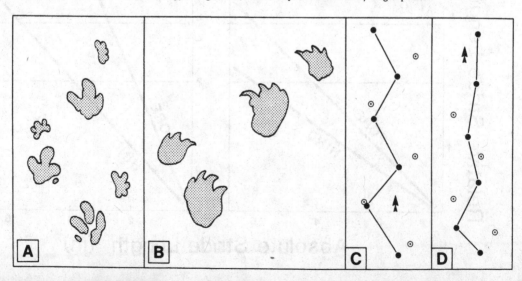

dinosaurs normally used the bipedal gait for relatively rapid progression.

In the quadrupedal gait the hindfeet took strides of moderate length; they produced a broad trackway with reduced pace angulation and consistent pace length (Fig. 5.12A,C). This consistency in pace length implies that the hindlimbs supported much of the trackmaker's body mass. By comparison the prints of the forefeet have an erratic distribution (Fig. 5.12A,C); the somewhat unpredictable placement of the forefeet implies that the forelimbs had no major supportive role. It is also noticeable that the forefeet left a wider trackway than the hindfeet (Fig. 5.12A–C) and that pace length for the forefeet often exceeds pace length for the hindfeet. Since the forelimbs were shorter than the hindlimbs this remarkably wide spacing of left and right forefeet implies that the shoulder region was depressed below the level of the hips: it is likely that head, neck and shoulders were carried close to the ground and that the elbows were stuck out sideways to some extent. This low–slung and somewhat sprawling gait (Fig. 5.13) was most probably used in slow progression; it is known to have been adopted by ornithopods and prosauropods (Fig. 5.12) and, perhaps, by some theropods (e.g., the trackmaker identified by Norman 1980, plate V, as *Iguanodon*; see comments of Thulborn 1984, p. 251).

Acknowledgments
Mary Wade spent many hours helping to compile measurements of trackways at the Lark Quarry site, Queensland, and James Farlow generously contributed data for Figure 5.5B. I am particularly grateful to David Gillette, David Hafner, Martin Lockley and Michael Morales for their unsparing efforts in helping me to attend the First International Symposium on Dinosaur Tracks and Traces.

References

Alexander, R. McN. 1976. Estimates of speeds of dinosaurs. *Nature* London 261:129–130.

1977. Mechanics and scaling of terrestrial locomotion. *In* Pedley, T. J. (ed.). *Scale Effects in Animal Locomotion* (Academic Press) pp. 93–110.

1985. Mechanics of posture and gait of some large dinosaurs. *Zool. Jour. Linn. Soc.* 83:1–25.

Alexander, R. McN., and Jayes, A. S. 1983. A dynamic similarity hypothesis for the gaits of quadrupedal mammals. *Jour. Zool. London* 201:135–152.

Avnimelech, M. A. 1966. Dinosaur tracks in the Judean Hills. *Proc. Israel Acad. Sciences and Humanities* (Science section) 1:1–19.

Baird, D. 1980. A prosauropod dinosaur trackway from the Navajo Sandstone (Lower Jurassic) of Arizona. *In* Jacobs, L. L. (ed.). *Aspects of Vertebrate History* (Museum of Northern Arizona Press) pp. 219–230.

Beaumont, G. de, and Demathieu, G. 1980. Remarques sur les extremités antérieures des Sauropodes (Reptiles, Saurischiens). *Comptes Rendus des Séances, Société d'Histoire Naturelle de Genève* (Nouvelle Série) 15:191–198.

Bernier, P. 1984. Les dinosaures sauteurs. *La Recherche* 15:1438–1440.

Bernier, P., Barale, G., Bourseau, J.-P., Buffetaut, E., Demathieu, G., Gaillard, C., and Gall, J.-C. 1982. Trace nouvelle de locomotion de chélonien et figures d'émersion associées dans les calcaires lithographiques de Cerin (Kimméridgien Supérieur, Ain, France). *Geobios* 15:447–467.

Bernier, P., Barale, G., Bourseau, J.-P., Buffetaut, E., Demathieu, G., Gaillard, C., Gall, J.-C., and Wenz, S. 1984. Découverte de pistes de dinosaures sauteurs dans les calcaires lithographiques de Cerin (Kimméridgien Supérieur, Ain, France): implications paléoécologiques. *Geobios, Mémoires Specials* 8:177–185.

Brown, J. C., and Yalden, D. W. 1973. The description of mammals — 2, limbs and locomotion of terrestrial mammals. *Mammal Review* 3:107–134.

Coombs, W. P. 1978. Theoretical aspects of cursorial adaptations in dinosaurs. *Quart. Rev. Biology* 53:393–418.

Currie, P. J. 1983. Hadrosaur trackways from the Lower Cretaceous of Canada. *Acta Palaeont. Polonica* 28:63–73.

Demathieu, G. 1970. Les empreintes de pas de vertébrés du Trias de la bordure nord–est du Massif Central. *Cahiers de Paléont.*, éditions du Centre National de la Recherche Scientifique 211 pp.

Figure 5.13. Restoration of an ornithopod dinosaur in its slow quadrupedal gait. Based on body proportions in the hadrosaur *Anatosaurus*.

1984. Utilisation de lois de la mécanique pour l'éstimation de la vitesse de locomotion des vertébrés tétrapodes du passé. *Geobios* 17:439-446.

Ellenberger, P. 1972. Contribution à la classification des pistes de vertébrés du Trias: les types du Stormberg d'Afrique du Sud (I). *Palaeovertebrata, Mém. Éxtraord.* 117 pp.

1974. Contribution à la classification des pistes de vertébrés du Trias: les types du Stormberg d'Afrique du Sud (IIème partie). *Palaeovertebrata, Mém. Extraord.* 147 pp.

Farlow, J. O. 1981. Estimates of dinosaur speeds from a new trackway site in Texas. *Nature*, London 294:747-748.

Galton, P. M. 1971. Manus movements of the coelurosaurian dinosaur *Syntarsus* and opposability of the theropod hallux. *Arnoldia (Rhodesia)* 5 (No. 15):1-8.

Hildebrand, M. 1965. Symmetrical gaits of horses. *Science* 150:701-708.

Ishigaki, S. 1985. Dinosaur footprints of the Atlas Mountains (1). *Nature Study* 31 (10):5-8. [in Japanese]

Kool, R. 1981. The walking speed of dinosaurs from the Peace River Canyon, British Columbia, Canada. *Canadian Jour. Earth Sci.* 18:823-825.

Kubiak, H. 1982. Morphological characters of the mammoth: an adaptation to the Arctic-steppe environment. *In* Hopkins, D. M., Matthews, J. V., Schweger, C. E., and Young, S. B. (eds.). *Paleoecology of Beringia* (Academic Press) pp. 281-289.

Lockley, M. G., Young, B. H., and Carpenter, K. 1983. Hadrosaur locomotion and herding behavior: evidence from footprints in the Mesaverde Formation, Grand Mesa coal field, Colorado. *Mountain Geol.* 20:5-14.

Muybridge, E. 1899. *Animals in Motion*. (Chapman and Hall) 264 pp.

Norman, D. B. 1980. On the ornithischian dinosaur *Iguanodon bernissartensis* of Bernissart (Belgium). *Memoires de l'Institut Royal des Sciences Naturelles de Belgique* 178:1-103.

Raath, M. A. 1972. First record of dinosaur footprints from Rhodesia. *Arnoldia (Rhodesia)* 5 (No. 27):1-5.

Russell, D. A., and Béland, P. 1976. Running dinosaurs. *Nature, London* 264:486.

Sanz, J. L., Moratalla, J. J., and Casanovas, M. L. 1985. Traza icnologica de un dinosaurio iguanodontido en el Cretacico Inferior de Cornago (La Rioja, España). *Estudios Geologicos* 41:85-91.

Thulborn, R. A. 1982. Speeds and gaits of dinosaurs. *Palaeogeogr., Palaeoclimat., Palaeoecol.* 38:227-256.

1984. Preferred gaits of bipedal dinosaurs. *Alcheringa* 8:243-252.

Thulborn, R. A., and Wade, M. 1979. Dinosaur stampede in the Cretaceous of Queensland. *Lethaia* 12:275-279.

1984. Dinosaur trackways in the Winton Formation (Mid-Cretaceous) of Queensland. *Mem. Queensland Museum* 21:413-517.

Tucker, M. E., and Burchette, T. P. 1977. Triassic dinosaur footprints from South Wales: their context and preservation. *Palaeogeogr., Palaeoclimat., Palaeoecol.* 22:195-208.

6 A Footprint as a History of Movement

RICHARD A. THULBORN AND
MARY WADE

Abstract

The foot of a moving animal interacts with the substrate in three phases: touch–down, weight–bearing and kick–off. We examine the morphological effects of these phases in footprints of Cretaceous dinosaurs (theropods and ornithopods) at the Lark Quarry Environmental Park, western Queensland, Australia.

Introduction

A footprint gives only an imperfect idea of a track-maker's foot structure. The footprint may be incomplete, because parts of the foot were not impressed into the substrate, or it may be obscured by adventitious or *extra-morphological* features (following terminology of Peabody 1948). Some of these deficiencies and extramorphological features result from local variations in the physical properties of the substrate; others originate from the dynamic inter-action between foot and substrate. That complex inter-action may be envisaged as a sequence of three phases: (i) touch–down (abbreviation T-phase), (ii) *weight–bearing* (W-phase), and (iii) *kick–off* (K-phase).

Touch–down is normally on the functional ball of the foot, with the tips of the toes turned slightly upwards. In the W-phase the trackmaker's center of gravity passes over the planted foot, which usually sinks to some degree into the substrate. The K-phase entails a backwards sweep of the toes. All three phases may be seen in living tetrapods, particularly in those such as kangaroos that move the hind-feet synchronously. Similar phases are evident in barefoot humans and probably occurred in dinosaurs too. In some instances a trackmaker's foot makes little impression on the substrate until the K-phase. Then, as the rear part of the foot is being lifted from the ground, the toes break through the surface to produce a foreshortened footprint. Such distinctive one–phase footprints, formed during a brief episode of breakthrough, are identified in our illustrations by the code–letter B.

We illustrate the morphological effects of these phases in a variety of footprints made by Cretaceous dinosaurs, both theropods and ornithopods. These footprints were found at a site known as Lark Quarry, in western Queensland, Australia (see Thulborn and Wade 1979, 1984; and Agnew et al. this volume).

Substrate and Paleoenvironment

The sediments at Lark Quarry are interbedded sand-stones and claystones. These accumulated in the floor of a broad creekbed that drained into a lake or waterhole to the southwest. Each sandstone/claystone couplet presum-ably originated from a single major influx of sediment–charged water. Dinosaur footprints occur at several horizons in the succession. At the single horizon discussed here near-ly 4,000 footprints were excavated in the form of natural molds (concave epireliefs). All these footprints were impressed in a single bedding plane which defines the top of a laminated claystone bed with a total thickness rang-ing between 6 cm and 12 cm. The claystone is overlain and underlain by finely cross–bedded arkosic sandstones.

Footprints

The Lark Quarry footprint horizon bears several generations of dinosaur tracks (Fig. 6.1), which are exam-ined in chronological order. All the footprints are original impressions; there is no indication that any of them are "underprints" transmitted through the overlying sediments.

First Generation

This category comprises a few scattered and poorly preserved footprints attributed to moderately large ornitho-pods. The tridactyl footprints have rather short, thick and bluntly–rounded digits and are randomly oriented (Fig. 6.1A). They were formed, and then eroded and filled with water–laid sediment, well before the substrate was exposed to the air and other dinosaurs traversed the area. These

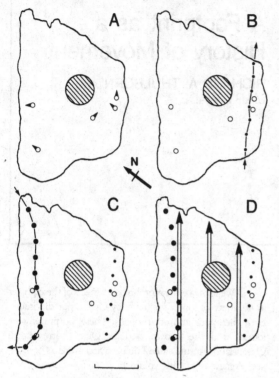

the area as it was draining free of water. This animal, estimated to have been about 1.6 m high at the hip, trotted across the southern part of the site from west–southwest to east–northeast (Fig. 6.1B). Some of its footprints were deeply impressed in saturated mud, which bulged up between the toes and behind the foot. This upwelling is most pronounced *behind* the footprint, indicating some backwards displacement of sediment at the end of the W–phase and during the K–phase. It is worth emphasizing that this displacement occurred in an almost horizontal substrate: evidently an upwelling of sediment behind a footprint does not necessarily indicate that the trackmaker was ascending a slope. Other footprints in the same trackway are not so deeply impressed nor so well preserved; they seem to have been formed in slightly lower–lying areas that were subject to scouring as the substrate finally drained free of water and became exposed to the air.

Third Generation

This category comprises the trackway of a single carnosaur (large theropod) that walked across the northern part of the site from northeast to southwest (Fig. 6.1C). The trackway comprises 11 tridactyl footprints, each with tapering or V–shaped digits and, in some cases, with clear indications of large and sharply–pointed claws. The foot-

Figure 6.1. Sequence in which dinosaur tracks were formed at the Lark Quarry site, Queensland. Each of the four charts shows outline of area excavated at the site; cross–hatched circle indicates approximate location of an outlier of undisturbed overburden; scale bar indicates approximately 5 m. **A,** randomly–oriented footprints produced by ornithopods crossing the area while it was still under water (not all examples illustrated). **B,** moderately large ornithopod traverses southern part of the area from west–southwest to east–northeast while it is draining free of water. **C,** carnosaur walks across northern part of the area from northeast to southwest. **D,** whole area is traversed by more than 160 small ornithopods and coelurosaurs running from southwest to northeast.

Figure 6.2. Variation in stride length and footprint size in the single carnosaur trackway (cf. *Tyrannosauropus*) at Lark Quarry. Footprint size index is the square root of footprint length x footprint width; it is plotted as the mean for the two footprints defining each stride. Data for each of the carnosaur's nine strides indicated by number. Note that the data fall into two distinct groups (strides 1 to 4 versus strides 5 to 9). Crosses indicate mean values plus and minus two standard deviations.

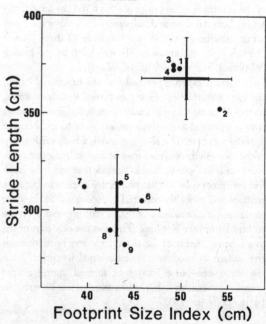

scattered prints were probably made by ornithopods that were swimming and only occasionally touching down on the bottom — much like the hadrosaur footprints described by Currie (1983) in the Cretaceous of Canada. The prints trapped some residual quartz sand, without appreciable feldspar, that had been winnowed across the subaqueously hardening mud, along with a few streaks of muddy sediment that was probably stirred up by dinosaurs venturing across the area before it drained free of water. The footprints represent fairly large animals; smaller dinosaurs might also have crossed the submerged area at about the same time, but these would not have left any traces if their legs were too short to have touched down.

Second Generation

At least one moderately large ornithopod traversed

prints resemble those in the ichnogenus *Tyrannosauropus* and were probably made by a carnosaur about 2.6 m high at the hip (Thulborn and Wade 1984).

The first part of the carnosaur's trackway is noticeably different from the second (Fig. 6.2). Initially (strides 1–4) the footprints are relatively large and the animal took long strides; thereafter (strides 5–9) the footprints are smaller and stride length is reduced. This weaving trackway ends with an abrupt turn to the right, which presumably reflects a change in the trackmaker's behavior. We suspect that the carnosaur might have been approaching or stalking a group of dinosaurs gathered around the waterhole to the southwest (Thulborn and Wade 1984).

The first part of the trackway comprises deep footprints, each surrounded by a prominent raised rim of sediment. Evidently the trackmaker's foot plunged right through the muddy surface layer and rested on the firmer sandy sediments beneath (Fig. 6.3W). In this W–phase the muddy sediment bulged between the toes and around the perimeter of the foot. Once again it is noteworthy that this upwelling of sediment also occurred *behind* the foot on a more or less horizontal substrate. In the second part of the trackway the foot sometimes made little impression on the substrate until the K–phase. Then, as the rear part of the foot lifted clear of the ground, the distal parts of the toes broke through the surface to produce a foreshortened footprint (Fig. 6.3B).

Fourth Generation

Finally a large number (160+) of small bipedal dinosaurs ran across the area from southwest to northeast. This unusual event has been interpreted as a dinosaur stampede, which was presumably triggered by the approach of the carnosaur (Thulborn and Wade 1979, 1984). By this time the muddy substrate had been exposed long enough to have achieved a firm plastic consistency. The substrate was not waterlogged, for none of the thousands of footprints slumped or collapsed; nor was it very tenacious, for it rarely adhered to a trackmaker's foot. The numerous small dinosaurs were of two types — ornithopods and coelurosaurs (small theropods) — each indicated by its own distinctive footprint morphology.

Ornithopod Footprints

The morphology of these blunt–toed tridactyl footprints, named *Wintonopus latomorum*, has been described elsewhere (Thulborn and Wade 1984). Well over 1,000 footprints are available for study; these show tremendous morphological variation, much of it related to events during the trackmaker's step–cycle (Figs. 6.4–6.7). At the start of this cycle (T–phase) the forwardly–extended foot was planted on the substrate with slight positive (inwards) rotation (Fig. 6.4T). Initially there would have been a very shallow footprint, or none at all. At mid-step (W–phase) the trackmaker's center of gravity passed forwards above the planted foot, which then sank deeper into the substrate. The degree to which the foot sank depended on many and variable factors, including the size and weight of the trackmaker, the attitude of its foot, and the local consistency of the substrate. Sometimes the foot sank up to, or beyond, the distal end of the metapodium, leaving a print that comprised three digital impressions joined together along a continuous posterior margin (Fig. 6.5C,D). At other times the foot did not sink so deeply, so that the three digital impressions are partly or completely separated and there is no

Figure 6.3. Two footprint morphologies in the carnosaur trackway at Lark Quarry. **W,** full-length footprint formed during weight-bearing phase of the step–cycle. Note that the foot penetrated the muddy surface sediment (c), which bulged up around the footprint, and came to rest on the firmer sandy sediment beneath (s). **B,** foreshortened footprint made by the foot breaking through the surface sediment during kick-off phase of the step–cycle. Approximate lengths of the two types of footprints indicated in cm.

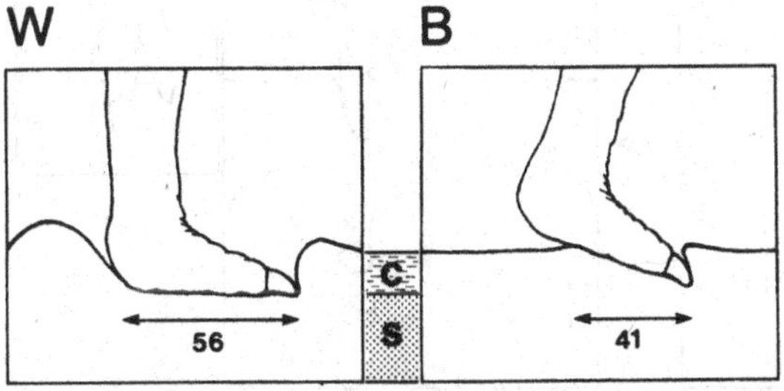

Figure 6.4. Morphological features of ornithopod footprints (*Wintonopus latomorum*) at Lark Quarry site, related to events during the trackmaker's step–cycle. Each diagram shows position of foot in profile (above) and corresponding plan view of the right footprint (below). **T,** touch-down phase; initial footprint (shaded) is shallow and inwardly rotated. **W,** Weight-bearing phase; the foot sinks deeper, and may rotate laterally and/or slip backwards (unshaded footprint). **Ka,** end of kick-off phase; tips of one or more toes drag through the front rim of the footprint, forming forwardly–directed scrape-marks. **Kb,** kick-off phase; tips of toes slip backwards, incising grooves in the floor of the footprint. **Kc,** kick-off phase (continued from Kb); toes continue to slip backwards, producing retro–scratches behind footprint.

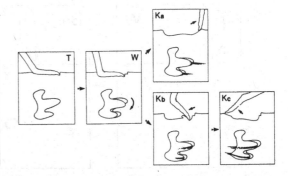

W   B

56   c s   41

continuous rear margin to the footprint (Fig. 6.5A,B). Other variations in footprint shape depended on the attitude of the foot during the T-phase and are examined elsewhere (see Thulborn and Wade 1984, Fig. 5); so too are size-related variations, such as the fact that the tips of the digits appear to have been distinctly sharper in small footprints. Variation in footprint shape with depth of tracks is a widespread phenomenon. In some sedimentary successions it may result in underprints which differ considerably from the true footprints.

Figure 6.5. Morphology of ornithopod footprints (*Wintonopus latomorum*) at Lark Quarry site, related to various depths to which the foot sank during the weight-bearing phase. Right foot is shown in anterior outline (center) with diagrams of corresponding right footprints at various depths. **A,** shallow footprint; only the undersurfaces of three separate digits are impressed. **B,** deeper footprint, with hypex (re-entrant) between digits III and IV impressed. **C,** still deeper footprint, with hypex between digits II and III also impressed. **D,** deep footprint, with extensive and continuous rear margin.

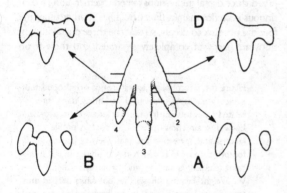

During the W-phase the foot sometimes slipped backwards as it sank deeper, so that the front margins of the digital impressions are distinctly stepped or terraced (Fig. 6.4W). In addition the foot sometimes rotated outwards, so that digit III pointed directly ahead; similar traces of rotation or sideways slippage of the foot were described by Soergel (1925) in *Chirotherium* footprints. Since the foot was planted into the substrate (T-phase) with inwards rotation, but was withdrawn (K-phase) in a different direction (forwards or antero-laterally), the tip of digit III may appear to be forked or Y-shaped (e.g., Fig. 6.4W).

During the K-phase the tips of the toes were the last parts of the foot to be withdrawn from the substrate. Sometimes the toes slipped backwards as they were being withdrawn, thus incising slots in the floor of the footprint (Fig. 6.4Kb). Occasionally the tips of the toes slipped back so far that they breached the rear wall of the footprint to leave retro-scratches (Fig. 6.4Kc). In some instances the tips of the toes did not slip backwards during the K-phase; instead they sometimes dragged through the front rim of the newly-formed footprint, to leave forwardly-directed scrape marks (Fig. 6.4Ka). Digit III was the longest in the trackmaker's foot and, for that reason, formed a scrape-mark most frequently; digit II was the shortest in the foot and rarely produced a scrape-mark.

In a few cases the foot made no impression during the T-phase and W-phase, but broke through the surface during the K-phase, when a considerable fraction of the animal's body mass was supported only by the distal parts of the toes. The tip of the long middle toe (III) sank into the substrate, anchoring itself in place (Fig. 6.6Ba), then, as the K-phase continued, the flanking toes (II and IV) swung forwards to touch down alongside the anchored digit III (Fig. 6.6Bb). The result was an extremely distinctive one-phase footprint consisting of three puncture-like markings arranged side-by-side. Such footprints are noticeably narrower than those "normal" footprints, where the trackmaker's toes splayed out sideways during the W-phase (cf. Figs. 6.4, 6.5, 6.7).

Figure 6.6. Unusual form of footprint formed by limited penetration of toes into substrate during the kick-off phase. **W,** weight-bearing phase; foot makes no impression on substrate. **Ba,** at kick-off phase the tip of digit III penetrates the substrate and is anchored in place. **Bb,** as the kick-off phase continues, digits II and IV swing forwards to touch down alongside the anchored digit III. Note that the digits are not splayed out (anterior view of foot, top right), so that the resulting footprint is distinctly narrower than a footprint formed during the W-phase (e.g., Figs. 6.5, 6.7, 6.8). Footprints of this type were produced by both ornithopods and coelurosaurs.

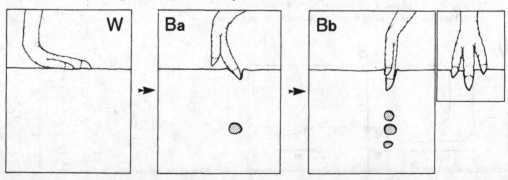

The axial digit (III) was probably the thickest in the trackmaker's foot and normally produced a correspondingly broad impression in the footprint (Fig. 6.7A,E–G). Sometimes the impact of this principal weight–bearing digit generated a thixotropic reaction in the underlying sediment; then, when the trackmaker's foot was withdrawn from the substrate, the fluid sediment flowed (or was sucked) inwards to leave a secondarily narrowed imprint of digit III (Fig. 6.7B–D).

Coelurosaur Footprints

These tridactyl and almost symmetrical footprints, named *Skartopus australis*, have also been described elsewhere (Thulborn and Wade 1984). They are readily distinguished from the ornithopod footprints on account of their narrow, tapering and sharply pointed digits. Many hundreds of footprints are available for study, but they show rather less morphological variation than the ornithopod footprints at the same site. Once again, the variations that do exist are best explained by reference to events during the trackmaker's step–cycle (Figs. 6.6, 6.8).

Figure 6.7. Ornithopod footprints *(Wintonopus latomorum)* showing variation in size and shape of the middle digit (3). All outlines are traced from photographs of footprints at Lark Quarry, with morphological features such as scrape-marks and retro-scratches omitted for sake of clarity. All examples are illustrated as left footprints (C, D and E reversed from right footprints), and each scale bar indicates 2 cm. Digit III was the stoutest in the trackmaker's foot and appears as such in many other footprints (A, E–G). In other footprints (B–D) the imprint of digit III was secondarily narrowed by inwards suction of fluid sediment during withdrawal of the foot.

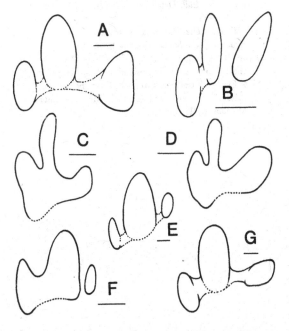

At the start of this cycle (T-phase) the forwardly-extended foot was planted on the substrate, producing a very shallow footprint or no perceptible footprint at all (Fig. 6.8T). At mid-step (W-phase) the trackmaker's center of gravity passed forwards above the planted foot, which in *some* cases sank deeper into the substrate (Fig. 6.8W, upper row). During the K-phase the tips of the toes pressed down and slightly backwards to leave imprints of the sharply-pointed claws (Fig. 6.8Ka). Sometimes the tips of the toes slipped backwards during the K-phase, incising grooves in the floor of the footprint (Fig. 6.8Kb). Occasionally the tips of the toes continued to slip backwards, forming retro-scratches that breach the rear wall of the footprint (cf. Fig. 6.8Kb). Retro-scratches are commoner than in the ornithopod footprints, possibly because the coelurosaurs had less flexible toes. There is no obvious indication that the coelurosaur foot rotated in its footprint during the W-phase (as it did in the ornithopods, cf. Fig. 6.4W) nor of forwardly directed scrape marks from the tips of the digits (cf. Fig. 6.4Ka).

In many cases the foot did *not* sink into the substrate during the T-phase or the W-phase (Fig. 6.8W, lower row). This seems to have happened very frequently, judging from the large number of "gaps" or "missing" footprints in the coelurosaur trackways at Lark Quarry. We suspect that the coelurosaurs had relatively large and broad-spreading feet that acted as analogues of snowshoes. Nevertheless the sharp tips of their claws did commonly break through the surface of the substrate during the K-phase, when the animal's body mass would have been supported by a small (and diminishing) area of the foot's undersurface (Fig. 6.8Ba). Then, quite frequently, the sunken tips of the toes lost their purchase in the muddy substrate and slithered backwards to leave a footprint consisting of superficial furrows or retro-scratches (Fig. 6.8Bb,Bc). Sometimes only one or two of the three digits broke through and furrowed the surface in this fashion. The coelurosaurs commonly produced distinctive one-phase footprints consisting of three puncture-like markings arranged side-by-side (Fig. 6.6).

Finally, a few of the coelurosaur footprints include a trace of the metapodium (Thulborn and Wade 1984, plates 12, 13B, 14B). This is a large subrectangular impression behind the three digit impressions; it is no wider than the maximum spread of the digits and is roughly as long as the imprint of digit III. Such footprints are rare and seem to occur randomly within otherwise normal *Skartopus* trackways; however, one section of trackway shows a sequence of four such footprints. Presumably these footprints originated when the normally digitigrade trackmakers used a "flat-footed" gait. Kuban (1986) described somewhat similar footprints from the Cretaceous of the Paluxy River, Texas, but these appear to differ in one respect from those at Lark Quarry: the metapodium imprint is *deeper* than the digit imprints in the Paluxy River examples whereas it is *shallower* than the digit impressions in the Lark Quarry footprints. The significance of this distinction remains unknown.

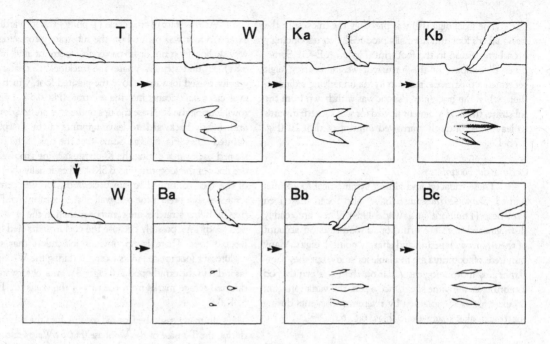

Figure 6.8. Morphological features of coelurosaur footprints *(Skartopus australis)* at Lark Quarry site, related to events during the trackmaker's step-cycle (see also Fig. 6.6). Each diagram shows position of foot in profile (above) and corresponding plan view of the right footprint (below). **T,** touch–down phase; the foot makes little or no impression on the substrate. **W** (upper row), weight–bearing phase; the foot sinks into the substrate. **Ka,** kick–off phase; as foot is lifted from substrate the tips of the toes produce sharp imprints of the claws. **Kb,** kick–off phase (frequently following Ka); tips of the toes slip backwards, incising grooves in floor of footprint. **W** (lower row), weight–bearing phase; the relatively large and broad-spreading foot fails to sink into the substrate. **Ba,** kick–off phase; only the tips of toes break through the surface of the substrate. **Bb, Bc** (frequently following Ba), kick–off phase; tips of the toes break through the surface of the substrate and slip backwards to form sub-parallel furrows.

References

Currie, P. J. 1983. Hadrosaur trackways from the Lower Cretaceous of Canada. *Acta Palaeont. Polonica* 28:63–73.

Kuban, G. J. 1986. Elongated dinosaur tracks. In Gillette, D. D. (ed.). *First International Symposium on Dinosaur Tracks and Traces, Albuquerque, New Mexico, 1986, Abstracts with Program.* New Mexico Museum of Natural History, pp. 17–18.

Peabody, F. E. 1948. Reptile and amphibian trackways from the Lower Triassic Moenkopi Formation of Arizona and Utah. *Bull. Dept. Geol. Sci. Univ. California* 27 (8):295–468.

Soergel, W. 1925. *Die Fährten der Chirotheria, eine paläobiologische Studie.* (Gustav Fischer) 92 pp.

Thulborn, R. A., and Wade, M. 1979. Dinosaur stampede in the Cretaceous of Queensland. *Lethaia* 12:275–279.

1984. Dinosaur trackways in the Winton Formation (Mid–Cretaceous) of Queensland. *Mem. Queensland Museum* 21:413–517.

7 Elongate Dinosaur Tracks

GLEN JAY KUBAN

Abstract

Elongate marks made by bipedal dinosaurs include tail impressions, foot slides, toe drags, and footprints with full or partial metatarsal impressions. This paper focuses on dinosaur trackways composed largely or entirely of metatarsal footprints, suggesting that some bipedal dinosaurs, at least at times, walked in a plantigrade or quasi–plantigrade manner.

Numerous elongate dinosaur tracks with apparent metatarsal impressions occur in the Paluxy River bed, near Glen Rose, Texas. These tracks vary somewhat in size, shape, and clarity, but are typically 55 to 70 cm long (including the metatarsal segment), and 25 to 40 cm wide (across the digits). In many cases the digit impressions on such tracks are indistinct, due to mud collapse, erosion, or a combination of factors — causing some to resemble giant human footprints, for which they have been mistaken by some creationists and local residents.

Some metatarsal tracks may have been made by dinosaurs whose anatomy was especially adapted to plantigrady. However, many metatarsal tracks may have been made occasionally by a wide variety of bipedal dinosaurs, perhaps during behaviors in which they walked low to the ground, as during foraging or prey stalking. Most elongate tracks near Glen Rose appear to have been made by moderate sized theropods, although metatarsal tracks at other sites throughout the world include theropod and ornithopod forms of varying size and shape.

Introduction

The track sites focused on in this paper occur in a section of the Paluxy River bed from 2.5 to 5 miles west of Glen Rose, Texas. The track beds are limestone and carbonate cemented sandstone layers near the base of the Glen Rose Formation, Lower Cretaceous (Langston 1979). The site locations are shown in Figure 7.1.

The Alfred West Site

This site was named for Alfred and Martha West, who lived near the site and discovered the tracks there in 1974. The site was visited in 1975 by Wann Langston, Jr., who described the elongate tracks there as problematic, and speculated that they may have been made by a dinosaur whose foot sank deeply in soft mud (Langston 1979). Creationist John Morris also visited the site (which he called "Shakey Springs") in the late 1970's. Morris, who identified similarly–shaped, indistinct tracks in other areas as human prints, stated that the elongate dinosaur tracks on the West site were unusual and deserve further study (Morris 1980). Between 1982 and 1985 I thoroughly studied and mapped the site. Much of the site is often under water, but in late summer the western half of the site usually becomes dry.

A few trackways on the West Site consist entirely of tridactyl, digitigrade tracks. These vary in size, clarity, and depth, but most are 25–45 cm in length and exhibit pace lengths and pace angles typical of other bipedal trails in Glen Rose. However, one trail (IIT, at far right of Fig.

Figure 7.1. Locations of the relevant track sites near Glen Rose, Somervell County, Texas. Glen Rose is about 70 miles southwest of Dallas, Texas.

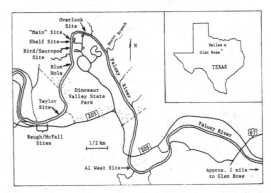

7.2) exhibits unusually long paces relative to track length (8.48 mean pace/track length ratio), indicating an exceptional speed of about 10 m/s, using the formula by Alexander (1976). Of more pertinence to this study are the many elongate tracks on the site, as well as some possible tail marks and a few problematic traces.

Metatarsal Tracks on the Al West Site

Elongate footprints with metatarsal impressions actually outnumber the digitigrade tracks on the West Site. In some trails elongate tracks are interspersed with nonelongate (digitigrade) tracks; other trails consist entirely or primarily of elongate tracks. Trails W and IIW, which occur on the western half of the site (Fig. 7.2) will be discussed as examples of each type; other trails with elongate tracks on the eastern half of the site (Fig. 7.3) will be summarized briefly.

Trail W occurs in an area of the site that is usually dry in summer, and where the rock surface is very coarse. The trackway includes both elongate and nonelongate forms alternating irregularly, with considerable variation in pace lengths and individual print features. Shorter pace lengths (less than 115 cm) predominate; however, some of the longer paces may be due to missing tracks (for example, between tracks W9 and 10 a print may be unrecorded, or may have been marred by IIW1, or may be under the bank). The elongate prints in Trail W average 55–65 cm in length; the nonelongate ones, 35–40 cm. The digit marks on most elongate tracks in Trail W are indistinct, evidently

due to mud collapse (slumping of mud back into the depressions), and later erosion. Note the superficially manlike shape of W2 (Fig. 7.4). Track W8 shows the outside digit (IV) more pronounced than the others — a common feature of elongate tracks in Glen Rose.

Trail IIW (Fig. 7.5) shows fairly long and consistent pace lengths (114–156 cm) relative to track size, and fairly large pace angles (140°–160°). All but one of the prints in Trail IIW (Fig. 7.5) are elongate, but they are slightly smaller than the elongate prints in Trail W (comparing corresponding areas of the prints). Many are indistinct, but a few are fairly well preserved. Track IIW2 shows digits splayed at a very wide angle (about 140° total divarication). Most other elongate tracks in Glen Rose also show wide, but less extreme, splaying — typically 75–90° total divarication. Several tracks in Trail IIW2 (and others in Glen Rose) show a narrow and/or raised area between the ball and heel areas, which might indicate a metatarsal "arch".

As Trail IIW approaches track IIW12 (and the area of the site usually under water), the track surface becomes smoother. Track IIW12 is very shallow and has short, blunt digit impressions (possibly due to a firm spot on the substrate), giving it an ornithopod–like appearance. However, most other tracks in the trail show more slender, pointed digits, strongly suggesting a theropod trackmaker. Track IIW13 (Fig. 7.5C), which is 53 cm long including the metatarsal segment, and 26 cm wide across the digits, is the best preserved track in Trail W, and one of the clearest on the site. The digit impressions, especially the center digit,

Figure 7.2. The Alfred West Site (western half).

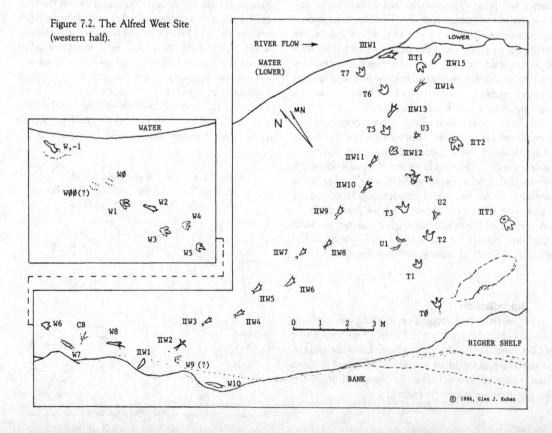

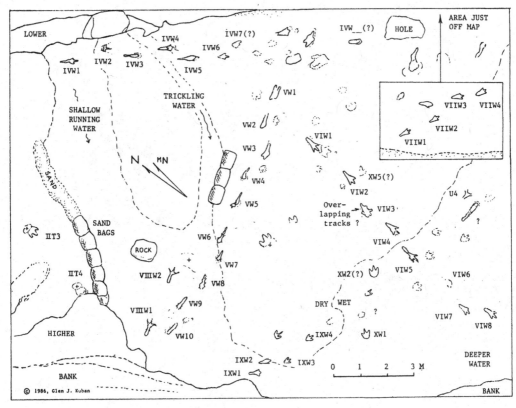

Figure 7.3. Eastern half of the Alfred West Site.

Figure 7.4. **A,** the W Trail on the Alfred West Site. Track W1 is at the lower right, followed by W2, W3, etc. **B,** track W2, showing elongate, man–like shape evidently due to metatarsal impression and mud collapse.

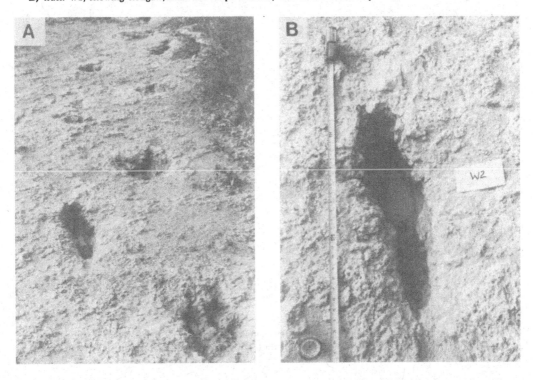

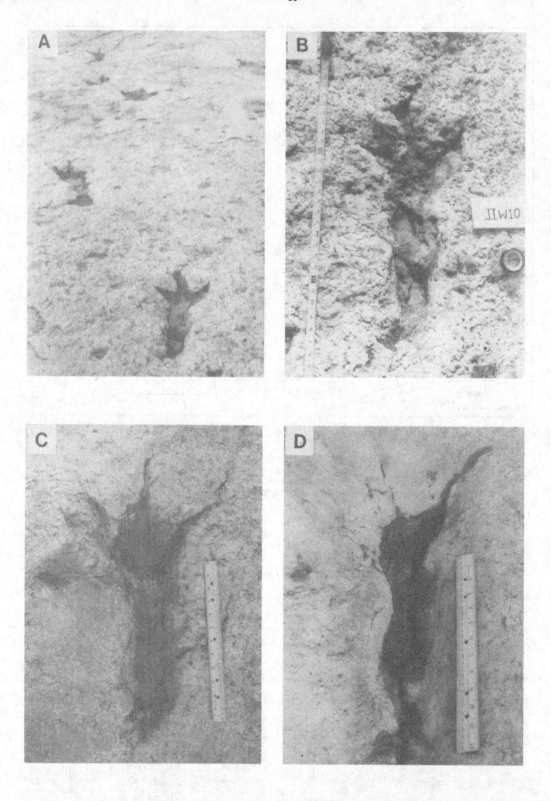

Figure 7.5. **A,** portion of the IIW Trail; track IIW10 at bottom. **B,** track IIW10. Tape measure at left is in inches. **C,** track IIW13, a well–preserved metatarsal track. **D,** track IIW14, a metatarsal track with mud collapsed digits.

are fairly narrow, apparently due to partial mud collapse. The "ball" area (metatarsophalangeal joint) is slightly more depresssed than the metatarsal and digit regions, which is typical of Glen Rose elongate tracks. The metatarsal segment is rounded at the posterior end (presumed to represent the tarsus), and narrows slightly at the center. Track IIW14 (Fig. 7.5D) is similar in shape to IIW13, but is somewhat narrower and has less distinct digits — features which may be attributed primarily to mud collapse. One can visualize how IIW13 and IIW14 would become very humanlike if the digit impressions were further obscured by erosion.

Trail IIIW is represented by a single metatarsal track near the broken north edge of the track layer, but originally may have been connected to Trail IVW. Trail IVW contains several deep elongate tracks with very indistinct (largely mud collapsed) digits, and one nonelongate track with three clear digit impressions (Fig. 7.6). This trail progresses into a very pockmarked area of the site. Trail VW contains only elongate tracks, some of which are very deep and distorted, and others of which are less deep and better preserved. Trails VIW and VIIW consist of several elongate tracks of variable clarity, but most show some indications of individual digits. Trail VIIIW comprises two well preserved metatarsal tracks, with a large pace (224 cm). Trail IXW begins with two elongate tracks near the southeast bank, then apparently becomes a digitigrade trail (the progression is ambiguous, since most of the tracks are indisinct).

Problematic Tracks and Possible Tail Marks on the West Site

Three Y–shaped depressions (U1, U2, U3) near Trail T are interpreted as possible tail marks. Each is situated about a half–meter from the midline of the trackway. The largest (U1) is about 65 cm across the longest dimension of the mark (Fig. 7.7). The Y–shape may be due to a double contact of a portion of the tail. Another small Y–shaped mark (U4) occurs in line with indistinct digitigrade tracks east of Trail VIW. Near U4 is an essentially straight elongate depression about 1 m long, and 8–13 cm wide. An apparent mud push–up on the wider end of the depression suggests that it may be a foot slide, but, since no digit marks are visible, it alternatively may be a tail mark. An unusually large trace of unknown origin is situated between tracks T0 and IIT3; this is an oblong, shallow, smooth–bottomed depression about 80 cm wide and 3 m long.

The Baugh/McFall Sites

The Baugh/McFall sites are a series of track exposures along a limestone ledge on the south bank of the Paluxy River, bordering the McFall property. The track surface is a coarse, friable limestone containing elongate and nonelongate tridactyl tracks, as well as some elongate marks of uncertain origin.

The far western end of the ledge constitutes the "Original McFall Site," a narrow exposure situated about a meter above the normal level of the river. The site features

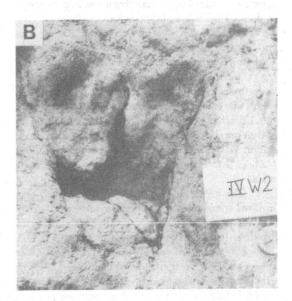

Figure 7.6. **A,** track IVW2, a mud–collapsed metatarsal dinosaur track with a somewhat humanlike shape, which occurred in the same trail with more obviously dinosaurian tracks, such as IVW2. **B,** track IVW2.

Figure 7.7. Depression #U1, a possible tail mark on the A1 West Side (see Figure 7.2 for position on site).

two long trails that contain both elongate and nonelongate tracks. The site was first studied about 1970 by a number of creationists, some of whom described the elongate tracks there as humanlike (Taylor 1971), although other creationists considered them dinosaurian (Neufeld 1975, Morris 1980). The tracks are eroded and indistinct, but most show indications of dinosaurian digits. The elongate tracks show apparent metatarsal impressions that are as deep, or almost as deep, as the ball and digit regions.

Other subsites along the Baugh/McFall Ledge have been excavated since 1982 by teams of creationists led by Carl Baugh, who claims to have found over 50 "man tracks" there. These sites were reviewed by a team of four mainstream scientists (hereafter referred to as the "C/E team") who refuted the "man track" claims (Cole and Godfrey 1985), but offered a few questionable interpretations of their own (discussed below). I began studying the Baugh sites in 1982, and in subsequent collaboration with Ron Hastings (a C/E team member) constructed detailed site maps.

The alleged human tracks on the Baugh sites involved several phenomena. Some were posterior extensions on definite dinosaur tracks, which Baugh interpreted as human tracks overlapping dinosaur tracks (Figs. 7.8, 7.9B). The extensions vary in length, but generally are smaller and narrower than those on the A1 West Site. The extensions were interpreted by the C/E team as hallux marks, but the blunt ends and direct posterior positions suggest they are more likely partial metatarsal impressions. Other elongate tracks occur on other areas of the ledge; most showed longer and more robust metatarsal segments (Fig. 7.10).

Other alleged "man tracks" on the Baugh sites included shallow, indistinct elongate marks near, but not in line with, dinosaur trackways. These marks, which are often slightly curved, are generally situated about 0.5 m

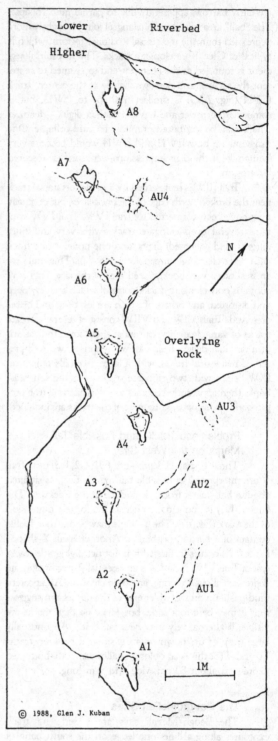

© 1988, Glen J. Kuban

Figure 7.8. Eastern section of the Baugh/McFall ledge, containing a trail of tridactyl tracks (A1–A6) with posterior extensions (partial metatarsal impressions?), and some indistinct elongate marks (AU1–AU4) that may be impressions of the dinosaur's tail or other body part.

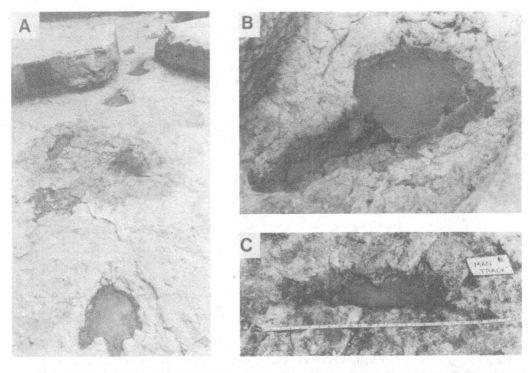

Figure 7.9. **A,** trail A on the Baugh/McFall ledge (compare to map, Figure 7.10). The oblong mark below the black plaque at the upper left is AU4, shown close-up in Figure 7.9C. **B,** track A5 from the same site. The posterior extension on this track was longer than most in the trail, but smaller than the metatarsal segments on many other elongate tracks in Glen Rose. **C,** AU-4, Creationist excavator Carl Baugh named this mark "Humanus bauanthropus," for the supposed humanoid that made it; but it shows no clear human features, and is not in a striding sequence. It may be an impression of the dinosaur's tail, toe, manus, or snout. Tape measure at bottom is marked in inches.

from the trackway midline (Figs. 7.8, 7.9A,C). Some were interpreted by the C/E team as tridactyl footprints whose side digits were poorly preserved, but the marks are not in stride with other tracks, and show no evidence of side digits. It seems more plausible that the marks were made by another part of the dinosaur's body, such as the tail, or possibly the snout or manus (the latter two possibly during food foraging). It is also possible that some of these marks may be plant impressions (recently a branch was found in such a position), or accidental gouges from excavation machinery.

Also identified by Baugh as "man tracks" were some vague, shallow depressions that appeared to be natural irregularities or erosional features on the rock surface. Some *Thalassinoides* burrows also were claimed by Baugh to be human "toes," and one abnormally–shaped "giant track" (over 60 cm long) was merely a carving in the firm marl that overlies the track surface.

Elongate Depressions in Dinosaur Valley State Park

Dinosaur Valley State Park is well known for its abundant tridactyl footprints and spectacular sauropod tracks. Not many elongate tracks occur in the park (Fig. 7.11A), but some indistinct ones usually under water occur

on the east side of the park, and may relate to claims of manlike tracks in this area (Taylor 1971, Morris 1980). A few other elongate tracks occur on the west side of the park, near the Blue Hole area. These also are usually near water; they are deeply impressed and show severe mud collapse (Fig. 7.11B).

Tail marks are rare in the park, but two possible tail marks occur near the "main" tourist area in the northwest portion of the park. One of these marks is a pronounced, curved impression about 1.5 m long and 10–15 cm wide (Fig. 7.12A). Overhanging the depression is what appears to be a ridge of mud pushed up by the tail, which partially slumped back into the depression. Along the bottom of the impression are shallow, parallel striations. Several tridactyl tracks and one sauropod manus/pes set are nearby, but it is difficult to determine which trackway, if any, is associated with the curved mark. Another possible trail mark (Fig. 7.12B) which occurs nearby is fairly straight, about 3.7 m long, and is straddled by two bipedal trackways. Overlapping the apparent tail drag is a footprint of one of these dinosaurs as well as a print of a third individual. Some shallow elongate grooves behind some sauropod pes tracks near Roland Bird's quarry site were interpreted by Fields (1980) as tail drags, but they are indistinct, and may be toe drags or river scours.

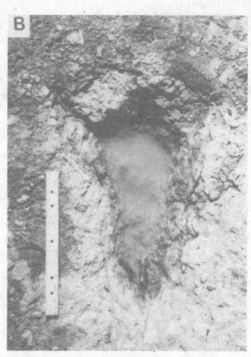

Figure 7.10. **A,** a trail on the Baugh/McFall ledge including some tracks with robust metatarsal impressions, and others with little or no metatarsal impression. **B,** close up of the elongate track at the upper left of Figure 7.10A. Note the indistinct digit marks and slightly humanlike shape. This particular track was not claimed to be human (since the preceding tracks were obviously dinosaurian), but similarly shaped tracks elsewhere on the ledge were called "man tracks" by some. Ruler at left is 30.5 cm (12 in.).

Figure 7.11. **A,** indistinct elongate tracks (partially mud–collapsed and eroded) on the east side of Dinosaur Valley State Park, near the Denio Branch. These tracks are usually under water. Nearby are additional tracks with varying degrees of elongation and clarity. **B,** severely mud–collapsed elongate track (under shallow water) on the west side of Dinosaur Valley State Park, just south of the Blue Hole.

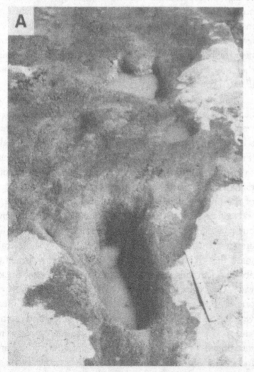

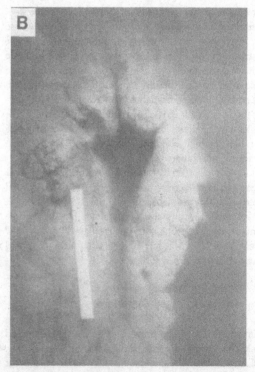

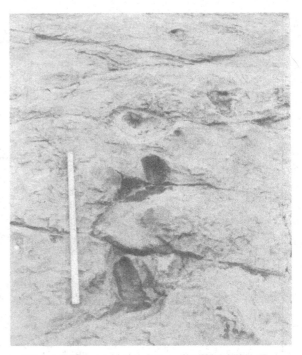

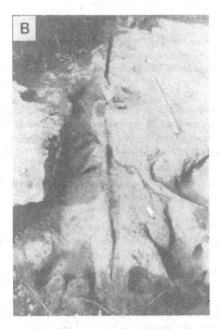

Figure 7.13. A portion of the Shelf Site in Dinosaur Valley State Park, showing erosional features (highlighted with water) claimed by some to be "man tracks."

Figure 7.12. Possible tail impressions at the "Main" Site in Dinosaur Valley State Park. **A,** a deep, curved depression situated among several tridactyl dinosaur footprints (one track is visible at the upper right). The overhanging lip of rock (mud pushed up by the tail?) covered the entire depression several years ago (when the mark appeared as a narrow slit) and gradually broke away in succeeding years, revealing shallow parallel striations on the bottom of the depression. **B,** a long, narrow depression (tail drag?) straddled by two bipedal dinosaur trails (note tracks visible at bottom of picture). A sauropod trail crosses the elongate mark at a right angle; part of a sauropod track is visible at the far right.

The "Shelf Site" in Dinosaur Valley State Park, situated above the "main track layer," was often claimed by creationists to contain many "man tracks." However, all the depressions there appear to be erosional markings, involving river scouring, karst dissolution, and weathering. None exhibit clear human features (Fig. 7.13).

The Taylor Site

For many years the Taylor Site (Figs. 7.14–7.16) was one of the most celebrated "man track" sites among creationists, since some authors claimed that at least four human trackways (named Taylor, Giant Run, Turnage, and Ryals Trails) occurred on the site (Taylor 1971, Morris 1980). Part of the site was originally excavated by Rev. Stanley Taylor and crew between 1968 and 1970. Taylor's subsequent film *Footprints in Stone* (Taylor 1973) helped popularize the "man track" claims. During the 1970's other creationists re-exposed the site, and most affirmed that the elongate tracks were human or humanlike (Beierle 1977, Fields 1980). However, other creationists disagreed and considered the tracks to be dinosaurian; they speculated that the elongate shapes might be due to erosion (Neufeld 1975).

Until the 1980's little study of the Taylor Site was made by noncreationists, apparently due in part to inaccessibility (the site is usually under water), and reluctance to treat the "man track" claims seriously. Some authors suggested the "man tracks" were erosion marks or carvings; others attributed them to single digit impressions of bipedal dinosaurs, or mud-collapsed typical (digitigrade)

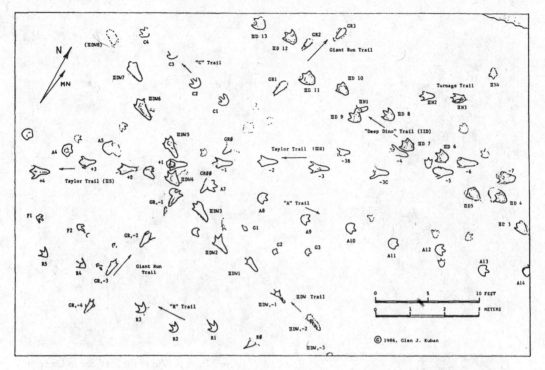

Figure 7.14. The Taylor Site (main section), based on fieldwork between 1980 and 1985. Track outlines indicate boundaries of color distinctions and/or relief differences from the surrounding substrate. The north bank of the Paluxy River occurs just past top of map. Additional tracks occur outside the map borders, including the Ryals Trail (Fig. 7.15).

dinosaur tracks. I have been studying the Taylor Site since 1980, and have concluded that the "man tracks" and other elongate tracks there are plantigrade dinosaur tracks.

The most renowned "man" trail on the site is the Taylor (IIS) Trail, containing 15 known tracks, most of which are large and elongate. The gait pattern is irregular, including some short, wide paces, and other long, narrow paces; most other trails on the site show more regular gait patterns (Figs. 7.14, 7.16). Most tracks in the Taylor Trail (Fig. 7.16) show a prominent metatarsal segment, slight mud push–ups along the sides, and a shallow anterior end with indications of a tridactyl pattern (often accentuated by color distinctions, explained below).

Other elongate tracks on the site typically show indistinct digit impressions; however, slight depressions and/or coloration features indicate dinosaurian digits on at least some tracks in each trail. The metatarsal segments are generally broad and long, with rounded heels. The ball and metatarsal sections of the track are often slightly deeper than the digit region, but most are shallow even at the deeper parts (2–5 cm in most cases). It was the basically oblong shape of the ball and metatarsal region that was often focused on as manlike, although this portion alone is typically over 35 cm long. Except for the IID Trail (comprised of deep digitigrade tracks), most nonelongate tracks on the site are also relatively shallow (some even show slight positive relief). Evidently the shallowness of the tracks, and the color distinctions mentioned above, are largely due to

a secondary sediment infilling of the original track depressions. In recent years many previously unknown tracks have been documented on the Taylor Site by virtue of the color distinctions, including the extension of the IIDW Trail, containing over 20 metatarsal tracks in sequence. The coloration phenomenon is discussed further in Chapter 50 in this volume.

Elongate Tracks in Other Areas

In the Connecticut Valley of New England, tridactyl tracks of the ichnogenus *Anomoepus*, which show metatarsal impressions (Fig. 7.17A), were described over a century ago (Hitchcock 1848). Hitchcock was uncertain as to what type of animal made the tracks (dinosaurs being little known at that time), and speculated that they may have been made by a froglike or kangaroolike creature. Later work clarified their dinosaurian origin and elucidated the locomotor behavior of the trackmaker (Lull 1953, Olsen 1986). Metatarsal *Anomoepus* tracks evidently occur as resting traces only; in striding trackways the animal assumed a digitigrade gait.

Apparent metatarsal dinosaur tracks in striding trails have been found at several sites besides those in Glen Rose; some of these are discussed below, with examples shown in Figure 7.17.

Trails of elongate tracks similar to those in Glen Rose have been observed at a Cretaceous site in Bandera County, Texas, by James Farlow (pers. comm. 1986), and by me and

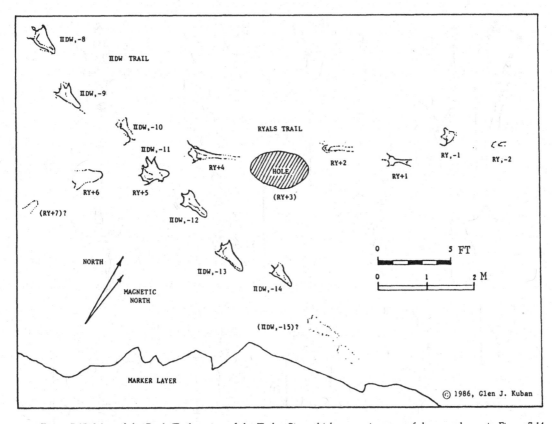

Figure 7.15. Map of the Ryals Trail section of the Taylor Site, which occurs just east of the area shown in Figure 7.14. Intersecting the Ryals (RY) Trail is the IIDW Trail, a long sequence of metatarsal tracks which traverses the entire riverbed, and intersects the Taylor Trail near the north bank (compare Fig. 7.14).

Figure 7.16. **A,** high overhead view of the main section of the Taylor Site, facing southwest (1984). The Taylor (IIS) Trail proceeds from bottom–center to upper left. It is crossed by the IID ("Deep Dino") Trail, proceeding from lower left to upper right. Other trails are visible in background (compare Fig. 7.14). **B,** Taylor Trail Track IIS,–2. Note faint indications of a tridactyl pattern at the anterior end. This photograph was taken in 1980, at which time the substrate was very dry, reducing the color contrasts later observed on many of the Taylor Site tracks. **C,** Taylor Trail Track IIS,+3. This track showed the unusually rounded anterior end (possibly due to the way the mud was pushed up by the middle digit), but still showed indications of a tridactyl pattern. This photograph was taken in 1984, when the color distinctions, ranging from blue–gray to rust–brown, were more vivid than in 1980, but less vivid than in 1985 (see paper on color distinctions, Chapter 50).

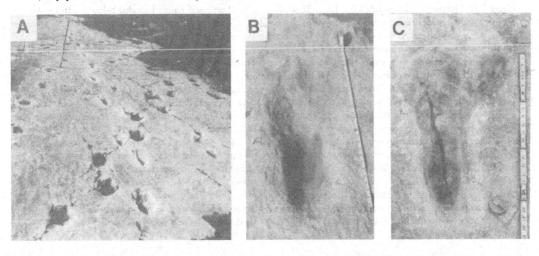

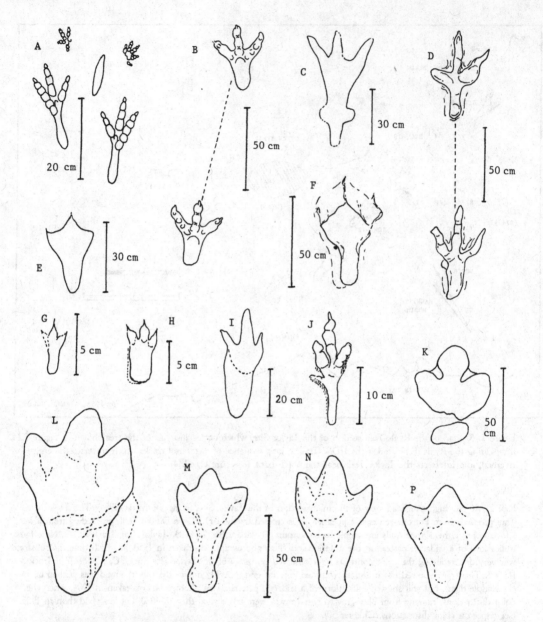

Figure 7.17. Elongate tracks outside Glen Rose, Texas. Note: the ichnological names included below are those assigned by the author(s) mentioned; no judgment is made concerning the validity of these track names. **A,** *Anomoepus scambus* tracks, made by a squatting dinosaur, from the Connecticut Valley, Lower Jurassic (redrawn from Lull 1953). **B,** *Moraesichnium barberenae*, from an Upper Jurassic/Early Cretaceous site in Paraiba, Brazil. Redrawn from Leonardi (1979). **C,** one of a series of elongate tracks from the Late Cretaceous of Spain, attributed by R. Brancas to *Ornitholestes*. Redrawn from Brancas (1979). **D,** trackway from the mid-Jurassic of Morocco. Redrawn from drawing by Shinobu Ishigaki (pers. comm. James Farlow 1986). **E.** one of a sequence of elongate tracks from Clayton Lake State Park (Lower Cretaceous), New Mexico, attributed by Gillette and Thomas (1985) to a web-footed theropod. **F,** one of a series of indistinct elongate tracks from the Schmidt Site at Hondo Creek, Bandera, Texas. Redrawn from photograph by James O. Farlow (pers. comm. 1986). **G,** one of a trail of small elongate tracks from Essex County, New Jersey, pictured in the *Audubon Guide to North American Fossils* (Thompson 1982). **H,** *Skartopus australis* (metapodial form), from the Winton Formation (mid-Cretaceous) of Australia. Redrawn from photograph in Thulborn and Wade (1984). **I,** *Dinosauriens theropodes*, from a site at Agadir, Morocco. Redrawn from Ambroggi and Lapparent (1954). **J,** *Jialinapus yuechiensis* from Yuechi, Sichuan (Upper Jurassic), redrawn from Zhen et al. (1983). **K,** unnamed, isolated track attributed to a hadrosaur. Redrawn from Langston (1960). The elliptical posterior depression was interpreted by Langston as a supporting pad under the metatarsus. **L,** *Dinosauropodes magrawii*, a huge elongate track from a Cretaceous coal mine in Utah. After Strevell (1940). **M–P,** other *Dinosauropodes* tracks from coal mines in Utah, after Strevell (1940).

others at another Cretaceous site in Comal County, Texas, although the tracks at both sites are indistinct (Fig. 7.17F).

Many large, blunt–toed tracks with apparent metatarsal impressions have been found in Cretaceous coal mine roofs in Utah (Fig. 7.17L–P); one such track was over 130 cm long (Strevell 1940). At the opposite end of the size spectrum, Thulborn and Wade (1984) described (5–10 cm long) tridactyl tracks of the ichnogenus *Skartopus* at a mid–Cretaceous Australian site and indicated that a few *Skartopus* tracks possessed what appeared to be metapodial impressions (Fig. 7.17H). Thulborn and Wade noted that metapodial *Skartopus* tracks usually occur as isolated prints, although one series of three such tracks was found.

Several trails of elongate dinosaur tracks occur at a Lower Cretaceous site at Clayton Lake State Park, New Mexico (Gillette and Thomas 1985). Gillette and Thomas interpreted some of these tracks as webbed–toed theropod tracks with metatarsal pads (Fig. 7.17E), and others as ornithopod tracks whose elongation was due to slippage in the mud.

Striding trails of tridactyl tracks with widely splayed digits and posterior extensions (Fig. 7.17B) were reported from sites in South America by Leonardi (1979), who tentatively interpreted them as plantigrade ornithopod tracks, although I consider them theropod tracks. Similar tracks (Fig. 7.17D) have been reported by Shinobu Ishigaki at a Jurassic site in Morocco (James Farlow pers. comm. 1986).

Interesting elongate tracks were reported from a Late Cretaceous site in Spain (Brancas et al. 1979). Near the track heels are curious lateral protrusions that may be "ankle" impressions (Fig. 7.17C).

The Plantigrade Interpretation vs. Alternate Explanations

Traditionally bipedal dinosaurs have been viewed as strict digitigrade walkers, since the vast majority of known tridactyl dinosaur tracks are unquestionably digitigrade, and the anatomy of most bipedal dinosaurs appears well suited to digitigrade locomotion. Perhaps for these reasons there has been a tendency to ascribe most elongate tracks, especially in Glen Rose, to phenomena other than metatarsal impressions, such as foot slides, erosion marks, hallux marks, or middle digit impressions. Although some elongate tracks may be due to these phenomena, and others are too indistinct to diagnose, there is much evidence that many elongate tracks are actually metatarsal footprints.

In many cases the length, shape, and position of the long posterior segment of elongate tracks rules out a hallux interpretation. A simple foot–slide explanation is also incompatible with the shape and details of many elongate prints, especially those with "pinched–in" centers. Further, although foot slides do occur, they generally show significant distortions, such as a mud "pile–up" at the anterior; whereas most mud push–ups, where present on elongate tracks in Glen Rose, usually are more pronounced at the sides than the front of the track. The above features also confirm that most elongate tracks in Glen Rose are not

simply eroded digitigrade tracks, especially since many are oriented contrary to river flow.

The idea that the posterior extensions may have been a thick "metatarsal pad" supporting an essentially digitigrade foot, as in elephants (or as suggested for an isolated hadrosaur track by Langston 1960), is inconsistent with the length and narrowness of many of the posterior extensions.

Roland Bird, who examined a limited number of elongate dinosaur tracks during his well–known sauropod studies in Glen Rose, proposed that such tracks may have been made when a dinosaur pressed its toes together upon withdrawing its foot from soft mud (Bird 1985). However, this would suggest considerable distortion within the track, especially at the back, which is inconsistent with appearance of the better–preserved elongate tracks, especially those which show a narrowing in the "arch" region, and a rounded tarsal element at the posterior.

A popular hypothesis in the past for elongate tracks in Glen Rose was that they were middle digit impressions. This concept has several variations, one of which holds that only the middle digit might register if the substrate were firm, or if the track were an overtrack or undertrack. However, on most elongate tracks one can see at least some indications of all three digits, and these digit marks are consistently located at the front of the track, not emanating from the back and sides as would be the case if the body of the print were a middle digit. Further, on well–preserved Glen Rose tracks the middle digit is about the same depth as the side digits.

Another variation of the middle–digit concept holds that in soft mud the side digits of a digitigrade track might collapse, leaving only the middle digit. However, this is contradicted by the same observations as those described above, as well as the mechanics of mud flow. When moist mud slumps back into a depression, it generally does so about equally from all sides, and therefore would not likely bury the side digits while leaving the middle digit intact (Fig. 7.18A). However, by the same principle, mud collapse would be likely to convert an already elongate (metatarsal) dinosaur track into an oblong — even "manlike" shape, since the relatively large metatarsal segment would be less completely buried than the smaller digit impressions, leaving a vague oblong depression (Fig. 7.18B). Such marks may be further subdued by erosion or other factors (Fig. 7.18C), fostering a "manlike" shape.

Although the above evidences establish that many elongate tracks are metatarsal tracks (recording at least part of the metatarsus), whether they are also "plantigrade" tracks raises a question of terminology. Although some workers may define a "plantigrade" track as one that records any of the metatarsus, others (including me) may wish to restrict the term "plantigrade" to tracks exhibiting complete or nearly complete metatarsal impressions oriented in a largely horizontal manner, or require that several such tracks occur in succession. In any case, some trails in Glen Rose and elsewhere do show several full metatarsal tracks

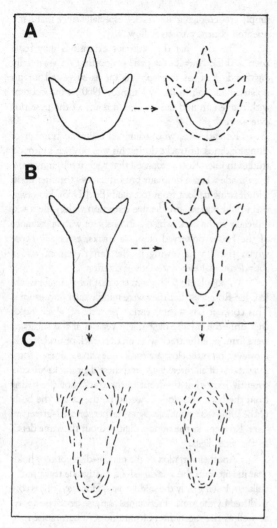

Figure 7.18. **A,** a typical tridactyl dinosaur track (left), and the common shape resulting from mud collapse (right). **B,** a typical metatarsal dinosaur track (left); and the common result of mud collapse: an elongate depression that somewhat resembles a large human footprint (right). **C** (left), the features of a metatarsal dinosaur track also may be subdued by erosion, a firm substrate, ghost impressions (over- and under-tracks), or a combination of factors, resulting in an indistinct elongate depression that roughly resembles a "giant man track." **C** (right), if a metatarsal dinosaur track is first mud collapsed, then less erosion is necessary to foster a humanlike shape, and even the size of the depression may be close to that of a normal human print.

in sequence, and therefore, even by strict definition, may be considered "plantigrade" or at least "quasi–plantigrade," the latter allowing for variations and uncertainties in the behaviors and environmental factors involved (discussed below).

Possible Causes of Metatarsal Impressions by Bipedal Dinosaurs

Although *Anomoepus* tracks record a resting, squatting posture, most other metatarsal tracks were clearly made during active locomotion. One may consider whether sediment consistency may have related to the making of such tracks (at least in some cases). One possibility is that even a foot held in a normal digitigrade fashion might incidentally record some or all of the metatarsus if the foot sank deeply in soft sediment. This may have occurred on some partially–elongate prints that show relatively short metatarsal segments or where the digit depression is much deeper than the metatarsal region. However, most elongate tracks show only a slightly deeper "ball" area, and rarely is there a steep angle toward the anterior end. In fact, some elongate tracks indicate an essentially horizontally impressed metatarsal segment. One might propose that some dinosaurs normally held their metatarsi at a low angle, but many deep digitigrade tracks are found, and indicate metatarsi held at a steep angle. Nevertheless, if the metatarsus were at times positioned at a low angle, a soft substrate would record more of the foot than a firm substrate; and sediment consistency may have contributed in other ways to metatarsal impressions (discussed below).

A soft and/or slippery sediment may have encouraged a dinosaur to lower its metatarsi onto the sediment in order to gain firmer footing. However, if this were a primary cause of metatarsal tracks, one might expect such prints to show relatively short paces, reflecting more cautious steps. In contrast, many elongate tracks in Glen Rose have moderate to long paces and pace angles. A few elongate trackways do show erratic gait patterns that might be construed as evidence that the trackmaker was contending with a slippery substrate, but similar gait patterns can be seen in some nonelongate trackways as well.

Another possibility, which might help explain the typically long paces found between elongate tracks, is that some dinosaurs may at times have traveled in a saltatory or "bouncy" manner, causing the tarsal joint to fold more than usual as the foot contacted the substrate (in a shock–absorbing function), bringing the metatarsus into contact with the sediment.

Perhaps the most plausible explanation, however, is that plantigrade tracks may have been made occasionally by a variety of bipedal dinosaurs whenever they walked low to the ground, which would decrease the angle between the metatarsus and the substrate (Fig. 7.19B). A low posture may have been assumed whenever a dinosaur foraged in mud flats or shallow water for small food items (such as molluscs, insects, crustaceans, amphibians, fish, eggs, or edible plant material); when stalking larger prey, or when approaching other dinosaurs.

That any pathology was involved in causing most metatarsal tracks seems unlikely in view of the relatively efficient stride patterns in most elongate trackways, and the great abundance of metatarsal tracks in some areas, such as Glen Rose.

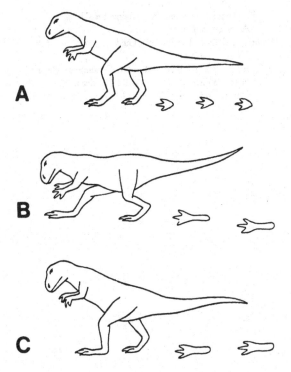

Figure 7.19. **A,** typical bipedal dinosaur walking in a digitigrade mode. **B,C,** possible postures assumed by bipedal dinosaurs during plantigrade locomotion.

It may be that some bipedal dinosaurs may have had leg anatomy more suited to plantigrady than others, although it is unlikely that any were obligatory plantigrade walkers, since most trails with elongate tracks also contain some nonelongate tracks. However, determining with certainty what caused such tracks is difficult, especially since more than one of the factors discussed above (and others) may have acted singly or in combination to foster metatarsal impressions.

Most metatarsal tracks near Glen Rose appear to have been made by moderate sized theropods; however, the variety of elongate tracks in Glen Rose suggests that more than one species in that area made them, and the additional varieties of metatarsal tracks found in other areas further suggests that plantigrade or quasi–plantigrade walking may have been a fairly widespread (though intermittent) behavior among both theropods and ornithopods.

Conclusions

Striding trackways composed partially or largely of elongate footprints suggest that some bipedal dinosaurs, at least at times, walked in a plantigrade or quasi–plantigrade manner. Some alleged "man tracks" in Glen Rose are indistinct metatarsal dinosaur tracks, whose digit impressions are obscured by mud collapse, erosion, or other factors. Other elongate depressions in Glen Rose include erosional features and possible tail marks, some of which also have been mistaken for human tracks.

Acknowledgments

I would like to acknowledge the field assistance of Ron J. Hastings, Tim Bartholomew and Alfred West, and students Dan Hastings, Marco Bonneti, Mike White, Alan Dougherty, Brian Sargent, and Tim Smith. I also would like to thank James O. Farlow for providing helpful comments and literature references.

References

Alexander, R. McN. 1976. Estimates of speeds of dinosaurs. *Nature* 261:129–130.

Ambroggi, R., and Lapparent, A. F. de. 1954. Les empreintes de pas fossils du maestrichtien d'Agadir. *Morocco Service Geologique* 10:43–57.

Beierle, F. 1977. *Man, Dinosaurs, and History* (Perfect Printing) 67 pp.

Bird, R. T. 1985. *Bones for Barnum Brown: Adventures of a Dinosaur Hunter*. Schreiber V. T. (ed.). (Texas Christian University Press) 225 pp.

Brancas, R., Blaschke, J., and Martinez, J. 1979. Huellas de dinosaurios en Enciso. *Unidad de Cultura de la Excma* 2:74–78.

Cole, J., and Godfrey, L. (eds.). 1985. The Paluxy River footprint mystery — solved. *Creation/evolution* 5 (1). (with articles by Cole, Godfrey, S. Schafersman, and R. Hastings.)

Fields, W. 1980. *Paluxy River Explorations*, revised edition. (Privately published) 48 pp.

Gillette, D. D., and Thomas, D. A. 1985. Dinosaur tracks in the Dakota Formation (Aptian–Albian) at Clayton Lake State Park, Union County, New Mexico. *New Mexico Geol. Soc. Guidebook, 36th Field Conf., Santa Rosa.* pp. 283–288.

Hitchcock, E. 1848. The fossil footmarks of the United Sates and the animals that made them. *Trans. Amer. Acad. Arts Sciences* Vol. 3, 128 pp.

Langston, W. Jr. 1960. A hadrosaurian ichnite. *Canada National Museum Nat. Hist. Paper* 4:1–9.

1979. Lower Cretaceous dinosaur tracks near Glen Rose, Texas. In *Lower Cretaceous Shallow Marine Environments in the Glen Rose Formation.* Amererican Association of Stratigraphic Palynologists. Field Trip Guide 12. Annual Meeting, pp. 39–55.

Leonardi, G. 1979. Nota preliminar sobre seis pistas de dinossauros Ornithischia da bacia do Rio do Peixe, en Sousa, Paraiba, Brasil. *Academia Brasileira de Ciencias* 51 (3):501–516.

Lull, R. S. 1953. Triassic life of the Connecticut Valley. *Connecticut Geol. Nat. Hist. Surv. Bull.* p. 81.

Morris, J. D. 1980. *Tracking Those Incredible Dinosaurs and the People Who Knew Them.* (Creation Life Publishers) 240 pp.

Neufeld, B. 1975. Dinosaur tracks and giant men. *Origins* 2 (10):64–76.

Olsen, P. E. 1986. How did the small Ornithischian trackmaker of *Anomoepus* walk? In Gillette, D. D. (ed.). *First International Symposium on Dinosaur Tracks and Traces, Abstracts with Program.* p. 21.

Zhen S., Li J., and Zhen B. 1983. Dinosaur footprints of Yuechi, Sichuan. *Mem. Beijing Nat. Hist. Museum* 25:1–20.

Strevell, C. N. 1940. *Story of the Strevell Museum.* (Salt Lake City Board of Education) pp. 7–15.

Taylor, S. E. 1971. The mystery tracks in Dinosaur Valley *Bible-Science Newsletter* 9 (4):1–7.

1973. *Footprints in Stone.* (film) (Films for Christ Association)

Thompson, I. 1982. *The Audubon Guide to North American Fossils.* pl. 472.

Thulborn, R. A., and Wade, M. 1984. Dinosaur trackways in the Winton Formation (Mid Cretaceous) of Queensland. *Mem. Queensland Museum* 21 (2):413–517.

8 The Stance of Dinosaurs and the Cossack Dancer Syndrome

MARY WADE

Abstract

The natural history of dinosaurs is a neglected field between an exact science and strongly visual popular material on which tradition holds so strong a grip that 100–year–old misinformation is still controlling what gets into most museum displays and book illustrations. This will continue unless scientists supply to their illustrators the parameters that controlled movement, and see that they are used.

Evidence of skeletal shapes and trackways are in broad agreement and consistent with the locomotory motions of ratites. The mental and physical abilities of dinosaurs lay between those of crocodiles and flightless birds, so that their hunting and escape abilities can be sharply defined. Modern animals make less use of defensive and aggressive armor than stegosaurs and ankylosaurs but African porcupines vigorously back to the attack, and the protective armor of a 20 cm tortoise may sustain little damage from a motor car running over it.

Introduction

The First International Symposium on Dinosaur Tracks and Traces drew together a large number of people who were well aware that almost all bipedal dinosaur trackways, and many quadrupedal trackways, have footprints that overlap a median line from left and right alternately (see Introduction, this volume). Yet when we turn to museum displays and recent illustrated texts on dinosaurs, we find the animals restored with their feet laterally separated, sometimes by a considerable distance, even when, as in a recent popular book, they are accompanied by trackways matched for size and probable kind (Charig 1979, p. 79, *Dilophosaurus/Eubrontes*). This book also has all but one moving bipedal dinosaur 'locomoting' with the head and shoulders held as high as if they had paused to look around, but without the concomitant shortening of the stride that is shown in trackways (e.g., Lockley 1986, pl. 2, top left). That shortening was related to keeping the center of gravity symmetrically between the feet as the stance altered. It is evident that something other than observation and documentation of facts is at work in the scientific mind, or the expression of what we know could not so greatly diverge from the established paleontological evidence.

Early Mounted Dinosaurs

The early attempts to restore dinosaurs were summarized, complete with period illustrations, by Norman (1985). This included the first whole mount of a *Hadrosaurus*, displayed in the Academy of Natural Sciences in Philadelphia in 1868. The knees were no wider spread than the hips, the tibia, fibula and metatarsals all sloped inward, the feet were close to the midline. In this position it was almost possible to fit the *Hadrosaurus* feet to a hadrosaur trackway, although the animal's back sloped upward at a mild angle now considered unusual, if not always incorrect. The thin end of its tail dragged on the ground. The notion that any reptile with a long tail must drag it, lizard–like, is still dying hard, in spite of the rarity of dinosaur tail dragmarks.

The astounding discovery of the many specimens of articulated *Iguanodon* at Bernissart (Dollo 1882–1884) profoundly impressed both scientific and lay thought. Norman (1980) redescribed the Bernissart iguanodons as bipeds, if relatively slight, to quadrupeds, if heavily built. In 1985 he published an 1878 photo of the work of mounting the first *Iguanodon*. It showed a hindquarters mount with a full-scale drawing of the front half, together with biped comparative material of "emu and wallaby". The emu was very badly mounted, sloping up from the rear of the pelvis to the base of the neck, femur at 90° to the ilium, and lower leg in rather a perching stance. This is roughly the leg position used for the dinosaur, but its planned vertebral column replicates the curves of the wallaby. We will probably never know how much the final pose owed to the desire to balance the heavy bones on a strong iron tripod supporting legs and tail. The Bernissart iguanodon mounts had a strong influence on the stance attributed to bipedal dinosaurs for the next hundred years. When workers gradually became conscious of the substantial lack of tail dragmarks with footprints, and the possession of calcified ligaments (or functionally equivalent structures in other groups), the lumbar to anterior tail regions of most dinosaur reconstructions were straightened and placed more nearly horizontal. Few scientists or scientific illustrators have faced the implications for stance of the axially overlapping footprints in bipedal trackways (Fig. 8.1). Even now it has remained customary for scientists to bow to the words of the tripod–maker: "But it will be more stable like this." Once the weight of the tail is transferred to the ground, the bipedal skeleton becomes front heavy and unbalanced specimens in tripod stance have a built–in tendency to fall forward.

The relatively heavy distal end of the pubis, possessed by many and various theropods, was probably thickened as a ballasting mechanism. It would have been as effective for bringing the center of gravity back and down as stones swallowed to ballast the hindquarters of a labyrinthodont (personally collected from shale), or the forequarters of crocodiles (Cott 1961). The gizzard stones of sauropods, however, are described as highly polished (Bird 1985) and were presumably primarily for grinding food, and only secondarily ballast. In prosauropods they may have been essential as ballast, for throughout prosauropod history the pubes retained a primitive broad, gently concave surface

to the main body cavity. A gizzard cannot be at the anterior of the gut, and it is logical to assume that it was placed far enough to the rear to rest on the pubes if the animal reared during feeding. The presence of stones may have been critical for bipedally balancing the relatively long and strongly–built fore–bodies.

The Cossack Dancer Syndrome

The two brilliant and influential illustrators, Charles Knight and Zdeněk Burian, worked during the period when scientists were most deeply imbued with the tripod stance. Since 1938, when Burian painted a pair of beautiful "*Trachodon*" (=*Edmontosaurus*) dithering in front of a *Tyrannosaurus*, dither has been the accepted norm for a duckbilled dinosaur's ability on land (Fig. 8.2). Any animal can get into a position from which it cannot quickly extricate itself, but unarmed and unarmored herbivores living in the wild cannot afford to make a habit of it. This interpretation of the bipedal duckbilled dinosaurs as habitual swimmers and clumsy, spread–legged, sub–erect to erect *walkers* caught on, though there is no reason why an animal thought to swim with its laterally flattened tail, and provided with finned forefeet to steer with, should automatically walk or run poorly. The great hind limbs were not obviously modified for swimming. They are very digitigrade, which usually implies speedy terrestrial locomotion. The swimming ability of hadrosaurs is still under debate.

The Royal Ontario Museum's *Kritosaurus* (Norman 1985, p. 121, as *Hadrosaurus*) shows the anterior of the tail lay above the rami of the ischia. It was not just included in the rear of the body proper, as in many Ornithischia, but so packed in calcified ligaments that it probably functioned like a rearward extension of the pelvis. This packing comes to an abrupt end, by failure to calcify more ligaments, only 5 or 6 vertebrae to the rear of the posterior symphysis of the ischia, approximately the end of the body cavity. The remaining three–quarters of the functional tail was apparently unconfined and normally supple in life. The neural spines, centra and chevron bones mimic the axial skeleton of a fish tail. If the hadrosaur tail could bend only 1° left and right at each vertebral joint, tail tip amplitude of swing would have been over 50°. There was also the possibility of additional power strokes from the hind legs, because the vertebrae bearing the caudofemoralis caudal attachments were stabilized.

Bird (1985, p. 18) shows an iguanodont wading in fairly deep water as a swimming animal. This painting by Paul at the Tyrrell Museum is brilliant, but the animal's hind feet, placed as in a trackway, make a normal walking contact with the substrate. The animal is not so deep that it uses a push and glide technique, though near to it. Swimming is the next gait. The more extreme "*Trachodon*" of Burian's pair has reappeared in many guises, as a direct copy, with other hadrosaur heads, or stripped to a skeleton. In the Dinosaur Tracks and Traces symposium logo, an exceptional trackway with a tail drag, no doubt accurately recorded by Gillette and Thomas (1985), was distorted by

Figure 8.1. Foot placement of a walking allosaur.

Figure 8.2. Sketch of Burian's dithering *"Trachodon"*.

Figure 8.3. The original Cossack dancer. Sketch of Michel's *Tarbosaurus* starting in a hurry from the "dithering duckbill stance".

the logo artist to match the tripod stance of whatever skeletal equivalent of Burian's pose was provided as model. In 1970, Burian painted a closely similar pose for a *Tarbosaurus* (Beneš and Burian 1979). But just how did a dinosaur move in or from that pose? Michel (1979) has obliged with a restoration of the first Cossack dancer — also *Tarbosaurus* (Fig. 8.3). As a reminder that it takes much more effort and deliberate muscular control to 'locomote' in an awkward stance than a good one, I suggest that the state of mind that maintains the tripod pose, and other offenses against observation, should be called "the Cossack dancer syndrome". It may be possible to laugh a good proportion of authors into common sense in the art work they accept. With rare exceptions, mostly in play, no wild creatures move awkwardly or inefficiently in their own milieu.

Facultative Gaits

For slow locomotion, kangaroos resort to plantigrade foot placement, which is rarely encountered in dinosaurs, but is the probable explanation of the long, narrow dinosaur footprints sometimes colloquially miscalled "man tracks" (Kuban, this volume). Unlike the dinosaurs, the kangaroos do place their tails as well as their hands on the ground during both alternating and symmetric slow gaits. The fact that there are changes in style of locomotion between browsing, "puttering about," and "going places" is the main similarity between these and comparative dinosaurian gaits. Happold (1967) vividly illustrated such changes in *Jerboa jerboa*.

Small front footprints look much further apart than the respective, much larger hind prints, but often the weight–bearing foci of the small and large feet are about the same distance from the axial plane (e.g., Bird 1985, pp. 166, 172, 179, 181). Some front feet, however, are significantly further apart (Lockley 1986, Fig. 9, 13A; Norman 1980). Sauropod trackways include evidence of overstepping, hind prints that blot out the front prints (Lockley 1986, Fig. 15 partim). This requires a change of gait to shift both feet on the same side at the same time. The action culminates in an amble, a slightly faster gait than a walk, as it utilizes one weightbearing interval for both feet. As a faster gait it sits well with the shape of sauropod bodies; there is no unsupported interval, the weight the feet usually bear is simply allotted to one foot of each pair more frequently than in walking.

The front feet of sauropods (other than brachiosaurs) carry about 20% of the weight while the percentage carried on an iguanodont hand is negligible (Alexander 1985), so it seems unlikely that ornithopods were ambling when the trackways show only hind prints. Any time locomotion was the main aim, ornithopods and hadrosaurs were likely to have been bipedal, for their hind legs were longer, their protractor muscles were well placed, they could see further, and most of their trackways are bipedal, whether the footprints are small or large. A certain amount of documentation of this is implicitly available from Alexander (1985) but he does not seem at home with the concept of facultative changes in style of locomotion. Even Thulborn (this volume) considered it "curious." Shipman (1986) used without comment a brilliant restoration (by Sibbick) of juvenile biped and adult quadrupedal *Iguanodon bernissartensis*, the adult with the hind feet wider spread than the front feet. It is the antithesis of Norman's data, on which she was commenting favorably, but is more consistent with the evidence provided by Lockley (1986) for quadrupedal ornithopod trackways.

Stance From Tracks

The rounded outlines of Early Cretaceous *Caririchnium* forefoot prints (Leonardi 1984, Lockley 1986) match the mummified Late Cretaceous *Edmontosaurus annectens* forefoot (Lull and Wright 1942, Lapparent and Lavocat 1955). Digit V is short and independent; IV and III occupy a finlike expansion of skin apparently covering connective tissue, for it did not compress very thin, the hoof of toe III breaks the outline of the flange, and an intermediate length of digit II has a small independent flange. The depth of the impression was as likely to decide whether digit II contributed to the track, as variation in morphology. The hind footprints attributed to hadrosaurs are large, often wider than long, with the toes widest at mid–length, rounded at the hooves, and no "heel." The skeletons are very digitigrade. This and narrow pigeon–toed trackways are in complete contrast to the typical "dithering duckbill" stance.

Figure 8.4 was sketched from a museum photograph of their own display, in their generally good leaflet on trackways. The carnosaur skeleton in Figure 8.4 shows that pubes and ischia slope inward and downward from their respective left and right hip–joints to adjoin in pairs. An almost almond–shaped vertical section to the body was thus defined in the pelvic area. If the skeleton was clad in appropriate muscles, the knees would have been closer together than the hips, and the feet, at furthest from the median plane, side by side. The center of gravity would have been in a plane passing between the feet. As the carnosaur moved forward, a slight sideward lean would have carried the center of gravity over the fixed foot (compare Norman 1985, p. 139, bottom right): the tracks show a typically pigeon–toed stance that would bring the advancing foot into the median plane. The momentarily rear foot, advancing in its turn, had to be swung outward to clear the rear of the newly fixed foot, and in again to the median plane. The support of the center of gravity is thus passed from foot to foot with very little lateral shift after start up, and then only about one half a foot's width. This is precisely what a moderately large ornithopod did (Lockley 1986, pl. 2, text–fig. 18). Humans, walking on open ground and turning their heads to look about, rarely walk straight even though they have stereoscopic vision, and dinosaurs, unless moving together, wandered repeatedly off course, perhaps as they turned their heads to do their forward looking with one eye or the other. The Lark Quarry carnosaur wandered rather more at a slow walk than a medium walk, but it was almost certainly slowing as it gauged the movements of intended prey animals (Thulborn and Wade 1984) and

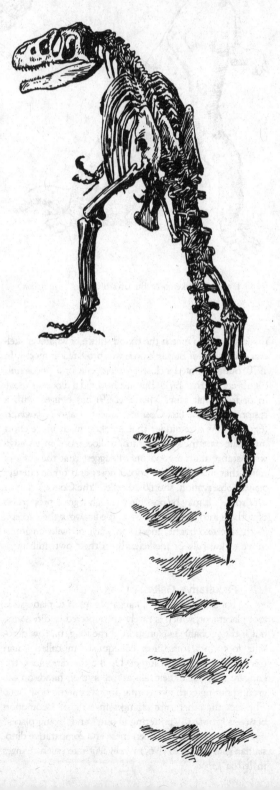

Figure 8.4. The normal, medially overlapping trackway of a carnosaur has had no power here to influence the foot placement of the skeletal mount placed at its end.

anticipating a direction change. At a foot size of 58 cm long x 50 cm wide it had presumably been hunting successfully for a number of years, time enough for any animal to have learned to recognize signs of other animals' intentions. The pressure on bipedal dinosaurs to place their feet strictly in the axial plane is greater for heavier and/or faster animals. At a faster speed the long paces give the advancing foot more distance to reach a median position. At the other extreme, Figure 8.5, *Skartopus*, records the flat–out running of a species whose adult size is tiny. The tracks lie about a footprint's width apart, like a silver gull (*Larus novohollandiae*) pattering steadily along a beach, supporting its light weight from the side without a waddle. The feet of the lightweight coelurosaurs were not restricted to the median plane but there is considerable individual variation in the lateral foot separation (cause of *pace angulation*). The angulation in a trackway depends on the kind of animal, not the size, for the *Wintonopus* individuals in growth stages as small as small *Skartopus* have the same trackway pattern as larger *Wintonopus*.

Locomotion, Defense and Attack

Looking at common modern tetrapods, it is quite noteworthy that plantigrade animals have rather free loco-motory hip–movements, and digitigrade animals tend to stabilize the hip and keep the greater movements to knee and ankle. The bipeds available for close comparison to the digitigrade bipedal dinosaurs are thus restricted to ground birds, particularly ratites, and kangaroos and other hoppers.

Ostriches, emus and kangaroos have relatively short femora and have delegated length to the tibia–fibula and the tarsometatarsus or metatarsals. The result is that the knees spend much of a stride quite high on the body and are nearly as widely spread as the hips, while the feet are brought close to, or into, the median plane of the body by the inward slope of the elongate bones. The various slow–walking emus (Fig. 8.6) show just how markedly tibia-fibula and tarsometatarsus must slope inward to achieve this, and how close to the ground the foot is when it is swung forward. A fine cover–photo of twenty running ostriches, taken by Jen and Des Bartlett, was used on the revised edition of *The Living Planet* (Attenborough 1985). Nine of the twenty are airborne, so that stride length (λ): hip height was just under 4. Whether the tibia is nearly vertical and the tarsometatarsus protracted for touch–down, or even weight–loaded, there is very little change in the position of the knee until the thrust of take–off began. Then the retractor muscles were placed under strain, a power stroke using the resistance of the ground to accelerate the animal's weight, and not until the hind foot was lifted did the femur reach its maximum backswing, vertical in 4 individuals, slightly more in one. The femur returned to its usual forward slope as the tarsometatarsus was protracted, so that it is angled forward for most of a stride, including the first half of the weight–bearing cycle. In other words, much of the support of the center of gravity has passed

Figure 8.5. *Skartopus* trackway, Lower Cenomanian, Winton. The lightweight adult animal had no need to place its feet below the center of gravity. Prints 7–12 of 24; 11 and 12 are elongated by retro-scratching at kickoff.

20 cm

from hip to knee with the minimizing of the tail. As the tarsometatarsus was swung forward, the tarsometaphalangeal joint passed the weight–bearing ankle at the same height above ground; the foot was thus kept well clear of the ground, here very flat. This is the chief difference between the running of these ostriches and a small group of emus I closely observed in the wild. The emus lifted their feet no higher from the ground to run at approximately 25 km/hr than to walk when the ground was bare, but crossing a growth of tough and tangled grass just over 30 cm high, they lifted their feet high enough to avoid entanglement without slackening their speed. They were running with plenty of acceleration available to them, and perhaps their gait would have been more similar to the ostriches' had they been similarly exerted. In running, a ratite body is, and presumably a dinosaur body was, rotated slightly upward at kick–off and more steeply down to land (Fig. 8.7). The very flat parabola is about 5° uptilt and 10° downtilt in the twenty ostriches. These attitudes may be observed in photographs of other creatures with unsupported intervals in their gaits, though in many mammals the back is also curved to add length to the strides (e.g., Archer 1985).

In debating how applicable a ratite stance is as a dinosaur model, it appears that even an "ostrich dinosaur" has a notably unshortened femur, relative to ostrich or emu, and, though the tibias of all are elongate, the metatarsal elongation is a lot less than the tarsometatarsal. The angle of rotation possible to the femur of a moving dinosaur must always have been dominated by the stretch and contraction and angle of attachment possible to the protractor muscles, most commonly the pubofemoralis, as the caudofemoralis is the longer in most dinosaurs. In squatting, the femur is pushed forward by the ground, plus tibia–fibula in the latter part of the movement, and the caudofemoralis must have had the stretch to accommodate the position.

Although the femoral protractors of Ornithischia were attached to the pelvis much higher than in most Theropoda, they could not have been used for pulling the femur nearly horizontal in continuous striding, as is often figured (e.g., Norman 1985, p. 98). At full speed it was essential that the protracted foot reached the ground with the knee fairly straight; otherwise the femur would have had to be lowered again to place the foot on the ground at maximum extension, which allowed the center of gravity to travel at virtually constant height. Bounding is a liga-

Figure 8.7. *Struthiomimus* restored after Russell (1972) and the action of *Struthio* running at approximately λ:h = 4. Most of the elevation of the foot in protraction was achieved by ankle movement. Protraction of the femur was achieved chiefly by the pubo(ischio)-femoralis externus, as the action of the internus has not modified the pelvis like it did in *Dromiceiomimus*. The pubes protruded about 35° from vertical and are the probable limit of protraction during striding in this and *Ornithomimus*.

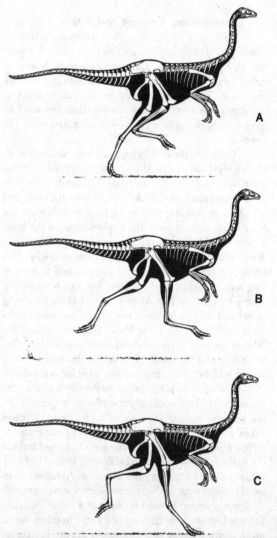

Figure 8.6. Walking emus take their weight on the advanced foot, with the femur oblique, before the unloaded femur droops greatly. The moving leg is rotated out to clear the fixed foot and in again to be placed medially. Drawn from photographs.

mentous action, never demonstrated by appropriate ligament scars on dinosaur limb bones (Russell 1972, Thulborn this volume). The ankle was able to provide most of the clearance for the protracted foot with much less effort than would have been used by heaving the thigh up to lift the foot. It did this by bending as acutely as was needed to lift tarsometatarsus and foot in ratites. Ornithopods could outpace the dangerous theropods of the same hip height and weight, but often only by a few percent per pace (Thulborn 1982, this volume). Measurement of scaled diagrams has accuracy limitations, but the theropod pelvis of *Struthiomimus* and *Ornithomimus* is unspecialized enough to allow the assumption that the pubo(ischio)femoralis externus was still the chief femoral protractor, and the femur can scarcely have been rotated further than the pubes, i.e., about 35° *maximum* forward rotation. *Dromiceiomimus* (Russell 1972) has a proportionately larger antilium than *Struthiomimus*, as does *Ornithomimus*, but a decidedly less protruded pubes than the other two, so the pubo(ischio)femoralis internus was the important femoral protractor, and femoral movements would have approximated those of a cursorial ornithopod. This combination of the proportionately longest legs, cursorially modified musculature, and moderate weight and size probably made it the fastest dinosaur ever, as Russell (1972) concluded with a different emphasis. He visualized a very free rotation of the hip for *Dromiceiomimus*, such as is not seen in any modern digitigrade animal. If the impressive pair of anterior and posterior iliofemoralis muscles which he has figured were mainly for swinging the leg forward and back, one would have to agree with him, but perhaps these muscles were employed in swinging the rear foot outward to clear the back of the front foot and in again. Not only would the outward and inward pull need to be balanced, as the muscle scars are, but the relevant muscles in a human thigh insert approximately where Romer (1923) suggested the tyrannosaurid iliofemoralis did, and suggest this use in theropods — perhaps in all biped dinosaurs. Run well or fight well; the alternative, hide well, is too difficult to recognize in fossils, although it is suspected in lizards and the more terrestrial of the crocodiles.

The ability to attain high speed swiftly is the most common and effective safety device employed by unarmored vertebrate herbivores, and even among fish. To launch successfully from the tripod pose to high speed would have required the muscular control and general aptitude of a Cossack dancer. To start slowly, optional bipeds could have dropped to all fours and drawn the hind legs inward to the plane defined by the lines of action of protractor and retractor muscles and the hipjoint, all of which would have taken time. A wild animal that habitually makes a slow start, yet has no notable protective devices, could only evolve or continue to exist in isolation.

Many people visualize carnivores as the fastest animals: "because they have to be to catch their food," and everyone knows the cheetah can reach 80 kph (but for a very short burst of speed). The ostrich may maintain this speed for 0.8 km (Howell 1944). In modern faunas the carnivores are fastest only for brief spurts, and rely on stalking, trapping or ambush, as apparently did the carnosaurs (Wade 1979, Thulborn and Wade 1984).

The fastest *sustained* speeds in the wild today, and those shown by Thulborn's calculations for dinosaurs, are the speeds of moderately large, unarmored herbivores, i.e., in dinosaurs, unarmored and/or unarmed Ornithischia and the relatively inoffensive ostrich dinosaurs. Thulborn (this volume) pointed out two particular oddities. One, the seemingly different λ:h attainable by the shorter front legs and larger hind legs of sauropods; the other, an even more exaggerated difference in λ:h for hind and fore legs of *Stegosaurus*. Norman (1985) remarked on the relatively short space available to femoral protractor muscles in sauropods (Romer 1923). It is easy to see a marriage of convenience between the short, muscle-limited hindleg paces and the height-limited foreleg paces. The height of the hind legs was not just inherited from bipedal forebears, it was also necessary to swing the long tails of sauropods. A grown sauropod's tail must have swung like a motile telegraph pole, so when considered with the herding evidence, now established (Bird 1941, 1944, 1985; Farlow 1981; Lockley 1986), they can hardly be considered defenseless, whether or not they could rear on the hind legs to attack as suggested by Norman (1985, p. 97) and others. Rearing or striking is not likely to have been a habitual defense, as it would have interfered with the operation of the all-important tail. It is far more likely that a conventional sauropod faced by one attacker would turn to keep the tail to the attacker, and, faced by attack from two directions, could strike with the neck as well as the tail. Mochi and MacClintoch (1973, p. 75) recorded a "thousand pound eland was tossed through the air to land with a broken shoulder" by "a powerful head swing" from a bull giraffe. So a deliberate sauropod swipe would have been formidable (Holland 1915). Until we recognize *Brachiosaurus* footprints unequivocally, we will not know what degree of freedom was possible to those huge front legs. The animals which today strike with a forefoot are very quick movers (e.g., horse, giraffe and some buck). They may not be good models for slow walkers. It would have made a very dramatic intraspecific ploy for display threatening, but inhibited use of the tail and neck to strike with could only be used against one attacker at a time, and at close quarters. The capability to raise or otherwise move the neck freely is assumed here, as there were extensive skeletal muscle and ligament attachments, and such a huge investment of material in construction had to be very useful. The later sauropods had the immense necks, and most probably they had muscular contraction of the arterial walls to shift blood up to the head (Dagg and Foster 1976, App. B); the mechanism is unknown but it is likely that they used their 3-story necks as observation towers and to convey food from the treetops. Alexander (1985) considered the ability of diplodocids to lift the neck with a dorsal muscle or ligament which filled or over-filled the median vertebral notch to be inconclusive; he also doubted the capacity of other muscles in the neck, but their sizes, arrangements and capacities are still unknown. A dorsal ligament was perhaps more likely to function to hold the neck in position once it was lifted. That would have freed the muscles for adjustor movements, as in many mammals.

In the "whiplash" sauropods the anteroposterior extension of the chevron bones formed a second row of bones, gaps alternating with intervertebral joints. It was probably chiefly for attaching ligaments and muscles to "splint" the vertebrae and thus help to carry the still thick mid–tail off the ground; the thin tip seems to have dispensed with chevron bones (Bird 1985, fig. on p. 56). The splinting effect of an alternating row of bones below the mid–tail vertebral joints must have strengthened the tail as a weapon, too. The anterior of the tail had unmodified, nearly transverse, chevron bones which did not reduce flexibility, so that the tail could still be swung hard against any misguided attacker that approched laterally or from the rear. Relatively short–tailed sauropods did not modify the distal chevron bones so much: enough to indicate that they attached homologous ligaments, but not enough to splint the joints. In the nodosaurs and ankylosaurs the chevron bones were extended anteriorly and posteriorly until they met, but did not fuse (Coombs 1978). Dorsally the neural arches completely roofed the neural canal. Thus the tail skeleton became effectively a triple row of bones, to which in ankylosaurs a club was added. In short, unless they were defensively aggressive, like those of stegosaurs and ankylosaurs, tails tended to be stabilized as balancing organs. A fine array of examples of modified tails can be viewed in Norman (1985). Illustrated nodosaurs crouching before carnosaurs look decidedly at risk, but on retrieving a run-over tortoise from a bitumen highway (100 km speed limit), I found its 20 cm shell had very few fractures; they were not displaced and did not hinder its subsequent activity. It had older damage, once more severe, almost fully repaired. Curved bone plates are very strong during life.

Stegosaurus in particular was not designed for escape, but for swivelling, and possibly backing to attack and retreating forward like an aggressive porcupine (Durrell 1964). As with mammalian carnivores, the carnosaurs probably learned young to leave the porcupine equivalents alone. With its head low and out of the way, its powerful shoulders and forelegs braced, and its large hind quarters equipped to step around a larger arc than the forequarters, *Stegosaurus* was adapted to swing its tail to meet any attack. It had no answer, however, to the attack of multiple carnivores. The patchy post–Jurassic occurrences of stegosaurs may reflect advances in the local carnivore hunting techniques or new carnivores. The last known stegosaur was the very Late Cretaceous *Dravidosaurus*, whose forebears presumably lived on the Indian plate during the Late Jurassic when seafloor spreading tore it from Africa. The group became extinct at the end of the Jurassic in North America, where probable pack–hunting theropods are described in Early Cretaceous. The group survived in Europe, Asia and perhaps Africa, at least until the middle of the Cretaceous.

The mainly solitary and frugivorous cassowary is the only hindfoot–fighter somewhat comparable to dromaeo-

saurs, even though its feet are much less specialized. It is still the inner claw, II, here straight, that is enlarged. The outer and axial toes are the more critical for balance. The cassowary leaps to the attack, kicking with one or both feet, and adds the weight of its dropping body to the kick, turning a limited spearthrust into a slash, and also using body to body collision in intraspecific fighting (Crome 1976). It is reputedly capable of defeating and driving off an aggressive feral boar. It is unlikely that dromaeosaurs would attack any but small prey without leaping for height, whether or not they closed with the ferocity of the pack illustrated by Sibbick in Norman (1985). The functional replacement of pubofemoralis with iliofemoralis dictates that the femur was operated wholly in advance of vertical, since 90° to a tangent at the dorsal edge of the acetabulum was the position of maximum stretch on the iliofemoralis. The forward inclination of the femur is complemented by the re-arrangements of balance indicated by the relatively heavy forequarters and the rearward twist of the pubes, which extended the adjacent end of the belly toward the rear. The Norman–Sibbick restorations, and some others, place the femur correctly.

Older illustrations that stem from Ostrom (1969) and Bakker's figure (1968) show *Deinonychus* standing on one foot to kick, and also show their heads and arms drawn back so that the kick can outreach them, else the heads and arms both outreach the feet. It thus is possible that *Deinonychus* attacks customarily showed a feline commitment of everything they had. At the same time, if Ostrom (1969) was correct in regarding the remains of three *Deinonychus* and *Tenontosaurus* as possible evidence of a pack kill, it must have been an attack on an unusually formidable prey. This loss rate is more than a pack can sustain for average kills. The supposition that carnivores habitually attacked the largest possible animals is suspect.

Although reptiles replace teeth throughout life, growth takes time, and young carnosaurs probably learned early to take no chances with their teeth. Favorite prey-sizes for mainline carnosaurs were probably small enough to be killed by a quick snap, hoist, and shake, like crocodiles often use to kill and to pulp the prey's bones before swallowing in one piece. Helpless prey of large size were doubtless eaten too. There is a continuous intergradation from fresh carcass to carrion, and the behavior of modern large carnivorous reptiles suggests that carnivorous dinosaurs would not be very particular about requiring freshly killed food.

Apart from the double crest that gave *Dilophosaurus* its name, it differs from all later carnosaurs in a relatively weak inner end to the premaxilla (where the nostril is so close to the mouth as to cause a diastema in the tooth row), and a relatively massive maxilla and maxillary teeth. The contrast is so great that it may have competed only for smallish live prey and carrion. Late Jurassic carnosaurs had all eliminated the weakness at the rear of the premaxilla. They were as ready for grasping larger prey as any crocodile and no doubt as cunning at stalking prey, though their habitats differed. A 4-m crocodile known to the author, although well fed, has a practice of stalking the man responsible for periodically cutting the vegetation in its spacious pen. Stalking and trapping faster prey than themselves is not beyond crocodiles and cannot be considered beyond the physical or mental ability of carnosaurs. Yet that supposition has been suggested by a number of scientists who have proposed alternative explanations to the Lark Quarry interpretation of active stalking.

The Dinosaur Heresies by Bakker (1986) was published while this paper was under review. Several of Bakker's observations require comment. *Megalania*, the Komodo dragon-like giant varanid, redescribed by Hecht (1975), reached a maximum length of 7 m and a corresponding weight of 600–620 kg. Bakker rounds out the weight to "half-ton" in his text, but draws an animal with legs approximately as long as its body (over 2 m) with the caption "Two-ton dragon lizard of ancient Australia. Fifteen feet long and as heavy as a bull rhinoceros", etc. This evocative journalese does not inspire confidence in his handling of less well-known fauna, *pace* Reid (1987).

Bakker argues strongly against the swimming ability of hadrosaurs, regarding the tail as virtually inflexible, and the finned forefoot as misinterpreted. He also emphasizes the swimming ability of emus which have no tail worth the name. As suggested above, I consider that he understates the freedom and size of that portion of the hadrosaur tail that is not included in the body proper. The illustrated contortions necessary to run hadrosaurs quadrupedally (Bakker 1986, p. 43; Paul in Bird 1985, p. 21) and the lack of quadrupedal trackways suggest to the impartial viewer that the animals ran bipedally.

Bakker's running *Ceratosaurus* legs (1986, p. 216) have the rear leg at maximum retraction (compare Fig. 8.7B herein), but the leading foot is weight-bearing, further into its cycle than as shown in Figure 8.7C. The *Ceratosaurus* is thus running much below λ = 4. This illustrates a slow to medium trot or jog. Thus Bakker's most detailed illustration of carnosaur locomotor ability tends to agree more with the speeds suggested by Thulborn (1982, this volume) than with Bakker's more lively illustrations, 6 or 8 of which have the femurs protracted ahead of the pubes.

The bounce and agility that Bakker attributes to carnosaurs in locomotion and their large, powerful limb joints were certainly needed to handle 2–7 tons on one foot at every pace. While sustained speed above a trot or jog seems unlikely based on known stride lengths, cadence could have been built up over 10 to 20 or 30 paces, even though it is very energy-expensive for long legs holding great weight. A carnosaur charge of 50–150 m is entirely possible when the 30+ m charges of large crocodiles are considered, but maximum attainable speed is not maximum sustainable speed. For that, the needs were good protractor muscle placement and/or elongate bones, found in varied proportions in Ornithischia, ostrich dinosaurs and other lightweight theropods.

Conclusions

Many of the problems in understanding the movements and general abilities of dinosaurs are due to failure to regard them as functional beings which had lives to keep and a living to make in the wild.

The main reason that there are so many bad poses seen in books and displays today is because scientists have not taken the time to insist on display artists using the known parameters. These are the structures and movements that have been preserved as skeletons or trackways, or that can be inferred from: (i) marks on the skeleton, muscle attachments, cranial casts, etc.; (ii) a heritage as crocodile/bird relatives; and (iii) accepting known controls (e.g., not protracting femora beyond the girdle attachment of the protractor muscles, or retracting them so far as to reverse the protractor function). A biped animal illustrated or mounted with both feet on the ground cannot imply a faster speed than a walk, and foot separation should not exceed hip height; preferably it is much less.

Whether the illustrations are impressions, sketches, finished paintings, or sculpture, only when we illustrate or supervise thoroughly will our specialized knowledge become generally available. From any implied criticism I exempt those artists who have delved into scientific literature in search of enough data to perfect their work.

Acknowledgments

I would like to thank all the hardworking organizers of the first Dinosaur Tracks and Traces Symposium for bringing the good idea of the meeting to fruition.

I owe thanks to Ralph Molnar and Tony Thulborn for encouragement, literature, and a modicum of criticism, which necessities have greatly extended my understanding of these fascinating animals. Laurie Beirne provided Figures 8.1 and 8.7, and the sketches of photographs and other illustrations.

References

Alexander, R. McN. 1985. Mechanics of posture and gait of some large dinosaurs. *Zool. Jour. Linn. Soc.* 83:1–25.

Archer, M. (with Flannery, T., and Grigg, G.). 1985. *The Kangaroo* (Australia: Weldons Pty. Ltd.) 263 pp.

Attenborough, D. 1985. *The Living Planet*, revised edition.

Bakker, R. T. 1968. The superiority of the dinosaurs. *Discovery* 3 (2):11–22.

1986. *The Dinosaur Heresies*. (New York: William Morrow & Co.) 482 pp.

Beneš, J., and Burian, Z. 1979. *Prehistoric Animals and Plants*. (Prague: Artia, Hamlyn)

Bird, R. T. 1941. A dinosaur walks into the museum. *Nat. Hist.* 47:74–81.

1944. Did *Brontosaurus* ever walk on land? *Nat. Hist.* 47:61–67.

1985. *Bones for Barnum Brown*. (Fort Worth: Texas Christian University Press) 225 pp.

Charig, A. 1979. *A New Look at the Dinosaurs*. (London: Heineman).

Coombs, W. P. 1978. The families of the ornithischian dinosaur order Ankylosauria. *Palaeontology* 21 (1):143–170.

Cott, H. B. 1961. Scientific results of an inquiry into the ecology and economic status of the Nile crocodile (*Crocodilus nilotica*) in Uganda and northern Rhodesia. *Trans. Zool. Soc. London* 39:212–356.

Crome, F. H. J. 1976. Some observations on the biology of the cassowary in northern Queensland. *Emu* 76 (1):8–14.

Dagg, A. I., and Foster, J. B. 1976. *The Giraffe: Its Biology, Behaviour and Ecology.* (Van Nostrand Reinhold Co.) 210 pp.

Dollo, L. 1882–1884. Première a cinquième note sur les Dinosauriens de Bernissart. *Bull. Mus. Hist. Nat. Belgique* I–III.

Durrell, G. 1964. *Menagarie Manor.* (Rupert Hart–Davies).

Farlow, J. O. 1981. Estimates of dinosaur speeds from a new trackway site in Texas. *Nature* 294:747–748.

Gillette, D. D., and Thomas, D. A. 1985. Dinosaur tracks n the Dakota Formation (Aptian–Albian) of Clayton Lake State Park, Union County, New Mexico. *In New Mexico Geol. Soc. Guidebook 36th Field Conf. Santa Rosa* pp. 283–288.

Happold, D. C. D. 1967. Biology of the jerboa, *Jaculus jaculus butleri* (Rodentia, Diplodidae), in the Sudan. *Jour. Zool.* (Proceedings of the Zoological Society, London) 151:257–275.

Hecht, M. K. 1975. The morphology and relationshps of the largest known terrestrial lizard, *Megalania prisca* Owen, from the Pleistocene of Australia. *Roy. Soc. Victoria Proc.* 87 (2):239–250.

Howell, A. B. 1944. *Speed in Animals.* (Chicago: University of Chicago Press).

Holland, W. J. 1915. Heads and tails: a few notes relating to the structure of the sauropod dinosaurs. *Ann. Carnegie Museum* 9 (3–4):272–278.

Lapparent, A. F. de, and Lavocat, R. 1955. Dinosauriens. *In* Piveteau, J. (ed.). *Traité de paléontologie. Amphibiens Reptiles Oiseaux.* 5. Maisson et Cie, 1113 pp.

Leonardi, G. 1984. Le impronte fossili di dinosauri. *In* Runzoni, Graggio and Guart (eds.). *Tulle orme dei Dinosauri* (Venice) pp. 165–186.

Lockley, M. G. 1986. *Dinosaur Tracksites.* University of Colorado Denver, Geology Department Magazine, Special Issue No. 1, 56 pp.

Lull, R. S., and Wright, N. E. 1942. Hadrosaurian dinosaurs of North America. *Geol. Soc. Amer. Spec. Paper* 40. 242 pp.

Michel, G. *In* Gilbert, J. 1979. *Dinosaurs Discovered.* (Hamlyn) 93 pp.

Mochi, U., and MacClintoch, D. 1973. *A Natural History of Giraffes.* (New York: Charles Scribners Sons) 134 pp.

Norman, D. B. 1980. On the ornithischian dinosaur *Iguanodon bernissartensis* of Bernissart, Belgium. *Memoirs de l'Institut Royal des Sciences Naturelles de Belgique* 178:1–103.

———. 1985. *The Illustrated Encyclopedia of Dinosaurs.* (Sydney: Hodder & Stroughton) 208 pp.

Ostrom, J. H. 1969. Osteology of *Deinonychus antirrhopus*, an unusual theropod from the Lower Cretaceous of Montana. *Peabody Museum Nat. Hist. Yale Univ. Bull.* 30.

Reid, R. E. H. 1987. The dinosaur heresies by R. T. Bakker. *Modern Geol.* 11:271–280.

Romer, A. S. 1923. The pelvic musculature of saurischian dinosaurs. *Bull. Amer. Museum Nat. Hist.* 48:605–517.

Russell, D. E. 1972. Ostrich dinosaurs from the Late Cretaceous of western Canada. *Canadian Jour. Earth Sci.* 9 (4):375–402.

Shipman, P. 1986. How a 125–million–year–old dinosaur evolved in 160 years. *Discover, British Museum (N. H.)* Oct. 1986:94–102.

Thulborn, R. A. 1982. Speeds and gaits of dinosaurs. *Palaeogeog., Palaeoclimat., Palaeoecol.* 38:227–256.

Thulborn, R. A., and Wade, M. 1984. Dinosaur trackways in the Winton Formation (Mid–Cretaceous) of Queensland. *Mem. Queensland Museum* 21 (2):413–517.

Wade, M. 1979. Tracking dinosaurs: the Winton excavation. *Australian Nat. Hist.* 19:286–291.

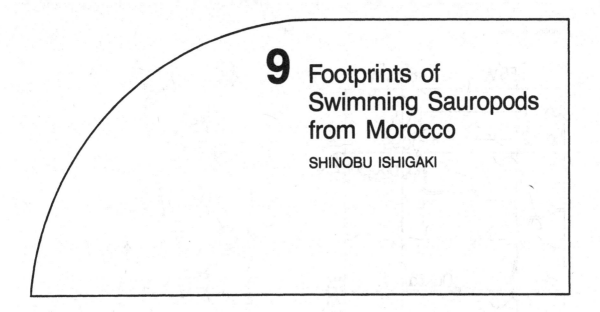

9 Footprints of Swimming Sauropods from Morocco

SHINOBU ISHIGAKI

Abstract

Four trackways consisting mainly of sauropod manus prints were discovered from the Middle Jurassic of the central High Atlas Mountains, Morocco. This supports R. T. Bird's theory of sauropod swimming ability.

Introduction

Bird (1944) reported a Cretaceous trackway consisting mainly of sauropod manus prints from the Mayan Ranch locality, Texas (U.S.A.). His interpretive sketch shows the trackmaking sauropod partially submerged, the forefeet pulling the animal forward by digging into the mud while the hindquarters float free of the bottom. There have been no other reports of manus–dominated sauropod trackways to test Bird's hypothesis. Herein is presented the discovery of four similar trackways from the Middle Jurassic of Morocco. These trackways were found by the writer during a survey of dinosaur footprints for the records of the National Earth Science Museum of Morocco.

The tracksites are in the Iouaridène Basin 10 km east of the town of Demnat in the central High Atlas Mountains. Iouaridène is one of the most important Middle Jurassic dinosaur tracksites in the world because of its excellent and abundant tridactyl dinosaur footprints. This site was discovered in 1937 and briefly described (Plateau et al. 1937) and has been frequently mentioned by later authors (e.g., Roch 1939; Bourcart et al. 1942; Lapparent 1942, 1945). Dutuit and Ouazzou (1980) were the first to report sauropod footprints from Iouaridène, a very large form described under the name *Breviparopus taghbaloutensis* (referred to below as Trackway Bre.). They also briefly reported sauropod manus prints near the village of Tirica, south of the Iouaridène Basin. The locality data are imprecise and the writer has been unable to relocate this site or study the Tirica prints as another instance of a swimming sauropod.

Geologic Setting and Tracksites

These trackways are preserved in association with many tridactyl and some normal sauropod footprints in the lower horizon of the Iouaridène Formation. This formation is composed of continental red beds, primarily red mudstones, and is distributed in the central Moroccan High Atlas Mountains. The footprints are preserved in red calcareous mudstone that also contains mudcracks and small ripple marks (wave length = 2–5 cm). Individual beds are a few centimeters thick with fine internal laminae. The depositional setting is interpreted as lagoonal or sebkha under semi–arid conditions, with intermittent temporary emergence. The age was initially thought to be Upper Lias, but a recent precise stratigraphy survey has redated the formation as Bathonian, Middle Jurassic (Jenny et al. 1981).

Localities where the writer found sauropod tracks are shown in Figure 9.1.

Locality no. 1 (31°44'8" N; 6°54'10" W): 300 m NNW from the hilltop (altitude 1095 m) of Taghbalout Village, northwest of the Iouaridène Basin. This is the locality where Dutuit and Ouazzou (1980) discovered an excellent trackway of a large sauropod (Trackway Bre.; part of this trackway is shown in Figure 9.5). About 25 m east from the eastern end of these footprints, two trackways (A and B) were found that are composed of sauropod manus prints only. These trackways trend S 50°W (Fig. 9.2).

Locality no. 2 (31°44'5" N; 6°54'18" W): 150 m SW from the western end of Trackway Bre. and about 250 m WSW of Trackways A and B. This single trackway (Trackway C) includes six manus prints that trend S 18°W (Fig. 9.3).

Locality no. 3 (31°42'36" N; 6°54'28" W): 350 m SW of Aghni Village, in the southwest of Iouaridène Basin. Trackway D at this location includes six manus prints that trend N 30°W (Fig. 9.4).

Description and Interpretation

All the individual prints of Trackways A–D are

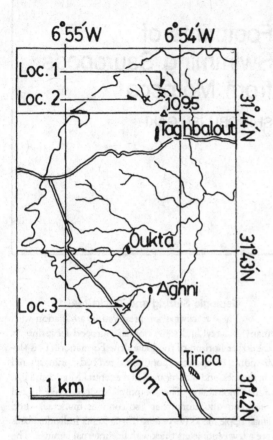

Figure 9.1. Map showing the tracksites (Locality nos. 1, 2, and 3).

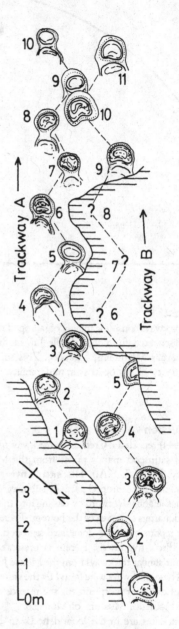

Figure 9.2. Diagram of part of Trackway A and B.

half–moon shaped, 45–65 cm in width and 30–50 cm in length. The central part of each half–moon is slightly raised which produces a horseshoe–shaped print. The footprints are very shallow (depth = 2–5 cm) and lack a sharply defined boundary which makes them very difficult to photograph. The bedding plane is not greatly disturbed by the footprints. Ripple marks and mudcracks run concordantly over both the surrounding bedding plane and the surface of the footprints.

Figure 9.5 shows part of the *Breviparopus taghbaloutensis* trackway (Trackway Bre.), and an enlarged drawing of a typical manus print from this trackway is shown in Figure 9.6. The size and shape of individual prints consisting Trackways A–D are almost the same as that of manus footprints of the Trackway Bre. However, Trackway Bre. differs from Trackways A–D in two important features: (1) footprints of Bre. destroy other surficial primary structures; and (2) the prints are deeper and more clearly defined.

Both ripple marks and mudcracks are obliterated by the prints of Trackway Bre. Individual prints are surrounded by small folded mounds of mud and a steep slope is present at the margin of the print. Sometimes, there are remnants of mud fragments in the footprint. It is likely that these adhered to the foot of the animal and fell into the print when it raised its foot.

The steady water current indicated by the ubiquitous ripple marks would erase small, upstanding surface irregularities such as mud mounds or the steep, well defined margins of the prints in Trackway Bre. Therefore this trackway was probably made with full aerial exposure under semi–dry conditions. The poorly defined, shallow manus-dominated Trackways A–D may have been partially erased by currents that formed the ripple marks, which supports the hypothesis that they were made underwater by a swimming animal. Mudcracks that developed after aerial exposure of the surface further degraded the quality of the prints. Although Lockley and Conrad (this volume) present an

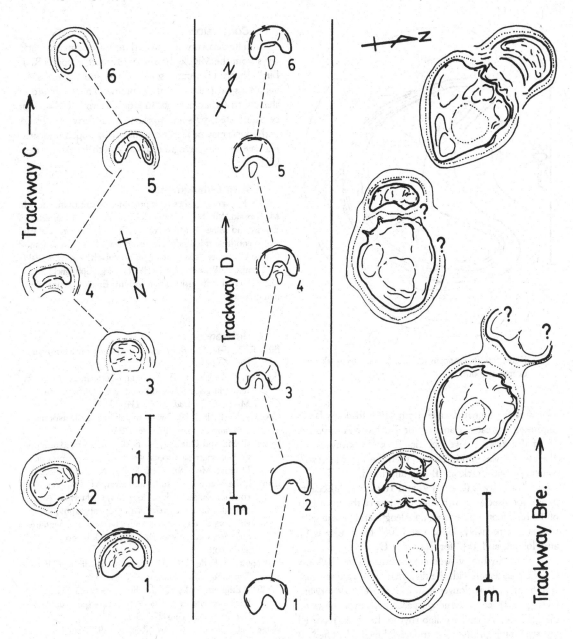

Figure 9.3. Diagram of Trackway C.　　Figure 9.4. Diagram of Trackway D.　　Figure 9.5. *Breviparopus taghbaloutensis.* Diagram of part of trackway.

alternative explanation, this author believes there is no possibility that these trackways are undertracks because no real footprints could be found on the surface of the covering bed which lies immediately above these tracks. An estimated hip height for the trackmaker can be computed from the pes prints of Trackway Bre. To float the hindquarters in the posture suggested by Bird (1944) would require a water depth of 3–4 m.

Trackways A and B extend almost parallel to each other for about 42 m. Figure 9.2 shows the last 15 m of these trackways. The average pace length and pace angulation are, respectively, 142 cm and 115° in Trackway A, and

172 cm and 114° in Trackway B. These trackways were not made at the same time by two sauropods swimming together. This conclusion is supported by the following observation. In Trackway A the front part of the hind foot always touches the bottom and makes faint imprints just behind the manus prints. These faint pes imprints are only intermittently present in Trackway B, and are missing from step nos. 4, 5, and 10, which suggests that the animal making Trackway B floated somewhat higher above the substrate than the animal that made Trackway A. However, the size of individual footprints of Trackway B indicates the trackmaker was larger than the maker of Trackway A.

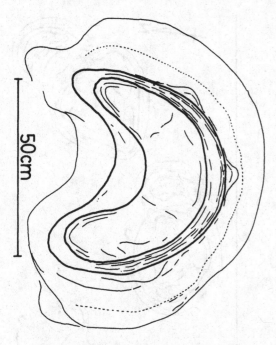

50cm

Figure 9.6. *Breviparopus taghbaloutensis*. Precise sketch of manus footprint.

Conclusion

The discovery of sauropod manus–dominated trackways from the Middle Jurassic of Morocco supports R. T. Bird's theory of swimming sauropods, and suggests some variations on their way of swimming. The prints are so shallow that they may go unnoticed during fieldwork or be interpreted, by some workers, as underprints. Tracks of this kind may be rather common at sauropod tracksites, but have not been recognized because of their indistinctness.

Acknowledgments

The writer wishes to express his sincere thanks to Mr. M. Bensaid, Mr. M. Dahamani, Mr. J-A. Jossen (Ministry of Energy and Mines in Morocco) and Dr. J. Jenny for their kind aid to continue this study. He also thanks Prof. Masae Omori (Azabu University, Tokyo, Japan) for helpful advice and Dr. Walter P. Coombs, Jr. (Western New England College, Springfield, Mass., U.S.A.) who kindly assisted him with the English version of this manuscript.

Therefore the estimated water depth when Trackway B was imprinted is greater than that for Trackway A. Step no. 10 of Trackway B overlaps no. 9 of Trackway A, so Trackway A was imprinted first, then the water level rose somewhat and Trackway B was imprinted.

Trackway C (Locality 2) and Trackway D (Locality 3) are not associated with faint imprints of the front part of the hind foot. The average pace length and pace angulation are, respectively, 133 cm and 106° for Trackway C, and 169 cm and 154° for Trackway D.

The rhythm of swimming is unusual in Trackway C. The pace length with the right manus forward (average = 156 cm) is always longer than with the left foot forward (average = 110 cm), indicating a gallop rhythm to the swimming strokes. A similar gallop rhythm has been reported for swimming prints of a tridactyl biped identified as *Anchisauripus* (Coombs 1980).

Trackway D has a larger pace angulation than that of the other trackways. Steps 3, 4, 5 and 6 are associated with small grooves just behind the prints that are interpreted as lines made by dragging toes.

References

Bird, R. T. 1944. Did *Brontosaurus* ever walk on land? *Nat. Hist.* 53:60–67.

Bourcart, J., Lapparent, A. F. de, and Termier, H. 1942. Un nouveau gisement de dinosauriens jurassiques au Maroc. *C.R. Acad. Sci.* t. 214:120.

Coombs, W. P., Jr. 1980. Swimming ability of carnivorous dinosaurs. *Science* 207:1198–1200.

Dutuit, J. M., and Ouazzou, A. 1980. Découverte d'une piste de Dinosaure sauropode sur le site d'empreintes de Demnat. *Mém. Soc. Géol. France*, N.S. 139:95–102.

Jenny, J., Le Marrec, A., and Monbaron, M. 1981. Les couches rouges du Jurassique moyen du Haut Atlas central (Maroc): correlations lithostratigraphiques, éléments de datations et cadre tectono-sédimentaire. *Bull. Soc. Géol. France Paris* (7), t. XXIII, no. 6:627–639.

Lapparent, A. F. de. 1942. Dinosauriens de Maroc. *C.R. Somm. Soc. Géol. France* 1942, fasc. 5:28.

———. 1945. Empreintes des pas de Dinosauriens du Maroc. Exposées dans la galerie de paléontologie. *Bull. Mus. Nat. Hist. Nat. Paris* 17:268–271.

Plateau, H., Giboulet, G., and Roch, E. 1937. Sur la présence d'empreintes de Dinosauriens dans la région de Demnat (Maroc). *C.R. Somm. Géol. France Paris* 1937:241–242.

Roch, E. 1939. Description géologique des montagnes à l'Est de Marrakech. *Mém. Service Mines et Carte Géol. du Maroc* 51:244.

IV Eggs and Nests

"No dinosaur eggs have ever been found, but these must be dinosaur eggs. They can't be anything else."

> W. Granger commenting on George
> Olsen's discovery of complete dinosaur
> eggs in Mongolia, in July 1923.
> Andrews 1932

Dinosaur egg remains were first discovered in Europe in the latter part of the nineteenth century, but it was not until 1923 that complete eggs were reported (quotation above) and assigned with some confidence to a dinosaurian egg layer. Even though many paleontologists have accepted that the Mongolian eggs were probably laid by *Protoceratops*, whose skeletal remains occur in abundance at the eggsites, this inference has not been proven because identifiable embryonic remains have not been found in the eggs (cf. Hirsch this volume). In fact it was not until very recently that Horner and Weishampel (1988) published absolute proof of the specific taxonomic affinity of any dinosaur egg by describing embryonic remains found inside.

Arising from the same, now classic studies in Montana, Horner and his colleagues have also convincingly demonstrated that dinosaurs nested in colonies over long periods of time. Such "site fidelity" is now being reported from several other regions including the Indian subcontinent (Jain this volume) and China (Mateer this volume). The unprecedented spate of egg and nestsite discoveries is fast exploding the myth that dinosaur eggs are rare. Progress in this field also promises to reveal more eggs with identifiable embryonic and hatchling remains. There are also indications that some nestsites or "hatcheries" (sensu Sahni this volume) are regionally extensive for tens of kilometers laterally, a phenomenon also recently noticed for certain trackbearing beds. Such evidence indicates that egg-bearing stratigraphic units are recurring, rather than exceptional features in Mesozoic successions.

References

Andrews, R. C. 1932. *The New Conquest of Central Asia*. (New York) 678 pp.

Horner, J. R., and Weishampel, D. B. 1988. A comparative embryological study of two ornithischian dinosaurs. *Nature* 332:256–257.

10 Interpretations of Cretaceous and Pre–Cretaceous Eggs and Shell Fragments

KARL F. HIRSCH

Abstract

Almost all eggs and eggshell fragments from pre-Cenozoic periods have been found in Upper Cretaceous rocks. All, as yet, positively identified fossil eggshell have a rigid shelled structure. The groups with soft and pliable shell are not represented, due to their poor fossilization potential. Thus the fossil record is biased.

The study of fossil eggshell is still in its early stages and neither universal descriptive terminology nor a systematic classification yet exists. To minimize the difficulties confronted in the study of eggshell material, all tools should be used for a complete description, and studies should be based on as much material as possible. We should distinguish clearly between structural and taxonomic types and should use a scale of reliability for both. Absolute identification at species or generic levels should be made only for eggs with embryonic remains or, as next best, for nests with hatchlings. Evolution studies should wait until sufficient descriptive literature is available and a universal classification exists.

Introduction

A survey of eggs and eggshells from the Cretaceous and pre–Cretaceous shows that almost all specimens have been found in Upper Cretaceous deposits; there are only a few localities with older eggshell. The first Mesozoic egg fragments were found in the Upper Cretaceous of France. Matheron (1869) attributed them to dinosaurs. These specimens fell into oblivion until the sensational finds of the American Museum of Natural History in Mongolia in 1923 stirred up new interest. This event was followed by many descriptions of dinosaur eggs and their shell structure. It also stimulated a more intense search for fossil eggs in other parts of the world (Jepsen 1931; Romer and Price 1939; Young 1954, 1959). In the last few decades, with the growing awareness of the paleontological value of eggshell, eggs, clutches and nesting sites, and the advancement and availability of modern techniques, localities too numerous to name have been found and many descriptive papers have been published (Young 1965; Jensen 1966, 1970; Sigé 1968; Sochava 1969, 1971, 1972; Erben 1970; Horner and Makela 1979; Zhao 1979a,b; Horner 1982, 1984; Sahni and Gupta 1982; Sahni et al. 1984). New structural eggshell types, including some with unusually thin eggshell, have been found in Cretaceous and pre-Cretaceous eggshell; thickness ranges from 0.05 to 4 mm. The abundance and variation of the eggs and eggshells now known, and the extinction of some egg–laying species intensify the problems of identification and classification.

Modern eggshell, composed of organic matter and crystalline calcium carbonate, can be divided into three rather loose groupings. In the first two groups (squamates, *Sphenodon,* and most turtles), the organic matter dominates; thus the eggshell is parchment-like or the outer calcareous layer is pliable. These shells have a poor chance of fossilization and no proven fossil record exists. In the third group the calcareous matter dominates. The well organized, interlocking crystalline structure makes the shell layer rigid, thus giving it a good chance of fossilizing. The documented fossil record of this group is good. In the rigid–shelled group we recognize structural types for avian, crocodilian, chelonian and gekko eggshell based on studies of modern eggshell (Hirsch 1979, 1983; Packard 1980). Gekko and crocodilian types have ben traced back into the Eocene, whereas avian and chelonian types extend back to the Late Cretaceous (Hirsch 1983, 1985; Hirsch and Packard 1987; Elzanowski 1981).

Notes and Abbreviations for Figures

In radial views, the outside of the eggshell is always up. The following abbreviations have been used: BM(NH), British Museum (Natural History); MCZ, Museum of Comparative Zoology; PU, Princeton University; UCM, University of Colorado Museum.

Review of Pre–Cenozoic Eggs and Eggshell
Permian

The oldest vertebrate "egg" (Fig. 10.1A) was described by Romer and Price (1939). It was found on the surface and is probably derived from the Admiral Formation. A re*evaluation (Hirsch 1979) showed that shape, size, eggshell–like fractures and only part of the chemistry are egg–like. The nodes in the outer, shell–like layer and the microstructure of this layer do not correspond structurally to those in rigid eggshells, and they even make the specimen a poor candidate for the pliable or the parchment–like group.

Triassic

Bonaparte and Vince (1979) reported a nest with several incomplete skeletons of juvenile Prosauropoda and two eggs from Patagonia, Argentina. However, the shell structure has not been described in detail. Kitching (1979) reported eggs (with embryonic remains) from South Africa. He described the eggshell as parchment–like. However, a recent preliminary study (Grine and Kitching 1987) shows that the South African eggs have a rigid eggshell structure.

Jurassic

Until recently, a limited number of eggshell fragments, but no eggs, had been found in the Morrison Formation of Colorado. These specimens, mainly from Fruita, are matched by a single eggshell fragment from the Garden Park area near Canyon City, Colorado. The eggshell is relatively thin and two different structural types (Fig. 10.1B,C) are represented; one of them (Fig. 10.1C) is probably chelonian (Hirsch unpublished). [More recently, abundant material (Hirsch et al. 1987), including complete eggs, was discovered in the Morrison Formation southeast of Grand Junction, and a single egg was reported from the Cleveland–Lloyd Quarry in Utah.]

Lower Cretaceous

A large number and variety of eggshell fragments, but no eggs, have been found in the Cedar Mountain and Kelvin Formations in Utah and Wyoming. Their macrofeatures have been described by Jensen (1970). However, he did not attempt to divide them into classes as he did with the Upper Cretaceous eggshells. A few new structural types have since been collected (Fig. 10.2G) but have not yet been described.

Upper Cretaceous

A wealth and large variety of egg material (clutches, eggs, shell fragments) are found world–wide. These localities can be grouped regionally.

France/Spain

This is the oldest known and also the most studied region. The egg fragments described by Matheron (1869) were, despite their chelonian–like shell structure, assigned to *Hypselosaurus* (based on their large size and some associated bones). Gervais (1877) and Roule (1885) con-

firmed this classification. These eggs have been studied often (Van Straelen and Denaeyer 1923; Van Straelen 1928; Lapparent 1947, 1957, 1958; Dughi and Sirugue 1957, 1966, 1976; Schwarz et al. 1961; Thaler 1965; Erben 1970; Erben and Newesely 1972; Erben et al. 1979; Kerourio 1981, 1982; Seymour 1979; Penner 1983, 1985). Because of their distinct structure (Fig. 10.1D), and their restriction to the European region, Young (1959), Jensen (1966), Sochava (1971) and others think that they are a group different from all other structural eggshell types. Embryonic remains or hatchlings have never been reported, in spite of the tremendous number of eggs.

In the last decades screen washing and a more intense search have produced eggshells with ornithoid–like structure from the same area. Beetschen et al. (1977) assigned the thin–shelled forms to avians; Kerourio (1982) questions this identification. He presumes that somewhat thicker eggshells of this same type found at a different locality could be dinosaurian.

Another peculiarity of this region are pathological eggshells (Fig. 10.1E). Multi–layered eggshells, which have so far only been reported from Upper Cretaceous beds at one other site (Jain this volume), have been discussed by Erben (1970) and Kerourio (1981). A thinning of eggshell towards the end of the Cretaceous, linked to the extinction of dinosaurs by Erben (1970), has been disputed by Kerourio (1981) and Penner (1985).

Mongolia

The sensational finds of eggs and clutches by the American Museum of Natural History expedition to Mongolia in 1923 have been described by Andrews (1927, 1932). A preliminary report by Van Straelen (1925) of two eggshell structures showed that they were considerably different from those reported from France. Sochava (1965, 1971) and Erben (1970) described these Mongolian eggshell types in detail. Embryonic remains found in an egg fragment were assigned to *Protoceratops* by Sochava (1972).

China

Chow (1951, 1954) and Young (1959, 1965) described this egg fauna; Zhao and Chiang (1974), Zhao and Ding (1976), Zeng and Zhang (1979) and Zhao (1979a,b) followed in their footsteps. Nesting areas, clutches and many eggs have been described (cf. Mateer this volume). However, no embryos or hatchlings have been reported. Linguistic and communication problems make it hard to evaluate and use these findings.

Western Interior of North America

Eggshell was reported first by Jepsen (1931) from Montana. Jensen (1966) described the macrofeatures of eggshell from Utah. Horner and Makela (1979) and Horner (1982, 1984) reported nesting sites, clutches, eggs, embryonic remains and hatchlings from Montana. Eggshell fragments, but no eggs, have also been found in the Lance and Kaiparowitz Formations. A systematic description of eggshell structure of this material has not yet been published.

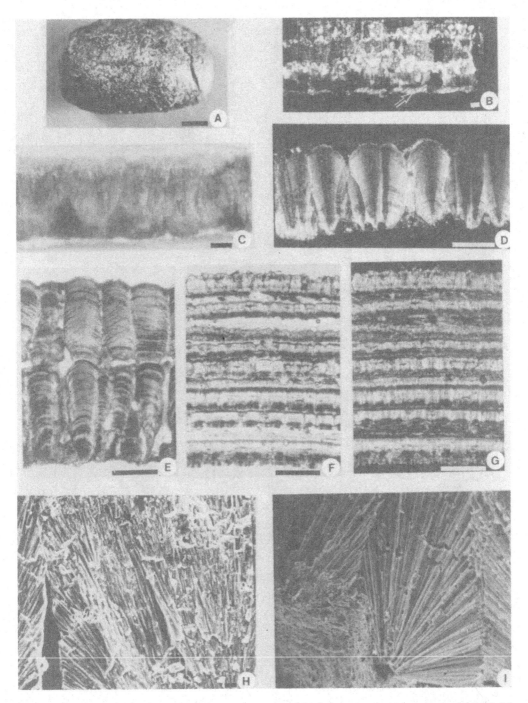

Figure 10.1. **A,** oldest vertebrate ?egg; MCZ 1107; Admiral Formation, Permian, Texas; bar = 10 mm. **B–C,** Radial thin sections of Jurassic eggshells; Morrison Fm., Colorado; viewed under polarized light. **B,** UCM 54471; columnar structure with extinction pattern. Arrow points to secondary deposits on inner surface of shell; bar = 100 m **C,** UCM 54322; chelonian–like structure with needle–like crystals; sweeping extinction pattern; bar = 100 m, **D,** radial thin section of Upper Cretaceous French/Spanish eggshell; UCM 54473; viewed under polarized light; chelonian–like shell structure, non–interlocking, tubocanaliculate type, no continuous horizontal layering; bar = 1 mm. **E–G,** radial thin sections of multi–layered eggshells, viewed under ordinary (E,F) and polarized light (G); bar = 1 mm. **E,** double–layered eggshell from Spain; UCM 54477. **F,G,** nine–layered *Geochelone elephantopus* eggshell; UCM OS1144. Note variations in size and shape of shell units in the various layers and the difference in the interspaces between these layers. **H–I,** micrographs of radial views of fossil and modern chelonian eggshell demonstrating the likeness of structure; bar = 10 m **H,** *Gopherus flavomarginatus*, UCM OS1128. **I,** fossil turtle from the Upper Cretaceous of England; BM(NH) 47208.

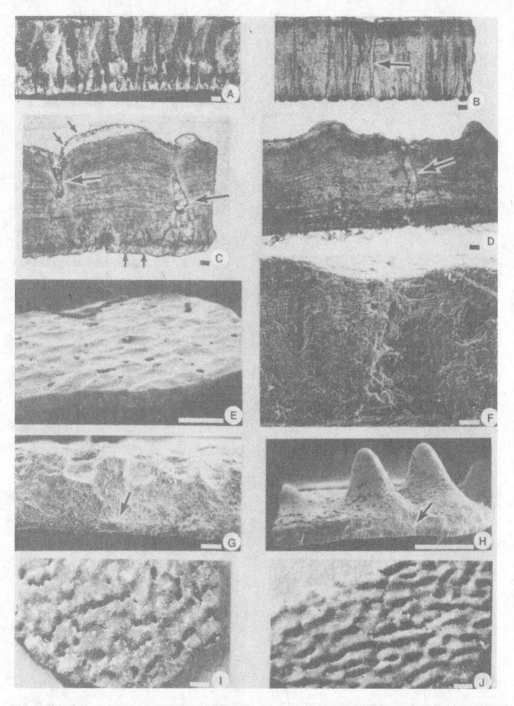

Figure 10.2. **A,** radial thin seciton of crane–like eggshell; UCM 47603; Eocene of Wyoming; viewed under polarized light; angusticanaliculate type ; bar = 100 m **B–D,** radial thin sections of dinosaur eggshells; Upper Cretaceous of Montana; viewed under ordinary light; bar = 100 m **B,** hypsilophodont eggshell; UCM 54474; angusticanaliculate type; smooth outer surface; continuous horizontal layering. Note likeness to avian shell structure. Arrow points to pore canal. **C,** hadrosaur eggshell; PU 22432; type prolatocanaliculate; outer surface sculptured with ridges; pronounced horizontal layering. Note irregular pore canals (single arrow) and thin secondary layer on inner and outer shell surface (double arrow). **D,** eggshell of unknown dinosaur; UCM 54320; angusticanaliculate type; nodose outer shell surface; pronounced horizontal layering. Arrow points to pore canal. **E–F,** micrographs of undescribed eggshell from a "?One–egg–site"; Upper Cretaceous of Utah; UCM 54319. Type not yet established. **E,** outer surface of shell; lightly ridged; edge of shell exhibits large shell units; bar = 1 mm. **F,** radial view; shell layer very loosely structured; the prounounced horizontal layering, especially in outer third of shell and the large interstices between units are

Other, Less Well Known Regions

In recent years a number of eggshell localities have been reported from India. They vary from sites having only eggshell fragments, some of which are extremely thin, to sites with eggs and complete clutches. This material is still in the early stages of being studied (Sahni and Gupta 1982, Sahni et al. 1984, Jain and Sahni 1985, Sahni et al. this volume).

In Peru only a limited amount of eggshell material has been found and two structural types have been described by Kerourio and Sigé (1984).

Classification
Problems in Identification and Classification

Work on fossil eggs is based on our knowledge of modern eggshell structure, which is still limited, especially among the squamates. In Cretaceous and older eggshell we deal also with extinct egg-laying species, some of which may have laid eggs with structures similar to modern eggshell, as is demonstrated by the avian-like eggshell of some dinosaurs. Others may have had structural forms not known in modern eggshell.

Modern eggshell is sensitive to the environment: temperature, humidity and gas exchange are critical for the developing embryo; thus the differences we see in eggshell structure may be due either to environmental or evolutionary structural malleability.

The organic matter of the eggshell decays, eliminating membrane, cuticle, pore covers and fibrous layers, all features used in the identification of modern eggs. The crystalline calcium carbonate of the shell layer has the best chance of being preserved as a fossil. However, diagenesis and recrystallization occasionally change the mineralogy and the microstructure of this layer partically or even completely. Thickness of the shell, appearance of the sculpturing of the outer shell surface, and the mammillae may change through weathering, abrade through transport, or partially dissolve through the acidity of the ground water. Secondary deposits, often hard to recognize, may cause the shell to appear thicker.

Pathological eggshells are another problem. The illustrated *Geochelone* egg (Fig. 10.1F,G), if fossilized and disintegrated into numerous shell fragments, would not only produce a multi-layered eggshell, but also a semblance of single-layered eggshells all very different from each other in shell thickness and in the shape of the shell units which compose the shell layer. Distributed within the sediment several "structural" types coming from a single egg would confront the unknowing finder. Erben (1970) described

pathologic, multi-layered eggshell from the France/Spain region in detail, assuming that they could have been one of the causes for extinction of dinosaurs. Kerourio (1981) disputed this assumption, quoting the small percentage of multi-layered shell he found within this region. However, he did not, and realistically he could not, take into consideration all the single-layered shell fragments produced by these pathological eggs.

The thinning of eggshell could be an evolutionary process or just be the effect of temporary change in the environment as studies of the condor eggshell have shown. Condor eggshell was only a fraction of its ordinary thickness from 1964 to 1983 during the time when DDT was used (Hirsch unpublished).

History

The desire to identify and classify eggs and eggshell is natural and arises with the discovery of local fauna. It is thus not strange that they are based on the specific characteristics of the specimens of the region and on the opinion of the researcher, sometimes ignoring the discovery of other regions. These classifications are discussed below:

China

Chow (1954) classified the eggs found in China by their shape into two forms: *Ooelithes spheroides* and *O. elongatus*. Young (1965), after new discoveries, added two new forms: *O. rugustus* and *O. nanhsiungensis*. Zhao (1979b) reclassified the Chinese eggs, based on shape, size, ornamentation of the outer surface and microstructure into three families: Faveoloolithidae, Elongatoolithidae and Spheroolithidae and added the French/Spanish eggs as Megaloolithidae. He hopes to establish a general scheme for the classification of all known dinosaur eggs. He is also concerned with the evolution of these eggs, based on the Chinese material. Regrettably his paper has not been translated into English (Zhao pers. comm. 1985).

Western Interior of North America

Jensen (1966) classified eggshell found in the North Horn Formation of Utah, based on the sculpturing of their outer surface into three classes: Class A with a smooth external surface; Class B with a nodose or nodular surface above the peripheral surface of the shell; Class C with a sculptured surface wherein the pattern is developed within the peripheral surface of the shell.

French/Spanish Region

Dughi and Sirugue (1958, 1964) and Schmidt (1967),

crocodilian-like features; bar = 100 m **G–H,** micrographs of eggshells with features unique and unknown in modern eggshell. **G,** undescribed eggshell; UCM 49387; Lower Cretaceous of Utah; deeply dimpled outer surface; narrow inner shell layer (arrow); bar = 100 m **H,** undescribed eggshell; UCM 54318; Upper Cretaceous of Utah; narrow inner shell layer (arrow); nodes are higher than eggshell is thick; bar = 21 mm. **I–J,** variations in the preservation of outer shell surfaces. Hadrosaur eggshell; UCM 54475; Upper Cretaceous of Montana; bar = 10 mm. **I,** well-preserved shell; Class B (sculpturing above peripheral surface). **J,** badly abraded eggshell; sculpturing seems to be Class C (below peripheral surface); pore holes and deep undercuts or depressions visible.

working only with the peculiar eggshell structure of the French/Spanish specimens, assumed that this eggshell structure (Fig. 10.1D) was typical for all dinosaurian and reptilian eggs. Dughi and Sirugue (1976) defended this assumption even at that time and classified all other structures as avian.

Mongolia

Sochava (1969) divided the eggs from Mongolia, based on their air canal structure, into three types: angusticanaliculate, prolatocanaliculate and multicanaliculate. In 1971 Sochava distinguished two taxonomic types of dinosaur eggshell: ornithoid, those with a shell structure similar to that of avian eggs and testudoid, those in which the shell structure is similar to the eggshell of modern turtles.

French/Spanish and Mongolian Region

Erben (1970) divided the eggshell from these regions by their microstructure into the types A, B, and C; the French/Spanish testudoid eggshell being type A and C, and the ornithoid eggshell from Mongolia as type B. In 1979 Erben summarized the work of Sochava (1969, 1971) and Erben (1970); and added the testudoid (French/Spanish) eggshell calling it tubocanaliculate to Sochava's (1969) classification. Erben also accepted with some reservation the taxonomic division into ornithoid and testudoid. With this publication (Erben et al. 1979), he also tried to establish a world–wide classification system.

The confusion which can arise from this "urge" to classify is best demonstrated on the French/Spanish eggs. They have been divided into varying numbers of types several times. Dughi and Sirugue (1958) suggested there were nine different types. Erben (1970) established two types, and later (Erben et al. 1979) one type. Williams et al. (1984) distinguished at least four types based on microstructure, porosity and shell thickness. A microstructural study of Penner (1985) recognizes at least three morphotypes. The newly discovered ornithoid eggshell (Beetshen et al. 1977, Kerourio 1982) is not included in these classifications.

Establishing of Types for the Fossil Record

Although useful descriptive features have been described, the approach to classification discussed above has been inadequate. The lack of a universal classification agreed upon by all workers attest to this. In part this is due to the early stage of the discipline, in part to incomplete studies of the egg or eggshell, and in part to a refusal to recognize the limits of knowledge confounding us at this time.

We can do nothing about the current state of the discipline, i.e., that it is still in an early descriptive stage and, in many cases, a new discovery can lead to a complete reevaluation of our current classification. However, we can attempt a complete description of the material we do have, and accept the constraints of what we know. Each structural type when possible should be described in terms of its size, shape, outer and inner surface sculpture, pore system, microstructure, and in radial and tangential sections or views under the PLM and SEM. As many specimens as feasible should be used in order to eliminate error due to population variations, stage of development, local environmental peculiarities and pathologies.

In terms of the constraint of the material, we should acknowledge that an absolute identification at the species or generic level can be done only for eggs with embryonic remains. The next best, in terms of low–level identification, is nests in which there are eggs and hatchlings. Until there is a fairly large set of genera and species in which the eggshell structure has been positively correlated with key remains, we will be unable to extend low–level identifications of eggshells that are not associated with bony remains.

The best results for establishing structural types come from studying whole eggs. Large numbers of like eggshell fragments, without complete eggs, are next in reliability. The least reliable are few eggshells of different structural types. On the other hand, we can now identify some eggs and eggshells to higher taxonomic levels based on comparison with modern–day forms (Fig. 10.1H,I). Each of the above–mentioned grades of reliability are discussed below in detail.

Eggs with Embryonic Remains

Elzanowski (1981) identified embryonic remains in egg fragments from Mongolia as avian. The eggshell of these fragments is regrettably too recrystallized to be compared with that of modern avian eggshell.

Clutches with eggs from Montana containing embryonic remains enabled Horner (1982) to identify and classify them as hypsilophodont dinosaurs. These eggs have an avian–like shell structure (angusticanaliculate) and the outer surface is smooth (Class A); however, we are not yet able to distinguish them from true avian eggshell (Fig. 10.2A,B).

Nests with Hatchlings

Eggshells from nests with the remains of hatchlings could be assigned to the hadrosaur *Maiasaura peeblesorum* (Horner and Makela 1979). This eggshell is different from that of the hypsilophodont (Fig. 10.2C,I,J); it is not as solid, has a prolatocanaliculate pore pattern and the outer surface is sculptured with ridges (Class B).

Eggs without Embryonic Remains

Other eggs from Montana are different from the two mentioned above and have no embryonic or hatchling remains. The outer surface of the eggs is nodose (Class B); the pore pattern seems to be angusticanaliculate (Fig. 10.2D). These eggs can not be assigned to a taxonomic group because they are not like any of the established fossil eggshell types.

Eggshell Fragments, But Not Complete Eggs

In the North Horn Formation of Utah, I found

about 400 similar eggshell fragments (Fig. 10.2E,F) concentrated in an area of about nine square meters, suggesting a deteriorated egg; this is a so–called "?one–egg–site". It is possible in situations like this to establish a structural type or fit them into an established one, as if they were a single egg. Again as many features as possible have to be studied. Measuring the radii of many eggshell fragments with the Geneva Lens Gauge may allow us to calculate size and shape of the egg.

Eggshell fragments with very unique and unknown features such as those from the Cedar Mountain (Fig. 10.2G) and North Horn Formations (Fig. 10.2H) may represent extinct eggshell types, in the latter case with a gekko-like structure.

When there are only a few eggshells or there are several types it is difficult, and a questionable undertaking, to establish structural types. Sculpture of the outer surface may vary to a degree within the same egg, or it may be weathered or abraded as the hadrosaur eggshell demonstrates (Fig. 10.2I,J).

Conclusions

Several areas in the Western Interior of North America have yielded a wealth and large variety of Upper Cretaceous and older eggshell. This material, when studied in detail, has to be compared and correlated to the eggshell material of the other regions of the world.

The study of fossil eggshell material is still in its early stages and the literature is spotty and regionally oriented. In describing an eggshell all of the tools for complete description should be used as far as possible. This, together with the development of a well–described comparative collection, will lead to a universal descriptive terminology and classification. Evolution of eggshell structure cannot be understood until sufficient descriptive literature from all regions and a worldwide classification are available.

In the study of eggshell we have to distinguish clearly between structural and taxonomic types. Structural types cannot be used for classification but can be used descriptively. Referring specimens to a lower or higher taxonomic group should be based on a scale of reliability according to the evidence found with the specimens.

Acknowledgments

My thanks to the people and institutions who furnished the specimens I used for this study, especially to the keepers in the Reptile House of the San Diego Zoo who surprised me with the "once in a lifetime" specimen of the nine-layered turtle egg.

I am deeply indebted to my mentor and friend Judith Harris, University of Colorado Museum. Without her encouragement, advice and help, I probably would not do these studies. My thanks are also extended to others who kindly lend their services, especially to John Drechsler and Richard Harding for their help to keep the SEM running and to the Geology Department, who generously let me use their facilities.

This study was supported by the University of Colorado, by my personal funds and, indirectly, by Rockwell International through their donations to the University of Colorado Museum.

References

Andrews, R. C. 1927. *Abenteuer und Entdeckungen dreier Expeditionen in die Mongolische Wueste.* (F. A. Brockhaus, Leipzig).

1932. *The New Conquest of Central Asia.* Natural History of Central Asia Vol I. (The American Museum of Natural History) 678 pp.

Beetschen, J-C., Dughi, R., and Sirugue, F. 1977. Sur la présence de coquilles d'oeufs d'Oiseaux dans la Crétacé supérieur des Corbières occidentales. *Compt. Rend. Acad. Sci.* 248:2491–2494.

Bonaparte, J. F., and Vince, M. 1979. El hallazgo del primer nido de dinosaurios Triasicos (Saurischia Prosauropoda), Triasicos superior de Patagonia, Argentina. *Ameghniana* 16:173–182.

Chow M. C. 1951. Notes on the Late Cretaceous dinosaurian remains and the fossil eggs from Laiyang, Shantung. *Bull. Geol. Soc. China* 31:89–96.

1954. Additional notes on the microstructure of the supposed dinosaurian eggshells from Laiyang, Shantung. *Acta Scientia Sinica* 3:523–525.

Dughi, R., and Sirugue, F. 1957. Les oeufs de Dinosauriens du bassin d'Aix-en-Provence. *Compt. Rend. Acad. Sci.* 245:707–710.

1958. Observation sur les oeufs de Dinosaures du bassin d'Aix-en-Provence, les oeufs a coquilles biostratifices. *Compt. Rend. Acad. Sci.* 246:2271–2274.

1964. Sur la structure des coquilles des oeufs des sauropsidés vivants ou fossiles du genre *Psammornis* Andrews. *Bull. Soc. Géol. France* 7:240–252.

1966. Sur la fossilisation des oeufs de Dinosaures. *Compt. Rend. Acad. Sci.* 262:2330.

1976. L'Extinction des dinosaures a la lumière des gisements d'oeufs du Crétacé terminal du sud de la France, principalement dans le bassin d'Aix-en-Provence. *Paléobiol. Continentale Montpellier* VII:1–39.

Elzanowski, A. 1981. Embryonic bird skeletons from the Late Cretaceous of Mongolia. *Palaeont. Polonica* 42:147–179.

Erben, H. K. 1970. Ultrastrukturen und Mineralisation rezenter und fossiler Eischalen bei Voegeln und Reptilien. *Biomineralisation* 1:1–66.

Erben, H. K., and Newesely, H. 1972. Kristalline Bausteine und Mineralbestand von kalkigen Eierschalen. *Biomineralisation* 6:32–48.

Erben, H. K., Hoefs, J., and Wedepohl, K. H. 1979. Paleobiological and isotopic studies of eggshells from a declining dinosaur species. *Paleobiology* 5:380–414.

Gervais, M. P. 1877. Structure des coquilles calcaires des oeufs et caracteres que l'on peut en tirer. *Jour. Zoologica* t. VI:88–96.

Grine, F. E., and Kitching, J. W. 1986. Early dinosaur eggshell structure: A comparison with other rigid sauropsid eggs. *Jour. Scan. Elect. Microscopy.* SEM Inc.

Hirsch, K. F. 1979. The oldest vertebrate egg? *Jour. Vert. Paleont.* 53:1068–1084.

1983. Contemporary and fossil chelonian eggshells. *Copeia* 2:382–397.

1985. Fossil crocodilian eggs from the Eocene of Colorado. *Jour. Paleont.* 59:531–542.

Hirsch, K. F., and Packard, M. J. 1987. Review of fossil eggs and their shell structure. *Scanning Microscopy* 1:383–400.

Hirsch, K. F., Young, R. Y., and Armstrong, H. J. 1987. Eggshell fragments from the Jurassic Morrison Formation of Colorado. *In* Averett, W. R. (ed.). *Paleontology and Geology of the Dinosaur Triangle*. (Grand Junction, CO: Museum of Western Colorado) pp. 79–84.

Horner, J. R. 1982. Evidence of colonial nesting and 'site fidelity' among ornithischian dinosaurs. *Nature* 297:675–676.

1984. The nesting behavior of dinosaurs. *Sci. Amer.* 250:130–137.

Horner, J. R., and Makela, R. 1979. Nest of juvenile provides evidence of family structure among dinosaurs. *Nature* 282:296–299.

Jain, S. L., and Sahni, A. 1985. Dinosaurian eggshell fragments from the Lameta Formation at Pisdura, Chandrapur District, Maharashtra. *Geoscience Jour.* VI:211–220.

Jepsen, G. L. 1931. Dinosaur eggshell fragments from Montana. *Science* 73:12–13.

Jensen, J. A. 1966. Dinosaur eggs from the Upper Cretaceous North Horn Formation of Central Utah. *Brigham Young Univ. Geol. Studies* 13:55–67.

1970. Fossil eggs in the Lower Cretaceous of Utah. *Brigham Young Univ. Geol. Studies* 17:51–65.

Kerourio, P. 1981. La distribution des "Coquilles d'oeufs de Dinosauriens multistratifieés" dans le Maestrichtien continental du Sud de la France. *Géobios* 14:533–536.

1982. Un nouveau type de coquille d'oeuf présumé dinosaurien dans le Campanian et Maestrichtien continental de Provence. *Palaeovertebrata Montpellier* 12:141–147.

Kerourio, P., and Sigé, B. 1984. L'apport des coquilles d'oeufs de dinosaures de Laguna Umayo a l'âge de la Formation Vilquechico (Pérou) et à la compréhensions de *Perutherium altiplanense*. *Newsletters Stratigraphy* 13:133–142.

Kitching, J. W. 1979. Preliminary report on a clutch of six dinosaurian eggs from the Upper Triassic Elliot Formation, northern Orange Free State. *Palaeont. Africana* 22:72–77.

Lapparent, A. F. de. 1947. Les dinosauriens du Crétacé Supérieur de Midi de la France. *Mém. Soc. Géol. France* N.S. 26:56.

1957. Les oeufs de dinosauriens fossiles de Rousset (Bouches du Rhone). *Comp. Rend. Acad. Sci.* 245:546.

1958. Decouverte d'un gisement d'oeufs de Dinosauriens dans la Crétacé supérieur du bassin de Tremp (Province de Lerida, Espagne). *Comp. Rend. Acad. Sci.* 247:1879–1886.

Matheron, M. P. 1869. Notice sur les Reptiles fossiles des dépots fluvio-lacustres Crétacés du bassin a lignite de Fuveau. *Mem. l'Academie Impériale des Sciences, Belles-Lettres et Arts de Marseille* 345–379.

Packard, G. C., and Packard, M. J. 1980. Evolution of the cleidoic egg among reptilian antecedents of birds. *Amer. Zool.* 20:351–362.

Penner, M. M. 1983. Contribution a l'étude de la microstructure des coquilles d'oeufs de dinosaures du Crétacé supérieur dans le bassin d'Aix-en-Provence: application biostratigraphique. Thèse de Doctorat 3eme cycle, Universite P. et M. Curie, Paris 6, pp. 1–234.

1985. The problem of dinosaur extinction. Contribution of the study of terminal Cretaceous eggshells from Southeast France. *Geobios Lyon* 18 (5):665–670.

Romer, A. S., and Price, L. I. 1939. The oldest vertebrate egg. *Amer. Jour. Sci.* 237:826–829.

Roule, L. 1885. Recherches sur le Terrain Fluvio-Lacustre inférieur de Provence. *Annales de Sciences Géologique* 18:2.

Sahni, A., and Gupta, V. J. 1982. Cretaceous eggshell fragments from the Lameta Formation, Jabalpur, India. *Bull. India Geol. Assoc.* 15:85–88.

Sahni, A., Rana, R. S., and Prasad, G. V. R. 1984. SEM studies of thin eggshell fragments from the Intertrappeans (Cretaceous–Tertiary transtion) of Nagpur and Asifabad, Peninsular India. *Jour. Palaeont. Soc. India* 29:26–33.

Schmidt, W. J. 1967. Struktur des Eischalenkalkes von Dinosauriern. *Zeitschrift fuer Zellforschung* 82:136–155.

Schwarz, L., Fehse, F., Mueller, G., Anderson, F., and Sieck, F. 1961. Untersuchungen an Dinosaurier-Eischalen von Aix en Provence und der Mongolei (Shabarakh Usu). *Zeitschrift fuer wissenschaftliche Zoologie* 165:344–379.

Seymour, R. S. 1979. Dinosaur eggs: gas conductane through the shell, water loss during incubation and clutch size. *Paleobiology* 5:1–11.

Sigé, B. 1968. Dents de Micromammifères et fragments de coquilles d'oeufs de Dinosauriens dans la faune de Vertébrés du Crétacé supérieur de Laguna Umayo (Andes péruviennes). *Compt. Rend. Acad. Sci.* 267:1495–1498.

Sochava, A. V. 1969. Dinosaur eggs from the Upper Cretaceous of the Gobi Desert. *Paleont. Jour.* 4:517–527.

1971. Two types of eggshell in Senonian dinosaurs. *Paleont. Jour.* 3:353–361.

1972. The skeleton of an embryo in a dinosaur egg. *Paleont. Jour.* 4:527–531.

Thaler, L. 1965. Les oeufs des dinosaures du Midi de la France livrent le secret de leur extinction. *Science Progres–La Nature* 45:41–48.

Van Straelen, V. 1925. The microstructure of the dinosaurian eggshells from the Cretaceous beds of Mongolia. *Amer. Mus. Novitates* 173:1–4.

1928. Les oeufs de reptiles fossile. *Palaeobiologica* 1:295–312.

Van Straelen, V., and Denaeyer, M. E. 1923. Sur les oeufs fossiles du Crétacé supérieur de Rognac en Provence. *Bull. Classe des Sciences de l'Acad. Royal de Belgique* 9:14–26.

Williams, D. L. G., Seymour, R. S., and Kerourio, P. 1984. Structure of fossil dinosaur eggshell from the Aix Basin, France. *Palaeogeog., Palaeoclimat., Palaeoecol.* 45:23–27.

Young C.-C. 1954. Fossil reptilian eggs from Laiyang, Shantung, China. *Scientia Sinica* 3:505–522.

1959. New fossil eggs from Laiyang, Schantung. *Vertebrata PalAsiatica* 3:34–35.

1965. Fossil eggs from Nanhsiung, Kwangtung and Kanchou, Kiangsi. *Vertebrata PalAsiatica* 9:141–170.

Zeng D. and Zhang J. 1979. On the dinosaur eggs from the western Dongting Basin, Hunan. *Vertebrata PalAsiatica* 9:141–170.

Zhao Z. 1979a. Discovery of the dinosaur eggs and footprint

from Neixiang County, Henan Province. *Vertebrata PalAsiatica* 17:304–309.

1979b. The advancement of researches on the dinosaurian eggs in China. *In South China Mesozoic and Cenozoic "Red Formation".* (Peking: Science Publishing Co.) pp. 329–340.

Zhao Z. and Chai Y–k. 1974. Microscopic studies on the dinosaurian egg–shells from Laiyang, Shantung Province. *Scientia Sinica* 17:73–83.

Zhao Z. and Ding S. 1976. Discovery of the dinosaurian egg–shells from Alxa, Ningxia and its stratigraphic significance. *Vertebrata PalAsiatica* 14:42–44.

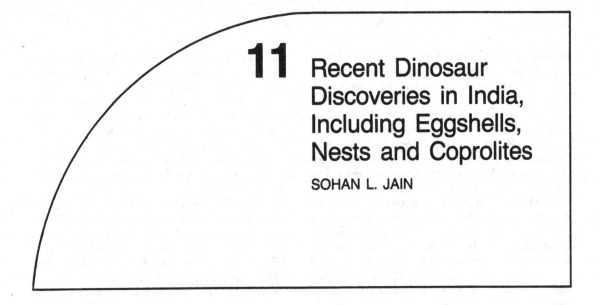

11 Recent Dinosaur Discoveries in India, Including Eggshells, Nests and Coprolites

SOHAN L. JAIN

Abstract

Although fragmentary skeletal remains of dinosaurs from India have been known for nearly a century, dinosaurian trace fossils are scarce. Since the 1980's several sites with definite eggshells, eggshell fragments, nests and clutches of Upper Cretaceous sauropod dinosaurs have been discovered. These include records of eggshell fragments of dinosaurian affinity at Bara Simla Hill, Jabalpur, discovered in 1982; dinosaurian eggshells from Infratrappean limestone in Kheda District Gujarat, discovered in 1983; nests and clutches of complete eggs in Gujarat, discovered in 1984, leading to detailed field and laboratory studies in 1986; possible pathologic dinosaurian eggshells in Gujarat, discovered in 1984; and dinosaur eggshell fragments from Lameta Formation in Chandrapur District, Maharastra, discovered in 1985. The most spectacular of these is an extensive hatchery encompassing an area of about 20 sq km. Nearly all eggs are spherical and seem unhatched. New sites of dinosaur eggshells were also discovered in districts Asifabad (Andhra Pradesh) and Nagpur (Maharastra) during 1984. A dinosaur footprint from an Upper Cretaceous locality in Gujarat, known for dinosaur eggs, was found in 1986.

Coprolites have been recorded from several Late Cretaceous and Lower Jurassic sites in India. Except for the large size and close proximity to the dinosaur remains, there is nothing characteristic in such coprolites to assign them to dinosaurs.

Introduction

This chapter provides basic information and reference to papers from the Indian subcontinent concerning discoveries of dinosaurs and their eggshells, complete eggs, nests and clutches, footprints and coprolites. This review is not comprehensive but attempts to include the more important references and compensate for the general paucity of contemporary literature on vertebrate fossils from India. World–wide overviews are handicapped due to the scattered nature of literature, most of which is inaccessible. This chapter attempts to piece together and summarize current knowledge. Among the general references, works by Steel (1970), Colbert (1974, 1977, 1984), Olshevsky (1978) and Halstead and Halstead (1981) have commented upon dinosaurs from the Indian subcontinent in a wider perspective.

Summary of Dinosaur Fossils from India

Skeletal remains of dinosaurs have been known from India for over a century. Most of the 19th century discoveries were, however, of very fragmentary material but gave a useful lead to further discoveries and description during the 20th century. Among the earliest workers, Lydekker (1877) identified posterior caudal vertebrae and an imperfect femur from the Lametas of Jabalpur and named it *Titanosaurus indicus*. Subsequently he also described some caudal vertebrae belonging to *Titanosaurus blanfordi* and a chevron of *Titanosaurus* sp. from Pisdura (Lydekker 1879), and *Megalosaurus* from the Ariyalur beds of Tiruchirapalli, based on a single tooth (Lydekker 1887).

The most extensive work on the Lameta beds at Jabalpur was done by the English geologist Matley (1921). His efforts led to the discovery of numerous dinosaur remains, mainly from the Main Lameta Limestone. Remains of an armored dinosaur (stegosaurian ankylosaur, *Lametasaurus indicus*) were also found by Matley (1923); these included a sacrum, a pair of ilia, a left tibia, two lateral spines and scutes. Huene and Matley (1933) made a detailed study of the fossiliferous horizons in the Lameta of Bara Simla Hill (Jabalpur) and recognized: (i) the Carnosaur bed, lying below the Main Limestone, with a large number of carnosaur, coelurosaur, stegosaur and sauropod bones; (ii) the Ossiferous conglomerate, lying on top of the Main Limestone, yielding some sauropod remains; and (iii) the Sauropod bed, lying 1.2 m above the conglomerate at the top of the Main Limestone. The last horizon has yielded the largest bulk of sauropod bones including *T. indicus* and

Antarctosaurus septentrionalis. Lametasaurus indicus (Matley 1923, Huene and Matley 1933), in the opinion of Chakravarthy (1934), was misidentified by Matley, though it was a carnosaur (megalosaur). However, Robinson (1967) strongly defended Matley's identification. The carnosaurs include: *Indosuchus raptorius, Indosaurus matleyi, Composuchus solus, Laevisuchus indicus, Jubbulpuria tenuis, Coeluroides largus, Dryptosauroides (?) grandis, Ornithomimoides mobilis* and *O. barasimlensis*. A faunal list of Cretaceous dinosaurs from India has been provided by Prasad (1968).

The age of the Jabalpur Lameta beds has been much debated. Huene and Matley (1933) considered the dinosaur fauna as a whole but Robinson (1967) pointed out that Lameta dinosaurs of Jabalpur do not all occur at the same horizon. Only the carnosaur pocket and the sauropod bed have provided a good yield of bones in a reasonably good state of preservation. Hence the age of the Jabalpur Lameta beds based on carnosaur pocket faunas is approximately Turonian (Robinson 1967). Chatterjee (1978), however, considered it younger than Turonian on the basis of study of Lameta tyrannosaurs and the sauropod bed (containing mainly titanosaurs), and made comparisons with the Maastrichtian of Patagonia. The Pisdura dinosaur assemblage is comparable to the latter. Sahni, Rana, Kumar and Loyal (1984) have recently suggested use of the term 'Lameta Formation' for Jabalpur Lameta as well as the Pisdura–Dongargaon beds.

Narayan-Rao et al. (1927) described vertebrae and some skeletal elements of sauropods from Kallamedu which have been recently reassessed by Yadagiri and Ayyasami (1987) as non–dinosaurian. Chakravarthy (1934) described a stegosaurian humerus from the Lameta beds of the Chota Simla Hills at Jabalpur. Walker (1964) commented on Cretaceous carnosaurs. Prasad and Verma (1967) described caudal vertebrae and hind limb elements of *T. indicus* and some elements of *Titanosaurus* sp. from Lameta beds near Umrer in Nagpur District. Prasad (1968) reviewed the Cretaceous dinosaurs of India. Yadagiri and Ayyasami (1979) described *Dravidosaurus blanfordi* (a stegosaur) from Upper Cretaceous rocks of the Trichinopoly Group in south India. The material includes a partial skull, a tooth, pelvic elements and armor plates. A Coniacian age has been assigned to the stegosaur, which is associated with brachiopods and ammonites. The description is very sketchy. Since there are extremely few records of Late Cretaceous stegosaurs, the new Indian material requires re–evaluation. A part of the dinosaur collection made by Barnum Brown during the 1920's from the Late Cretaceous Lameta Formation, lodged in the American Museum of Natural History, New York, was recently examined by Chatterjee (1978). He recognized two carnosaurs in the Lameta Group: *Indosaurus matleyi*, a megalosaur; and *Indosuchus raptorius*, a tyrannosaur. Berman and Jain (1982) described the braincase of a small sauropod dinosaur from the Lameta Group at Dongargaon. The skeletal material collected from the site, especially the vertebrae, resembles that of *Titanosaurus indicus*. The braincase has been compared to the only other dinosaur braincase known from the Lameta Formation,

described by Huene and Matley (1933) as *Antarctosaurus septentrionalis*, although the former is half the size of the latter. A new collection of skeletal material from Dongargaon is in hand and is being prepared for study by the author. Mathur and Pant (1986) recently described some skeletal elements (humeri) of *Antarctosaurus* sp. from a Late Cretaceous site near Balasinor in Kheda District, Gujarat. Mohabey (1987) has described a juvenile sauropod dinosaur (genus et species indet.) from the Lameta Formation in Gujarat.

A major breakthrough in dinosaur discoveries in India came to light when remains of a Lower Jurassic sauropod dinosaur were found in the Kota Formation in the Pranhita–Godavari Valley in central India (Jain et al. 1962). The skeletal elements represented several individuals of different sizes. Except for the skull, most of the dissociated parts of the skelton were found. This early sauropod dinosaur, *Barapasaurus tagorei* (Jain et al. 1975), was found to have a number of unique characteristics (Jain et al. 1979). Because of these peculiarities, Halstead and Halstead (1981) have raised the genus *Barapasaurus* to the familial level as Barapasauridae, although Bonaparte (1988) has considered it in the family Cetiosauridae.

Knowledge of Triassic dinosaurs from India is sketchy. Kutty (1969) recognized an Upper Triassic fauna in the Gondwana formations of the Pranhita–Godavari Valley, as distinct from the Maleri fauna. He named it the Dharmaram fauna, characterized by the presence of a plateosaurid and a thecodontosaurid prosauropod, co–eval to the Knollenmergel and Rhaetsandstein of Germany. The plateosaurid material includes "a partial, though disjointed skeleton found in situ." The skull is not represented and the size of the animal is comparable to *Plateosaurus*. The thecodontosaurid prosauropod is inferred from elongate cervical vertebrae and some limb bones. To date, however, no description of the material has been published. Chatterjee (1987) announced the discovery of the earliest dinosaur from Asia. He found the fossils in the Upper Triassic Maleri Formation and erected the new taxon *Walkeria maleriensis*. It is a small, podokesaurid theropod, very similar to *Procompsognathus* of Germany and *Coelophysis* of North America (Table 11.1).

In addition to the general works mentioned earlier, dinosaurs from India have also been commented upon, in different contexts, by Lapparent (1957), Walker (1964), Prasad (1968) and Rozhdestvensky (1977). Colbert (1984) and Bonaparte (1986), however, have dealt with the relationships of Indian dinosaurs in greater detail.

Dinosaur Eggshell Records from India

The earliest report of a fossil egg is from the marine Cenomanian Uttatur Formation (Sahni 1957) and is associated with ammonites and belemnites. It is assigned to marine turtles. Discoveries of dinosaurian eggshell fragments, complete eggs, clutches and nests have been made during the 1980's.

Table 11.1

DINOSAURS FROM THE INDIAN PENINSULA

		Formation	Age	Material
CRETACEOUS	**SAUROPODA**			Caudal vertebrae dorsal v, part femur braincase caudal v, ribs, part scapula, part femur, humerus, vertebrae.
1,2,3,4,5	Titanosaurus indicus Lyd.	(1) Lameta Fm. Jabalpur.		
2	Titanosaurus blanfordi Lyd.			
1	Antarctosaurus septentrionalis Huene	(2) Lameta Fm. Pisdura.		
2,6	Antarctosaurus sp.	(3) Lameta Fm. Umrer.		
2	Laplatosaurus madagascariensis (Déperet)	(4) Lameta Gr. Ariyalur.	Maestrichian	
	THEROPODA	(5) Lameta Gr. Trichy.	Turonian (Robinson) Younger Than Turonian (Chatterjee)	Part skull, braincase
	Indosaurus matleyi Huene	(6) Lameta Gr. Balasinor.		Part skull roof
	Indosuchus raptorius Huene	Lameta Fm. Jabalpur.		Cervical vertebra
	Compsosuchus solus Huene			Cervical vertebra
	Laevisuchus indicus Huene			Dorsal vertebra
	Jubbulpuria tenuis Huene			Dorsal vertebra
	Coeluroides largus Huene			Dorsal vertebra
	Dryptosauroides grandis Huene			Dorsal vertebra
	Ornithomimoides mobilis Huene			Dorsal vertebra
	Ornithomimoides barasimlensis Huene			Dorsal vertebra
	Massospondylus rawesi Lyd.	Takli Fm. Nagpur	Maestrichian	tooth
	Megalosaurus sp.	Lameta Gr. Ariyalur.	,,	Vertebrae
	ORNITHISCHIA			
	Lametasaurus indicus Matley	Lameta Fm. Jabalpur	,,	Sacrum, tibia, ilia humerus
	Brachypodosaurus gravis chakravarthi	,,		
	Dravidosaurus blanfordi Yadagiri & Ayyasami	Trichinopoly Gr.	Coniacian	Part skull, scute, tooth, sacrum, ilium
JURASSIC	**Sauropoda** Barapasaurus tagorei Jain, Kutty, Roychowdhury & Chatterjee	Kota Fm.	Liassic	most skeletal elements except skull and footbones
TRIASSIC	{ a plateosaurid { a thecodontosaurid prosauropod	Dharmaram Fm.	Norian Rhaetian	partial skeleton cervical v & limb bones
	Walkeria maleriensis Chatterjee	Maleri Fm	Carnian	part skull

Fossil eggshell localities in India are currently restricted to rocks of Upper Cretaceous age. In some cases dinosaur skeletal remains have also been found near eggshell sites, but it is difficult to determine a definite relationship between the type of eggshell and the specific dinosaur group. At present all the known eggshell localities are in four states (Fig. 11.1) as described below:

(a) *Madhya Pradesh:* Eggshells, fragmented eggs and clutches found at the Bara Simla Hills in District Jabalpur (Sahni and Gupta 1982, Tiwari 1986, Vianey–Liaud et al. 1987).

(b) *Gujarat:* Eggs, clutches and nests found in District Kheda (localities at Balasinor, Rohioli, Khempur and Khevariya) encompassing an area of about 20 sq km (Mohabey 1983, 1984a,b; Srivastava et al. 1986).

(c) *Maharastra:* Eggshell fragments found near Pisdura in District Chandrapur (Jain and Sahni 1985); Takli Formation in District Nagpur has also yielded eggshells having an affinity towards dinosaurs, lizards or avians (Rana 1984; Sahni, Rana, and Prasad 1984;

Vianey–Liaud et al. 1987).

(d) *Andhra Pradesh:* Eggshell fragments found in the Intertrappeans near Asifabad, District Adilabad (Sahni, Rana and Prasad 1984; Prasad 1985). The eggshells are thin, of uncertain affinities, but structural features show a resemblance to dinosaur eggs.

Morphological Features of Indian Dinosaur Eggshell Material

The finding of dinosaur eggshell material from India has naturally attracted attention from many quarters. In view of scattered reports it is useful to categorize the published information in three parts: (i) eggs, nests and clutches, (ii) eggshell fragments, and (iii) microstructure studies.

Eggs, Nests and Clutches. The Kheda localities in the Lameta Formation (Gujarat) are the most spectacular in India; they contain complete eggs in clutches. Among the three localities, one near Balasinor Quarry has only a clutch containing 6 eggs and another near Khempur has been recorded with the stray occurrence of incomplete eggs, but

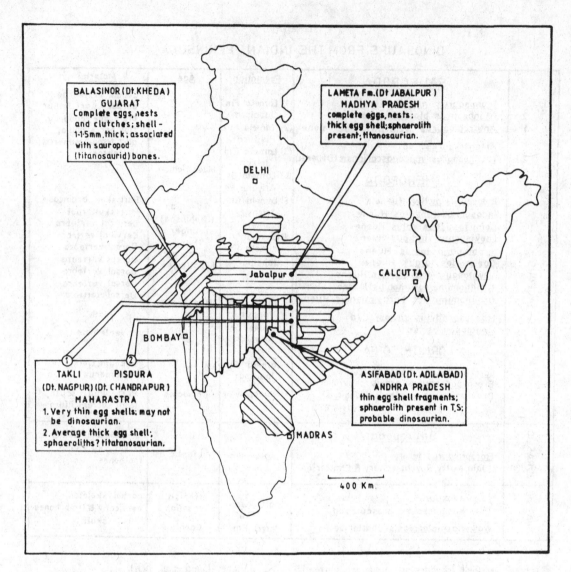

Figure 11.1. Map of India showing dinosaur eggshell, eggs and nest localities.

a third locality (Locality 2 of Mohabey 1984a) near Rahioli village has at least 7 complete clutches of eggs in an area of about 700 sq m (Fig. 11.2A). The area of each clutch is less than one sq m and the clutches occur in shallow pits with less than 20 cm depth. The eggs are spherical with a diameter ranging from 11 to 15 cm. Mohabey suggested that they represent the original nesting ground of the dinosaurs. He also suggested that the dinosaurs probably dug shallow pits, laid the eggs and covered them with available dug out material. The egg clutches are very close to each other and some are within a distance of less than one meter. Each clutch contains 3 to 6 eggs; some of the eggs in the clutches are crushed and shell fragments are scattered. The clutches occur in the same stratigraphic horizon. The shell is chocolate brown and the thickness of the shell is slightly variable from locality to locality. Srivastava et al. (1986) suggested that at least three eggs were laid in one oviposition by one animal and that in

the Rahioli area the same site was used for oviposition again and again. A similar situation has been noted earlier by Horner (1982) among some ornithischian dinosaurs.

Another dinosaur eggsite has been recently discovered by Prof. A. Sahni (Tiwari 1986). The clutches have been found embedded in the Main or Lower Limestone horizon of the Lameta Formation at Bara Simla Hill. Estimated diameter of each egg ranges from 10 to 20 cm, although no complete eggs have been recovered. The largest clutch contains as many as 8 eggs, while the smallest has one or two (Fig. 11.2B). Individual clutches are separated by 2.6 to 9 m but some are as far as 92 m apart. The eggs within the individual clutches are very close to each other and sometimes superimpose one another. Each clutch occupies a maximum area of about one meter.

Eggshell Fragments. Fossil eggshell fragments of dinosaurs were first reported by Sahni and Gupta (1982) from the

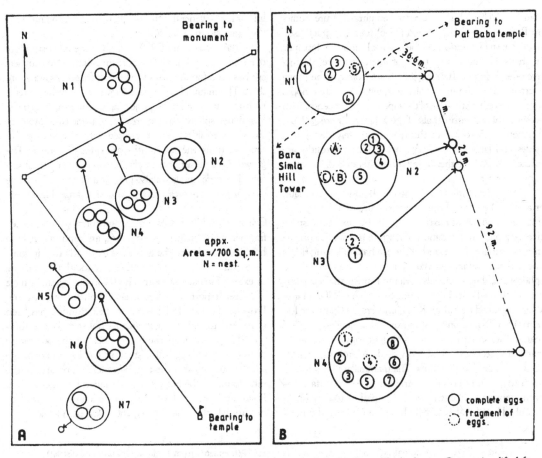

Figure 11.2. **A,** distribution of dinosaur egg clutches at a site in Rahioli village, Kheda district, Gujarat (modified from Mohabey 1984a). **B,** distribution of dinosaur egg clutches in Limestone horizon of Lameta Formation at Bara Simla Hill, Jabalpur district, Madhya Pradesh (modified from Tiwari 1986 MS).

Lameta Formation of Bara Simla Hill, Jabalpur District. The thickness of the eggshell ranges from 1 to 1.5 mm. Mohabey (1983) recovered 6 spherical dinosaurian fossil eggs from Infratrappean limestone in the area north of Balasinor in Kheda District, Gujarat. Srivastava et al. (1986) distinguished Kheda Type 'A' eggs having a thickness of about 1 mm with a dimaeter of 14–20 cm from Type 'B' being over 2 mm in thickness with 12–16 cm diameters. More dinosaur eggs have also been reported from near Rahioli, about 5 km north–northeast of Balasinor. The structural characteristics of these eggs have been found to be different from the eggs reported at Jabalpur. The Takli Formation of Nagpur has also yielded fragments of eggshells having an affinity with either dinosaurs, lizards or avians (Rana 1984; Sahni, Rana, and Prasad 1984; Vianey-Liaud et al. 1987). Intertrappean deposits of Asifabad, Adilabad District (A.P.) have also yielded thin eggshell fragments of uncertain affinity. Sahni, Rana, and Prasad (1984) claimed that their structural features show some similarity with dinosaur eggs. At Pisdura (Lameta Formation) in Chandrapur District, Maharastra, Jain and Sahni (1985) have reported eggshell fragments while wet sieving the white ossiferous quartzitic sandstone for microvertebrates (Fig. 11.3A). Most of the

fragments average 1 sq cm in size and vary from 1 to 2 mm in thickness, similar to the Jabalpur material. The Rahioli eggshells are, however, generally thicker, averaging slightly over 3 mm. The pathological forms (Mohabey 1984b) have an even greater thickness. The shell fragments from the Takli Formation at Nagpur and Asifabad, in the majority of cases, are extremely thin, averaging from 130μ to 10μ.

Microstructure Studies. Thin section and SEM (Scanning Electron Microscope) studies have been carried out on all eggshell material from different localities and horizons in India. In fact the authentic identification of dinosaur eggshell material largely depends upon diagnostic characteristics such as the presence of nodose external features, and spheroliths with canaliculae seen in thin sections. These features have been identified in nearly all cases.

Mohabey (1984a), while describing Balasinor eggshell material, differentiated the shell into two types: (1) comparable to angusticanaliculate type (Sochava 1970), and (2) comparable to tubocanaliculate type (Sochava 1970). In both cases mammillary knobs are present on the external and internal surfaces. Srivastava et al. (1986) and Vianey-

Liaud et al. (1987) have made detailed ultrastructure studies on Balasinor eggshells. Tiwari (1986) found characteristic nodular ornamentation on the external surface of Jabalpur eggshells. He found two types of microstructure, distinctive of each group. Type 1 eggshells are from clutches; they average 0.6 to 1.8 mm in thickness, with distinctive nodes (not coalesced), with spheroliths tapering at both ends, with presence of aeration canals. Type 2 eggshells are isolated fragments, 2.5 to 3 mm thick, with sparsely distributed nodes and large mammillary knobs. The spheroliths are cylindrical and neither fused nor interlocked; aeration canals are rare.

Sahni, Rahni, and Prasad (1984) and Vianey-Liaud et al. (1987) described the microstructure of Takli eggshell fragments, which are extremely thin but exhibit distinctive external ornamentation of discrete mammillae on the internal surface. The external surface has coàlescing ridges and isolated solitary tubercules. There are no well defined spheroliths, but vertically oriented prismatic elements extend throughout the shell thickness. The Asifabad eggshells described by Sahni, Rana, and Prasad (1984) are less than 0.5 mm thick and possess external ornamentation and mammillary structure similar to Nagpur eggshell. However, transverse sections reveal individual spheroliths, characteristic of sauropod dinosaurs. The mammillary layer is confined to the internal surface and the external surface has prominent node-like structures. The pathological eggshells described by Mohabey (1984b) have double layered sphero-liths and are generally thicker than other fragments from the same site (Fig. 11.3C).

Jain and Sahni (1985) described the microstructure of Pisdura eggshells from Chandrapur District. No complete egg has been found at this locality but fragments are abundant. The mammillary surface shows well developed relief with mammillary knobs projecting as columns internally. Distinctive spheroliths are present; in some cases the individual spheroliths are partly fused with one another. In some sections the spheroliths are quite indistinct. The growth lines persist and are distinct up to the external surface. The pore canals are also present and external; minute circular pores are present for gaseous exchange (Fig. 11.3B).

Comment. While discussing the morphology of eggshell fragments and their microstructure, an important feature has been noted, viz., the eggshell is extremely thin in some cases and relatively thick in others, in addition to varying in external ornamentation. Thin dinosaur eggshells have also been reported by Erben et al. (1979) from Aix-en-Provence Locality in France and their thinness has been attributed to physiological and pathological causes (hormonal imbalance). At the French locality, the percentage of thin to thick eggshell is small in the basal part of the section but, towards the top, becomes comparable to the ratio found in the Nagpur and Asifabad localities (Sahni, Rana, and Prasad 1984). At present it is not possible to be sure whether the thin eggshells are pathologic.

Figure 11.3. **A, A',** external morphology of dinosaurian eggshell fragments from Pisdura, Chandrapur district, Maharastra (x 10). **B,** microstructure (SEM) of eggshell from above (from Jain and Sahni 1985). **C,** microstructure of thick eggshell from Kheda district, Gujarat, showing double-layered spheroliths (from Mohabey 1984b).

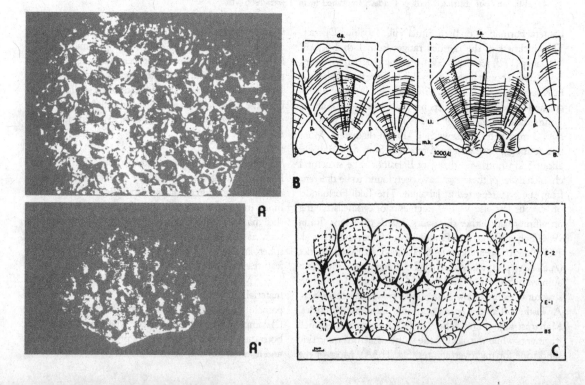

Association of Eggshells and
Dinosaur Skeletal Elements from India

It is somewhat odd that nearly all the discoveries of dinosaur eggshells anywhere in the world are restricted to the Late Cretaceous (Senonian–Maastrichtian), as noted by Sochava (1971). Until recently the dinosaur egg–bearing localities were restricted to the continental deposits in central Asia, southwestern Europe and North America. Recently Indian finds have added to the ever increasing documentation of such material (Sahni and Gupta 1982; Mohabey 1983, 1984a,b; Sahni, Rana, and Prasad 1984; Jain and Sahni 1985; Srivastava et al. 1986; Vianey–Liaud et al. 1987). Mohabey (1987) claimed to have found "the partial remains of a juvenile sauropod" from "one of the dinosaur egg clutches" in the Upper Cretaceous Lameta Formation of the Panchmahals district (Gujarat). The material, however, requires reassessment since there are some apparent inconsistencies. For example it is stated that: (i) cervical vertebrae have longer neural spines than dorsals, (ii) dorsal and caudal vertebrae have zygosphene–zygantra articulation, (iii) caudal vertebrae are amphicoelous to slightly amphiplatyan and (iv) there is no fusion between centrum and neural arch. The material mostly consists of a series of vertebrae, in addition to questionably identified skeletal elements. My quick look at the specimen in November 1985 suggested that it resembles a ?booid snake. In fact such snakes are well documented from a number of Upper Cretaceous Lameta localities in India (Jain and Sahni 1983, Sahni et al. 1982).

Several attempts have been made at identifying a specific dinosaur genus from skeletal material found near dinosaur eggs, or based on the indirect evidence of egg morphology. Tiwari (1986) has recently made a detailed SEM study of eggshell found at sites near Jabalpur. He pointed out that, in size and external and internal morphology, these bear features identical to those from Aix–en–Provence of southern France (Maastrichtian), ascribed to the sauropod genus *Hypselosaurus*. Among the skeletal remains at Jabalpur, *Titanosaurus* remains are by far the most common in the sauropod bed. It may, therefore, suggest a possible relationship between sauropod eggshell and *Titanosaurus* skeletal remains at Jabalpur.

The Gujarat localities in Kheda District are the most spectacular in which complete eggs in clutches have been found. In fact a hatchery spread over an area of 20 sq km has been identified. Mohabey (1984a) indicated the occurrence of sauropod dinosaur bones at the Kheda beds in association with dinosaur eggs in these beds (calcareous sandstones) as well as those lying below (pebbly sandstones and conglomerate). Mathur and Pant (1986) reported several well preserved humeri, estimated length 70–85 cm, belonging to sauropod dinosaurs, from the same locality. Because the Kheda and Jabalpur dinosaur eggshell material is somewhat similar, the Kheda specimens are assignable to *Titanosaurus*.

In Maharastra, eggshells reported by Sahni, Rana, and Prasad (1984) are from the Takli Formation near Nagpur. The precise age of the Takli Formation is a matter

of some debate but it may be between Uppermost Cretaceous and Paleocene (Vianey–Liaud et al. 1987). The eggshell fragments are found in both the green clay and the sandy marl beds. There is a rich invertebrate fauna among the fossils (*Paludina*, *Valvata*, *Limnaea* and *Physa*) and a number of ostracods; vertebrates are represented by fishes (*Dasyatis*, *Lepidotes*, *Lepisosteus*, *Pycnodus*, *Enchodus*, *Belonostomus*), pelobatid frog, turtles, scincomorph lizard, booid snake and *Crocodylus*. A carnosaur tooth (*Massospondylus rawesi*) was reported by Lydekker (1890) and several serrated tooth fragments have been recently found by the Indo–French team of research scientists (Sahni et al. 1982, Vianey–Liaud et al. 1987). Sahni, Rana, and Prasad (1984) pointed out that extremely thin shell, far thinner than commonly known for dinosaurs, is the main obstacle in assigning the Nagpur eggshell to dinosaurs.

A second site of dinosaur eggshell fragments in Maharastra, at Pisdura, was described by Jain and Sahni (1985). The eggshells were found in white ossiferous sandstone and were obtained by wet screen sieving of matrix. The microvertebrate assemblage, remarkable for its diversity, is represented by an admixture of freshwater and brackish to marine forms (Jain and Sahni 1983). The freshwater component includes *Lepisosteus*, pelobatid frog, booid snake, pelomedusid turtle, crocodile and dinosaurs; while the brackish to marine taxa include the fishes *Dasyatis*, *Rhombodus*, *Igdabatis*, *Rhinoptera*, pycnodonts, *Enchodus*, tetraodont fishes and *Arius*. The mixed facies assemblage resembles the vertebrate asseemblage near Nagpur. The Pisdura dinosaur fauna includes the sauropods *Titanosaurus*, cf. *T. indicus*, *Antarctosaurus* and *Laplatosaurus* (Huene and Matley 1933, Berman and Jain 1982). The Pisdura eggshells reveal a structure which encompasses the morphological variation observed in tubocanaliculate and angusticanaliculate eggshell types (Erben et al. 1979). Jain and Sahni (1985) eliminated turtles as the egg layers on the basis of external sculpturing and the fusion of spheroliths; instead they suggested sauropod affinity. It appears that all dinosaur eggshell material so far discussed is titanosaurid, even though there are variations in thickness of shell and detailed microstructural features.

Lastly, the Asifabad locality in Andhra Pradesh (Sahni, Rana, and Prasad 1984) yielded extremely thin (less than 0.5 mm thick) eggshells. Though the external ornamentation and mammillary structure is somewhat similar to the Nagpur eggshells, the microstructure is different. Individual spheroliths are distinct and similar to the structure seen in sauropod dinosaurs. At Asifabad, Intertrappean deposits are thicker than at Nagpur and the eggshells have been recovered from both the Upper and Lower horizons which represent brackish water as well as freshwater conditions. The associated microfossils consist of charophytes, cyprid ostracods, unionid pelecypods and pulmonate gastropods. The presence of dinosaurs was first reported by Rao and Yadagiri (1981) in the Intertrappeans of Asifabad. So far only a few limb bones having probable affinities to dinosaurs have been reported. Sahni, Rana, and Prasad (1984), while commenting on Asifabad eggshells, pointed out that (i) in external sculpturing, structure of spheroliths and mammillary surface, the eggshells are similar to those of sauropods, and (ii) except for the thinness of the eggshells, no other feature mitigates against assigning them to the Dinosauria.

Mohabey (1986) briefly described an isolated footprint from Infratrappean limestone of the Jetholi Talao area near Balasinor in Kheda district, Gujarat. The footprint was found near a dinosaur egg clutch and was assigned to a sauropod dinosaur. It comprises three digits with the longest, middle digit flanked by shorter ones. It is an important first record from India, although the photograph is not very convincing as to its authenticity. The shape (3 digit) is more suggestive of a theropod or ornithopod than a sauropod.

Coprolites

Coprolites occur in various localities in India, including those where dinosaur bones have been found. The earliest reports of the occurrence of coprolites together with vertebrate remains are by Hislop (1860, p. 163; 1864, p. 281). Subsequently Hughes (1877), Blanford (1878), King (1881) and Aiyenger (1937) mentioned collections of coprolites. The most detailed papers on coprolites from India are those by Matley (1939a,b), who described the coprolites from Maleri beds (attributed to Ceratodus, a lungfish) and from Pisdura, attributed to reptiles. Coprolites from Maleri beds have been recently commented upon by Sohn and Chatterjee (1979) and Jain (1983). These are not associated with any dinosaur remains. However, coprolites from Pisdura (and those recently collected from the Dongargaon locality near Pisdura by the author) have a close association with dinosaur skeletal remains. These will be discussed in detail. Chiplonkar (1980) reviewed the Indian literature on coprolites.

The locality near Pisdura (south of Nagpur) has been known for dinosaur bones for over a century (Lydekker 1879); recently fragments of eggshells assignable to sauropod dinosaurs were reported (Jain and Sahni 1985). Among other vertebrate remains from Pisdura are a pelomedusid turtle (Jain 1977, 1986) and fish, amphibian and reptile (booid snake and Crocodylus) fossils (Jain and Sahni 1983). Matley (1939b) had at his disposal 600 specimens of coprolites from which he selected 350 for further examination. Matley's classification of various coprolites was mainly based on (i) the presence or absence of longitudinal ribbing and (ii) size: large, medium and small. Some 60 specimens in his collection range in weight from 1 to 2 kg with a diameter of about 10 cm each. These forms do not have ribbing on their external surface. The coprolites are rather elongated or hemispherical, often with wrinkled surface, elliptical in transverse section. In section, they reveal fine-grained matrix, without any spiral features; some have contraction cracks, filled with calcite or silica. As to the producers of these coprolites, Matley (p. 540) suggested: "The great size of these coprolites makes it clear that they came from very large animals and they must therefore be assigned to the titanosaurs whose remains are associated with them."

A second group (Group B and Ba of Matley) consist of coprolites with ribbing impressed on the surface. Smaller in size than the first group, length up to 9 cm and breadth 4–6 cm, these coprolites have a diameter up to 64 mm. They also exhibit sun-cracks and similar features of subaerial drying. Matley suggested that these coprolites represent droppings on land, in contrast to the earlier group which appear to have been deposited in water. The ribbing is a characteristic feature (10 to 20 ribs may be counted), running generally longitudinally but in some cases obliquely or spirally. It has been suggested that ribbing has been "caused by pressure on the plastic material during the opening and distension of the anal or cloacal aperture under the action of the sphincter muscle". Matley assigned these coprolites to the tortoise since "modern Chelone has longitudinal rugae in its great intestine and also can contract its intestine so as to produce temporary corrugations."

One last group of coprolites from Pisdura (Group C of Matley) represents some 130 specimens ranging in size from 12 to 68 mm with a diameter of 16–34 mm. These are generally sausage shaped with a rounded, terminal end; the surface is generally smooth, occasionally with muscular marks and wrinkles. Ribbing is absent. Matley assigned these coprolites to the chelonians or possibly small dinosaurs, as it is known that a large variety of Carnosauria, including Coelurosauria, inhabited India in Upper Cretaceous time.

A large collection of coprolites in hand with the author is currently being studied. Pending completion of these studies, a few remarks may be made. A collection of coprolites from the excavation site of the Lower Jurassic sauropod Barapasaurus tagorei has revealed the presence of large coprolites with desiccation marks. These are either flattened, spheroid, ovoid or ellipsoidal. There is no ribbing nor any spiral coiling. No other large animal is associated with the fossils except the sauropod dinosaur. As such, the conclusion to associate them together is very inviting. Some 25 specimens of coprolites are in hand from the Pisdura site but these do not add any significant information not already given by Matley (1939b). Lastly a collection of coprolites has just been brought in from the Dongargaon site (about 16 km south of Pisdura) where titanosaurian skeletal remains were found in situ. The large coprolites occur in the green clays and sandstone along with bones, occasionally exhibiting a faint groove or two, but no ribbing and very few sun-cracks. Some coprolites are as large as 20 cm long with about 10 cm diameter. Others are about half this size. These are often covered with a thin, sandy matrix. On the basis of the above discussion, these are designated as belonging to titanosaurid dinosaurs. A few small coprolites have also been found which possess ribbing similar to the Pisdura coprolites. Since tortoise remains are quite common in Dongargaon, the latter are assigned to tortoises.

Summary

The documentation of trace fossils to supplement skeletal evidence of the dinosaurs is valuable in better understanding their behavior. It also often gives clues to paleoenvironment. However paleoichnology is also a neglected area of research. The skeletal remains of dinosaurs from the Indian peninsula have been known for over a century but a footprint has been recorded recently (Mohabey 1986). Coprolites have been known even for a longer period but no serious attempts to study them have been made, except sporadically.

While attempting to review dinosaur trace fossils from India, it was necessary to include a brief review of the skeletal remains as well. The reason is two-fold: (i) any evaluation of dinosaur trace fossils would naturally attempt to locate possible skeletal remains in the same horizon/locality, and (ii) in the absence of a recent review (not available otherwise) readers would not connect the two. The discoveries of eggshells, eggs, nests and clutches from India attributed to dinosaurs are a welcome addition to the ever increasing documentation from different parts of the world. SEM studies of some Indian eggshell material shows similarities with material from localities in southern France. Much remains to be done concerning (i) location of new sites, (ii) development of a uniform grouping of ultrastructure (SEM) studies of eggshell to identify different dinosaur groups, distinctive from other reptiles, and (iii) locating dinosaur tracks in India.

The study of coprolites has not attracted many investigators because comparative data on fecal matter of living reptiles is very rarely available. Moreover, neither the external morphology nor the internal features of coprolites reveal any useful clues. The large size of the coprolites, general absence of ribbing on the external surface and proximity with dinosaurs (skeletal material) is the only reason to associate them.

References

Aiyenger, K. N. 1937. A note on the Maleri beds of Hyderabad State (Deccan) and the Tiki beds of South Rewa. *Rec. Geol. Surv. India* 71 (4):401–406.

Berman, D. S., and Jain, S. L. 1982. The braincase of a small sauropod dinosaur (Reptilia: Saurischia) from the Upper Cretaceous Lameta Group, central India, with a review of Lameta Group localities. *Ann. Carnegie Museum Pittsburgh* 51 (21):405–422.

Blanford, W. T. 1878. On the stratigraphy and homotaxis of the Kota–Maledi (Maleri) deposits. *Pal. Indica* Ser. IV, 1 (2):17–25.

Bonaparte, J. F. 1986. The early radiation and phylogenetic relationships of the Jurassic sauropod dinosaurs, based on vertebral anatomy. *In* Padian, K. (ed.). *The Beginning of the Age of Dinosaurs* (Cambridge University Press) pp. 247–258.

Chakravarthy, D. K. 1934. On a stegosaurian humerus from the Lameta beds of Jubbulpore. *Quart. Jour. Geol. Min. Met. Soc. India* 6:75–82.

Chatterjee, S. 1978. *Indosuchus* and *Indosaurus*, Cretaceous carnosaurs from India. *Jour. Paleont.* 52 (3):570–580.

1987. A new theropod dinosaur from India with remarks on Gondwana–Laurasia connection in the Upper Triassic. Gondwana Six: Stratigraphy, Sedimentology and Paleontology. McKenzie, G. D. (ed.). *Geophysical Monograph* 41 (Washington: A.G.U.) pp. 183–189.

Chiplonkar, G. W. 1980. First twelve years of Palichnology

in India (presidential address). *Proc. 3rd Ind. Geol. Congr. Poona* pp. 1–38.

Colbert, E. H. 1974. *Wandering Lands and Animals*. (London: Hutchinson) pp. 1–223.

1977. Mesozoioc tetrapods and the northward migration of India. *Jour. Palaeont. Soc. India* 20:138–145.

1984. Mesozoic reptiles, India and Gondwanaland. *Ind. Jour. Earth Sci.* 11 (1):25–37.

Erben, H. K. 1970. Ultrastrucktur und mineralisation rezenter und fossiler Eischalen bei Vogeln und Reptilien *Biominer. Res. Rep.* 1:1.66.

Erben, H. K., Hoefs, J., and Wedepohl, H. K. 1979. Palaeobiological and isotopic studies of egg shells from a declining dinosaur species. *Palaeobiology* 5 (4):380–414.

Halstead, L. B., and Halstead, J. 1981. *Dinosaurs*. (Poole (Dorset), U.K.: Blanford Press) pp. 1–170.

Hislop, R. S. 1860. On the Tertiary deposits associated with Trap rocks in the East Indies. *Quart. Jour. Geol. Soc. London* 16:163.

1864. Further discovery of fossil teeth and bones of reptiles in central India (extracts from letters). *Proc. Geol. Soc. London* 20:280–282.

Horner, J. R. 1982. Evidence of colonial nesting and site fidelity among ornithischian dinosaurs. *Nature* 297:675–676.

Huene, F. v., and Matley, C. A. 1933. The Cretaceous Saurischia and Ornithischia of the central provinces of India. *Palaeont. Indica* (Geol. Surv. India) 21 (1):1–74.

Hughes, T. W. H. 1877. The Wardha Valley coalfield. *Mem. Geol. Surv. India* 13:1–154.

Jain, S. L. 1977. A new fossil pelomedusid turtle from the Upper Cretaceous sediments, central India. *Jour. Palaeont. Soc. India* 20:360–365.

1983. Spirally coiled 'coprolites' from the Upper Triassic Maleri Formation, India. *Palaeontology* 26 (4):813–829.

1986. New pelomedusid turtle (Pleurodira: Chelonia) remains from Lameta Formation (Maastrichtian) at Dongargaon, central India, and a review of pelomedusids from India. *Jour. Palaeont. Soc. India* 31:63–75.

Jain, S. L., Robinson, P. L., and Roychowdhury, T. K. 1962. A new vertebrate fauna from the early Jurassic of the Deccan, India. *Nature* 194:755–757.

Jain, S. L., Kutty, T. S., Roychowdhury, T., and Chatterjee, S. 1975. The sauropod dinosaur from the Lower Jurassic Kota Formation of India. *Proc. Roy. Soc. London* A, 188:221–228.

1979. Some characteristics of *Barapasaurus tagorei*, a sauropod dinosaur from the Lower Jurassic of Deccan, India. *Fourth Int. Gondwana Sym. Proc.* (Calcutta, 1977) 1:204–216.

Jain, S. L., and Sahni, A. 1983. Upper Cretaceous vertebrates from central India and their palaeogeographical implications. *In* Maheswari, H. K. (ed.). *Cretaceous of India* (Lucknow: Indian Assoc. Palyn.) pp. 66–83.

1985. Dinosaurian egg shell fragments from the Lameta Formation at Pisdura, Chandrapur District, Maharastra. *Geosci. Jour.* 6 (2):211–220.

King, W. 1881. The geology of the Pranhita–Godavari Valley. *Mem. Geol. Surv. India* 18:151–131 (reprint, 1931).

Kutty, T. S. 1969. Some contributions to the stratigraphy of the Upper Gondwana Formations of the Pranhita-Godavari Valley, central India. *Jour. Geol. Soc. India* 10 (1):33–48.

Lapparent, A. F. 1957. The Cretaceous dinosaurs of India and Africa. *Jour. Palaeont. Soc. India* 2:109–112.

Lydekker, R. 1877. Notices of new and other vertebrates from Tertiary and secondaries of India. *Rec. Geol. Surv. India* 10:30–43.

    1879. Fossil Reptilia and Batrachia. *Pal. Indica* (Geol. Surv. Ind.) 1:1–36.

    1887. The fossil vertebrata of India. *Rec. Geol. Surv. India* 20 (2):51–80.

    1890. Note on certain vertebrate remains from the Nagpur District. Pt. II. Part of the chelonian plastron from Pisdura. *Rec. Geol. Surv. India* 25 (6):329–337.

Matley, C. A. 1921. Stratigraphy, fossils and geological relationships of the Lameta beds of Jubbulpore. *Rec. Geol. Surv. India* 53 (2):142–164.

    1923. Note on an armoured dinosaur from the Lameta beds of Jubbulpore. *Rec. Geol. Surv. India* 55:105–109.

    1939a. On some coprolites from the Maleri beds of India. *Rec. Geol. Surv. India* 74:530–534.

    1939b. The coprolites of Pijdura, Central Provinces. *Rec. Geol. Surv. India* 74:535–547.

Mathur, U. B., and Pant, S. C. 1986. Sauropod dinosaur humeri from Lameta Group (Upper Cretaceous – ?Palaeocene) of Kheda District, Gujarat. *Jour. Palaeont. Soc. India* 31:22–25.

Mohabey, D. M. 1983. Note on the occurrence of dinosaurian eggs from the Infratrappean limestone in Kheda District, Gujarat. *Curr. Sci.* 52 (24):1194.

    1984a. The study of dinosaurian egg shells from the Infratrappean Limestone of Kheda District, Gujarat. *Jour. Geol. Soc. India* 25 (6):329–337.

    1984b. Pathologic dinosaurian egg shells from Kheda District, Gujarat. *Curr. Sci.* 53 (13):701–703.

    1986. Note on dinosaur footprint from Kheda District, Gujarat. *Jour. Geol. Soc. India* 27 (5):456–459.

    1987. Juvenile sauropod dinosaur from Upper Cretaceous Lameta Formation of Panchmahals District, Gujarat, India. *Jour. Geol. Soc. India* 30 (3):210–216.

Narayan-Rao, C. R., and Seshachar, B. R. 1927. A short note on certain fossils taken in the Ariyalur area, south India. *Jour. Mysore Uni.* 1 (2):144–152.

Olshevsky, G. 1978. The archosaurian taxa (excluding the Crocodylia). *In Mesozoic Meanderings*, I. (Illinois: Stevens Publishing Co.) pp. 1–48. (Supplements 1, 2)

Prasad, G. V. R. 1985. Microvertebrates and associated microfossils from the sedimentaries associated with Deccan traps of the Asifabad region, Asifabad District, Andhra Pradesh. Ph.D. thesis (MS), Panjab University, Chandigarh, pp. 1–320.

Prasad, K. N. 1968. Some observations on the Cretaceous dinosaurs in India. *In Cretaceous–Tertiary Formation of South India. Geol. Soc. India Mem.* 2:248–255.

Prasad, K. N., and Verma, K. K. 1967. Occurrence of dinosaurian remains from the Lameta beds of Umrer, Nagpur District, Maharastra. *Curr. Sci.* 36 (20):547–548.

Rana, R. S. 1984. Microvertebrate palaeontology and biostratigraphy of the Infra- and Intertrappean beds of Nagpur, Maharastra. Ph.D. thesis (MS), Panjab University, Chandigarh, pp. 1–234.

Rao, B. R. J., and Yadagiri, P. 1981. Cretaceous Inter-

trappean beds from Andhra Pradesh and their stratigraphic significance. *Geol. Soc. India Mem.* 3:287–291.

Robinson, P. L. 1967. The Indian Gondwana Formations — a review. *Internat. Gondwana Symp.* (Strat. and Pal.), Mar del Plata 201–268.

Rozhdestvensky, A. K. 1977. The study of dinosaurs in Asia. *Jour. Palaeont. Soc. India* 20:102–119.

Sahni, A., and Gupta, V. J. 1982. Cretaceous egg shell fragments from the Lameta Formation, Jabalpur. *Bull. Ind. Geol. Assoc.* 15 (1):85–88.

Sahni, A., Kumar, K., Hartenberger, J. L., Jaeger, J. J., Rage, J. C., Sudre, J., and Vianey-Liaud, M. 1982. Discovery of terrestrial microvertebrates in the Palaeocene of the Deccan Traps, India. Geodynamic implications. *Bull. Geol. Soc. France* 24 (5/6):1093–1099. [in French]

Sahni, A., Rana, R. S., Kumar, K., and Loyal, R. S. 1984. New stratigraphic nomenclature for the Intertrappean beds of the Nagpur reegion, India. *Geosci. Jour.* 5 (1):55–58.

Sahni, A., Rana, R. S., and Prasad, G. V. R. 1984. SEM studies of thin egg shell fragments from the Inter-trappean (Cretaceous – Tertiary transition) of Nagpur and Asifabad, peninsular India. *Jour. Palaeont. Soc. India* 29:26–33.

Sahni, M. R. 1957. The fossil reptilian egg from Uttatur (Cenomanian) of south India, being the first record of vertebrate fossil egg in India. *Rec. Geol. Surv. India* 87 (4):671–674.

Sochava, A. V. 1970. Microtexture of dinosaur egg shells from the Lower Cretaceous of the Northern Gobi. *Dokl. Akad. Nauk. SSR.* 192 (5):1137–1140. [English translation]

    1971. Two types of egg shells in Cenomanian dinosaurs. *Palaeo. Jour.* 2 (2):1–13 [English translation]

Sohn, I. G., and Chatterjee, S. 1979. Freshwater ostracodes from Late Triassic coprolites in central India. *Jour. Paleont.* 53 (3):578–586.

Srivastava, S., Mohabey, D. M., Sahni, A., and Pant, S.C. 1986. Upper Cretaceous dinosaur egg clutches from Kheda District (Gujarat, India). *Palaeontographica Abt. A* 193:219–233.

Steel, R. 1970. Saurischia. *In Kuhn, O. (ed.). Encyclopaedia of Paleoherpetology* 14:1–87.

Tiwari, A. 1986. Biostratigraphy, palaeoecology and dinosaur egg shell ultrastructure of the Lameta at Jabalpur, Madhya Pradesh. MS thesis (MS), Panjab University, Chandigarh pp. 1–129.

Vianey-Liaud, M., Jain, S. L., and Sahni, A. 1987. Dinosaur eggshells (Saurischia) from the Late Cretaceous Inter-trappean and Lameta Formations (Deccan, India). *Jour. Vert. Paleont.* 7 (4):408–424.

Walker, A. D. 1964. Triassic reptiles from the Elgin area: *Ornithosuchus* and the origin of carnosaurs. *Phil. Trans. Roy. Soc. London B*, 248 (744):53–134.

Yadagiri, P., and Ayyasami, K. 1979. A new stegosaurian dinosaur from Upper Cretaceous sediments of south India. *Jour. Geol. Soc. India* 20 (11):521–530.

    1987. A carnosaurian dinosaur from the Kallamedu Formation (Maestrictian Horizon), Tamil Nadu *In Geol. Surv. India, Spec. Publ. No. 11, Three Decades of Developments in Palaeontology and Stratigraphy in India* Vol. I, Precambrian to Mesozoic, pp. 523–528.

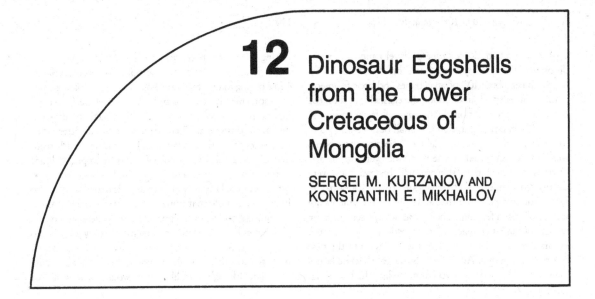

12 Dinosaur Eggshells from the Lower Cretaceous of Mongolia

SERGEI M. KURZANOV AND
KONSTANTIN E. MIKHAILOV

Abstract

Eggshells are reported from a probable nest site at the Builjasutuin–Khuduk locality in Mongolia. This is the first report of eggshell material of Lower Cretaceous age from this region. The site also yields evidence which permits inferences about the behavior of carnivorous dinosaurs at nest sites.

Introduction

Mongolia is rich with localities of Upper Cretaceous dinosaur eggs and eggshells including Bayn-Dzak (Shabarakh–Usu), Khermin–Tsav, Tugrikiin–Shireh, and Guriliin–Tsav, which are the largest sites. The age of the known eggshell locality at Olgoi–Ulan–Tsav is uncertain. Almost all of these localities are located in the southern Gobi districts. No eggs or eggshell, up to now, had been found in the numerous Lower Cretaceous localities which had produced an abundance of other fossil remains. Lower Cretaceous eggshell was only known from Utah, USA (Jensen 1970).

In 1985 the East–Gobian group of the Joint–Soviet–Mongolian paleontological expedition visited the known Lower Cretaceous dinosaur locality Builjasutuin–Khuduk, 20 miles northeast from Barun–Bajan–Ula, Ubur–Khangai Aimak. The locality is dated as Upper Neocomian–Aptian by coniferous pollen, fresh water molluscs, ostracods and also shells of turtles and ornithopod dinosaur skeletons, *Psittacosaurus mongoliensis*.

The remains of dinosaurs are found predominately in the medial part of the exposure. The character of these deposits suggests a limnological origin, which is also confirmed by analysis of the fauna, crossbedding, lithological composition and the close position of these deposits to the Paleozoic massif (Shuvalov 1975). There are very few complete skeletons; heads and limbs are always separated. Skeletal fragments composed of articulated bones occur often. Single bones and badly tumbled bone fragments are numerous. All this suggests that the *Psittacosaurus* carcasses disintegrated on the ground (?sub–aerial) and were then buried by deposits of temporary streams. The possibility exists that other places in the basin were not flooded and were used by dinosaurs to lay their eggs. In these southern parts of the locality, in the limnological deposits, the bones of *Psittacosaurus* are quite rare.

The Builjasutuin–Khuduk Site

The main collecting was done in horizons corresponding to the lower part of layer 2 (Shuvalov 1975) or layer 5 (Bratseva and Novodvorskaya 1976). In this 3 m thick layer of dark clay with large flattened lime concretions is a lens of lighter grayish–brown clay with green and blue spots and small lime concretions. In the lens is also an inclusion of blue, pale purple–spotted aleurite. Some part of this inclusion (which has a diameter of 32–35 cm) eroded to a thickness of 11 cm. The inclusion is filled with eggshell fragments which do not vary much in size and have a maximum size of 2–2.5 cm. The eggshell is scattered throughout the whole thickness of the inclusion rather uniformly, without any obvious concentrations; however, some small horizontal compressions were noticed. The eggshell is not oriented but points in all directions. Eggshell groups which could indicate the form and size of eggs were not found.

It is very likely that this accumulation of eggshell represents the remains of a nest. This assumption is supported by the high concentration of eggshells (over 30%) and the observation that eggshells are only found in the oval shaped, nest–like inclusion but not in the surrounding sediments. The eggshell found downslope from the nest derived from the eroded part of one side of the inclusion (nest). The large amount of eggshell found indicates that the nest contained many eggs, maybe even too many for one laying cycle. However, since egg size and shape are not known, we do not have enough evidence to say that this nesting site had been used repeatedly.

The 'nest' also suggests that the hatchlings occupied it for a longer time and that they were fed by their parents, thus indicating an altricial type of development for some dinosaurs. This assumption is based on the presence of a large number (about 100) of small bones and bone fragments of different reptiles found among the eggshells, and on the character of preservation of the eggshell fragments.

The first thought, that the small bones belonged to the embryos of the eggs, was discarded because these bones were not concentrated in one area, and were not in any obvious relationship to the eggshell. Instead they were distributed uniformly throughout the inclusion. A systematic study showed that all bones and fragments are very small, just a few mm. They are all well preserved, show no sign of tumbling, maceration or deforming which often occurs during burial; apparently they all got into the nest before it was buried. All these facts suggest that the bones may be food remains, leftovers from feeding the hatchlings

during their stay in the nest.

The combination of the bones is rather variable. Most belong to the small ornithopod dinosaur *Psittacosaurus*, consisting mainly of dorsal and caudal vertebrae. Two teeth belong to a sauropod and four other teeth to different predatory dinosaurs. There are also many undetermined fragments of other dinosaurs. A jaw fragment with teeth and a few cervical vertebrae of a scincomorphous lizard (*Paranecellodidae*) were also found. Such mixed bone concentration could hardly be formed accidentally. It is also important to note that none of these bones went through the process of digestion: coprolites and pellets are absent. The bones have no teeth or gnaw marks, they are just too small for this. These small, often tiny, bones evidently belonged to small animals on which the adults also could have fed, although most likely they would have swallowed

Figure 12.1. **A–D,** eggshell with "ornithoid" structure from Builjasutuin-Khuduk, Mongolia, Lower Cretaceous. **A,B,** outer surface of eggshells; sculpture of heteromorphous hillocks and short ridges in longitudinal (long axis of egg) oriented. Scale bars in mm. **C,** SEM micrograph of radial fracture of eggshell showing cones, spongy layer with squamous pattern and vertical striations in hillock. Bar = 150 mkm. **D,** SEM micrograph of inner eggshell surface showing mammillary tips with craters of resorption of calcium carbonate by embryo during incubation. x 220.

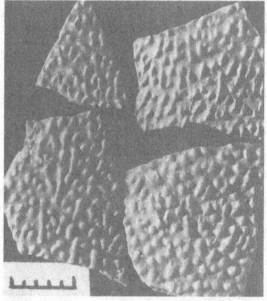

A

B

C

D

them whole. However, it is more possible that these small animals were brought in to feed the young.

Paleoecological Interpretation

The preservation and size of the eggshell fragments suggests that the hatchlings stayed for some time in the nest and during this time they reduced the eggshells to such small size. The variation of size within all the fragments is minimal (Fig. 12.1A,B). The eggshell looks crushed and abraded. If the young leave the nest right after hatching, the eggshell will vary in size considerably and often even egg fragments are preserved.

Judging by the distribution of the eggshell fragments, which are more concentrated in the rougher matrix of the lower part of the inclusion, we get an idea about the probable nest structure. It was an open nest type, a shallow hole in the ground with eggs in more than one layer or circle, separated by plant material. After hatching, plant material and eggshell got thoroughly mixed, with the eggshell almost uniformly distributed through the nest.

This corresponds well to other known fossil nests and the nests of the mound–building birds (*Megapodiidae*). In some of the fossil nests in the Upper Cretaceous, elongated eggs have been found in 2–3 concentric circles, sometimes the eggs of the inner circle overlying those of the external circle. In other nests these eggs were found standing vertically in their natural position, a position similar to that found in the nests of mound-building megapodian birds. Here the eggs are oriented with their broad top up to facilitate embryonic air–exchange and other peculiarities of embryogenesis (Bolotnikov et al. 1985). If eggs that stand vertically and in circles in plant material do not hatch for some reason, they will fall over when the plant material decays, producing the horizontal position described above.

Our chain of assumption — nest of eggs; nest for raising young, feeding young — is speculative but nevertheless is probably similar to the nest building and care for the young found in recent reptiles and crocodiles (Basu and Bustard 1981) and also in fossil forms as described by Horner and Makela (1979) for hadrosaurs.

Eggshell Morphology

The thickness of the eggshell (tubercles included) varies from 0.8 to 1.3 mm, with the thicker shells being rather rare. The outer surface is ornamented with heteromorphic, pointed tubercles (Fig. 12.1B). Sometimes 2–3 tubercles form a short rib of 3 mm length. All ribs and tubercles are mostly oriented in the same direction, forming a network of diagonal and longitudinal rows. Some eggshell has only single tubercles without any orientation. In a few fragments, a vortexed structure typical for egg poles (Mikhailov 1986) can be seen; this eggshell also has the most curvature. Less curved and almost flat fragments are typical for the equatorial region of elongated eggs. It is possible that the eggs of the described nest had an elongated form with ornamentation oriented along the longitudinal axis

Figure 12.2. **A–E,** stylistic drawings of dinosaur eggshell structure based on SEM studies (radial view); shell thicknesses are disregarded. **A–C,** "spherolithic" type; exospherite (crystalline shell unit) consists of thin radiating wedges. **A,** "prolatocanaliculate" pore canal; horizontal growth lines; exospherites have no discrete boundaries in upper 1/3–1/2 of eggshell. Hadrosaurs, Upper Cretaceous of Mongolia, China and Montana, USA. **B,** "tubocanaliculate" pore canal; exospherites have discrete boundaries; semi-concentric growth lines. Sauropods; Upper Cretaceous of France and India. **C,** "multicanaliculate" pore canal; in upper part of eggshell thin wedges are strongly recrystallized. ?Sauropods; Upper Cretaceous of Mongolia and China. **D,** "spherolitho–prismatic" type; in upper part of exospherite thin wedges are fused into large vertical prisms; both zones consist of the same type of structure (tabular); "angusticanaliculate" pore canal. Protoceratopsians; Upper Cretaceous of Mongolia. Hypsilophodont eggshell from Montana, USA similar type. **E,** "ornithoid" type with "angusticanaliculate" pore canal; exospherite is discrete morphological unit only in lower 1/3–1/2; upper part (spongy layer with squamous structure) is a continuous layer; growth lines undulating. Theropods; Lower Cretaceous of Mongolia; Upper Cretaceous of Mongolia, China, Kazakhstan, and Montana, USA.

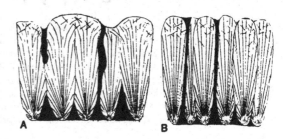

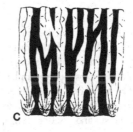

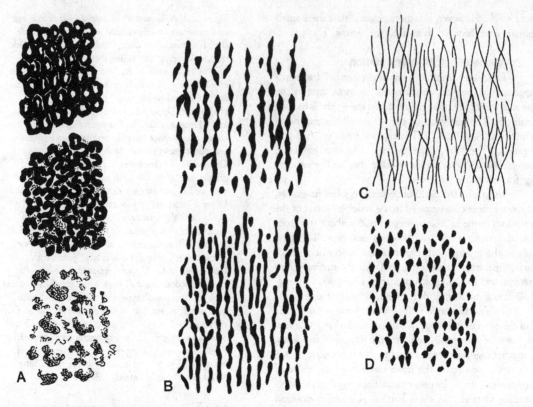

Figure 12.3. **A–D,** external surface of dinosaur eggshells, sculpture pattern in the equatorial region of the eggs.
A, round or lightly ellipsoid eggs; spherolitic type. Various reticulate patterns of nodes. Hadrosaurs; Upper Cretaceous of Mongolia, China and Montana, USA. **B,** elongated symmetrical eggs with blunt ends; "ornithoid" type. Various patterns of ridges and hillocks, oriented longitudinal. Theropods; Upper Cretaceous of Mongolia, China, Kazakhstan. **C,** elongated eggs (slightly asymmetrical) with a blunt and a pointed end; "spherolitho-prismatic" type. Pattern of long fine ridges, oriented longitudinal; very often shell has smooth surface; no sculpture on poles of egg. Protoceratopsians; Upper Cretaceous of Mongolia. **D,** eggshell, perhaps elongated egg; ornithoid type. Heteromorphous hillocks and short ridges, oriented longitudinal. Lower Cretaceous at Builjasutuin–Khuduk, Mongolia.

and no orientation on the egg poles. Eggshells of similar microstructures are widespread in the Upper Cretaceous of Mongolia (Fig. 12.3B–D).

Study of thin sections and examination of shell fragments under the scanning electron microscope shows a typical "ornithoid" structure (Fig. 12.1C) with the lower mammillary and the upper (3 to 5 times thicker) spongy layer. No vertical dividing lines of the exospherites can be observed in the spongy layer. The horizontal growth lines are wavy. The pore canals are straight (angustocanaliculate [Sochava 1969]), of a constant diameter and not branching.

It is known that the embryo withdraws calcium carbonate from the inner part (mammillae) of the eggshell. The described eggshell has resorption craters on the mammillae (Fig. 12.1D), indicating that the eggs of this nest had been incubated.

Ornithoid eggshell structure is probably characteristic only for theropods, whereas "spherolite" structure is found in hadrosaurs, ceratopsians and sauropods. Spherolite-eggshell has vertical divisions in the shell layer and the "spongy" layer is not continuous. In the exospherite, clear

radial lines and horizontal or semi-concentric growth lines can be observed. Pore canals are prolatocanaliculate or multicanaliculate (Fig. 12.2A–E).

Conclusions

Dinosaur eggshell and nests give us information on the peculiarities of dinosaur behavior: (i) The bones, interpreted as food remains, indicate the feeding of young with meat, and the feeding took place in the nest. (ii) The nest probably belonged to a small carnivorous dinosaur. It must have been small, because all the bones belong to tiny animals that were fed to the young. A big predator, for example, *Tarbosaurus* size or even smaller, would hardly catch such small animals for feeding its young or even itself. Thus the parents had to be such a size that they could catch these small agile lizards or tiny dinosaurs. Small predators, like deinonychosaurs whose behavior was more lively than that of the larger ones, fed their nestlings for a while (Ostrom 1980). (iii) The incubated eggshell shows that the nest at the Builjasutuin–Khuduk locality was once inhabited by hatchlings. The absence of embryonic carnivorous dinosaur bones confirms this fact too. The teeth found at this site belonged to different theropods that were brought in as food for the young.

Acknowledgments

We thank Karl Hirsch for his help in improving the English of the original version of this manuscript.

References

Basu, D., and Bustard, H. 1981. Maternal behavior in the Gharial. *Jour. Bombay Nat. Hist. Soc.* 78 (2):390–392.

Bolotnikov, A. M., Dobrinski, L. N., Kamenski, J. N., and Shurakov, A. I. 1985. *Ekologiya rannego ontogenesa ptits.* (Sverdlovsk: Ural. nauchniy tsentr) 228 s.

Bratseva, G. M., and Novodvorskaia, I. M. 1976. Ob ussloviyah zakhoroneniya rannemelovoy fauny i flory v mestonakhogdeniyakh Builjasutuin–Khuduk i Anda–Khuduk. V kn: *Paleontologiya i biostratigrafiya Mongolii.* M., *Nauka* s. 285–290 (Tr. SSMPE, vyp. 3).

Horner, J. R., and Makela, R. 1979. Nest of juveniles provides evidence of family structure among dinosaurs. *Nature* 282:296–298.

Jensen, J. A. 1970. Fossil eggs in the Lower Cretaceous of Utah. *Brigham Young Univ. Geol. Studies* 13:55–67.

Mikhailov, K. E. 1986. Porovye kompleksy skorlupi yaits beskilevikh ptits i mekhanism formirovaniya por. *Paleont. Zhurn.* 3:84–93.

Ostrom, J. H. 1980. The evidence for endothermy in dinosaurs. *In A Cold–blooded Look at the Warm-blooded Dinosaurs.* (Boulder, Colorado) pp. 15–60.

Shuvalov, V. F. 1975. Novoye mestonakhozhdenie rannemelovykh dinozavrov Builjasutuin–Khuduk. V kn.: *Iskopaemaya fauna i flora Mongolii.* M., Nauka, s. 210–213. (Tr. SSMPE, vyp. 2).

Sochava, A. V. 1969. Yaitsa dinozavrov is verkhnego mela Gobi. *Paleont. Zhurn.* 4:76–78.

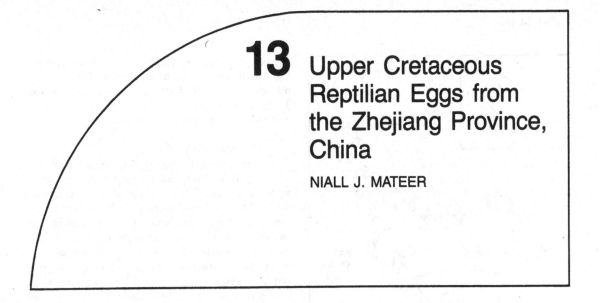

13 Upper Cretaceous Reptilian Eggs from the Zhejiang Province, China

NIALL J. MATEER

Abstract

Two dinosaur (?) egg localities in the Upper Cretaceous of Zhejiang Province, eastern China, are described. The first locality (Jinhua Basin) reveals several clutches of spherical and semi–spherical eggs over a two–hectare area. The second locality (Tian Tai Basin) is in a vertical sequence where semi–spherical eggs were found through a 10 m range, indicating site use through a considerable period.

Geologic Setting and Paleontology

This note reports a brief visit to two reptilian egg localities in Zhejiang Province, eastern China. Both localities, Tian Tai and Quxian, occur within Upper Cretaceous intermontane basins which were formed during Mid to Late Cretaceous Yanshanian orogenic episodes (Fig. 13.1). These basins constitute thick clastic wedges injected with abundant volcanics and volcanoclastics (Fig. 13.2). The sedimentary clastic sequence comprises a thick sequence of gray siltstones interbedded with conglomerates and medium– to coarse–grained sandstones. Facies are typically lacustrine (finely laminated extensive siltstones with occasional bioturbation) and alluvial plain (channel sandstones, upward–fining point–bar sequences, fanglomerates).

The Jurassic–Cretaceous stratigraphy of these basins is problematic owing to differing interpretations of diagnostic fossils which can be suitably correlated with marine sequences elsewhere in China (Gu 1985, Chen and Mateer 1986). However, the age of the Laija Formation, in which both egg localities occur, is generally believed to be Upper Cretaceous by virtue of the ostracod assemblage of *Candoniella, Ziziphocypris, Cypridea, Metacypris, Eucypris, Darwinula*, which are well–known throughout China (Chen 1983, Chen et al. 1985a). The biostratigraphic resolution of this assemblage, however, is not such as to permit further breakdown of the age of the egg localities. No vertebrate fossils have been reported from the Quxian Formation,

Figure 13.1. Generalized stratigraphy of the Cretaceous of Zhejiang Province, China. After Chen and Mateer (1986).

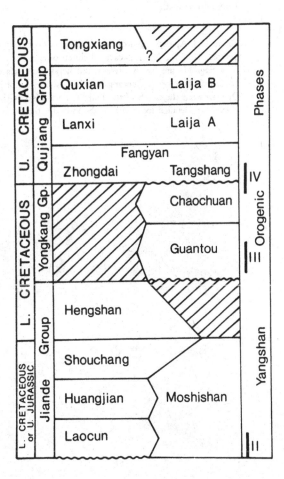

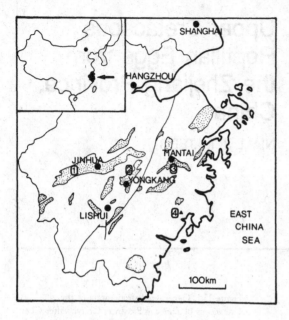

although the theropod *Chiliantaisaurus zhejiangensis* has been found in this region from the slightly older Tangshan Formation. Reptilian bone fragments were found at the Tian Tai locality, but identification was not possible. Further exploration of this area is underway.

Egg Localities

The Quxian Formation egg locality lies in the Jinhua–Quxian Basin, approximately 110 km west-southwest of Jinhua City in west–central Zhejiang Province (Fig. 13.2) at the village of Gaotangshi. The exposed Quxian Formation comprises gray–brown to brick red thick–bedded massive fine– to medium–grained sandstone with occasional gravel conglomerate channel beds. This formation is approximately 700 m thick (Zhen 1985). Owing to the poor vertical exposure of this formation, an exact stratigraphc level of the eggs was not determined. The eggs from this locality appeared to be of two types: small and spherical (c. 11 cm), and large and sub–spherical (c. 16 cm long) (Fig. 13.3b), with each type occurring in clutches (up to nine eggs) or nests, indicating two different parental origins. The distribution of these clutches was estimated to be from 10 m to 50 m apart over an area of approximately two hectares (this distribution was estimated from photographs [Fig. 13.3c] where local villagers assembled around each site pro-

Figure 13.2. Outline of the Cretaceous fault-bounded basins of Zhejiang Province, China. Egg locality 1 is at the "1", and egg locality 2 is at the dot marking Tian Tai (see text).

Figure 13.3. **A,** two eggs from the Laija B Formation north of Tian Tai town. Now in the collections of the Guoqing Temple, Tian Tai. **B,** an excavated clutch of eggs from the Quxian Formation, Gaotangshi village southwest of Jinhua City. Eggshells are mostly present, but no sign of surface patterning is noted. Bar marks 10 cm. **C,** the Quxian Formation outcrop at Gaotangshi village, southwest of Jinhua City. Arrows (and people) mark some major egg sites. **D,** view from the north of the Laija A Formation at the type locality, southwest of Tian Tai town (see Chen et al. [1985a] for detailed location). Arrow indicates the main sandstone of basal unit 4 and the volcanoclastics of unit 3.

viding a dramatic visual, if not very precise, distribution pattern). Some clutches, mainly the smaller eggs, were mold impressions in the sandstone, while the larger tended to be preserved as casts, clearly showing a thin shell infilled with sandstone. There is no positive evidence that the smaller eggs were dinosaurian in origin, being, for example, of a similar size to those of modern crocodiles. The larger eggs presumably were lain by animals larger than crocodiles; thus a dinosaurian origin is more probable.

The second egg locality lies in the Tian Tai Basin of east–central Zhejiang Province about two km north of the town of Tian Tai (Fig. 13.2), by the Guoqing Temple. They occur within the Laija B Formation, generally thought to be coeval with the Quxian Formation further to the west (Chen et al. 1985b). The eggs have been known for some time by local residents, monks at the Guoqing Temple in particular, who have made a considerable collection. A road metal quarry is the principal source of the eggs, which appear in a vertical quarry wall section. Unlike the Quxian Basin egg locality, spatial distribution was impossible to estimate; however, the stratigraphic distribution was noted to range some 10 m at the lower part of the Laija B Formation (Fig. 13.4). This may indicate 'nest fidelity' as espoused by Horner (1982); that is, the continued use of a nest area through a significant period. The lithology is primarily a chocolate to gray–brown coarse sandstone intercalated with discontinuous conglomeratic beds. No clutches were seen in situ, although several lie in the Guoqing Temple collection and, like those from Quxing, they number nine or less, far fewer than the 24 noted by Horner (1982) or from sites in Mongolia (Van Straelen 1925). All eggs seen were identical to the large sub–spherical type noted from the Quxian Basin, preserved as casts with the thin shell extant, although no surface texture was identified (Fig. 13.3a). The stratigraphic position of these eggs (unit 5, Fig. 13.4) in the Laija B Formation lower sandstone conglomerate set indicates that nesting took place in an alluvial fan environment at the edge of the intermontane basin. Above this facies set is a thick sequence of fine–grained interior basin clastics including lacustrine facies. Bone fragments were noted in the coarse sandstone at the base of unit 4 (Laija A Formation) and in the maroon siltstones of unit 1 where eggshell fragments were also found (Figs. 13.3d, 13.4). According to Chen et al. (1985b), eggs assigned to Faveoloolithinae have been found in the sandstone at the base of unit 2 (Laija A Formation).

Visits to these localities were brief, but it is hoped that future finds of reptilian egg localities will be more thoroughly studied with a view to their environmental context. No indication of nest mounds or any particular distribution within the clutch was noted at either locality such as is apparent from Horner's (1982) localities. Poor knowledge of vertebrate remains from these horizons do not permit any good indication of the potential identity of the egg–layers. Moreover, only direct association of embryos and/or juveniles with eggs could indicate that theses eggs are in fact dinosaurian.

Dinosaur eggs are not uncommon in China, having been noted from the Upper Cretaceous of Shandong, Jiangxi, Guandong, Hunan, Henan, Hubei, Anhui, Ningxi, Xinjiang, and now Zhejiang provinces (see Zhen et al. 1986 for a brief review). In a number of papers, Zhao (e.g., 1979; see other references in Zhen et al. 1986) has developed a classification of reptilian eggs involving seven generic names. No attempt is made here to use this classification in view of the brevity of inspection of the Zhejiang eggs.

Acknowledgments

I am indebted to the Palaeontological Society of China, the Chinese Petroleum Society, and the Zhejiang Petroleum Geological Brigade for arranging visits to these localities. Drs. Chen Pei–ji and Shen Yanbin were of invaluable help in the field. The Stillwater Foundation provided financial aid for this work. This is a contribution to IGCP-245, "Nonmarine Cretaceous Correlations".

References

Chen P–j. 1983. A survey of the non–marine Cretaceous in China. *Cretaceous Resarch* 4:123–143.

Figure 13.4. Generalized stratigraphic section of the Upper Cretaceous Laija Formation measured at the type section southwest of Tian Tai town.

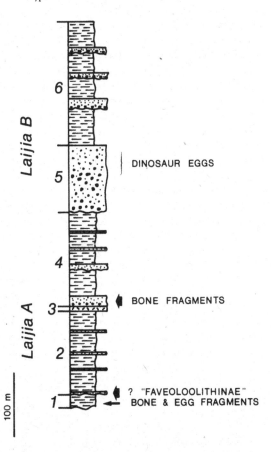

Chen P-j., Li L-t., Zhen J-s., and Shen Y-b. 1985a. Upper Cretaceous of Zhejiang. *In: Excursion Guide to the Cretaceous of Zhejiang Province* (Palaeontological Society of China, Chinese Petroleum Society, and Zhejiang Petroleum Geological Brigade) pp. 6–12.

Chen P-j., Zhen J-s., Li L-t., and Sha J-g. 1985b. Tiantai–Linhai region. *In: Excursion Guide to the Cretaceous of Zhejiang Province* (Palaeontological Society of China, Chinese Petroleum Society, and Zhejiang Petroleum Geological Brigade) pp. 21–24.

Chen P-j. and Mateer, N. J. 1986. Cretaceous of eastern and southern China. *Episodes* 9:41–42.

Gu Z-w. 1985. Brief accounts of the Cretaceous in Zhejiang. *In: Excursion Guide to the Cretaceous of Zhejiang Province* (Palaeontological Society of China, Chinese Petroleum Society, and Zhejiang Petroleum Geological Brigade) pp. 1–6.

Horner, J. R. 1982. Evidence of colonial nesting and 'site fidelity' among ornithischian dinosaurs. *Nature* 297:675–676.

Van Straelen, V. 1925. The microstructure of the dinosaurian eggshells from the Cretaceous beds of Mongolia. *Amer. Museum Novitates* 173:1–4.

Zhao Z. 1979. Advances in the studies of fossil dinosaur eggs in China. *In: Mesozoic and Cenozoic Redbeds in Southern China* (Beijing: Institute of Vertebrate Palaeontology and Palaeoanthropology, and Nanjiang Geology and Palaeontology Institute Science Press) pp. 330–340. [In Chinese]

Zhen J-s. 1985. Jinhua–Quxian region. *In: Excursion Guide to the Cretaceous of Zhejiang Province* (Palaeontological Society of China, Chinese Petroleum Society, and Zhejiang Petroleum Geological Brigade) pp. 15–17.

Zhen S., Zhen B., Mateer, N. J., and Lucas, S. G. 1986. The Mesozoic reptiles of China. *In: Lucas, S. G., and Mateer, N. J. (ed.). Studies of Chinese Fossil Vertebrates. Bull. Geol. Institutions Univ. Uppsala*, n.s. 11:133–150.

V Paleoecological, Paleoenvironmental and Regional Synthesis

"[T]rails ... are most effective in establishing population numbers or individual size ... every scale on the underside of particularly the larger animals' feet can be counted."
Affenberg 1981 p. 139

Because tracks are the in situ evidence of animal activity, they should provide useful ecological data on animal populations and communities. This has been established in modern field studies of Komodo dragons (Auffenberg 1981) and elephants (Western et al. 1983). In the latter study there was "close agreement of the age structure derived from footprint measurements with the known age structure" of the populations observed.

The utility of tracks in modern population studies suggests some potential for fossil tracks. Although this has not been attempted to any degree with dinosaurs, whose growth rates and hence age structure have not been estimated accurately, ichnologists have begun to recognize that tracks are useful as indicators of the composition of paleocommunities. Leonardi (this volume) and Lockley and Conrad (this volume) have attempted preliminary continental and regional scale syntheses, showing that ichnofaunas are controlled to some degree by facies distributions, paleogeography, paleolatitude and preservational biases. These syntheses suggest that tracks and traces can only be fully understood in the context of the rock successions from which they originate. This is illustrated by Pittman's synthesis of Cretaceous Gulf Coastal Plain data and his elegant demonstration that tracks predominate in definable "zones" which fit into classic depositional environment models (Figs. 15.1–15.3). Prince and Lockley (this volume) present similar data for lacustrine subenvironments and Sahni (this volume) demonstrates that it is now possible to place important dinosaur nest sites on the Indian subcontinent in their proper paleoenvironmental context.

Emerging from Pittman's study, as well as others in this volume, is evidence of the widespread distribution of certain track- and egg-bearing beds. As outlined in the summary chapter, these may extend for tens or even hundreds of kilometers, indicating trace bearing deposits of regional extent.

References

Affenberg, W. 1981. *The Behavioral Ecology of the Komodo Monitor.* (University Presses of Florida) 406 pp.

Western, D., Moss, C., and Georgiadis, N. 1983. Age estimation and population age structure of elephants from footprint dimensions. *Jour. Wildlife Management* 47:1192–1197.

14

The Paleoenvironmental Context, Preservation and Paleoecological Significance of Dinosaur Tracksites in the Western USA

MARTIN LOCKLEY AND
KELLY CONRAD

Abstract

A survey of Mesozoic dinosaur tracksites in the western USA shows that they occur in a broad spectrum of sedimentary facies representing many different continental depositional environments. Synthesis of the available data is best undertaken using a sedimentological classification of depositional environments. Different paleoenvironments were characterized by distinct ichnofaunas which varied in composition and diversity. In general the ichnofaunas are very good indicators of the paleobiological and paleoenvironmental conditions known or inferred to have existed for these subenvironments, and as such are an essential part of any paleontological assessment.

Local paleoenvironmental conditions are important in contributing to different modes of preservation which can significantly bias the ichnofaunas and complicate their interpretation. In general however, it is clear that tracks and traces are much more widespread than previously recognized, and of considerable importance in facilitating paleoecological and paleobiogeographical interpretations.

Because they are generally larger, dinosaur tracks appear to dominate those of other vertebrates in most cases; however this bias is similar to that encountered in the body fossil record in many cases.

Introduction

"Behind the history of every sedimentary rock there lurks an ecosystem, but what one sees first is an environment of deposition" Edward S. Deevey.

It has recently been demonstrated that the western USA is rich in dinosaur tracksites throughout the Mesozoic (Upper Triassic to Late Cretaceous: Lockley 1986a,b; Gillette 1986). This is largely a reflection of the widespread temporal and spatial distribution of terrestrial sediments (Fig. 14.1). Consequently it is useful to look at the distribution of dinosaur and other ichnofaunas, in the context of depositional environments which range from desert dune fields through arid and humid fluvial systems to carbonate and clastic lakes, to coastal plains, marine shorelines and delta complexes. The ichnofaunas vary widely in composition and diversity but are generally a reliable indicator of biotic diversity, as inferred from other body fossil evidence. Moreover, the context and preservation of the tracks is so much an integral part of the sedimentology that it tells us much about biotic activity and diversity within the short time frame of specific depositional events such as storms and floods. With this overview in mind we present a discussion of the context of the most important ichnofaunas in the western USA, with relevant comparisons to other areas.

Context and Preservation

It is logical to differentiate between "context and preservation" (*sensu* Tucker and Burchette [1977]) and the broader depositional environment setting, even though individual examples of given modes of preservation may occur in many different environments. Although little published information exists, "context and preservation" defines the occurrence of tracks in lithostratigraphic successions (Fig. 14.2A), which in turn has clear implications for interpretation of the ancient depositional environment (Fig. 14.2B). For example, Tucker and Burchette's report from Wales is typical of Triassic fluvial systems and can be compared with our example from the western USA, in which trackbearing horizons appear at the top of flood deposited sequences. Similar occurrences are noted elsewhere by Buffard and Demathieu (1969) and Visser (1984). In a broader sense, a majority of trackbearing planes represent animal activity following specific depositional events or at particular instances during depositional cycles and hiatuses.

Clay drapes are important in track preservation because they provide the ideal medium for preserving impressions and underprints (Fig. 14.2D), and facilitate subsequent parting between lithified bedded units. Clay drapes essentially denote diastemic breaks in the stratigraphic record, and represent waning flood/storm condi-

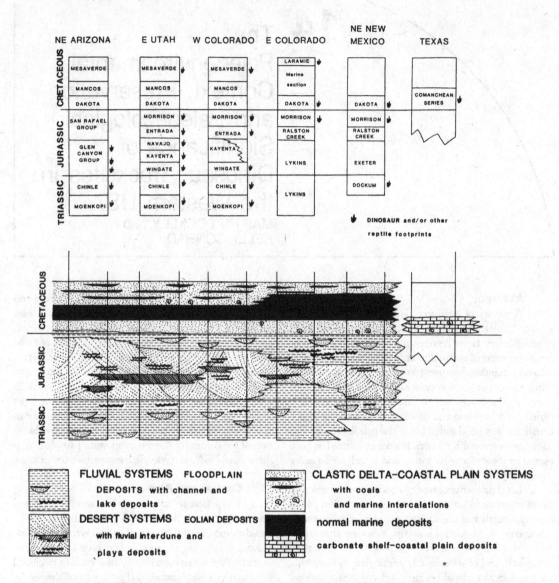

Figure 14.1. Simplified Mesozoic stratigraphic successions of the Colorado Plateau (above) after Lockley (1986b), with corresponding broad paleoenvironmental categories (depositional systems) below. Details given in text.

tions, slack water deposition or other hiatus phases that expose new sheets of sediment suitable for trampling or bioturbation by terrestrial animals. It is a common observation that good tracks are seen after storms, rains, floods and fresh snowfalls. Tracks at clay drape horizons were reported from Triassic fluvial sediments as early as 1838 (Tomlinson *in* Sarjeant 1974), but the drapes themselves are often overlooked because they are so susceptible to weathering. For example, slabs of fine sandstone from the Popo Agie Formation of Colorado reveal clay partings only when split, and these have already dissolved back several centimeters from the rock edge. In cases where no clay drape exists, the bedding surface may exhibit accumulations of heavy mineral grains (Stokes and Madsen 1979). Exposure itself leads to physical and chemical changes (dessication and oxidation) which alter the nature of the sediment,

regardless of the lithological similarity of the subsequent deposit. McKee (1944) suggested that dew may facilitate the preservation of tracks made in sand dunes.

As clay drapes frequently represent a hiatus in deposition, it is important to ask how much time is represented by such diastems or disconformities, a question which should be of considerable interest to sedimentologists. The problem can be approached by developing a bioturbation index similar to that proposed by Droser and Bottjer (1986). In this case however we can refer to a "dinoturbation" index using the term coined by Dodson et al. (1980)*. The basic difference between the two schemes is that ours measures footprint density on a two dimensional planar surface rather than in cross section. Exposures available for study may be categorized as lightly, moderately or heavily trampled (0–33%, 34–66% and 67–100% of surface area respectively).

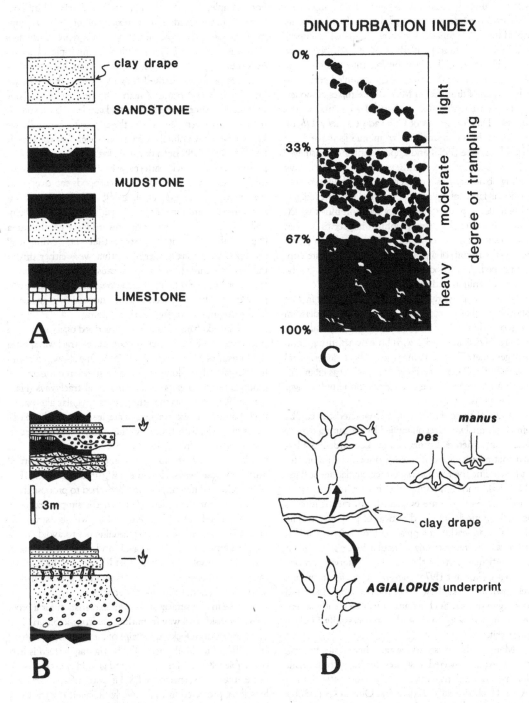

Figure 14.2. The context and preservation of dinosaur tracks in stratigraphic (**A & B**) and plan view (**C**) as explained in text. **A,** simplified examples of tracks in cross section at various lithologic boundaries. **B,** (upper) after Tucker and Burchette (1977), lower after Lockley (1986b). The dinoturbation index (**C**) is modeled after the flashcard scheme of Droser and Bottjer (1986). **D** shows the underprinting phenomenon and its effect on track preservation, using *Agialopus* from the Popo Agie Formation as an example of a trackmaker which bore most of its weight on its hind feet.

For example, the four trackbearing Morrison Formation beds at the Purgatoire site (Lockley et al. 1986) range from lightly through moderately trampled (beds 1–3) to heavily trampled (bed 4). Similarly, we know that the Davenport Ranch herd of 23 sauropods (Bird 1985) disturbed up to 29% of the substrate (Lockley 1987a), thus transforming it from a virgin to almost moderately trampled condition in a short span of time. Other Mesozoic examples of moderate to heavy trampling (dinoturbation) are noted in the Cretaceous Dakota Group of Colorado (Lockley 1987b). Examples of heavy trampling by mammals have been reported by Loope (1986) and Laporte and Behrensmeyer (1980) from the Cenozoic and Recent respectively. Heavy trampling obviously has a greater effect on the internal (3 dimensional) fabric of a particular bed and it is theoretically possible to develop a vertebrate bioturbation index which measures the vertical or stratigraphical impact in a given rock succession. Degree of "dinoturbation" reflects: (i) the duration of exposure of a given bed, (ii) the moisture content of the bed, (iii) the local population density, and (iv) the frequency of animal activity in an area.

By assessing the relative importance of these factors we should be able to constrain our estimates of diastem and disconformity duration. As shown below, a diverse assemblage of delicate tracks, with little overprinting, indicates a genuinely diverse fauna and a short depositional hiatus, as in the Popo Agie Formation, whereas examples of beds with more than one track generation imply longer hiatus durations.

Many of these dinoturbation characteristics can be systematically analyzed and quantified by measuring density and depth parameters. Demathieu (in Leonardi 1987) has shown that relative footprint depth is a crucial parameter in understanding the mechanics of track emplacement (Figs. 14.2D, 14.3), and estimating the center of gravity of trackmakers. As tracks are emplaced, downward force is translated anteriorly, in the direction of progression (Fig. 14.3E). This can lead to the preservation of deeper underprint impressions representing particular feet, e.g., the front feet, or particular parts of the feet, e.g., anterior portions of the digits. Sarjeant (1975) suggested that, as an animal moves faster, its tracks show less of the heel impression and its toes dig in deeper. Such a pattern will be greatly accentuated in undertracks, and need not necessarily imply a running gait.

Many of the questions we raise herein are not frequently asked or answered in studies dealing with dinosaur tracks, nor are modern experimental studies often undertaken (see Hitchcock 1858, Padian and Olsen 1984 and this volume for examples of the much needed empirical approach). Hitchcock's work is exemplary in many respects and much can be learned from his observations. He addressed the potential for preservation of actual tracks, undertracks and casts in successive layers, under a variety of situations, including subaerial and subaqueous settings (Hitchcock 1858, p. 31–33).

Systematic and thoughtful consideration of these preservation phenomena may suggest the need for radical reinterpretation of trackway data, at least in some instances. For example, consider the famous case described by Bird (1944) from the Cretaceous of Texas, interpreted as a swimming sauropod, because all but one track represent manus impressions (Fig. 14.3A). Using a multiple working hypothesis approach, at least two other plausible interpretations are possible. Firstly, following Hitchcock (1861), we know that overprinting of manus by pes tracks does not necessarily obliterate the former, and can lead to the manus track being driven deeper into the substrate. Secondly we know that some animals carry more weight on their front feet (Manter 1938). Both of these alternative scenarios can produce similar track patterns, whether considered separately or together. We know that sauropods can overprint, pes on manus (Lockley et al. 1986), and that their front feet were very small in comparison with the pes, thus potentially exerting much greater downward force per unit area (Fig. 14.3E). Thus it is possible that the "swimming" sauropod tracks are underprints that were either driven more deeply into the substrate because of greater force on the front feet, or by being driven downward through overprinting by the hind feet. Further discussion is included in the summary chapter of this volume.

Ishigaki (this volume) has described other sauropod trackways consisting of manus–dominated trackways, from the Jurassic of Morocco (Fig. 14.3b,c). One shows an asymmetric gait (short–long–short–long steps) reminiscent of group–intergroup gaits seen in mammal trackways (Halfpenny 1986), whereas the other gait is a regular alternating walk pattern. These trackways are associated, like Bird's Texas example, with trackways that show partial pes tracks consisting of the front part of tracks. In all cases the manus tracks are very shallow in comparison with normal sauropod trackways. Because all three examples (Fig. 14.3A–C) are different, it seems farfetched to propose three different swimming styles. We prefer the simpler explanation of differential underprint depths. We also stress the need to explore these preservational issues to avoid erroneous interpretation. If Ishigaki and Bird are correct, vertebrate ichnologists will in effect be reviving the "aquatic sauropod" hypothesis. Our interpretation is more in line with the terrestrial sauropod scenario.

We infer a similar underprint explanation for a probable therapsid trackway from the Late Triassic Sloan Canyon Formation (Dockum Group) of New Mexico (Conrad et al. 1987, Fig. 14.3D herein). The zig–zag pattern is irreconcilable with any known tetrapod gait. However, as soon as the intervening spaces are filled in with inferred pes track locations, projected from an overlying layer, the trackway conforms to a predicted normal alternating therapsid gait.

The Depositional Environment Setting of Western Tracksites

Sedimentological classification of depositional environments traditionally follow proximal (near source) to distal transects (e.g., Reading 1986). We have conformed to this tradition in our classification of continental sediments by broadly defining: (i) Alluvial/Fluvial Systems,

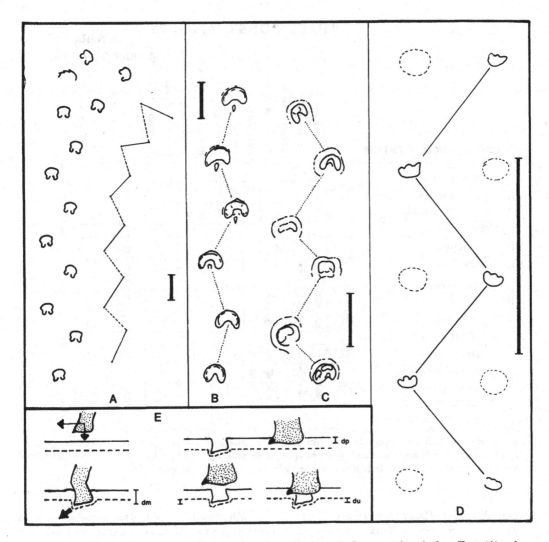

Figure 14.3. Unusual incomplete trackways explained as underprints. **A–C**, sauropod tracks from Texas (A) and Morocco (B–C), and therapsid tracks from New Mexico (**D**) with inferred position of missing footprints. Inset (**E**) model of differential depth of manus tracks (dm), pes tracks (dp) and underprints (du) for a hypothetical sauropod, with and without overprinting. Bar scale = 1 m.

(ii) Desert Systems, (iii) Lacustrine Systems, and (iv) Deltaic–Coastal Plain Systems. When dealt with in this order, our survey of trackbearing deposits follows the general Mesozoic succession of the Colorado Plateau region (Fig. 14.1). Similar trends in the "evolution" of sedimentary environments are noted elsewhere, as in southern Africa (Visser 1984). Each of these broad categories can be divided into at least two more specific paleoenvironments (Fig. 14.4) known to yield distinctive ichnofaunas.

When known ichnofaunas are plotted according to their depositional environment, various patterns emerge, the most obvious of which is a predictable correlation between paleoecological diversity deduced from the ichnofauna and the estimated biotic diversity based on body fossil documentation. This correspondence does much to establish faith in the usefulness of trace fossil assemblages, and

suggests that single ichnofaunas can provide a reliable ecological census. In stressing this point, it is worth emphasizing that trace fossils provide a dynamic record of in situ animal activity in a given paleoenvironment.

In each of the following sections we present a brief explanation of the criteria used in differentiating subenvironments before giving examples of named deposits, the context and preservation of the ichnofaunas, and inferred paleoecological and paleoenvironmental significance.

I. Alluvial/Fluvial Systems

Alluvial sediments include proximal (coarse grained) alluvial fan type deposits often attributed to braided stream deposition, as well as more distal (fine grained) flood plain deposits attributed to meander belt and overbank deposition (Reading 1986).

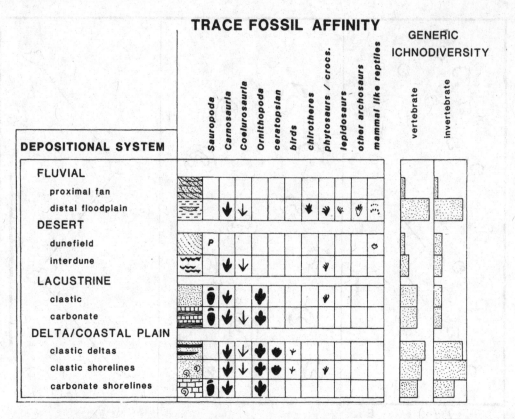

Figure 14.4. The distribution of Mesozoic vertebrate traces in western North America according to the environments of deposition in which they commonly occur. (All dubious or doubtful occurrences are omitted.) Ichnodiversity, both vertebrate and invertebrate, is estimated on a semi-quantitative scale (right). A rigorous scale can not be used in the absence of formal ichnogeneric designations for many ichnofaunas. P = prosauropod.

A. Proximal Alluvial Fan, Fluvial Environments

We know of few reports of tracks from coarse-grained, proximal fluvial deposits. This presumably reflects low preservation potential in coarse grained facies assemblages susceptible to reworking and scour. Such environments are known in the Mesozoic of the Colorado Plateau (e.g., Peterson 1984 for the Morrison Formation) but rarely reveal extensive trackway evidence. The Triassic track example cited above (Tucker and Burchette 1977), like the Colorado example (Fig. 14.2B) shows that coarse grained (> 2 mm) trackbearing fluvial sediments often represent localized channel and sheet flood deposits in otherwise predominantly fine grained successions. It is therefore convenient to discuss such occurrences in the context of the more distal flood plain successions they incise.

B. Distal Fluvial Flood Plain Environments

Lowland fluvial environments are particularly well represented in the Late Triassic Popo Agie, Chinle and Dockum Formations where humid settings predominate (for example with associated lake deposits). By contrast, late Jurassic fluvial environments, like those of the Morrison Formation, indicate less humid conditions (Dodson et al. 1980).

All three late Triassic ichnofaunas are described elsewhere (Lockley 1986b, Conrad et al. 1987) and can be summarized as follows:

The POPO AGIE ichnofauna is preserved in thinly bedded, clay draped, fine grained fluvio-lacustrine siltstones which comprise thin (> 0.5 m) laterally impersistent beds in a predominantly fine clay-silt succession. Such facies assemblages suggest a lowland fluvio-lacustrine floodplain (High and Picard 1969). This interpretation is supported by the faunal evidence which includes seven named taxa of large vertebrates (4 terrestrial and 3 aquatic).

The ichnofauna is diverse, comprising at least 5 named invertebrate ichnogenera and 3 named small vertebrate ichnogenera together with at least two other unnamed but distinctive ichnotaxa (total minimal ichnodiversity = 10) (see Lockley and Conrad [1987]). This ichnofauna more than doubles the paleontologic [systematic] record represented by body fossils (excluding plants). Although the new paleontologic evidence is entirely different from the existing record, as shown below, it is consistent with what can be expected in this paleoenvironment.

The CHINLE ichnofaunal occurrences include at least four footprint bearing sites spanning the greater part of the formation. Records of isolated small coelurosaurian and chirotherian track specimens from the Monitor Butte Member in northwest New Mexico and *?Rhynchosauroides* from the Petrified Forest Member in the type area cannot be placed in their exact depositional environment context. However, it is known that the New Mexico specimens are natural casts from predominantly fluvial successions (Stewart et al. 1972). The context of the third Upper Chinle tracksite, near Gateway, Colorado, is illustrated in Figure 14.2B.

The fourth trackbearing horizon in the upper Church Rock Member of eastern Utah is an extensive tracksite (Lockley 1986b) which yields a large ichnocoenosis comprising three named vertebrate ichnogenera and other traces. The footprints are impressed on the upper surface of an extensive sand sheet which grades up from ripple-marked medium sandstone to thinly bedded, well sorted fine sandstone containing plant debris and a few partially disintegrated carnivore coprolites. Although badly eroded, the top of the sandstone bed, in contact with the overlying silty claystones, exhibits small traces (cf. *Rhynchosauroides*) reminiscent of ichnofacies B (Fig. 14.5). We have suggested (Lockley 1986a,b) that the deposit represents a major depositional flood event, possibly in a point bar environment. This interpretation is consistent with the fluviatile interpretation of this unit by O'Sullivan (1970) and others.

The overall picture of Chinle paleoecology based on ichnofaunas (about 8 vertebrate and at least two named invertebrate ichnotaxa) is consistent with the body fossil evidence, but has biostratigraphic implications which suggest that the Chinle fauna was more evenly distributed in space and time than previously supposed. This may reflect relatively stable paleoenvironmental conditions.

The DOCKUM ichnofauna described by Conrad et al. (1987) is based mainly on the large Peacock Canyon

Figure 14.5. Schematic cross section of the Peacock Canyon Tracksite (Dockum Group), New Mexico, showing different ichnofacies in the channel and overbank subenvironments. Note that large vertebrate tracks occur in ichnofacies B, but small vertebrate tracks are not characteristic of ichnofacies A.

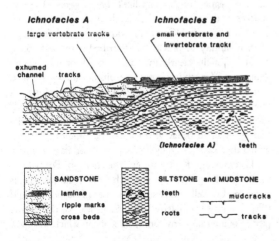

site which comprises a sandy fluvial channel complex incising mottled red and green floodplain siltstones and mudstones (Fig. 14.5). Unlike the other examples described above, the Peacock Canyon site comprises a series of multiple trackbearing horizons (at least 12) which begin with impressions in mud-cracked siltstones (beds 0–5) and are overlain by well-sorted, ripple-marked sandstones (beds 6–11) constituting the main channel sand body (Conrad et al. 1987, Fig. 3). We infer that the lower trackbearing beds, which dip towards the channel axis, probably represented a small pond, tributary or backwater that was subject to only limited sediment influx during normal flooding. The upper well-sorted sand sequence represents a larger flood deposit. It is cross-bedded in the main channel axis and thinly bedded where it spills out into the overbank areas. The paleoenvironmental interpretation (Lockley 1987a Fig. 3) is similar to that proposed by Johnson (1986 Fig. 5) for a Tertiary fluvial complex.

The Dockum ichnofauna comprises about ten named ichnotaxa (7 vertebrate and 3 invertebrate) which, like the Chinle trace fossil assemblage, show a reasonably good degree of correspondence with the body fossil record. It is important to note that smaller tracks and invertebrate traces are preserved only in the finer, thinly bedded overbank sandstones, not in the ripple-marked facies associated with the channel. This preservational difference is schematically illustrated in Figure 14.5 where two ichnofacies are proposed (A & B).

A synthesis of the ichnofaunas from the Popo Agie, Chinle and Dockum reveals assemblages from similar paleoenvironmental settings with similar overall character, namely high diversity and a mix of both invertebrate and vertebrate traces (Fig. 14.4). The Popo Agie differs from the other two in lacking large vertebrate traces and bearing little correspondence to the body fossil record. The difference probably primarily reflects preservational conditions, but may also reflect differences in age, paleogeography and paleoecology. Specifically we infer that the Popo Agie assemblage is reminiscent of Dockum ichnofacies B whereas the Chinle assemblage generally resembles ichnofacies A in that it is dominated by large tracks.

Elsewhere we have outlined the utility of Late Triassic ichnofaunas in providing useful paleoecological census data (Lockley and Conrad 1987).

POST TRIASSIC LOWLAND FLUVIAL SYSTEMS from the western USA generally lack well documented ichnofaunas. However, it is instructive to compare Late Jurassic Morrison fluvial facies with the above-mentioned Triassic systems because the ichnofaunas apparently reflect different paleoenvironmental conditions. In short, we note that except in lacustrine settings Morrison ichnodiversity rarely exceeds a single vertebrate or invertebrate ichnotaxon. The difference in diversity is entirely consistent with evidence of a somewhat restricted fauna, particularly of small vertebrates, related to the short supply of water (cf. Dodson et al. 1980, Prince and Lockley this volume). Again the ichnofauna mirrors what is generally known of the paleoecosystem in the sense that small vertebrate traces are rare, especially in comparison with Late Triassic fluvial systems. However, recent discoveries of small vertebrates in the Morrison suggest that the lack of traces may be a preservational artifact of the type that we are learning to expect (i.e., large tracks predominate over small in most cases).

II. Desert Systems

In the Mesozoic of western North America, a large number of ancient desert (or arid paleoenvironment) deposits exist in the Jurassic successions (Glen Canyon and San Rafael Groups). Despite Stokes' (1978) reference to "a desert as a place almost barren of life", a significant number of dinosaur tracks have been reported.

Desert paleoenvironments may be divided into sand-seas or dune fields (see Kocarek 1981 for erg interpretation of the Entrada Formation) and various fluvial, playa or other interdune settings. Blakey and Middleton (1983) recognized "at least six facies within the main body of the Kayenta and the intertonguing interval with the Navajo." Although documentation is sparse, it is possible to demonstrate that the majority of tracks in desert systems are those of theropods (both large carnosaurs and small coelurosaurs) (Fig. 14.6) and that most tracks occur in the interdune fluvial and playa deposits susceptible to wetting, rather than in paleoenvironments representing actual dune fields. Therefore the evidence suggests that dune and interdune ichnofaunas are different. This is probably also in part due to different preservational potential.

A. Desert dune fields

Tracks preserved in desert dunefield environments are often recognized because of the steeply dipping foresets where downslope sand crescents form below the tracks (Leonardi and Godoy 1980). A good example from the western USA is the prosauropod trackway *Navahopus* described by Baird (1980). Peterson (pers. comm. 1986) has also observed this type of occurrence in the Entrada Formation and we have observed it in the Kayenta Formation of Utah. Despite the Permian work of McKee (1947, 1949), who concluded that clear tracks were formed in dry sand, the mode of preservation of tracks in such environments is incompletely understood. Albers (1975) made the interesting observation that ?Nugget Formation tracks are mainly associated with fine laminae. However, like most authors, she does not discuss the possibility of underprinting. Instead she follows McKee in suggesting that, in order to preserve the tracks, the sand would have to be dampened, then covered with dry loose sand, allowing for the development of a parting plane. Although this mechanism is supported by the inferences of Stokes and Madsen (1979) for a Navajo Formation tracksite, a minority of authors (e.g., Brand 1979) suggest that the tracks were made underwater. Such an inference is highly unlikely in the case of the preservation of scorpionoid–like tracks reported by Albers (1975) and Faul and Roberts (1951). The occurrence of scorpionoid tracks (c.f. *Paleohelcura*) in similar eolian facies

of Permian and Jurassic age suggests that this was their preferred habitat. We suggest here that vertebrate tracks impressed, through superficial dry (or wet) layers, onto moist or well–laminated underlayers, have a relatively good preservation potential. Because invertebrates are so light, they will not leave many undertraces unless they burrow in the substrate.

B. Desert interdune environments

Desert interdune paleoenvironments which yield tracks include settings which have been characterized as fluvial (Middleton and Blakey 1983), soil (Morales and Colbert 1986, in Gillette) and playa lake (Van Dijk et al. 1978). Considered collectively in our classification (Fig. 14.4), they may all be viewed as ephemeral deposits which indicate periodic wetting of interdune substrates. Sedimentological support for such interpretations comes from reports of fresh water limestones (Brady 1959), sun or dessication cracks (Gregory 1917, Middleton and Blakey 1983) and clay or mud partings (Bunker 1957, Welles 1971). We have recently investigated a Kayenta Formation tracksite in eastern Utah which consists of a localized limestone lens comprised of theropod trackbearing carbonates and stromatolites (Museum of Western Colorado specimens 180–181). Whereas tracks are abundant in the playa facies, the surrounding desert sandstone facies lacks dinosaur tracks. Evidently water or "the shallow margin of an ephemeral lake was probably attractive to dinosaurs in an arid environment" (Van Dijk et al. 1978). The Navajo Sandstone exhibits many limestone lenses "replete with three-toed dinosaur trackways" (Gilland 1979).

Also, Marzolf (1970) reported localized structureless beds of limited lateral extent in the Navajo Formation of Utah. Such facies are often associated with trackbearing beds, and we infer that they represent localized dino-turbated facies of the type referred to above.

Most dinosaur tracks from interdune settings are attributable to theropods, ranging in size from 4 to 56 cm (foot length); see Madsen (1986 in Gillette, Lockley 1986b) and Gregory (1917) for representative dimensions. Many of the larger forms (foot length 30–40 cm) have been casually assigned to ichnotaxa like *Eubrontes giganteus*, *Anchisauripus minisculus* or described under new names (*Dilophosauripus* and *Kayentipus*) without detailed discussion of their affinity to previously described ichnotaxa. Such a state of affairs clearly indicates a need for a thorough analysis of Western tracksites for fruitful comparisons with ichnofaunas from the eastern USA and other regions. "A complete study ... would well repay the effort involved" (Welles 1971, p. 27). Much needs to be learned about the affinity of small tridactyl trackmakers. Brady (1959) suggested that *Segisaurus* may have been responsible for small, 12–13 cm long footprints in the Navajo Formation. Although the possibility of ornithopod trackmakers was suggested by Welles' 1971 interpretation of *Hopiichnus*, and Olsen's report (in Gillette 1986) of *Anomoepus*, the overall picture of interdune environments is one in which theropod

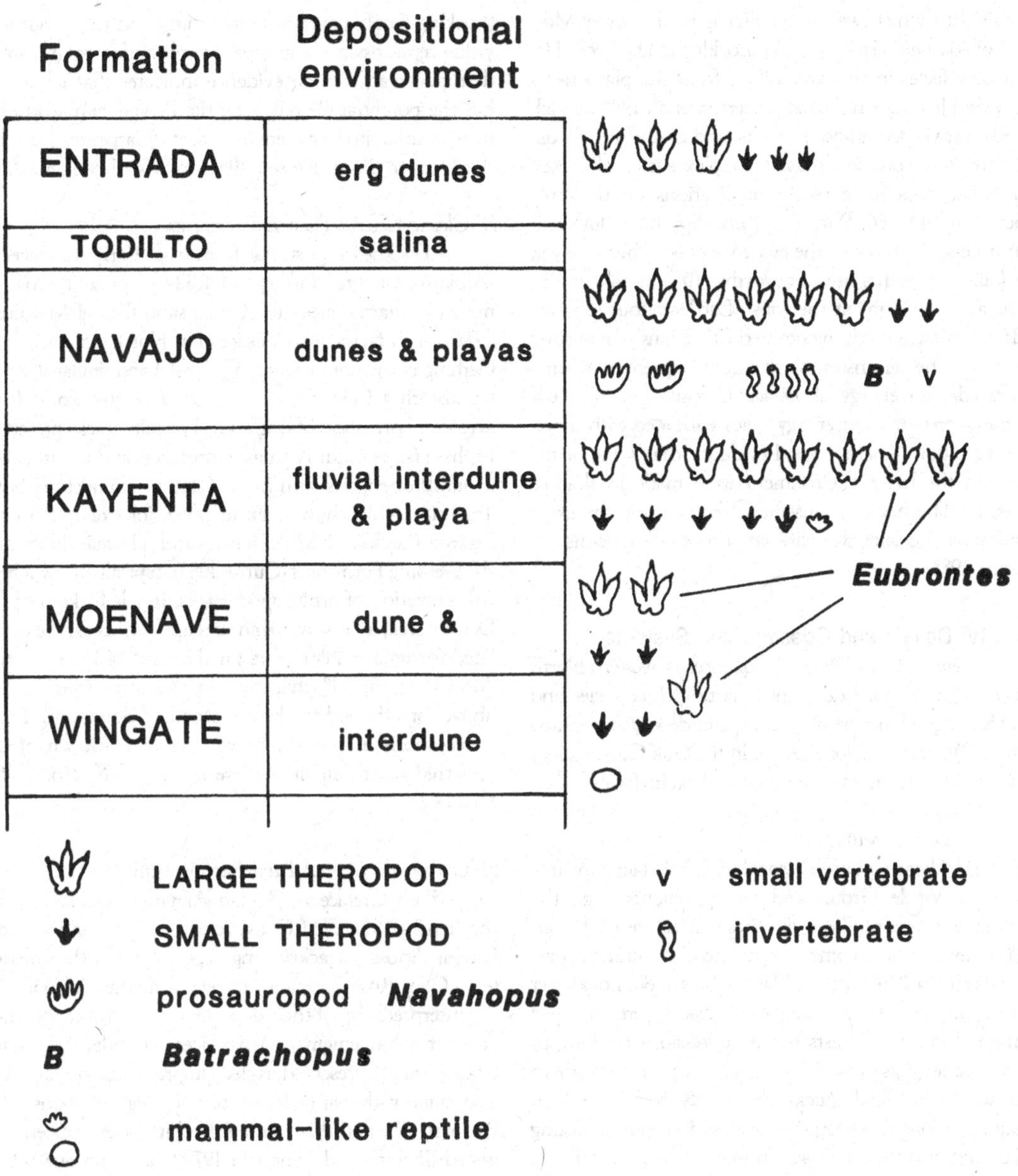

Figure 14.6. The frequency of dinosaur and other animal tracks in desert subenvironments, based on the Jurassic section of Colorado, Utah and Arizona. Note that the majority of tracks occur in playa and interdune deposits; see text for details. Some tracks in the Wingate–Kayenta interval have been named *Eubrontes*.

activity predominated over that of other dinosaurs (Fig. 14.6). This observation is consistent with the work of Leonardi (this volume), who concluded that, in South America, coelurosaurs show a preference for arid regions, and that true sauropod tracks are unknown in desert environments. To date, our observations in the western USA support these paleoecological inferences.

III. Lacustrine systems

Lakes may form in many different paleogeographic settings, and can be depocenters for both clastic and carbonate/chemical deposition (Fig. 14.4). Fluvial systems may grade into fluvio–lacustrine complexes, as in the Popo Agie,

or, in desert systems, playa lakes may characterize enclosed interdune drainage systems. In this section we comment only on large lake systems which have not been considered in the context of other facies assemblages. Mesozoic lakes are not well known, although the Triassic rift lake complexes of eastern North America have been increasingly well documented in recent years (Olsen et al. 1978). In the western USA however, where Tertiary lakes are well known, there are only a few preliminary studies dealing with Mesozoic lakes (Ash 1978; Lockley et al. 1986; Brown and Wilkinson 1981, respectively for Triassic, Jurassic and Cretaceous examples).

The most intensively studied trackbearing lacustrine

deposit with which we are familiar is in the lower Morrison of southeastern Colorado (Lockley et al. 1986). The carbonate facies in this area differs from the playa facies of the San Juan Basin (Turner–Peterson et al. 1986), which has not yet yielded dinosaur tracks, and is also entirely different from the proximal fluvial complexes in Utah (Peterson 1984). Because of recent publications on this area (Lockley et al. 1986, Prince and Lockley this volume), it is unnecessary to discuss the evidence at length. However, the following points are noteworthy: (i) Dinosaur tracks occur as impressions at the tops of at least four separate beds which range from micrite to oolitic grainstone in composition. The grainstones probably represent storm-generated/high-energy shoreface deposits, whereas the micrites represent lower energy facies associated with shoreline vegetation. There is good evidence that some of the tracks in the latter environment were made in shallow water. (ii) The tracks are useful in determining the exact position of the lake shoreline at successive horizons (cf. Alonso 1985).

IV. Deltaic and Coastal Plain Systems

Cretaceous trackbearing sequences in western North America include well developed clastic deltaic facies and various marginal marine or coastal plain deposits associated with the Western Interior Seaway. In the Gulf Coast region, tracks are found in carbonate coastal plain facies.

A. *Clastic Deltaic Settings*

Trackbearing Late Cretaceous deltaic facies include the Mesa Verde Group and its equivalents (e.g., the Laramie Formation). Typically these facies are coal bearing and many tracks come to light through mining operations (Peterson 1924, Brown 1938, Balsley 1980, Lockley et al. 1983). In all the studies cited above, tracks were preserved as natural sandstone casts filling impressions in the tops of coal seams. Balsley (1980) suggested that tracks made in peat would be filled quickly by sandy bed load from flooding, before the spongy peat could rebound. Young (1976) suggested that dinosaurs trampled the peat, actually contributing to the coal formation process. If this was true, then coals may, in part, represent "heavily trampled" beds which only preserve discrete tracks at the upper contacts. The Laramie Formation near Denver exhibits some "heavily trampled" beds at the interface between kaolinitic clays and overlying sands, and the excellent plant impressions are in part due to the plant material having been trampled into the fine underlying muds (cf. Lockley et al. 1986 for further evidence of dinosaurs trampling plants and clams).

In general tracks from the Mesa Verde Group are predominantly those of large herbivores (hadrosaurs) (see Lockley et al. [1983] and Balsley [1980]), with only a few tracks attributable to theropods. The trackway data is consistent with what is known of the paleoecosystem. Several reports refer to the abundance of herbivores' tracks in what were obviously heavily vegetated lowland environments. An ichnofauna from the Laramie Formation near Denver

(Lockley 1986b) contains large ornithopod tracks but also yields ceratopsian tracks and a variety of large and small theropod tracks. Such evidence indicates that ichnology has the potential for reflecting the diverse faunas known to have inhabited such environments (Carpenter 1985, for the Denver Basin; Russell 1984, for North America).

B. *Clastic Coastal Plain Environments*

Trackbearing strata from the Early Cretaceous (Albian) Dakota Group typically represent coastal, marginal–marine facies (e.g., tidal sand flats of McKenzie 1972). Such facies are widespread in North America (e.g., Gething Formation of western Canada) and similar to other trackbearing facies from other parts of the world (e.g., Wealden Formation of England). Typically tracks preserved in these facies occur both as impressions and natural casts in sandstone facies with intercalated fine mudstone beds. Tracksites which have been mapped, for example, in the Dakota (Lockley 1987b, Gillette and Thomas 1985) and the Gething Formation (Currie 1983), typically show a high concentration of ornithopod tracks. In the Dakota of the Denver area, at least one thin package of beds can be correlated for tens of kilometers on the basis of its distinctive "heavily trampled" character. In the three studies cited above, ornithopod tracks are generally dominant. These observations are consistent with what is known of the preferred environment of these herbivores (Ostrom 1964, Fig. 14.7).

C. *Carbonate, Coastal Plain Environments*

Early Cretaceous (Aptian–Albian) coastal plain environments of the Gulf Coast region are among the most famous dinosaur trackbearing deposits in North America (e.g., Glen Rose Formation). Yet, from the viewpoint of the interpretation of their depositional environment, they are somewhat enigmatic. In places, abundant dinosaur tracks, usually preserved as deep impressions, are found in association with marine faunas of reefal origin (Perkins 1968, Pittman 1984 and this volume), and it has even been suggested (Perkins and Langston 1979) that dinosaur tracks were made when tropical storms caused waters to recede from back reef lagoons, thereby exposing mud flats. Certainly the evidence for lagoonal environments is well substantiated, and in some areas extensive back reef evaporites accumulated. In such environments, dinosaur tracks are often very abundant (Pittman 1984, this volume), and there is evidence that tracks are mainly associated with lowstand diastems. Detailed stratigraphic studies suggest that the Glen Rose and its equivalents represent a complex alternating sequence of marginal marine deposits ranging from sub- to supra-tidal. Depositional units are thin and laterally discontinuous, while dinosaur tracks are deep and liable to impact on several layers beneath the one on which the animals actually walked (Perkins and Langston 1979).

The ichnofaunas of the Glen Rose and its equivalents include a significant proportion of sauropod tracks (Bird 1985, Pittman 1984 and this volume) in association with a large number of tridactylous forms mainly attributable to theropods (Farlow et al. this volume). However, ornithopod tracks are uncommon. This is in stark contrast to the coeval Dakota Group and its equivalents which is ornithopod–dominated and currently lacks any evidence of sauropods. Hence we conclude that there were considerable paleoecological differences between the dinosaurian faunas of clastic and carbonate coastal plain environments, at least in these two areas.

A preliminary survey of known Cretaceous dinosaur ichnofaunas indicates a clear facies fauna pattern (Fig. 14.7, taken from Lockley in press). Ornithopod tracks dominate in mid- to high-latitude paleoenvironments characterized by clastic lowland coastal plain deposits with coals and other evidence of abundant vegetation. In such settings, sauropod tracks are currently unknown. By contrast, in low-latitude lacustrine and carbonate evaporite settings, theropod-sauropod ichnofaunas generally dominate. Such a pattern appears to indicate that distinct facies faunas can be identified from footprint assemblages, at least in the Late Mesozoic (Late Jurassic through Cretaceous, Prince and Lockley this volume, Lockley in press).

Conclusions

1. Dinosaur tracks and other traces occur throughout the well–developed continental successions in the Mesozoic of the western United States, making the area one of the best in the world for a regional ichnological synthesis.

2. Tracks occur in *all* representative continental facies including fluvial, lacustrine, desert and coastal deposits.

3. In all these different paleoenvironmental settings, tracks are liable to occur repeatedly at successive stratigraphic intervals representing depositional hiatuses of varying frequency and duration.

4. Intensity of dinosaur (and other animal) activity can be measured on a "dinoturbation" scale ranging from lightly to heavily trampled. At the intense end of the scale, considerable impact on sedimentary layering is implied.

5. Serious study of differential track depth and the importance of underprints has been neglected. Consequently certain well–known but problematic trackways need reevaluation.

6. Ichnofaunal lists (or ichnocoenoses) for various formations or localities provide a very useful indication of biotic activity in extinct animal communities, and in many instances add much to our knowledge of former biotas.

7. In a majority of cases, ichnofaunas are consistent with the body fossil evidence, particularly after preservational biases have been taken into account.

8. Preservational biases generally favor the preservation of large vertebrate tracks relative to small tracks. However, this bias is not fundamentally different from the size biases encountered in the study of body fossil prservation.

9. Ichnofaunas are facies controlled to a significant degree just like other paleontological assemblages. Mesozoic track assemblages show patterns of differentiation which

UPPER CRETACEOUS

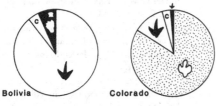

Bolivia   Colorado

Figure 14.7. Relative proportions or ratios of ornithopod, sauropod and theropod tracks estimated from Cretaceous tracksites where large ichnofaunas are documented. Accuracy of estimates depends on quality of available data. Ichnofaunas with a sauropod component are all at Cretaceous low-latitude sites. With the probable exception of the Korean sites, all the ornithopod dominated ichnofaunas represent mid-high latitude clastic coal-bearing facies. (Modified after Lockley, in press)

LOWER CRETACEOUS

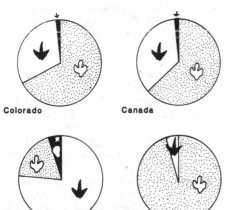

Colorado   Canada

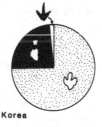

Korea   Texas

Brazil   England

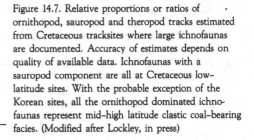

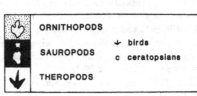

ORNITHOPODS	
SAUROPODS	↓ birds
	c ceratopsians
THEROPODS	

relate to paleogeography and paleoenvironment.

Each track, trackway and assemblage of traces is a legacy of animal activity in a particular environment. As data on animal distributions begins to emerge for different Mesozoic deposits, we often see patterns which are consistent with our knowledge of the various ecosystems under consideration. In many cases the ichnofaunas add significantly to our knowledge of the taxonomic composition of particular faunas or at least the relative frequency of forms frequenting specific subenvironments. In other instances the traces enhance our understanding of the configuration of the paleoenvironment and the depositional events responsible for the accumulation and preservation of strata. Until attempts are made to adequately understand the ichnology of continental deposits, paleontologists and sedimentologists cannot expect to fully appreciate the evidence of ancient ecosystems which lurk therein.

Acknowledgments

Dr. Giuseppe Leonardi is the only ichnologist to have previously attempted a continental scale synthesis of dinosaur footprint data. His preliminary data for South America suggested the potential for a synthesis on this scale. We have also benefited from useful discussion with Jeff Pittman and many of the other ichnologists contributing to this volume. This work has been supported in part by National Science Foundation Grant EAR–8618206.

References

Albers, S. H. 1975. Paleoenvironment of the Upper Triassic – Lower Jurassic (?) Nugget (?) Sandstone near Heber, Utah. M.S. thesis, University of Utah.

Alonso, R. 1985. Ichnitas de Aves como control de niveles boratiferos. *Soc. Cient. Norteste Argentino.* 1 (4):37–42.

Ash, S. R. (ed.) 1978. *Geology and Paleontology of a Late Triassic Lake, Western New Mexico. Brigham Young Univ. Geol. Studies* 23:

Baird, D. 1980. A prosauropod dinosaur trackway from the Navajo Sandstone (Lower Jurassic) of Arizona. *In* Jacobs, L. L. (ed.) *Aspects of Vertebrate History.* (Museum of Northern Arizona Press).

Balsley, J. K. 1980. Cretaceous wave dominated delta systems, Book Cliffs, eastern central Utah. *AAPG Field Seminar Guidebook.*

Bird, R. T. 1944. Did *Brontosaurus* ever walk on land? *Nat. Hist.* 53:61–67.

1985. *Bones for Barnum Brown.* (Fort Worth: Texas Christian University Press).

Blakey, R. C., and Middleton, L. T. 1983. Evolution of early Mesozoic tectonostratigraphic environments. *Southwestern Colorado Plateau to Southern Inyo Mountains, Utah Geolog. and Mineral Surv.* Special Studies 60. pp. 33–39.

Brady, L. F. 1959. Dinosaur tracks from the Navajo and Wingate sandstones. *Plateau* 32:81–82.

Brand, L. 1979. Field and laboratory studies on the Coconino Sandstone (Permian) vertebrate footprints and their paleoecological implications. *Paleogeogr., Paleoclimat., Paleoecol.* 28:25–38.

Brown, B. 1938. The mystery dinosaur. *Nat. Hist.* 41:190–202, 235.

Brown, R. E., and Wilkinson, B. H. 1981. The Draney Limestone: Early Cretaceous lacustrine carbonate deposition in western Wyoming and southeastern Idaho. *Contrib. Geol. Univ. Wyoming* 20 (1):23–31.

Buffard, R. G., and Demathieu, G. 1969. Mise in evidence de deux niveaux biens individualises a empreintes theromorphorides, lacertoides et crocodiloides danc le gres bigarre de Haute Saone Annales Scientifiques de L'universite de Besancon. 3. Serie. Geologie: 13–20.

Bunker, C. M. 1957. Theropod saurischian footprint discovery in the Wingate (Triassic) Formation. *Jour. Paleont.* 31:973.

Carpenter, K. 1985. Late Cretaceous non-marine vertebrates of the Denver Basin. *Geol. Soc. Amer. Abs. Prog.* 17:212.

Conrad, K., Lockley, M. G., and Prince, N. K. 1987. Triassic and Jurassic vertebrate dominated trace fossil assemblages of the Cimarron Valley region: implications for paleoecology and biostratigraphy. *New Mexico Geol. Soc. Guidebook, 38th Field Conf.* pp. 127–138.

Currie, P. J. 1983. Hadrosaur trackways from the Lower Cretaceous of Canada. *Acta Paleont. Polonica* 28:63–73.

Dodson, P., Behrensmeyer, A. K., Bakker, R. T., and McIntosh, J. S. 1980. Taphonomy and paleoecology of the dinosaur beds of the Jurassic Morrison Formation. *Paleobiology* 6:208–232.

Droser, M., and Bottjer, D. 1986. A semi-quantitative field classification of ichnofabric. *Jour. Sedim. Petrol.* 56:558–559.

Faul, H., and Roberts, W. A. 1951. New fossil footprints from the Navajo (?) Sandstone of Colorado. *Jour. Paleont.* 25:266–274.

Gilland, J. K. 1979. Paleoenvironments of a carbonate lens in the Lower Navajo Sandstone near Moab, Utah. *Utah Geol.* 6:29–38.

Gillette, D. D. (ed.). 1986. *Abstracts with Program. First International Symposium on Dinosaur Tracks and Traces.* (Albuquerque: New Mexico Museum of Natural History) 31 pp.

Gillette, D. D., and Thomas, D. A. 1985. Dinosaur tracks in the Dakota Formation (Aptian–Albian) of Clayton Lake State Park, Union County, New Mexico. *New Mexico Geol. Soc. Guidebook 36th Field Conf., Santa Rosa* pp. 283–288.

Gregory, H. E. 1917. Geology of the Navajo Country. *U.S. Geol. Surv. Prof. Paper* 93.

Halfpenny, J. 1986. *A Field Guide to Mammal Tracking in Western America.* (Boulder, Colorado: Johnson Books) 163 pp.

High, L. R., and Picard, M. D. 1969. Stratigraphic relations within Upper Chugwater Group (Triassic Wyoming). *Amer. Assoc. Petrol. Geol. Bull.* 53:1091–1104.

Hitchcock, E. 1858. A report on the sandstone of the Connecticut Valley and its fossil footmarks. (Boston: W. White; reprinted by Arno Press 1974) 220 pp. 6 pls.

1861. Remarks upon certain points in ichnology. *Proc. Amer. Soc. Adv. Sci.* 144–156.

Johnson, K. R. 1986. Palaeoecene bird and amphibian tracks from the Fort Union Formation, Bighorn Basin, Wyoming. *Contrib. Geol. Univ. Wyoming* 24 (1):1–10.

Kocorek, G. 1981. Erg reconstruction: the Entrada Sandstone (Jurassic) of northern Utah and Colorado. *Paleogeog., Paleoclimat., Paleoecol.* 36:125–153.

Laporte, L. F., and Behrensmeyer, A. K. 1980. Tracks and substrate reworking by terrestrial vertebrates in quaternary sediments of Kenya. *Jour. Sedim. Petrol.* 50:1337–1346.

Leonardi, G. (ed.) 1987. *Glossary and Manual of Tetrapod Footprint Palaeoichnology.* (Departamento Nacional du Producao Mineral, Brasil) 75 pp.

Leonardi, G., and Godoy, L. C. 1980. Novas pistas de tetrapodes de formacao Botucatu no estado de Sao Paulo. *Anais. do 31 Congr. de Brasil. de Geol., Santa Catarina* 5:3080–3089.

Lockley, M. G. 1985. Vanishing tracks along Alameda Parkway. *In* Chamberlain, C. K. (ed.). *Soc. Econ. Paleont. Miner. Field Guide 2nd Midyear Ann. Meeting, Golden, Colorado* p.3:131–142.

——— 1986a. The paleobiological and paleoenvironmental importance of dinosaur footprints. *Palaios* 1:37–47.

——— 1986b. *A Guide to Dinosaur Tracksites of the Colorado Plateau and American Southwest.* University of Colorado Denver, Geology Department Magazine Special Issue 1. (published in conjunction with the First International Symposium on Dinosaur Tracks and Traces. Albuquerque, New Mexico) 56 pp.

——— 1987a. Dinosaur trackways and their importance in paleontological reconstruction. *In* Czerkas, S., and Olsen, E. C. (ed.). *Dinosaurs Past and Present Symposium.* (Los Angeles County Museum) pp. 80–95.

——— 1987b. Dinosaur footprints from the Dakota Group of Eastern Colorado. *Mountain Geol.* 24:107–122.

——— in press. Cretaceous dinosaur-dominated footprint assemblages: their stratigraphic and paleoecological potential. *In* Mateer, N. J., and Chen, P. J. (ed.). *Proceedings of the First International Symposium on Non-Marine Cretaceous Correlations.* (Beijing: China Ocean Press)

Lockley, M. G., and Conrad, K. 1987. Mesozoic tetrapod tracksites and their application in paleoecological cenuss studies. *In* Currie, P. M., and Koster, E. H. (eds.). *4th Symposium on Mesozoic Terrestrial Ecosystems.* pp. 144–149.

Lockley, M. G., Young, B. H., and Carpenter, K. 1983. Hadrosaur locomotion and herding behavior. *Mountain Geol.* 20:5–13.

Lockley, M. G., Houck, K., and Prince, N. K. 1986. North America's largest dinosaur tracksite: Implications for Morrison Formation paleoecology. *Geol. Soc. Amer. Bull.* 57:1163–1176.

Loope, D. E. 1986. Recognizing and utilizing vertebrate tracks in cross section: Ceonozoic hoofprints from Nebraska. *Palaios* 1:141–152.

Manter, J. T. 1938. The dynamics of quadrupedal walking. *Jour. Experim. Biol.* 15:522–540.

Marzolf, J. E. 1970. Evidence of changing depositional environments in the Navajo Sandstone, Utah. Ph.D. thesis, Geology, University of California at Los Angeles.

McKee, E. D. 1944. Tracks that go uphill. *Plateau* 16:61–72.

——— 1947. Experiments on the development of tracks in fine cross-bedded sand. *Jour. Sedim. Petrol.* 17:23–28.

——— (ed.). 1979. *A Study of Global Sand Seas.* U.S. Geol. Surv. Prof. Paper 1052.

McKenzie, D. B. 1972. Tidal sand flat deposits in lower Cretaceous, Dakota Group near Denver, Colorado. *Mountain Geol.* 9:269–277.

Middleton, L. T., and Blakey, R. C. 1983. Processes and controls on the intertonguing of the Kayenta and Navajo Formations, northern Arizona: eolian fluvial interactions. *In* Brookfield, M. E., and Ahlbrandt, T. S. (ed.). *Developments in Sedimentology 38: Eolian Sediments and Processes.* (Elsevier) pp. 613–634.

Morales, M., and Colbert, E. H. 1986. (in Gillette)

Olsen, P. E., Remington, C. L., and Cornet, B. 1978. Cyclic change in Late Triassic lacustrine communities. *Science* 201:729–732.

Ostrom, J. H. 1964. A reconsideration of the paleoecology of hadrosaurian dinosaurs. *Amer. Jour. Sci.* 262:975–997.

O'Sullivan, R. G. 1970. *The Upper Triassic Chinle Formation and Related Rocks, Southeastern Utah and Adjacent Areas.* U.S. Geol. Surv. Prof. Paper 664E. 22 pp.

Padian, K., and Olsen, P. E. 1984. The fossil trackway *Pteraichnus:* not a pterosaurian, but crocodilian. *Jour. Paleont.* 58:178–184.

Perkins, B. F. 1968. Geology of a rudist reef complex (abst.). *Geol. Soc. Amer. Abs. Prog.* p. 233.

Perkins, B. F., and Langston, W. Jr. 1979. Lower Cretaceous. Shallow marine environments in the Glen Rose Formation. Dinosaur tracks and plants. *Amer. Assoc. Stratigraphic Palynologists.* Field Trip Guide 12th Annual Meeting.

Peterson, F. 1984. Fluvial sedimentation on a quivering craton: Influence of slight crustal movements on fluvial processes, Upper Jurassic Morrison Formation, Western Colorado Plateau. *Sedim. Geol.* 38:21–49.

Peterson, W. 1924. Dinosaur tracks in the roofs of coal mines. *Nat. Hist.* 24:388–391.

Pittman, J. G. 1984. Geology of the De Queen Formation of Arkansas. *Gulf Coast Assoc. Geol. Soc. Trans.* 34:201–209.

Reading, H. G. 1986. *Sedimentary Environments and Facies.* (2nd ed.). (New York: Elsevier) 561 pp.

Russell, D. A. 1984. *A Checklist of Families and Genera of North American Dinosaurs.* Syllogeus 53. National Museum of Canada.

Sarjeant, W. A. S. 1974. A history and bibliography of fossil footprints in the British Isles. *Paleogeog., Paleoclimat., Paleoecol.* 16:265–378.

——— 1975. Fossil tracks and impressions of vertebrates. *In* Frey, R. (ed.). *Trace Fossils.* (Springer Verlag) pp. 283–324.

Stewart, J. H., Poole, F. G., and Wilson, R. F. 1972. *Stratigraphy and Origin of the Chinle Formation and Related Upper Triassic Strata in the Colorado Plateau Region.* U. S. Geol. Surv. Prof. Paper 690. 195 pp.

Stokes, W. L. 1978. Animal tracks in the Navajo-Nugget Sandstone. *Contrib. Geol. Wyoming* 16:103–107.

Stokes, W. L., and Madsen, J. H. Jr. 1979. The environmental significance of pterosaur tracks in the Navajo Sandstone (Jurassic), Grand County, Utah. *Brigham Young Univ. Geol. Studies* 26:21–26.

Tucker, N. E., and Burchette, T. P. 1977. Triassic dinosaur footprints from South Wales: their context and preservation. *Palaeogeog., Paleoclimat., Paleoecol.* 22:195–208.

Turner-Peterson, C. E., Santos, E. S., and Fishman, N. S. (ed.). 1986. A basin analysis case study: The

Morrison Formation, Grants Uranium Region, New Mexico. *Amer. Assoc. Petrol. Geol.* Studies in Geology 22. 391 pp.

van Dijk, D. E., Hobday, D. K., and Tankard, A. J. 1978. Permo–Triassic lacustrine deposits in the eastern Karoo Basin, Natal, South Africa. *Spec. Publ. Int. Assoc. Sediment.* 2:225–239.

Visser, J. N. J. 1984. A review of the Stormberg Group and Drakensberg volcanics in southern Africa. *Paleont. Africana* 25:5–27.

Welles, S. P. 1971. Dinosaur footprints from the Kayenta Formation of northern Arizona. *Plateau* 44:27–38.

Young, R. G. 1976. Genesis of western Book Cliff coals. *Brigham Young Univ.* Geol. Studies 22:3–14.

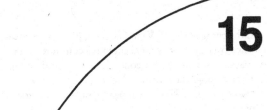

15 Stratigraphy, Lithology, Depositional Environment, and Track Type of Dinosaur Track–bearing Beds of the Gulf Coastal Plain

JEFFREY G. PITTMAN

Abstract

Age of the beds at dinosaur tracksites in the Gulf Coastal Plain ranges from Late Aptian or Early Albian (Twin Mountains and Glen Rose Formations) to Cenomanian (Woodbine Formation). Tracks at all sites except one (the Cenomanian site) occur on strata within the Trinity and Fredericksburg Groups of the provincial Comanche Series. Most sites occur on Early to Middle Albian beds of the Glen Rose Formation of the Trinity Group. Numerous sites on strata lying very near the top of the Glen Rose Formation constitute a dinosaur track zone.

The majority of tridactyl tracks were made by small to large theropods. Definite ornithopod tracks occur at only three sites. Sauropod tracks are known at nine sites, occurring with theropod tracks at four of those nine sites and with ornithopod tracks at one site. Parallel orientation of sauropod trackways occurs at five of those nine sites.

The tracks occur on beds of lithologies including fine-grained sandstone, algal-laminated micrite, micrite, biomicrite, pelsparite, and rudist biolithite. Allochems in these limestones include common miliolid foraminifera, ostracods, pelecypod fragments, algal fragments, pellets, and occasional intraclasts and silt. The most common lithology is pure micrite or limestone with some allochem present, but dominantly micrite. Many of these limestones have been dolomitized.

Substrate consistency indicated by depth and preservation of tracks varies from "soupy" to hard. Most tracks were made in exposed or very shallow water settings. Exposed or supratidal conditions are indicated by mud cracks and fenestral bird's eye cavities and laminae. Intertidal conditions are indicated by algal lamination and oscillation ripples in several micrites. Exposed wet conditions are perhaps indicated by a lack of the above, in beds where deep tracks occur. Theropod tracks at one site occur atop a rudist mound that was perhaps exposed when the tracks were made.

Introduction

Most published reports of dinosaur tracks of the Gulf Coastal Plain have been brief descriptions of specific sites. Several authors (Adkins 1932, Getzendaner 1956, Langston 1974) have listed tracksites known to them at the time. However, the stratigraphic position, lithology, and track type of track–bearing beds exposed at the numerous sites have not been compiled. In addition, since Langston's (1974) review, many more sites have been discovered. In this paper I discuss the stratigraphy, depositional environment, and track type of previously reported and recently discovered dinosaur track–bearing beds of the Gulf Coastal Plain.

History of Dinosaur Discoveries and Reports

The first discovery of dinosaur remains in the region was made by Robert T. Hill, probably in the late 1870's (Langston 1974). Hill found a tooth (probably carnosaur) in Lower Cretaceous sandstone near Hamilton, Texas. Since Hill's discovery, reports of dinosaur tracks and body fossils in the region have been infrequent. However, this has been the result of sporadic study; much more exists in Cretaceous rocks of the province in the way of a dinosaur fossil record.

Dinosaur Track Discoveries and Reports

A Mr. Herndon of Kinney County, near Uvalde, may have been the first discoverer of dinosaur tracks in the region. He noticed three-toed dinosaur tracks near his ranch on limestone ledges exposed along Live Oak Creek before the turn of this century (Elmer Herndon, son, pers. comm.). The first published report of dinosaur tracks in the region was made by Shuler (1917) of tracks exposed in the bed of the Paluxy River west of Glen Rose, Somervell County, Texas. Since then a number of papers describing specific sites and several general review papers have been published: Wrather 1922; Gould 1929; Adkins 1932; Houston 1933; Shuler 1937; Bird 1939, 1941, 1944, 1954, 1985; Albritton 1942; Booth 1956; Getzendaner 1956; Moore 1964; Perkins and Stewart 1971; Stricklin et al. 1971;

Davis 1974; Langston 1974, 1986; Perkins 1974; Stricklin and Amesbury 1974; Skinner and Blome 1975; Perkins and Langston 1979; Farlow 1981, 1986, 1987; Sams 1982; Pittman 1984, 1986; Kuban 1986; Herrin et al. 1986; Hawthorne 1987; Langston and Pittman 1987; Pittman and Gillette this volume; Farlow, Pittman, and Hawthorne this volume; Kuban this volume.

The work of Roland T. Bird (1939 and later) made famous the tracks of Texas sauropods and carnosaurs by describing, mapping, and excavating tracks and trackways in the bed of the Paluxy River near Glen Rose, Somervell County, and in West Verde Creek, Bandera County, Texas. He also described the "swimming sauropod" trackway in the bed of the Medina River at Bandera, Medina County, Texas. His work has been cited in discussions of sauropod habits and habitats, notably in Bakker (1971), Ostrom (1972), Coombs (1975), and Lockley (1986).

Dinosaur Body Fossils of the Region

Dinosaurs recovered from Lower Cretaceous rocks of the Gulf Coastal Plain include a carnosaur, *Acrocanthosaurus*; a sauropod, perhaps *Pleurocoelus*; an ornithomimid; the ornithopod *Tenontosaurus*; a small hypsilophodontid; and a large ornithopod, probably an iguanodontid.

Acrocanthosaurus, named for two Oklahoma specimens described by Stovall and Langston (1950), is known from remains in Texas, Oklahoma, and Arkansas. A recently discovered specimen from Oklahoma includes the complete skull and jaws, and much of the postcranial skeleton, including most of the right pes. This specimen is in private hands and is not currently available for study. A foot of an ornithomimid, not yet described, was recovered from Lower Cretaceous sandstone in Arkansas (Dan Chure pers. comm.).

The first sauropod fossil discovered in the region was a coracoid from Oklahoma (Larkin 1910). Langston (1974) listed the following other sauropod remains, which he referred to *Pleurocoelus*: a partial braincase and other scanty remains from the lee flank of a rudist reef in the middle Glen Rose Formation; a larger, more complete skeleton from the Paluxy Sand; a partial hind leg and foot (described by Gallup 1975); and teeth. New Texas material is being collected from a mass accumulation of sauropod bones in the Bluffdale Sandstone (Lower Glen Rose equivalent). This new material supports Langston's consideration of the Texas material as "brachiosaurid".

The ornithopod *Tenontosaurus* is known in the region from material collected in Texas and Oklahoma. A recent discovery in north-central Texas was made of a mass accumulation of small hypsilophodontids (Phillip Murray pers. comm.). Another important recent discovery was made of scanty remains of a large ornithopod, probably an iguanodontid (Langston pers. comm., Wilcox and Rohr [1986]), from a Lower Cretaceous sandstone in West Texas.

Methods

In addition to the reports of tracks listed above, I used the personal files of Wann Langston, Jr., who has accumulated over the years a number of undocumented reports of tracks. This study was concentrated on finding and "authenticating" these reports, locating the sites described in the literature, and documenting new sites found through "door-to-door" questioning.

Each site was recorded on county, topographic, and geologic maps. The geologic maps, from which the stratigraphic position of each site was taken, are sheet maps (scale: 1:250,000) of the Geologic Atlas of Texas compiled by the Bureau of Economic Geology, Austin, Texas.

During visits to each tracksite, samples of the track-bearing bed were taken at the tracked interface. Thin sections were made of each sample and stained for dolomite (TMM #s [Texas Memorial Museum] in Table 15.1 refer to these thin sections). Lithologic descriptions are also listed in Table 15.1.

Use of landowner names for site names was avoided. Since most sites occur in the beds of rivers and creeks, the name of the creek or river was given to the tracksite. In the case of more than one site on a given creek or river, the tracksites were arbitrarily assigned numbers, e.g. Cowhouse Creek #1, Cowhouse Creek #2, etc. Names of those tracksites not occurring in stream beds were taken from the nearest town or permanent landmark.

Regional Geologic Setting

Much has been learned during this century about the Cretaceous System of the Gulf Coastal Plain through inland drilling from Florida to Texas and in Mexico. The Cretaceous outcrop, which marks the landward boundary of the province, has also been thoroughly studied. Outcrop studies date back to the second half of the 19th century with the work of Römer (1846, 1852), the Shumards (G.G. Shumard 1853, B. F Shumard 1860), Hill (1887 and later), Taff (1892), and others.

Cretaceous Outcrop of the Gulf Coastal Plain

The outcrop of Upper and Middle Cretaceous strata along the interior margin of the Gulf Coastal Plain is laterally more extensive than that of Lower Cretaceous rocks. Nevertheless, the Lower Cretaceous outcrop is areally very extensive, mainly because of the wide band of Lower Cretaceous outcrop in the Lampasas Cutplain and Edwards Plateau regions of Texas (Fig. 15.1). Outcrop of Lower Cretaceous strata in the Gulf Coastal Plain is limited to Texas, where it is most widespread, southeastern Oklahoma, and southwestern Arkansas. East of southwestern Arkansas, along the interior margin of the Gulf Coastal Plain, Upper Cretaceous and younger sedimentary rocks crop out, covering Lower Cretaceous and older strata below. In far western Texas the Lower Cretaceous outcrop is limited.

Geographic Distribution of Tracksites

Tracksites occur in Arkansas and in the following Texas Provinces: Edwards Plateau; the area of outcrop skirt-

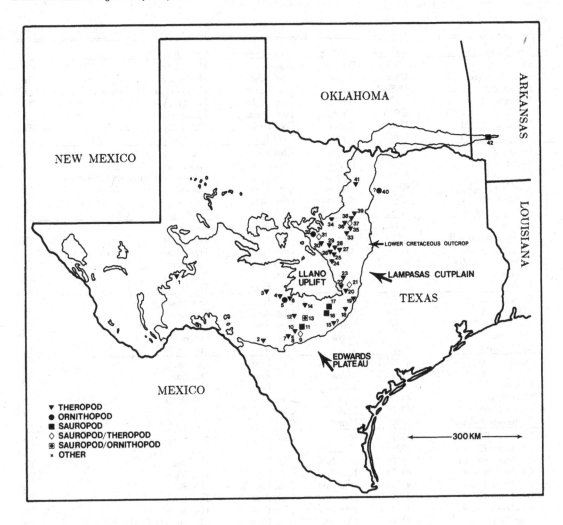

Figure 15.1. Dinosaur tracksites of the Gulf Coastal Plain. The outcrop of Lower Cretaceous strata is shown. See Table 15.1 for the list of tracksites and site numbers.

ing the Llano Uplift; Lampasas Cutplain; and north–central Texas (Fig. 15.1). Exposure of track–bearing beds at these sites is the result of a combination between the gentle dips of well bedded limestones, the erosional regime, low stream gradients, and climate of the region. Most naturally exposed tracks occur in stream beds floored by gently dipping Lower Cretaceous limestones.

The tracksite located farthest west is the Girvin Site, west of Girvin, Pecos County, Texas, and the farthest east is the Briar Site in southwestern Arkansas. A continuous scattering of tracksites occurs across the Edwards Plateau, around the Llano Uplift of central Texas, on the Lampasas Cutplain, to north–central Texas. No tracksites are known from the area between the West Trinity River Site in north Texas to the site in Arkansas, although Getzendaner (1956) listed a site in the Antlers Sand of Oklahoma with no locality information.

I list 54 tracksites in this report (Table 15.1). In Table 15.1 sites which occur in very close proximity are given the same site number; many of these "sites" are only a few hundred meters apart (e.g., sites along the Paluxy River). The stratigraphic position, lithology, specific location, and track type of 22 tracksites are reported here for the first time. I am aware of a number of other reports, many of which will probably be authenticated in the future.

Specific locality information, thin sections, and some photographs and casts from the sites are retained in the files of the Vertebrate Paleontology Laboratory, Texas Memorial Museum.

Stratigraphy

All but one site (the Grapevine Lake site in the Cenomanian Woodbine Formation) occur on strata within the Fredericksburg and Trinity Groups of the provincial

Table 15.1.

No.	Site Name	TMM #	County	Formation	Identification			
1	Girvin	42987	Pecos	Segovia	This Study			Th
2	Live Oak Creek	42988	Kinney	Glen Rose	This Study			Th
3	McKegan Draw #1a	42989-1	Kimble	Fort Terrett	Gould (unpublished photo)			Th
3	McKegan Draw #1b	42989-2	Kimble	Fort Terrett	This Study			Th
3	Middle Copperas Ck.	42990	Kimble	Fort Terrett	This Study			Th
4	Dry Cedar Creek	42991	Kimble	Fort Terrett	This Study			Th
5	Johnson Fork Ck.	42992	Kimble	Hensel	This Study	O		
6	James River	42993	Kimble	Glen Rose	This Study			Th
7	Sabinal River	42994	Uvalde	Glen Rose	This Study			Th
8	Seco Creek	42996	Medina	Glen Rose	Stricklin and Amsbury (1974)			
9	West Verde Ck.	40637-3	Medina	Glen Rose	This Study		S	Th
10	Hondo Creek #1	42996	Bandera	Glen Rose	This Study		S	Th
10	Hondo Creek #2	42997	Bandera	Glen Rose	Perkins (1974) [photo]			Th
11	Medina River	42998	Bandera	Glen Rose	This Study		S	
12	South Guadalupe R.	42999	Kerr	Glen Rose	This Study			Th
13	Guadalupe River	43000	Kerr	Glen Rose	This Study	O	S	
14	Dittmar Creek #1a	43001-1	Gillespie	Glen Rose	This Study			Th
14	Dittmar Creek #1b	43001-2	Gillespie	Glen Rose	This Study			Th
15	Sattler	43002	Comal	Glen Rose	This Study			Th?
16	Blanco River	43003	Blanco	Glen Rose	This Study		S	
17	Miller Creek	43004	Blanco	Glen Rose	This Study		S	
18	Bear Creek	43005	Hays	Glen Rose	This Study			Th
19	Bull Creek	43006	Travis	Glen Rose	Langston (pers. comm.)			Th
20	Sandy Creek	43007	Travis	Glen Rose	This Study			Th
21	South San Gabriel #1a	41988-2	Williamson	Glen Rose	This Study			Th
21	South San Gabriel #1b	41988-3	Williamson	Glen Rose	This Study		S	Th
22	Oatmeal	43008	Burnet	Glen Rose	This Study			Th
23	Olive Branch	43009	Burnet	Glen Rose	This Study			Th
24	Simms Creek	43010	Lampasas	Glen Rose	This Study			Th
25	Cottonwood Creek	43011	Hamilton	Glen Rose	Wrather (1922) [sketch]			Th
26	Lampasas River	43012	Hamilton	Glen Rose	This Study			Th
27	Cowhouse Ck. #4	43018	Hamilton	Glen Rose	Davis (1974) [photo]			Th
28	Gholson Draw #1	43013	Hamilton	Glen Rose	This Study			Th
28	Gholson Draw #2	43014	Hamilton	Glen Rose	This Study			Th
29	Cowhouse Ck. #1	43015	Hamilton	Glen Rose/Paluxy	This Study			Th
29	Cowhouse Ck. #2	43016	Hamilton	Glen Rose	This Study			Th
29	Cowhouse Ck. #3	43017	Hamilton	Glen Rose	This Study			Th
30	Lake Eanes	43019	Comanche	Glen Rose	Albritton (1942) [sketch]			Th?
31	Sidney	43020	Comanche	Glen Rose	This Study		S?	Th
32	Jimmys Creek	43021	Comanche	Glen Rose	This Study	0		
33	North Bosque R.	43022	Bosque	Glen Rose	Local newspaper photos			Th
34	Huckabay	43023	Erath	Glen Rose/Paluxy	This Study			Th
35	Cross Branch	43024	Somervell	Glen Rose	This Study			Th
36	Rough Creek	43025	Somervell	Glen Rose/Paluxy	Booth (1956) [photo]			Th
37	Paluxy River #A1	43026	Somervell	Glen Rose	This Study			Th
37	Paluxy River #A2	43027	Somervell	Glen Rose	This Study		S	Th
37	Paluxy River #A3	40638	Somervell	Glen Rose	This Study		S	Th
37	Paluxy River #A4	43028	Somervell	Glen Rose	This Study		S	Th
37	Palxuy River #B1	43029	Somervell	Glen Rose	Shuler (1937) [photo]			Th
37	Paluxy River #C1	43030	Somervell	Glen Rose	This Study			Th
37	Paluxy River #C2	43031	Somervell	Glen Rose	Kuban (1986) [photo, sketch]			Th
37	Paluxy River #D1	43032	Somervell	Glen Rose	Shuler (1917) [photo]			Th
37	Paluxy River #E1	43033	Somervell	Glen Rose	This Study			Th
37	Paluxy River #E2	43034	Somervell	Glen Rose	This Study			Th
37	Paluxy River #F1	43035	Somervell	Glen Rose	This Study			Th
38	Squaw Creek #1	43036	Somervell	Glen Rose	This Study			Th
38	Squaw Creek #2	43037	Somervell	Glen Rose	This Study			Th
39	Lake Granbury	43038	Hood	Glen Rose	This Study			Th
40	Grapevine Lake	43039	Tarrant	Woodbine	Gillette (pers. comm.)	O?		
41	West Trinity River	43040	Wise	Glen Rose/Paluxy	This Study			Th
42	Briar (Arkansas)	43041-1	Howard	Glen Rose equiv.	This Study		S	
42	Briar (Arkansas)	43041-2	Howard	Glen Rose equiv.	This Study		S	

Alternate Name	Lithology	Components	No.
	miliolid micrite/wackestone		1
	dolomicrite/mudstone		2
	dolomitic pelmicrite/mudstone		3
	pelsparite/packstone		3
F6 Ranch	pelmicrite/mudstone	miliolids, ostracods present	3
	miliolid micrite/mudstone	occ. pelecypods, pellets	4
	pelsparite/wackestone	rare miliolids, pelecypods; occ. quartz	5
Garner Ranch	dolomicrite/mudstone		6
	dolomicrite/mudstone		7
	No sample		8
Davenport Ranch	dolomicrite/mudstone	common quartz	9
	dolomicrite/mudstone		10
	rudist biloithite (Perkins 1974)		10
Mayan Ranch	micrite/mudstone		11
	dolomicrite/mudstone		12
	micrite/mudstone	rare miliolids, ostracodsd; occ. pelecypods	13
	dolomicrite/mudstone		14
	dolomicrite/mudstone		14
Thayer	micrite/mudstone	occ. miliolids, ostracods	15
	micrite/mudstone	algal lamination	16
	micrite/mudstone	occ. ostracods	17
	dolomicrite/mudstone	laminoid fenestrae, bird's eyes	18
	No sample		19
	dolomicritic miliolid micrite/mudstone		20
	dolomicrite/mudstone	birds' eyes, clotted micrite	21
	dolomicrite/mudstone		21
	dolomicrite/mudstone	common quartz	22
	miliolid micrite/mudstone	occ. ostracods	23
	micrite/mudstone	occ. miliolids, ostracods, pellets	24
	No sample		25
	dolomicrite/mudstone	common quartz	26
	dolomite (Davis 974)		27
	dolomicrite/mudstone		28
	dolomicrite/mudstone		28
	calcite–cemented sandstone	vfg sand, coarse silt	29
	arenaceous dolomicrite/mudstone		29
	arenaceous micrite/mudstone		29
	No sample		30
	calcite–cemented sandstone	fine to medium grained	31
	dolomite–cemented sandstone	fine to medium grained	32
	No sample		33
	calcite–cemented sandstone	fine to medium grained	34
	dolomitic micrite/wackestone	rare ostracods, pelecypods	35
	sandstone (Booth, 1956)		36
DSVP–Denio Branch Mouth	dolomicrite/mudstone		37
DVSP–Overlook	arenaceous dolomicrite/mudstone		37
DVSP–Bird Excavation	arenaceous dolomicrite/mudstone		37
DVSP–Possum Creek	arenaceous dolomicrite/mudstone		37
4th Crossing	No sample		37
2nd Crossing–Lancaster	dolomicrite/mudstone	rare quartz	37
Hall	No sample		37
3rd Crossing–Loc.?	No sample		37
Taylor	No sample		37
McFall	pelemicrite/packstone	c. miliolids, pelecypods, algae, occ. quartz	37
McFall	dolomitic pelsparite/packstone	occ. algae, pelecypods, ostracods, miliolids, qtz.	37
	micrite/mudstone	occ. miliolids, ostracods, pellets, quartz	38
	micrite/mudstone	c. miliolids, pelecypods; rare ostracods	38
	dolomicrite/mudstone	common ostracods, pelecypods	39
	No sample		40
Paradise	dolomite–cemented sandstone	medium grained	41
	arenaceous pelsparite/packstone		42
	arenaceous micrite/mudstone		42

Comanche Series (Fig. 15.2). Most sites occur on rocks of the Lower to Middle Albian Glen Rose Formation of the Trinity Group. Within the Glen Rose Formation, the large number of tracksites occurring in strata lying very near the top of the formation constitutes a dinosaur track zone. There is a less-well-defined zone in the middle part of the formation at and near the *Corbula* Bed and equivalent strata.

Trinity Group Tracksites

The tracksites at the lowest stratigraphic position may be those in the Twin Mountains Formation in Comanche County, Texas (Sidney and Jimmys Creek). However, the precise stratigraphic position of these sites is not known, because the contact between the Twin Mountains and Glen Rose climbs toward the west and north (Fig. 15.3). The Johnson Creek Site, occurring in the Hensel Formation on the Edwards Plateau, is also difficult to correlate. Its stratigraphic position is not precisely known because the top of the Hensel also climbs to the north and west. However, the beds on which the tracks occur lie so near the base of the overlying Fredericksburg Group, it is likely that they are equivalent to beds of the uppermost Glen Rose Formation (see Stricklin et al. 1971 Fig. 14).

Within the middle part of the Glen Rose Formation tracksites occur in strata lying adjacent to the *Corbula* Bed, a thin regional marker bed between the Lower and Upper Glen Rose. The *Corbula* Bed (actually there are several, but only one is widespread and used as a regional marker) consists of very abundant shell debris of the minute pelecypod *Carycorbula martinae*. It is plotted on geologic maps of the region. At the following sites on the Edwards Plateau dinosaur tracks occur on beds lying near the *Corbula* Bed, in the middle part of the Glen Rose: Sabinal River, Hondo Creek, Medina River, South Guadalupe River, Guadalupe River, Blanco River, and Miller Creek (Fig. 15.2).

"*Corbula* beds" also occur in the Glen Rose sequence exposed in the Paluxy River valley, Somervell County, Texas, far to the north (about 175 km north of the northernmost known extent of the *Corbula* Bed proper). Sauropod and carnosaur tracks at Dinosaur Valley State

Figure 15.2. Correlation chart, Lower and Middle Cretaceous strata of the Gulf Coastal Plain (modified from stratigraphic charts on sheet maps of the Geologic Atlas of Texas listed across the top). Sites are listed left-right in their general order from southwest to northeast as they occur on Figure 15.1. The geologic maps listed across the top of the chart correspond to the stratigraphy presented below them, but tracksites do not necessarily occur on the map which they are listed under. Symbols as in Figure 15.1. Note concentration of tracksites at top of Glen Rose (see Fig. 15.3 also).

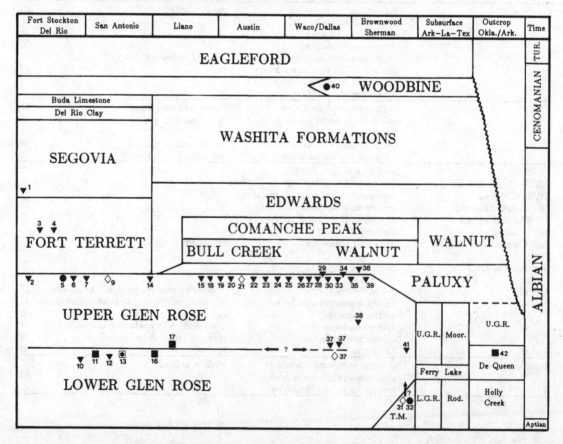

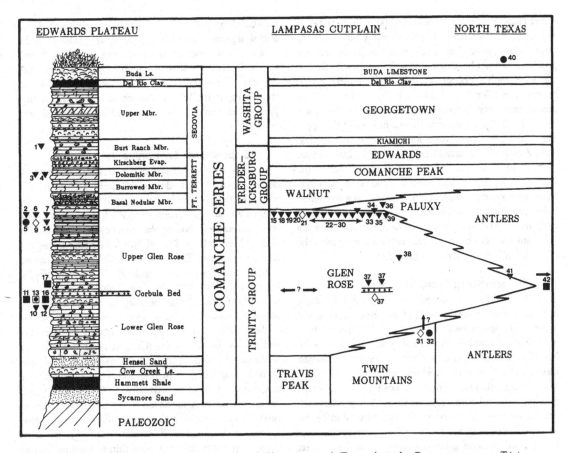

Figure 15.3. Stratigraphic chart from the Edwards Plateau to north Texas, along the Cretaceous outcrop. Trinity Group profile in Edwards Plateau from Perkins (1974); Fredericksburg Group and younger profile from Rose (1972). Symbols as in Figure 15.1.

Park on the Paluxy River occur on beds a few meters above and below one such "*Corbula* bed". Whitney (1952) correlated the *Corbula* Bed of the Edwards Plateau with the Paluxy River "*Corbula* bed". However, later workers (see Nagle 1968) questioned this correlation because in both areas there are, in fact, several "*Corbula* beds".

This problem of the correlation of the trackbearing strata in the middle part of the Glen Rose of the Edwards Plateau and of the Paluxy River Valley relates directly to the issue of the stratigraphic position of trackbearing beds in the De Queen Formation of southwestern Arkansas (Glen Rose equivalent). Tracks exposed in the quarry at the Briar Site in Arkansas occur on beds lying only 10 m above a regionally extensive sequence of anhydrite beds (Pittman 1984, 1985). These outcropping beds of the Ferry Lake Anhydrite may be traced into the subsurface across southern Arkansas and eastern Texas to near the outcrop in the Paluxy River Valley. The trackbeds in the Glen Rose in the Paluxy River Valley occur below a distinctive, 10 m thick limestone bed referred to as the Thorp Springs. Beds equivalent to the Ferry Lake Anhydrite occur below the Thorp Springs in the subsurface near the outcrop in this area. (This is based on my own correlations, but see also cross-sections of Fisher and Rodda [1966] and

Hayward and Brown [1967] and the work of Nagle [1968].) On the basis of this correlation, it is apparent that tracks in Arkansas and in the Paluxy River Valley lie on beds of very near the same age. Tracks in these two areas occur on beds lying at about the stratigraphic position of the thin stringers of anhydrite lying above the Ferry Lake Anhydrite in the western Gulf Coastal Plain (see Pittman 1985).

In the subsurface near the outcrop of the Glen Rose Formation on the Edwards Plateau, thin gypsum beds occur a few meters above the *Corbula* Bed. These gypsum beds may be traced through the subsurface of the Edwards Plateau, and through laterally equivalent limestone beds around the Llano Uplift to a stratigraphic position lying in the middle part of the Glen Rose section above the Ferry Lake Anhydrite. Trackbearing beds in the middle part of the Glen Rose on the Edwards Plateau, in the Paluxy River Valley, and in Arkansas are very near the same age. At the time of this writing the resolution of subsurface control does not allow the specific correlation of the *Corbula* Bed of the Edwards Plateau with the "*Corbula* bed" in the Paluxy River Valley (although it is not ruled out). If they are not specifically equivalent, they are very near the same age.

Theropod tracks at the Comanche Peak Nuclear

Power Plant (Squaw Creek Sites) lie about 10–20 m below the top of the Glen Rose. The West Trinity Site lies at the outcrop limit of the Glen Rose Formation. Its stratigraphic position could not be determined with accuracy. Therefore, the stratigraphic positon of this track–bearing bed is presented as Glen Rose/Paluxy transition (Fig. 15.3). No precise vertical correlation within the Glen Rose is intended.

Tracksites are most numerous in beds of the upper half of the Glen Rose Formation, very near the top (Fig. 15.2). Across the Edwards Plateau, around the Llano Uplift, and on the Lampasas Cutplain these upper beds of the Glen Rose are thin–bedded, micritic, and often dolomitic. Many sites occur at the upper boundary or just a few meters below, constituting a dinosaur track zone. Many more tracksites may be discovered in the future by systematically searching the beds of rivers and creeks flowing across the upper Glen Rose formational boundary.

Fredericksburg Group Tracksites

Rose (1972) summarized the stratigraphy, lithology, and depositional history of rocks of the Fredericksburg and Washita Groups exposed on the Edwards Plateau. Three tracksites occur on beds of this age in this area, two in the Fort Terrett Formation (Dry Cedar Creek and McKegan Draw/Middle Copperas Creek) and one in the Segovia Formation (Girvin). The stratigraphic position of the Girvin Site was estimated from a geologic map. More precise stratigraphic control is provided for the other two sites by the work of Rose (1972). His Junction composite section (section J1) was made of good exposures of the Fort Terrett Formation along Dry Cedar Creek. At the Dry Cedar Site, the tracks lie above the well exposed Burrowed Member, on miliolid–rich beds in the Dolomitic Member. Rocks at the McKegan Draw/Middle Copperas Creek Site, a few kilometers west of Junction, exhibit a similar lithology and features. On this evidence and through observation of the geologic map is the correlation made of these sites with the Dolomitic Member of the Fort Terrett Formation.

The Rough Creek Site occurs on sandstone near the top of the Paluxy (Booth 1956, Keith Young pers. comm.). Although the track–bearing beds at the Cowhouse Creek #1 and Huckabay Sites, lying at the base of the Paluxy, are sandstone, the time equivalent of these beds may be the uppermost part of the Glen Rose. The contact between the Glen Rose and Paluxy is gradational, the limestones of the Glen Rose passing upward into more arenaceous beds (Owen 1979). For this reason, and because of poor exposures at the Cowhouse Creek #1 and Huckabay sites, the stratigraphic position of these beds is presented here as Glen Rose/Paluxy boundary (Fig. 15.3).

The Outlier: The Grapevine Lake Site

The tracksite in the Woodbine Formation, exposed at the spillway of Grapevine Lake, Tarrant County, Texas, is a stratigraphic outlier. It lies in Cenomanian strata, much younger than other sites of the Gulf Coastal Plain. However, when the lithology of outcropping Cretaceous rocks of the region is considered, its occurrence is not quite so unusual. Many of the units of the upper Edwards/Fredericksburg Group and Washita Group are "marine" in the area of the outcrop. Lying above these strata, the Woodbine Formation consists predominantly of sand and clay deposited in nearshore and terrestrial settings. In addition, the outcrop of the Woodbine Formation is limited to that part of Texas and is best developed where the tracksite occurs.

Depositional Environments

The Gulf of Mexico region was situated near 30 degrees north latitude during the Early Cretaceous (see reconstruction by Ziegler et al. 1983). During the later Aptian, Albian, and Cenomanian Ages of the Early Cretaceous, extensive carbonate platforms developed around the Gulf of Mexico (Figs. 15.4, 15.5). Major clastic influx was limited to northern Texas/southern Oklahoma and the area south of the Appalachian highlands. Wide areas (50–250 km from shelf to shoreline) of clear, shallow, warm water induced carbonate production on the shelves in offshore areas and away from these areas of major clastic influx (Wilson 1975). Most of the six evolutionary stocks of rudists had evolved by the Aptian (Coogan 1977) and were key to the development of these platforms. Rudist–coral reefs formed along the shelf margin and rudist buildups were common in backreef areas (Perkins 1974). Behind the reefs along the shelf margin, shallow water carbonates accumulated. Oolitic–bioclastic grainstone was deposited in shallow, agitated areas on the shelf. Typical back reef rocks, aside from the rudist buildups, include thin to medium bedded micrite often rich in miliolid foraminifera, some miliolid grainstone, and bioturbated mollusc wackestone. Periodically widespread subaqueous gypsum precipitation (Ferry Lake Anhydrite) occurred in this lagoonal area when it became isolated from the open Gulf across the shelf margin (Forgotson 1963, Loucks and Longman 1982, Pittman 1985). Typical rock types deposited in nearshore settings include micrite, often dolomitized, algal laminated in intertidal zones, and fenestrated in supratidal settings, and mollusc–miliolid dominated wackestone. In nearshore areas where clastic sediment influx was higher, claystones and arenaceous limestones were deposited, breaking the sequence of carbonate beds. Current–deposited beds in the nearshore sequence are often rich in mollusc fragments. Particularly significant, in regard to stratigraphy, are the thin current–deposited beds rich in the very small clam *Carycorbula martinae*. Tracksites occur in beds of the varying lithologies outlined above, deposited near the shore in this lagoonal area.

Lower Glen Rose and Units Below

Tracks at the Johnson Creek Site, in the Hensel Formation, are in pelsparite. This lithology, the lateral facies equivalents, and the geographic location of site indicate a very nearshore environment.

As in the case of the Johnson Creek Site in the Hensel Sand, the stratigraphic position of beds at the two sites in the Twin Mountains Formation (Sidney and Jimmys

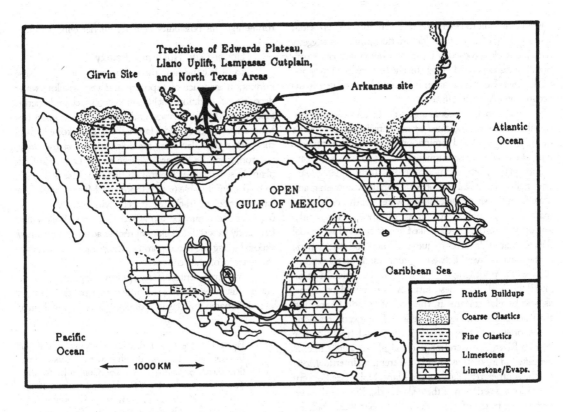

Figure 15.4. Facies map of Comanchean age strata of the Gulf of Mexico region (modified from Bryant et al. [1969], Cook and Bally [1975], and Wilson [1975]).

Figure 15.5. Comanche Shelf sediments, Gulf of Mexico region (modified from Wilson [1975] and McFarlan [1977]), showing paleoenvironmental context of most trackbearing sediments.

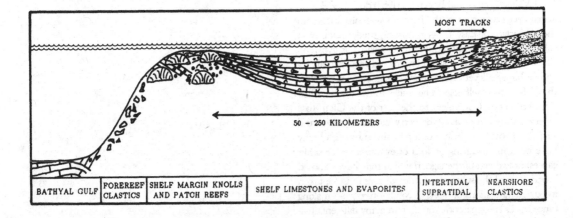

Creek) is difficult to determine with certainty. Nevertheless, the lithology of both beds (dolomitic sandstone) suggests that the tracks exposed at these sites were made on transitional sands deposited along the shoreline while limestone of the Glen Rose Formation (probably Lower Glen Rose) accumulated farther offshore.

Evidence of exposure and very shallow water depositional conditions is indicated by mud cracks, fenestral bird's eye cavities, and algal lamination in rocks of the middle part of the Glen Rose Formation of the Edwards Plateau, Lampasas Cutplain, and north–central Texas. The Comanche shelf was a broad platform with shallow bottom gradients; from the shoreline offshore, water depth probably remained very shallow for great distances. Apparently, sauropods and theropods walked across exposed supratidal flats and intertidal areas frequently. Theropod tracks made at the Sabinal River Site were certainly made on an exposed surface (see photograph of mudcracks in Stricklin et al. [1971] Pl. XIII, B). These surfaces were bored by clams, as were similar surfaces in the bed of Hondo Creek and the track bed at the Squaw Creek #2 Site (exposures along the shoreline of the reservoir). Sauropod tracks at the Blanco River Site were made on algal–laminated micrite deposited in a northeast–southwest trending intertidal zone on the southeastern flank of the Llano Uplift (see Behrens [1965] for a description of these algal beds). Bird (1954) interpreted that tracks of the Medina River Site were made by a "swimming sauropod". See Lockley and Conrad (this volume) for an alternate interpretation of this trackway. Sauropod and ornithopod tracks occur at the Guadalupe River Site on a limestone bed lying below the *Corbula* Bed (the ornithopod trackway in Fig. 15.6 is from the Guadalupe River Site). Hundreds of theropod tracks occur in the bed of the South Guadalupe River in Kerr County. These tracks are deep and many are poorly preserved. They were certainly made in deep mud, either exposed or in shallow water. Theropod tracks occur atop a rudist mound at the Hondo Creek #2 Site (Perkins 1974). This reef, however, was not a great distance offshore and shows evidence of frequent exposure (Perkins 1974).

Nagle (1968) described a cyclical alternation of subtidal through supratidal facies in the thin track–bearing sequence of the Glen Rose Formation exposed in the Paluxy River Valley. Loucks and Longman (1982) described depositional cycles in a core of the Ferry Lake Anhydrite from the subsurface of East Texas consisting predominantly of subtidal carbonates and anhydrite (deposited as gypsum). These cycles "shallow" upward and record repeated salinity and perhaps sea level variations in the lagoonal area behind the reefs at the shelf edge. The tracks in the middle Glen Rose Formation (equivalent to the part of the Glen Rose wherein anhydrite beds of the Ferry Lake Anhydrite and stringers above and below occur in the subsurface) may have been made during periods of evaporative (or other causes) drawdown of the lagoon which may have exposed broad areas of the substrate. Although care must be taken in our interpretations of "special" conditions of trackmaking (since tracks are apparently common in many different envi-

ronments), this possibility deserves further consideration.

Upper Glen Rose and Paluxy

In beds lying near the top of the upper Glen Rose Formation evidence of exposure and very shallow water depositional conditions include mud cracks, clotted micrite, fenestral bird's eye cavities, and pholad clam borings (see Moore 1964, Moore and Martin 1966). Limestone beds of this part of the Glen Rose are dolomitic across the broad outcrop, from the Edwards Plateau to the Lampasas Cutplain. In several areas the top of the Glen Rose is thought to be disconformable (e.g., Moore and Martin 1966). The large number of tracksites in beds lying near the top of the Glen Rose may be attributed to a very shallow and frequently exposed depositional regime across the area now exposed along the outcrop and to the erosion and weathering pattern of the sequence.

In north–central Texas, during deposition of the Albian Paluxy Sandstone, the nearshore depositional system was dominated by clastic sediment. Moore and Martin

Figure 15.6. Sections of dinosaur trackways of the Gulf Coastal Plain. **A,** single ornithopod track (drawn from field sketch of Jimmys Creek Site, Comanche County, Texas). 1 square = 1 m². **B,** theropod trackway from Huckabay Site, Erath County, Texas. **C,** ornithopod trackway from Guadalupe River Site, Kerr County, Texas. Corner brackets mark the boundary of the photograph in Figure 15.11. **D,** sauropod trackway from Briar Site, Howard County, Arkansas (Trackway #1 — see Pittman and Gillette this volume, Fig. 15.17.).

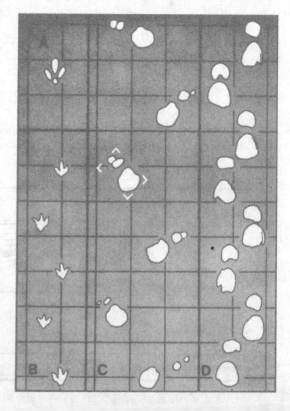

(1966) described a tongue of the Paluxy extending across the Lampasas Cutplain southward from the main body of the formation, between the top of the Glen Rose and the base of the Fredericksburg Group (Fig. 15.3). They believed these sands were deposited by longshore currents, and modified shoreward by wave swash and tidal action in a nearshore environment. Farther north, in north–central Texas, similar depositional conditions are indicated for the Paluxy (Atlee 1962, Caughey 1977, Owen 1979). Theropod tracks at the Cowhouse Creek #1, Rough Creek, and Huckabay sites were made on such transitional nearshore sands.

Fredericksburg Group

Tracks at the Dry Cedar Creek and McKegan Draw/Middle Copperas Creek sites occur on mud–cracked, thin bedded limestones, probably in the Dolomitic Member of the Fort Terrett Formation. Rose (1972, p. 34) gave the following depositional summary for these beds: "stromatolitic hard crusts, root marks, mud cracks, and birdseye structure indicating tidal flat deposits.... Ripple marks, current streaks, fine planar cross beds, and flat mud clasts are scattered throughout the member. Evidence of restriction and high salinity is provided by indications that the dolomitic sediments once contained gypsum." The tracks at these sites were probably made on exposed flats.

Tracks at the Girvin Site occur on strata in the lower part of the Segovia Formation. Rose (1972, p. 35) described the Burt Ranch Member of the Segovia, in which the track beds probably occur (or its stratigraphic equivalent), as consisting generally of "marl, marly micrite, miliolid biosparite, and rudist biosparite, with a few scattered beds of soft massive dolomite". Lozo and Smith (1964) and Moore (1964) had earlier noted a disconformity at the top of the Fort Terrett Formation, marked by iron staining and bored surfaces. Rose (1972) noted similar surfaces within the Burt Ranch Member of the Segovia, and concluded they were the result of submarine lithification, instead of frequent exposure. He believed the unit was deposited on positive areas where wave agitation was high enough to inhibit clay accumulation. The similarity of these bored surfaces to bedding surfaces containing dinosaur tracks in the Glen Rose and Fort Terrett Formations is striking. Perhaps these surfaces in the Burt Ranch Member also indicate exposure. In either case, the tracks at the Girvin Site were probably made in an exposed setting.

Description and Rationale for Identification of Tracks

The majority of tridactyl tracks in the region were made by theropod dinosaurs (contrary to the opinion presented in Pittman [1986] wherein ornithopod tracks were said to be common). Farlow (1986, 1987) has independently come to this conclusion. The three sites at which ornithopod tracks definitely occur are first reported here (Jimmys Creek [Fig. 15.6a], Guadalupe River [Figs. 15.6c, 15.10, 15.11], and Johnsons Fork Creek Sites).

Theropod Tracks

Typical theropod trackways (Figs. 15.6b, 15.7) are rather narrow and pes prints are consistently "toed–in". Characteristics of tridactyl tracks which I attribute to theropods include (Figs. 15.7d, 15.8d): (i) an indentation along the medial margin of the track, along the medial edge of digit II, just distal to the apex of the "heel"; (ii) a medially directed (away from the longitudinal axis of the track along digit III) distal termination of digit II (the distal termination of digit IV is less often outwardly directed [laterally]); (iii) an inturned (medially directed) distal termination of digit III; (iv) a topographic separation in the floor of the track of digits III and IV from II (if water were poured into a typical track, a pool would form along digit II and another pool would form in digits III and IV, the two pools separated by a gentle topographic ridge along the margin

Figure 15.7. Photograph of Simms Creek theropod trackway. The first left print is very poorly preserved. The first right (at arrow; same track in Figs. 15.8a, 15.9a) and other tracks are well-preserved. The narrow gauge of this trackway and "toed–in" prints is characteristic of Texas theropod trackways. Data for the "left–right–left–right" sequence visible in this photograph (in centimeters and degrees): Strides: left–302, right–302. Paces: left/right–155; right/left–158; left/right–155. Bearings: left–11°; right–5°; left–12°; right–4°. The rotation of the bearing of a particular print from one step to the next in this trackway (through about 7°) is typical.

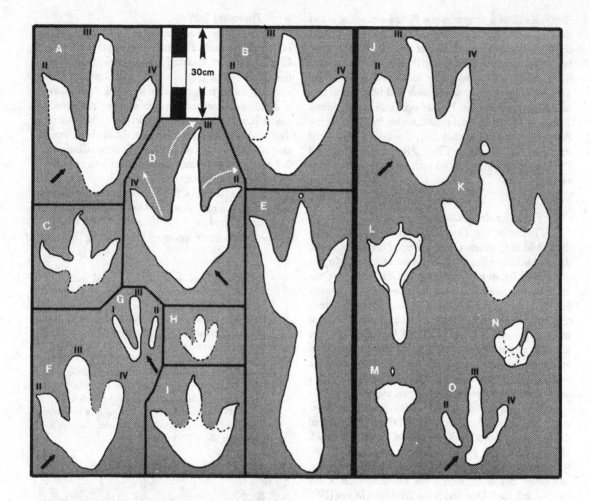

Figure 15.8. Variation in theropod track as illustrated by selected examples. The left side of this figure is an overlay sketch of the photographs in Figure 15.9. All tracks drawn to same scale. **A,** Simms Creek (the first right print in the trackway in Fig. 15.7) — Arrow marks the indentation along the medial edge of the digit II impression. Note the undulating lateral edge of the digit IV impression (metatarsal–phalangeal and phalangeal articulations?; see also O).

B, Huckabay (right) — The abrupt, oblique boundary at the distal termination of digit III distinguishes several Huckabay prints. The demarcation of the digit II impression from that of the digit III/IV impression in the floor of the track is clearly defined in this track. **C,** Sidney (?right) — The tracks at the Sidney Site lie atop a cross-bedded sandstone stratum. Slight deformation of the tracks by migration of small bed forms apparently occurred. **D,** Squaw Creek #2 (left) — Black arrow marks the indentation along the medial edge of the digit II impression. White arrows indicate the orientation of the digital impressions in this print, which is typical of theropod tracks of the region. **E,** Cowhouse Creek #3 (side unknown) — The very elongate posterior extension from this track is atypical. See text for discussion. **F,** Middle Copperas Creek (right) — Arrow marks the indentation along the medial edge of the digit II impression. Note the rounded distal termination of digits III and IV. A claw mark is present at the distal termination of digit II — its rounded appearance is due to fracturing of the rock filling the track. **G,** Middle Copperas Creek (left) — Ghost print exhibiting clear demarcation of the digit II impression from that of digits III/IV. **H,** Sidney (side unknown) — Very small print (approximately 16 x 14 cm). **I,** Lampasas River (?left) — See text for discussion. **J,** West Verde Creek (right) — Arrow marks the indentation along the medial edge of the digit II impression. The distal termination of the digit III impression is sharply inturned (as in other prints in this trackway). **K,** Sandy Creek (?right) — A separate claw mark at the end of the digit II impression is present in this track. **L,** Hondo Creek #2 (side unknown) — The elongate "heel" region of this track is of a more typical dimension than E. **M,** Cowhouse Creek #3 (side unknown) — This print occurs within the same sequence of trackways as E. **N,** Dry Cedar Creek (side unknown) — Very small print of similar form and dimensions as H. **O,** Dry Cedar Creek (right) — Another example of a ghost print in which the digit II impression is distinctly separated from the impression of digits III/IV.

between digits II and III); and (v) rather narrow digital impressions (as compared to the broad digital impressions of ornithopod tracks). Claw marks of all three digits are present in many tracks. Langston (1974) referred the larger tracks of this type from the Texas Comanche Series to *Irenesauripus* Sternberg (1932), considering Shuler's (1917) earlier assignment to *Eubrontes* inappropriate.

In tracks at the West Verde Creek Site an unusually large claw mark of digit III turns more abruptly and to a greater extent medially than in most typical tracks (Fig. 15.8j). The distal termination of digit III in some tracks at the Huckabay Site is formed by an abrupt, oblique boundary (Figs. 15.8g, 15.9g). A separate claw mark of the end of the ungual of digit III is present in several tracks in the region (e.g., the Sandy Creek Site, Fig. 15.8k). Similar marks are present in tracks at the Bear Creek Site (excavated and partially destroyed by vandals).

The surface outline of the track at the Lampasas River Site shown in Figures 15.8i and 15.9i is slightly smaller than the actual foot of the trackmaker; cavities formed by impression of the digits into soft mud extend downward and forward. This is a very common characteristic of deeper tracks in the region. Initial contact of the foot with the soft substrate was made while the foot was travelling downward and forward (in a sense, the toes are "jabbed" into the mud). The weight of the animal increasingly forces the foot farther into the mud until weight–bearing shifts to the next foot and step. Retraction of the digits is accomplished by contraction and coadunation of the digits, as in extant theropods. Thereby an impression less than the size of the foot is made.

Tracks at the Sidney Site (Figs. 15.8c, 15.9c) and those at the Cowhouse Creek #1 Site were made in soft sand. The digits now appear somewhat short. This is due to distortion and deformation after trackmaking by active sand deposition. The sandstone at these sites is crossbedded.

At some sites (Sattler, Rough Creek, some larger tracks at the Middle Copperas Site [Figs. 15.8f, 15.9f]) tracks are of the same form as the previous examples, but the digital terminations are rounded. Many of these tracks are ghost prints. The deeper tracks of this form at the Rough Creek (see photograph in Booth 1956) and other sites are perhaps the result of retraction of the unguals during locomotion.

In some tracks (Middle Copperas, Figs. 15.8g, 15.9g; Dry Credar Creek, Fig. 15.8o) the topographic ridge in the floor of the track between digit II and III/IV is most pronounced (suggesting ghost prints), resulting in their total separation. Farlow (pers. comm.) recalls a track, probably a ghost print, at the Middle Copperas Creek Site in which only impressions of digits III and IV occur.

Elongate dinosaur tracks at several Paluxy River sites have been interpreted as metatarsal impressions (Kuban 1986). There are also elongate tracks at the Hondo Creek #1 and Cowhouse Creek #3 sites. In the elongated tracks at the Hondo Creek #1 Site (Fig. 15.8l) an elongate "heel" region extends from a poorly preserved digital region, probably the result of impression in very deep mud during which

either the "heel" and/or the metatarsals cut through the mud as the foot was very deeply impresssed (the animal was probably not "walking on the metatarsals" — the "heel" impression is the result of locomotion in deep mud). In one elongate track at the Cowhouse Creek #3 Site (Figs. 15.8e, 15.9c) the "heel" region is very elongate. In this track the digital region is more well preserved than usual — in this same trackway, and in adjacent trackways, other tracks consist only of an elongate groove or a deep digital region and an elongate "heel" (Fig. 15.8m). Experimental study of birds walking and running through deep mud may offer an explanation for these tracks.

Very small prints (one is 16x14 cm) at the Dry Cedar Creek (Fig. 15.8n) and Sidney (Figs. 15.8h, 15.9h) sites occur with larger tracks more definitely attributed to theropods. Identification of these tracks as having been made by theropods is based only on this circumstantial evidence.

Larger tracks of the type shown in Figures 15.7, 15.8, and 15.9 may be attributed to *Acrocanthosaurus*, the large carnosaur which lived in the region during the Early Cretaceous. Some bones of the pes and other skeletal material of two individuals were earlier recovered from sands in Oklahoma (Stovall and Langston 1950). Recently, a more complete specimen, also from Oklahoma, including much of a right pes, was discovered. These bones exhibit a structure very similar to that of *Allosaurus*, described by Madsen (1976). Phalanges from both *Allosaurus* and *Acrocanthosaurus* can be arranged to fit these large tridactyl tracks described above. Articulations support an arrangement where the ungual of digit II turns away from the axis of the track, where the ungual of digit III turns inward, and where the ungual and distal phalanges of digit IV are directed forward or turned away from the axis of the track (digit IV appears to have been the most "laterally flexible"). The "heel" of the foot must have consisted of a cartilaginous "pad" coming off metatarsals and proximal phalanges of digits III and IV; hence the indentation distal to the "heel" impression along the medial margin of the track.

Ornithopod Tracks

Definite ornithopod tracks occur at only three sites, the Jimmys Creek, Johnson Fork Creek, and Guadalupe River sites (Fig. 15.6a,c). The tracks at the Jimmys Creek Site are large prints consisting of three broad rounded impressions of digits II, III, and IV and a distinct "heel" impression (Fig. 15.6a). These tracks were perhaps made by an iguanodontid ornithopod. Tracks of similar size and form occur at the Johnson Fork Creek Site. The trackway at the Guadalupe River Site, complete with manus prints (Figs. 15.6c, 15.10, 15.11) certainly records the walk of a large ornithopod. Possible ornithopod tracks occur at the Grapevine Lake Site in the Cenomanian Woodbine Formation.

The Guadalupe River Ornithopod Trackway

In addition to the ornithopod trackway, several sauropod trackways and an unidentified tridactyl trackway occur at the Guadalupe River Site (all tracks at this site

Figure 15.9. Photograph of theropod prints shown in outline in Figure 15.8. See Figure 15.8 caption for discussion.

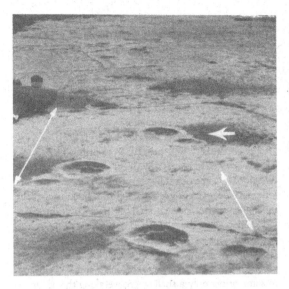

Figure 15.10. Photograph of the Guadalupe River ornithopod trackway (travel direction toward viewer). Short arrow indicates trackset shown in the photograph in Figure 15.11. The distance between the lateral edges of manus prints, as indicated by long arrow, is about 2.4 m.

Figure 15.11. Photograph of a trackset from the Guadalupe River ornithopod trackway. Boundaries of this photograph are shown by corner brackets in Figure 15.6. Arrow indicates travel direction. One meter is reeled out of the tape measure (not including the case).

are ghost prints). The ornithopod trackway deserves a brief description. Pes prints in this trackway are "toed out" slightly (the second right pes print has been partially deformed by a joint in the rock). The outline of three broad digital impressions may be discerned through close observation of pes prints. Two impressions comprise each manus print: a larger, medial impression, and a smaller, lateral impression. The large size of the larger impression suggests that it was made by "bundled" digits II and III. If this interpretation is correct, the smaller impression would have been made by digit IV. The orientation of these two impressions, relative to one another, is slightly varied along the trackway. Manus impressions lie well outside pes prints, relative to the direction of travel. The breadth of this trackway from the lateral edge of manus impressions on the left and right sides is about 2.4 m.

The ornithopod tracks at the three sites discussed above were not known to Langston (1974). Langston did consider some Texas tracks to have been made by ornithopods, describing them as "generally smaller than most tracks ascribed to *Irenesauripus*, and differ from them in having a more rounded 'heel' and relatively shorter, blunter, and more widely divergent digits. The tips of the toe marks may terminate bluntly or acutely, but there is less suggestion of sharp claws...." I suspect that these are poorly preserved theropod prints, or theropod prints lacking claw marks (see Figs. 15.8f, 15.9f; Farlow [1987 Fig. 13g]; Booth 1956). Regardless, the tracks at the Guadalupe River, Jimmys Creek, and Johnson Fork Creek sites I consider unquestionably to have been made by ornithopods.

Sauropod Tracks

Texas sauropod tracks have been described by Bird (1939 and later), Langston (1974), and Farlow (1987). Langston attributed the tracks to the sauropod knwon from body fossils of the region, perhaps the genus *Pleurocoelus*. Descriptions of sauropod tracks in the region are given by Farlow et al. (this volume) and Pittman and Gillette (this volume). A brief description of a typical trackway from the Briar Site in Arkansas (Fig. 15.6d) and a very well–preserved track set excavated by R. T. Bird in 1940 from the West Verde Creek Site (TMM 40637-1, -2, Fig. 15.12) are presented here. The pattern of alternating left and right manus–pes sets distinguishes most sauropod trackways in the region (e.g., Fig. 15.6d), although exceptions do exist (see Farlow et al., and Pittman and Gillette this volume). Broadly U-shaped manus impressions consistently lie slightly closer to the midline than the larger pes prints. Indentations from the sidewall and rear margin of manus prints separate impressions of digits I, II–III–IV, and V. The axis of symmetry of manus prints, bisecting the track

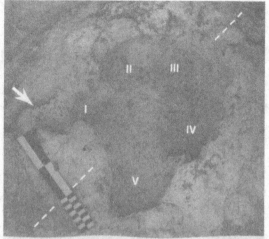

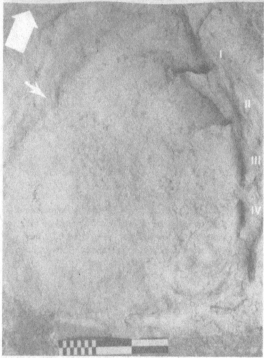

through the apparent location of the digit III impression, is usually rotated outward relative to the direction of travel (e.g., Fig. 15.12, although this manus print is turned outward more than usual). No claw marks are present in manus prints, although the impression to which the arrow points in Figure 15.12 may be a mark made by the ungual of digit I. Other examples of this unique feature are needed to substantiate this interpretation.

Impressions of the unguals of digits I, II, III, and IV are present in very well–preserved pes prints (Fig. 15.12). These claw marks are usually directed in a posterolateral direction. In some cases these claw marks are more anteriorly directed, evidently for traction or a braking or turning maneuver (see Pittman and Gillette this volume). An indentation along the medial wall of some pes prints occurs at the apparent location of the metatarsal–astragalar articulation (at the arrow in Fig. 15.12). In less well–preserved pes prints only impressions of unguals I and II occur and in some prints only a shallow groove along the apparent position of unguals II, III, and IV. Even in very poorly preserved pes prints (ghost prints), the medial half of the track is deeper than the lateral half.

Other Ichnites

Stricklin and Amsbury (1974) reported an interesting sidelight to the ichnology of Texas dinosaur tracksites. They described "scratch and claw marks" and "wingtip scoop marks ... possibly of a pterosaur" atop a limestome stratum in the bed of Seco Creek, Medina County, Texas (photographs are presented in Stricklin and Amsbury [1974, Textfigs. 12, 13] and in Langston [1974, Pl. 4, Fig. 3]). These ichnites have reportedly been destroyed (Langston 1974).

Paleoecology

The habits and habitats of sauropod dinosaurs have long been debated. Their great mass led to ideas of obligate life in the water. Later trackway discoveries and osteological studies provided evidence that sauropods could walk on land. The trackways in sediments deposited along the shoreline of the Cretaceous Gulf of Mexico show that sauropods walked across exposed mudflats or waded in the shallows. Sauropod body fossils have been recovered from fluvial sandstones deposited by streams and rivers not far from the shoreline and even from the lee flank of a rudist reef. Altogether this information suggests that the Gulf Coast sauropods frequented the watercourses and the shoreline and may have ventured into the waters of the broad shallow lagoon.

Carnosaur tracks occur in isolated trackways at some sites, in multiple trackways at others, and with sauropod tracks at a few sites. Their occurrence with sauropod prints has been interpreted as a suggestion of a predator/prey relationship (Bird 1944, 1954; Langston 1974; Farlow 1987). The abundance of single and multiple theropod trackways of both small and large individuals indicates that these animals may have regularly hunted in numbers along the shoreline.

Parallel orientation of multiple trackways of both

Figure 15.12. Photograph of a track set from the West Verde Creek Site excavated by R. T. Bird and crew (manus = TMM40637–1; pes = TMM40637–2). The block containing these prints has been sawed along the white line. This track set is apparently the first right set in Farlow (1987, Figure 5b). The "fat" arrow indicates the travel direction, as determined from this photograph in Farlow (1987). The roman numerals refer to apparent digital impressions in the manus print and claw marks in the pes print. The dotted line bisects the manus print along the apparent position of the digit III impression. Features indicated by the small arrows are discussed in the text.

theropods and sauropods indicates that these animals occasionally traveled in groups. Although interpretations of the timing of trackmaking are difficult, the great number of tracks at many sites suggests that these animals occurred in large numbers in the region.

Undoubted ornithopod tracks occur at only a few sites in the Comanche Cretaceous. It is therefore not possible to construct a well-informed interpretation about their habits. At the Guadalupe River Site, ornithopod and sauropod tracks occur together atop a limestone stratum.

Acknowledgments

Special thanks go to Wann Langston, Jr., who allowed me free access to his personal files and made many helpful suggestions. James Farlow and Phillip Murray provided information about the location of specific sites, as did numerous citizens. James Farlow, Wann Langston, Jr., Timothy Rowe, and Keith Young read versions of the manuscript and made helpful suggestions. My sincere thanks go to them all. Financial support provided for this study by the Geology Foundation, Department of Geological Sciences, University of Texas at Austin, and the Gulf Coast Association of Geological Societies is greatly appreciated.

References

Adkins, W. S. 1932. The Mesozoic System in Texas. *In* Sellards, E. H., Adkins, W. S., and Plummer, F. B. (eds.). *The Geology of Texas, Vol. 1, Stratigraphy. Univ. Texas Bur. Econ. Geol. Bull.* 3232:1–1007.

Albritton, C. C. 1942. Dinosaur tracks near Comanche, Texas. *Field and Laboratory* 1:160–181.

Atlee, W. A. 1962. The Lower Cretaceous Paluxy Sand in central Texas. *Baylor Geol. Studies Bull.* 2:1–26.

Bakker, R. T. 1971. Ecology of the brontosaurs. *Nature* 229:172–174.

Behrens, E. W. 1965. Environment reconstruction for a part of the Glen Rose Limestone, central Texas. *Sedimentology* 4:65–111.

Bird, R. T. 1939. Thunder in his footsteps. *Nat. Hist.* 43:254–261.

——— 1941. A dinosaur walks into the museum. *Nat. Hist.* 47:74–81.

——— 1944. Did *Brontosaurus* ever walk on land? *Nat. Hist.* 53:61–67.

——— 1954. We captured a "live" brontosaur. *National Geographic* 105:707–722.

——— 1985. *Bones for Barnum Brown: Adventures of a Dinosaur Hunter.* Schreiber, V. T. (ed.). (Texas Christian University Press).

Booth, C. C. 1956. Geology of the Chalk Mountain quadrangle, Bosque, Erath, Hamilton, and Somervell counties, Texas. Master's thesis, University of Texas at Austin.

Bryant, W. R., Meyerhoff, A. A., Brown, N. P., Furrer, M. A., Pyle, T. E., and Antoine, J. W. 1969. Escarpments, reef trends, and diapiric structures, eastern Gulf of Mexico. *Amer. Assoc. Petr. Geol. Soc. Bull.* 53:2506–2542.

Bureau of Economic Geology, University of Texas at Austin, Geologic Atlas of Texas Sheet Maps. Scale: 1:250,000.

Caughey, C. A. 1977. Depositional systems in the Paluxy Formation (Lower Cretaceous), northeast Texas — oil,

gas, and groundwater resources. *Univ. Texas Bur. Econ. Geol. Geological Circular* 77–8:1–59.

Coogan, A. H. 1977. Early and Middle Cretaceous Hippuritacea (rudists) of the Gulf Coast. *In Cretaceous Carbonates of Texas and Mexico, Applications to Subsurface Exploration. Univ. Texas Bur. Econ. Geol. Report of Investigations* 89:32–68.

Cook, T. D., and Bally, A. W. (ed.). 1975. *Stratigraphic atlas of North and Central America.* (Princeton, New Jersey: Princeton University Press) 272 pp.

Coombs, W. P., Jr. 1975. Sauropod habits and habitats. *Paleogeog., Paleoclimat., Paleoecol.* 17:1–33.

Davis, K. W. 1974. Stratigraphy and depositional environments of the Glen Rose Formation, north-central Texas. *Baylor Geol. Studies Bull.* 26:1–43.

Dunham, R. J. 1962. Classification of carbonate rocks according to depositional texture. *In* Ham, W. E. (ed.). *Classification of Carbonate Rocks. Amer. Assoc. Petrol. Geol. Mem.* 1:108–121.

Farlow, J. O. 1981. Estimates of dinosaur speeds from a new trackway site in Texas. *Nature* 294:747–748.

——— 1986. *In* Gillette, D. D. (ed.). *First International Symposium on Dinosaur Tracks and Traces, Abstracts with Program.* 31 pp.

——— 1987. *A Guide to Lower Cretaceous Dinosaur Footprints and Tracksites of the Paluxy River Valley, Somervell County, Texas.* Field Trip Guide, South Central Section, Geology Society of America Annual Meeting, 1987. 50 pp.

Farlow, J. O., Pittman, J. G., and Hawthorne, M. Chapter 42, this volume. *Brontopodus birdi*, Lower Cretaceous sauropod footprints from the Gulf Coastal Plain of North America.

Fisher, W. L., and Rodda, P. U. 1966. Nomenclature revision of basal Cretaceous rocks between the Colorado and Red rivers. *Univ. Texas Bur. Econ. Geol. Report of Investigation* 58:1–20.

Forgotson, J. M., Jr. 1963. Depositional history and paleotectonic framework of Comanchean Cretaceous Trinity Stage, Gulf Coast area. *Amer. Assoc. Petr. Geol. Bull.* 47:69–103.

Folk, R. L. 1959. Practical petrographic classification of limestones. *Amer. Assoc. Petr. Geol. Bull.* 43:1–38.

Gallup, M. R. 1975. Lower Cretaceous dinosaurs and associated vertebrates from north-central Texas in the Field Museum of Natural History. Master's thesis, University of Texas at Austin. 159 pp.

Getzendaner, F. M. 1956. Geology of the Trinity Division of the Lower Cretaceous in Texas and adjacent states with special reference to its stratigraphy and the environment of deposition. *San Angelo Geol. Soc. Four Provinces Field Trip Guidebook* pp. 77–105.

Gould, C. N. 1929. Comanchean reptiles from Kansas, Oklahoma and Texas. *Geol. Soc. Amer. Bull.* 40:457–462.

Hawthorne, M. 1987. The stratigraphy and depositional environments of Lower Cretaceous track-bearing strata in the Edwards Plateau and Lampasas Cut Plain physiographic provinces of Texas. Master's thesis, Baylor University.

Hayward, O. T., and Brown, L. F., Jr. 1967. Comanchean (Cretaceous) rocks of central Texas. *In Comanchean (Lower Cretaceous) Stratigraphy and Paleontology of Texas.*

Permian Basin Section Soc. Econ. Paleont. Mineral. pp. 31–47.

Herrin, T., Gillette, D. D., Gillette, J. L., and Campbell, J. 1986. Dinosaur tracks in Paradise. In Gillette, D. D. (ed.). First International Symposium on Dinosaur Tracks and Traces, Abstracts with Program. 31 pp.

Hill, R. T. 1887. The Texas section of the American Cretaceous. Amer. Jour. Sci. 134:287–309.

Houston, 1933. Fossil footprints in Comanchean limestone beds, Bandera County, Texas. Jour. Geol. 41:650–653.

Kuban, G. J. 1986. Elongated dinosaur tracks. In Gillette, D. D. (ed.). First International Symposium on Dinosaur Tracks and Traces, Abstracts with Program p. 17.

1988. Elongate dinosaur tracks. this volume.

Langston, W., Jr. 1974. Nonmammalian Comanchean tetrapods. Geoscience and Man 8:77–102.

1986. Stacked dinosaur tracks from the Lower Cretaceous of Texas — A caution for ichnologists. In Gillette, D. D. (ed.). First International Symposium on Dinosaur Tracks and Traces, Abstracts with Program p. 18.

Langston, W., Jr., and Pittman, J. G. 1987. Lower Cretaceous dinosaur tracks near Glen Rose, Texas. In Lower Cretaceous Shallow Marine Environments in the Glen Rose Formation: dinosaur tracks and plants. Amer. Assoc. Petr. Geol. Southwest Section 1987 Meeting Field Trip Guidebook pp. 39–69.

Larkin, P. 1910. The occurrence of a sauropod dinosaur in the Trinity Cretaceous of Oklahoma. Jour. Geol. 28:93–98.

Lockley, M. 1986. A guide to dinosaur tracksites of the Colorado Plateau and American Southwest. Univ. Colorado Denver, Geol. Dept. Publ. 56 pp.

Loucks, R. G., and Longman, M. W. 1982. Lower Cretaceous Ferry Lake Anhydrite, Fairway Field, East Texas: product of shallow–subtidal deposition. In Depositional and Diagenetic Species of Evaporites — A Core Workshop. Soc. Econ. Paleont. Mineral. Core Workshop 3:130–173.

Lozo, F. E., Jr., and Smith, C. I. 1964. Revision of Comanche Cretaceous stratigraphic nomenclature, southern Edwards Plateau, southwest Texas. Gulf Coast Assoc. Geol. Soc. Trans. 14:285–307.

McFarlan, E., Jr. 1977. Lower Cretaceous sedimentary facies and sea level changes, U.S. Gulf Coast. In Cretaceous Carbonates of Texas and Mexico, Applications to Subsurface Exploration. Univ. Texas Bur. Econ. Geol. Report of Investigations 89:5–10.

Madsen, J. H., Jr. 1976. Allosaurus fragilis: a revised osteology. Utah Geol. Miner. Surv. Bull. 109:1–163.

Moore, C. H., Jr. 1964. Stratigraphy of the Fredericksburg Division, southcentral Texas. Univ. Texas Bur. Econ. Geol. Publ. 52:1–48.

Moore, C. H., Jr., and Martin, K. G. 1966. Comparison of quartz and carbonate shallow marine sandstones, Fredericksburg Cretaceous, central Texas. Amer. Assoc. Petr. Geol. Bull. 50 (5):981–1000.

Nagle, J. S. 1968. Glen Rose cycles and facies, Paluxy River Valley, Somervell County, Texas. Univ. Texas Bur. Econ. Geol. Geological Circular 68 (1):1–25.

Ostrom, J. H. 1972. Were some dinosaurs gregarious? Paleogeog., Paleoclimat., Paleoecol. 11:287–301.

Owen, M. T. 1979. The Paluxy Sand in north–central Texas. Baylor Geol. Studies 36:1–36.

Perkins, B. F. 1974. Paleoecology of a rudist reef complex in the Comanche Cretaceous Glen Rose Limestone of central Texas. Geoscience and Man 8:131–173.

Perkins, B. F., and Langston, W., Jr. 1979. Lower Cretaceous shallow marine environments in the Glen Rose Formation: dinosaur tracks and plants. Amer. Assoc. Stratigraphic Palynologists, 12th Annual Meeting, Field Trip Guide 1–55.

Perkins, B. F., and Stewart, C. L. 1971. Dinosaur Valley State Park. In Perkins, B. F. (ed.). Trace Fossils, a Field Guide to Selected Localities in Pennsylvania, Permian, Cretaceous, and Tertiary Rocks of Texas and Selected Papers. Louisiana State University Misc. Publ. 71 (1):56–59.

Pittman, J. G. 1984. Geology of the De Queen Formation of Arkansas. Gulf Coast Assoc. Geol. Soc. Trans. 34:201–209.

1985. Correlation of beds within the Ferry Lake Anhydrite of the Gulf Coastal Plain. Gulf Coast Assoc. Geol. Soc. Trans. 35:251–260.

1986. Correlation of dinosaur trackway horizons in the Cretaceous of the Gulf Coastal Plain of North America. In Gillette, D. D. (ed.). First International Symposium on Dinosaur Tracks and Traces, Abstracts with Program p. 23.

Pittman, J. G., and Gillette, D. D. Chapter 34, this volume. The Briar Site: a new sauropod dinosaur trackway in Lower Cretaceous beds of Arkansas.

Römer, F. 1846. A sketch of the geology of Texas. Amer. Jour. Sci. 2 (2):358–385.

1852. Die Kreidebildungen von Texas, and ihre organischen Einschlusse. (Bonn: Adolph Marcus) 100 p.

Rose, P. R. 1972. Edwards Group, surface and subsurface, central Texas. Univ. Texas Bureau Econ. Geol. Report of Investigations 74:1–198.

Sams, R. H. 1982. Newly discovered dinosaur tracks, Comal County, Texas. South Texas Geol. Soc. Bull. 23:19–23.

Shuler, E. W. 1917. Dinosaur track mounted in the band stand at Glen Rose, Texas. Amer. Jour. Sci. 4:294–297.

1937. Dinosaur tracks at the Fourth Crossing of the Paluxy River near Glen Rose, Texas. Field and Laboratory 5 (2):33–36.

Shumard, B. F. 1860. Observations upon the Cretaceous strata of Texas. Trans. St. Louis Acad. Sci. 1:582–590.

Shumard, G. G. 1853. Remarks upon the general geology of the country passed over by exploring expedition to the sources of the Red River, under command of Capt. R. B. March, USA. In March, R. B. Exploration of the Red River of Louisiana, in the year 1852, U.S., 32nd Congress, 2nd Session, S. Ex. Doc. 54, 8:179–195.

Skinner, S. A., and Blome, C. 1975. Dinosaur track discovery at Comanche Peak Steam Electric Station, Somervell County, Texas. Southern Methodist Univ. Contrib. Anthropology 13:1–16.

Stovall, J. W., and Langston, W., Jr. 1950. Acrocanthosaurus atokaensis, a new genus and species of Lower Cretaceous Theropoda from Oklahoma. Amer. Midland Naturalist 43:696–728.

Stricklin, F. L., and Amsbury, D. L. 1974. Depositional environments on a low–relief carbonate shelf, Middle Glen Rose Limestone, central Texas. Geoscience and Man 8:53–66.

Stricklin, F. L., Smith, C. I., and Lozo, F. E., Jr. 1971. Stratigraphy of Lower Cretaceous Trinity deposits of central Texas. Univ. Texas Bur. Econ. Geol. Report of

Investigations 71:1–63.

Taff, J. A. 1892. Reports on the Cretaceous area north of the Colorado River. *Texas Geol. Surv. Third Annual Report* pp. 269–379.

Whitney, M. I. 1952. Some new pelecypoda from the Glen Rose Formation of Texas. *Jour. Paleont.* 26 (5):697–707.

Wilcox, R. E., and Rohr, D. M. 1986. The stratigraphy and depositional systems of the Bissett Formation. *Geol. Soc. Amer., Abst. Prog. 99th Ann. Meeting.*

Wilson, J. L. 1975. *Carbonate Facies in Geologic History.* (Springer–Verlag) 471 pp.

Wrather, W. E. 1922. Dinosaur tracks in Hamilton County, Texas. *Jour. Geol.* 3:354–360.

Ziegler, A. M., Scotese, C. R., and Barrett, A. F. 1983. Mesozoic and Cenozoic paleogeographic maps. In Broche, P., and Sunderman, J. (ed.). *Tidal Friction and the Earth's Rotation.* (New York: Springer) pp. 240–252.

16 The Sedimentology of the Purgatoire Tracksite Region, Morrison Formation of Southeastern Colorado

NANCY K. PRINCE AND
MARTIN G. LOCKLEY

Abstract

Localized bedding plane exposures of the Morrison Formation along the Purgatoire River in southeastern Colorado have been mapped to reveal over 1,300 dinosaur footprints representing at least 100 different trackways, some ranging up to 215 m in length. The tracks indicate a sauropod–theropod fauna.

Preliminary study of complete Morrison sections within a 40 km radius indicates that these trackways are in a package of cyclic lacustrine sediments near the top of the generally shallowing–upward lower Morrison sequence. The presence of fresh–water flora, vertebrate and invertebrate skeletal remains as well as the vertebrate traces support the lacustrine origin of the strata at the tracksite.

The Morrison Formation in southeast Colorado can be subdivided into 4 units. Unit A, calcareous shales and mudstones; Unit B, which contains the Purgatoire tracksite exposures; Unit C, sandy limestones interbedded with mudstones; and Unit D, mudstones interbedded with thin sandstones. Sedimentary facies of Unit B include gray fissile mudstones indicative of relatively quiescent shallow–water deposition; higher energy micritic shoreface limestones containing the footprints; and minor quartzose sandstones with crystal casts. Subaerial exposure features such as mud cracks and ripple marks and evidence of rooting and paleosol development are found throughout the sequence.

The dinosaur traces provide valuable data that combine with sedimentologic evidence to help develop a reconstruction of the paleoshoreline and a complete picture of the paleoenvironment of "Dinosaur Lake".

Introduction

Detailed study of the six meters of limestone and shale exposed at the Purgatoire River tracksite has resulted in an understanding of the Jurassic paleoenvironment at this locality (Lockley et al. 1986). Here on the Muddy Creek Monocline (Fig. 16.1), the section is partially covered by colluvium and exposures are mainly confined to the river bed. The purpose of this report is to include data from measured sections and oil wells on the Black Hills Monocline that help establish the regional stratigraphy and extend the depositional model southwestward beyond the tracksite (Fig. 16.2). We also present some revised interpretations of the ichnofauna to advance the inferences published in our preliminary report (Lockley et al. 1986).

Stratigraphy

The oldest rocks exposed in the area of the Black Hills Monocline are orange–red sandstones and siltstones below algal laminated dolomites interbedded with red sandstones and shales of probable Permo–Triassic age (Scott 1968). Deep maroon siltstones and sandstones with occasional thin limestones equivalent to the Triassic Dockum Group are followed by the salmon–colored massive crossbedded sandstones of the Exeter/Entrada Formation (Heaton 1939).

Discontinuous light–red thin crossbedded sandstone, siltstone and conglomerate above this surface contain possible Middle Jurassic fossils (Kauffman et al. 1985). Large root–like trace fossils and paleosols characterize this unit. Thin beds of clastic gypsum interbedded with brown and grayish green siltstone and mudstone, occasional thin sandstones and minor crystalline calcite appear to be a facies variation of the stratigraphically equivalent Bell Ranch Formation of east central New Mexico and "brown silt member" of the Morrison (Lucas et al. 1985) in northeast New Mexico and western Oklahoma.

A 10 to 30 cm thick band of blue chert is taken as the basal Morrison contact since it marks the end of silt– and sand–size quartz grain and gypsum deposition and beginning of clay and micritic limestone deposition in the Purgatoire Uplift.

The Morrison Formation in the study area can be informally subdivided, in ascending order, into Unit A,

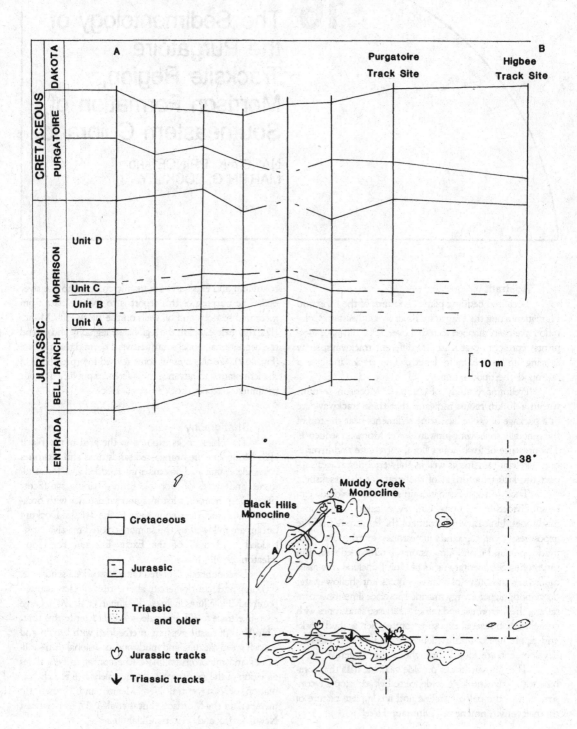

Figure 16.1. Correlation of Jurassic and Lower Cretaceous strata exposed along the Purgatoire River and Chaquaqau Creeks and encountered in nearby oil wells of southeastern Colorado. Inset (below) shows inliers of Jurassic and Triassic strata in Purgatoire and Cimarron River valleys, southeast Colorado, northeast New Mexico and northwest Oklahoma. (Map modified from King and Beikman 1974.)

Figure 16.2. Aerial view of a part of the Purgatoire River tracksite showing prominent trackways in Bed 2.

largely calcareous shales and mudstones; Unit B, interbedded claystones, mudstones and limestones; Unit C, sandy limestones and mudstones; and Unit D, mudstones, siltstones and sandstones. To the south along the Cimarron River in New Mexico and east around Two Buttes Reservoir in Baca County, Units A and B thin and become indistinguishable and the sandstones in Unit C become more abundant.

The massive sandstones of the Cretaceous Lytle Member of the Purgatoire Formation overlie the Morrison, followed by the marine shales of the Glencairn Member. The Dakota Sandstone forms the resistant cap which rims most of the canyons in the study area.

Sedimentology
Unit A

Unit A consists of calcareous, laminated- to thin-bedded, fissile, light gray to green clayshales, argillaceous micrites and mudstones (Fig. 16.3). Unfossiliferous clayshales predominate and are interbedded with clayshales containing ostracods, conchostrachans, rare *Chara* oogonia and fish debris; and occasional resistant claystone beds with thin (‡ 1 cm thick) lenses of rounded sand-size quartz and carbonate grains. The laminated sediments were deposited in a relatively quiet environment punctuated by infrequent pulses of higher energy. The high clay content creates a characteristic friable covered slope in outcrop and a distinctive highly radioactive gamma-ray trace on geophysical well logs (Fig. 16.4) which assist with placing the lower Morrison contact on a regional basis.

Unit B

Argillaceous and arenaceous micrite interbedded with clayshale in a cyclic pattern of quiet, below wave-base deposition followed by shoreline facies and exposure prior to the beginning of the next cycle are typical of Unit B (Fig. 16.5).

The clayshales at the tracksite contain tiny lymaneid and valvatid gastropods, *Chara* oogonia, ostracods and conchostrachans (Lockley et al. 1986, Fig. 5). They contain varying amounts of quartz silt, are structureless and appear to have been deposited below wave base in a shallow lake.

At the tracksite the vertebrate tracks are at the top of the shoreline facies beds. Footprint beds 1 and 4 are heavily trampled (see Lockley and Conrad this volume),

Figure 16.3. Typical Unit A sequence of laminated mudstones, clayshales and limestones as exposed at Goat Ranch, T 30 S, R 56 E, section 4, two meters above the Morrison/Bell Ranch equivalent contact.

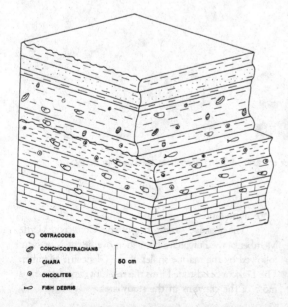

OSTRACODES
CONCHCOSTRACHANS
CHARA
ONCOLITES
FISH DEBRIS

50 cm

argillaceous micrites with largely untransported biota including unionid clams, *Equisitum* stem impressions and partially disarticulated sauropod bone. Beds 2 and 3 are less heavily trampled and contain imbricate lithoclasts at the base with minor quartz grains, ooids and transported biota including chara, ostracods, fish debris and transported unionid shells in a sparry micritic matrix, and fine upward to micrites with ripple marks and mudcracks, representing higher energy shoreline conditions indicative of flood/storm events (Lockley 1987). Thin rippled sandstones and algal mats are other, less abundant shoreline facies.

Horizontal root mats in the clay drapes above the micrites indicate a high water table while vertical, spar-filled tubes within the micrites show the water table was lower at times. The beginnings of paleosol development during dry spells are suggested by calcareous and siliceous nodules interbedded with orange- and yellow-stained siltstone. Dessication features such as crystal casts of unknown salts are common as are small mudcracks. Large, meter-scale polygons (Fig. 16.6) are further evidence that at times the climate was very dry.

The cyclic pattern of deposition, similar sedimentary structures and biota, and relationship to overlying units observed in both the Black Hills and Muddy Creek Monocline exposures suggest that the track horizons are in Unit B. Several of the micritic limestones in the Black Hills are highly disturbed, probably "dinoturbated". No dinosaur tracks have been discovered however, because there are no bedding plane exposures of these limestones.

There is some evidence that the paleoenvironment in the Muddy Creek Monocline and Black Hills area was slightly different. Abundant fish and mollusc remains, present at the tracksite in the former area, are very sparse, or missing, from Unit B in the Black Hills area. However, other biota, such as ostracods and charophytes, are present in both areas. Abundant subaerial exposure surfaces which dominate the measured section between the two areas suggest two separate water bodies with different chemistries and biota during this time.

Unit C

The maroon and green mudstones and tan-weathering sandy limestones of Unit C interfinger with the micritic limestones of Unit B below. Crystalline calcite forms the matrix for the abundant fine-grained angular quartz. The limestones weather into rounded "shepherds bread" boulders. Sedimentary structures include penecontemporaneous soft-sediment deformation — possibly dinoturbation, shallow crossbedding and burrows, though some beds lack bedding entirely due to bioturbation and diagenesis.

Dinosaur bone fragments are common in thin conglomerates near the base of this unit, which is possibly correlative to Stovall's bone quarries in northwestern Oklahoma (West 1978).

The intimate association of the sandy limestones and adjacent floodplain deposits suggest nearshore deposition, such as on a bench or bar as the lake basin filled and base level dropped.

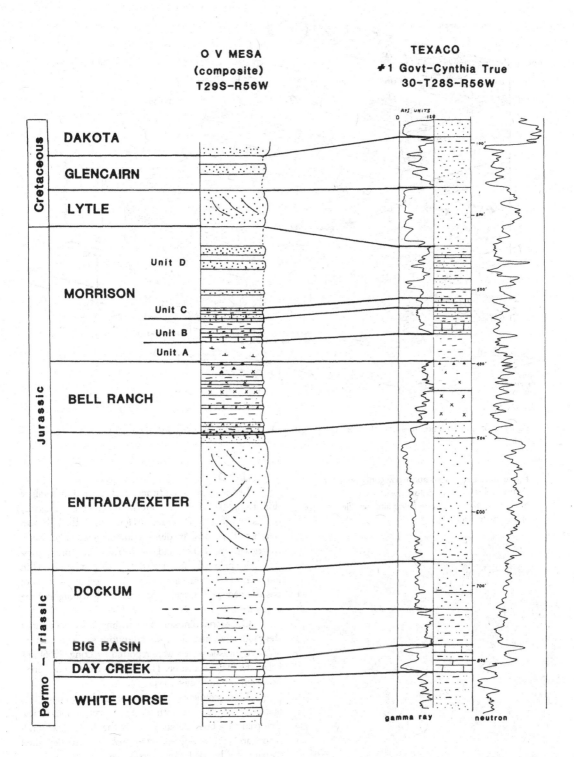

Figure 16.4. Comparison of subsurface oil well data with measured outcrop. The subsurface lithologies are based on well cuttings and comparision with log characteristics.

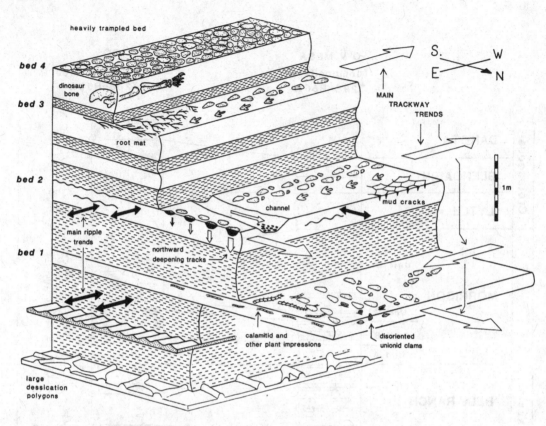

Figure 16.5. Unit B limestones and shales at the Purgatoire River tracksite showing features in all four track–bearing beds.

Figure 16.6. Large dessication polygon features in positive relief in strata below the track–bearing horizons at the Purgatoire River site (see Fig. 16.5 for approximate scale).

Unit D

Red and green mudstones and covered intervals of drab to lightly colored mudstones make up the majority of this unit (Fig. 16.7). Three- to four–meter thick, brown weathered, porous, medium–grained crossbedded sandstones with lag of chert and clay pebbles often form a resistant ledge below the massive white Lytle sandstones. Thin limestone beds and calcareous and siliceous nodules of possible pedogenic origin appear randomly throughout the sequence.

Occasional dinosaur bone fragments weather out of the mudstones and thin, discontinuous sandstone lenses and late Jurassic pollen were reported by Long (1966) just below the Lytle sandstone. The brightly colored mudstones, however, are unfossiliferous.

These sediments are interpreted as fluvial flood plain and overbank deposits, with some soil formation and a few thin palustrine beds. Because some of the thick, upper sandstones are quite widespread, they probably were deposited as part of a braided stream sequence.

The Dinosaur Tracks

A summary and preliminary synthesis of the dinosaurian ichnofauna at the Purgatoire tracksite has been presented elsewhere (Lockley et al. 1986). There it was shown that the assemblage consisted of the trackways of about 100 animals, of which 40 were quadrupeds (sauropods) and 60 represented tridactyl bipeds. Details of footprint and step dimensions were also presented together with a summary and interpretation of trackway orientations and sauropod track depth data.

In that report we differentiated tridactyl tracks on the basis of the presence or absence of claw impressions in order to infer theropod or ornithopod affinities respectively. A subsequent reexamination of the tracks and the published data (Lockley et al. 1986 Table 1) suggests that other criteria, including stride length and footprint length/width ratio are important in making the theropod-ornithopod distinction. The results of this reappraisal suggest that the majority of the tridactyl tracks may represent theropods, *not* ornithopods, as initially inferred by early workers and ourselves. When taking preservational problems into account the lack of claw impressions is not a decisive factor that outweighs the evidence of foot shape and stride length.

Support for the interpretation of a theropod-sauropod dominated fauna is suggested by a recent synthesis of Late Mesozoic (Cretaceous) ichnofaunal data (Lockley and Conrad this volume) which indicates a repeat pattern of theropod-sauropod associations in low latitude paleoenvironments characterized by seasonal climatic regimes. A global overview of dinosaurian ichnofaunas, and dinosaur assemblages in general, also indicates that ornithopod-dominated assemblages did not become established until the Cretaceous, whereas theropod-sauropod faunas were well-established in the Jurassic.

Distribution of Jurassic Dinosaur Tracksites in Southeastern Colorado and Surrounding Areas

In addition to the Purgatoire River tracksite described in this report, dinosaur tracks and trackways have been studied in other Jurassic localities in the region. Tridactyl trackways near Kenton, Oklahoma (Lockley 1986 p. 38) occur in a thin sandstone of the "brown silt member" (possible Bell Ranch equivalent). Tridactyl (theropod) trackways near Higbee, Colorado (Lockley 1986, Conrad et al. 1987) are found in a clay-rich sandstone which is probably at the transition of Unit B to Unit C. A heavily trampled micritic horizon has also been observed in a canyon along the southeastern rim of Mesa de Maya near the Colorado/New Mexico state line. A sauropod tracksite in the Morrison Formation of the Cimarron Valley is also known (Conrad et al. 1987, Langston 1974, Schoff and Stovall 1943).

Probable archosaurian tracks of non-dinosaurian affinity have been reported from regions bordering the study area (Fig. 16.8). These have been briefly reviewed by Conrad et al. (1987) who concluded that the so-called pterodactyl tracks from the Cimarron Valley (West 1978) are

Figure 16.8. Crocodilian tracks from the Morrison Formation and elsewhere. **A,** from Cimarron Valley after West (1978) and Conrad et al. (1987). **B,** *Pteraichnus saltwashensis* (Stokes, 1957). **C,** modern track after Padian and Olsen (1984). **D,** track from Alcova, Wyoming after Logue (1977). **E,** *Purbekopus pentadactylus* (Delair, 1963); see text for details.

Figure 16.7. Usually largely covered, Unit D is shown in the composite section which include data from O V Mesa, T 29 S, R 56 E, section 22, and Colbert Canyon, T 34 S, R 54 E, section 35.

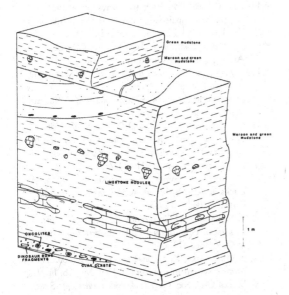

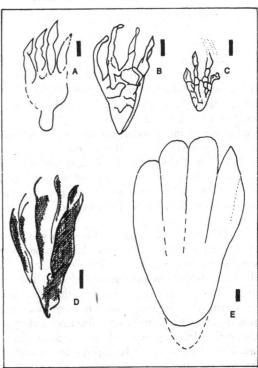

probably crocodilian (cf. Padian and Olsen 1984). The same interpretation may hold for a tetradactyl track reported by Hatcher (1903) from the Morrison Formation near Canon City, Colorado. In addition to these two occurrences we know of only two other such tracks from the Morrison Formation. These are the original account of *Pteraichnus saltwashensis* (Stokes 1957) from Utah and a report of the same ichnospecies from Alcova, Wyoming (Logue 1977).

When excluding the Hatcher specimen, for which insufficient information exists, the three other Morrison specimens are remarkably similar in both size and morphology. It is also interesting to note that they represent small animals (crocodiles). According to Dodson et al. (1980), aquatic vertebrates from the Morrison are characteristically small, indicating environments in which water was in short supply. Thus the footprint evidence is consistent with the skeletal and sedimentologic record.

Although we have not undertaken a systematic global search for comparable footprints, we note that *Purbeckopus pentadactylus* (Delair 1963) resembles *Pteraichnus saltwashensis* and originates from the Purbeck beds which are traditionally considered coeval with the Morrison. In our opinion the main difference between the ichnospecies is size, with the Purbeck fauna being larger. The common Morrison crocodile, *Goniopholis*, is also known from the Purbeck, which has yielded a more diverse crocodilian fauna than the Morrison. We consider that *P. pentadactylus* is a misnomer since Delair himself (1963, p. 92) diagnosed the inferred digit I as "diminutive" and illustrated a type specimen which could easily be interpreted as tetradactyl. Delair acknowledged the possibility that *Purbeckopus* might be attributable to a crocodile and he is probably right.

Conclusions

Unit A of the Morrison Formation was deposited in a tectonically stable lake basin following Middle Jurassic clastic evaporitic deposition. Unit B, a cyclic unit with trackbearing beds represents quiet water deposits interbedded with micrites showing high energy and exposure features, deposited in a seasonal semi-arid to arid climate (cf. Dodson et al. 1980). Complete filling of the basin and initiation of flood plain deposits represented by Units C and D and the Lytle Formation occurred prior to transgression of the Glencairn Sea from the north during the Cretaceous.

The dinosaur tracks are found in at least four layers which form part of a cyclic, climatically controlled sequence which accumulated during the middle portion (Unit B) of Morrison deposition in southeastern Colorado. There is no conclusive evidence of tracks at any other horizon in the formation in this region. Therefore we conclude that there is a correlation between track frequency and cyclic lake deposits of the type found in Unit B which are characterized by shallow-water lake deposits and abundant shoreface sediments.

Acknowledgments

This research has been supported in part by NSF Grant EAR-8618206.

References

Baldwin, B., and Meuhlberger, W. R. 1959. Geologic Studies of Union County, New Mexico. *New Mexico Bur. Mines Bull.* 63, 171 pp.

Conrad, K., Lockley, M. G., and Prince, N. K. 1987. *Triassic and Jurassic Micrite-Dominated Trace Fossil Assemblages of the Cimarron Valley Region: Implications for paleoecology and biostratigraphy in New Mexico*. New Mexico Geol. Soc. Guidebook 38th Field Conf. Northeastern New Mexico.

Delair, J. B. 1963. Notes on Purbeck fossil footprints, with descriptions of two hitherto unknown forms from Dorset. *Proc. Dorset Nat. Hist. Archaeological Soc.* 84:92-100.

Dodson, P., Behrensmeyer, A. K., Bakker, R. T., and McIntosh, J. S. 1980. Taphonomy and paleoecology of the dinosaur beds of the Jurassic Morrison Formation. *Paleobiology* 6 (2):208-232.

Hatcher, J. B. 1903. Osteology of *Haplocanthosaurus* with description of a new species and remarks on the probable habits of sauropods and the age and origin of the *Atlantosaurus* beds. *Carnegie Museum* Memoir 2, pp. 1-72.

Heaton, R. L. 1939. Contribution to Jurassic stratigraphy of Rocky Mountain region. *Amer. Assoc. Petrol. Geol. Bull.* 23 (8):1153-1177.

Kauffman, E. G. 1985. Geological and paleontological site analysis of the Pinon Canyon Maneuver site, Las Animas County, Colorado, a report submitted to the United States Army under .contract DAKF06-85-C-0168, 132 pp.

King, P. B., and Beikman, H. M. 1974. Geologic Map of the United States. *U. S. Geol. Surv.*

Langston, W. Jr. 1974. Nonmammalian Comanchean tetrapods. *Geoscience and Man* 8:77-102.

Lockley, M. G. 1987. Dinosaur trackways and their importance in paleoenvironmental reconstruction. In Czerkas, S., and Olsen, E. C. (eds.). *Dinosaurs Past and Present Symposium*. (Los Angeles, CA: Los Angeles County Museum) pp. 81-95.

Lockley, M. G. 1986. *A Guide to Dinosaur Tracksites of the Colorado Plateau and American Southwest*. University of Colorado at Denver Geology Department Magazine Special Issue No. 1. 56 pp.

Lockley, M. G., Houck, K. J., and Prince, N. K. 1986. North America's largest dinosaur trackway site: Implications for Morrison Formation paleoecology. *Geol. Soc. Amer. Bull.* 97 (10):1163-1176.

Logue, T. J. 1977. Preliminary investigation of pterodactyl tracks at Alcova, Wyoming. *Wyoming Geol. Assoc. Earth Sci. Bull.* 10:29-30.

Long, C. S. 1966. Basal Cretaceous strata, southeastern Colorado. Unpublished Ph.D. dissertation, University of Colorado. 479 pp.

Lucas, S. G., Kietzke, K. K., and Hunt, A. P. 1985. The Jurassic system in east-central New Mexico. *In New Mexico Geol. Soc. Guidebook 36th Field Conf.* Santa

Rosa. pp. 213–242.

Padian, K., and Olsen, P. E. 1984. The fossil trackway *Pteraichnus*: not pterosaurian, but crocodilian. *Jour. Paleont.* 58:178–184.

Schoff, S. L., and Stovall, J. W. 1943. Geology and ground-water resources of Cimarron County, Oklahoma. *Oklahoma Geol. Surv. Bull.* 64, 317 pp., 217 figs., 23 pls.

Scott, G. R. 1968. Geologic and structure contour map of the La Junta Quadrangle, Colorado, and Kansas. *U.S. Geol. Surv. Misc. Invest.* I-560.

Stokes, W. L. 1957. Pterodactyl tracks from the Morrison Formation. *Jour. Paleont.* 31:952–954.

Stovall, J. W. 1938. The Morrison Formation of Oklahoma and its dinosaurs. *Jour. Geol.* 46:583–600.

West, E. S. 1978. Biostratigraphy and paleoecology of the lower Morrison Formation of Cimarron County, Oklahoma. Unpublished Ph.D. dissertation. 61 pp.

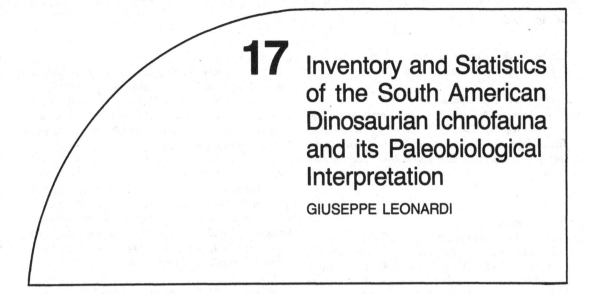

17 Inventory and Statistics of the South American Dinosaurian Ichnofauna and its Paleobiological Interpretation

GIUSEPPE LEONARDI

Abstract

Data concerning dinosaur tracks from 38 tracksites of South America are summarized and studied statistically. The distribution of dinosaurs by period, region, latitude, climate, environments, behavior and direction is examined. A number of conclusions and working hypotheses are presented. The following are worthy of special mention: (i) dinosaurs were rare in the Triassic in South America and were mainly bipedal; (ii) thereafter, South America possessed a dinosaurian fauna that was abundant and varied. (iii) Dinosaurs increased in individual and species number in the Jurassic and even more so in the Cretaceous. (iv) Theropod tracks were consistently the most abundant. (v) The ratio Sauropoda/Ornithopoda = 1/2 is probably due to a different ethology and ecology. (vi) Coelurosaurs decreased proportionately from the Triassic to the Cretaceous whereas carnosaurs increased. The former were better adapted to arid and temperate regions than were the other dinosaur groups. (vii) Ornithopods were more abundant in the south than in the north. Though the great majority of South American dinosaurs were not gregarious, some interesting cases of gregarious behavior are documented. The different groups of dinosaurs preferred different environments. The dinosaurs were strongly influenced in their movements by local or regional landscape, in particular by bodies of water.

Introduction

This work summarizes, in the form of a preliminary inventory, the data concerning dinosaur trackways from 38 known tracksites in South America (Fig. 17.1) and assembles this information to improve knowledge of dinosaur associations and their distribution in time, space and environment.

The material studied furnishes a record of 615 dinosaurs. The author has personally examined, on site or at the repositories where specimens are lodged, 594 individual trackways or footprints (97% of the total). The classification herein presented does not always correspond with that given by their finders or by the describers of the material. This is just a partial and preliminary synthesis from an annotated Atlas on more than 100 local tetrapod ichnofaunas from South America which will be published in the near future (Leonardi ms).

The twelve years of field, laboratory and library research which led to the collection of these data and its interpretation were financed and encouraged by the Brazilian Conselho Nacional de Desenvolvimento Científico e Tecnológico (CNPq).

Abbreviations Used. AM = American Museum, New York, USA; DNPM–RJ = Secção de Paleont., Departamento Nacional de Produção Mineral, Rio de Janeiro, Brazil; FL = Fundación Lillo, Tucumán, Argentina; MACN = Museo Argentino de Ciencias Naturales, Buenos Aires, Argentina; MCC = Museu Câmara Cascudo, UFRN, Natal–RN, Brazil; MN–RJ = Museu Nacional, Rio de Janeiro, Brazil; MIJ = Museo de Ingeniero Jacobacci, Río Negro, Argentina; MLP = Museo de La Plata, La Plata, Argentina; UFRGS = Dept. Paleont. and Stratigr. of Universidade Federal do Rio Grande do Sul, Porto Alegre–RS, Brazil; UNESP–RC = Dept. of Geol. of UNESP, Rio Claro–SP, Brazil; UNS = Dept. Geol. de Universidad Nacional de Salta, Salta, Argentina.

Inventory

Numbers refer to map localities as shown in Figure 17.1.

Triassic

Local Fauna of Quebrada de Ischichuca (1)

Argentina, La Rioja Province, Quebrada de Ischichuca, 36 km SSW from Villa Unión. Sedimentary basin of Ischigualasto–Villa Unión. Ischichuca Formation, Agua de la Peña Group, Middle Triassic. A single small footprint, with narrow digits; possibly a coelurosaur, or perhaps a pre–dinosaurian thecodont, for instance *Lagosuchus* or

Lagerpeton whose bones are found at the same level (J. F. Bonaparte pers. comm. 1981).

Local Fauna of Bajo Caracoles (2)

Argentina, NW section of Santa Cruz province, Bajo Caracoles region. "Complejo Porfirico" *auctorum*, Upper Triassic. Footprint of a coelurosaur: the exact source of the specimen is unknown. Repository: MLP (Casamiquela 1964).

Jurassic

Local Fauna of Araraquara (3)

Brazil, state of São Paulo, Araraquara, Ouro District, from the quarries São Bento, Cerrito Velho, Cerrito Novo and others and from sidewalks in the city of Araraquara and in other towns of the state of São Paulo. Aeolian sandstones of the Botucatu Formation (*sensu stricto*), São Bento Group. Age believed as Lower to Middle Jurassic by this author, based on study of the ichnological material and comparison with material from Argentina. About 8 or 9 forms of dinosaur trackways: coelurosaurs, small carnosaurs and (especially the larger forms) ornithopods; all biped and tridactyl. The known dinosaurian trackways represent about 40 theropod individuals, most of them coelurosaurs, and around 12 ornithopods. These tracks are associated with those of mammals (predominant) and Tritylodontoidea. Repositories: DNPM–RJ; MN–RJ; UNESP–RC (Leonardi 1980a, Leonardi and Godoy 1980, Leonardi and Sarjeant 1986, Leonardi unpublished data).

Local Fauna of Jurucê (Brodósqui) (4)

Brazil, state of São Paulo, Brodósqui, road from Brodósqui to the village of Jurucê. Visconde quarry. Aeolian sandstone as above. The footprints were found not in the quarry, which is closed and filled in, but in the sidewalks of Franca (São Paulo). Formation and group as above. Two bipedal trackways probably attributable to coelurosaurs. Association with Tritylodontoidea tracks and mammals, as above. "In situ" in the sidewalks of Franca – SP, Brazil (Leonardi and Godoy 1980).

Local Fauna of Rifaina (5)

Brazil, state of São Paulo, Rifaina, "Chave do Calixto" quarry. Aeolian sandstone. Formation and group as above. The footprints were found in the sidewalks of the city of Rifaina. Two trackways attributed to coelurosaurs. Association as above. "In situ" in the sidewalks of Rifaina – SP, Brazil (Leonardi and Godoy 1980).

Local Fauna of Botucatu Range (6)

Brazil, state of São Paulo, highway SP-300, on the Serra de Botucatu slope, 11 km E of Botucatu, 793 m above sea level. Sandstone, formation and group as above. Two footprints of two bipedal dinosaurs, tridactyl; probably Ornithopoda. Repository: DNPM–RJ (Leonardi and Godoy 1980).

Local Fauna of Tramandaí (7)

Brazil, state of Rio Grande do Sul. Sidewalks of the town of Tramandaí. The stone pavement, according to local sources, came from the Santa Cruz do Sul (RS) area. Sandstone, formation and group as above. One uncertain footprint attributable, with some probability, to a theropod (Leonardi and Sarjeant 1986).

Local Fauna of Santa Cruz do Sul (8)

Brazil, Rio Grande do Sul state, Santa Cruz do Sul, quarry in the "Sete Curvas" area. The rocky paving–stone with the track, however, was found in a town sidewalk. Sandstone, formation and group as above. One biped trackway probably from a very small dinosaur. Repository: UFRGS (Leonardi and Sarjeant 1986).

Local Fauna of Estância Laguna Manantiales (9)

Argentina, Santa Cruz Province, southern Patagonia. Estancia Laguna Manantiales. Tuffaceous sandstone, from the middle section of the "Complejo Porfirico". Jurassic; Callovian to Oxfordian, most probably Low to Middle Callovian. Arid environment. Several trackways were found (52 dinosaurian trackways known):

a) *Wildeichnus navesi* Casamiquela 1964. Tridactyl footprints with thin toes, bipedal and narrow trackway; attributed to a small coelurosaur. 23 specimens known.

b) *Sarmientichnus scagliai* Casamiquela 1964. Narrow footprints of a bipedal animal, attributed to a coelurosaur that was highly specialized as a desert runner. 10 individuals known.

c) *Delatorrichnus goyenechei* Casamiquela 1964. Quadrupedal trackway, with small forefeet and larger hindfeet, all tridactyl. Trackway attributed to a theropod; Casamiquela considered this to be coelurosaurian. 18 specimens known.

d) Another unnamed form, attributed to the Coelurosauria by Casamiquela. One specimen.

The four dinosaur forms are accompanied by a preponderance of mammal trackways (*Ameghinichnus patagonicus* Casamiquela 1964) and tracks of Coleoptera. The environment, as mentioned, was arid, but the footprints, in many cases, were made when the sand (of pyroclastic origin) was wet, probably in transitory river beds. Repositories: MLP, MACN, FL, MIJ (Casamiquela 1964, Leonardi unpublished data).

Local Fauna of Chicarilla (10)

Chile, Tarapacá, Chacarilla, Quebrada de Chacarilla. Sandstone of the Chacarilla Formation, Upper Jurassic, probably Oxfordian. At least 5 different outcrops in the area yield trackways. Indeed, there are hundreds of trackways and single footprints, still to be studied. Among them have been registered (Dingman and Galli 1965, Casamiquela and Fasola 1968) at least four carnosaurs and two sauropods; perhaps one ornithopod; and one trackway, probably quadruped and tridactyl, attributed to a stegosaur.

Marine platform environment. In situ.

Cretaceous

11 – Local Fauna of São Domingos

Brazil, state of Goiás, Itaguatins, São Domingos District. Coarse sandstone: Corda Formation, Lower Cretaceous; pre Aptian; *non* Jurassic. Fluvial environment. Seven sauropod trackways, of which at least five are sub-parallel, all of them (except one) large. These were previously classified as trackways of large ornithopods, owing to their bipedal aspect (by total primary overlap) and the poor quality of the material (Leonardi 1980b). In situ.

Local Fauna of Las Quijadas (12)

Argentina, San Luis Province, Sierra de Las Quijadas, 90 to 100 km NW of the town of San Luis. Fine reddish sandstone, Neocomian (Lower Cretaceous). The material was originally considered as Middle or Upper Triassic (Lull 1942). Two individual footprints of *Anchisauripus australis* Lull 1942, attributed to Coelurosauria and associated with a crocodilian footprint (*Batrachopus argentina* Lull 1942). Repository: AM.

Local Fauna of Cianorte (13)

Brazil, state of Paraná, Cianorte, headwaters of Catingueiro Creek. Deep pink medium-grained sandstone, Caiuá Formation, Bauru Group, Lower Cretaceous. Probably of arid environment, with dunes. Two poor quality trackways of bipedal dinosaurs, probably theropods (perhaps one carnosaur and one coelurosaur). Repository: ?carnosaurian footprints at DNPM-RJ; coelurosaurian trackway in situ (Leonardi 1977).

Local Fauna of Indianópolis (14)

Brazil, state of Paraná, Indianópolis, Dois Coqueiros. Sandstone, formation, group, age and environment as above. One poor quality footprint, probably attributable to a coelurosaur (Leonardi 1977). Not collected.

Local Fauna of Baños del Flaco (15)

Chile, Colchagua, Baños del Flaco. Calcareous sandstone of the Baños del Flaco Formation, "Formaciones Basales". Lowest Cretaceous (Berriasian). Coastal swamp environment. Seven trackways attributed to ornithopods (Iguanodontidae) of two different forms: 3 individuals of *Iguanodonichnus frenkii* Casamiquela 1968 and 4 individuals of *Camptosaurichnus fasolae* Casamiquela 1968. In situ (Casamiquela and Fasola 1968).

Local Faunas of the Rio do Peixe Basin (16–31)

Brazil, Paraíba State, Sousa and Antenor Navarro Municipalities. Rio do Peixe Group, Lower Cretaceous, pre Aptian. From bottom to top:

a) Antenor Navarro Formation. Coarse sandstone, yellowish or gray colored. Fluvial environment. Three localities (Serrote do Lotreiro and Serrote do Pimenta, in Sousa Municipality; Aroeira, in Antenor Navarro Municipality) with 48 trackways of medium to large-sized carnosaur; three ornithopod trackways (one *Iguanodon* sp.; one probable bipedal Iguanodontid; one trackway attributable to a quadrupedal Iguanodontid, perhaps *Ouranosaurus* Taquet 1976, classified as *Caririchnium magnificum* Leonardi 1984; one uncertain trackway, probably of an undetermined quadrupedal ornithischian; six sauropod trackways in herd; and short trackways or single bipedal footprints of large dimension of at least 8 individual dinosaurs, unclassifiable.

b) Sousa Formation. Eleven localities (Passagem das Pedras, Piau, Matadouro, Piedade, Pedregulho, in the Municipality of Sousa; Engenho Novo, Juazeirinho, Cabra Assada, Zoador, Barreira do Domício, Poço da Volta in the Municipality of Antenor Navarro) with trackways or footprints of 22 Theropoda individuals; 24 coelurosaurs; 182 carnosaurs; 21 ornithopods, bipedal or semi-bipedal; 9 individuals probably attributable to the Sauropoda and 1 quadrupedal ornithischian trackway, unclassified. From the Ornithopoda the following taxa were established: *Sousaichnium pricei* Leonardi 1979; *Moraesichnium barberenae* Leonardi 1979; *Staurichnium diogenis* Leonardi 1979. Some theropod footprints were attributed to *Eubrontes* cf. *E. platypus* Lull 1904, and to the ichnogenus *Grallator* E. Hitchcock 1858 *vel Eubrontes* E. Hitchcock 1845. Lacustrine environment, semi-arid, flood plain.

c) Piranhas Formation. Fluvial environment. Coarse sandstone, white or pink colored. Six carnosaur trackways or single footprints and the same number of ornithopods.

Repositories: generally in situ; some pieces at DNPM-RJ and MCC etc. (Huene 1931; Leonardi 1979a,b, 1984; Godoy and Leonardi 1985; Leonardi and Sarjeant 1986; Moraes 1924; Leonardi unpublished data).

Local Fauna of São Romão (32)

Brazil, state of Ceará, Orós, district of São Romão; Lima Campos basin. Yellow, medium-grained sandstone; Antenor Navarro Formation, Rio do Peixe Group, Lower Cretaceous (pre Aptian). Fluvial environment. Trackways and single footprints, on a single rocky pavement, of at least 14 individual dinosaurs, among them seven carnosaurs; one ornithopod; three small bipedal dinosaurs (perhaps ornithopods); and three bipedal trackways and some single footprints which are not classifiable. In situ. (Leonardi and Muniz in press).

Local Fauna of Cabeça de Negro (33)

Brazil, state of Ceará, Orós, Palestina Basin (called also Malhada Vermelha Basin), Cabeça de Negro District. Fine-grained purple sandstone; Sousa Formation, Rio do Peixe Group. Lower Cretaceous. One single carnosaur footprint. Environment: transitional between fluvial to lacustrine. In situ (Leonardi and Muniz in press).

Local Fauna of Plottier (34)

Argentina, Neuquén Province. Red sandstone, Upper Cretaceous (Upper Senonian) redbed environment. One right footprint attributed to a coelurosaur. Repository: MLP (Huene 1931).

Local Fauna of El Chocón (35)

Argentina, Neuquén Province, Nueva Península, El Chocón Reservoir. Variegated sandstones belonging to the Huicol Member, Rio Limay Formation, Neuquén Group. Upper Cretaceous (post–Albian and probably pre–Maastrichtian). Semi–arid continental environment. Several trackways, among them one very large single footprint of the *Grallator* type, commonly attributed to the Coelurosauria; one trackway of large dimensions, with at least four footprints, probably attributable to the Carnosauria (Abelisauridae Bonaparte & Novas 1985); one set of at least two large footprints, attributed to the Ornithopoda (Hadrosauridae). Repository: in situ. Casts kept at MLP (Gasparini and Leonardi unpublished data).

Local Fauna of Quebrada de la Escalera (36)

Argentina, Salta Province, San Carlos Department, Quebrada de la Escalera, west tributary of the Río Tonco. Calcarenite of the Yacoraite Formation, Salta Group, Upper Cretaceous (Maastrichtian). Marine platform environment. Seven trackways (at least) classified as *Hadrosaurichnus australis* Alonso 1980. Although these were classified as Hadrosauridae by the mentioned author, an examination of the photographs leads us to the assumption that at least some were theropods (Carnosauria). For statistical purposes they were considered with 0.5 points in both systematic positions. Repository: in situ; casts at UNS (Alonso 1980).

Local Fauna of Toro–Toro (37)

Bolivia, Potosí Department, Charcas Province, near the village of Toro–Toro. Yellow to red colored sandstones of the El Molino Formation, Pilcomayo Group (ex Puca Group). Upper Cretaceous (Maastrichtian). Marine platform environment.

a) Approximately 60 parallel carnosaur trackways, medium to large sized, associated with two large trackways of quadrupedal ornithischians (perhaps Ceratopsians and certainly not sauropods) classified as *Ligabueichnium bolivianum* Leonardi 1984.

b) At least 32 carnosaurian trackways associated with, and parallel to, eight sauropod trackways (six adults and two young).

c) Several other carnosaur trackways.

In situ. (Branisa 1968, Leonardi 1984).

Local Fauna of Parotani (38)

Bolivia, Department of Cochabamba, Quillacollo Province, near Parotani. White and yellow colored sandstone of the Santa Lucia Formation, Potosí Group, Upper Cretaceous (Maastrichtian). Marine platform environment. Six trackways of bipedal dinosaurs, of poor quality, without morphological details, attributed either to carnosaurs or, less probably, to hadrosaurs. In situ. A recent landslip has almost completely destroyed the material (Leonardi 1981).

Statistical Analysis
General Consideration of the Data

The South American dinosaurian ichnofauna is interesting for its size: 615 is a substantial number of dinosaurs, even when compared to the total number of individual dinosaurs that are known by fossil bones from this continent.

However, the distribution of the material by periods is uneven: 2 possible dinosaurs in the Triassic (0.33%); 119 in the Jurassic (19%)* and 494 in the Cretaceous (80%) (see Fig. 17.2A). The distribution is also very irregular if considered by region and period. The Triassic, for example, is not represented in Brazil.

Overview (Fig. 17.2B, 17.2C)

The large number of theropods (Fig. 17.2B), and consequently of carnivores (82%) in relation to herbivores (16%), is a proportion which does not correspond to the osseous record from the continent. This disparity seems to contradict the biomass pyramid theory, which maintains that a large herbivorous mass is necessary to feed a small carnivorous mass. We must naturally take into account that carnivorous dinosaurs did not limit their diet to herbivorous dinosaurs. More than a few of them must have been piscivorous; others probably ate other reptiles besides dinosaurs and also mammals; still others must have eaten eggs; some, small sized and mainly from arid regions, must have been insectivorous. We should also remember that the problem is not simply one of numbers, but also of mass: herbivores frequently weighed many times more than the carnivores.

Nevertheless we should consider in particular the aspect of the different metabolism and activity levels. The herbivores, especially the quadrupeds, lived as browsers in restricted environments, while the carnivores must have accomplished wide–ranging and continuous hunting expeditions, leading to proportionally more footprints. It is also evident that the latter were excellent pedestrians and fast runners, consuming ample energy reserves. We can conclude also, from the ichnological data, that the majority of theropods were active hunters.

One is surprised next by the 1:2 ratio of sauropods to ornithopods. This certainly does not correspond to the ratio of skeleton findings on the continent, where Sauropoda are common and Ornithopoda rare. The phenomenon might be attributed, among other causes, to the greater activity of the Ornithopoda, which were better walkers and runners than is usually thought. The South American ornithopod trackways generally present a very high step angle and are very narrow, indicating that their legs were well beneath their bodies and that their walk was fast and efficient.

*In general, the percentages, in the text, have been rounded off to nearest unit figure.

In the whole sample, the quadrupedal ornithischians (0.8%) are poorly represented. This fact will be commented on later.

The frequency of the Carnosauria in the sample is 70% (Fig. 17.2C). It is of course possible that classification mistakes exist, but there is no doubt that large carnivores dominated.

Rio do Peixe Basin (Fig. 17.2D–17.2G)

One part of the material deserves particular comment from the outset, owing to its uncommon parameters. This is the great set of trackways, representing 330 dinosaurs (54% of the whole) originating from about sixteen localities in the Rio do Peixe Basin in the northeast of Brazil, systematically explored and studied by the writer during the last eleven years.

The size of this sample might easily compromise the results of the whole statistical study. However, we observe that there is not a significant difference between the Rio do Peixe Basin sample and the set of other ichnofaunas. Perhaps due to its age — Lower Cretaceous, intermediate between the Jurassic (19%) and the other Cretaceous faunas which belong in general to the Upper Cretaceous (27%) — the Rio do Peixe ichnofauna is well balanced.

The calculation of some interesting ratios (made using percentages) offers the following results:

Herbivorous/carnivorous: Rio do Peixe 1/6; other basins 1/4.

Quadrupeds/bipeds: Rio do Peixe 1/18; other basins 1/13.

Group by group, we note that the greatest difference is among the theropods, which are 5% more abundant in the Rio do Peixe and among which there is a notable preponderance of carnosaurs (70% of the Rio do Peixe theropods, compared to 53% in other basins). The coelurosaurs are evidently less abundant in the Rio do Peixe sample. The values related to the herbivores are surprisingly close:

Sauropoda: 5% in Rio do Peixe; 6% in the other basins.

Ornithopoda: 9% in Rio do Peixe; 12% in the other basins.

Quadrupedal Ornithischia: ‡ 1% in Rio do Peixe; 1% in the other basins.

Distribution by Period (Fig. 17.3)

Triassic (Fig. 17.3A,B)

Only the Theropoda (Coelurosauria) are represented. Although the sample is very small (two trackways), it corresponds well to reality. Among Triassic dinosaurian trackways, in the world as a whole, the great majority of footprints belong to the Theropoda, especially to the Coelurosauria.

Jurassic (Fig. 17.3C,D)

Theropoda maintains an absolute majority (84%), while the Sauropoda (2%), the Ornithopoda (13%) and one quadrupedal ornithischian, a stegosaur (1%), are minor components.

Cretaceous (Fig. 17.3E,F)

The frequency of Theropoda decreases slightly (81%). The Sauropoda increase to 6%; the Ornithopoda decrease to 10%; and the quadrupedal Ornithischia maintain low frequency (below 1%). Following the evolution of the several groups in time, we observe that the Theropoda, which represent the totality of dinosaurian tracks in the Triassic, start to decrease relatively and to increase absolutely. Thus in the Triassic we have, in percentages, 100% theropods; in the Jurassic, 15% herbivores compared to 84% carnivores (Theropoda); in the Cretaceous, 17% compared to 81%. In absolute numbers, however, we have in the Triassic two theropods; in the Jurassic, 100; in the Cretaceous, 399.

The much lower number of quadruped individuals (18%) compared to bipeds (82%) in the Jurassic, with 7% compared to 93% in the Cretaceous, does not seem to correspond to the reality of dinosaurian associations as evidenced by fossil bones, although there is a lack of statistics in this regard. If it is true that the quadrupedal Ornithischia were always rare in this continent, as appears also by bone records, it seems that, by the same records, the sauropods were always very numerous. The rarity of quadrupedal trackways in the Jurassic seems to indicate not so much a reduced number, but a low activity level for these heavy animals, which would probably remain in their feeding grounds for long periods of time.

The increase of the Sauropoda from 2% in the Jurassic (Fig. 17.3C) to 6% in the Cretaceous (Fig. 17.3E) is due to the fact that the Jurassic sample comes almost totally from arid environments. Since, during the Jurassic, much of the continent was covered by deserts and arid zones, it is probable that in general these large herbivorous dinosaurs lived only in isolated areas, with adequate vegetation.

Among the theropods (Fig. 17.3B,D,F), the Coelurosauria, which occupy all the small Triassic sample, decrease to slightly more than 1/3 of the population (37%) in the Jurassic — even with the abundance of arid zones during this period, where coelurosaurs were, apparently, in their optimal environment. The coelurosaurs decrease in abundance in the Cretaceous (only 7%), although their absolute numbers remain high (N = 37 in the Jurassic, N = 29 in the Cretaceous). We should note that, among the large sample of Theropoda tracks in the Jurassic, many may eventually prove to be Coelurosauria. In the Botucatu Formation, in particular, where most of the theropod footprints are small and of bad quality, it becomes difficult or perhaps impossible to distinguish between the two groups.

In this case, it does not seem that we can invoke a lower activity as explanation for the predominance of the carnosaurs over the coelurosaurs. The anatomy, as well as the trackway characteristics, of the latter leads us to think that they were active animals. So the relationship between the abundance of the trackways ought to provide a valid basis for determining the relative abundance of individuals of both infraorders. Similar frequencies for the ichnofaunas of Europe and North America lend support to this hypothesis.

The Carnosauria are not recognized in the Triassic, but account for 4% of the theropods in the Jurassic and increase to 87% in the Cretaceous when these medium to large sized theropods were dominant.

Distribution by Region, Latitudes and Climates (Fig. 17.4)

In the two graphs representing the southern region, including Argentina and Chile (Fig. 17.4C,D), and the two representing the northern portion of the continent, corresponding to Bolivia and Brazil (Fig. 17.4A,B), the numerical and percentage statistics are similar. The theropod percentage is almost equal: 79% south and 82% north. However, there is a large variation in the composition of the theropod samples. In the south the Coelurosauria are predominant (59% compared to 12% Carnosauria); in the north, the Carnosauria are overwhelmingly predominant (79% compared to 7%). The difference is due, in lesser part, to the absence of Triassic material in Brazil and Bolivia and, somewhat more perhaps, to the fact that the great majority of individuals in the northern region (89%) belong to the Cretaceous Period when the Carnosauria prevailed.

Another possible explanation is that the coelurosaurs were more adaptable to temperate climates while the carnosaurs, in general of larger size, tended to be more frequent in regions close to the equator. This argument might seem satisfactory if we consider that, among modern reptiles, larger individuals and overall abundance (in contrast to mammals, and indeed to many invertebrates) are concentrated in intertropical regions. These patterns, however, cannot be confirmed either from the distribution of theropod footprints in the boreal continents or from the general distribution of dinosaur skeletons. On the other hand, the Cretaceous climate was warmer than at present, with temperate regions extending to the poles and with a climatic zonation linked to latitude, much less marked than at present. Consequently, the concentration of the Carnosauria, as large theropods, in the intertropical regions cannot be satisfactorily explained in terms of a distribution related to latitudinal zonation.

The Ornithopoda are more widespread in the south (16%) than in the north (10%), which corresponds also to osseous findings. We do not have accurate numerical data on the number of individual sauropods represented by osseous remains in the two regions under consideration, but it does not seem that the pattern of osseous findings corresponds with the proportions registered by the footprints — that sauropods are more frequent in the north (6%) than in the south (2%). In fact, the ratios herbivores/carnivores are, as expected, reasonably close: approximately 1:4 in the south and 1:5 in the north.

Finally, among the quadrupedal Ornithischians the larger percentage in the south (1.2% with only one specimen) than in the north (0.8%, with four specimens) corresponds to the relative frequency of osseous material. Bones and footprints are exceptionally rare on this continent, where it seems these reptiles were never abundant. How-ever, the frequency of tracks made by quadrupedal Ornithischians is always rare, even on other continents where bones are relatively common (as, for instance, in North America).

A more obvious influence of the climate is on the number of morphological types. As in modern reptiles, the more meridional (more temperate climate) faunas are more monotonous. In the same arid environment, for instance, at the Estância Laguna Manantiales, we find only three forms (maybe four) — all Theropoda, and perhaps all Coelurosauria. In contrast, in the Botucatu Formation, not very far away in geological time and 26° (latitude) farther to the north, there are eight or nine forms, all bipeds but probably belonging to the Coelurosauria, the Carnosauria, and perhaps the Ornithopoda.

In general, the localities with the larger numbers of forms and diversified groups are concentrated in the northern (intertropical) region, starting approximately at 20°S. The localities which are richest and most diversified, with dozens of forms, are those of the Rio do Peixe Basin, near the equator (approximately 6°S latitude) today and not very far from this latitude even in the Lower Cretaceous.

Distribution by Environments Figs. 17.4E,F,G,H)

More surprising, and difficult to interpret, are the data when organized by environments. We should take into account that the data herein synthesized have been collected from several sources written at different times; consequently, they are not equally reliable or current. A similar synthesis for western North America is presented by Lockley and Conrad (1987 and this volume).

Arid–Desert Environment (Fig. 17.4E)

Limitations in our study regarding this environment are (i) the difficulty in classifying the Brazilian Jurassic material of the Botucatu Formation, and (ii) the degree of aridity of the environment. We should remember, however, the importance of the ichnological record, even with these heavy limitations, since osteological findings within a desert paleoclimate and in highlands are extremely rare. Only biped dinosaurian trackways occur in the arid environments. All the tracks observed in the Botucatu are of small dimensions, pointing to a dwarf fauna. All dinosaurs were very small at the Estancia Laguna Manatiales and almost all were small in the Botucatu Formation; even the rare "giants", judged as Ornithopoda, could reach only a length of 3 m (maximum) and a height of 1.5 m.

The frequency of Theropoda in the arid environment is high (87%) with a predominance of coelurosaurs (proved in some cases, probable in others). We must note also the absence of sauropods, which is surely connected with the aridity of the environment.

Semi–arid Alluvial Environment (Fig. 17.4G)

Among environments with semi–arid climates and large rivers during the rainy seasons, we note: (i) a relatively small number of theropods (67%), which is the minimum for all environments, regions and periods represented here; (ii) a large proportion of sauropods and ornithopods, which is indeed the highest among all environments considered; (iii) the large proportion of carnosaurs in relation to the coelurosaurs (98% of the Theropoda); and (iv) the extraordinary abundance of herbivores (30%).

As the localities with this environment are almost exclusively in the northern region, the characteristics of the fauna are in part influenced by this latitudinal factor.

Semi–arid Lacustrine Environment (Fig. 17.4F)

In the semi–arid climate in a region of flood plains, with large lakes and lagoons in the rainy season, theropod frequency is high (88%) and the abundance of herbivores is low.

It is clear that the sampling of this environment is strongly influenced by the enormous occurrence in the Rio do Peixe Basin (Sousa Formation). We should note, however, that the dominance of theropods over herbivores, already visible in the Rio do Peixe Basin (Fig. 17.2D,E), is even more pronounced in this environment.

Coastal Belt Environment (Fig. 17.4H)

This environment generally encompasses the intertidal belt and/or the marine shelf. In both cases we have coastal swamps and the influence of the sea. It is necessary to emphasize that: (i) the sample is surprisingly similar to the pattern of the whole (Fig. 17.2B,C); (ii) the quadruped Ornithischia are more numerous (almost three times greater) than the continental mean; (iii) the Sauropoda are more abundant than the continental mean (30% more) but not as much as we might expect since the majority of the sauropod trackways worldwide come from this environment; (iv) the Carnosauria occupy 100% of the theropod sample; (v) the quadrupeds are considerably more abundant than the continental mean (63% more); and (vi) the herbivores in general are also more abundant.

If the abundance of the footprints is related to activity, we should conclude that in the coastal environment the herbivores and, among them, mainly the quadrupeds, were more active than average and that the carnivores were less active than average.

Distribution by Behavior

Among 38 localities, 12 do not provide sufficient data to analyze behavior. Of the other 26 localities, 22 present isolated individual tracks, which denote individual and not gregarious behavior. In only four cases are there signs of social behavior and, in three of these, alongside the organized groups, there are other trackways not similarly organized.

As expected, since trackways should depend on local landscape for their directions, there are no predominant directions on a continental level.

The four sites that demonstrate "social" organization are:

(i) São Domingos (Itaguatins, Goiás, Brazil) — Seven sauropod trackways, of which the majority are subparallel, in a 50 m front (approximately 8 m per animal). Some of these, at least, must represent the passage of a herd.

(ii) Serrote do Pimenta (Sousa, Paraíba, Brazil) — Among many independent theropod and ornithopod trackways, there are six parallel trackways attributable to Sauropoda, which go forward in a front of about 20 m (an average of about 4 m per animal, considering that two trackways are superposed). Here we have probable gregarious behavior of a herbivore herd.

(iii) Piau (Sousa, Paraíba, Brazil) — The almost two hundred trackways from this site, belonging mainly to Theropoda, do not indicate social behavior. There is, however, a set of trackways of large dimensions (up to 1 m diameter) on layer 13/3, which may represent the passage of a herd of seven large sauropods. In this case, the herd would be advancing in a front of about 18 m (2.5 m per animal, which indicates that they could hardly be walking side by side).

(iv) Toro–Toro (Potosí, Bolivia) — Many sets of trackways at this locality demonstrate gregarious behavior: (a) In the main rocky pavement are two parallel trackways of large dimensions attributed dubiously to Ceratopsia. They leave the sea, where the argilliferous sand was more plastic, and proceed across the shore, where the sand was dry and hard, so that the trackways, from deep, turn to shallow and disappear. The midlines of the trackways are 5 m apart. (b) In the same rocky pavement are 60 parallel trackways of Carnosauria. It is questionable whether all trackways are contemporary and whether they represent the passage of a giant pack. It is more probable, owing to differences in depth and clarity of the tracks, that various packs passed at different times (perhaps also on different layers, in which case some of the trackways are underprints). The 60 trackways occupy an 85 m front: 1.42 m per animal. Although the trackways are not superimposed, it seems improbable that the Theropoda were walking all in the same front. (c) In another rocky pavement, about 300 m downstream, are eight parallel trackways of large sauropods (six adults, two juveniles) which advanced along the shore line in a front of about 200 m, an average of 25 m per animal; however, both juveniles moved together, next to an adult. They were followed, undoubtedly shortly afterwards, by a pack of at least 32 medium–sized carnosaurs, which sometimes stepped into the sauropod footprints and were moving in the same direction and on the same front (an average of 6.25 m per animal).

Thus, the cases of gregariousness, with herds and packs, are relatively rare: 15% of the localities present cases of social behavior and only 20% of the trackways are in organized groups, if the 60 Theropoda trackways of Toro–Toro are interpreted in this way.

The most gregarious are the Sauropoda (28 among 32 sauropodian individuals: 87%). Next come the quadruped Ornithischia (2 out of 5: 40%) and the Theropoda (92 out of 501: 13% or 32 out of 501: 6%, depending on how we consider the Toro–Toro site). Last come the Ornithopoda, of which no gregarious trackways have been found on the continent (0 out of 64: 0%). The Ceratopsia would have a percentage of 100% if the 2 Toro–Toro trackways are attributed to them, but the sample is very small and localized.

Another aspect of behavior that should be considered is the direction of the tracks. Of the 38 localities, we do not have sufficient information from about 13, as they involve only one or two individuals. Data were not collected from two other localities while from 15 localities abundant data has been collected but has still not been studied in detail. We can comment on only the remaining eight cases:

1) Araraquara

The animals — not only the small dinosaurs but also their other companions — preferred certain directions. Indeed, 76% of examined trackways are oriented to directions within the 2nd and 4th quadrants (1st quadrant: 12%; 2nd quadrant: 29%; 3rd quadrant: 12%; 4th quadrant: 47%). The fact is even more evident if, instead of dividing the field into quadrants by N–S and E–W axes, we divide it by arbitrary N13°E–S13°W and S77°E–N77°W axes. In this case we find that 88% of the tracks go towards points in the 2nd and 4th quadrants (1st quadrant: 6%; 2nd quadrant: 29%; 3rd quadrant: 6%; 4th quadrant: 59%). As a whole the animals have crossed most frequently (and diagonally) a large transverse dune. The motives that explain this special direction are not known at present.

2) São Domingos (Itaguatins, Goiás, Brasil)

This case is in all probability a dinosaur herd, at least in part (as mentioned above). The animals, or at least those from which we have morphological details of the footprints, were probably going towards the mean direction S7°E (or in both ways along this direction).

3) Baños del Flaco (Colchagua, Chile)

The trackways do not reveal preferential directions.

4) Rio do Peixe (Sousa, Paraíba, Brazil)

Only the Piau trackways were studied in detail (Godoy and Leonardi 1985). The curve that is obtained by placing the directional data relative to 24 levels in rose diagrams and histograms is tetramodal, with two main modes in the 1st and 3rd quadrants and two secondary modes in the others. The lowest layer, however, provides orientation different from the other 24, with modes in the 75° and 340° azimuth sectors. The directions of these dinosaurs were mainly parallel or subparallel to the wave ripple crests and therefore also to the waterline of lakes or other bodies of water. The dominant direction NE–SW seems to be a general choice also in the other localities of the basin, where a detailed study was not made (see Lockley 1986 Fig. 5A for summary).

5) São Romão (Orós, Ceará, Brazil)

We observe the presence of a preferential corridor in the same direction as the preceding case.

6) Parotani (Cochabamba, Bolívia)

Of the six trackways (four in one level and two in an adjacent level), four — three in one level and one in another — are going approximately along a direction N15°E and two others (one in each level) along a direction N70°E. It is not known in which way the animals were moving along this direction.

7) La Escalera (Salta, Argentina)

We do not observe preferential directions here.

8) Toro–Toro (Potosí, Bolívia)

On the main pavement, the theropodian trackways are moving to a sector between N65°W and N18°W (76% of the tracks to N40°W–N26°W). The mean of the azimuths is 325°; the mode by degrees is 320° (N = 6); the mode by 5° sectors falls on sector 320°–324° (N = 15). The other preferential sectors are contiguous, to the right: 325°–329° (N = 9) and 330°–334° (N = 7). The two contiguous sectors on the left have low frequencies (N = 2). The median is the same as the mean (325°). A small group of trackways (5) goes towards 340°–344°. The standard deviation of the trackway direction is j = 9.17. Calculating using only 76% of the trackways from sector N40°W–N26°W, we obtain j = 4.25. The two parallel quadruped trackways in the same pavement are directed to N45°E and so are at an angle of almost 90° to the preceding.

On the other large pavement, all the footprint fauna (that is, the sauropod and theropod populations) is directed north (approximately). The great majority of the Toro–Toro fauna moved along a preferential corridor parallel to the shore, with routes generally between N and N65°W. In this case we have the rare phenomenon that, within this sector, almost all trackways go in the same way (approximately N), and none in the opposite way (S). The two *Ligabueichnium* are the only exceptions to the general trend.

Results and Working Hypotheses

1 The above mentioned 38 ichnofaunas show that South America possessed a dinosaurian fauna, from both dinosaurian orders, that was abundant and varied. This fact confirms what is known from osseous findings and contrasts with the vacuum we note on our continent on maps of the world distribution of dinosaurs proposed by the paleontologists of "Laurasia" maps, that seem many times to contain, in South America, the medieval cartographer's phrase *"Hic sunt leones"* — loosely translated as "wilderness" or *"terra incognita."*

2 There was a great increase in dinosaurian associations and populations in the Jurassic, and even more so in the Cretaceous.

3 The large number of theropod footprints should be attributed more to the greater activity (exploration and hunting) of these saurischians than to their real abundance.

4   The Sauropoda/Ornithopoda ratio (1/2) may also be the result of differential ethology and ecology.

5   In some regions, as in the Rio do Peixe Basin and Toro-Toro (respectively Lower and Upper Cretaceous), tetrapods were practically represented only by dinosaurs.

6   The coelurosaurs decreased in proportionate numbers from the Triassic to the Cretaceous on the continent, whereas the Carnosauria increased.

7   The quadrupedal ornithischians were present in South America, though always rare. The especial rarity of their footprints might be attributed in part, however, to their probable slowness and low activity level. They were more abundant in the south than in the north of the continent.

8   It is probable that, under the particular climatic situation of the continent in the Jurassic, the sauropods lived concentrated in somewhat restricted areas, particularly humid and green and close to large bodies of water.

9   The coelurosaurs, at least in South America, adapted themselves better to arid and desert regions than did the other dinosaurian groups.

10 The coelurosaurs in South America were more abundant in more temperate regions, the carnosaurs in hotter regions. It is necessary to observe that the more arid regions of the continent, mainly in the Jurassic, corresponded to the temperate belts and not to the hot regions, as is more usual today.

11 The ichnofaunas that are more rich in individuals and types occur in the intertropical regions.

12 The Ornithopoda were more abundant in the south than in the north of South America. Perhaps there was a phenomenon of a climatic and latitudinal vicariousness between the Ornithopoda and Sauropoda.

13 The desert environment supported, at least in the Jurassic, a relatively varied dinosaur fauna (along with other reptiles and mammals). This was a dwarf fauna, almost completely of bipeds and consisting probably of Coelurosauria, Carnosauria and Ornithopoda, with a predominance of theropods that was above the continental mean. It is probable that the larger-sized dinosaurs of this environment were ornithopods.

14 The theropods were not particularly abundant in the riverside environment in semi-arid climates. Instead, they preferred the sea and lake shores and the highlands and/or arid environments. The coelurosaurs are the group least represented in this semi-arid fluvial environment.

15 The sauropods and ornithopods, in contrast, were well adapted to this environment. The sauropods especially preferred the seashore and platform environments.

16 The great majority of South American dinosaurs were not gregarious, with the exception of the sauropods which were almost always gregarious. The theropods rarely operated in packs. In order of decreasing gregariousness, the data presents itself in this order form: Sauropoda, quadrupedal Ornithischia, Theropoda, Ornithopoda.

17 The dinosaurian populations in general were strongly influenced by the local and regional landscape, in particular by the bodies of water (water-lines) and, indirectly, by the region's tectonic texture in their movements, so travelling along preferential corridors.

18 Even in the desert there were preferential directions followed by the dinosaurs (along with the associated tetrapods), perhaps "caravan tracks" between oases. Some of the statements above are better classed as working hypotheses. It would be important to compare these data and interpretations with those from other continental plates, and also with pertinent osseous material, in a systematic fashion.

Acknowledgments

I wish to express my deepest thanks to Professor Dr. José F. Bonaparte (Buenos Aires) and to Professor Dr. William A. S. Sarjeant (Saskatoon, Canada) for innumerable observations and suggestions, for their critical revision of the manuscript, and for their constant and kind support.

References

Alonso, R. N. 1980. Icnitas de Dinosaurios (Ornithopoda, Hadrosauridae) en el Cretácico Superior del Norte de Argentina. *Acta Geológica Lilloana.* 15 (2):55-63.

Branisa, L. 1968. Hallazgo del amonite *Neolobites* en la Caliza Miraflores y de huellas de dinosaurios en la formación El Molino y su significado para la determinación de la edad del "Grupo Puca". *Bol. del Instituto Boliviano del Petróleo.* 8 (1):16-29.

Casamiquela, R. M. 1964. Estúdios Icnológicos. *Buenos Aires, Colegio Industrial.* Pio X, 229 pp.

Casamiquela, R. M.; and Fasola, A. 1968. Sobre pisadas de Dinosaurios del Cretácico inferior de Colchagua (Chile). *Publ. del Departamento de Geología de la Universidad de Chile, Santiago.* 30:1-24.

Frakes, L. A. 1979. *Climates throughout Time.* (Elsevier) 310 pp.

Dingman, R. J., and Galli O., C. 1965. Geology and groundwater resources of the Pica Area, Tarapaca Province, Chile. *U.S. Geol. Surv. Bull.* No. 1189.

Godoy, L. C., and Leonardi, G. 1985. Direções e comportamento dos dinossauros de localidade de Piau, Sousa, Paraíba (Brasil), formação Sousa (Cretáceo inferior). *In* DNPM, Coletânea de Trabalhos paleontológicos. *Série "Geologia"* 27:65-73.

Huene, F. v. 1931. Verschiedene mesozoische Wirbeltierreste aus Súdamerika. *Neuen Jahrbuch für Mineral., Geol. und Paläont.* 66 (B):181-198.

Leonardi, G. 1977. Two new ichnofaunas (vertebrates and invertebrates) in the Eolian Cretaceous sandstones of the Caiuá Formation in northwest Paraná. *Actas I Simpósio de Geologia Regional.* SBG-SP:112-128.

1979a. Nota preliminar Sobre Seis Pistas de Dinosauros Ornithischia da Bacia do Rio do Peixe (Cretáceo Inferior) em Sousa, Paraíba, Brasil. *Anais Acad. brasileira de Ciências.* 51 (3):501-516.

1979b. New archosaurian trackways from the Rio do Peixe Basin, Paraíba, Brasil. *Annali dell'Università di Ferrara.* (N.S.) IX 5 (14):239-249.

1980a. On the discovery of an abundant ichnofauna (vertebrates and invertebrates) in the Botucatu Formation s.s. in Araraquara, São Paulo, Brasil. *Anais*

Acad. brasileira de Ciências. 52 (3):559–567.

1980b. Ornithischian trackways of the Corda Formation (Jurassic), Goiás, Brazil. *Actas I Congreso Latinoamericano de Paleont.* 1:215–222.

1981. As localidades com rastros fósseis de Tetrápodes na América Latina. *Anais II Congresso Latino-americano de Paleont.* 2:929–940.

1984. Le impronte fossili di dinosauri. *In* Bonaparte, J. F. *et al.* Sulle Orme dei dinosauri. (Venezia: Erizzo) 333 pp.

Leonardi, G., and Godoy, L. C. 1980. Novas pistas de Tetrápodes de Formação Botucatu no Estado de São Paulo. *Anais do XXXI Congresso brasileiro de Geologia.* 5:3080–3089.

Leonardi, G., and Muniz, G. B. C. in press. Ichnological Notes (Dinosaurs and Invertebrates) in the Cretaceous Continental of the Ceará, Brasil, with Reference to Fresh-Water Fossil Molluscs.

Leonardi, G., and Sarjeant, W. A. S. 1986. Footprints representing a new Mesozoic vertebrate fauna from Brazil. *Modern Geology.* 10:73–84.

Lockley, M. G. 1987. The paleobiological and paleoenvironmental importance of dinosaur footprints. *Palaios* 1:37–47.

Lockley, M. G., and Conrad, K. 1987. Mesozoic tetrapod tracksites and their application in paleoecological census studies. *In* Currie, P. M., and Koster, K. (eds.). *4th Symposium on Mesozoic Terrestrial Ecosystems* (Tyrell Museum of Paleontology) pp. 144–149.

Lull, R. S. 1942. Triassic footprints from Argentina. *Amer. Jour. Sci.* 240:421–425.

Moraes, L. J. 1924. Serras e montanhas do nordeste. Brasil, Inspectoria Obras contra Seccas. *Publicação Série* I.D. 28, 2 vol. xi + 122 + 122 pp.

Smith, A. G. et al. 1981. *Phanerozoic Paleocontinental World Maps.* (Cambridge: Cambridge University Press) 102 pp.

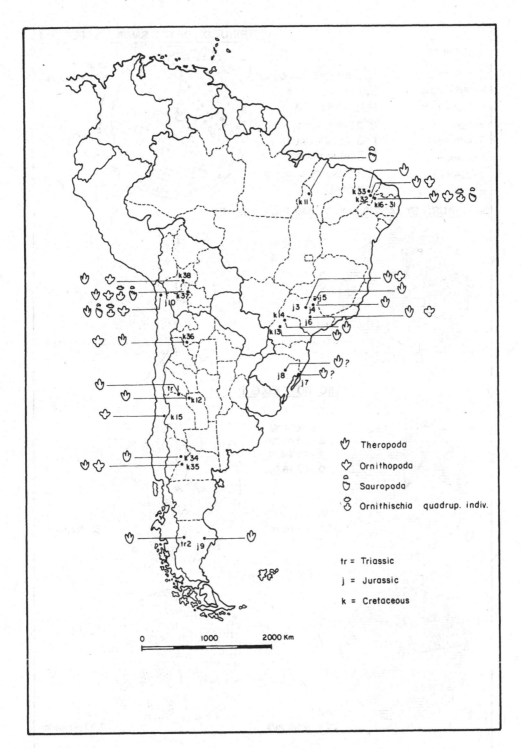

Figure 17.1. Dinosaur tracksites in South America. The numerical order of the sites in the map, from 1 to 38, is the same as in the text and it corresponds to the probable chronological order, from Late Triassic to Late Cretaceous. 1 = Ischichuca; 2 = Bajo Caracoles; 3 = Araraquara; 4 = Jurecê; 5 = Rifaina; 6 = Botucatu Range; 7 = Tramandaí; 8 = Santa Cruz do Sol; 9 = Estancia Laguna Manatiale; 10 = Chacarilla; 11 = São Domingos; 12 = Las Quijadas; 13 = Cianorte; 14 = Indianópolis; 15 = Baños del Flaco; 16–31 = localities of the Rio do Peixe Basin; 32 = São Romão; 33 = Cabeça de Negro; 34 = Plottier; 35 = El Chocón; 36 = Quebrada de La Escalera; 37 = Toro-Toro; 38 = Parotani.

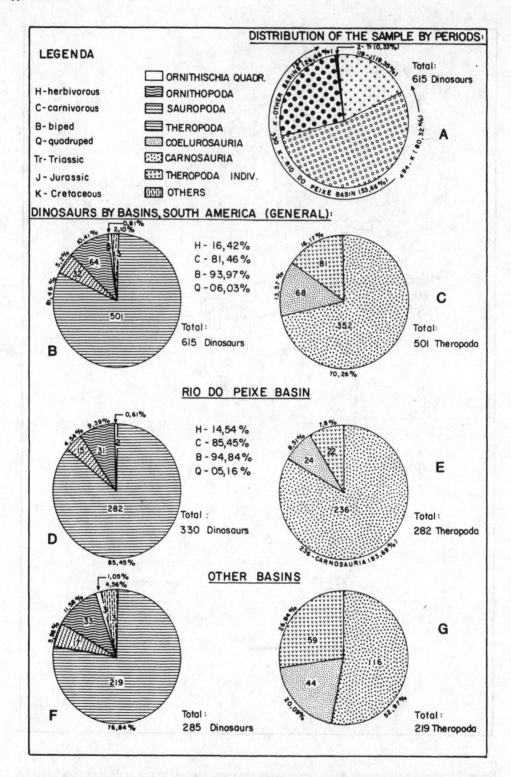

Figure 17.2. Distribution of South American dinosaurs (N = 615) by periods. The graphic symbols on A are different from those on B–G. Cretaceous: 494 individuals (80%); Jurassic: 119 individuals (19%); Triassic: 2 individuals (0.33%).

The graphic symbols of the legend apply to B–G.

B–G, distribution of the South American dinosaurs and their groups by basins, according to their tracks. Left: general; right: Theropoda. **B,C,** general distribution in South America. **D,E,** their distribution in the Rio do Peixe Basin. **F,G,** distribution in the other South American basins.

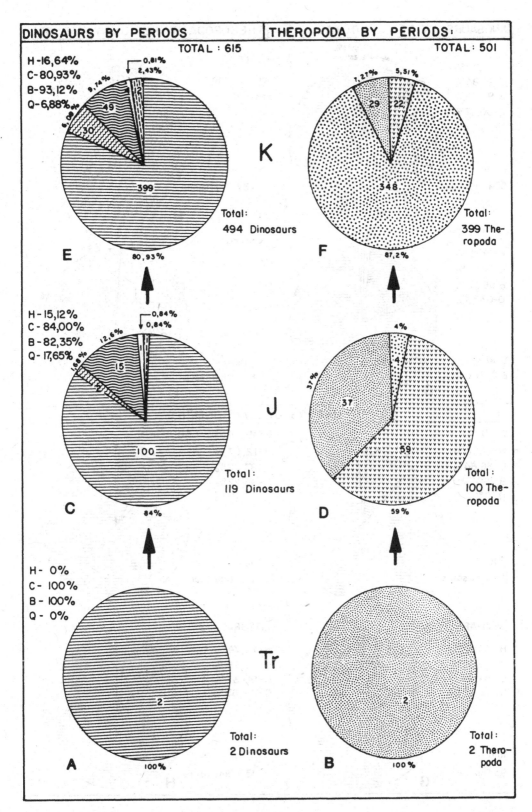

Figure 17.3. Distribution of the South American dinosaurs and their groups by periods, according to their tracks. Left: general; right: Theropoda. **A,B,** Triassic; **C,D,** Jurassic; **E,F,** Cretaceous. Arrows point to the evolution of the ichnofaunas.

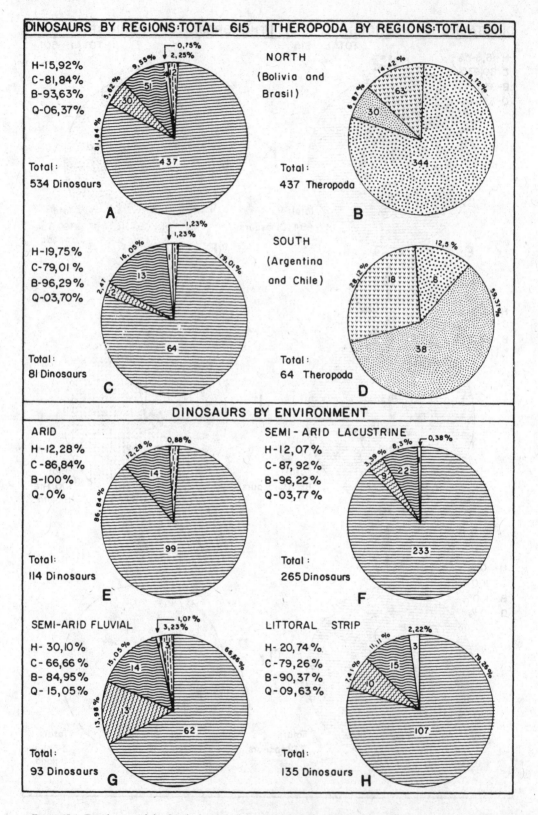

Figure 17.4. Distribution of the South America dinosaur ichnofaunas by regions (north and south), and environment. Left: general; right: Theropoda. **A,B,** distribution of the dinosaurs in the north of the continent (Bolivia and Brazil); **C,D,** distribution in the south (Argentina and Chile); **E,F,** distribution by environments.

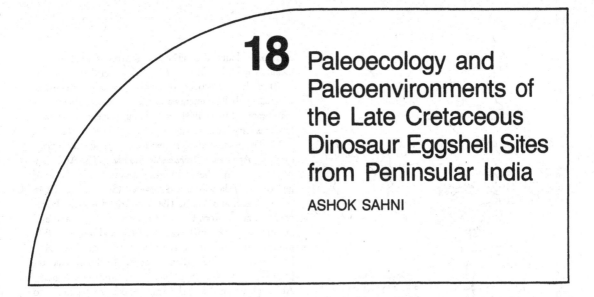

18 Paleoecology and Paleoenvironments of the Late Cretaceous Dinosaur Eggshell Sites from Peninsular India

ASHOK SAHNI

Abstract

Late Cretaceous dinosaur localities of the Indian subcontinent are confined to the thin (less than 40 m) but widespread Lameta Formation underlying the Deccan basalts of central peninsular India. The Lameta dinosaurs are represented by titanosaurids, theropods including a primitive tyrannosaurid, a possible stegosaur and other poorly documented ornithischians. The Lameta Type Section at Jabalpur was previously considered to represent an inter-tidal, stable shelf environment on the basis of sedimentological criteria (glauconite, algal mats in limestones, *Thalassinoides* crab burrows). Based on bulk screening, present findings of abundant cyprid ostracods, charophytes, pulmonate gastropods, freshwater fishes and frogs clearly demonstrate freshwater, fluvial conditions of deposition with well developed paleosol horizons, for the greater part of the Lameta sequence. At Kheda and at Jabalpur, in situ nests of sauropod dinosaurs are found in massive lacustrine limestones. At Kheda, dinosaur tracks are associated with sauropod eggs. Only in one locality in southern India, at Kallamendu, do dinosaurs occur in beds intercalated with a predominantly marine benthonic fauna.

The latest record of dinosaurs in India (terminal Cretaceous) is from the "Intertrappean beds" which represent thin (less than 3 m) fluvio-lagoonal, clastic sediments interbedded with basalt.

The Maestrichtian record of Indian dinosaurs indicates the presence of large populations living in coastal plain regions at a time when the initial Deccan volcanics had started to erupt.

Introduction

One of the earliest recorded dinosaurs from Asia was from the Upper Cretaceous Lameta Formation of Jabalpur, central peninsular India and was collected by Captain Sleeman in 1828, who later earned fame as the suppressor of the "Thuggee Cult". Though a century and a half has elapsed since the first findings, our knowledge of Late Cretaceous dinosaurs of the Indian subcontinent and their paleoenvironmental setting has progressed at a slower rate than discoveries in other parts of the world. During the 1920's and 1930's there was resurgence of interest in Indian dinosaurs and a number of new taxa, localities and horizons were discovered and recorded (Matley 1921, 1923, 1929; Huene and Matley 1933). During the last decade, probably as a result of global interest in Cretaceous dinosaurs, attention has again been focused on the taxonomic status, paleogeographical implications and traces of the Lameta dinosaurs (Chatterjee 1978; Colbert 1977, 1984; Sahni and Gupta 1982; Sahni et al. 1984). The most significant finding in recent years has been the discovery of two extensive dinosaur hatcheries of Late Cretaceous (Maastrichtian) age in Gujarat, western India (Mohabey 1984a,b; Srivastava et al. 1986) and at Jabalpur (Sahni et al. 1984). In addition, isolated dinosaur remains and traces have been reported from several other peninsular Indian localities (Jain and Sahni 1983, 1985). The only dinosaur track recorded from India is from a horizon yielding dinosaur eggshells at Kheda (Mohabey 1986). The footprint occurs over a dinosaur egg clutch. It has three digits, of which the central is the largest, and represents the left front foot (Mohabey 1987). The length is 22.5 cm while the maximum width is 16.5 cm.

The distribution of Late Cretaceous dinosaur localities is given in Figure 18.1. The westernmost locality in peninsular India has recently been discovered by field parties of the Geological Survey of India (Mohabey 1984a,b; Srivastava et al. 1986). At Kheda, sauropod and theropod remains are found in great abundance in calcareous grits and conglomerates overlying the Godhra Granite. The bone-yielding horizon is overlain by finer clastics, marls and dark colored shales bearing nests and isolated fragments of eggshells. The classic locality of Bara Simla Hill near Jabalpur Cantonment has yielded a primitive tyrannosaurid, megalosaurids, titanosaurid sauropods (*Antarctosaurus, Laplatasaurus* and *Titanosaurus*), possible stegosaurians

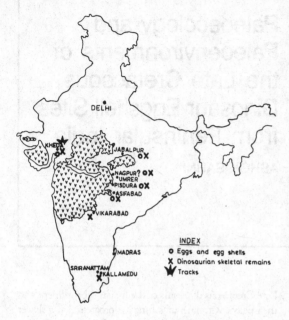

Figure 18.1. Late Cretaceous localities of peninsular India showing the distribution of dinosaur skeletal remains, eggs and tracks.

and some doubtful ornithischians. Recent work in this locality (by D. D. Gillette, P. A. Murry and A. Sahni in 1982) resulted in recovery of a large ostracode fauna, some isolated eggshell fragments and a few unidentifiable microvertebrates. Since 1984, additional vertebrate–yielding horizons have been recognized in the same section (Tripathi 1986). The associated microvertebrates include *Lepisosteus*, *Pycnodus*, *Apateodus*, *Stephanodus* and *Phareodus*. An extensive dinosaur hatchery yielding the same types of eggs as found at Kheda was also discovered (Sahni et al. 1984).

Prasad and Verma (1967) described isolated bones of titanosaurids from Umrer. At nearby Dongargaon and Pisdura, after the initial work of Huene and Matley (1933), intensive studies have been undertaken by the Geological Studies Unit of the Indian Statistical Institute, Calcutta (Berman and Jain 1982; Jain 1977 and in progress). Although fragmentary, the fauna appear to be similar to the assemblages described from other Lameta exposures (Fig. 18.2).

As yet, no dinosaur elements are known from the Lameta Formation of Asifabad, Andhra Pradesh, although fragments of large sauropod bones have been recovered from the Intertrappeans (Rao and Yadagiri 1979, Sahni 1984). Dental elements tentatively assigned to hypsilophodonts have also been described (Prasad 1985). Some of the thicker eggshell fragments (thickness about 1 mm) may also belong to dinosaurs, although at present there is no way of confirming this. Lydekker (1890) had reported a theropod tooth from the Nagpur Intertrappeans (Takli Formation).

Figure 18.2. Stratigraphic sections giving the position of the dinosaur–yielding horizons of the Lameta and Takli Formations (and equivalents), **a**, Kheda, **b**, Jabalpur, **c**, Nagpur, **d**, Pisdura, **e**, Asifabad.

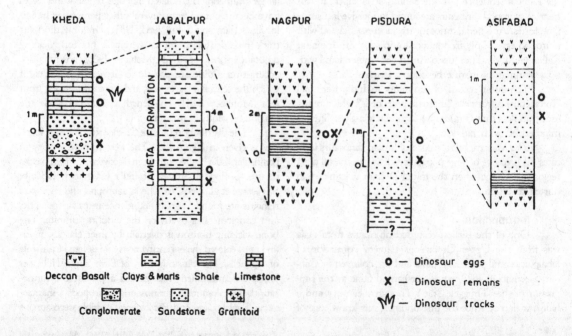

The southernmost locality from which dinosaurs of Maastrichtian age have been recovered is from the Ariyalur District, near Madras. Dinosaurs were first noticed in the area by Blanford in the years 1857–1860. Later, Matley (1929) updated and revised the poorly preserved assemblage. The Kallamendu locality is significant as it is intercalated in a foraminifera- and ammonite–zoned sequence permitting a fairly accurate determination of its age and also allowing for a correlation with Maastrichtian marine benthonic assemblages of Madagascar (Sahni 1983).

The dating of other dinosaur localities mentioned above, although they contain essentially the same generic components as the Ariyalur assemblage, has been a subject of much controversy. The problem rests essentially on the designation of European Stage names for the Indian horizons in spite of the hazards raised by the lack of congruity of correlatable fossils and paleogeographical constraints for the drifting Indian Plate. One view which remained unchallenged until very recently was that of Huene and Matley (1933), who stated that the Jabalpur dinosaurs are mainly of Turonian age. This viewpoint is supported by a majority of workers, who believe that the Jabalpur Lametas are eastern extensions of the better–dated Bagh Group/Formation of the Western Narbada axis (Singh 1981, Singh and Srivastava 1981). However, the recognition of a primitive tyrannosaurid at Bara Simla Hill (Chatterjee 1978) and the views of Colbert (1977, 1984) and Sahni (1984) have reopened the issue. The preliminary results of the Jabalpur microfossil assemblages currently being studied (Tripathi 1986) indicate that there was no significant time gap between the Lameta Formation and the Takli Formation on the basis of microvertebrates, ostracods, pulmonate gastropods and charophytes. The Jabalpur Lametas are probably of Maastrichtian age.

Lameta Paleoenvironments

Interpretations of basinal paleoenvironments of the Lameta Formation were initially based on the presence of terrestrial dinosaurs (Matley 1921). Later, mainly as a result of sedimentological studies of the type and other reference sections at Jabalpur, Chanda (1963, 1964) proposed a shallow–water marine environment for the Lameta. The conflict in the interpretation of the Lameta paleoenvironment as it stands today is mainly a result of the opinions of a marine origin advocated by the sedimentologists and of a freshwater nature as suggested by the paleontological evidence. The relationship of lithology to the type of occurrence of dinosaur remains is shown in Figure 18.3.

There are at least four major criteria for paleoenvironmental reconstruction which have been differently interpreted by the sedimentologists and the paleontologists. These are i) the presence of glauconite; ii) nature of the limestones, whether marine or lacustrine; iii) nature of tubular structures, whether these represent (marine) crab burrows of the *Thalassinoides* type or whether the tubules are root casts; and iv) the paleontological evidence.

(i) The Presence of Glauconite

By far the strongest evidence in support of a "stable-shelf environment" is the supposed presence of glauconite in the basal unit of the Lameta Formation, namely the Green Sandstone. This mineral was principally diagnosed on the basis of high K_2O percentages as well as some optical and X–ray diffraction studies (Chanda and Bhattacharya 1967). Since then, this evidence has been widely quoted (Singh 1981) but the original determination has never been seriously questioned or tested. Recently, as a result of the recovery of a predominantly freshwater assemblage and reinterpretation of the sedimentological history of the Lameta (Sahni and Brookfield 1986, Brookfield and Sahni 1987), it became essential to reanalyze the "glauconite" in the Green Sandstone to verify its exact nature. Detailed examination of the clay minerals in the Green Sandstone by a variety of modern techniques (flame photometry, X–ray diffraction and mass balance studies) shows that the clay mineral in question is not glauconite, but celadonite. Celadonite is a rare mineral which is formed by the hydrothermal alteration of basalts, probably redeposited before and after Trap eruptions by groundwaters. The field evidence also shows that, at a number of localities, the Green Sandstone is not green in color but is white orthoquartzite that is extensive over tens of kilometers.

Figure 18.3. Relationships of sediment type and preservation of different kinds of dinosaurian skeletal elements.

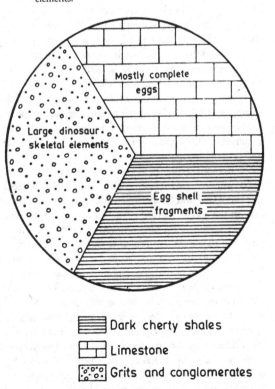

Dark cherty shales

Limestone

Grits and conglomerates

(ii) Lameta Limestones

The next important sedimentological evidence comes from the nature of the Lameta Limestone, which is also an extensive lithounit found in most sections wherever the Lametas have been recognized. Because of the purity of the limestone (it is exploited commercially at Balasinor, Gujarat for cement manufacture), its thickness (2–3 m) and the supposed occurrence of algal mats and crab burrows, this unit was considered to be intertidal, marine in origin, by Singh (1981). Recent work at the Vertebrate Paleontological Laboratory has shown that the limestone is of variable thickness with varying additions of silt and sand-sized grains, so that at places (for instance at the Bara Simla Hill Section), the limestone is an impure sandy marlstone. In fact, in a green marly lens in between the limestone of Bara Simla Hill, an abundant assemblage of freshwater ostracods, charophytes and microvertebrates were recovered (Sahni 1984, Tripathi 1986).

The limestone is a typical semi-arid pedogenic calcrete deposit. In other localities in the world where a similar assemblage of microvertebrates and dinosaur eggs have been reported (de Muizon et al. 1983, Erben et al. 1979), associated limestones have been interpreted as non-marine.

(iii) Nature of Tubular Structures

Until the 1970's, when Kumar and Tandon (1977, 1978, 1979) first observed the occurrence of the tubular burrows in the Mottled Nodular Bed, there was no previous record of these abundant and extensive structures which have now been reported from all the five main lithological units of the type locality of the Lameta Formation. From the time of their initial description as crab burrows belonging to the *Thalassinoides* type of traces, workers in the area have accepted this identification by consensus. Singh (1981) dealt comprehensively with the structure, distribution of the crab burrows and their facies relationships. He pointed to their occurrence in the limestone units as well, thereby giving him additional evidence to support his viewpoint on the intertidal marine nature of the sediments. His well documented descriptions of the *Thalassinoides* burrows is still accepted by many workers.

Recently, however, as a sequel to field mapping and paleoenvironmental studies by Sahni and Brookfield (1986) and Brookfield and Sahni (1987), doubts have again been cast on the specific nature of the tubular structures, the majority of which are now believed to be calcrete structures and root casts. According to Sahni and Brookfield (1986), the tubular structures represent undoubted rootlet horizons associated with pedogenic carbonate structures comprising calcareous pedosols and calcretes. These pedotubules have a simple or branching tubular form. Most of the rootlet types are slender and tapering and can be ascribed to shrub-type vegetation. As a general rule, it is often difficult to distinguish root casts from invertebrate burrows. The main criteria are the downward branching of the roots, an irregularly tubular form which tapers unilaterally, and a sedimentary facies showing activity of groundwater solutions. The tubules of the Lameta Formation are characterized by the above mentioned features, although there is evidence that biogenic animal burrows were also present. Burrows by themselves are not particularly helpful in demarcating continental or marine environments unless a large number of diversified forms are found. Such is not the case at Jabalpur where the predominant structures are in the shape of pedotubules.

(iv) The Paleontological Evidence

The strongest evidence for the depositional environment of the Lametas comes from the fossil assemblages recovered from various horizons, but principally from the basal units comprising the Green Sandstone, Lower Limestone and associated conglomerates and the Variegated Shale immediately overlying the Lower Limestone. Until the present work on microfossils was undertaken, the main thrust of the paleontological evidence rested on the abundant dinosaur skeletal elements recovered in the 1920's and 1930's. The data of the large dinosaur bones were equivocally interpreted by both marine and non-marine protagonists. Matley (1921) considered the dinosaurs to be a good indicator of freshwater environments, although, in the interpretation of Singh (1981), this very same evidence served as an indicator of river-transported material into the tidal zone. As dinosaur remains can also be found as transported fragments in marine environments, the presence of large bones was not of much use in deciding the issue. The finds of a diverse and abundant microfossil assemblage from horizons which have produced the large dinosaurs as well as from others just above the main dinosaurian-yielding horizons has strengthened the view advocating a mainly non-marine origin for the Lametas (Sahni 1984, Sahni 1985, Tripathi 1986).

The fossil evidence is represented by the existence of pulmonate gastropods and unionid pelecypods, cyprid and other freshwater ostracods, *Chara* gyrogonites, osteoglossid fishes, frogs, delicately preserved limb bones and teeth of smaller dinosaurs, and in situ nests of sauropods (Tripathi 1986).

The presence of pulmonate gastropods referrable to *Paludina (Vivipara) normalis* were reported from the Mottled Nodular Bed at the Chui Hill Quarry section (Sahni and Mehrotra 1974). Recently, from the green marlstone band intercalated in the Lower Limestone, other gastropods of freshwater habitat have been recovered: *Physa*, *Lymnaea* and *Planorbis*. Ostracods have been referred to the following genera: *Candona*, *Paracypretta*, *Cyprois*, *Metacyprois* etc. This assemblage is also non-marine and so far no definite marine ostracods have been recorded (Tripathi 1986). The remains of charophytes are not very common but still constitute a small and fairly diversified component of the Lameta biotas. They include the taxa *Platychara*, *Peckichara* and *Microchara*, all of which are abundant in the Intertrappeans of peninsular India.

Microvertebrates have been recovered by bulk screening and washing of the sandy marl horizons intercalated with the Lower Limestone or occurring just above it. Of significance is the presence of the osteoglossid *Phareodus*, which is one of the most common components of the Cretaceous–Eocene assemblages of both peninsular India and the Himalayan belt of the subcontinent. Osteoglossids are typically freshwater and constitute one of the largest known Recent non–marine fish groups of South America. The other genera of fishes (eotrigonodontids, *Apateodus* and the holostean *Lepisosteus*) are found in other assemblages from the Intertrappeans. These taxa have been reported from the Takli Formation of Nagpur (Gayet et al. 1985).

Dinosaur Eggs and Paleoenvironments

Of all the dinosaur traces, the finds of complete nests of eggs at Kheda, Gujarat (Mohabey 1983, 1984a,b) and at Jabalpur, Madhya Pradesh, constitute one of the best lines of evidence for the reconstruction of paleoenvironments (Brookfield and Sahni 1987). There is considerable morphological and microstructural diversity among Late Cretaceous environments found in the Lameta amd Takli Formations of India (Sahni and Gupta 1982, Srivastava et al. 1986, Jain and Sahni 1985, Sahni et al. 1984, Vianey–Liaud et al. 1987). The positive identification of eggs to taxa known on the basis of cranial or skeletal remains is a difficult task in most instances unless embryonic or juvenile skeletons are recognized in the nesting sites. On the basis of hatchling skeletons, eggshell structure has been recognized for hypsilophodonts and hadrosaurs (Hirsch and Packard 1987). In India as well, the recent description of a juvenile skeleton found in large (diamater 16 cm) spherical, nodose–ornamented eggs from the Lameta Formation of Kheda District is the best evidence at present to relate such eggs to the titanosaurid sauropod (Mohabey 1987, summary this volume). Earlier, large eggs of similar size and microstructure were referred to titanosaurids (Srivastava et al. 1986) on the basis of their similarity to eggs referred to *Hypselosaurus* by Erben et al. (1979). This assignment was mainly based on size considerations and on the association of *Hypselosaurus* remains in the egg–bearing localities.

The record of saurischian eggs in general is poor and at present is best represented by the Late Cretaceous localities of southern France and India. Vianey–Liaud et al. (1987) have demonstrated a close correspondence between microstructural types in which the individual spheroliths are distinct, elongated with highly arched growth increments truncated at spherolith boundaries (Kheda Type B) to eggshells earlier referred to sauropods (Erben et al. 1979). Other Late Cretaceous eggshells, particularly those that are extremely thin (0.02 to 0.05 mm), are of uncertain taxonomic affinity (Sahni et al. 1984). Some of these may be dinosaurian or even gekkonid in nature, as recently suggested by Hirsch and Packard (1987).

At Kheda and Jabalpur, there are at least two morphologically distinct types of eggs. These have been designated as Type A and Type B (Fig. 18.4) for the Kheda localities (Srivastava et al. 1986). Both these types can also be recognized at Jabalpur, although the frequency of occur-

rence of Type B eggs at Jabalpur appears to be lower than that for Kheda. Paleoenvironmental factors influencing oxygen isotopes of the Kheda eggs have recently been studied (Sarkar et al. 1986).

Type A Eggs

These are by far the most common type found and have also been reported in isolated fragmentary conditions from Pisdura (Jain and Sahni 1985). The shells are characterized by thicknesses varying between 1 to 2 mm, with an external ornamentation consisting of isolated, coalescing tubercles with pore openings usually occurring between the adjacent tubercles. The spheroliths themselves are broadly oval, tapering proximally, and bluntly rounded externally, corresponding to the structure of the external tubercle. The spheroliths show considerable variability in their relationship to their neighboring units: the spheroliths may retain their respective boundaries or they may be confluent with the adjacent units so that the increment lines of growth do not terminate at the boundaries of the spheroliths themselves but are stratified across the thickness of the shell.

Type B Eggs

These eggs are characterized by thicker walls (2–3 mm) with an external ornamentation that consists of discrete tubules or nodes, not as a rule joined together. The spheroliths are high, cylindrical with parallel boundaries, clearly distinct from their neighbors. The growth increments are highly arched and terminate at the margins of the individual spheroliths. This type resembles the tubocanaliculate structure.

Concluding Remarks

The dinosaur faunas from the Late Cretaceous of India are dominated by the titanosaurid sauropods. Nearly all the large spherical eggs and nests recorded so far from India can apparently be ascribed to this group. Theropods, including megalosaurids and a primitive tyrannosaurid, are less common. No theropod eggs have so far been reported. The ornithischian component is presently too poorly documented to be of any significance. Associated vertebrates from the dinosaur–yielding horizons are even more

Figure 18.4. Eggshell structure of large spherical eggs assigned to the sauropods: **a,** Type A. **b,** Type B.

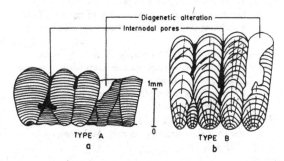

poorly documented. Crocodiles are represented by isolated teeth and scutes and are indistinguishable at the familial level. A single vertebra assignable to the Dryosauridae is currently being described by Rana (1987) from the Lametas near Vicarabad. Turtles are represented mainly by the Pelomedusidae (Jain 1977). The fishes are generically identical to those reported from the Intertrappeans.

Paleobotanical information (both megafloral and palynofloral) is meager, preventing interpretations of paleclimatic conditions. Most of the earlier studies on Indian Late Cretaceous floras indicate a close similarity to those from northern Africa and Europe (Srivastava 1983; Nandi 1982).

Future work is needed to obtain additional information on dinosaur tracks and traces in India. A beginning has been made and it is hoped that more data will be forthcoming on dinosaur paleoecology and paleoclimates from India.

References

Berman, D., and Jain, S. L. 1982. The braincase of small sauropod dinosaur (Reptilia–Saurischia) from the Upper Cretaceous Lameta Group, Central India, with a review of Lameta Group localities. *Annals Carnegie Museum Nat. Hist.* 51:403–422.

Brookfield, M. E., and Sahni, A. 1987. Palaeoenvironment of the Lameta Beds (Late Cretaceous) at Jabalpur, Madhya Pradesh, India: Soil and biotas of a semi-arid alluvial plain. *Cretaceous Research* 8:1–14.

Chanda, S. K. 1963. Cementation and diagenesis of Lameta Beds. Lametaghat, M.P. India. *Jour. Sedim. Petrol.* 33:728–738.

1964. Petrography and origin of the Lameta Sandstone, Lametaghat, M.P. *Proc. National Institute of Science, India* 29 (5):578–587.

Chanda, S. K., and Bhattacharya, A. 1966. A re-evaluation of the stratigraphy of the Lameta–Jabalpur contact around Jabalpur, M.P. *Jour. Geol. Soc. India* 7:92–93.

Chatterjee, S. 1978. *Indosuchus* and *Indosaurus*, Cretaceous carnosaurs from India. *Jour. Paleont.* 53 (3):570–580.

Colbert, E. H. 1977. Mesozoic tetrapods and the northward migration of India. *Jour. Palaeont. Soc. India* 20:136–145.

1984. Mesozoic reptiles, India and Gondwanaland. *Indian Jour. Earth Sci.* 11 (1):25–37.

de Muizon, C., Gayet, M., Lavenu, A. I., Marshall, L. C., Sigé, B., and Villareel, C. 1985. Late Cretaceous vertebrates, including mammals from Tiupampa, southcentral Bolivia. *Geobios* 16 (6):747–753.

Erben, H., Hoefs, K. J., and Wedepohl, H. K. 1979. Paleobiological and isotopic studies of egg shells from a declining dinosaur species. *Paleobiology* 55 (4):380–414.

Gayet, M., Rage, J. C., and Rana, R. S. 1985. Nouvelles ichthyofaune et herpetofaune de Gitti Khadan le plus ancien gisement connu du Deccan (Cretace/Paleocene) a Microvertebres. Implication paleogeographiques. *Mem. Soc. Geol. France* 147:55–65.

Hirsch, K. F., and Packard, M. J. 1987. Review of fossil eggs and their shell structure. *Scanning Microscopy* 1:383–400.

Huene, F. v., and Matley, C. A. 1933. The Cretaceous Saurischia and Ornithischia of the central province of India. *Mem. Geol. Surv. India Palaeont. Indica N. S.* 21:1–72.

Jain, S. L. 1977. A new fossil pelomedusid turtle from the Upper Cretaceous Pisdura sediments, central India. *Jour. Palaeont. Soc. India* 20:360–365.

Jain, S. L., and Sahni, A. 1983. Upper Cretaceous vertebrates from central India and their palaeogeographical implications. Symposium Cretaceous of India. *Proc. Indian Assoc. Palynostratigraphers Lucknow* pp. 66–83.

1985. Dinosaur egg shell fragments from the Lameta Formation at Pisdura, Chandrapur District, Maharashtra. *Geosci. Jour.* 2:211–220.

Kumar, S., and Tandon, K. K. 1977. A note on the bioturbation in the Lameta beds, Jabalpur area, M.P. *Geophytology* 7:135–138.

1978. *Thalassinoides* in the mottled nodular beds, Jabalpur, M.P. *Current Science* 47:52–53.

1979. Trace fossils and environment of deposition of sedimentary succession of Jabalpur, Madhya Pradesh. *Jour. Geol. Soc. India* 20:103–106.

Lydekker, R. 1890. Note on certain vertebrate remains from the Nagpur District, II Part of Chelonian plastron from Pisdura. *Record Geol. Surv. India* 23:22–23.

Matley, C. A. 1921. On the stratigraphy, fossil and geological relationship of Lameta beds of Jabbulpore. *Record Geol. Surv. India* 55:105–108.

1923. Note on an armoured dinosaur from the Lameta Beds of Jabalpur. *Record Geol. Surv. India* 55:105–108.

1929. The Cretaceous dinosaurs of the Trichinapoly District. *Record Geol. Surv. India* 61:337–349.

Mohabey, D. M. 1983. Note on the occurrence of dinosaurian fossil eggs from Infratrappean limestone in Kheda District, Gujarat. *Current Science* 52 (24):1194.

1984a. The study of dinosaurian eggs from Infratrappean limestone in Kheda District, Gujarat. *Jour. Geol. Soc. India* 25 (5):329–337.

1984b. Pathologic dinosaurian egg shells from Kheda District, Gujarat. *Current Science* 53 (13):701–702.

1986. Note on dinosaur foot print from Kheda District, Gujarat. *Jour. Geol. Soc. India* 27 (5):456–459.

1987. Juvenile sauropod dinosaur from Upper Cretaceous Lameta Formation of Panchmahals District, Gujarat, India. *Jour. Geol. Soc. India* 30:210–216.

Nandi, B. 1982. Palynostratigraphy of the Gumaghat Formation, Meghalaya, India, with special reference to the significance of the *Normapollea* group. In *Evolutionary Botany and Biostratigraphy*, A.K. Ghosh Volume pp. 521–540.

Prasad, G. V. R. 1985. Microvertebrates and associated microfossils from the sedimentaries associated with Deccan traps of Asifabad region, Adilabad District, Andhra Pradesh. Unpublished Ph.D. thesis, P. U. Chandigarh. 320 pp.

Prasad, K. N., and Verma, K. K. 1967. Occurrence of dinosaurian remains from Lameta beds of Umrer, Nagpur District, Maharashtra. *Current Science* 36 (20):547–548.

Rana, R. S. 1987. Dryosaurid crocodiles (Mesosuchia) from the Infratrappean beds, Hyderabad District, A. P.

Current Science 56:532–533.

Rao, B. R. J., and Yadagiri, P. 1979. Cretaceous Intertrappean beds from Andhra Pradesh and their stratigraphic significance. *Geol. Soc. India Mem.* 3:287–291.

Sahni, A. 1983. Upper Cretaceous palaeobiogeography of peninsular India and the Cretaceous–Palaeocene transition: the vertebrate evidence. *Proc. Indian Assoc. Palynostratigraphers, Lucknow:* pp. 128–140.

——— 1984. Cretaceous–Palaeocene terrestrial fauna of India. Lack of endemism during drifting of Indian–Plate. *Science* 226:441–443.

——— 1985. Upper Cretaceous–Early Palaeocene palaeobiostratigraphy of India based on terrestrial vertebrate fauna. *Mem. Soc. Geol. France* 147:125–137.

Sahni, A., and Brookfield, M. E. 1986. Re-assessment of Lameta palaeoenvironments, Jabalpur (M.P.) *Abstract VI Conv. Indian Assoc. Sedimentologists, Dehra Dun* pp. 83–84.

Sahni, A., and Gupta, V. J. 1982. Cretaceous egg shell fragments from Lameta Formation, Jabalpur, India. *Indian Geol. Assoc. Bull.* 15 (1):85–88.

Sahni, A., and Mehrotra, D. K. 1974. Turonian terrestrial communities of India. *Geophytology* 4 (1):102–105.

Sahni, A., Rana, R. S., and Prasad, G. V. R. 1984. SEM studies of thin egg shell fragments from the Intertrappean (Cretaceous–Tertiary transition) of Nagpur and Asifabad, peninsular India. *Jour. Palaeont. Soc. India* 29:26–33.

Sarkar, A., Bhattacharya, S. K., and Mohabey, D. M. 1986.

Interclutch $_6O^{18}$ variations in dinosaurian egg shells and its palaeoclimatic implications. *Symp. Isotope Based Studies on Problem of Indian Geology (Calcutta) (Abst.)*, pp. 20–22.

Singh, I. B. 1981. Palaeoenvironment and palaeogeography of Lameta Group sediments (Late Cretaceous) in Jabalpur area, India. *Jour. Palaeont. Soc. India* 25:38–53.

Singh, S. K., and Srivastava, H. K. 1981. Lithostratigraphy of Bagh Beds and its correlation with Lameta Beds. *Jour. Palaeont. Soc. India* 26:77–85.

Srivastava, S. K. 1983. Cretaceous phytogeoprovince and palaeogeography of the Indian plate based on palynological data. *Proc. Indian Assoc. Palynostratigraphers, Lucknow:* pp. 141–157.

Srivastava, S. K., Mohabey, D. M., Sahni, A., and Pant, S. C. 1986. Upper Cretaceous dinosaur egg clutches from Kheda District (Gujarat, India), their distribution, shell ultrastructure and palaeoecology. *Palaeontographica A* 193:219–233.

Tripathi, A. 1986. Biostratigraphy, palaeoecology and dinosaur egg shell ultrastructure of the Lameta Formation at Jabalpur, Madhya Pradesh. Unpublished M.Phil. thesis, Panjab University, Chandigarh. 128 pp.

Vianey-Liaud, M., Jain, S. L, and Sahni, A. 1987. Dinosaur eggshells (Saurischia) from the Late Cretaceous Intertrappean and Lameta Formation (Deccan, India). *Jour. Vert. Paleont.* 7 (4):408–424.

19 A Review of Dinosaur Footprints in China

ZHEN SHOUNAN, LI JIANJUN,
RAO CHENGGANG,
NIALL J. MATEER AND
MARTIN G. LOCKLEY

Abstract

All dinosaur footprints from China named in the literature are briefly reviewed. The first dinosaur footprints were found in 1929; since then 16 ichnogenera belonging mainly to theropods and ornithopods have been identified. Very few Triassic footprints have been found, but Jurassic footprints are common in Sichuan and Yunnan provinces, and northeast China, and afford some useful correlations with other areas. Cretaceous footprints are known from Hunan and Sichuan provinces and, recently, from Nei–Monggol and Xizang (Tibet). The ichnogenus *Jeholosauripus* is reassigned to *Grallator*, and Kuangyuangpus is identified as *Batrachopus* (a non–dinosaurian ichnogenus).

Historical Overview

Although the history of dinosaur footprints is some 160 years old, it was not until 1929 that the first dinosaur footprints in China were discovered. These were found by Teilhard de Chardin and Young (1929) but their discovery was not made known to the scientific world until some 30 years later when they were given the scientific name *Sinoichnites youngi* by Kuhn (1958).

This first discovery was followed 11 years later by the finding of another locality at Chaoyang, in Liaoning province (Fig. 19.1), where over 4000 footprints were noted, all belonging to a single species, *Jeoholosauripus s-satoi* (Yabe et al. 1940).

In 1943, *Kuangyuanpus szechuanensis* was found in Sichuan and was the first footprint described by a Chinese paleontologist (Young 1943). Herein we reassign the ichnospecies to the genus *Batrachopus* and attribute it to a crocodilian. Young (1960) summarized fossil footprints in China and described three new taxa: *Changpeipus carbonicus*, *Yangtzepus yipingensis*, and *Laiyangpus liui*. Following this review, Young published two further dinosaur footprint papers in which he erected the taxa *Shensipus tungchuanensis* and *Changpeipus luanpingeris* (Young 1966 and 1979, respectively). As the second Chinese paleontologist to study

dinosaur footprints, Zeng (1982) erected two new genera and three species from the Upper Cretaceous in Xiangxi, Hunan province: *Xiangxipus chenxiensis*, *X. youngi*, and *Hunanpus jiuguwanensis*.

Perhaps the best preserved fossil footprints found in China are those described by Zhen et al. (1983) named *Jialingpus yuechensis* in which the pads, and creases between the pads, are clearly visible. They are also the first Late Jurassic tracks reported with distinctive metatarsal impressions (Fig. 0.3, Lockley and Gillette this volume). They resemble *Anomoepus* (Fig. 19.2D,E).

In 1983, in Jinning, Yunnan province, hundreds of dinosaur footprints were found which belong to six genera and species, four genera and five species of which are new (Zhen et al. 1986). Many footprints from this locality have yet to be described as work continues. In addition to the above mentioned footprints, thousands of dinosaur footprints have been found in Nei–Monggol (Li 1985) and Sichuan provinces, but they too remain unstudied. These recent discoveries suggest a considerable ichnological potential in China.

Systematic Descriptions

Suborder Theropoda
  Infraorder Coelurosauria
    Ichnofamily Anchisauripodidae
      Ichnogenus *Grallator* Hitchcock 1858 (Lull 1904)
        Grallator limnosus Zhen, Li, and Rhao 1986
          1986 *Grallator limnosus* n.sp. Zhen, Li, and Rao, p. 16.

Diagnosis. Bipedal tridactyl tracks, tail trace absent. Claws accuminate, moderately developed. The lengths of the pes and pace are in a ratio of about 1:3.6. The step angle is 147°. The angles between the toes are II–20°–III–15°–IV. Length of footprints is 27.0–28.0 cm and the width is 17.0–20.0 cm. The pace is 95.0–103.5 cm, and the stride is 190.0 cm.

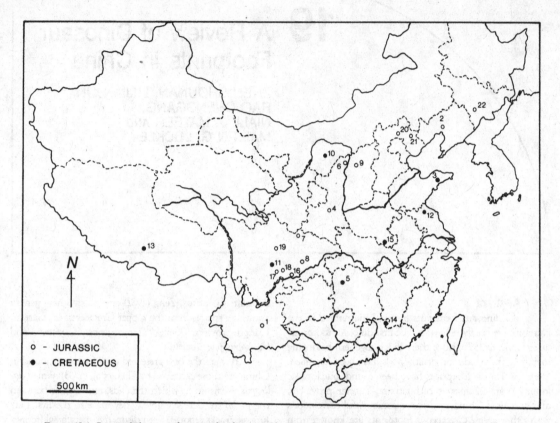

Figure 19.1. Principal dinosaur footprint localities in China. **1.** Jinning, Yunnan province. **2.** Chaoyang, Liaoning province. **3.** Laiyang, Shandong province. **4.**Tungchuan, Shaanxi province. **5.** Xiangxi, Hunan province. **6.** Shenmu, Shaanxi province. **7.** Guangyuan, Sichuan province. **8.** Yuechi, Sichuan province. **9.** Shanxi province. **10.** Chabu, Nei–Monggol. **11.** Emei Shan, Sichuan province. **12.** Lianyungang City, Jiangsu province. **13.** Xigaze, Xizang Autonomous Region. **14.** Nanxiong, Guandong province. **15.** Xiaguan, Henan province. **16.** Yiping, Sichuan province. **17.** Sichuan province. **18.** Sichuan province. **19.** Sichaun province. **20.** Hebei province. **21.** Chengde, Hebei province. **22.** Huinan, Jilin province.

Material. (Fig. 19.3A) Part of a trackway consisting of three successive footprints (Beijing Museum of Natural History, No. BPV–FP1).

Horizon and Locality. Lower Fengjiahe Formation, Lower Jurassic of Jinning, Yunnan province (Fig. 19.1, locality 1).

Discussion. The characters of this specimen are similar to those of *Grallator*, of which only *G. maximus* is comparable in size; other species are much smaller. Digit III of *G. maximus* is more than 50% longer than lateral digits, but BPV–FP1 is only about 30% longer than its lateral digits. In *G. maximus*, the angle betwen digits II and III is smaller than that between digits III and IV, the opposite of BPV–FP1.

Grallator s-satoi (Yabe et al. 1940) comb. nov.
  1940 *Jeholosauripus s-satoi* n.g., n.sp. Yabe et al., p. 560–563.
  1942 *Jeholosauripus* Shikama, p. 21–31.
  1954 *Jeholosauripus* Baird, p. 182.
  1957 *Jeholosauripus s-satoi* Baird, p. 471–472.

  1958 *Jeholosauripus s-satoi* Kuhn, p. 23, 27.
  1960 *Jeholosauripus s-satoi* Young, p. 54.
  1963 *Jeholosauripus s-satoi* Kuhn, p. 100–101.
  1971 *Jeholosauripus s-satoi* Haubold, p. 67–68.
  1983 *Jeholosauripus s-satoi* Zhen, Li, and Zhen, p. 8.

Diagnosis. Plantigrade, tridactyl, but no trace of hallux, lateral digits, nor caudal impression. General outline deltoid. Proximal end of III starting much anterior to that of II and IV. Size rather small: length of footprints is 7.0–12.0 cm and the width is 5.0–8.5 cm. The angles between the digits are II–14°–III–13°–IV.

Material. (Figs. 19.3, 19.4B) The type material is no longer in China, transferred to the Institute of Geology and Palaeontology, Tohoku Imperial University, Sendai (No. 61677). The description is based on that of Young (1960).

Horizon. Rhaetic–Lias, Chaoyang, Liaoning province, and at Chengde, Hebei province (Fig. 19.1, localities 2 and 21, respectively).

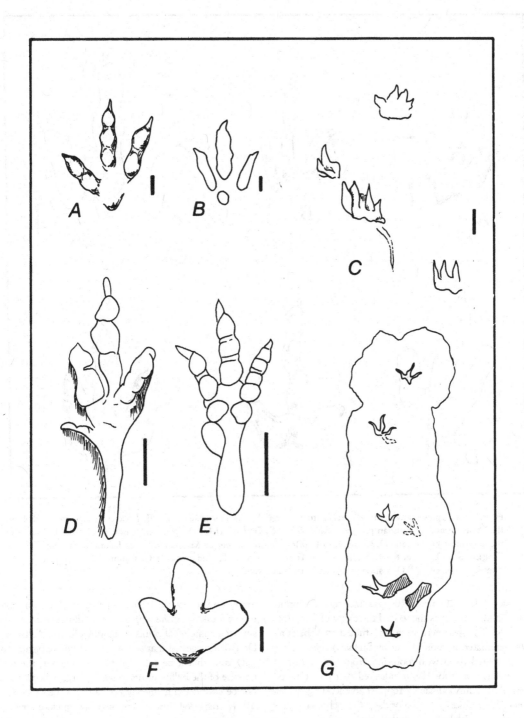

Figure 19.2. Representative dinosaur tracks and trackways from China. **A,** *Schizograllator* from the Lower Jurassic of Yunnan province with *Kayentipus* (**B**) from the Lower Jurassic of Arizona for comparison. **C,** *Batrochopus* (probably from a pseudosuchian-like crocodile and not a dinosaur) from the Middle Jurassic of Sichuan province.
D, *Jialingpus* from the Upper Jurassic of Sichuan province with *Anoemoepus intermedius* (**E**) (redrawn after Lull 1953, Fig. 63) for comparison. **F,** *Sinoichthys* from the Upper Jurassic of Shaanxi province.
G, *Xianxipus chenxiensis* trackway from the Upper Cretaceous of Hunan province. Bar scales are cm.

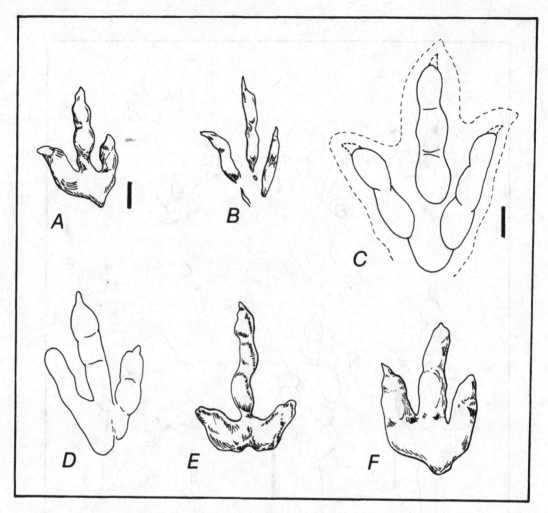

Figure 19.3. Representative dinosaur tracks from China. **A,** *Grallator limnosus* and **B,** *Paracoelurosaurichnus*, both from the Lower Jurassic of Yunnan province. Bar is 5 cm. **C,** *Grallator (Jeholosauripus) s-satoi* nov. comb. (bar is 2 cm) from the lowermost Jurassic and **D,** *Changpeipus* (x 0.16) from the Lower or Middle Jurassic of Liaoning province. **E,** *Zhengichnus* (x 0.2) from the Lower Jurassic of Yunnan province. **F,** *Youngichnus* (x 0.23) from the Lower Jurassic of Yunnan province. A,B,D–F are approximately the same scale.

Discussion. It has been nearly 50 years since these footprints were found, which was the second discovery of footprints in China. The specimens were first studied by Yabe et al. (1940) and later, in more detail, by Shikama (1942). Having compared them with footprints from the Triassic of North America, Yabe (1940) indicated that they differed from Anchisauripodidae, Otouphepodidae, Gigandipodidae, Eubrontidae, Grallatoridae, Selenchnidae, and Anoemoepodidae by virtue of their shape and smaller size. He noted that the Anchisauripodidae and Gigantipodidae were tetradactyl, unlike this specimen, and that the Otouphepodidae and Eubrontidae were much broader and shorter. The size of the Chinese footprints is similar to Anchisauripodidae, hence Haubold's (1971) assignment of them to this ichnofamily. Moreover, Baird (1957) considered them to be synonymous with *Anchisauripus*. However, after carefully studying more than 70 specimens in 1960, Young

found no trace of the hallux impression which was commonly found in *Anchisauripus*; neither did Yabe et al. after studying some 4000 prints. Young noted that "although Dr. Baird suggested that the absence of hallux prints may not prove that it be excluded from this genus, the total absence of the hallux prints in such a large collection can not be considered as merely accidental" (Young 1960, p. 63). He regarded these footprints as more closely related to *Grallator* than to *Anchisauripus*, in contrast to Shikama (1942) who found the only difference being the divergence between the third and fourth digits of less than 14°. Shikama thought that in *Grallator* this angle consistently exceeded 14°; however, Lull (1904, 1953) described some species of *Grallator* where the angle measured only 12°.

The digit III to IV angle can now be dismissed as insignificant; thus the difference between *Grallator* and *Jeholosauripus* is negated, and *Jeholosauripus* is considered

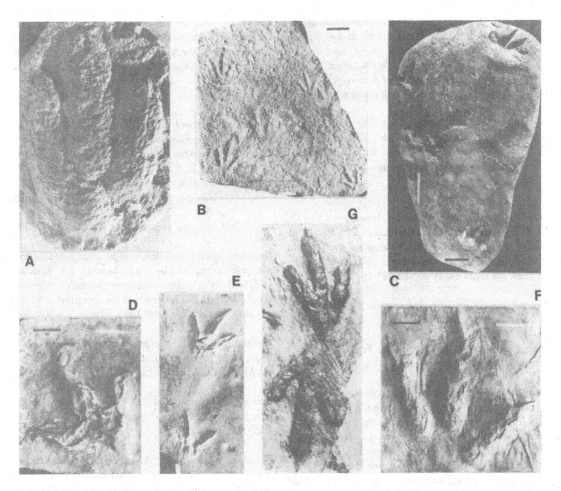

Figure 19.4.**A,** *Yangtzepus yipingensis* Zhen, Li, and Rao (1986). Lower Upper Jurassic (Chiating Series), Sichuan province. (No scale.) **B,** *Grallator (Jeholosauripus) s–satoi* Yabe, Inai, and Shikama (1940). ?Lower Jurassic, Liaoning province. (From Young 1960.) **C,** *Kuangyuangpus szechuanensis* Young (1943). Middle Jurassic, Sichuan province. Bar is 5 cm long. **D,** *Xiangxipus chenxiensis* Zeng (1982). Upper Cretaceous (Jinjiang Formation), Hunan province. Bar is 5 cm long. **E,** *Shensipus tungchuanensis* Young (1966). Middle Jurassic (Lower Chiloo Group), Shaanxi province. (No scale.) **F,** *Schizograllator xiaohebaensis* Zhen, Li, and Rao (1986). Lower Jurassic (Lower Fenjiahe Formation), Yunnan province. Bar is 5 cm long. **G,** *Changpeipus carbonicus* Young (1960). ?Lower-Middle Jurassic, Liaoning province. (No scale.)

a junior synonym of *Grallator*; the name *Jeholosauripus* should thus be rejected. The footprints described appear to differ from other *Grallator* species; thus we suggest that a separate specific assignment be retained, using the original specific name *s–satoi* with *Grallator* in a new combination. Ichnogenus *Schizograllator* Zhen, Li, and Rao 1986

　　Schizograllator xiaohebaensis Zhen, Li, and Rao 1986
　　　　1986 *Schizograllator xiaohebaensis* n.g., n.sp. Zhen, Li, and Rao, p. 16–17.

Diagnosis. Bipedal, digitigrade with three clawed toes (II, III, IV). Neither impression of hallux, toe V, nor tail are present. The relative lengths of pes and pace are in a ratio of about 1:4.3. Angles between toes are large:

II–30°–III–45°–IV. The phalangeal pads are elliptic. The creases are wide. The length of the type specimen, a track of the left pes, is 28.0 cm, and 30.0 cm wide. The pace is 120 cm.

Material. (Figs. 19.2A, 19.4F) Part of a trackway consisting of eleven footprints (casts in the Beijing Museum of Natural History, No. BPV–FP4).

Horizon and Locality. Lower Fenjiahe Formation, Lower Jurassic of Jinning, Yunnan province (Fig. 19.1, locality 1).

Discussion. According to Baird's (1957) classification of theropod footprints, the skeletal characters of BPV–FP4 are very similar to those of *Gigandipus caudatus*. However, *Gigandipus* is much larger than BPV–FP4 and has a hallux and tail trail. This specimen differs from *Anchisauripus* in that it has four digits, and from *Grallator* in that digit III is more than 50% longer than the lateral digits.

Schizograllator is very similar to *Kayentipus* (Welles 1971) from the Kayenta Formation of Arizona (Fig. 19.2B). Haubold (1984, p. 48) interpreted this genus as a theropod track.

Ichnogenus *Paracoelurosaurichnus* Zhen, Li, and Rao 1986
  Paracoelurosaurichnus monax Zhen, Li, and Rao 1986
    1986 *Paracoelurosaurichnus monax* n.g., n.sp. Zhen, Li, and Rao, p. 16.

Diagnosis. Bipedal, digitigrade, with three clawed toes (II, III, IV). Impressions of toe V and hallux are absent. The outline of the foot is compact. The length of the footprint is 25.5 cm (digit III) and the width is 19.0 cm. The angles between the digits are II–30°–III–10°–IV.

Material. (Fig. 19.3B) Two sole footprints (Beijing Museum of Natural History, No. BPV–FP2).

Horizon and Locality. Lower Fengjiahe Formation, Lower Jurassic of Jinning, Yunnan province (Fig. 19.1, locality 1).

Discussion. The characters of this specimen are similar to Anchisauripodidae (Lull 1904), which has nine genera, but BPV–FP2 differs from these genera in the following ways. *Anchisauripus* usually has a hallux trace; *Grallator* is much smaller in size; *Saltopoides* has a lower ratio of pace to pes length, digits II and III are crooked, and there is no trace at the center of the footprints; *Stenonyx* is considerably smaller; *Swinnertonichnus* is plantigrade with a web trace; *Talmontopus* is also plantigrade; *Wildeichnus* has a hallux trace; *Coelurosaurichnus* is smaller in size and digits significantly thicker.

Coelurosauria indet.
  Ichnogenus *Laiyangpus* Young 1960.
    Laiyangpus liui Young 1960.
      Laiyangpus liui n.g., n.sp. Young, p. 64–65.
      Laiyangpus liui Kuhn, p. 101.
      Laiyangpus liui Haubold, p. 101, 119.

Diagnosis. Small footprints. Manus with three digits and pes with four digits. Apparently the first digit of the pes and first and fifth of the manus are not preserved. Digits slender and sharply pointed, roughly parallel. Tail impression present. The manus is 1.2–1.9 cm long and 1.7–2.3 cm wide. The pes is 1.8–2.2 cm long and 2.1–2.6 cm. wide.

Material. A slab with numerous prints (Institute of Vertebrate Palaeontology and Palaeanthropology, No. V2471).

Horizon and Locality. Laiyang Series, Upper Jurassic of Laiyang, Shandong province (Fig. 19.1, locality 3).

Discussion. The general structure is very similar to *Batrachopus* (*Kuangyuanpus*) particularly with regard to the parallel position, the sharpness and slenderness of the digits. Hence a non–dinosaurian affinity is suggested. *Laiyangpus* is only 30% the size of *Batrachopus* (*Kuangyuanpus*) *szechuanensis*.

Ichnogenus *Shensipus* Young 1966.
  Shensipus tungchuanensis Young 1966.
    1966 *Shensipus tungchuanensis* n.g., n.sp. Young, p. 68–71.
    1971 *Shensipus tungchuanensis* Haubold, p. 73.
    1983 *Shensipus tungchuanensis* Zhen, Li, and Zhen, p. 8.
    1984 *Shensipus tungchuanensis* Haubold, p. 49, 184.

Diagnosis. Tridactyl with slender digits. Digit II and IV considerably divergent. The angles between the digits are II–32°–III–58°–IV. Tips of digits with distinct pellet–like claws. Heel small. The length of the footprints is 8.1–9.8 cm, and the width is 8.9–9.1 cm. The pace is 9.7 cm.

Material. (Fig. 19.4E) Two print impressions on a slab of fine–grained sandstone (Institute of Vertebrate Palaeontology and Palaeanthropology, No. V3229).

Horizon and Locality. Lower Chiloo Group, Middle Jurassic, near Jiaoping Coal Mine, Tungchuan in Shaanxi province (Fig. 19.1, locality 4).

Discussion. *Shensipus* is similar to *Talmontopus* but is much smaller and has no web traces. Haubold (1984, p. 9) interpreted this specimen as a theropod track. Lockley et al. (1986) compared this ichnogenus to a trackway in the Morrison Formation in Colorado.

General Discussion of Anchisauripodidae. We recognize that the six ichnospecies discussed above may be fruitfully compared with a number of other known ichnotaxa. Such comparisons and the synonymies that might result are considered premature at this stage. Many of the classic North American ichnotaxa require revision first.

Infraorder Carnosauria
  Ichnofamily Eubrontidae
    Ichnogenus *Eubrontes* Hitchcock 1845
      Eubrontes platypus Lull 1904
        1858 *Amblonyx giganteus* Hitchcock, p. 71
        1904 *Eubrontes platypus* Lull, p. 512.
        1915 *Eubrontes platypus* Lull, p. 199.
        1953 *Eubrontes platypus* Lull, p. 181.
        1963 *Eubrontes platypus* Kuhn, p. 79, 83.
        1971 *Eubrontes platypus* Haubold, p. 75.
        1986 *Eubrontes platypus* Zhen, Li, and Rao, p. 9.

Diagnosis. Angles between pes digits: II–12°/17°–III; III–20°/22°–IV. Maximum length is 27 cm. II is 14–16 cm; III is 17–17 cm; IV is 20–21.5 cm. Length of pace is 100–120 cm.

Material. Nine footprints in one trackway (Beijing Museum of Natural History, No. BPV-FP5).

Horizon and Locality. Lower Fenjiahe Formation, Lower Jurassic of Xiyan, Jinning, Yunnan province (in Beijing Natural History Museum) (Fig. 19.1, locality 1).
Ichnogenus *Youngichnus* Zhen, Li, and Rao 1986
  Youngichnus xiyangensis Zhen, Li, and Rao 1986.

Diagnosis. Bipedal, digitigrade with three clawed toes (II, III, IV). Impressions of hallux, toe V, and tail are absent. Relative lengths of pes and pace are approximately 1:4.3. The step angle is about 156°. Digit III is more slender than II and IV. Digit IV closely parallels III. The phalangeal pads of toe II and IV are elliptical, with the width greater than the length. The length is 26.0–27.0 cm and the width is 16.0–19.0 cm. The pace is 110.0–114.0 cm and the stride is 219.0 cm. The angles between the digits are II–10°–III–11°–IV.

Material. (Fig. 19.3F) Part of a trackway consisting of three successive prints (Beijing Museum of Natural History, No. BPV-FP6).

Horizon and Locality. Lower Fengjiahe Formation, Lower Jurassic of Jinning, Yunnan province (Fig. 19.1, locality 1).

Discussion. Although all the pads of traces PBV-FP6 are not clear, the pad traces of lateral digits are, nonetheless, visible. It appears similar to the Eubrontidae, but digit III is thinner, crooked, and without pad traces.

Carnosauria indet.
  Ichnogenus *Changpeipus* Young 1960
    Changpeipus carbonicus Young 1960
      1960 *Changpeipus carbonicus* n.g., n. sp. Young, p. 59.
      1963 *Changpeipus carbonicus* Kuhn, p. 101.
      1971 *Changpeipus carbonicus* Haubold, p. 79, 119.
      1983 *Changpeipus carbonicus* Zhen, Li, and Zhen, p. 8
      1984 *Changpeipus carbonicus* Haubold, p. 51, 184

Diagnosis. Quadrupedal, tridactyl, semi-plantigrade. No trace of the hallux. About three to four times larger than *Grallator* (*Jeholosauripus*) *s-satoi*. The pes length is 29.2–38.3 cm and the width is 19.3–23.4 cm. The angles between the digits are II–26°–III–36°–IV. General outline is deltoid. Fourth digit projects farther than the second. Manus print present with digits I–III. Digit III is rudimentary. The manus length is 13.3 cm and the width is 7.0 cm. The angles between the digits are I–35°–II–18°–III.

Material. (Figs. 19.3D, 19.4G) Three prints on a slab (Institute of Vertebrate Palaeontology and Palaeanthropology, No. V2472, V2470).

Horizon and Locality. Lower or Middle Jurassic of Hebei, Liaoning and Jilin provinces (Fig. 19.1, localities 20, 2, 22, respectively).

Discussion. The structure of *Changpeipus* is similar to that of *Grallator*, apart from its larger size. *Changpeipus* also has a well-developed heel and a greater digit divergence. It is also larger than *Anchisauripus* and differs from *Eubrontes* by having a more compact heel and narrower digits. Haubold (1984, p. 51) considers *Changpeipus* a carnosaur track. The inferred manus and overall trackway configuration requires further study.

Ichnogenus *Xiangxipus* Zeng 1982
  Xiangxipus chenxiensis Zeng 1982
    1982 *Xiangxipus chenxiensis* n.g., n.sp., Zeng, p. 485.
    1983 *Xiangxipus chenxiensis* Zhen, Li, and Zhen, p. 8.

Diagnosis. Medium-sized and tridactyl. The length is 22.0 cm and the width is 21.5 cm. Length of digit II is equal to that of digit IV. Digit II and IV meet at a right-angle. Claws strong and sickle-shaped.

Material. (Figs. 19.2G, 19.4D) Five prints comprising a single trackway on a slab of sandstone (Hunan Geological Museum, Changsha, No. HV003-4).

Horizon and Locality. Jinjiang Formation, Upper Cretaceous of Xiangxi, Hunan province (Fig. 19.1, locality 5).

Discussion. An unusual trackway characterized by slender, sinuous digits in individual tracks. One footprint is anomalous because it appears tetradactyl, which is probably the result of a single digit (II) being moved after it first made contact with the ground, suggesting a very flexible digit (Fig. 19.2G).

Xiangxipus youngi Zeng 1982
    1982 *Xiangxipus youngi* n.g., n.sp. Zeng, p. 487.

Diagnosis. Similar to X. *chenxiensis* but smaller in size (12.0 cm long and 12.5 cm wide), and digit II longer than digit IV, and the hallux is closer.

Material. Two prints together with X. *chenxiensis* on the same slab (Hunan Geological Museum, Changsha, No. HV003-1, 2).

Horizon and Locality. See X. *chenxiensis*.

Ichnogenus *Hunanpus* Zeng 1982
  Hunanpus jiuguwanensis Zeng 1982.
    1982 *Hunanpus jiuguwanensis* n.g., n.sp., Zeng, p. 488.

1983 *Hunanpus jiuguwanensis* Zhen, Li, and Zhen, p. 8.

Diagnosis. Large tridactyl print. Proximal ends of toes are thick, thinning distally. Angle between II and IV is approximately 60°. The length is 33.2 cm and is 19.2 cm wide. Pellet–like claws.

Material. A print preserved together with *Xiangxipus* (Hunan Geological Museum, Changsha, No. HV003-8).

Horizon and Locality. See *Xiangxipus chenxiensis.*

Discussion. This specimen is preserved on the same slab as *Xiangxipus chenxiensis,* but the stride direction is at a 120° angle to the latter.

Hunanpus closely resembles *Platysauropus* (Ellenberger 1972), but is distinguished by sharper tips to the digits and the pellet–like claws. Furthermore, *Platysauropus* is Early Jurassic, thus considerably older than *Hunanpus.*

Suborder Ornithopoda
  Ichnofamily Iguanodontidae
    Ichnogenus *Sinoichnites* Kuhn 1958
      Sinoichnites youngi Kuhn 1958
        1929 Not named. Teilhard and Young, p. 132
        1958 *Sinoichnites youngi* n.g., n.sp. Kuhn, p. 24.
        1960 *Sinoichnites youngi* Young, p. 53.
        1963 *Sinoichnites youngi* Kuhn, p. 92, 101.
        1971 *Sinoichnites youngi* Haubold, p. 86.
        1983 *Sinoichnites youngi* Zhen, Li, and Zhen, p. 8.
        1984 *Sinoichnites youngi* Haubold, p. 56, 184.

Diagnosis. Digitigrade, bipedal, tridactyl print. Toe impressions are wide. Total length is 30 cm and width is 33 cm.

Material. (Fig. 19.2F) This, the first footprint found in China is now lost.

Horizon and Locality. Upper Jurassic of Shenmu, Shaanxi Province (Fig. 19.1, locality 6).

Discussion. This specimen was named 29 years after its discovery by Kuhn (1958) who regarded it as iguanodontid, as did Haubold (1984, p. 56).

Ichnofamily Anoemoepodidae Lull 1904
  Ichnogenus *Jialingpus* Zhen, Li, and Zhen 1983
    Jialingpus yuechiensis Zhen, Li, and Zhen 1983
      1983 *Jialingpus yuechiensis* n.g., n.sp. Zhen, Li, and Zhen p. 14, 16.

Diagnosis. Bipedal, semi–plantigrade tracks with occasional tail trace. Relative lengths of pes and pace are approximately 1:3.6. Pes tetradactyl with elongated metatarsal segment,

generally touching the ground with three clawed toes (II, III, IV). Hallux occasionally present, rotating sideways. Pad formula is 2–2–3–3. The metatarso–phalangeal pad of III is the longest of the three pads. The angles between digits are: I–40°/50°–II; II–20°/24°–III; III–24°/28°–IV. Claw impressions are weak. The length of the type specimen, a trace of the left pes, is 20.7 cm, and the distance between the tips of the lateral digits is 11.1 cm. There is also a 13.0 cm trace of a metatarsal behind the footprint.

Material. (Fig. 19.2D) A slab with 38 prints (Beijing Museum of Natural History, No. SCFP24).

Horizon and Locality. Upper Jurassic Penglaizhen Formation of Yuechi, Sichuan province (Fig. 19.1, locality 8).

Discussion. This is the first dinosaur track reported from China with a metatarsal impression. Although similar to *Anoemoepus* (Zhen et al. 1983, p. 15) (Fig. 19.2E) from the Lower Jurassic of the U.S.A., this form is the first of its type reported from the Upper Jurassic. Metatarsal impressions are generally rare in dinosaur tracks, but with this record they are now known from the Upper Triassic to Upper Cretaceous (see Fig. 0.3, this volume). The trackway configuration requires further study.

Ornithopoda indet.
  Ichnogenus *Yangtzepus* Young 1960
    Yangtzepus yipingensis Young 1960
      1960 *Yangtzepus yipingensis* n.g., n.sp. Young, p. 62.
      1963 *Yangtzepus yipingensis* Kuhn, p. 101.
      1971 *Yangtzepus yipingensis* Haubold, p. 89, 119.
      1983 *Yangtzepus yipingensis* Zhen, Li, and Zhen, p. 8.
      1984 *Yangtzepus yipingensis* Haubold, p. 51, 184.

Diagnosis. Plantigrade tridactyl print with three closely spaced digits. The manus length is 14.1 cm and the width is 10.2 cm. The angle between digits II and III is 75°. The lateral digits of the manus dinstinctly converge. The pes length is 29.0 cm and the width is 15.5 cm. The angle between digits II and IV is 81°. Lateral pes digits rather long but of subequal length. Digit III well separated from the hallux. Pad number in II is two, in III and IV is three. Possible skin impressions present showing coarse granulations.

Material. (Fig. 19.4A) Three prints (one manus and two pes) (Institute of Vertebrate Palaeontology and Palaeoanthropology, No. V2473).

Horizon and Locality. Lower part of the Upper Jurassic Chiating Series of the Kuanyin district, Yiping, Sichuan province (Fig. 19.1, locality 16).

Discussion. Young (1960), who first studied these tracks,

noted their peculiar structure and could not find comparable tracks. The pes print is similar to *Eubrontes* and *Otouphepus*, but it differs from the former by its quadrupedal posture and the parallel position of the digits. It differs from *Otouphepus* by the absence of claws and a web–like expansion between the digits. The trackway configuration requires further study.

Crocodylia indet.
    Ichnogenus *Batrochopus* Hitchcock 1845.
        Batrochopus szechuanensis comb. nov.
            1943 *Kuangyuangpus szechuanensis* n.g., n.sp.
                Young, p. 151–154.
            1958 *Kuangyuangpus szechuanensis* Kuhn, p. 25.
            1960 *Kuangyuangpus szechuanensis* Young, p. 54.
            1963 *Kuangyuangpus szechuanensis* Kuhn, p. 100.
            1971 *Kuangyuangpus szechuanensis* Haubold,
                p. 101, 119.
            1984 *Kuangyuangpus szechuanensis* Haubold,
                p. 184.

Diagnosis. Quadrupedal gait, semi–plantigrade, manus with three parallel arranged digits. Digit lengths: II, 2.7–3.6 cm; III, 3.8–4.8 cm; IV, 4.5–5.4 cm. Pes tetradactyl, toes parallel. Length of pes digits: I to IV, 4.5, 7.5, 8.4, and 8.4 (about one–third larger than manus). Depression is weak.

Material. (Figs. 19.2C, 19.4C) Slab with four prints (East China Geological Exhibition Hall, Nanjing, No. C161).

Horizon and Locality. Middle Jurassic, Guangyuan in Sichuan province (Fig. 19.1, locality 7).

Discussion. There is no doubt that this ichnospecies is referable to the ichnogenus *Batrochopus* Hitchcock 1845 which is well known from Lower and Middle Jurassic ichnofaunas in North America. Olsen and Padian (1986) have revised the ichnogenus *Batrochopus* and suggested that the trackmaker was probably a *Pseudosuchus*–like crocodilian which has been reported from the Lufeng fauna of Yunnan Province.

Reptilia indet.
    Ichnogenus *Zhengichnus* Zhen, Li, and Rao 1986
        Zhengichnus jinningensis Zhen, Li, and Rao 1986
            1986 *Zhengichnus jinningensis* n.g., n.sp. Zhen, Li,
                and Rao, p. 11, 17.

Diagnosis. Digitigrade pes with three digits (II, III, IV). Digit III is extremely long (approximately 20 cm). The impression of the posterior part of digit III is very narrow. Digits II and IV are more robust and shorter. The length is 28.0 cm and the width is 19.0 cm. The angles between the phalanx are large: II–50°–III–53°–IV. The outline of the print is similar to an inverted letter "T". No tail impression.

Material. (Fig. 19.3E) Single print (Beijing Museum of

Natural History, No. BPV–FP7).

Horizon and Locality. Lower Jurassic Lower Fengjiahe Formation of Jinning, Yunnan Province (Fig. 19.1, locality 1).

Discussion. This specimen resembles a gracile coelurosaur track with very short digits II and IV. Delair (1963) described a T–shaped track from the Purbeck (Upper Jurassic) of England, but, like the Chinese specimen, it is described from a single specimen.

Stratigraphic and Geographic Distribution of Footprints

Of the 30 provinces of China, 21 have yielded dinosaur fossils, and 14 have yielded dinosaur footprints, most of which come from Jurassic beds (Fig. 19.1). The oldest footprints, not yet formally described, have been found in the Xujiahe Formation (Upper Triassic), Sichuan, where two tetradactyl footprints, 33.8 cm long and 29.7 cm wide, have been found. An abundance of later, Jurassic dinosaur remains and footprints are known from Sichuan.

The greatest abundance of dinosaur footprints in China come from the Lower Jurassic, particularly from Chaoyang, Liaoning province (Fig. 19.1, locality 2), where some 4000 prints of *Grallator (Jeholosaurus) s–satoi* have been found (Yabe et al. 1940). The Lower Jurassic of Yunnan province has revealed several hundred footprints within three horizons of the Fengjiahe Formation including *Grallator limosus*, *Schizograllator xiaohebaensis*, *Paracoelurosaurichnus monax*, *Youngichnus xiyangensis*, *Zhengichnus jinningensis* (Zhen et al. 1986). *Eubrontes platypus* (Lull 1904) has also been found in this region. This ichnofauna is comparable with that of the Stormberg of South Africa and the Glen Canyon and Newark Groups in the U.S.A.

Middle Jurassic footprints are not common but are represented by *Batrochopus (Kuangyuangpus) szechuanensis* from Sichuan (Young 1943), *Shensipus tungchuanensis* from Shaanxi (Young 1966), and *Changpeipus carbonicus* from Jilin and Shanxi provinces (Young 1960) (Fig. 19.1, localities 7, 4, 22, 9).

The Upper Jurassic beds also do not have a prolific dinosaur ichnofauna, in contrast to the more abundant dinosaur bones that have been found. Taxa include *Laiyangpus liui* from Shandong (Young 1960), *Changpeipus luanpingeris* from Hebei (Young 1979), *Sinoichnites youngi* from Shaanxi (Kuhn 1958), *Jialingpus yuechiensis* from Sichuan (Zhen et al. 1983), and *Yangtzepus yipingensis* (Young 1960), also from Sichuan.

In 1984, footprints similar to *Changpeipus* were found in the Lower Cretaceous beds of Chabu, Nei–Monggol; later more than a thousand footprints were discovered at this locality (Li 1985) (Fig. 19.1, locality 10). Several types and sizes of footprints were found at this locality, including numerous tail impressions. Lower Cretaceous footprints have also been found in abundance in a fine–grained sandstone near Mt. Emei, Sichuan province (Zhen et al. 1987)

(Fig. 19.1, locality 11). This ichnofauna is interesting because it contains bird tracks (cf. *Aquatilavipes* Currie 1981) and a didactyl footprint provisionally attributed to a deinonychid (*Deinonychosaurichnus*, Zhen et al. 1987). These prints were recovered from the Jiaguan Formation which, according to Hao and Guan (1984), is Aptian–Albian in age; Chen (1983), however, regards this formation as early Late Cretaceous (Cenomanian to ?Senonian). Thus, these bird tracks are probably as old as those reported by Currie (1981), which are Albian and the oldest known.

Upper Cretaceous footprints are common, notably in Hunan province (Fig. 19.1, locality 5), where prints of *Xiangxipus chenxiensis*, *X. youngi*, and *Hunanpus jiguwanensis* have been found (Zeng 1982). Footprints have also been discovered, together with dinosaur eggs, in the Nanxiong Formation, Guangdong province (Fig. 19.1, locality 14), but have not yet been described. Possible mammal tracks have recently been reported from the Upper Cretaceous of Linhai County, Zhejiang province (Rao and Zhang 1987). The largest footprint known from China, and as yet unstudied (approximately 76 cm long), was discovered in the Upper Cretaceous Wangshi Formation in Jiangsu province (Fig. 1, locality 12). Small (10.0 cm long, 19.0 cm wide) Upper Cretaceous footprints have been reported in the Quwu Formation of Xizang (Tibet) (Fig. 19.1, locality 13). From sketches, these prints appear similar to an unnamed tridactyl footprint from a Mesozoic erratic block from Caithness, Scotland, reported by Sarjeant (1974, p. 282).

A single footprint from the Upper Cretaceous of the Xiaguan basin, Henan province (Fig. 19.1, locality 15), has been discovered in direct association with dinosaur eggs, and it is thought likely that the same animal was responsible for both the eggs and the footprint (Zhao 1979).

Conclusions

Traditionally, unique Chinese names (ichnogenera) have been erected for ichnotaxa even when they closely resemble those described in the Western literature. Recognition of the similarity between Chinese tracks and those from other parts of the world facilitates some preliminary correlation as suggested here. Many ichnogenera accommodate only a single species, and in future some may prove to be junior synonyms of existing ichnospecies. A significant number of ichnogenera appear to represent facultative bipeds which left occasional manus tracks. These require further study.

Tracks are now known for about 30 localities in China, including some large sites discovered recently. The potential for further ichnological studies is thus considerable.

Our preliminary survey suggests a preponderance of theropod tracks (12 ichnotaxa reported here) and a notable lack of sauropod and, to a lesser extent, ornithopod tracks (only three ichnotaxa reported), especially in Cretaceous rocks where they are commonly found elsewhere.

References

Baird, D. 1954. *Chirotherium lulli*, a pseudosuchian reptile from New Jersey. *Bull. Mus. Comp. Zool.* 3 (4):163–192.

1957. Triassic reptile footprint faunules from Milford, New Jersey. *Bull. Mus. Comp. Zool.* 117:449–520.

Chen, P. 1983. A survey of the non-marine Cretaceous in China. *Cretaceous Research* 4:123–143.

Currie, P. J. 1981. Bird footprints from the Gething Formation (Aptian, Lower Cretaceous) of Northeastern British Columbia, Canada. *Jour. Vert. Paleont.* 1:257–264.

Delair, J. B. 1963. Notes on Purbeck fossil footprints, with descriptions of two hitherto unknown forms from Dorset. *Proc. Dorset Nat. Hist. Archaeological Soc.* 84:92–100.

Ellenberger, P. 1972. Contribution à la classification des pistes de vertébrés du Trias: Les types du Stormberg d'Afri-que du Sud (1) *Paleovertebrata Mem. Extraord.* 1972:1–117.

Hao Y. and Guan S. 1984. The Lower–Upper Cretaceous and Cretaceous–Tertiary boundaries in China. *Bull. Geol. Soc. Denmark* 33:129–138.

Haubold, H. 1971. Ichnia amphibiorum et reptiliorum fossilium. *Handbuch der Palaoherpetologie* part 18 (Stuttgart: Gustav Fischer Verlag) 124 p.

1984. *Saurierfahrten.* (Wittenberg: Ziemsen Verlag) 229 pp.

Hitchcock, E. 1845. An attempt to name, classify and describe the animals that made the fossil footmarks of New England. *Proc. 6th Mtg. Amer. Assoc. Geol. Naturalists.* New Haven, Conn. pp. 23–25.

1858. *Ichnology of New England.* A report on the sandstone of the Connecticut Valley, especially its footmarks. (Boston: W. White) 220 pp.

Kuhn, D. 1958. *Die Fahrten der vorzeitlichen Amphibien und Reptilien.* (Bamberg Verslagshaus) 64 pp.

1963. Ichnia Tetrapodium. *Fossilim Catalogus I.*

Lockley, M. G., Houck, K., and Prince, N. K. 1986. North America's largest dinosaur trackway site: implications for Morrison Formation paleoecology. *Bull. Geol. Soc. Amer.* 97:1163–1176.

Li R. 1985. The discovery of a large expanse of dinosaur footprints on the Ordos plateau. *Vertebrata PalAsiatica.*

Lull, R. S. 1904. Fossil footprints of the Jura–Trias of North America. *Mem. Boston Soc. Nat. Hist.* 5:461–557.

1915. Triassic life of the Connecticut Valley *Connecticut Geol. Nat. Hist. Surv. Bull.* 24:1–285.

1953. Triassic life of the Connecticut Valley. Revised edition. *Connecticut Geol. Nat. Hist. Surv. Bull.* 81:1–336.

Olsen, P. E., and Padian, K. 1986. Earliest records of *Batrachopus* from the southwestern United States, and a revision of some early Mesozoic crocodylomorph ichnogenera. In Padian, K. (ed.). *The Beginning of the Age of Dinosaurs* (Cambridge University Press) pp. 259–273.

Rao C. and Zhang B. 1987. Footprints from Upper Cretaceous Tangshang Formation in Linhai, Zhejiang. *First Internat. Symp. Nonmarine Cretaceous Correlations* Abstracts, pp. 38–39.

Sarjeant, W. A. S. 1974. A history and bibliography of the study of fossil vertebrate footprints in the British Isles. *Palaeogeog., Palaeoclimat., Palaeoecol.* 16:265–378.

Shikama, T. 1942. Footprints from Chinchou, Manchoukuo, of *Jeholosauripus*, the eomesozoic dinosaur. *Central Natural Museum Manchoukou* 3:21–31.

Teilhard de Chardin, P., and Young C.C. 1929. On some traces of vertebrate life in the Jurassic and Triassic beds of Shansi and Shensi. *Bull. Geol. Soc. China* 8:131–133.

Welles, S. P. 1971. Dinosaur footprints from the Kayenta Formation of northern Arizona. *Plateau* 44:27–38.

Yabe, H., Inai, Y., and Shikama, T. 1940. Discovery of dinosaurian footprints from the Cretaceous (?) of Yangshan, Chinchou: preliminary note. *Proc. Imperial Academy Japan* 16:560–563.

Young C.C. 1943. Note on some fossil footprints in China. *Bull. Geol. Soc. China* 13:151–154.

1960. Fossil footprints in China. *Vertebrata PalAsiatic* 4:53–66.

1966. Two footprints from the Jiaoping Coal Mine of Tungchuan, Shensi. *Vertebrata PalAsiatica* 10:68–71.

1979. Footprints from Luanping, Hebei. *Vertebrata PalAsiatica* 17:116–117.

Zeng X. 1982. *Fossil handbook of Hunan Province.* (Geology Bureau of Hunan Province) [in Chinese].

Zhao Z. 1979. Discovery of the dinosauria eggs and footprints from Neixiang county, Henan Province. *Vertebrata PalAsiatica* 304–309.

Zhen S., Li J., and Zhen B. 1983. Dinosaur footprints of Yuechi, Sichuan. *Mem. Beijing Nat. Hist. Museum* 25:1–19.

Zhen S., Zhen B., and Rao C. 1986. Dinosaur footprints of Jinning, Yunnan. *Mem. Beijing Nat. Hist. Museum* 33:1–18.

Zhen S., Zhen B., Zhang B., Chen W., and Zhu S. 1987. Bird and dinosaur footprints from the Lower Cretaceous of Emei County, Sichuan. *First Internat. Symp. Nonmarine Cretaceous Correlations* Abstracts pp. 37–38.

VI Biostratigraphy

"La stratigraphie, également, peut utiliser les ichnites des vertébrés pours les datations relatives de plusieurs gisements, grâce aus ichnogenres et aux ichnospèces et à leurs fréquences relatives."

Courel, Demathieu and Gand
(1982 p. 313)

Although still not widely recognized by geologists and paleontologists as stratigraphic tools, vertebrate tracks have considerable biostratigraphic utility, and hold much promise of further potential in this area. The above quotation was translated, in the original publication, as follows: "Stratigraphy (also) uses ichnites for dating some layers with ichnogenera and ichnospecies and their ratios."

Until recently there was a perception that tracks were mainly useful when they occurred in strata otherwise devoid of biostratigraphically diagnostic zone fossils. As indicated in the Introduction, tracks are much more than a last paleontological resort. Like any evidence, they always add some significant information.

In ascending stratigraphic order, we can now cite a number of important "ichnostratigraphic" studies. Demathieu (this volume) and his colleagues have shown the considerable potential of the European Triassic sequences in shedding light on the origin of dinosaurs and the evolution of contemporaneous reptile faunas (Demathieu and Haubold 1978, Courel et al. 1982).

Subsequent work by Olsen and Galton (1984) has demonstrated the utility of tracks in elucidating the rise of Early Jurassic faunas worldwide. Similarly, the emergence of sauropod faunas can be traced through the Early and Middle Jurassic track record in North Africa (Ishigaki this volume) and Monbaron et al. (1985 ref. in Introduction).

The ichnological transition from Jurassic to Cretaceous is, as yet, poorly documented. However, Currie (this volume) has observed some distinct changes at this time, and the rise of large ornithopod trackmakers is clearly evident in Early Cretaceous ichnofaunas. Similarly it is possible to distinguish different ichnofaunas in Lower and Upper Cretaceous deposits. Studies by Parker and others (Chapter 39 and 40) for example indicate the considerable potential of Late Cretaceous ichnofaunas for systematic study. In most cases stratigraphically useful tracks occur as components of larger footprint assemblages, and need to be understood in their paleoecological context (Section V) as well as from a purely biostratigraphic viewpoint.

References

Courel, L., Demathieu, G., and Gand, G. 1982. Contribution apportée par l'ichnologie a quelques disciplines de la géologie. Extrait de livre jubilaire Gabriel Lucas, géologie sédimentaire. *Mem. Geol. Univ. Dijon, Inst. Sci. de la Terre* 515 pp.

Demathieu, G., and Haubold, H. 1978. Du probleme de l'origine des dinosauriens d'apres les donnees de l'ichnologie du Trias. *Géobios* 11 (3):409–412.

Olsen, P. E., and Galton, P. M. 1984. Review of the reptile and amphibian assemblages from the Stormberg of southern Africa, with special emphasis on the footprints and the age of the Stormberg. *Paleont. Afr.* 25:87–110.

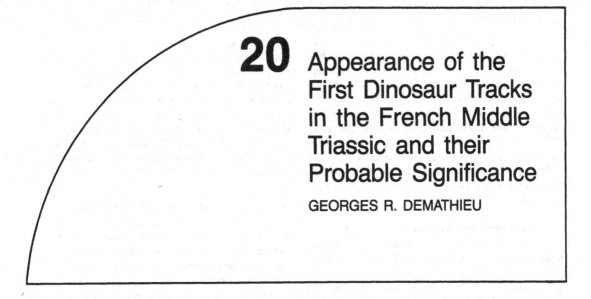

20 Appearance of the First Dinosaur Tracks in the French Middle Triassic and their Probable Significance

GEORGES R. DEMATHIEU

Abstract

The dinosauroid tracks described here come from the 'lower sandstones' unit of the Triassic eastern border of the French Massif Central.

These tridactyl footprints mark at least the functional reduction of the toes I and V. Very scarce in the oldest southern part of this border (Anisian), they become more common in the northern younger part (Lower Ladinian). This increasing frequency of tridactyl footprints through time and other paleontological considerations enable us to infer that their trackmakers were effectively dinosaurs of rather small size. These dinosaurs were not the ruling reptiles of this time; they were only 3 to 5% of the whole population, where the greatest part was formed by thecodonts, generally of larger size, and by numerous small lepidosaurs.

As they were active and swift predators, the dinosaurs continued to exist and even increase. They lived in a marginal biotope at the shore of a sea or lagoon that shows numerous primary sedimentary structures and invertebrate tracks or trails. If this interpretation is accurate, one can infer that dinosaurs (probably coelurosaurs) appeared in the Triassic at least as early as the Middle Anisian.

Brief Geological Setting

The dinosauroid tracks discussed herein come from the Triassic of the eastern and southeastern border of the French Massif Central. The outcrops form a discontinuous narrow zone located in the Saône–Rhône Valley from Dijon (Côte d'Or) to Alès (Gard) (Fig. 20.1) about 450 km long. Several horizons bearing ichnofaunules occur in the lower sandstones, the basal unit of the Triassic sequence (Demathieu 1985) referred to the Middle Triassic (e.g., Courel 1973; Demathieu 1984a,b).

The footprints are preserved as natural casts on the undersurface of gray or pink quartz sandstones with car-

bonate or siliceous cement, or as impressions on the upper surface of the strata. In the southern region, the thickness of this unit can be relatively high (c. 100 m). Conglomeratic at the bottom, these sandstones have locally a cross–bedded structure, and become horizontally bedded and fine grained toward the top. Footprints appear in these latter strata (cf. Lockley and Conrad, Fig. 14.2, this volume). Northward, where their thickness does not exceed 20 m, they show a general upward decrease in grain size and in bedding thickness. The footprint–bearing slabs are horizontally bedded and exhibit a peculiar graded bedding (Demathieu 1985).

Figure 20.1. Location of Middle Triassic deposits along the eastern and southeastern border of the Massif Central (France).

In each locality the distinct footprint horizons bear similar ichnofaunules consisting of vertebrate trackways, invertebrate trails, with primary sedimentary structures as well as plant impressions. The simultaneous presence of all these tracks together on the same levels assist with the interpretation of the depositional environments (Demathieu 1985). No body fossil has been discovered in these strata but the overlying carbonate unit has yielded some microfossil faunules and invertebrates (Roman 1926, Ricour 1962, Courel 1973, Doubinger and Adloff 1977, Adloff and Doubinger 1979).

The Vertebrate Footprints

The lower sandstones have furnished numerous fossil trackways belonging to (i) Ichnofamily Chirotheriidae (Abel 1935) with the ichnogenera *Synaptichnium* Nopcsa 1923, *Chirotherium* Kaup 1835, *Brachychirotherium* Beurlen 1950, *Isochirotherium* Haubold 1971, and *Sphingopus* Demathieu 1966 (Fig. 20.2), an ichnogenus characteristic of this region which suggests the functional reduction of toes I and V; (ii) Ichnofamily Rotodactylidae (Peabody 1948); (iii) Ichnofamily Rhynchosauroidae (Haubold 1966); and (iv) Ichnofamily Grallatoridae (= Anchisauripodidae) (Lull 1904). These families were probably (i) thecodont, (ii) thecodont, (iii) lepidosaurian, and (iv) dinosaurian, respectively.

Figure 20.2. *Sphingopus ferox* Demathieu 1966 shows a partial reduction of the toes I and V; cast of a right pes; the manus near the 2nd toe is poorly preserved. Slab 2.1.1. BQ, Chasselay, Rhône. (x 0.95)

The dinosauroid tracks are very scarce in the south and more common to the north. Their frequency with other paleontological considerations has allowed construction of a relative time scale between the different layers: Middle Anisian in Alès to Lower Ladinian in the north in Autun (Courel, Demathieu and Buffard 1968, Demathieu 1970, Courel and Demathieu 1976). Palynology studies have corroborated this point of view (Doubinger and Adloff 1977, Adloff and Doubinger 1979).

Dinosauroid Tracks and their Interpretation

The term 'dinosauroid' is used to avoid an a priori paleontological attribution of tridactyl tracks. These tridactyl footprints are traces where only the toes II, III, IV have impressed the sediment. Sometimes the imprint of the hallux is present. But a tridactyl footprint may be an incomplete pentadactyl track as seen among Chirotheriidae, an underprint of a pentadactyl trackmaker, or it may have been made by a reptile other than a dinosaur. In the case in point, undertracks are ruled out because of (i) the thickness of the footprint–bearing or overlying layers, or (ii) the presence of other different imprints above or below the footprint surfaces, and (iii) the sharp intersection between the surface of the imprint and that of the sediment. A true dinosaur footprint indicates a functional anatomical reduction of toes I and V, even if the hallux is poorly impressed as in the case of *Grallator* and *Gigandipus*. This characteristic reduction is only observed among the archosaurs, essentially the dinosaurs. Nevertheless, some crocodiles like *Trialestes* and thecodonts like *Saltoposuchus* (?) could have made such footprints.

Our problem is to know whether the Middle Triassic dinosauroid footprints were made by dinosaurs: thecodonts like *Hesperosuchus* and *Ornithosuchus* have been figured in a bipedal motion with only three toes making their imprints in the soil. But I think that these reptiles may have been the trackmakers of the ichnogenus *Sphingopus*. *Trialestes*, as far as it is known, seems to have been a very peculiar crocodile. On the other hand, ornithosuchids and crocodiles like *Trialestes* appear during the Upper Triassic, whereas some dinosaur skeletons are known as early as the end of the Middle Triassic (Bonaparte 1978). Besides, we must take into account the regular and continuous increase of dinosauroid tracks from the Middle Anisian to the top of the Triassic.

All the considerations stated above lead me to infer that the dinosauroid tracks of these areas were made by dinosaurs. The fact that no dinosaur skeleton has been discovered in lower and median Middle Triassic is another problem. On the one hand, as far as I know, no tetrapod skeleton has been discovered in the trackway areas, and, on the other hand, the conditions of formation and preservation of the footprints are very different from those of the preservation of vertebrate carcasses.

The Dinosaur Trackways

The known dinosaur tracks of this region (21 track-

ways, 80 measured footprints) belong to two ichnogenera
Coelurosaurichnus Huene 1941 (Figs. 20.3, 20.4); *Grallator*
Hitchcock 1858 (= *Anchisauripus* Lull 1904) (Figs. 20.5, 20.6
Others not very numerous and poorly known have no
been named. The ichnogenus *Grallator* is well represente
by numerous small footprints and contains here only on
species *Grallator* (= *Anchisauripus*) *bibractensis*, relativel
common in the north and scarce in the south (Demathie
1970, 1977; Demathieu and Gand 1972).

The difference between the *Coelurosaurichnus* an
Grallator ichnogenera can lie in the fact that digits II an
IV are more important relative to digit III i
Coelurosaurichnus than in *Grallator*, i.e., the prolongatio
of the third digit beyond the tips of digits II and IV is les
in *Coelurosaurichnus* than in *Grallator*. Furthermore, th
manus is sometimes present on *Coelurosaurichnus* trackway
(Demathieu and Gand 1981, Demathieu 1985). To thes
two ichnogenera must be added asymmetric tridactyl foo
prints (Fig. 20.7) which are rare and known only in th
Lyon–Mâcon region. These traces are comparable to thos
of the dinosaur footprints of Wallis, Switzerland (Demathie
and Weidmann 1982) from the top of the Ladinian or th
base of the Carnian. Other dinosaur footprints of the Lowe
Ladinian of Nottinghamshire have been described by Sa

jeant (1967). We can see with these examples how difficult
it can be when the ichnologist has to deal with tridactyl
dinosaur footprints. They show few quantitative or
morphologic characters, much fewer than in chirotheroid
pentadactyl footprints where the manus furnishes additional
information.

Paleoecological Inferences

One of the characteristics of these dinosaur footprints lies in their small size. The biggest is 10.3 cm long
and 8.5 cm wide. The trackmakers of *Coelurosaurichnus* and
Grallator were probably coelurosaurs of small and very small
size, although no coelurosaur body fossils are known at this
age. Prosauropods could not be the trackmakers because
their feet do not fit these tracks. The *Coelurosaurichnus* and
Grallator trackmakers were probably carnivores and
predators. Their minority status is based on the trackway
census in areas of Middle Triassic outcrop where they comprise 3 to 5% of the whole set of imprints. Most numerous
were the lepidosaurs of the ichnogenus *Rhynchosauroides* and
small thecodonts represented by *Rotodactylus*, followed by
large thecodonts of the ichnogenus *Isochirotherium*, which
crossed in small herds. Around these gregarious and herbivorous potential prey lived other thecodonts, represented
by the ichnogenera *Synaptichnium* and *Brachychirotherium*.
These could have belonged to the Aetosauria for the former
and Sphenosuchia for the latter. Perhaps they were omniv-
orous or predators of small animals only. The ornitho-

Figure 20.3. *Coelurosaurichnus perriauxi* (Demathieu
and Gand 1972); cast of a right pes; the manus, near
the 4th toe, is poorly impressed. Slab 2.1.8. AQ,
Autun, Saône-et-Loire (x 0.88)

Figure 20.4. *Coelurosaurichnus largentierensis* (Courel
and Demathieu 1976); trackway. Largentière mines,
Ardèche. (Scale: 50 cm)

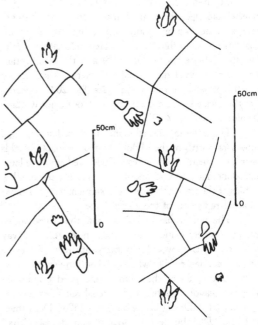

Figure 20.5. *Anchisauripus bibractensis* (Demathieu 1971); cast of a left pes. Slab 2.1.17. C, Largentière, Ardèche. (x 1.2)

suchian trackmakers of *Sphingopus* were in competition with the dinosaurs. The whole set of predators represents about 10%, the so-called omnivores 15%, herbivores 65%, and microvertebrates 5%. Undetermined animals constitute the remaining 5% (Demathieu 1985).

The numerous primary sedimentary structures and invertebrate lebensspuren demonstrate vast beaches between water, sea or lagoon, and dry land covered with vegetative zones in these trackway areas (Demathieu 1985). The invertebrate tracks (Fig. 20.8) and the sedimentary structures give evidence of shallow waters at their border and the presence of salt crystal casts (Fig. 20.9) indicates that these waters were probably brackish or marine (Courel, Demathieu and Gall 1979).

The existence of small numbers of medium or small dinosaurs as early as the Middle Anisian probably heralds their future increase and spread. They lived among a large thecodont fauna with advanced characters, like the ornithosuchids or the trackmakers of *Isochirotherium*, which allowed a partial reduction of the external toe.

The numerous trackways show that these widespread biotopes, including sea or lagoon, shores, trackway areas and dry land, would have been particularly attractive for archosaurian reptiles whose locomotion was easy and swift. Among them, thecodonts were good quadruped walkers, occasional runners with bipedal gait (*Chirotherium*, *Isochirotherium*) (Demathieu and Haubold 1982, Demathieu and Leitz 1982), but in this mode bipedal dinosaurs had

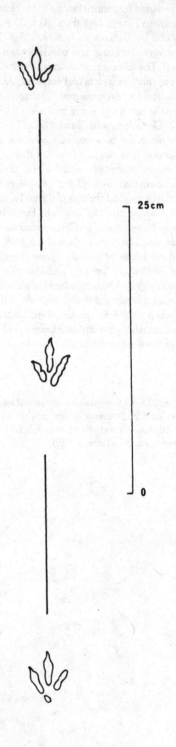

Figure 20.6. *Anchisauripus bibractensis* (Demathieu 1971); trackway. Autun, Saône-et-Loire. (Scale: 25 cm)

Figure 20.7. Dinosauroid asymmetrical footprint, concave epirelief. Slab 2.1.1. JO, Chasselay, Rhône (x 0.5)

Figure 20.8. Invertebrate trail; *Planolites* with the cast of a right pes *Isochirotherium coureli*. Slab 2.1.1. FJ, Chasselay, Rhône. (x 0.5)

Figure 20.9. Salt crystal casts. Slab 2.1.2. XA20, Autun, Saône–et–Loire. (x 0.45)

a higher efficiency. This explains the decrease in frequency of chirotheroid tracks from the Lower Middle to the Upper Triassic and the complementary increase in frequency of dinosauroid tracks during this time. The competition between the two groups of reptiles must have been relatively great and, at the end, thecodonts vanished from these landscapes. The part played by locomotion was prominent in this rivalry for the conquest of these territories.

Ichnological information is different from the osteological data. The skeletal record shows that the number of thecodont taxa, as well as that of dinosaurs, increases in the Upper Triassic. But it must not be forgotten that archosaur footprints give evidence of life in particular biotopes, whereas skeletons furnish generally little or no information about life environments because the carcasses could have been transported before fossilization (Lockley 1986).

Conclusion

I have written (Demathieu 1981) that the ichnological information does not have the same chronological significance as the information provided by vertebrate body fossils, because the formation and preservation of tracks and traces depends essentially on the nature and physical state of the sediment, whereas the presence of skeletons is essentially dependent more on the number in living populations. It is also possible to know animals by their footprints before (or after) knowing them by their skeletons.

The rise of dinosaurs, according to ichnological information, seems to have been relatively swift. Very scarce in the Middle Anisian, their tracks increase in proportion during the median Middle Triassic and exceed the number of chirotheroid trackways as early as the base of the Upper Triassic. Later, they become more numerous, and dinosaurs are the ruling reptiles of these biotopes.

The appearance of tridactyl bipeds, the carnivorous coelurosaurian dinosaurs, as early as the lower Mid-Triassic, with well marked, specialized features, shows that these reptiles already had a long evolution behind them, and it is possible that some chirotheroid tracks from the Lower Triassic could have been made by dinosaurs or predinosaurs (Demathieu and Haubold 1978). Consequently the characters common to dinosaurs and thecodonts lead one to conclude that these two groups had a common origin rather than dinosaurs evolving from thecodonts, an origin which could be among the eosuchians.

Acknowledgments

I am very much indebted to the editors for their helpful contribution to the final writing of this paper, especially to David D. Gillette, who has kindly accomplished a thoroughly thought-out revision of the original manuscript; my sincere thanks to all. Many thanks also to Liliane Gallet-Blanchard, University of Paris, for her friendly participation; to Annie Bussière for the drafting, and Alain Godon for preparing some photographs, Centre des Sciences de la Terre, Dijon; finally to Germaine J. Demathieu-Mallet, not only for her typing but also for her significantly constructive comments and suggestions.

References

Abel, O. 1935. *Vorzeitliche Lebensspuren.* (G. Fischer Verlag) 644 pp.

Adloff, M.-C., and Doubinger, J. 1979. Etude palynologique dans le Mesozoique de base de la bordure N.E. du Massif Central français. 7e R. A. S. T., *Société Géologique de France*, ed., 1 p.

Beurlen, K. 1950. Neue Fährten Funde aus dem Fränkischen Trias. *Neues Jahrb. Geol. Paläont.* 308-320.

Bonaparte, J. F. 1978. El mesozoico de America del sur y sus Tetrapodos. *Opera Lilloana* 26: 596 pp.

Courel, L. 1973. Modalités de la transgression mésozoïque; Trias et Rhétien de la bordure Nord et Est du Massif Central français. *Société Géol. France* Mémoire 118, LII, 152 pp.

Courel, L., Demathieu, G. R., and Buffard, R. 1966. Empreintes de pas de vertébrés et stratigraphie du Trias. *Société Géol. France* X:275-281.

Courel, L., and Demathieu, G. R. 1976. Une ichnofaune reptilienne remarquable dans le grès triasiques de Largentière (Ardèche, France). *Palaeontographica* Abt. A, 151:195-216.

Courel, L., Demathieu, G. R., and Gall, J.-C. 1979. Figures sédimentaires et traces d'origine biologique du Trias moyen de la bordure orientale du Massif Central. Signification sédimentologique et paléoécologique.

Géobios 12 (3):379-397.

Demathieu, G. R. 1966. *Rhynchosauroides petri* et *Sphingopus ferox*. Nouvelles empreintes des grès triasiques de la bordure Nord-Est du Massif Central. *Académie des Sciences*, Paris D 263:483-486.

1970. Les empreintes de pas de vertébrés du Trias de la bordure N.E. du Massif Central. *Cahiers Paléont.* C.N.R.S. ed., 211 pp.

1971. Cinq nouvelles espèces d'empreintes de reptiles du Trias de la bordure N.E. du Massif Central. *Académie des Sciences Paris* D 272, 812-814.

1977. La Palichnologie des vertébrés. Développement récent et rôle dans la Stratigraphie du Trias. *Bull. Bur. Recherches Géolog. Minières* 2e série, section IV, 3:269-278.

1981. Comparaison des informations fournies par l'ichnologie des vertébrés et par la paléontologie ostéologique dans le domaine de la chronologie. *Bull. Scientifique Bourgogne* 34:5-12.

1984a. Les pistes de vertébrés du Trias du Sud-Est de la France. Apport à la stratigraphie. *Bull. Bureau Recherches Géolog. Minières* 1-2:175-177.

1984b. Paleoichnologie. In Synthèse Géologique du Sud-Est de la France. *Mémoire Bureau Recherches Géolog. Minières* n° 125:63-64.

1985. Trace fossil assemblages in Middle Triassic marginal marine deposits. Eastern border of the Massif Central, France. In Curran, H. A. (ed.). *Biogenic Structures. Their use in interpreting depositional environments.* S.E.P.M. special publication n°35:53-66.

Demathieu, G. R., and Gand, G. 1972. Les pistes dinosauroides du Trias moyen du Plateau d'Antully et leur signification paléozoologique. *Bull. Société Histoire Naturelle d'Autun* 62:2-18.

1981. Interprétation paléoécologique de traces d'origine biologique et mécanique observées dans la carrière traisique de Pont d'Argent, Saône-et-Loire (France). *Bull. Société Histoire Naturelle d'Autun* 98:3-22 and 99:19-34.

Demathieu, G. R., and Haubold, H. 1978. Du problème de l'origine des dinosauriens d'après les données de l'ichnologie du Trias. *Géobios* 11 (3):409-412.

1982. Reptilfährten aus dem mittleren Buntsandstein von Hessen (B.D.R.). *Hallesches Jahrb. Geowissenschaften* 7:97-110.

Demathieu, G. R., and Leitz, F. 1982. Wirbeltier-Fährten aus dem Röt von Kronach (Trias, Nord-Ost-Bayern). *Mitteilungen Bayerischen Staatssammlung Paläont.* 22:63-89.

Demathieu, G. R., and Weidmann, M. 1982. Les empreintes de pas de reptiles dans le Trias du Vieux Emosson (Finhaut, Valais, Suisse). *Ecologae Helveticae* 75 (3):721-757.

Doubinger, J., and Adloff, M.-C. 1977. Etudes palynologiques dans le Trias de la bordure S-E. du Massif Central français (Bassin de Largentière, Ardèche). *Bull. Sciences Géolog. Strasbourg* 30 (1):59-74.

Haubold, H. 1966. Therapsiden- und Rhynchocephalen-Fährten aus dem Butsandstein Südthuringens. *Hercynia* 3 (2):147-183.

1971. Ichnia Amphibiorum et Reptiliorum fossilium. *Hdbuch Paläoherpet.* Teil 18 (G. Fischer Verlag) 124 pp.

Hitchcock, E. 1858. *Ichnology of New England. A report on the* sandstone of the Connecticut Valley, especially its fossil footmarks. (Boston: William White). 220 pp.

Huene, F. v. 1941. Eine Fährtenplatte aus dem Stubensandstein der Tübinger Gegend. *Zentralblatt Mineral., Geolog., Paleont.* B:138–141.

Kaup, J. J. 1835. Mitteilung über Tierfährten von Hildburghausen. *Neues Jahrb. Geognosis, Geologie, Petrographie* 327–328.

Lockley, M. G. 1986. The paleobiological and paleo-environmental importance of dinosaur footprints. *Palaios* 1:37–47.

Lull, R. S. 1904. Fossil footprints of the Jura-Trias of North America. *Mem. Boston Soc. Nat. Hist.* 5:461–557.

Nopsca, F. v. 1923. Die Familien de Reptilien. *Fortschritte Geol. Palaeont.* 2:1–210.

Peabody, F. E. 1848. Reptile and amphibian trackways from the Lower Triassic Moenkopi Formation of Arizona and Utah. *Univ. California Bull. Dept. Geol. Sciences* 27 (8):295–468.

Ricour, J. 1962. *Contribution à une révision du Trias français. Mémoire pour servir à l'explication de la carte geologique de France.* (Imprimerie Nationale Paris ed.) 471 pp.

Roman, F. 1926. *Géologie lyonnaise.* (Presses Universitaires de France ed.) 356 pp.

Sarjeant, W. A. S. 1967. Fossil footprints from the Middle Triassic of Nottinghamshire and Derbyshire. *The Mercian Geologist* 2 (3):327–341.

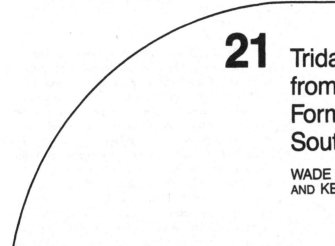

21 Tridactyl Trackways from the Moenave Formation of Southwestern Utah

WADE E. MILLER, BROOKS B. BRITT
AND KENNETH L. STADTMAN

Abstract

A previously unreported dinosaur trackway site was discovered in Warner Valley of southwesternmost Utah in 1982. 161 separate prints representing 23 trackways occur on the upper parting surfaces of three separate fine-grained, well-sorted sandstone units in the Dinosaur Canyon Member of the Moenave Formation. Recent literature indicates that the age of this formation (Late Triassic and/or Early Jurassic) has not yet been resolved, although the latter time appears most probable.

Many of the tracks are distinct. Some show pad impressions and claw marks, suggesting the dinosaurs were walking on damp sediment. A few prints, especially those of the easternmost trackway-bearing horizon, imply that sediments were water saturated at the time they were made. While the trackways lead in several directions, a faint trend is to the southwest. At the main site two bipedal dinosaurs are indicated: *Grallator*, a coelurosaurid which might represent *Coelophysis*; and *Eubrontes*, a possible plateosaurid. Four trackways can be attributed to the latter and 15 to the former. No tail drag marks for either are in evidence. A single complete track and part of another, representing a third type of dinosaur, probably a megalosaurid, were also found in the same formation. They, however, lie about 100 m northwest of the main site and 12 m higher stratigraphically.

Introduction

In the spring of 1982 dinosaur footprints were discovered in southwestern Utah by Mr. Gary Delsignore of Cedar City. He contacted Dr. Blair Maxfield, a geologist at Southern Utah State College, who in turn informed one of us (WEM) of the discovery. In May, 1982, the senior author and one of his graduate students, Samuel Webb, investigated the site making preliminary evaluations. Since that time, the authors have made several trips to the locality in an attempt to gather all relevant data relating to the tracks, as well as to obtain geologic and paleoenvironmental information.

Until recently tracks and trackways have not received much serious scientific attention, especially those in the western United States (Lockley 1986). Although dinosaur tracksites in the eastern half of Utah are not uncommon, very few have been reported from the western half of the state. The site discussed in this paper, from the Moenave Formation, appears to be the oldest exclusively dinosaur tracksite thus far identified in Utah. Of the very limited dinosaur tracks in the western part of the state, Stokes and Bruhn (1960) identified two tracksites from the Kayenta Formation of the Zion National Park area. Other tracksites in this region have been discovered, but thus far remain unreported or unstudied. This paper presents a significant dinosaur trackways find, and includes descriptions of exposed track types, the geological and paleoenvironmental setting and interpretations regarding the dinosaurs which made them. However, it is beyond the scope of the present article to delve deeply into ichnological taxonomy.

Location

The tracksite discussed in this paper occurs in the southwestern corner of Utah in Washington County (Fig. 21.1). It is situated in Warner Valley about 18 km south-southwest of the town of Hurricane and 2.5 km north of the Arizona border in the NW 1/4, section 30, T43S, R13W at an elevation of 963 m. The Hurricane Cliffs, a recent fault scarp of major magnitude, lies approximately 5.5 km to the east. With the exception of two isolated tracks, the majority of trackways exist in the bottom of a wash in a sandstone unit at the southern base of a cuesta named Sand Mountain. The lower two exposed trackway horizons cover an area about 60 m long (east–west) by 12 m wide (north–south). Unexposed tracks, which undoubtedly exist, would lie mostly north and south of the exposed

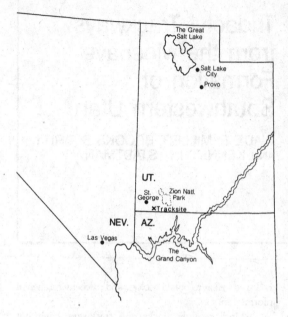

Figure 21.1. Locality map. The x indicates the location of the trackway site in the southwestern corner of Utah.

ones where the fossil-bearing units are covered. The two isolated tracks occur about 100 m up the same wash but on its north upper bank rather than at the bottom (SW 1/4, section 19, T43S, R13W).

Geologic Setting and Age

All the dinosaur tracks and trackways composing the present locality exist within the lowest member (Dinosaur Canyon member) of the Moenave Formation. Along with other formations in the area, mostly of the Glen Canyon Group (Fig. 21.2), it constitutes part of a major structural feature in this region of Utah, the Virgin Anticline. The main tracksite exists in the bottom of a wash (Fig. 21.3) at the base of a cuesta (Fig. 21.4). Trackways here lie approximately 25 m above the top of the unconformably underlying Chinle Formation. The closest contact between the Moenave and Chinle can be seen approximately 1 km southwest of the tracksite.

Strata exposed on the southern (steepest) face of the Sand Mountain cuesta show a north to northwest dip ranging from 6° to 12°. The wash bearing the tracksite extends northwest up the face of the cuesta from the track locality, and has been developed along the trace of a normal fault.

Originally the Dinosaur Canyon member of the Moenave was named as a separate formation (Dinosaur Canyon Sandstone) based on an area just east of Cameron, Arizona (Colbert and Mook 1951). Later Harshbarger, Repenning and Irwin (1957) named the Moenave Formation from a site immediately east of Tuba City, Arizona, and included the Dinosaur Canyon Sandstone as the lowest member of their new formation. This assignment has generally been followed to the present time.

At the Warner Valley tracksite locality the stratigraphy of the Dinosaur Canyon member consists of alternating layers (25–50 cm) of fine-grained, well-sorted sandstone, siltstone and shale/mudstone. Mostly the units are reddish-brown, but some light-colored layers also exist. The unit which contains the majority of footprints consists of a fine-grained, well-sorted reddish sandstone 25 to 30 cm thick. This sandy layer exhibits micro-crossbedding and a mottling of white inclusions high in $CaCO_3$. The layers immediately above and beneath the track-bearing unit are each reddish brown shale/mudstone. Neither of these show any footprints. Two layers beneath the main dinosaur track layer is a 70 to 100 cm thick, fine-grained, well-sorted sandstone, light gray to white in color. A limited exposure of this bed (about 11 m²) in the wash a few meters east of the main trackways displays at least four short trackways and several isolated footprints. The two isolated tracks that come from the bank of the wash about 100 m upstream lie 12 m stratigraphically above the main tracksite. Repeating the condition of the other tracks, these also are impressed on the top of a fine-grained, well-sorted sandstone, reddish-brown in color, and overlain by a shale/mudstone unit of nearly the same color. The sandstone measures 40 to 45 cm in thickness. It contains a few abraded bone chips. All the sediment layers show slightly undulating surfaces with some minor cut and fill structures. Apparently they represent unconformities involving minor breaks in the depositional record.

The Moenave Formation was originally considered to be *questionably* of Late Triassic age (Harshbarger, Repenning and Irwin 1957). Others, though, have designated it unquestionably as Late Triassic (e.g., Galton 1971, Hintze 1973). And as observed by common usage, both practices have continued until very recently. However, Welles (1954) cast doubt on a Late Triassic age assignment which had been made for the younger Kayenta Formation (see Fig. 21.2 for relationship to the Moenave Formation). He thought that the relatively advanced stage of a dinosaur (*Dilophosaurus wetherilli* = *Megalosaurus wetherilli*) he collected 38.4 m (125 ft) above the upper surface of the Chinle Formation, within beds he considered referable to the Kayenta Formation, was of Early Jurassic age. Few authors followed his lead initially. But Olsen and Galton (1977), on the basis of a holostean fish, *Semionotus kanabensis*, and a palynoflora from the Moenave Formation of northern Arizona, also inferred an Early Jurassic rather than a Late Triassic age. At present more authors seem to be considering the Moenave as an Early Jurassic deposit (e.g., Lockley 1986, Madsen 1986). The relatively advanced stage of footprints from Warner Valley also appear suggestive of an Early Jurassic age for the Moenave Formation.

Paleoenvironment

As indicated above, the Dinosaur Canyon member of the Moenave Formation in the area of study contains intercalated units of fine-grained, well-sorted sands, silts and mud. Accumulated thicknesses of this member's units

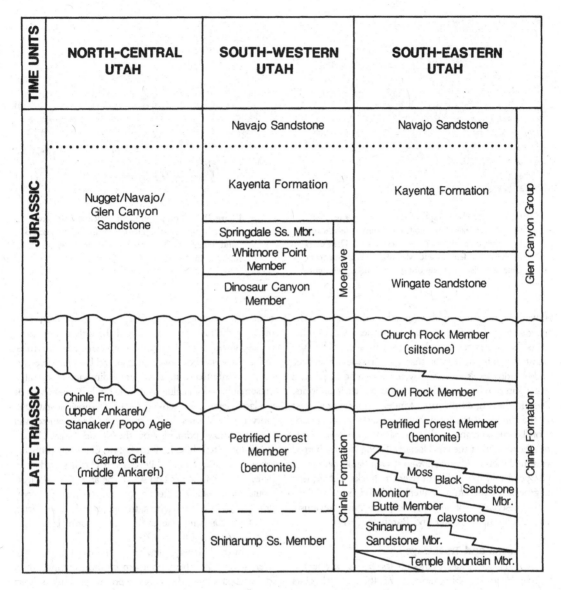

Figure 21.2. Stratigraphic columns of Late Triassic and Lower Jurassic lithostratigraphic units in Utah pertinent to this study. The trackways are in the Dinosaur Canyon Member of the Moenave Formation.

in the region can exceed 100 m. The nature and extent of these fine-grained sediments, including micro-crossbedding of the sands, imply that lower floodplain conditions (i.e., those not far from the river's mouth) existed here during Dinosaur Canyon time. Fragments of petrified wood, probably conifer, found in the vicinity of the dinosaur tracks within this member suggest that wooded areas were not far away. No evidence, though, exists for in situ vegetation. In a study of the paleoenvironments of the Moenave Formation at St. George, Utah, approximately 20 km to the northwest, Davis (1977) determined that the Dinosaur Canyon member sediments exposed there were indicative of tidal flat deposition. Based on bimodality of paleocurrent trends, he interpreted that much

of the sedimentation was occurring in the subaqueous subtidal zone. However, Paul D. Proctor (pers. comm. 1986) considers the beds Davis studied to possibly belong to the Kayenta Formation rather than the Moenave. Day (1967) also studied the Moenave Formation in southwestern Utah. His conclusion was that a broad, low-lying floodplain existed in this region during that time and that the source area was in highlands to the east, possibly as far away as western Colorado. Our limited study on the environment of deposition supports a lower floodplain model.

    Since tracks have been discovered only on the upper surfaces of fine-grained, well-sorted sandstones overlain by mudstones, special conditions probably existed to permit preservation. Clarity of many prints and the thicknesses

Figure 21.3. Photograph of the trackway site, looking to the northeast. The wash containing the trackways is in the lower right of the photograph. The cuesta in the upper left is Sand Mountain. The Hurricane Cliffs are visible in the distance of the upper right.

Figure 21.4. Photograph of the trackway site area, looking to the north. The cuesta in the distance is Sand Mountain. The arrow points to a person standing in the wash.

of overlying mudstone units suggest that the tracks on the upper surface of the fine–grained sandstones are not underprints. No footprints have been seen in any level of the mudstones. It is our opinion that, in each case of the three track–bearing levels, dinosaurs were walking on moist sand (water saturated in one instance) and left their imprints. Dried in the sun, a thin crust of modest resistance formed on the surface of the sand. Shortly thereafter significant rainfall occurred higher in the watershed; no evidence exists for a heavy rain at the fossil locality. Rising waters charged with mud then gently overflowed stream banks depositing a drape of very fine sediment over the tracks. Subsequent diagenesis left the latter unaltered. Mechanisms involved in the preservation of dinosaur footprints have previously been discussed by Tucker and Burchette (1977).

Tracks and Trackways

Although isolated dinosaur footprints exist at the Warner Valley tracksite, most of the 161 exposed prints comprise components of trackways (Fig. 21.5). Overprinting is rare despite the relatively high concentration of tracks. In all 23 trackways (two or more footprints made by one individual) have been counted. Unfortunately, heavier than usual rains in southwestern Utah in 1983 and 1984 caused erosion to completely or partially destroy a number of tracks previously observed, this without exposing new ones.

No strong directional tendency appears for the trackways or for isolated prints (Figs. 21.5, 21.6). It seems that the Warner Valley tracksite does not represent a narrow shoreline. Based on regional geology, the ocean shore was probably not far distant, perhaps less than 5 km to the south or west. It is possible that other bodies of water were nearby. Day (1967) reported that a lacustrine facies of the Dinosaur Canyon member outcrops throughout most of southwestern Utah.

All dinosaur prints at the present site show that the three genera recognized were functional tridactyl bipeds,

none of which dragged their tails. The most numerous footprints at the main track level, and the only ones present on the lowest level, belong to the smallest of the three forms. At least 19 trackways and several isolated footprints can be attributed to it. The type of dinosaur represented presumably was a coelurosaur, possibly like *Coelophysis* or a related genus. Two ichnogenera that show closest similarity to the Warner Valley tracks are *Anchisauripus* and *Grallator*, known primarily from the Newark Basin. Colbert (1963) indicated that the form–genus, *Grallator*, probably represents *Coelophysis*, while Ostrom (1968) stated that, based on foot structure and size of the animal, *Coelophysis* would have made a perfect match for *Anchisauripus* footprints (these two ichnogenera, though, may be congeneric, with *Grallator* being the earlier named form (Hitchcock 1858, Olsen and Galton 1984)

In the Warner Valley tracks the third digit is relatively shorter and the entire footprint more compact than in described species; they lack the pronounced sulcus Baird (1957) considered diagnostic of *G. sulcatus*, which had a shorter digit III than previously named species of this genus. No distinct metatarsal pads show in the many footprints; however, a very faint, nearly circular depression seems to exist just behind the toe marks on a few prints (Fig. 21.7a). Size of prints vary in greatest length from 12 to 18 cm and 9 to 10 cm in greatest width. The pace varies from 68 cm to 104 cm, and does not seem to correlate with foot size. As Lull (1953) observed, differences in trackways due to the sex and age of the individual are difficult if not impossible to determine. Toe divarication shows considerable variation. Baird (1957) stated such measurements have a very limited value in differentiation of theropod genera.

Four trackways and no isolated footprints, all on the main track–bearing horizon, represent the largest dinosaur at the presently described site (Fig. 21.7b). Individual tracks exhibit large phalangeal pads but not a

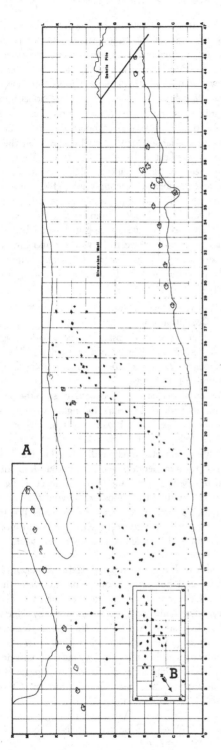

Figure 21.5. Maps of prints and trackways. **A,** map of the main trackway–bearing horizon with both *Eubrontes*, the large prints, and *Grallator*, the small prints, being represented; **B** (inset), map of the lowermost trackway horizon which contains the most deeply impressed prints; all tracks on this horizon are *Grallator*. Note the short, erratic strides evident of the upper trackway of B. The side of each square is 1 m.

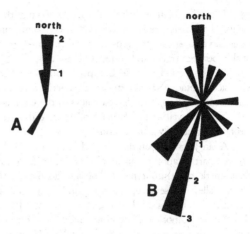

Figure 21.6. Rose diagrams of trackway orientations. **A,** rose diagram of lowermost trackway–bearing horizon. **B,** rose diagram of main trackway horizon. The smallest unit radiating from the center represents one trackway.

separate metatarsal one. Sizes vary moderately. Greatest lengths of prints of different individuals range from 40.2 cm to 48.0 cm, and greatest widths from 24.3 cm to 36.3 cm. Within each trackway the dimensions are surprisingly constant. The pace measurements of the four individuals making the trackways also show constancy, for a given trackway and to each other. These vary only from 124 cm to 138 cm for the smallest to the largest animal respectively. Toe divarication displays only modest variation. Although larger, these footprints look nearly identical to the ichnogenus *Eubrontes*, a well-known form from the Newark Basin. However, the actual animal represented still remains equivocal. Lull (1953) listed it as a carnosaur, but inferred that the trackmaker was not strictly carnivorous. Colbert (1963) indicated that some of the tracks in the Connecticut Valley might be prosauropods but that Baird recognized only theropod dinosaurs there. Ostrom (1968) allowed that prosauropods could have left three–toed prints, but believed *Eubrontes* was probably a carnosaur. Earlier, Bock (1952) made a case supporting *Eubrontes* as a prosauropod. He said there is no doubt that the large tracks of *E. giganteus* from the Connecticut Valley could match ones made by *Plateosaurus*, a prosauropod. However, Baird (1980) described a distinctive quadrupedal trackway, *Navahopus*, from the Navajo Formatiaon of Arizona, as that of a plateosaurid prosauropod. In our opinion, *Eubrontes*, and more particularly the type which made the large prints at Warner Valley, Utah, probably represents a plateosaurid prosauropod and so we follow Bock (1952) rather than Baird (1980). We compared photographs and illustrations of the pes of *Plateosaurus* and related forms which to us indicated they could have made prints similar to those from the present site. The very broad phalangeal pads strongly suggest a heavy animal, and the short, blunt claw marks do not seem to represent a carnivore. These footprints represent a dinosaur much larger than any North American theropod known at this time based on skeletal remains. The only known dinosaurs of this size and age are the prosauropods. A similar relationship between tracks and bones of dinosaurs and other reptiles of this period exists in Europe, a continent then in physical contact with eastern North America. We also infer that the makers of the largest tracks in the Connecticut Valley, as well as in Warner Valley, were the prosauropods, specifically plateosaurids.

A single good footprint (Fig. 21.7c) and a poorly preserved part of another, apparently of a different individual, mark the third type of dinosaur discovered at Warner Valley. These lie up-wash about 100 m, and 12 m stratigraphically higher than the main trackway level, still within the Dinosaur Canyon member. For the complete print, reliable measurements were taken. However, some collapsing of the sand into the print just after it was made obscured details, such as the exact length and shape of the third digit and its claw. In greatest length the footprint measures about 28 cm and in greatest width 29.5 cm. No impression of a metatarsal pad shows. The indication of relatively stong and sharp claws suggests that the represented animal was a theropod. This track appears to resemble a type named by Welles (1971) as *Kayentapus hopii*, from the Kayenta Formation of northern Arizona. Omitting the metatarsal pad print shown in Welles's figure of *Kayentapus*, the sizes of the footprints of both forms are nearly the same. Divarication between toes II and III is 33°, and 36° between III and IV on the Warner Valley footprint. This falls well within the range for Welles' ichnotaxon (1971, pp. 34–35). Three lines of evidence suggest that the trackmaker was moving in a gait faster than a walk: (i) lack of a metatarsal pad print, (ii) much deeper impressions of the toes at their distal ends, and (iii) push-up ridges of sand behind each toe. Unfortunately no successive prints are available to help substantiate this conjecture. The moderately large carnivore represented may have been generically similar to the Arizona theropod described by Welles. Although the former is older, the age difference is not great.

With one exception, the Warner Valley trackways show no unusual markings attesting to some specific behavioral activity of the dinosaurs represented other than minor directional changes by individuals. The easternmost trackway, at the lowest track horizon, is of a dinosaur whose stride definitely changed (Fig. 21.5A). Individual prints of this animal are the deepest relative to their small size, yet least distinct due to collapse. Although the size of footprints is among the largest for the small coelurosaur, the pace measures significantly less than the others. We interpret this to mean the individual was walking on saturated sediment and possibly in standing water, which necessitated shorter steps. The widening of the gait starting between prints five and six (heading from northeast to southwest) might have been an attempt to gain a better balance. A much longer pace between tracks eight and nine could represent the regaining of balance following a stumble. No markings in the sediment here indicate stepping over an object.

Conclusions

Three types of dinosaurs of Early Jurassic age are represented at the Warner Valley tracksite. The most numerous is a small coelurosaur, which could possibly be indicative of *Coelophysis*. It shows similarity to the ichnogenus *Grallator*. Four trackways attest to the presence of four individual dinosaurs belonging to one taxon, *Eubrontes*, whose large, broad prints and short blunt clawmarks suggest a prosauropod, possibly a plateosaurid. A single complete track indicates the third form, *?Kayentapus*, a moderately large carnivorous dinosaur that could represent a carnosaur. Based on local stratigraphy and sedimentation, the Dinosaur Canyon member of the Moenave Formation here represents a lower floodplain, probably coastal, environment where the ocean shoreline was no more than a few kilometers to the west.

Editorial Note: It is worth emphasizing that ichnofaunas with *Grallator* and Eubrontes, now being discovered in the western USA, are very similar to the classic Connecticut Valley ichnofaunas, originally described by Hitchcock (1858). These are known to be Liassic (Early Jurassic), *not* Late Triassic, thus underscoring their usefulness in establishing an Early Jurassic age. A similar ichnofauna with *Eubrontes* and *Grallator* is found in the Wingate Formation at Cactus Park near Grand Junction, Colorado (Eriksen 1979, Lockley in prep.).

Eriksen, L. 1979. Tell Tale Tracks. *Museum Notes Museum Western Colorado* 5:2, 5–7.

Acknowledgments
We wish especially to acknowledge and thank Mr. Gary Delsignore of Cedar City, Utah for finding and reporting the dinosaur tracks discussed in this paper. Dr. Blair Maxfield, geologist at Southern Utah State College, to whom the find was reported, informed us. We express our appreciation to him. Dr.

Figure 21.7. **A,** *Grallator.* **B,** *Eubrontes.* **C,** *Kayentapus.* A and B are composite drawings based on several prints of each form. C is based on a single print. Dashed lines indicate approximated outline. The scales each represent 10 cm.

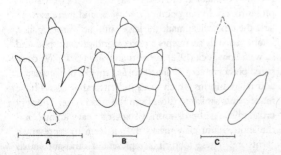

Samuel Welles of the University of California at Berkeley visited the site and discovered the track of a third type of dinosaur. For this and useful comments we extend our thanks. Mr. Samuel Webb, a former graduate student at Brigham Young University (BYU), helped in some of the initial trackway preparation and measurements. Drs. W. Kenneth Hamblin and Paul D. Proctor, both of BYU, shared knowledge of the geology they have obtained in their separate research studies of nearby areas. We have benefited from their useful comments.

References

Baird, D. 1957. Triassic reptile footprint faunules from Milford, New Jersey. *Harvard Mus. Comp. Zool. Bull.* 117 (5):449–520.

——— 1980. A prosauropod trackway from the Navajo Sandstone (Lower Jurassic) of Arizona. In Jacobs, L. L. (ed.). *Aspects of Vertebrate History.* (Flagstaff: Museum of Northern Arizona Press) pp. 219–230.

Bock, W. 1952. Triassic reptilian tracks and trends of locomotive evolution. *Jour. Paleont.* 26 (3):395–433.

Colbert, E. H. 1963. Fossils of the Connecticut Valley: The age of dinosaurs begins. *State Geol. Nat. Hist. Surv. Connecticut Bull.* 96:1–31.

Colbert, E. H., and Mook, C. C. 1951. The ancestral crocodilian *Protosuchus. Amer. Mus. Nat. Hist. Bull.* 97 (3):149–182.

Davis, J. D. 1977. Paleoenvironments of the Moenave Formation, St. George, Utah. *Brigham Young Univ. Geol. Studies* 24 (2):17–31.

Day, B. S. 1967. Stratigraphy of the Upper Triassic (?) Moenave Formation of southwestern Utah. University of Utah, Department of Geology Masters' Thesis, 58 pp.

Galton, P. M. 1971. The prosauropod dinosaur *Ammosaurus,* the crocodile *Protosuchus,* and their bearing on the age of the Navajo Sandstone of northeastern Arizona. *Jour. Paleont.* 45 (5):781–795.

Harshbarger, J. W., Repenning, C. A., and Irwin, J. H. 1957. Stratigraphy of the uppermost Triassic and the Jurassic rocks of the Navajo Country. *U.S. Geol. Surv. Prof. Paper* 291, 74 pp.

Hintze, L. F. 1973. Geologic history of Utah. *Brigham Young Univ. Geol. Studies* 20 (3):1–181.

Hitchcock, E. 1858. Ichnology of New England — a report on the sandstone of the Connecticut Valley, especially its fossil footmarks. (Boston: W. White, Printer to the State) 220 pp.

Lockley, M. G. 1986. Dinosaur tracksites. *Univ. Colorado Denver, Geol. Dept. Magazine* Special Issue 1, 56 pp.

Lull, R. S. 1953. Triassic life of the Connecticut Valley. *State Geol. Nat. Hist. Surv. Connecticut Bull.* 81:1–331.

Madsen, S. 1986. The rediscovery of dinosaur tracks near Cameron, Arizona. In Gillette, D. D. (ed.). *First International Symposium on Dinosaur Tracks and Traces.* (New Mexico Museum of Natural History) p. 20.

Olsen, P. E., and Galton, P. M. 1977. Triassic–Jurassic tetrapod extinctions: are they real? *Science* 197 (4307):983–986.

——— 1984. Review of the reptile and amphibian assemblages from the Stormberg of Southern Africa, with special emphasis on the footprints and the age of the Stormberg. *Paleont. Africana* 25:87–110.

Ostrom, J. H. 1968. The Rocky Hill dinosaurs. Guidebook for fieldtrips in Connecticut: New England Intercollegiate Geological Conference 60th annual meeting. *State Geol. Nat. Hist. Surv. Connecticut Guidebook* 2 (C–3):1–12.

Stokes, W. L., and Bruhn, A. F. 1960. Dinosaur tracks from Zion National Park and vicinity, Utah. *Utah Acad. Proc.* 37:75–76.

Tucker, M. E., and Burchette, T. P. 1977. Triassic dinosaur footprints from South Wales: Their context and preservation. *Palaeogeog., Palaeoclimat., Palaeoecol.* 22:195–208.

Welles, S. P. 1954. New Jurassic dinosaur from the Kayenta Formation of Arizona. *Geol. Soc. Amer. Bull.* 65:591–598.

Welles, S. P. 1971. Dinosaur footprints from the Kayenta Formation of northern Arizona. *Plateau* 44:27–38.

22 Stratigraphy and Age of Cretaceous Dinosaur Footprints in Northeastern New Mexico and Northwestern Oklahoma

SPENCER G. LUCAS,
ADRIAN P. HUNT AND
KENNETH K. KIETZKE

Abstract

Footprints of theropod and ornithopod dinosaurs occur in the uppermost Mesa Rica Sandstone and/or Pajarito Formation at three localities in northeastern New Mexico and northwestern Oklahoma: Mosquero Creek (Harding County) and Clayton Lake (Union County), both New Mexico, and South Carrizo Creek (Cimarron County), Oklahoma. Marine invertebrate fossils and palynology indicate that these dinosaur–footprint occurrences are of late Albian (Early Cretaceous) age.

Introduction

Dinosaur footprints are known from three localities in northeastern New Mexico and northwestern Oklahoma (Fig. 22.1) in strata of Cretaceous age. However, the stratigraphic position and age relationships of these footprints have not been well documented. In this paper, we present lithostratigraphic and biostratigraphic data to elucidate the stratigraphic position and precise age of the dinosaur–footprint localities in northeastern New Mexico and northwestern Oklahoma.

Regional Stratigraphic Context

Cretaceous strata in northeastern New Mexico and northwestern Oklahoma that contain dinosaur footprints are part of a sequence deposited near the end of the Skull Creek cyclotherm (late Albian) and just before the onset of the Greenhorn cyclotherm (early Cenomanian) (Fig. 22.2). In Quay, Guadalupe and Harding Counties, New Mexico, the basal marine deposits of the Skull Creek cyclotherm are termed Tucumcari Shale and contain an invertebrate mega- and microfauna of late Albian (Duck Creek) equivalent age (Kietzke 1985, Kues et al. 1985). In Union County, New Mexico and Cimarron County, Oklahoma, the basal marine deposits are termed Glencairn Shale and also contain a late Albian marine–invertebrate fauna (Stovall 1943, Scott 1970). Throughout northeastern New Mexico and northwestern Oklahoma, the Tucumcari

and Glencairn are overlain by a tripartite sequence of sandstone – "shale" – sandstone. This sequence is overlain by the early Cenomanian Graneros Shale. In Quay and Guadalupe Counties, New Mexico this tripartite sequence, often referred to as the Dakota Group, Formation or Sandstone (e.g., Mateer 1985 and references cited therein), has been divided into a basal Mesa Rica Sandstone, medial Pajarito Formation and upper "Dakota Sandstone" (Dobrovolny et al. 1946, Wanek 1962, Lucas and Kues 1985). In Union County, New Mexico and Cimarron County, Oklahoma, this tripartite sequence has been referred to only as Dakota Sandstone, Formation or Group (Stovall 1943, Baldwin and Muehlberger 1959). However, northward extension of the Quay–Guadalupe Counties nomenclature for this stratigraphic interval is clear and has been justified elsewhere (Lucas et al. 1986).

The medial, "shaley" unit of the tripartite "Dakota" sequence was named Pajarito Shale by Dobrovolny et al. (1946). However, this unit contains little shale and is as much as 33 m of interbedded siltstone, silty shale and bioturbated sandstone. Here, we refer to this unit as Pajarito Formation. The Pajarito Formation represents a delta plain that prograded eastward over the deltaic Mesa Rica Sandstone during the regression of the Tucumcari–Glencairn seaway (Fig. 22.2). All dinosaur footprints known from northeastern New Mexico and northwestern Oklahoma are either in the Pajarito Formation or the top of the underlying Mesa Rica Sandstone.

Footprint Localities and Lithostratigraphy

The three known occurrences of dinosaur footprints in northeastern New Mexico and northwestern Oklahoma are at Mosquero Creek, Clayton Lake and South Carrizo Creek (Fig. 22.1).

The dinosaur footprints at Mosquero Creek are on the point of the escarpment south of the creek, 6.8 km west of NM Highway 39 in the SW 1/4 NE 1/4 SW 1/4 section 23 T17N, R29E (Fig. 22.1). On this point, the Mesa

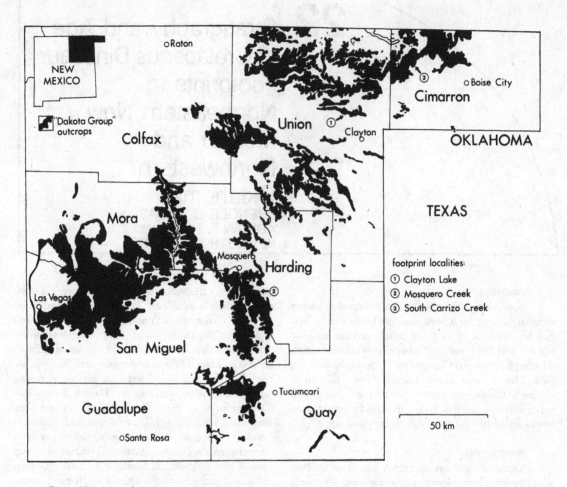

Figure 22.1. Map of northeastern New Mexico and part of northwestern Oklahoma showing distribution of Lower Cretaceous strata (after Dane and Bachman, 1965) and Cretaceous dinosaur-footprint localities.

Figure 22.2. Cross-section through Albian–Cenomanian strata in east-central New Mexico from the northwest (San Miguel County) to the southeast (Quay County). Age relationships, correlative stratigraphic units in western Texas and stratigraphic position of dinosaur footprints also are indicated.

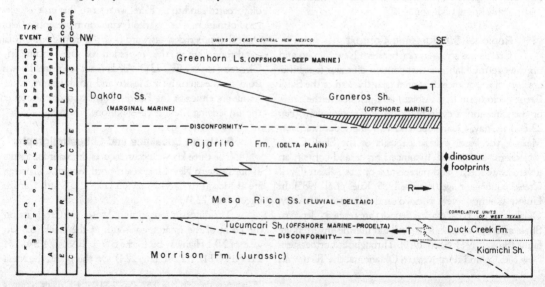

Rica Sandstone underlies 2.1 m of light brownish gray silty shale that we identify as the basal unit of the Pajarito Formation (Fig. 22.3). The sandstone above the shale exposes about 114 dinosaur footprints over an area of 557 m^2 and is overlain by carbonaceous shale (Figs. 22.3, 22.4c). This sandstone is 30 cm of grayish orange to yellowish brown, fine- to medium-grained, subangular to rounded and moderately well sorted quartzarenite. Its lower 20 cm preserve sinuous- and straight-crested ripples, low angle trough crossbeds, horizontal laminae and tool marks. The upper 10 cm of the footprint-bearing sandstone are intensely bioturbated by both the dinosaur footprints and numerous small, smooth-walled horizontal tubes. The footprints, described by Lucas et al. (1987), pertain to ornithopod dinosaurs, ichnogenus *Amblydactylus* (Fig. 22.4d).

At Clayton Lake approximately 500 footprints are exposed in the dam spillway in the SW 1/4 NE 1/4 section 15 T27N, R34E (Figs. 22.1, 22.4a). Gillette and Thomas (1985) described and illustrated the footprints which are of theropod (Fig. 22.4b) and ornithopod (including ichnogenus *Amblydactylus*) dinosaurs. Lucas et al. (1986) discussed the stratigraphic context of the footprints and concluded that they occur at the top of the Mesa Rica Sandstone and at two horizons in the lower part of the Pajarito Formation (Fig. 22.3).

In the east-facing bank of South Carrizo Creek, Cimarron County, Oklahoma, three dinosaur footprints are present in the NE 1/4, NE 1/4, NW 1/4 section 7, T4N, R2E (Figs. 22.1, 22.4e). These footprints are of ornithopod dinosaurs (ichnogenus *Amblydactylus*) (Fig. 22.4f) and occur at a single stratigraphic level in the upper part of the Pajarito Formation (Fig. 22.3). The footprint-bearing unit is a clayey, bioturbated quartzarenite that is well sorted, well rounded and ranges in color from very pale orange to pale yellowish brown.

Age of the Footprints

Lucas et al. (1986) revealed biostratigraphic data relevant to the age of the dinosaur footprints at Clayton Lake. Foraminifera from the upper part of the Glencairn Shale at Clayton Lake are: *Ammobaculoides plummerae, A. ?phaulus, Ammodiscus ganitinus, Ammobaculites subcretaceus, Qunqueloculina nanna, Reophax* sp., *Dentalina* cf. *D. cylindroides* and *Nodosaria* sp. These taxa indicate a late (but not latest) Albian age and thus set a maximum age for the dinosaur footprints at Clayton Lake. Foraminifera from the lower part of the Graneros Shale at Clayton Lake set a minimum age of early Cenomanian for the dinosaur footprints. These Foraminifera are: *Bigenerina hastata, Haplophragmium arenatus, Trochammina* sp., *Heterohelix globulosa* and *Guembelitria harrisi*. No other direct evidence at the footprint localities is now available to determine their age.

However, there is regional evidence that suggests the age of the uppermost Mesa Rica Sandstone and the Pajarito Formation and the dinosaur footprints they contain are late Albian. First, invertebrate fossils from the lower

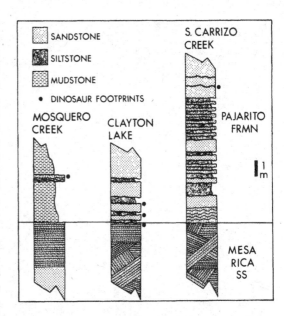

Figure 22.3. Measured stratigraphic sections at the three Cretaceous dinosaur-footprint localities known in northeastern New Mexico and northwestern Oklahoma.

and middle parts of the Mesa Rica Sandstone include the ammonoid *Mortoniceras equidistans*, indicative of a late Albian age (Kues et al. 1985). Second, palynomorphs from the Pajarito Formation at Montezuma Gap near Las Vegas, San Miguel County, New Mexico are of late Albian age (Elmer H. Baltz, Jr. oral communication 1986). Provided that the Pajarito does not significantly transgress time from west to east, these palynomorphs are the first direct evidence of the late Albian age of the Pajarito Formation.

The dinosaur footprints from the Cretaceous of northeastern New Mexico and northwestern Oklahoma thus are approximately the same age as footprints from the Dakota Group of Colorado described by Lockley (1985, 1987) who determined a Late Albian to Early Cenomanian age. It is also interesting to note that there are several similarities in the ichnofauna which is dominated by ornithopod tracks including *Amblydactylus* and *Caririchnium* in both areas (Lockley 1987, pers. comm. 1988). Other forms from both areas include a slender-toed theropod (Fig. 22.4b).

Acknowledgements

We thank S. Hayden, J. Holbrook, J. Hunley, B. Kues, P. Reser, R. Sullivan and D. Wolberg for assistance in the field. E. Baltz, D. Gillette, D. Thomas and R. Tschudy shared information with us. M. Kisucky drafted Figure 22.2.

References

Baldwin, B., and Muehlberger, W. R. 1959. Geologic studies of Union County, New Mexico. *New Mexico Bur. Mines Min. Res. Bull.* 63.

Figure 22.4. Cretaceous dinosaur-footprint localities in northeastern New Mexico and northwestern Oklahoma. **a,** overview of footprint locality at Clayton Lake. **b,** theropod footprint at Clayton Lake. **c,** overview of footprint locality at Mosquero Creek. **d,** *Amblydactylus* footprint at Mosquero Creek. **e,** overview of footprint locality at South Carrizo Creek (arrow indicates footprint-bearing sandstone). **f,** *Amblydactylus* footprint at South Carrizo Creek.

Dane, C. H., and Bachman, G. O. 1963. Geologic Map of New Mexico. *U.S. Geol. Surv.* scale 1:500,000.

Dobrovolny, E., Bates, R. L., and Summerson, C. H. 1946. Geology of northwestern Quay County. New Mexico. *U.S. Geol. Surv. Oil Gas Inv. Prelim. Map* OM-62, scale 1:62,500.

Gillette, D. D., and Thomas, D. A. 1985. Dinosaur tracks in the Dakota Formation (Aptian–Albian) at Clayton Lake State Park, Union County, New Mexico. *In*

Lucas, S. G., and Zidek, J. (eds.). 1985. *Santa Rosa-Tucumcari Region. New Mexico Geol. Soc. Guidebook 36th Field Conf.* (University of New Mexico Press) pp. 261–281.

Kietzke, K. K. 1985. Microfauna of the Tucumcari Shale, Lower Cretaceous of east-central New Mexico. *In* Lucas, S. G., and Zidek, J. (eds.). *Santa Rosa - Tucumcari Region. New Mexico Geol. Soc. Guidebook to 36th Field Conf.* (University of New Mexico Press) pp.

247–260.

Kues, B. S., Lucas, S. G., Kietzke, K. K., and Mateer, N. J. 1985. Synopsis of Tucumcari Shale, Mesa Rica Sandstone and Pajarito Shale paleontology, Cretaceous of east-central New Mexico. *In* Lucas, S. G., and Zidek, J. (eds.). *Santa Rosa–Tucumcari Region. New Mexico Geol. Soc. Guidebook to 36th Field Conf.* (University of New Mexico Press) pp. 261–281.

Lockley, M. G. 1985. Vanishing tracks along Alameda Parkway implications for Cretaceous dinosaurian paleobiology from the Dakota Group, Colorado. *In* Chamberlain, C. K., (ed.). 1985. *A Field Guide to Environments of Deposition (and Trace Fossils) of Cretaceous Sandstones of the Western Interior.* (Society of Economic Paleontologists) pp. 131–142.

——— 1987. Dinosaur footprints from the Dakota Group of eastern Colorado. *Mountain Geol.* 24:107–122.

Lucas, S. G., and Kues, B. S. 1985. Stratigraphic nomenclature and correlation chart for east-central New Mexico. *In* Lucas, S. G., and Zidek, J. (eds.). 1985. *Santa Rosa–Tucumcari Region. New Mexico Geol. Soc. Guidebook 36th Field Conf.* (University of New Mexico Press) pp. 243–246.

Lucas, S. G., Hunt, A. P., Kietzke, K. K., and Wolberg, D. L. 1986. Cretaceous stratigraphy and biostratigraphy, Clayton Lake State Park, Union County, New Mexico. *New Mexico Geol.* 8:60–65.

Lucas, S. G., Holbrook, J., Sullivan, R. M., and Hayden, S. N. 1987. Dinosaur footprints from the Cretaceous Pajarito Formation, Harding County, New Mexico. *In* Lucas, S. G., and Hunt, A. P., (eds.). 1987. *Northeastern New Mexico. New Mexico Geol. Soc. Guidebook 38th Field Conf. (University of New Mexico Press)* pp. 31–33.

Mateer, N. J. 1985. Pre-Graneros Cretaceous stratigraphy of northeastern New Mexico. *In* Lucas, S. G., and Zidek J. (eds.). 1985. *Santa Rosa–Tucumcari Region. New Mexico Geol. Soc. Guidebook 36th Field Conf.* (University of New Mexico Press) pp. 243–246.

Scott, R. W. 1970. Stratigraphy and sedimentary environments of Lower Cretaceous rocks, southern Western Interior. *Amer. Assoc. Petrol. Geol. Bull.* 54:1225–1244.

Stovall, J. W. 1943. Stratigraphy of the Cimarron Valley (Mesozoic rocks). *Oklahoma Geol. Surv. Bull.* 64:43–132.

Wanek, A. 1962. Reconnaissance geologic map of parts of Harding, San Miguel, and Mora Counties, New Mexico. *U.S. Geol. Surv. Oil Gas Inv.* Map OM–208, scale 1:96,000.

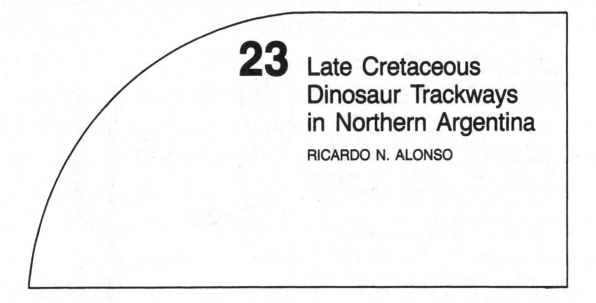

23 Late Cretaceous Dinosaur Trackways in Northern Argentina

RICARDO N. ALONSO

Abstract

Presently the only known region with dinosaur tracks in northern Argentina occurs in the Tonco Valley (Salta Province). In this site two ichnological localities have been found in the Quebrada de la Escalera and Quebrada del Tapon in Cretaceous limestones of the Yacoraite Formation (Maastrichtian). They are tridactyl and belong to bipedal dinosaurs. One set of the tracks corresponds to imprints of ornithopods (probably hadrosaurians), the other to carnosaurians. In the Quebrada del Tapon bird tracks are also found.

Introduction

The study of dinosaur tracks is still a young field of research in northern Argentina. In 1968, M. Raskovsky mentioned dinosaur tracks in the Tonco Valley in an unpublished stratigraphic report of the National University of Salta. Alonso (1978, 1980) studied these tracks in further detail. At the end of 1985 another track locality was discovered in the same valley. The outcrop is about 25 km southeast of the previously known locality (Alonso and Marquillas 1986).

The Tonco Valley is located about 150 km southwest of the city of Salta, within the Andean Eastern Cordillera at about 25°35' S and 66°00' W, at 2500 m altitude (Fig. 23.1). Access to the fossil track localities is via national routes 9 and 59 and secondary roads that lead to a recently abandoned uranium mine (Don Otto) of the National Atomic Energy Commission (CNEA).

Geology

The area (Fig. 23.2) is underlain by metasediments of the Late Precambrian Puncoviscana Formation, which is overlain and in angular unconformity with the continental sediments of the Upper Cretaceous/Lower Tertiary Salta Group. The Salta Group is divided into the Pirgua, Balbuena (Lecho and Yacoraite Formations) and Santa Bar-

bara subgroups (Fig. 23.3). These are covered by continental deposits of the Upper Tertiary Oran Group. All sedimentary units are folded into a large syncline with a north–south trend (Fig. 23.2). The eastern limb is affected by a large reverse fault of regional importance. Geologic investigations have primarily been conducted under the direc-

Figure 23.1. Location map of study area (specific track locations shown in Figure 23.3).

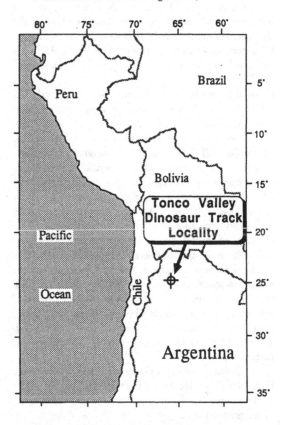

tion of the CNEA (Raskovsky 1968a,b; Fugueroa 1978; Antonietti et al. 1984).

The ichnites occur in the Yacoraite Formation, which is characterized by yellowish calcarenites (Figs. 23.2, 23.3). The beds have been rotated to a nearly vertical orientation and the tracks are exposed on a vertical face. Toward its top the Yacoraite Formation is conformable with the Paleocene red sandstones of the Santa Barbara Subgroup. At the base, the Yacoraite Formation grades into a white sandstone of the Lecho Formation. Due to its particular yellowish color and composition, the Yacoraite Formation is easily recognized in the field. The formation is 140 m thick and corresponds to a shoreline depositional environment of a water body whose marine or possible lacustrine origin is still unclear.

Specific studies concerning the Yacoraite Formation in northern Argentina have been carried out by Lencinas and Salfity (1973), Marquillas et al. (1984), and Marquillas (1985).

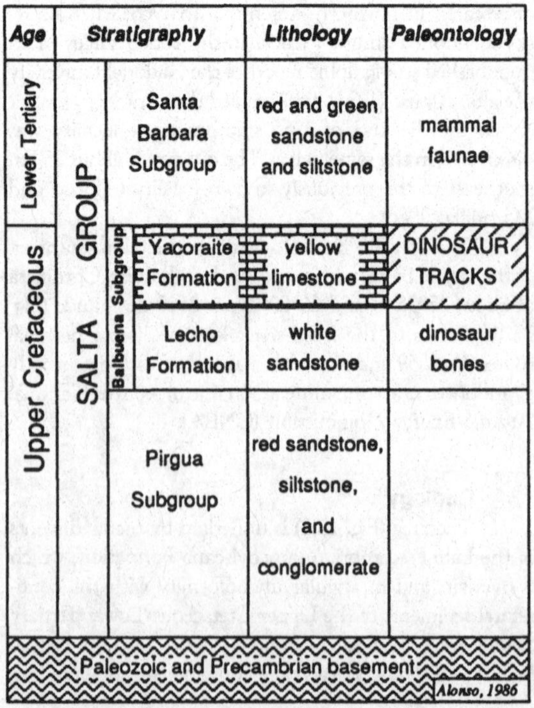

Ichnological Localities
Quebrada de la Escalera

This locality (Fig. 23.2) is known from the Los Berthos uranium deposits (Raskovsky 1968), and occurs within the western limb of the syncline. The terrain is extremely rugged and the fossil site can only be reached with two ladders (each of 15 m height) that were installed on the nearly vertical Yacoraite beds. The tracks occur in an oolitic calcarenite about 10 m above the base of the Yacoraite Formation (Fig. 23.3). Both overlying and underlying strata are characterized by mud cracks, ripple marks, gastropod coquina and intraformational conglomerates.

Paleontology. Seven trackways, each 2–5 m in length and with more than 30 poorly to well preserved tridactyl footprints, occur in the Quebrada de la Escalera (Fig. 23.4). The dimensions of the tracks are shown in Figure 23.5. Six trackways (1, 2, 4, 5, 6, 7) have the same morphology: a narrow imprint with the III digit parallel to the midline of the trackways, indicating an agile and erect gait (Fig. 23.6). The footprints are slightly longer than their width and it appears that the digits were connected by a web. The digits terminate in a blunt, triangular nail. This character is very well expressed in one footprint of trackway 1. The general aspect of thse tracks suggests ornithopods of the hadrosaurian group and has been designated *Hadrosaurichnus australis* (Alonso 1980), although Leonardi (this volume) believes a proportion may be theropod trackways. The track number 3 has a different morphology with V-shaped digits and pronounced claw imprints (Fig. 23.5). These tracks are interpreted to belong to carnosaurians.

In this locality other dinosaur tracks are observed. One track is imprinted on a stromatolitic bed. All tracks are poorly preserved and are difficult to measure because of dangerous access.

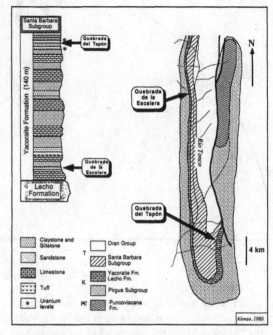

Figure 23.2. Stratigraphic column of the Yacoraite Formation and geologic map of the Tonco Valley showing location of fossil tracks discussed in the text.

Figure 23.3. Stratigraphy, lithology and paleontology of Mesozoic and Lower Cretaceous rocks of northern Argentina.

Age	Stratigraphy		Lithology	Paleontology
Lower Tertiary		Santa Barbara Subgroup	red and green sandstone and siltstone	mammal faunae
Upper Cretaceous	SALTA GROUP / Balbuena Subgroup	Yacoraite Formation	yellow limestone	DINOSAUR TRACKS
		Lecho Formation	white sandstone	dinosaur bones
		Pirgua Subgroup	red sandstone, siltstone, and conglomerate	
	Paleozoic and Precambrian basement			

Alonso, 1986

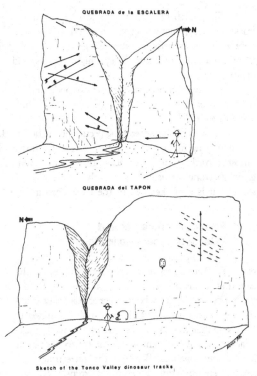

QUEBRADA de la ESCALERA

QUEBRADA del TAPON

Sketch of the Tonco Valley dinosaur tracks

Figure 23.4. Sketch of Quebrada de la Escalera and Quebrada del Tapon dinosaur tracks in the Tonco Valley showing spatial relationships and ripple crest trends (dashed lines).

Quebrada del Tapon

The Quebrada del Tapon locality is in the vicinity of the major uranium deposit of the Don Otto mine in the eastern limb of the Tonco syncline (Fig. 23.2). The ichnites occur in a very fine, oxidized yellowish–green sandstone of the Yacoraite Formation which is further characterized by green clay laminae, ripple marks and mud cracks (Figs. 23.4, 23.6). The sandstone underlies an important layer with mineralized uranium, and occurs about 10 m below red sandstones of the Paleocene Mealla Formation (Santa Barbara Subgroup) (Fig. 23.2).

Paleontology. Four different types of footprints can be observed in Figure 23.7. The first two types occur as isolated footprints: one of them occurs on ripple marks (A), the other on incipient mud cracks (B). Both of them are characterized by short digits, absence of claws and the existnce of an apparent web. The first type (A) has two U–shaped digits without an expression of digit II. Convex ridges in the sediment parallel the outline of the digits. The second type (B) has the general characteristics of a "web-foot", reminiscent of duck footprints. Both types of prints are attributed to ornithopods and possibly hadrosaurids.

The third type (C) is defined by six moderately to well preserved tridactyl footprints that belong to a bipedal dinosaur. The digits are generally V–shaped and in addition have pronounced claws. Based on these morphologic characteristics it is attributed to a carnosaurian. The track dimensions are shown in Figure 23.7.

Figure 23.5. Dimensions and features of dinosaur trackways in the Tonco Valley.

Trackway Number	Preservation	Trackway Length (cm)	Trackway Direction	Stride Length (cm)	Pace Angulation	No. of Footprints	Footprint Length (cm)	Footprint Width (cm)
QUEBRADA DE LA ESCALERA								
1	Good	190	N -> S	198	160°	3	37	28
2	Mod.	240	NNE->SSE	200	140°	3	34	31
3	Mod.	220	NE ->SW	193	167°	3	30	22
4	Good	564	SW -> NE	150	160°	9	25	17
5	Good	543	NW -> SE	170	160°-164°	8	27	24
6	Poor	292	NW - SE	120	160°-164°	8	27	11
7	Poor	362	SE -> NW	120	160°-164°	8	19	7
QUEBRADA DEL TAPON								
--	Good	--	--	--	--	1	52	39 ?
--	Good	--	--	--	--	1	57	24
1	Mod.	460	W -> E	164	100°	6	40	38

Alonso, 1986

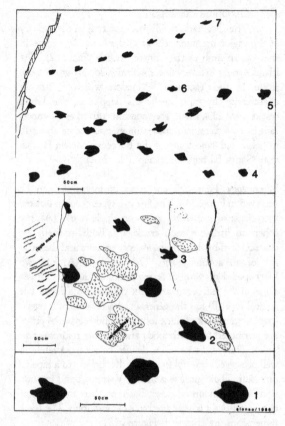

Figure 23.6. Sketch drawn from photographs of Quebrada de la Escalera dinosaur tracks. Numbers refer to descriptions in the text. Dimensions of tracks are given in Figure 23.5.

Figure 23.7. Sketch drawn from photographs of Quebrada del Tapon dinosaur tracks. Dimensions in Figure 23.5.

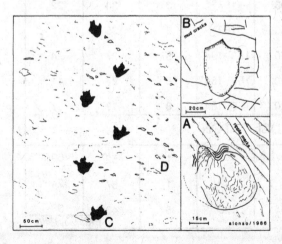

The fourth type (D) occurs with the carnosaurian tracks. Although many trackway directions can be observed here, their preservation is extremely poor, such that their identification is difficult. They might correspond to ornithopods.

The distinction between ornithopods and carnosaurians is based on the approach suggested by Thulborn and Wade (1984).

There is also an interesting occurrence of bird tracks (8 cm in length) in a layer 1 m above a horizon with dinosaur tracks. The bird tracks are well preserved in a green claystone that is in contact with a white sugary sandstone that is a marker bed in the Don Otto mine.

Paleoenvironmental and Paleoecological Implications

The paleogeographic and sedimentological aspects of the Yacoraite Formation in the Tonco Valley clearly indicate that it was deposited along the western margin of a water–filled sedimentary basin. Higher relief existed towards the west and a large but shallow basin extended to the east. Whether the basin was part of a shallow marine or a lacustrine environment is still subject to discussion and further investigation. Between the terrestrial and subaqueous environment, there was an extensive coastal plain with a very low gradient that was subject to periodic submergence due to cyclic and periodic changes in water level. In the transition to the terrestrial environment in the southwestern part of the Tonco Valley occur fluvial facies associated with indicators of soil development (Marquillas 1985). Subaqueous facies are increasingly found towards the east. They are characterized by stromatolite mats, gastropod coquina, and layers of green and black shale which contain abundant fish fossils.

The dinosaur tracks in the profile of the Quebrada de la Escalera occur within oolitic calcarenites which were deposited over stromatolites that covered calcarenite beds that are characterized by well defined ripple marks. In a few cases the dinosaur prints are superimposed on the stromatolite mats. Several decimeters below the stromatolite beds occur more strata with ripple marks that overlie an intraformational breccia, which may indicate an event with higher water energy, perhaps the influence of storms. The preservation of the dinosaur tracks was made possible by the deposition of a pelitic layer above the oolitic calcarenite. Above the pelite follow more calcarenite beds with ripple marks which are overlain by a siltstone with pronounced dessication cracks.

In the Quebrada del Tapon locality a similar situation is found, except that breccias and siltstones with dessication cracks do not occur. The dinosaur imprint in Figure 23.7A is associated with ripple marks. At the time the footprint was made, the sediment must have been saturated with water or at least very moist, as indicated by compressive convex ridges around the tracks, whereas the tracks in Figure 23.7B were generated on incipient mud cracks. Above the fossil track–bearing beds occurs a massive white "sugar" sandstone which indicates a period of higher water levels with intense reworking of shoreline sediments. A thin cap of green claystone with bird tracks on top of the sugar-sandstone shows a period of inundation followed by exposure, accompanied by formation of bird tracks in the muds.

In the Quebrada del Tapon sequence it is further possible to observe numerous beds of green to gray uranium bearing pelites ("Seccion Verde" local name) which are undoubtedly related to a local strong reducing environment. This is a major contrast with the Quebrada de la Escalera locality, where light colored calcarenites are dominant. The Quebrada del Tapon sequence might be related to a coastal environment, characterized by lagoons with restricted circulation behind sand barriers.

It is possible that the herbivorous dinosaurs (ornithopods) with apparent web-feet could correspond to swimming hadrosaurians that lived close to the coastal environment for nourishment. The coeval presence of hadrosaurians and the terrestrial carnosaurians in the coastal environment suggest that carnosaurians were the predators of hadrosaurians or other dinosaurs. It is also possible that both groups came to the shoreline environment to drink, and, in the case of the hadrosaurians, it may be even possible that they fed on the stromatolite mats.

Furthermore, the Quebrada del Tapon tracks may be of stratigraphic importance because they lie very near the horizon recognized locally as the Cretaceous–Tertiary boundary, which is conformable. This might be an excellent site for a geochemical study to examine the cosmogenic hypothesis of catastrophic extinction.

Discussion and Conclusions

In northern Argentina the sedimentary depositional cycle began during the Late Cretaceous and terminated during the Eocene without any unconformity between Upper Cretaceous and Lower Paleocene strata (Fig. 23.3). In general, the occurrence of dinosaur fossils is rather limited, except in the Upper Cretaceous Lecho and Yacoraite Formations. The Lecho Formation is known for its dinosaur bones (Bonaparte et al. 1977, Bonaparte and Powell 1980), whereas the Yacoraite Formation is characterized by dinosaur tracks (Fig. 23.3). These tracks are a very important element in paleofaunistic reconstructions and aid in the evaluation of the abundance of Cretaceous reptilian fauna in the region.

All the discovered dinosaur tracks are bipedal and the majority are imprints of ornithopods which are attributed to hadrosaurians; the remainder belong to carnosaurians. One would prefer not to assign these tracks to a specific family of dinosaurs; they could correspond to native groups of dinosaurs as suggested by Casamiquela and Fasola (1968) and Leonardi (1979). This concept has recently been supported by the works of J. Bonaparte (Museo "Bernardino Rivadavia", Buenos Aires), who has recently found in Patagonia numerous forms exclusive to South America (pers. comm.). There are very few occurrences of dinosaur tracks of the Upper Cretaceous in South America and these are limited to Argentina, Chile and Bolivia (Leonardi 1981 and this volume).

The study of vertebrate ichnology of northern Argentina is still in a preliminary state and with further investigations new discoveries are to be expected. For example other sites with possible tracks have been mentioned informally but have not been carefully studied. The search for dinosaur tracks is facilitated by the geologic and physiographic setting of the area. The region is characterized by excellent outcrops due to the arid climate and the intense tectonic movements within the Eastern Cordillera that have produced steeply inclined sedimentary strata which may permit widespread scale reconnaissance of ichnologic features in northern Argentina.

Acknowledgments

I wish to thank the following persons for their assistance at various stages of this work: R. Marquillas and A. Tapia of the Universidad Nacional de Salta; R. Bustos, R. Figueroa and S. Gorustovich of the Comision Nacional de Energia Atomica; G. Bossi of the Universidad Nacional de Tucuman, and Mr. M. Sutti, all of them from Argentina. In the USA I thank D. Gillette, formerly of the New Mexico Museum of Natural History, now State Paleontologist of Utah, and R. Allmendinger, T. Jordan, M. Strecker, J. Beer and P. Flemmings of Cornell University, Ithaca, New York.

References

Alonso, R. 1978. Icnitas de dinosaurios de la Formacion Yacoraite (Cretacico superior). Su importancia paleozoogeografica y cronoestratigrafica. Facultad de Ciencias Naturales, Universidad Nacional de Salta. Tesis Profesional. Unpublished. Salta.

—— 1980. Icnitas de dinosaurios (Ornithopoda, Hadrosauridae) en el Cretacico superior del Norte de Argentina. *Acta Geologica Lilloana* 15 (2):55–63. S.M. de Tucuman.

Alonso, R., and Marquillas, R. 1986. Nueva localidad con huellas de dinosaurios y primer hallazgo de huellas de aves en la Formacion Yacoraite (Maastrichtiano) del norte argentino. *IV Congress. Argentino de Paleontologia y Biostratigrafia* Actas 2, pp. 33–41.

Antonietti, C., Gorustovich, S., Valdiviezo, A., Benitez, A., and Saucedo, P. 1984. Geologia y Metalogenesis de los depositos uraniferos de Argentina. In *Geology and Metallogenesis of Uranium Deposits of South America.* (Vienna: International Atomic Energy Agency)

Bonaparte, J., Salfity, J., Bossi, G., and Powell, J. 1977. Hallazgo de dinosaurios y de aves cretacicas en la Formacion Lecho de El Brete (Salta), proximo al limite con Tucuman. *Acta Geologica Lilloana* 14:5–17. S.M. de Tucuman.

Bonaparte, J., and Powell, J. 1980. A continental assemblage of tetrapods from the Upper Cretaceous beds of El Brete, northwestern Argentina (Sauropoda, Coelurosauria, Carnosauria, Aves). *Mem. Soc. Geol. Fr. N.S.* 139:19–28.

Casamiquela, R., and Fasola, A. 1968. Sobre pisadas de dinosaurios del Cretacico inferior de Colchagua

(Chile). *Universidad de Chile, Departamento de Geologia* Publicacion No. 30, 24 pp. Santiago.

Lencinas, A., and Salfity, J. 1973. Algunas caracteristicas de la Formacion Yacoraite en el oeste de la cuenca andina, provincias de Salta y Jujuy, Argentina. V *Congreso Geologico Argentino* Actas III:253–267. Buenos Aires.

Leonardi, G. 1979. Nota preliminar sobre seis pistas de dinossauros Ornitischia da Bacia do rio Peixe, em Sousa, Paraiba, Brasil. *An. Acad. Brasil.* 51 (3):501–516.

1981. Glossario Comparado da Icnologia dos Vertebrados. *Universidad Estadual de Ponta Grossa. Cuadernos Universitarios* No. 17. Parana.

Marquillas, R. 1985. Estratigrafia, sedimentologia y paleoambientes de la Formacion Yacoraite (Cretacico Superior) en el tramo austral de la cuenca, norte

argentino. Tesis Doctoral. Universidad Nacional de Salta. Unpublished. Salta.

Marquillas, R., Bosso, M., and Salfity, J. 1984. La Formacion Yacoraite (Cretacico Superior) en el norte argentino al sur del paralelo 24. *Noveno Congreso Geologico Argentino* II:300–310. Buenos Aires.

Raskovsky, M. 1968. Estudio estructural y radimetrico al norte del Yacimiento Don Otto. Seminario I. Facultad de Ciencias Naturales. Universidad Nacional de Salta. Unpublished. Salta.

1968. Relevamiento geologico en el sector sur del yacimiento Los Berthos. Facultad de Ciencias Naturales. Universidad Nacional de Salta. Seminario II. Unpublished. Salta.

Thulborn, R., and Wade, M. 1984. Dinosaur trackways in the Winton Formation (Mid–Cretaceous) of Queensland. *Mem. Queensland Museum* 2 (2):413–516.

VII Experimentation and Functional Morphology

"Experimental work on trackways, coupled with considerations of limb kinematics and substrate conditions, will permit the most robust inferences about paleoichnologic trackmakers, and will thus maximize the utility of fossil footprint data."

Padian and Olsen (1984 p. 178)

It is difficult to fully understand modern tracks without observing the trackmakers and understanding their functional morphology to some degree. It therefore follows that an adequate understanding of fossil tracks is even harder to achieve without observing trackways of the modern analogs of fossil trackmakers and studying their functional morphology. As pointed out by Padian and Olsen (1984 p. 183), "there have been relatively few studies of the relationships between patterns of locomotion and footprints of living reptiles." Although this may account for the lack of experimental studies conducted by paleontologists interested in fossil tetrapods, there is no doubt that a number of living groups including lizards, crocodiles, birds, and even some mammals provide good analogs for Mesozoic trackmakers.

Padian and Olsen have broken new ground in this field, first with their study of the tracks of the Komodo dragon (1983, see also Auffenberg ref. p. 119 herein), then with their controversial but convincing study of *Pteraichnus* tracks (cf. Chapters 16, 27 and 36 this volume) and now with their study of modern ratites (Chapter 24), living analogs of Mesozoic theropods.

In the past, vertebrate ichnologists have sometimes overlooked available anatomical evidence which could have facilitated trackmaker identification. Although complete manus and pes skeletons are not known for a number of fossil groups, sufficient information exists to constrain many interpretations of trackmaker affinity. This is essentially the predictive approach implied and advocated by Parrish (Chapter 16) and Unwin (Chapter 27). Such approaches are most effective when well integrated with fossil track studies to establish correspondence between body and trace fossils. Because the extent to which manus and pes skeletal morphology reflects the shape of tracks varies with the amount of soft tissue or flesh found in the feet of different vertebrate groups (Chapter 50), caution must be used in the predictive approach. Fossil feet will not always fit the trackways, and unequivocally prove or disprove the affinity of footprints.

References

Padian, K., and Olsen, P. E. 1983. Footprints of the Komodo dragon and the tracks of fossil reptiles. *Copeia* 1984 (3):662–671.

1984. The fossil trackway *Pteraichnus*: not pterosaurian, but crocodilian. *Jour. Paleont.* 58:178–184.

24 Ratite Footprints and the Stance and Gait of Mesozoic Theropods

KEVIN PADIAN AND
PAUL E. OLSEN

Abstract

Footprints of the rhea (*Rhea americana*) are identical in several diagnostic features to tridactyl footprints of the Mesozoic Era attributed to theropod dinosaurs. Of particular interest, (i) the rhea's feet are placed very close to its body midline as it walks, so that it virtually places one foot in front of the other; (ii) its middle toe (digit III), the central weight–bearing axis, is directed slightly inward under normal conditions; and (iii) the feet are very deliberately placed on the substrate, and the toes and claws leave no drag marks. These are all characteristic of Mesozoic theropod (and ornithopod) trackways, and invite extended comparison of fossil and recent theropods. Modern ratites and Mesozoic theropods are essentially identical in bone morphology and in joint structure and articulations. Their trackways are similar because the structure and function of the hindlimbs of the two groups are also essentially identical. These similarities are to be regarded as homologies because birds are descended from Mesozoic theropods, and the ratites merely retain characters plesiomorphic for the group since the Late Triassic. Mesozoic theropods had fully erect stance and parasagittal posture, as both bone structure and articulation, and footprints reveal. Hypotheses of semi–erect posture based on hypothetical muscle reconstructions are not supported by the available evidence.

Introduction

Dinosaurs were unknown as a group when Edward Hitchcock described gigantic three–toed trackways from the redbeds of the Connecticut River Valley in the early 1830s. Hitchcock (1836) noted that they were first regarded as prints made by giant birds, including "Noah's raven." He named the prints "*Ornithichnites*" to reflect their origin and thereby differentiate them from the tracks of reptiles, or "*Saurichnites,*" found in the same beds. Sir Richard Owen named the Dinosauria in 1842, but on such fragmentary material that the Connecticut Valley tracks could not be associated with the osseous remains of dinosaurs until the

late 1800s (Colbert 1968, Desmond 1975), when relatively complete skeletons of carnivorous and herbivorous dinosaurs became known from places as diverse as Belgium and the western United States. By the time Lull published the first edition of *Triassic Life of the Connecticut Valley* (1915), footprints could be referred not only to dinosaurs but to other archosaurian groups, often at the family or even genus levels. The parataxonomy of fossil footprints is still preserved, but in specific cases the inference that certain bones and trackways may have been left by the same animals has been of great use in reconstructing stance, gait, and functional morphology of extinct tetrapods, particularly reptiles (Haubold 1971).

Experimental studies (e.g., Schaeffer 1941, Peabody 1959, Padian and Olsen 1984a,b) of the trackways of living tetrapods have been able to shed light on the process of trackmaking and how it relates to the structure of the foot, the kinematics of the limb, and the competence of the substrata (Baird 1954, 1957; Padian and Olsen 1984a). In many cases, the form of a footprint can reveal not only the identity of the trackmaker and the condition of the surface, but also the animal's stance and gait, which may vary with behavioral and environmental circumstances. The experimental approach is especially powerful when comparing fossil and living members of a single phylogenetically restricted group. For example, the earliest known crocodilian trackways (*Batrachopus*, reviewed in Olsen and Padian 1986) differ in no appreciable way from those of living crocodiles (Padian and Olsen 1984a), which suggests that crocodilian locomotory trends have remained conservative over nearly 200 million years.

The present study is an attempt to test alternative hypotheses of the stance and gait of Mesozoic theropod dinosaurs, using the evidence of fossil and recent footprints. No living vertebrates have precisely the same pelvic structure as Mesozoic ornithischian and saurischian dinosaurs. Some lines of evidence, such as the shapes of pelvic and hindlimb joints and the inferred angles of articulation of their bones, have led paleontologists to conclude that Mesozoic dinosaurs walked in fully upright position (e.g., Ostrom 1970, Charig 1972). However, certain problems associated with reconstruction of muscles on these dinosaurs have suggested to others (e.g., Tarsitano 1983) that fully erect posture was impossible, and a more crocodilian, semi-erect stance was more likely.

We proceed from the premises that (i) similarities of bone shapes and joint articulations between two tetrapods suggest similar gaits: that is, if no functionally significant difference is demonstrated, none can be assumed; and (ii) functional similarity between animals is correlated with degree of phylogenetic relationship, given the caveat listed in (i). If the bones and articulations of two animals show no significant functional differences, and if their footprints match in all respects related to kinematics of the limbs, then the inference that those kinematics are fundamentally similar is a strong one. We realize that, except in certain cases, fossil tracks cannot be assigned to a trackmaker at a generic or specific level, and we will content ourselves with similarities and differences pertinent to the "family" level or above. In the case of Mesozoic theropods, as Olsen (1980) has demonstrated, differences that characterize ichnogenera can often reduce to a single allometric transformation series; that is, *Grallator*, *Anchisauripus*, and other theropodan footprint taxa intergrade statistically according to size alone, and it is impossible to know whether a given footprint was made by an adult of a small species or a juvenile of a large species.

Previous Work

For 150 years, the resemblance of footprints of living ratite birds to those of Mesozoic dinosaur tracks has been noted — even before the latter were recognized as dinosaurian. Hitchcock (1858, Plate LV, Fig. 1) figured what he described as a track of a "South American Ostrich" (i.e., a rhea) to compare with fossil tracks of the Connecticut Valley. Ratites are the largest of living birds as well as the most primitive, and so they were natural models for comparison with the large petrified tracks that were being found in New England, Europe, and elsewhere.

Because living ratites are confined to the Southern Hemisphere, most records of ratite footprints, both living and fossil, come from southern continents. Dr. Donald Baird has kindly provided us with his file of some references to these records, most of which are likely to be overlooked. They include records of footprints of the extinct moa (*Dinornis*) in New Zealand (Gillies 1872, Williams 1872, Owen 1879, Voy 1880, Benham 1913, Wilson 1913, Lambrecht 1933 p. 174–175, Archey 1941) and a Tertiary record of a "kiwi-like bird" from New Zealand (Hutton 1899). Heilmann, who was extremely thorough in his paleo-ornithological investigations, figured cassowary tracks in *The Origin of Birds* (1926 p. 180–181) along with tridactyl dinosaur tracks from the Connecticut Valley; the latter photos were taken from the work of Beebe (1906 p. 396–397). Dr. Baird notes two other treatments of ratite

tracks, this time of the emu, in Jaeger (1948 p. 215–218) and Mountford (1946). Finally, Rich and Green (1974) compared the footprints of living cassowaries and emus with those of fossil dromornithine footprints from Tasmania.

The footprints of ratites should be of special interest to dinosaurian paleontologists because birds are living dinosaurs. Their origin from Mesozoic coelurosaurian theropods is now beyond reasonable dispute (Ostrom 1985; Gauthier and Padian 1985; Gauthier 1984, 1986). By cladistic convention, birds must be classified as theropod dinosaurs because they evolved from theropod dinosaurs. Because Mesozoic theropods are so closely related to living birds, it ought to prove interesting to compare their tracks. Differences in limb proportions (especially the general elongation of the metatarsus), changes in pelvic structure, and the loss of the tail have been salient features in the evolution of birds from non-avian theropods. Trackways, as indicators of stance and gait, ought to shed light on whether any changes in these functional features have occurred as a result of the morphological evolution mentioned above. Yet we know of no study in the published literature that has compared trackways of living and extinct dinosaurs in an effort to address these questions.

Dr. W. A. S. Sarjeant has kindly drawn our attention to two studies of tridactyl footprints from the Triassic conglomerate of Wales, a set of five theropod prints that Sollas (1879) and artist T. H. Thomas (1879) compared to footprints of the living emu and to the feet of stuffed cassowary and rhea. Sollas' study, of which we will repeat some details below, concludes with the following passage:

> So complete is the agreement in all essential points between the footprints in the Triassic conglomerate and those of the living Emu, that, leaving all other considerations out of the question, one would not feel much hesitation in declaring for the Avian and, indeed, Ratitous character of the animal which produced the former. But the other considerations are too important to be overlooked. Although the remains of fossil vertebrates have in several instances been discovered in the Triassic deposits of S.W. England, yet none have hitherto been referred or referable to Birds; on the other hand many of them are true Reptiles, though with extraordinarily strong ornithic affinities. The existence of Dinosaurs during the Trias gives, indeed, a strong *prima facie* probability to the supposition that these associated bird-like footprints were really produced by some form of Ornithic Reptile.
>
> The occurrence of *Thecodontosaurus* and *Palaeosaurus* in the magnesian conglomerate of Durdham Down, Bristol, which is on the same parallel of latitude as Newton Nottage, and only 45 miles distant, is very suggestive; and I cannot help thinking that one or other of the animals which possessed the bones must have been a near relation to that which has left its footprints in the magnesian conglomerate of S. Wales.

The Discussion at the end of Sollas' paper records: "Professor Hull pointed out that Prof. Marsh had suggested that the supposed footprints of birds in the Connecticut Valley may probably have been made by Dinosaurs." From these discussions it is clear that the similarities of bird and dinosaur tracks were being considered, but in the absence at that time of a viable hypothesis of ancestry of birds from dinosaurs (apart from Huxley's [1868, 1870] studies, which had by no means received full acceptance), there was no indication that the resemblances of ratite tracks to those of the presumed "Ornithic Reptiles" was anything more than coincidental.

Materials and Methods

We decided to record the footprints of the rhea (*Rhea americana*) because (unlike the ostrich) it conserves all three pedal digits primitive to theropods, it is relatively docile, and it was readily available at the Oakland Zoo. The rhea that we used was subadult and weighed about 25 kg. These birds are not known for their intelligence, but this one was relatively accustomed to human contact.

Our methods were similar to those we used in taking the footprints of the Komodo monitor (Padian and Olsen 1984b). On a base of plywood 120 cm by 240 cm we laid slabs of potters' clay 2 cm thick, smoothed the surface, and sprayed it occasionally with water to prevent drying. The rhea was placed at the end of the trackbed and encouraged to walk across it. At first it avoided the clay but eventually accepted it. Because its first trail was uncertain, we had the rhea walk across the clay a second time. We examined and photographed the trackbed and then took molds of the entire clay slab with ordinary casting plaster. All measurements and observations used in the study were taken from the plaster molds.

Results

Figure 24.1 is a diagram of the rhea's trackways; it represents the plaster mold of the clay trackbed. It should be remembered that the topographic "left" foot of the plaster mold was really made by the rhea's right foot; however, for the sake of clarity we will refer to the topographic left foot of the plaster mold as the track of the left foot.

In Figure 24.1 the first trackway made by the rhea proceeds down the page, and the second trackway proceeds up the page. The impressions of the digits II–IV measure 8, 13, and 7.5 cm respectively. Of particular interest are the following features.

1. The rhea's phalangeal formula is 3–4–5 for digits II–IV. In general, the proximal phalanx of the second digit and the proximal two phalanges of the fourth digit were not impressed. All four phalanges of the third toe are visible except the proximal part of the first phalanx, which is separated by a slight gap from the round central metatarsal pad. Similar spaces separate the other toes from this pad. (Thomas [1879] noted that this feature in the rhea resulted from the thick horny boss at the end of the metatarsal, but that the separation was not complete in the first digit of the emu.) It should be remembered that the pads of sauropsids (including birds) cover the interphalangeal joints, not the phalanges themselves as in the grasping hands of mammals. A reconstruction of the rhea's skeletal

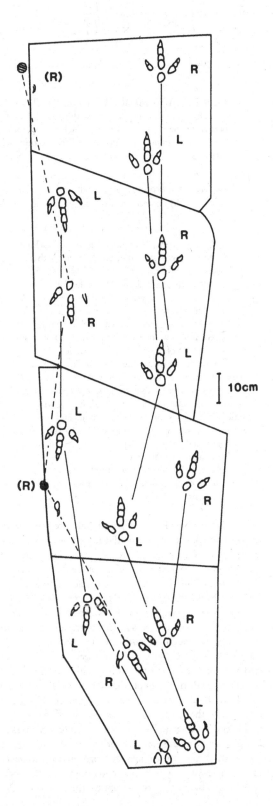

Figure 24.1. Diagram of *Rhea americana* trackways (UCMP 131683), plaster mold of clay trackbed.

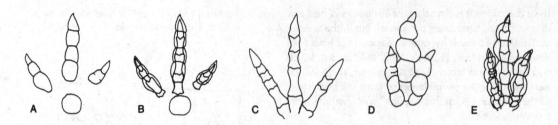

Figure 24.2. **A.** Footprint of *Rhea americana* (UCMP 131683). **B.** Reconstruction of foot skeleton based on preserved phalangeal impressions. Note how this differs from **C**, The foot skeleton of a real rhea, by virtue of the functional pattern of the rhea's gait (see Fig. 24.3 and discussion in Padian and Olsen, 1984a). **D**, *Anchisauripus* sp. (A.C. 49/1) from slab diagrammed in Figure 24.4. **E**, Reconstruction of the theropod foot skeleton that made this footprint. All drawings of left foot; not to scale.

foot, based on its footprints, is given in Figure 24.2.

Sollas (1879) noted in his study of the emu's footprints:

> On comparing the regions of the sole of the Emu's foot with its skeletal structure, one is struck with their wide divergence in details, which clearly shows the futility of too closely arguing in all cases from the skeletal structure of a foot to the impression it might make on the surface of a sedimentary deposit. Thus, while the feet of most of the Ratitae possess a prominent heel, the end of the tarso–metatarsal bone, on the other hand, does not appear to reach the level of the ground; so too, while the articulations of the phalanges are the most swollen parts of the digital skeleton, on the sole of the foot they are the least so, owing to the excessive development of tissue over the middle of the phalangeal bones; and, finally, while the inner toe possesses three phalanges in all, and the outer toe as many as five, yet the imprints left by these digits on the ground show only two depressions in each case — one a mere pit indicating the nail, the other a long groove representing all the rest of the phalanges. Moreover the number of phalanges indicated varies with the way the foot is set on the ground....

Our observations are in general agreement with those of Sollas, except that he did not recognize that the interphalangeal articulations are the most swollen parts of the sole of the foot, and therefore of the footprint.

Sollas' remark about the incomplete impression of the proximal phalanges, which we have noted in our results, also applies to the footprint figured by Hitchcock (1858 Plate LV, Fig. 1). This seems to be a general characteristic of avian trackways. Although the outline of the posterior part of the foot is visible, the proximal phalangeal impressions are seldom clear. We suspect that this is related to the longer metatarsus and short toes of these birds, and to the angle at which the metatarsus meets the ground. The third metatarsal is the longest, the second and fourth are shorter and laterally divergent. This was true for theropods plesiomorphically, but birds have longer metatarsals than other theropods. Because these bones are longer, in order to maintain the relative position of the center of mass of the body (represented by, but not equivalent to, the position of the

acetabulum over the metatarsal–phalangeal joint), they may have been angled more vertically than those of other theropods (Fig. 24.3). Another possibility is that the metatarsals of birds elongated to move the center of support farther forward to compensate for (1) the increased pectoral muscle mass of birds, which make up one-fourth to one-third of their muscle mass, and (2) the loss of the long, fleshy tail of their dinosaurian ancestors.

If this generalization is true, then we would expect a similar pattern in the trackways of ornithomimid theropods, which had exceptionally long metatarsals and foreshortened toes. Unfortunately, it is difficult to verify that particular footprints were made by ornithomimids. Sternberg (1926) named "*Ornithomimipus angustus*" without diagnosis as the footprint of *Ornithomimus*, because the latter was "the only animal which would make tracks similar to those here preserved." Like the rhea tracks, Sternberg's show a gap where the proximal part of the second toe and distal part of its metatarsal would have been expected to have been impressed. Significantly, metatarsal II is shorter than III and IV so Sternberg's inference about the trackmaker may be correct. Avnimelech (1966) assigned some mid-Cretaceous (upper Albian or lower Cenomanian) tracks from Israel to ornithomimids, but most are too poor for diagnosis. A few (his Fig. 2, bottom row; 'Type A') show both the circular impression of the metatarsal 'boss' and gaps where the proximal parts of digits II–IV would have been impressed. Unfortunately Avnimelech gave no measurements and only figured isolated tracks, not trackways, so it is impossible to evaluate the variation.

2. The angle between digits II–III ranged from 33° to 47°, with a mean value of 40°. The angle between digits III–IV ranged from 45° to 54°, with a mean value of 52°. The uniformly wider angle between digits III–IV is characteristic of birds and facilitates assignment of isolated tracks to left or right. The angle between digits II–IV ranged from 84° to 99°, with a mean value of 91°. Most of the variation in this latter measurement was due to the variation in the angle between digits II–III. The average stride length was 67 cm, with a range of 54 to 78 cm, but stride length should not be considered characteristic, because the rhea walked slowly and cautiously as it tested the trackbed.

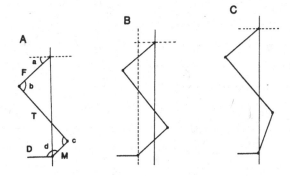

Figure 24.3. Why are the rhea's proximal phalanges not apparent in its footprints? **A** represents a schematic dinosaur hindlimb, with the center of gravity (vertical line) arbitrarily passing through the hip joint and ground contact point. In **B,** the metatarsus has elongated, but no other change has been made. This forces the center of gravity backwards. In **C,** to maintain the original alignment, the orientation of the metatarsus has changed to a more vertical position (but angles *a* and *b* have remained constant). This pulls the metatarsal-phalangeal joints off the ground — a possible answer to the question. Abbreviations: D = digits, F = femur, M = metatarsus, T = tibia.

3. In those portions of the trackbed in which the rhea walked in a straight line and without hesitation, the trackways show two important characteristics: first, one foot is placed almost directly in front of the other; and, second, the third toe (the main weight–bearing axis of the foot) is directed straight ahead or even toed slightly inward. These observations are corroborated by the trackways of birds figured by Hitchcock (1858 Pls. VII, XXXI, LIV), Deane (1861 Pl. III), and later workers.

Comparison of the Tracks of the Rhea with Those of Mesozoic Dinosaurs

In addition to examining the Mesozoic dinosaur trackways in the literature cited below, we photographed and mapped a slab in the Hitchcock Ichnological Collection of the Pratt Museum at Amherst College (A.C. 49/1). It was collected in 1862 from the south side of Turner's Falls, Massachusetts (Early Jurassic, probably Portland Formation), but was never previously figured. Hitchcock (1865 p. 82) referred the numerous tracks on this slab to *Brontozoum validum, B. exsertum,* and *B. sillimanium,* which we regard as synonymous with *Anchisauripus* sp. following Olsen (1980). We have mapped two *Anchisauripus* trackways in Figure 24.4, to compare with those of the rhea. Trackway A, proceeding down the right side of the slab, consists of four prints separated by 82, 82, and 84 cm respectively. Trackway B, proceeding up the left side, comprises three prints separated by 113 and 109 cm respectively. Lengths of digits II–IV are approximately 10, 15, and 10 cm respectively for both trackways. The disparity in pace length sug-

gests that the animal that made trackway B was traveling faster.

Compared to the rhea trackway, A.C. 49/1 shows some similarities and some differences. The angles between the digits are much smaller than in the rhea, averaging 20° (range 19°–21°) and 16° (range 14–17°), respectively, for angles II–III and III–IV. Like the rhea, one foot is placed almost directly in front of the other. This can be seen in other theropod trackways (e.g., Hitchcock 1858 Pls. XXXIII, XXXVIII, XXXIX, XL, XLI, XLII, XLVII Figs. 3 and 6, and LIII Fig. 5; Lull 1953; Thulborn and Wade 1984 Fig. 3; etc.). In the tracks on A.C. 49/1, unlke those Mesozoic tracks just cited, the toes are directed just slightly outward, as they are in some living bird tracks (e.g., Hitchcock 1858 Pl. XXXII Fig. 2, Pl. LIV Fig. 3). (The reasons for this bear further investigation, but are outside our present study.) We illustrate A.C. 49/1 to show this variation in fossil theropod trackways, but the majority of them are directed inward, as our cited references indicate. Several of these are reproduced diagrammatically in Figure 24.5.

We have stressed the similarity between the trackways of the rhea and those of Mesozoic theropods because of their close phylogenetic relationship. However, we should point out that some of the characteristics of all the trackways mentioned above also apply to the trackways of ornithisichian dinosaurs: for instance, *Anomoepus* trackways typically show one foot planted directly in front of the other (or nearly so) and are toed inward. We have given an example of this (A.C. 52/10, after Hitchcock 1865 Pl. XV) in Figure 24.6. The evidence suggests that these features were present in the common ancestor of saurischians and ornithischians, as we will explain in the next section.

Hindlimb Anatomy and Kinematics

We argue that the footprints of rheas and Mesozoic theropods are similar in these derived respects because their basic patterns of stance and gait have not changed since the Late Triassic. To support this, we must show (*contra* Tarsitano 1983) that the hindlimb anatomy and kinematics of these two groups have the same patterns.

The anatomy and kinematics of the rhea pelvis and hindlimb are straightforward. The femur is subhorizontally and slightly laterally oriented and its distal end moves up and down during locomotion, while the tibiotarsus moves parasagittally. The long tarsometatarsus also moves parasagittally because the metatarsal ankle, like the knee, allows movement in only the fore–and–aft plane.

The recent discovery of the pelvis and hindlimb of a primitive theropod dinosaur (cf. *Coelophysis*) in the Petrified Forest National Park, Arizona (Chinle Formation, Late Triassic, Late Carnian or Early Norian), allows some unusual insight into anatomical and kinematic patterns of early theropods. This specimen (University of California, Museum of Paleontology V82250/129618) has recently been described by Padian (1986); readers are refered to this paper for further details and illustrations. Here, only a few func-

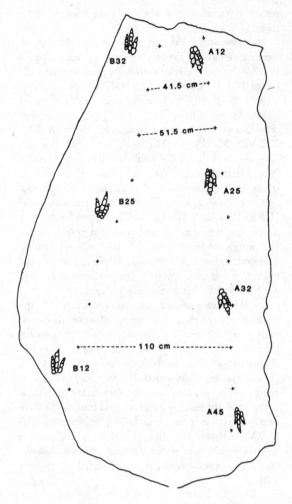

Figure 24.4. Diagram of *Anchisauripus* sp. trackway (A.C. 49/1).

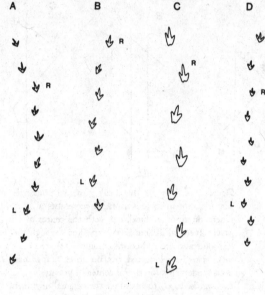

Figure 24.5. Toed-in theropod trackways, with successive footprints very nearly in a straight line. **A,D,** after Hitchcock 1858 Pl. XLI Fig. 2. **B,C,** *ibid.* Pl. XLII Fig. 2. L = left; R = right.

tional considerations will be provided.

The pelvis of UCMP 129618 has a long, low ilium with a pronounced posteroventral arch behind the ischial peduncle, an overhanging supraacetabular crest, a corresponding crest on the inside of the acetabulum, and a pronounced bifaceted antitrochanter on the posterior border of the acetabulum at the junction of the ilium and ischium (Fig. 24.7). The proximal end of the femur is flat, and the head is offset 90° from the shaft. The flat proximal end of the femur abuts against the antitrochanter when the hip is articulated. The broad head can fit into the ovoid acetabulum only when the femur is subhorizontal to horizontal. If it is held more vertically, it snaps off the supraacetabular crest and will not fit in the acetabulum. Furthermore, when the proximal end of the femur is properly fitted against the antitrochanter, the head angles upward at approximately 45° into the acetabulum (Fig. 24.7). In this orientation, as Padian (1986) showed, it is more correct to regard the dinosaurian femur as bowed in two

planes (the shaft dorsoventrally, the distal end laterally), not "sigmoid."

These features demonstrate the restrictions on the femoral position in even the earliest theropods. Martin (1983) correctly noted this orientation of the femur in the acetabulum of *Archaeopteryx*, which we conclude is a synapomorphy of theropods and probably of dinosaurs in general. Martin also argued that the partial internal closure on the inside of the acetabulum in *Archaeopteryx* was primitive for archosaurs, but not found in saurischian dinosaurs. However, it is present in UCMP 129618, in *Syntarsus* (Raath 1969), and in *Dilophosaurus* (Welles 1984), to name only a few primitive theropods, and Gauthier (1984) has shown that this feature is synapomorphic at the level of Dinosauria. These features are probably not an adaptation for jumping in *Archaeopteryx* as Martin (1983) suggested, but for upright posture. In order for the acetabulum to accommodate the head of the femur, the distal end must be abducted slightly from the parasagittal plane, thus enabling the hindlimb to clear the abdominal region (birds have retained this structural complex).

As Figure 24.7 also shows, the condyles of the distal end of the femur are subterminal. The lateral condyle fits between the tibia and fibula, as in birds. The available range of movement therefore occurs around the natural position of a 90° bend in the knee. Both the knee and ankle move parasagittally, as in birds (Fig. 24.8), and this follows from the features of hingelike, nonrotary joints, parallel surfaces, bilateral symmetry, and straight shafts (Schaeffer 1941,

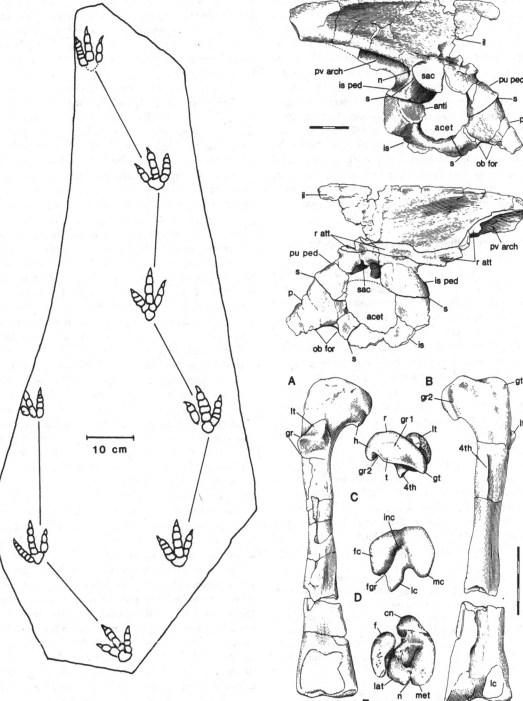

Figure 24.6. Diagram of the ornithischian trackway *Anomoepus* (A.C. 52/10), after Hitchcock 1865 Pl. XV.

Figure 24.7. Pelvis and hindlimb bones of a small theropod, cf. *Coelophysis bauri*, UCMP 129618 (Late Triassic: Chinle Formation). Above: right pelvis in lateral (upper) and medial (lower) views. Below: **A–D**, right femur in dorsal (A), ventral (B), proximal (C), and distal (D) views. **E**, left tibia and fibula in proximal view. Scale divisions = 1 cm. After Padian (1986).

Coombs 1978, Brinkman 1980, Padian 1983). These features also appear to apply to early sauropodomorphs (e.g., *Plateosaurus*), ornithischians (e.g., *Heterodontosaurus*), and dinosaurs outside the saurischian–ornithischian dichotomy (e.g., *Herrerasaurus*), and hence characterize the ancestors of dinosaurs. Padian (1983) has shown that these also apply to pterosaurs, and therefore most probably to the common ancestor of pterosaurs and dinosaurs. We therefore suggest that theropod hindlimb articulation and kinematics have not changed in any substantial details since the Late Triassic. This conclusion appears to be supported by the evidence of Early Mesozoic theropod footprints detailed above. It would be falsified if it were demonstrated that we have incorrectly assembled and manipulated the pelvis and hindlimb, or that the footprint patterns have been misinterpreted.

Crocodilian vs. Avian Paradigms for Mesozoic Theropods

The need to learn more about dinosaurian posture and gait has long led paleontologists to dissect the structures of living reptiles and compare them to those of extinct dinosaurs; Romer's doctoral dissertation work (1923a,b, 1927) is a classic case in point. This tradition has continued to the present day, but one problem common to all studies of paleobiological structure and function has yet to be solved: how can the effectiveness of a paradigm be assessed, and how can the limits of an analogy be recognized? The answer seems to be that homology is generally better than analogy, and evidence better than models. We will demonstrate some consequences of these differing approaches.

Tarsitano (1983) recently assessed the stance and gait of theropod dinosaurs of the Mesozoic. Because he disagreed with the evidence that shows that birds evolved from small coelurosaurs of the Mesozoic (e.g., Ostrom 1976, Gauthier and Padian 1985, Gauthier 1986), he had to regard most features shared by theropods and birds as convergent. In his 1983 paper, Tarsitano claimed that in birds

the tibiotarsus bone–muscle complex is the primary system of locomotion. In thecodonts, crocodilians and dinosaurs it is the tail–femoral bone–muscle complex which is most important in locomotion. Thus, in order to interpret the osteology and muscle scars of theropods, it is better to compare theropods to crocodilians which have the same morphology as the pseudosuchian predecessors of theropods.

In other words, according to Tarsitano, theropods evolved from "pseudosuchians," which are allegedly very much like crocodilians, so it is legitimate to base restorations of theropod musculature and functional morphology on the model of a crocodile. This overlooks the obvious difference in anatomy and kinematics between crocodiles and theropods. However, the real problem with the terms of this comparison is the lack of definition of the groups with which theropods are to be compared. Tarsitano (1983

p. 255) asserted "It is apparent that all saurischian dinosaurs have evolved from a pseudosuchian ancestry since the remnants of the crocodilian tarsus is [sic] to be seen in theropods, sauropods and prosauropods. The ischia and pubes of pseudosuchians are decidedly saurischian and not crocodilian."

In these and other passages, Tarsitano did not supply the membership of the "pseudosuchians" or tell us how they differ from crocodilians or other "thecodonts". The difficulty with this, as Gauthier and Padian (1985) pointed out, is that the groups commonly called "thecodonts" and "pseudosuchians" are not defined or united by any shared derived evolutionary features and therefore cannot be diagnosed. As a result, it is difficult to generalize about most aspects of their evolution, including stance and gait.

Gauthier (1984, 1986) suggested that the name "thecodont" be discarded and the name "pseudosuchian" be restricted to crocodilians and all archosaurs closer to crocodiles than they are to birds. He proposed the name "ornithosuchian" for all archosaurs closer to birds than to crocodiles. If, in the above passages, this concept of "ornithosuchian" is substituted for the word "pseudosuchian," the statements are now true in a phylogenetic sense but not in an anatomical sense, because crocodilians and ornithosuchians have different ankle joints ("crocodile-normal" versus "crocodile-reversed"). Morever, in advanced ornithosuchians (*Lagosuchus*, *Lagerpeton*, Pterosauria, and Dinosauria) a metatarsal ankle evolved, with many concomitant changes in the pelvis, femur, and tibia–fibula (Gauthier 1984, 1986; Padian 1984). Theropods should not be compared to crocodiles, a different evolutionary lineage, but to other ornithosuchians. Birds are living theropods and living ornithosuchians, and retain more "remnants" of the plesiomorphic dinosaurian structure.

The consequence of these different views for locomotion and trackway studies can be seen by contrasting Figure 24.8 with Figure 24.9, which is an adaptation of Tarsitano's (1983) view of how the theropod hindlimb was articulated, using the example of *Tyrannosaurus rex*. Note that in his figure the femur moves in a 90° arc about the vertical plane, with the knee joint nearly straight. According to our conclusions described above, in Tarsitano's left figure the hip joint is severely dislocated and the knee is nearly hyperextended. It is probable that *Tyrannosaurus*, like other very large dinosaurs, secondarily adapted to some graviportal characteristics, including a more columnar stance. But even large theropods made tracks (e.g., *Eubrontes*) like smaller ones (see Thulborn and Wade 1984).

If, as Tarsitano seems to be suggesting, crocodiles are a better model for theropodan stance and gait than are birds, theropod trackways should be considerably broader, with lower step angles, than is actually the case. We do not expect that Tarsitano's crocodilian theropod could have made the tracks in the fossil record ascribed to theropods, because the distal end of the femur would have to have been positioned considerably lateral to the proximal end (as in birds) in order to clear the body cavity. In this posi-

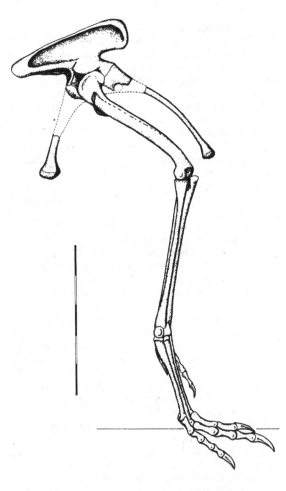

Figure 24.8. Reconstruction of pelvis and hindlimb of *Coelophysis bauri*, partly based on UCMP 129618. The femur is drawn slightly forward of its actual articulation in the acetabulum in order to show the position of the head with respect to the trochanter. Also, the position of the femur as shown is near maximum retraction, toward the end of the propulsive phase of the hindlimb. Scale division = 5 cm.

Figure 24.9. Tarsitano's (1983) reconstruction of hindlimb kinematics in *Tyrannosaurus*, with restored muscles removed for clarity.

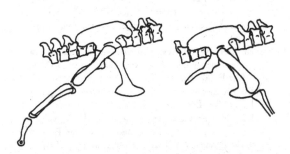

tion the tibia articulates in columnar fashion against the distal end of the femur, not its asymmetrical subterminal condyles. As positioned, it is unlikely that the tibia could have sloped back underneath the body to produce trackways so close to the body midline, and even toed slightly inward. Birds can do this because the subterminal distal femoral condyles are asymmetrical, and when the knee articulates at a 90° angle, the disparity in the position of the condyles angles the tibia–fibula medially underneath the body. This seems to have been the plesiomorphic condition for dinosaurs.

We agree with Tarsitano on a different point: that the vertebral column in Mesozoic non-avian theropods was probably held at an angle of about 20°, rather than the 0° or 50° models advocated by other authors. But we conclude on the basis of comparative anatomy and kinematics of theropods, living and extinct, that there were no substantial differences in hindlimb function, despite sweeping changes in the pelvis, tail, and mode of life that occurred during the evolution of birds. Our independent evidence for this conclusion is the ichnological record, which demonstrates no change in trackmaking pattern since the Early Mesozoic. If we are correctly interpreting Tarsitano's ideas about the stance and gait of theropods, then we have no choice but to conclude that, if he were right, then the Mesozoic footprints ascribed to theropods must be misidentified. It seems more likely that a crocodile is not a good model for theropod stance and gait.

Conclusions

The first conclusion that we wish to emphasize is the importance of phylogeny in understanding function. From the above considerations, it is unlikely that Mesozoic theropods differed from living birds in their mode of walking, except in individual details (for example, in large forms such as *Tyrannosaurus*, which are clearly secondarily derived). It is a precept of comparative anatomy that functional inferences are most likely to be correct when the structural similarities on which they are based are phylogenetically homologous. As the phylogenetic level becomes more specific, the functional inference becomes more powerful. Modern birds are descended from Mesozoic theropods, and crocodilians represent an outgroup that retains many primitive archosaurian patterns, as well as its own derived ones. With this phylogenetic paradigm as an independent body of evidence, some reasons emerge for the functional similarities of living birds and Mesozoic theropods.

The second conclusion is that footprints can serve as an independent test of functional or ecological hypotheses. On the basis of comparative anatomy alone, we could have hypothesized that the locomotory patterns of the rhea and other birds were essentially unchanged from those of *Coelophysis*. A comparison of their footprints was perhaps the only possible independent test of this hypothesis. In our view, the hypothesis has proven robust. We expect that this conclusion will not surprise most workers; we will be content if it helps to lay to rest some arguments about the paleobiology of Mesozoic dinosaurs that are not supported by several independent lines of evidence.

Acknowledgments

For allowing us access to their rhea and for much help in taking its footprints, we thank Director of Research Joel Parrott, D.V.M., and the staff of the Oakland Zoo. We also thank Ken Warheit for assistance in taking the footprints, and for useful discussion. Dr. Edward Belt kindly provided access to the Hitchcock Ichnological Collection of the Pratt Museum at Amherst College. Drs. Don Baird, W. A. S. Sarjeant, and several anonymous reviewers kindly provided many helpful references and good critical comments on the manuscript. We are especially grateful to Steven M. Gatesy, who offered many good insights and criticisms based on his work in progress on hindlimb kinematics of living birds and other reptiles. Figure 24.7 (from Padian [1986], with Figure 24.8 courtesy Cambridge University Press) was drawn by Jaime Pat Lufkin of UCMP, which also supported the costs of making the rhea trackways. Mr. Howard Schorn (UCMP) provided his usual good and timely photographic help.

References

Archey, G. 1941. The moa: a study of the Dinornithiformes. *Auckland Institute Museum Bull.* 1:1–119, XV Plates, Tables A–O.

Avnimelech, M. A. 1966. Dinosaur tracks in the Judean hills. *Proc. Israel Acad. Sci. Humanities, Section Sci.* 1:1–19.

Baird, D. 1954. *Chirotherium lulli*, a pseudosuchian reptile from New Jersey. *Bull. Museum Comparative Zool., Harvard Univ.* 111:165–192.

1957. Triassic reptile footprint faunules from Milford, New Jersey. *Bull. Museum Comparative Zool., Harvard Univ.* 117:449–520.

Beebe. 1906. *The Bird: Its Form and Function.* (Henry Holt & Co.) 496 pp.

Benham, W. B. 1913. Notes on footprints of the moa. [Note in paper by K. Wilson.] *Trans. New Zealand Institute* 45:211.

Brinkman, D. 1980. The hindlimb step cycle of *Caiman sclerops* and the mechanics of the crocodile's tarsus and metatarsus. *Canadian Jour. Zool.* 58 (12):2187–2200.

Charig, A. J. 1972. The archosaur pelvis and hindlimb: an explanation in functional terms. *In* Joysey, K. A., and Kemp, T. S. (ed.). *Studies in Vertebrate Evolution* (Winchester Press) pp. 121–156.

Colbert, E. H. 1968. *Men and Dinosaurs: The search in field and laboratory.* (E. H. Dutton Company) 283 pp.

Coombs, W. P. 1978. Theoretical aspects of cursorial adaptations in dinosaurs. *Quart. Rev. Biology* 53:393–418.

Deane, J. 1861. *Ichnographs from the Sandstone of the Connecticut River.* (Little, Brown & Co.) 91 pp.

Desmond, A. J. 1975. *The Hot-Blooded Dinosaurs: A revolution in paleontology.* (Blond and Briggs) 238 pp.

Gauthier, J. A. 1984. A cladistic analysis of the higher systematic categories of the Diapsida. Ph.D. thesis, Department of Paleontology, University of California, Berkeley. Dissertation #85–12825, University

Microfilms, Ann Arbor, Michigan.

1986. Saurischian monophyly and the origin of birds. *In* Padian, K. (ed.). *The Origin of Birds and the Evolution of Flight. Mem. California Acad. Sci.* 8:1–55.

Gauthier, J. A., and Padian, K. 1985. Phylogenetic, functional, and aerodynamnic analyses of the origin of birds. *In* Hecht, M. K., Ostrom, J. H., Viohl, G., and Wellnhofer, P. (ed.). *The Beginnings of Birds.* (JuraMuseum) pp. 185–197.

Gillies, T. B. 1872. On the occurrence of footprints of the moa at Poverty Bay. *New Zealand Institute Proc. Trans.* 4:127–128.

Haubold, H. 1971. *Ichnia Amphibiorum et Reptiliorum fossilium. Handbuch der Palaeontologie, Teil 18.* (Gustav Fischer) 124 pp.

Heilmann, G. 1926. *The Origin of Birds.* (Appleton) 210 pp.

Hitchcock, E. 1836. Ornithichnology. Description of the footmarks of birds (*Ornithichnites*) on New Red Sandstone in Massachusetts. *Amer. Jour. Sci.* 29:307–340.

1858. *Ichnology of New England. A report on the sandstone of the Connecticut Valley, especially its fossil footmarks.* (William White) 199 pp. + LX pls.

1865. *Supplement to the Ichnology of New England.* (Wright and Potter) 96 pp. + XX pls.

Hutton, F. W. 1899. On the footprint of a kiwi-like bird from Manaroa. *Trans. and Proc. New Zealand Institute* 31:486, Pl. XLV.

Huxley, T. H. 1868. On the animals which are most nearly intermediate between the birds and reptiles. *Annals and Magazine Nat. Hist.* 2 (4):66–75.

1870. Further evidence of the affinity between the dinosaurian reptiles and birds. *Quart. Jour. Geol. Soc. London* 26:12–31.

Jaeger, E. 1948. *Tracks and Trailcraft.* (Macmillan) 381 pp.

Lambrecht, K. 1933. *Handbuch der Palaeornithologie.* (Gebrüder Borntraeger Verlag) 1024 pp.

Lull, R. S. 1915. Triassic life of the Connecticut Valley. *Connecticut State Geol. Nat. Hist. Surv. Bull.* 24:1–285.

1953. Triassic life of the Connecticut Valley (revised). *Connecticut State Geol. Nat. Hist. Surv. Bull.* 81:1–331.

Martin, L. D. 1983. The origin of birds and of avian flight. *In:* Johnston, R. (ed.). *Current Ornithology* 1:105–129.

Mountford, C. P. 1946. *Spinifex* town. *Nat. Hist.* 55 (1):62–68.

Olsen, P. E. 1980. Fossil great lakes of the Newark Supergroup in New Jersey. *In* Manspeizer, W. (ed.). *Field Studies of New Jersey Geology and Guide to Field Trips, 52nd Annual Meeting, New York State Geological Association. Geology Dept., Rutgers Univ.* pp. 352–398.

Olsen, P. E., and Padian, K. 1986. Earliest records of *Batrachopus* from the southwestern United States, and a revision of some Early Mesozoic crocodylomorph ichnogenera. *In* Padian, K. (ed.). *The Beginning of the Age of Dinosaurs.* (Cambridge University Press) pp. 259–276.

Ostrom, J. H. 1970. Terrestrial vertebrates as indicators of Mesozoic climates. *Proc. North Amer. Paleont. Conv.* Part D:347–376.

1976. *Archaeopteryx* and the origin of birds. *Biological Jour. Linn. Soc. London* 8:91–182.

1985. The meaning of *Archaeopteryx*. *In* Hecht, M. K., Ostrom, J. H., Viohl, G., and Wellnhofer, P. (ed.). *The Beginnings of Birds* (JuraMuseum) pp. 161–176.

Owen, R. 1842. Report on British fossil reptiles, part II. *Report British Assoc. Adv. Sci. 11th Meeting Plymouth* 1841:60–294.

——— 1879. *Extinct Wingless Birds of New Zealand; with an Appendix on Those of England, Australia, Newfoundland, Mauritius, and Rodriguez.* (John Van Voorst) pp. 451–453, pl. CXVI.

Padian, K. 1983. A functional analysis of flying and walking in pterosaurs. *Paleobiology* 9 (3):218–239.

——— 1984. The origin on pterosaurs. *In:* Reif, W.-E., and Westphal, F. (eds.). *Third Symposium on Mesozoic Terrestrial Ecosystems* (Attempto Verlag) pp. 163–168.

——— 1986. On the type material of *Coelophysis* Cope (Saurischia: Theropoda), and a new specimen from the Petrified Forest of Arizona (Late Triassic: Chinle Formation). *In* Padian, K. (ed.). *The Beginning of the Age of Dinosaurs: Faunal change across the Triassic-Jurassic boundary.* (Cambridge University Press) pp. 45–60.

Padian, K., and Olsen, P. E. 1984a. The track of *Pteraichnus*: not pterosaurian, but crocodilian. *Jour. Paleont.* 58:178–184.

——— 1983b. Footprints of the Komodo dragon and the trackways of fossil reptiles. *Copeia* 1984 (3):662–671.

Peabody, F. E. 1959. Trackways of living and fossil salamanders. *Univ. California Publ. Zool.* 63 (1):1–72.

Raath, M. A. 1969. A new coelurosaurian dinosaur from the Forest Sandstone of Rhodesia. *Arnoldia (Rhodesia)* 4:1–25.

Rich, P. V., and Green, R. H. 1974. Footprints of birds at South Mt. Cameron, Tasmania. *The Emu* 74 (4):245–248.

Romer, A. S. 1923a. Crocodilian pelvic muscles and their avian and reptilian homologues. *Bull. Amer. Museum Nat. Hist.* 48:533–552.

——— 1923b. The pelvic musculature of saurischian dinosaurs. *Bull. Amer. Museum Nat. Hist.* 48:605–617.

——— 1927. The pelvic musculature of ornithischian dinosaurs. *Acta Zoologica* 8:225–275.

Schaeffer, B. 1941. The morphological and functional evolution of the tarsus in amphibians and reptiles. *Bull. Amer. Museum Nat. Hist.* 78:395–472.

Sollas, W. J. 1879. On some three-toed footprints from the Triassic Conglomerate of South Wales. *Quart. Jour. Geol. Soc. London* 35:511–517.

Sternberg, C. M. 1926. Dinosaur tracks from the Edmonton Formation of Alberta. *Canada Dept. Mines Geol. Surv. Bull.* 44 (Geological Series 46):85–87, 134–135.

Tarsitano, S. 1983. Stance and gait in theropod dinosaurs. *Acta Palaeont. Polonica* 28 (1–2):251–264.

Thomas, T. H. 1879. Tridactyl uniserial ichnolites in the Trias at Newton Nottage, near Porthcawl, Glamorganshire. *Cardiff Naturalists Soc. Reports Trans.* 10:72–91, pl. II.

Thulborn, R. A., and Wade, M. 1984. Dinosaur trackways in the Winton Formation (Mid–Cretaceous) of Queensland. *Mem. Queensland Museum* 21 (2):413–517.

Voy, C. D. 1880. On the occurrences of footprints of *Dinornis* at Poverty Bay, New Zealand. *Amer. Naturalist* 14:682–684.

Welles, S. P. 1984. *Dilophosaurus wetherilli* (Dinosauria, Theropoda): osteology and comparisons. *Palaeontographica* 185A:85–180.

Williams, W. L. 1872. On the occurrence of footprints of a large bird, found at Turanganui, Poverty Bay. *New Zealand Inst. Proc. Trans.* 4:124–2127.

Wilson, K. 1913. Footprints of the moa. *Trans. New Zealand Inst.* 45:211.

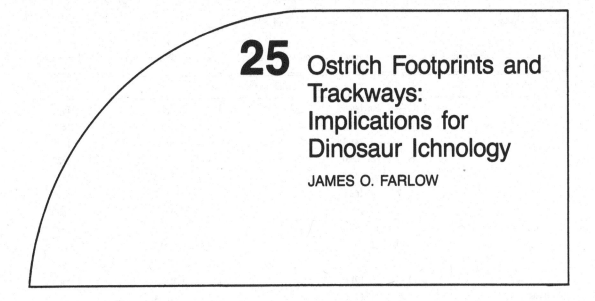

25 Ostrich Footprints and Trackways: Implications for Dinosaur Ichnology

JAMES O. FARLOW

Abstract

A fully-grown male ostrich (*Struthio camelus*) was allowed to walk and trot over a plowed, smoothed, and moistened dirt surface at the Fort Wayne, Indiana Children's Zoo. The ostrich left footprints about 22 cm long and took strides ranging 239–424 cm (stride/footprint length ratio = 11–19); an ostrich running at top speed probably takes strides in excess of 800 cm (stride/footprint length approaching 40). Step angles for these ostrich trails ranged from 133° to about 180°, with most values falling in the range of 140°–170°, as in bipedal dinosaur trackways. Unlike the trails of most bipedal dinosaurs, individual ostrich footprints angle outward (positive rotation) with respect to the animal's direction of travel. Some variations in footprint form among the tracks made by my bird can be related to differences in gait when the footprints were emplaced.

Introduction and Methods

Although large ratites are structurally very similar to bipedal dinosaurs, rather little has been done to compare the trackways of the two groups of animals (see Padian and Olsen this volume). In July, 1983 I had the opportunity to examine footprints and trails made by a one–year–old, fully grown (but not sexually mature) male ostrich (*Struthio camelus*; named "Redlegs") at the Fort Wayne, Indiana Children's Zoo.

Zoo personnel under my direction used a tractor to smooth the dirt surface of a large animal holding pen. Water sprinklers were used to wet the soil, after which the ostrich was permitted to walk and trot about the enclosure (the pen was too small to permit the bird to run at top speed). After the ostrich was removed from the pen, I measured and photographed selected portions of its trackways. Casts (negative copies or artificial convex hyporeliefs) of a few footprints were then taken with plaster of Paris.

Trackway paces and strides were measured with the anterior end of the large inner toe (digit III) as the reference point. I then used these values to calculate trackway step angles (pace angulations) by the law of cosines.

Results

Strides of the ostrich ranged 238.8–424.4 cm (Table 25.1). The footprint length for this bird was about 22 cm, so the stride/footprint length ratio ranged about 11–19. Step angles ranged 133°–180°, with most values in the range 140°–170°.

Although I made no systematic effort to investigate ostrich footprint morphology as a function of gait or substrate conditions, I did make a few observations. Three of the footprints that I copied in plaster were made when the ostrich was accelerating. Tracks 3 (Fig. 25.1) and 4 of the north–northeast trail represent an interval in which the bird increased its pace length by 23% (pace 3–4) and then 30% (pace 4–5) over the length of pace 2–3. The large inner toe (III) is more deeply impressed than the small outer toe (IV), and is most deeply impressed at the toetip. Track 3 of the south trail was made during an interval in which the ostrich increased its pace length by 51% over the previous pace (2–3), but here the inner toe is impressed rather evenly over its length. Track 8 (Fig. 25.2) of the south trail was made as the ostrich changed its direction of travel by angling off to the left, perhaps decelerating slightly as it did so. There are three conspicuous slide marks at the rear of the footprint, made by the posterior portions of the two toes and (probably) the distal end of the tarsometatarsus. The ungual "hoof" of the inner toe is the deepest part of the track; the plantar surface of this toe is most deeply impressed along its medial margin, and is markedly shallower along its lateral edge. In all of the footprints the track margins are conspicuously grooved, presumably by fringe scales of the ostrich's toes.

Discussion

For ostriches running at top speed, the stride length and the stride/footprint length ratio are undoubtedly much

Table 25.1. *Measurements of trackways of an adult male ostrich ("Redlegs") at the Fort Wayne, Indiana Children's Zoo.*

Trackway	Pace (cm)	Stride (cm)	Step Angle
South	Track 1 (L) — 2 (R): 144.8	Track 1 — 3: 302.3	165°
	2 — 3: 160.0	2 — 4: 401.3	180°
	3 — 4: 241.3	3 — 5: 424.2	162°
	4 — 5: 188.0	4 — 6: 330.2	162°
	6 — 7: 167.6	6 — 8: 330.2	180°
	7 — 8: 162.6	7 — 9: 315.0	155°A
	8 — 9: 160.0	8 — 10: 325.1	166°
	9 — 10: 167.6	9 — 11: 320.0	155°
	10 — 11: 160.0	10 — 12: 299.7	
		11 — 13: 279.4	
North–	Track 1 (L) — 2 (R): 101.6	Track 1 — 3: 238.8	156°
Northeast	2 — 3: 142.2	2 — 4: 312.4	159°
	3 — 4: 185.4	3 — 5: 358.1	166°
	4 — 5: 185.4	4 — 6: 369.6	171°
	5 — 6: 185.4	5 — 7: 368.3	167°
	6 — 7: 185.4	6 — 8: 354.3	163°
	7 — 8: 172.7	7 — 9: 335.3	133°
	8 — 9: 193.0		
Boardwalk	Track 1 (L) — 2 (R): 137.2	Track 1 — 3: 297.2	180°
	2 — 3: 160.0	2 — 4: 330.2	180°B
	3 — 4: 165.1		
Southwest	Track 1 (L) — 2 (R): 160.0		

AThe ostrich changed direction at footprint #8 (a right), veering to the left.

BThe measured stride length is greater than that of the combined pace lengths. This is probably due to inaccuracy of measurement, exacerbated by the conversion from English units (the units of my tape measure) to their metric equivalents.

greater than the values reported here. For ostriches in southern Africa, the British missionary–explorer David Livingstone wrote:

> When the ostrich is feeding his pace is from twenty to twenty–two inches; when walking, but not feeding, it is twenty–six inches; and when terrified, as in the case noticed, it is from eleven and a half to thirteen and even fourteen feet in length. Only in one case was I at all satisfied of being able to count the rate of speed by a stop-watch, and, if I am not mistaken, there were thirty [sic – steps?] in ten seconds.... If we take the above number, and twelve feet stride as the average pace, we have a speed of twenty–six miles an hour (Livingstone 1857:172).

Calculated correctly, the speed of Livingstone's running ostrich would actually be closer to 25 miles/hour, or 39 km/hr. Alexander et al. (1979) estimated speeds of running ostriches as approximately 12–14 m/sec., or 43–50 km/hr, in reasonably good agreement with Livingstone's much earlier estimate.

Livingstone's estimate of the "pace" of a walking,

non–feeding ostrich is 66 cm, considerably less than the pace values reported here for the Fort Wayne bird. This suggests first that Livingstone was using the term "pace" in the same way that I am (distance from left to right foot-print, or right to left; cf. Leonardi 1987), and second that the Fort Wayne ratite was somewhat agitated, and not com-pletely calm. Assuming that Livingstone's ostriches had foot lengths comparable to that of Redlegs, the pace/footprint length ratio of Livingstone's walking birds would have been close to that reported for other ratites; the range of values of the pace/footprint length ratio in the Fort Wayne ostrich trails overlaps that observed in emu trackways (Table 25.2). In bipedal dinosaur trackways from the Lower Cretaceous of Texas, the stride/footprint length ratio (roughly twice the pace/footprint length ratio) usually ranges 5–7 in animals thought to have been walking (Farlow 1987), within the range of expected values of the stride/footprint length ratio in slowly–moving ratites.

If my interpretation of Livingstone's use of the term "pace" is correct, his observations indicate that an ostrich running at top speed might have a stride length approach-

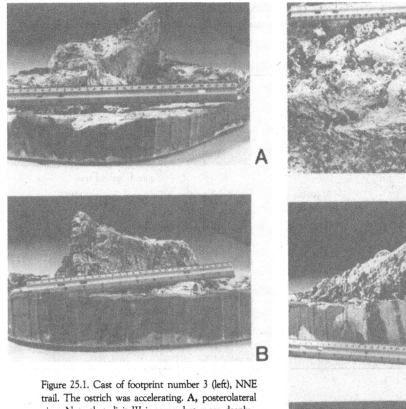

Figure 25.1. Cast of footprint number 3 (left), NNE trail. The ostrich was accelerating. **A,** posterolateral view. Note that digit III is somewhat more deeply impressed along its medial than its lateral edge, especially posteriorly. **B,** medial view.

ing 850 cm (Saharan ostriches were said to take "strides" of 22–28 feet [671-853 cm]; Smit 1963), and a stride/footprint length ratio approaching 40, for a bird leaving a footprint 22 cm long. Such a ratio is considerably greater than values commonly reported for bipedal dinosaur trackways; the stride/footprint length ratio is 15–20 in Texas trackways believed to have been made by running dinosaurs (Farlow 1987). The only dinosaurian stride/footprint length ratio comparable to that estimated for a rapidly running ostrich is that given for the *Hopiichnus shingi* trail (c. 38 – Welles 1971, see below). Given the exaggerated lengths of the ostrich's limb bones (Coombs 1978), however, its stride/footprint length ratio is probably not directly comparable with ratios of at least most bipedal dinosaurs, the above-mentioned similarities in the pace/footprint length ratio notwithstanding.

A 41.5 kg female ostrich dissected by Alexander et al. (1979) had a femur length of 24.9 cm, tibiotarsus length of 44.4 cm, tarsometatarsus length of 41.8 cm, and a length of the longest toe (III, not II as stated by Alexander et al. [1979]) of 17 cm. The toe length may be slightly longer than the length of a footprint the ostrich would make; as the bird's center of mass passes over its foot, "the metatarso-phalangeal joint is off the ground but the first inter-

Figure 25.2. Cast of footprint number 8 (right), south trail. The ostrich angled to the left as it made this trace.
A, overhead view. **B,** medial view. **C,** lateral view.

phalangeal joint and the part of the toe distal to it are on the ground" (Alexander et al. 1979). Ostriches run with less movement in the vertical plane about the hip joint than mammals (Alexander et al. 1979); consequently the hip height of the 41.5 kg female would have been between 86.2 cm (combined lengths of the tibiotarsus and tarsometatarsus) and 111.1 cm (combined lengths of tarsometatarsus, tibiotarsus, and femur), and Alexander et al. (1979) estimated the bird's hip height at 100 cm. This yields a hip height/footprint length ratio of about 6 (100/17). The male ostriches in the Fort Wayne Zoo collection are similar in size to each other, with hip heights a little over 115 cm.

Table 25.2. *Comparisons of trackways of modern and extinct ratites.*

Known or Probable Trackmaker	Footprint Length (cm)	Pace Length (cm)	Pace/Footprint Length Ratio	Source of Data
Struthio camelus	22	66 (walking, not feeding)	3.0	Livingstone (1857); footprint length from the present study
		102–141 (walking or trotting)	4.6–11	this study; Fort Wayne Zoo
		427 (running)	19	Livingstone (1857); footprint length from the present study
Rhea americana	19	c.31–39 (walking)	2.4–5.9	Padian and Olsen (this volume: Figure 1, second trail)
Dromaius novaehollandiae	22	52–130 (walking)	2.4–5.9	Rich and Green (1974)[a]
		120–145 ("running")	5.5–6.6	
	23	64–77 (walking)	2.8–3.3	Rich and Green (1974)[a]
		94–130 ("running")	4.1–5.7	
Dromornithids	15–24	52–63	c.3	Rich and Green (1974)[1]
Moas	20	c.34	c.2	Gillies (1872)
	20	51	3	Williams (1872)
	20	48	2	Voy (1880)
	38	66	1.7	Hill (1895)
	31	76	2.5	Wilson (1913); possibly a continuation of the trail seen by Hill (1895)

[a]I assume that the authors' "stride" = pace.

For these birds the hip height/footprint length ratio would be somewhat over 5 (115/22).

For a large adult ostrich like Redlegs taking 850 cm strides, the stride/hip height ratio would be about 7 (850/115), or a little less. Alexander et al. (1979) estimated that some of the ostriches they filmed took strides of about 500 cm; if a bird with a hip height of 100 cm can take strides of that length, its stride/hip height ratio would be 5.

These values are equal to or slightly higher than those inferred for rapidly running small bipedal dinosaurs and the suggested upper limit (5) of the stride/hip height ratio of a bipedal animal (Thulborn and Wade 1984) — again, with the possible exception of *Hopiichnus*. Welles (1971) believed *Hopiichnus* to have been made by a long-legged dinosaur that was walking, rather than running; Thulborn and Wade (1984) observed that this would require the trackmaker to have had hindlimbs three times as long as those of ornithomimids of similar foot length. Alternatively, if the *Hopiichnus* trackmaker is assumed to have had ornithomimid-like proportions, Thulborn and Wade (1984) calculated that it would have had a stride/hip height ratio of 7, and an estimated speed of 47 km/hr, similar to values estimated here for ostriches running at top speed; if the *Hopiichnus* dinosaur is assumed to have been more like "typical" coelurosaurs than ornithomimids in its proportions, the stride/hip height ratio becomes 9, and its estimated speed 58 km/hr. It is difficult to believe that so small a dinosaur (footprint length = 10 cm) could have outrun an ostrich, and the interpretation of *Hopiichnus* (a trackway consisting of only three footprints – Welles 1971) remains open to question. [Ed. Note: *Another simple explanation of the* Hopiichnus *trackway is that it contains "missing" footprints, thus implying shorter steps than inferred by Welles.*]

Because my reference point for measuring paces and strides was the tip of digit III, step angles calculated from pace and stride data by the law of cosines may be different from values calculated from pace and stride; data based on other reference points, such as the base of digit III or the midpoint along the length of the footprint. In general, if the footprint angles outward (positive rotation) with respect to the trackmaker's direction of travel, the step angle computed from toetip–based pace and stride lengths will be less than that computed from track midpoint–based stride and pace lengths; in the case of trails with negative rotation, toetip–based step angles will be greater than footprint midpoint–based step angles. The difference in computed step angles will depend on the magnitude of the positive or negative rotation of the footprint from the trackmaker's direction of travel, and on the linear distance between alternative reference points on the footprint, relative to the stride length. For the simplest case, in which both the pace length and the angle between the footprint's long axis and the animal's direction of travel remain constant along the trackway, use the formulae given in Figure 25.3.

To investigate the magnitude of the effect of alterna-

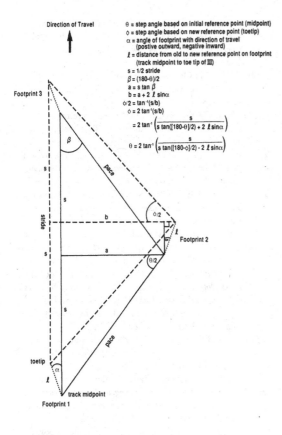

Direction of Travel

θ = step angle based on initial reference point (midpoint)
ϕ = step angle based on new reference point (toetip)
α = angle of footprint with direction of travel
(positive outward, negative inward)
ℓ = distance from old to new reference point on footprint
(track midpoint to toe tip of III)
s = 1/2 stride
$\beta = (180-\theta)/2$
$a = s \tan \beta$
$b = a + 2 \ell \sin\alpha$
$\phi/2 = \tan^{-1}(s/b)$
$\phi = 2 \tan^{-1}(s/b)$

$$= 2 \tan^{-1}\left(\frac{s}{s \tan([180-\theta]/2) + 2 \ell \sin\alpha}\right)$$

$$\theta = 2 \tan^{-1}\left(\frac{s}{s \tan([180-\phi]/2) - 2 \ell \sin\alpha}\right)$$

Footprint 3

Footprint 2

Footprint 1

track midpoint

toetip

pace

stride

Figure 25.3. Analysis of the effects of alternative reference points for measuring paces and strides on the computation of the step angle by the law of cosines. Two paces and their associated stride define a triangle; in this simple model, the two paces are assumed to be equal. The angle formed by the long axis of a footprint and the trackmaker's direction of travel is assumed to remain constant along the trackway. (N.B. Equations do not work where step angle is 180° *and* footprint rotation = 0°.)

Figure 25.4. Overhead view of footprints 2–4 of the south trackway. The ostrich increased its speed between footprints 3 (a left footprint; adjacent to the meter stick) and 4. Note outward angling (positive rotation) of footprints with respect to the ostrich's direction of travel.

tive reference points on calculated step angles, I measured paces and strides of a theropod dinosaur trackway from the Lower Cretaceous of Texas using the toetip and the "base" of digit III (the latter defined as the midpoint of a line segment connecting hypexes [Leonardi 1987] II–III and III–IV) as alternative reference points. Computed step angles based on the alternative reference points are within 5°–10° of each other. Thus the divergence of results by the two methods is not great for relatively long-striding animals. Furthermore, I used the same reference point for measuring paces and strides of the Fort Wayne ostrich that I used for Lower Cretaceous bipedal dinosaur trails from Texas (Farlow 1987), and so the computed dinosaurian and ostrich step angles should be directly comparable.

Step angles calculated for the Fort Wayne ostrich (Table 25.1) are similar to those I have calculated for the

bipedal dinosaur trackways from Texas (typically 140°–180°, Farlow 1987). In contrast to the footprints of most bipedal dinosaurs, which angle inward (negative rotation) with respect to the animal's direction of travel (cf. Farlow 1987), ostrich footprints toe outward (Fig. 25.4). This may be due to the relatively wide hips of ostriches, or an adaptation to facilitate forward kicking (D. Baird pers. comm.). Alternatively, or in addition to these explanations, fast running in ostriches may involve a controlled, diagonally inward forward falling, the inner toe having been lost, and the foot turned outward, to enhance this manner of rapid locomotion.

My observations on the three–dimensional shape of ostrich footprints in different portions of Redlegs' trails suggest that a careful examination of the influences of gait and substrate on track morphology in ratites would be of great help in reconstructing gaits of Mesozoic dinosaurs from their footprints (cf. Dollo 1905, Ballerstedt 1914, Lehmann 1978). The ostrich, with its peculiar didactyl foot, would probably not be the best candidate for such a study, however. Emus, cassowaries, and rheas, with foot structures more like those of most bipedal dinosaurs, would probably be more instructive experimental animals.

[Ed. Note: At least one didactyl dinosaurian trackmaker has been reported; namely Deinonychosaurichnus from the Lower Cretaceous of China (Zhen et al. 1987). This form resembles Wildeichnus from the Jurassic of South America, another unusual track type (Casamiquela 1964). Both forms resemble ostrich tracks except that they are considerably smaller.]

Acknowledgments

I thank Earl Wells for permission to conduct my study at the Fort Wayne Children's Zoo, and Mark Weldon and his staff for assistance in carrying it out. Kevin Padian made useful comments on an early version of the manuscript, and Donald Baird offered insights into the peculiarities of ostrich gaits. Raymond Pippert advised me on trigonometric arguments. Illustrations were prepared with the assistance of the Learning Resource Center, and this research was made possible by a grant from the Office of Sponsored Research, both of Indiana University–Purdue University at Fort Wayne.

References

Alexander, R. Mc.N. 1983. On the massive legs of a moa (Pachyornis elephantopus, Dinornithes). Jour. Zool. London 201:363–376.

Alexander, R. McN., Maloiy, G. M. O., Njau, R., and Jayes, A. S. 1979. Mechanics of running of the ostrich (Struthio camelus). Jour. Zool. 187:169–178.

Ballerstedt, M. 1914. Bemerkungen zu den alteren Berichten über Saurierfährten im Wealdensandstein und Behandlung einer neuen, aus 5 Fussabdrucken bestehenden Spur. Zentralblatt für Mineralogie, Geologie, and Paläontologie 1914:48–64.

Casamiquela, R. M. 1964. Estudious Ichnològicos. Buenos Aires, Colegio Industrial. Pio X, 229 pp.

Coombs, W. P., Jr. 1978. Theoretical aspects of cursorial adaptations in dinosaurs. Quart. Rev. Biol. 53:393–418.

Dollo, L. 1905. Les allures des iguanodons, d'après les empreintes des pieds et de la queue. Bull. Sci. de la France et de la Belgique 40:1–12.

Farlow, J. O. 1987. A Guide to Lower Cretaceous Dinosaur Foot-prints and Tracksites of the Paluxy River Valley, Somervell County, Texas. Field trip guidebook, South-Central Section, Geol. Soc. Amer. (Waco, Texas: Baylor University) 50 pp.

Gillies, T. B. 1872. On the occurrence of footprints of the moa at Poverty Bay. Proc. Trans. New Zealand Institute 4:127–128.

Hill, H. 1895. On the occurrence of moa–footprints in the bed of the Manawatu River, near Palmerston North. Trans. Roy. Soc. New Zealand Institute 27:476–477.

Lehmann, U. 1978. Eine Platte mit Fährten von Iguanodon aus dem Obernkirchener Sandstein (Wealden). Mitteilungen Geologische Paläontologisches Institut Hamburg 48:101–114.

Leonardi, G. (ed.) 1987. Glossary and Manual of Tetrapod Footprint Palaeoichnology. (República Federativa do Brasil, Ministério das Minas e Energia, Departamento Nacional de prodção Mineral) 75 pp.

Livingstone, D. 1857. Missionary Travels and Researches in South Africa; including a Sketch of Sixteen Years' Residence in the Interior of Africa, and a Journey from the Cape of Good Hope to Loanda on the West Coast; thence across the Continent, down the River Zambezi, to the Eastern Ocean. (Freeport, New York: Books for Libraries Press) Reprint of 1857 edition, 1972. 732 pp.

Padian, K., and Olsen, P. E. This volume. Ratite footprints and the stance and gait of Mesozoic theropds.

Rich, P. V., and Green, R. H. 1974. Footprints of birds at South Mt. Cameron, Tasmania. Emu 74:245–248.

Smit, D. J. v. Z. 1963. Ostrich Farming in the Little Karoo, Republic of South Africa. Department of Agricultural Technical Services, Bull. 358. 103 pages.

Thulborn, R. A., and Wade, M. 1984. Dinosaur trackways in the Winton Formation (Mid-Cretaceous) of Queensland. Mem. Queensland Mus. 21:413–517.

Voy, C. D. 1880. On the occurrence of footprints of Dinornis at Poverty Bay, New Zealand. Amer. Naturalist 14:682–684.

Welles, S. P. 1971. Dinosaur footprints from the Kayenta Formation of northern Arizona. Plateau 44:27–38.

Williams, W. L. 1872. On the occurrence of footprints of a large bird, found at Turanganui, Poverty Bay. Proc. Trans. New Zealand Institute 4:124–127.

Wilson, K. 1913. Footprints of the moa. Proc. Trans. New Zealand Institute 45:211–212.

Zhen Z., Li J., Zheng B., Chen W. and Zhu Z. 1987. Bird and dinosaur footprints from the Lower Cretaceousof Emei County, Sichuan. (Abst.) 1st Internat. Symp. Non-marine Cretaceous Correlations, Urumqi, China pp. 37–38.

26 Phylogenetic Patterns in the Manus and Pes of Early Mesozoic Archosauromorpha

J. MICHAEL PARRISH

Abstract

One of the biggest problems in the study of fossil reptile trackways is the establishment of correspondence between ichnofossils and skeletal remains. Patterns of manus and pes morphology are evaluated throughout the Triassic archosaurs and related early diapsids in order to determine trends throughout the Archosauromorpha. Patterns of digital length, metapodial length, and digital robustness are emphasized, as are phalangeal formulae and evidence for digital divarication.

The plesiomorphic condition for diapsids is for digit IV to be longest in both manus and pes, with digits II and III shorter and subequal in length, with digits I and V shorter still and subequal in length. The plesiomorphic diapsid phalangeal formulae are 2-3-4-5-3 (manus) and 2-3-4-5-4 (pes). This basic pattern is retained in archosauromorphs such as *Prolacerta*, *Trilophosaurus*, and rhynchosaurs, and is also present in the proterosuchid archosaurs.

A trend towards symmetry of both manus and pes occurs in derived members of both the crocodile–normal and crocodile–reverse archosaurs, with reduction of external digits taking place in the more cursorial members of both groups. Anterior pedal alignment, compression of metapodials, and increased robusticity of manual digit I occurs in the erect members of each lineage. Phalangeal formulae tend to be conservative, with changes in relative phalangeal length the usual cause of increased symmetry in more derived archosaurs. Phalanges are lost in pedal digit V in dinosaurs, poposaurids, and crocodylomorphs, and in manual digits V and IV in dinosaurs.

Footprint evidence demonstrates earlier occurrences of archosaurs with symmetrical pedes and functionally tridactyl feet than are known from the skeletal record. Used in conjunction with skeletal evidence, footprints can also illustrate trends in archosauromorph locomotor evolution.

Introduction

In the Triassic, the diapsid amniotes underwent an extensive radiation, becoming the most diverse group of terrestrial tetrapods by the end of the period. The large diapsid reptiles of the Triassic comprised the archosaurs, including the early dinosaurs, crocodylomorph reptiles, and pterosaurs, along with the primitive archosaurs commonly grouped as "thecodonts". The archosaurs, along with the rhynchosaurs, trilophosaurs, protorosaurs, and tanystropheids, comprise the Archosauromorpha (Gauthier 1984, Benton 1985). Renewed study of the Triassic archosaurs in recent years has permitted some refining of hypotheses concerning archosaur phylogeny. Over the last several years, a consensus has begun to emerge regarding the phylogeny of the major groups of archosaurs (e.g., Brinkman 1981b; Chatterjee 1982; Gauthier 1984; Parrish 1983, 1986b), although dissenting views are still expressed (e.g., Cruickshank 1979, Thulborn 1980, Cruickshank and Benton 1985).

The purposes of this paper are to survey the morphological patterns present in the forefoot (manus) and hindfoot (pes) of Triassic archosauromorphs, and to consider the implications of these patterns for the phylogeny, functional morphology, and paleoecology of Triassic archosaurs. Because of the diversity of archosauromorphs of similar size sharing the same basic pedal pattern in the Triassic, establishment of correspondence between skeletal fossils and fossil footprints has been a difficult problem. By surveying the diversity of morphological patterns of the manus and pes among the Triassic forms in a phylogenetic context, this study may permit more ready delineation of the possible taxonomic affinity of Triassic footprints than was previously possible.

Archosauromorph Phylogeny

The phylogeny of the Diapsida has recently been considered by Gauthier (1984) and Benton (1985). They concur on the Diapsids comprising two divergent mono-phyletic lineages, called the Archosauromorpha and Lepido-sauromorpha in each study. In the Triassic, the Lepido-sauromorpha tend to consist of small forms that will not be considered here. The most primitive diapsid, *Petrolaco-saurus*, has no derived characters that define either the Archosauromorpha or Lepidosauromorpha, and will be considered as the outgroup for the archosauromorphs here.

The base of the archosauromorph radiation consists of the archosaurs plus the following groups: Protorosauria, Trilophosauria, Rhynchosauria, and Tanystropheidae. Gauthier (1984) and Benton (1985) disagree about the relationships of these groups, but both consider them all archosauromorphs that are sister groups of the Archosauria (Archosauriformes of Gauthier &1984é, considered here as Archosauria for convenience). For the purposes of this paper, these groups will be grouped as an unresolved tetratomy that is the sister group of the Archosauria (Fig. 26.1).

Proterosuchus is commonly considered the most plesiomorphic archosaur (e.g., Cruickshank 1972; Parrish 1983, 1986b; Gauthier 1984; Benton 1984). Most workers consider *Erythrosuchus* a primitive archosaur that is the sister group of the more derived forms (e.g., Romer 1972, Brinkman 1981b, Parrish 1986b, Gauthier 1984, Benton 1984), but Cruickshank (1979) considers *Erythrosuchus* to be the derived sister group of *Euparkeria*.

Many workers (Brinkman 1981b; Chatterjee 1982; Parrish 1983, 1986b; Gauthier 1984) have recognized two divergent lineages among the later archosaurs, as follows:
1. the crocodile–reversed archosaurs (Ornithosuchia of Gauthier 1984), comprising the Euparkeriidae, Ornithosuchidae, Pterosauria, and Dinosauria (including birds);
2. the crocodile–normal archosaurs, comprising the Phytosauria, Aetosauria, Rauisuchia, and Croco-dylomorpha (including crocodilians). This division, erected on the basis of different tarsal patterns derived from the primitive archosauromorph type, is also supported by other postcranial and cranial character distributions (Gauthier 1984). Although divergent opinions have been presented (e.g., Cruick-shank 1979, Thulborn 1980, Cruickshank and Benton 1985), this bipartite scheme will be employed in this study, and tested by its results.

Within the crocodile–reversed lineage, dinosaurs plus birds will be considered the most derived group, with pterosaurs, ornithosuchids, and euparkeriids successively distant sister groups. Within the crocodile–normal lineage, crocodylomorphs will be considered the most derived group, with rauisuchians plus aetosaurs (unresolved dichotomy) and phytosaurs successively more distant outgroups. Phytosaurs are probably the sister group of the crocodile–normal archosaurs, but have an incompletely developed crocodile–normal tarsus (Parrish 1986a). A cladogram reflecting the relationship employed in this paper, synthe-sized from Chatterjee (1982), Gauthier (1984), and Parrish (1986b) is presented as Figure 26.1. Several nodes have not

been resolved as dichotomies, and their resolution is beyond the scope of this study.

Patterns of Manus and Pes Morphology

In this section, manus and pes morphology is sur-veyed among the Triassic archosauromorphs. Basic characters considered are number of digits, phalangeal formulae, relative digital length, and morphological evidence for digital divarication. The metric data are presented superimposed on cladograms in Figures 26.2 and 26.3.

Early Diapsids and Non–Archosaurian Archosauromorphs

The basic archosauromorph manus and pes patterns can also be seen in the earliest diapsid, *Petrolacosaurus* (Fig. 26.4; Peabody 1952; Reisz 1981). In *Petrolacosaurus*, digit IV is longest in both manus and pes, digit III slightly shorter, digits II and V still shorter and usually roughly equal in length, and digit I shortest. The phalangeal formulae are 2–3–4–5–3 for the manus, and 2–3–4–5–4 for the pes. In most of the non–archosaur archosauromorphs, the basic pattern seen in *Petrolacosaurus* is retained (Fig. 26.4), differ-ing from the primitive condition mainly in the develop-ment of a medially directed neck on the fifth metatarsal. This pattern is shared by both *Noteosuchus*, the earliest rhynchosaur (but with a manual phalangeal formula of 2–3–4–4–3) and by *Trilophosaurus*. In the tanystropheids

Figure 26.1. Cladogram illustrating hypothesized rela-tionships among the Archosauromorpha. Based on Chatterjee (1982), Parrish (1983, 1986b), and Gauthier (1984).

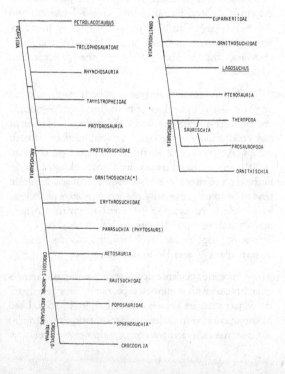

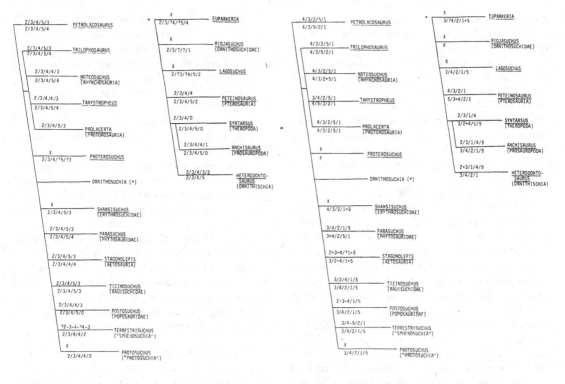

Figure 26.2. Phalangeal formulae for the manus (above line) and pes (below line) for representative Archosauromorpha. Phalangeal formulae based on composite or incomplete digits are indicated by a question mark. An X indicates that the manus or pes is not sufficiently known to determine phalangeal formula.

Figure 26.3. Relative digital lengths for manus (above line) and pes (below line) for representative Archosauromorpha. Relative lengths estimated from fragmentary digits are indicated by a question mark. An X indicates that the manus or pes is not sufficiently known to determine relative digital lengths.

Figure 26.4. Manus (top) and pes (bottom) of the primitive diapsid *Petrolacosaurus* and non-archosaur archosauromorphs. Scale = 1 cm unless noted otherwise. **A,** *Petrolacosaurus.* After Reisz (1981). **B,** *Trilophosaurus.* After Gregory (1945). **C,** the rhynchosaur *Noteosuchus.* After Carroll (1976). **D,** the protorosaur *Prolacerta.* After Gow (1975). **E,** *Tanystropheus.* After Wild (1973).

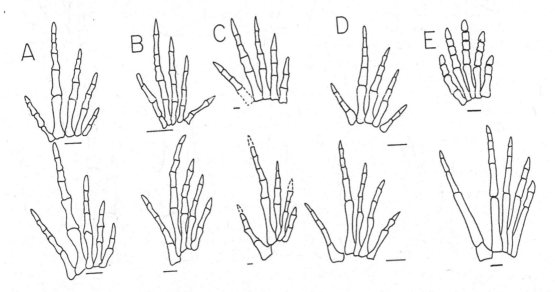

Tanystropheus and *Tanytrachelos*, the plesiomorphic pedal digital formula is retained, but the fifth digit is hypertrophied by elongation of the proximal phalanx with the result that digit V is as long (*Tanytrachelos*; Olsen 1979, Fig. 4a) or longer (*Tanystropheus*; Wild 1973, his Figs. 75–77) than digit III. The manus in *Tanystropheus* has a phalangeal formula of 2–3–4–4–2, and is notable in being symmetrical about the third digit, with digits decreasing in length through the series 3–4–2–5–1. The manus is not fully described in *Tanytrachelos*, but appears to be of the more typical early archosauromorph pattern (Olsen 1979).

Primitive Archosaurs (Fig. 26.5)

The feet are incompletely known in the earliest archosaurs, the Proterosuchidae. However, the available material of *Proterosuchus* from South Africa and *Chasmatosaurus* from China is consistent with the primitive archosauromorph pattern. The manus and pes are somewhat better known in erythrosuchids, but still from abundant disassociated material (Young 1964, Cruickshank 1978). Other than having relatively shorter phalanges that presumably represent a graviportal adaptation, the foot patterns are basically the same as the primitive diapsid pattern. The phytosaur manus and pes are known from complete material from India (Chatterjee 1978) and from the southwestern U.S. (Parrish 1986a). The pes differs from the primitive pattern in having subequal digits IV and III, and manual digit III is longer than digit IV.

Crocodile–Normal Archosaurs (Figs. 26.6, 26.7)

The crocodile–normal archosaurs principally differ from primitive archosauromorphs in having symmetrical manus and pes, with digits II, III, and IV nearly equal in length, and digits I and V shorter and subequal in length. The primitive phalangeal formulae are retained in most cases; digit IV is characterized by much shorter phalanges than in the primitive condition. The manus and pes of aetosaurs are otherwise of the primitive archosaur pattern (Fig. 26.6).

The partial pes known from the rauisuchid *Saurosuchus* is of the pattern seen in aetosaurs. In the rauisuchid *Ticinosuchus* and the poposaurid *Postosuchus*, the medially directed process on the fifth metatarsal is reduced relative to that in aetosaurs. In the poposaurid *Postosuchus* digit V is reduced to a metatarsal splint. The manus is very similar in form among aetosaurs and rauisuchians (Fig. 26.7).

Early crocodylomorph feet (Fig. 26.7) share the basic pattern seen in *Postosuchus*. The sphenosuchian *Terrestrisuchus* retains two phalanges on the fifth pedal digit. The pedal phalangeal formula in most other crocodylomorphs (including the "modern" Eusuchia) is 2–3–4–4–0.

Ornithosuchia (Fig. 26.8)

Feet are not well known in the early ornithosuchians. Little can be said about the manus; it is known only from proximal parts in *Euparkeria* (Ewer 1965) and *Riojasuchus* (Bonaparte 1971). The pes in *Euparkeria* is similar

Figure 26.5. Manus (top) and pes (bottom) of primitive archosaurs. Scale = 1 cm. **A,** *Proterosuchus*. After Cruickshank (1972). **B,** the erythrosuchid *Shansisuchus*. After Young (1964). **C,** the phytosaur *Parasuchus*. After Chatterjee (1978).

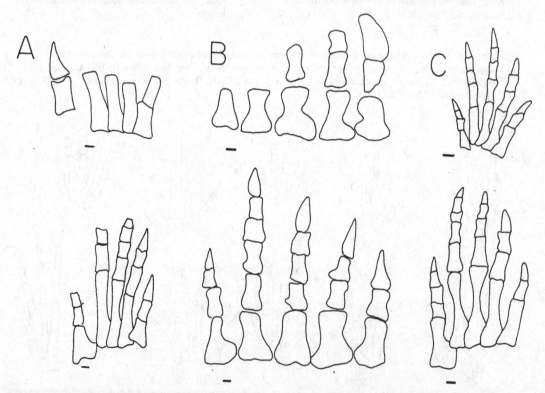

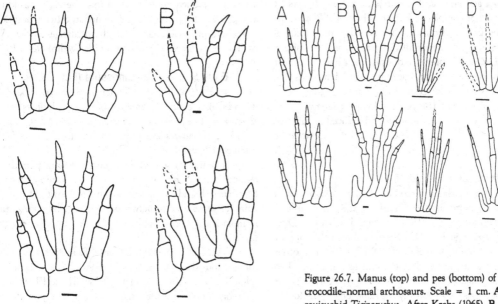

Figure 26.6. Manus (top) and pes (bottom) of aetosaurs. Scale = 1 cm. **A,** *Stagonolepis.* After Walker (1961). **B,** *"Typothorax" meadei.* After Sawin (1947).

Figure 26.7. Manus (top) and pes (bottom) of crocodile–normal archosaurs. Scale = 1 cm. **A,** the rauisuchid *Ticinosuchus.* After Krebs (1965). **B,** the poposaurid *Postosuchus.* After Chatterjee (1985). **C,** the "sphenosuchian" crocodylomorph *Terrestrisuchus.* After Crush (1984). **D,** the "protosuchian" crocodile *Protosuchus.* After Colbert and Mook (1951).

Figure 26.8. Manus (top) and pes (bottom) of primitive ornithosuchians and pterosaurs. Scale = 1 cm. **A,** *Euparkeria.* After Ewer (1965). **B,** the ornithosuchid *Riojasuchus.* After Bonaparte (1971). **C,** *Lagosuchus.* After Bonaparte (1975). **D,** the pterosaur *Eudimorphodon.* After Wild (1978). **E,** the pterosaur *Peteinosaurus.* After Wild (1978).

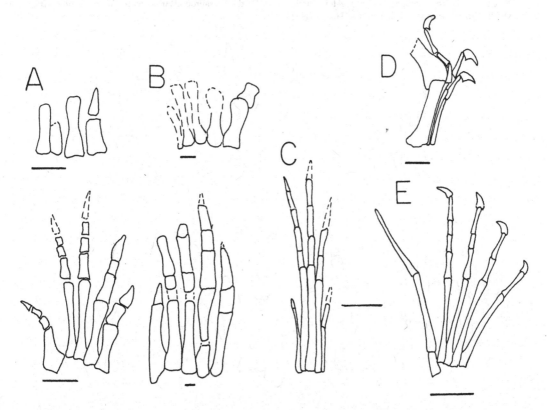

to those of aetosaurs and rauisuchids, although digits III and IV are not completely known. The pes of *Riojasuchus* is mainly known from the metatarsus and proximal phalanges; the only apparent difference from the pattern in *Euparkeria* is the presence of only one phalanx in digit V. As opposed to the condition in crocodylomorphs, the fifth metatarsal is not reduced in relative length.

In *Lagosuchus*, the pes is strongly bilaterally symmetrical, with only two phalanges and reduced metatarsals in both digits I and V. The manus of *Lagosuchus* is unknown.

Pterosauria (Fig. 26.8)

In the earliest pterosaurs, *Eudimorphodon* and *Peteinosaurus*, from the Late Triassic of Italy, the manus is greatly altered from the primitive diapsid condition. Digit V is lost, and digit IV is hypertrophied into the support structure for the wing membrane (Wild 1978). The other digits are much shorter, decreasing in length from III to I. Manual phalangeal formula is 2-3-4-4. Although the four metacarpals are of roughly the same length, the fourth metacarpal is stoutest, with the more medial digits much less robust.

The pes of *Peteinosaurus* has a phalangeal formula of 2-3-4-5-2, with the lateral digits somewhat longer than I and II. Digit V is of an unusual pattern; the metatarsal is very short and splint-like, whereas the two phalanges

are hypertrophied in length such that the digit is as long as digit IV. This pattern is maintained in other rhamphorhynchoid pterosaurs, although the phalanges of digit V are lost or reduced in pterodactyloids (Wellnhofer 1978).

Early Saurischian Dinosaurs (Fig. 26.9)

In the primitive saurischian *Herrerasaurus* (Bennedetto 1973), five manual digits are present, with digits I and V stoutest. The phalangeal formula of the first three digits is 2-3-4; phalanges are unknown in digits IV and V (Bennedetto 1973).

The metatarsus of *Herrerasaurus* is characterized by reduction in length and breadth of the external digits. Metatarsals II and IV are subequal in length with IV somewhat longer; all are equally robust. The pedal phalangeal formula is not known, although digit V has only a single phalanx.

The Triassic theropod *Syntarsus* (Raath 1977) has only the first four digits, with digit IV reduced to a metacarpal splint. Digit I is shortest, and metatarsals II and III are considerably longer and subequal in length. The phalangeal formula is 2-3-4-0.

The pes of *Syntarsus* is also reduced externally; digits II–IV are the most robust, with digit III slightly longer than II and III. Metatarsals I and V are reduced and splintlike. The phalangeal formula is 2-3-4-5-0. Similar pedal patterns are seen in the early theropods *Procompsognathus* (Ostrom

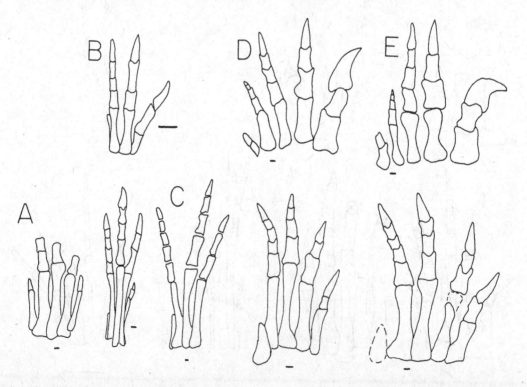

Figure 26.9. Manus (top) and pes (bottom) of primitive saurischian dinosaurs. Scale = 1 cm. **A,** *Herrerasaurus*. After Reig (1963). **B,** the theropod *Syntarsus*. After Raath (1977). **C,** the prosauropod *Anchisaurus*. After Galton (1976). **D,** the prosauropod *Plateosaurus*. After Huene (1932).

1982) and *Dilophosaurus* (Welles 1984), although no phalanges are known on digit I in *Dilophosaurus*.

Podial structures in prosauropod dinosaurs are quite uniform (Galton 1976, Huene 1932). The manus has five digits, with digit I hypertrophied and the more lateral digits reduced in breadth. The manual phalangeal formula is 2–3–4–4–2. The claw on digit I is enlarged and trenchant. Digit II is longest and digits IV and V are greatly reduced in length.

The pes is of the basic pattern seen in theropods, although digit I is not reduced to the extent seen in *Syntarsus*. The metatarsus is also not elongated to the extent that it is in the theropods, although relative reduction of metatarsal V is greater than in the Triassic theropods.

Ornithischia (Fig. 26.10)

In the earliest ornithischian *Pisanosaurus* (Bonaparte 1976), two digits of the pes are known, III and IV. The metatarsus is elongate, apparently bilaterally symmetrical, and four phalanges were probably present on digit III, and five on digit IV.

In the somewhat younger ornithischian *Heterodontosaurus*, the manus is reduced laterally as in saurischians. Digits I–III are considerably more robust than IV and V, and the digits increase in length in the following sequence: V, IV, I, II, III. The phalangeal formula is 2–3–4(?)–2–2.

The pes shares with theropods the elongation and lateral compression of the metatarsus. The foot is basically symmetrical about digit III, although digit V is lost entirely. The phalangeal formula is 2–3–4–5. The same basic pattern is present in the incompletely preserved pes of *Scutellosaurus* (Colbert 1981).

Implications for Locomotor Patterns

The basic diapsid manus and pes patterns are retained throughout the archosaurs, although modified to differing degrees in the various lineages. Because of the central importance of locomotor performance to adaptation in tetrapods, most podial modifications appear to reflect directly differing types of locomotor behavior.

The primitive diapsid condition is retained with little modification in *Sphenodon* and many lizards, all of which utilize a sprawling gait similar to that hypothesized for the early archosauromorphs. At least in the pes, the elongation of the lateral digits may be related to the rotation of the foot about the long axis of the crus during the latter parts of the step cycle in the sprawling gait (Brinkman 1981a; Parrish 1986a,b).

The development of a symmetrical pes in both ornithosuchians and crocodile–normal archosaurs correlates with the shift to a more erect gait in both lineages (Parrish 1983, 1986b). Particularly important is the decreased lateral splay of the fifth digit, effected in part by reduction of the medially directed neck of the fifth metatarsal. In poposaurids, crocodylomorphs, and dinosaurs, loss of lateral phalanges and development in some forms of a bilaterally symmetrical pes with an elongate, compact metatarsus can

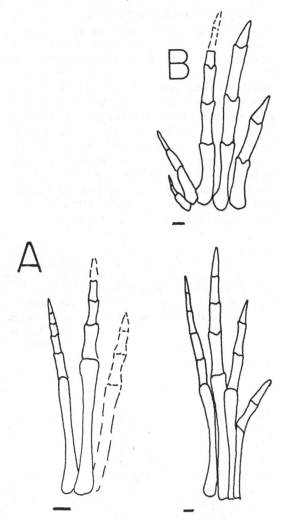

Figure 26.10. Manus (top) and pes (bottom) of primitive ornithischian dinosaurs. Scale = 1 cm.
A, *Pisanosaurus*. After Bonaparte (1976).
B, *Heterodontosaurus*. After Santa Luca (1980).

probably be related to attainment of more cursorial habits, as has been widely noted previously (e.g., Charig 1972, Romer 1956). The pes of pterosaurs is not strongly modified towards cursoriality, and the presence of the aberrant elongate fifth digit seems most strongly to reflect attachment of a membrane of some sort.

In the manus, the tendency towards more symmetrical form in the crocodile–normal archosaurs is also probably related to a shift to a more erect gait. No good samples of primitive ornithosuchian forefeet are known. Reduction of lateral digits and enlargement and specialization of the medial digits in early dinosaurs probably relate to utilization of the manus for grasping in such apparently bipedal forms as theropods, prosauropods, and early ornithischians.

Correlation with Trackways

Studies of phylogenetic patterns of podial structure and of fossil trackways are mutually complementary. In some cases, very close correspondence between given skeletal fossils and particular ichnotaxa can be demonstrated, such as between *Chirotherium barthii·* and *Ticinosuchus* grade rauisuchians (Krebs 1965), between Liassic prosauropods and *Navahopus falcipollex* (Baird 1980), or between *Tanytrachelos* and *Gwynnedichnium* (Baird 1986). More often than not, the footprint record is more complete than that of the skeletal manus and pes. For example, we have only a sketchy idea of what a primitive crocodile-reverse archosaur (euparkeriid or ornithosuchid) footprint should look like, owing to incomplete skeletal records of these taxa. Trackways of these forms may be preserved as

existing ichnotaxa, but proving correspondence would be difficult at this time.

The footprint record is useful in elucidating the timing of trends in archosaurian pedal evolution. For example, the presence in the Lower Triassic of the ichnotaxa *Isochirotherium, Brachychirotherium,* and *Chirotherium* all demonstrate a relatively symmetrical pes (Haubold 1971, 1984), and indicate that archosaurs with feet specialized relative to the plesiomorphic condition were present at that time, although no skeletal evidence of this is known. Similarly, the presence of tridactyl, theropod–like pedal tracks have been recorded in the Middle Triassic of France (e.g., Gand 1975) and South Africa (e.g., Demathieu and Haubold 1978, Demathieu this volume) well before the appearance of skeletal fossils with such feet, which are

Figure 26.11. Stratigraphic trends in chirotheriid footprint types. **A–E,** Lower and Middle Triassic forms with everted digit V and lateral rotation of print relative to line of march. **A,** *Chirotherium barthii.* **B,** *Chirotherium moquinense.* **C,** *Isochirotherium soergeli.* **D,** *Isochirotherium herculis.* **E,** *Isochirotherium marshalli.* **F–H,** Middle and Upper Triassic forms with anteriorly directed digit V and/or pes aligned parallel to line of march. **F,** *Brachychirotherium circaparvum.* **G,** *Brachychirotherium eyermani.* **H,** *Brachychirotherium parvum.* **I,J,** Upper Triassic forms with everted digit V (I) and pes everted relative to the line of march (J). **I,** *Brachychirotherium thuringiacum.* **J,** *Chirotherium* sp. Scale = 10 cm. A–I redrawn after figures in Haubold (1971). J redrawn after Conrad et al. (1987).

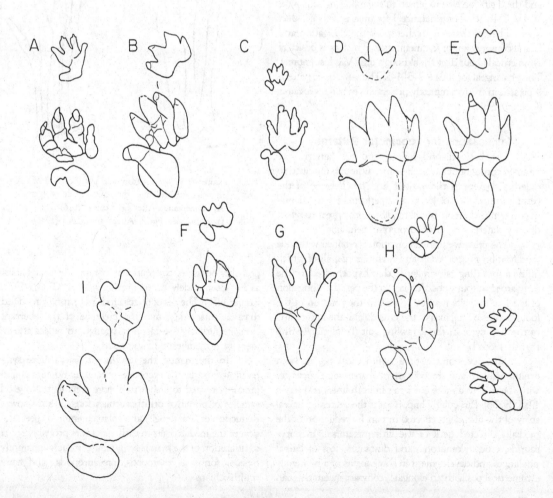

known no earlier than the Carnian.

Footprints can also demonstrate trends in locomotor evolution. A good example is the diversity in form of large "thecodont" footprints that are pentadactyl and are generally symmetrical. In the Lower and Lower Middle Triassic, pes prints of such "chirotheroid" ichnogenera (e.g., *Chirotherium barthii, C. moquiensis, Isochirotherium soergeli, I. herculis, I. marshalli,* and *Brachychirotherium tinanti* [Figures in Haubold 1971, 1984, 1986]) have digit V strongly recurved posterolaterally, a character that is a correlate of sprawling gaits in tetrapods (e.g., Parrish 1986b). Starting in the Middle Triassic (e.g., *Brachychirotherium circaparvum* [Haubold 1971]), and particularly in the Upper Triassic (e.g., *B. eyermani, B. parvum* [Baird 1957, Haubold 1971]) lateral splay of digit V is reduced and the pes is directed forwards relative to the line of march, correlates of a more erect gait. Other Late Triassic "chirotheriids" appear to represent animals walking with a more sprawling gait (e.g., *Brachychirotherium thuringiacium* [Haubold 1971]; *Chirotherium* sp. of Conrad et al. [1987]) or at least had an everted fifth digit (e.g., *C. wondrai* [Heller 1952]).

Biostratigraphic evidence of fossil footprints can also reinforce arguments made on the basis of skeletal fossil evidence. For example, the disappearance of "chirotheriid" footprints at the Triassic/Jurassic boundary has been well documented by Olsen and Galton (1977, 1984).

Conclusions

The patterns of digital modification of the manus and pes throughout the archosauromorphs are basically consistent with the current notions of diapsid evolution (Fig. 26.11, Gauthier 1984, Benton 1985). The podial structures in a few taxa are of phylogenetic significance.

1. The retention of the basic primitive archosaur pattern in phytosaurs is further evidence of their being primitive sister groups of the more derived crocodile–normal archosaurs.

2. The combined reduction of the medial and lateral pedal digits and of the lateral manual digits in dinosaurs are further evidence for their monophyly (e.g., Bakker and Galton 1974, Gauthier 1984). The dinosaurian condition is rather different from the reduction pattern in the more cursorial crocodile–normal archosaurs, where pedal digit V is reduced but pedal digit I and the manus are unmodified. Interestingly, modern crocodilians do exhibit reduction of the lateral manual digits as in dinosaurs, but this pattern is not present in protosuchian crocodilians for which the manus is known (e.g., Colbert and Mook 1951).

In cases where both skeletal and ichnofossil material is complete and well–preserved, correspondence between the two can be inferred with a fair degree of confidence. However, the trackway record includes several records of particular foot types at stratigraphic levels that precede skeletal evidence of such feet. Trends in locomotor evolution, such as the shift from sprawling to more erect gaits

in archosaurs, can be monitored both stratigraphically and morphologically using a combination of skeletal material and fossil footprints.

Acknowledgments

This work was supported in part by NSF Grants BSR–860748 to JMP and EAR–8618206 to Martin G. Lockley and JMP. I would like to thank three anonymous reviewers for helpful comments on this manuscript. This work has also benefited from discussions with Don Baird, Kelly Conrad, Martin Lockley, Paul Olsen, and Kevin Padian.

References

Baird, D. 1957. Triassic footprint faunules from Milford, New Jersey. *Bull. Museum Comparative Zool.* 111:165–192.

——— 1980. A prosauropod dinosaur trackway from the Navajo Sandstone (Lower Jurassic) of Arizona. In Jacobs, L. L. (ed.). *Aspects of Vertebrate History* (Museum of Northern Arizona Press) pp. 219–230.

——— 1986. Some Upper Triassic reptiles, footprints, and an amphibian from New Jersey. *The Mosasaur* 3:125–153.

Bakker, R. T., and Galton, P. M. 1974. Dinosaur monophyly and a new class of vertebrates. *Nature* 248:168–172.

Bennedetto, J. L. 1973. Herrerasauridae, nueva familia de la saurisquios triasicos. *Ameghiana* 10:89–102.

Benton, M. J. 1985. The Triassic reptile *Hyperdapedon* from Elgin: functional morphology and relationships. *Phil. Trans. Roy. Soc. London, Series B* 302:605–720.

Bonaparte, J. F. 1971. Los tetrapodos del sector superior de la Formacion Los Colorados, La Rioja, Argentina (Triasico Superior). *Opera Lilloana* 22:1–182.

——— 1975. Nuevos materiales de *Lagosuchus tamalpayensis* Romer (Thecodontia – Pseudosuchia) y su significado en el origen de los Saurischia. Chañarense inferior, Triasico Medio de Argentina. *Acta Geologica Lilloana* 13:5–90.

——— 1976. *Pisanosaurus mertii* Casamiquela and the origin of the Ornithischia. *Jour. Paleont.* 503:805–820.

Brinkman, D. 1981a. The hindlimb step cycle of *Iguana* and other primitive reptiles. *Jour. Zool.* 181:91–103.

——— 1981b. The origin of the crocodiloid tarsi and the interrelationships of thecodontian archosaurs. *Breviora* 464:1–23.

Carroll, R. L. 1976. *Noteosuchus* – the oldest known rhynchosaur. *Ann. South African Museum* 72:35–57.

Charig, A. J. 1972. The evolution of the archosaur pelvis and hindlimb, an explanation in functional terms. In Joysey, K. A., and Kemp, T. S. (eds.). *Studies in Vertebrate Evolution* (Oliver and Boyd) pp. 121–151.

Chatterjee, S. 1978. A primitive parasuchid (phytosaur) from the Upper Triassic Maleri Formation of India. *Palaeontology* 21:83–127.

——— 1982. Phylogeny and classification of thecodontian reptiles. *Nature* 295:317–320.

——— 1985. *Postosuchus*, a new thecodontian reptile from the Triassic of Texas and the origin of tyrannosaurs. *Phil. Trans. Roy. Soc. London, Series B* 309:395–460.

Colbert, E. H. 1981. A primitive ornithischian dinosaur from the Kayenta Formation of Arizona. *Museum Northern Arizona Bull.* 53:1–61.

Colbert, E. H., and Mook, C. C. 1951. The ancestral crocodilian *Protosuchus*. *Bull. Amer. Museum Nat. Hist.* 97:143–182.

Conrad, K., Lockley, M. G., and Prince, N. K. 1987. Triassic and Jurassic vertebrate-dominated trace fossil assemblages of the Cimarron Valley region: implications for paleoecology and biostratigraphy. In Lucas, S. G. (ed.). New Mexico Geological Society, 38th Field Conference, Northeastern New Mexico. pp. 127–138.

Cruickshank, A. R. I. 1972. The proterosuchian thecodonts. In Joysey, K. A., and Kemp, T. S. (eds.). Studies in Vertebrate Evolution (Oliver and Boyd) pp. 89–117.

1978. The pes of Erythrosuchus africanus Broom. Zool. Jour. Linn. Soc. London 62:161–177.

1979. The ankle joint in some early archosaurs. South African Jour. Sci. 77:307–308.

Cruickshank, A. R. I., and Benton, M. J. 1985. Archosaur ankles and the relationships of the thecodontian and dinosaurian reptiles. Nature 317:715–717.

Crush, P. J. 1984. A late Upper Triassic sphenosuchid crocodilian from Wales. Palaeontology 27:131–157.

Demathieu, G., and Haubold, H. 1978. Du probleme de l'origiene des dinosauriens d'apres les donnees de l'ichnologe du Trias. Geobios 11:409–412.

Ewer, R. F. 1965. The anatomy of the thecodont reptile. Euparkeria capensis Broom. Phil. Trans. Roy. Soc. London, Series B 248:379–435.

Galton, P. M. 1976. Prosauropod dinosaurs (Reptilia; Saurischia) of North America. Postilla 169:1–98.

Gand, G. 1975. Sur quelques emprientes de pas de vertébrés recoltées dans de la Stephanien moyen de Montceau-les-Mines (Saone-et-Loire France). Bulletin Society Histoire Naturale Musee d'Autun 74:9–31.

Gauthier, J. A. 1984. A cladistic analysis of the higher categories of Diapsida. Ph.D. dissertation, University of California, Berkeley.

Gow, C. E. 1975. The morphology and relationships of Youngina capensis Broom and Prolacerta broomi Parrington. Pal. Africana 18:89–131.

Gregory, J. T. 1945. Osteology and relationships of Trilophosaurus. Univ. Texas Publ. 4401:273–359.

Haubold, H. 1971. Handbuch der Paläoherpetologie. Teil 18. Ichnia Amphibiorum et Reptilorum Fossilum. (Stuttgart: Fischer Verlag) 124 pp.

1984. Saurierfährten. (Wittenberg: Ziemsen Verlag) 231 pp.

1986. Archosaur footprints at the terrestrial Triassic-Jurassic transition. In Padian, K. (ed.). The Beginning of the Age of Dinosaurs: Faunal change across the Triassic-Jurassic boundary. (Cambridge University Press) pp. 189–202.

Heller, H. 1952. Reptilfährten-funde aus dem Ansbacher Sandstein des Mittleren Keupers von Franken. Geol. Blätt. NO Bayern 2:129–141.

Huene, F. v. 1932. Die fossile reptile-ordnung Saurischia, ihre entwicklung und geschischte. Monatschefte Geologische Palaontologica 4:1–361.

Krebs, B. 1965. Ticinosuchus ferox nov. gen. nov. sp. Ein neuer Pseudosuchier aus der Trias der monte San Giorgio. Schweiz Palaontologische Abhandlungen 81:1–140.

Olsen, P. E. 1979. A new aquatic eosuchian from the Newark Supergroup (Late Triassic–Early Jurassic) of North Carolina and Virginia. Postilla 176:1–14.

Olsen, P. E., and Galton, P. M. 1977. Triassic–Jurassic extinctions: are they real? Science 197:983–986.

1984. A review of the reptile and amphibian assemblages from the Stormberg of South Africa, with special emphasis on the footprints and on the age of the Stormberg. Paleont. Africana 25:87–110.

Ostrom, J. H. 1982. Procompsognathus: theropod or thecodont? Paleontographica (A) 175:179–195.

Parrish, J. M. 1983. Locomotor adaptations in the pelvis and hindlimb of the Thecondontia (Reptilia: Archosauria). Ph.D. dissertation, University of Chicago.

1986a. Structure and function of the tarsus in the phytosaurs (Reptilia: Archosauria). In Padian, K. (ed.). The Beginning of the Age of Dinosaurs: Faunal change across the Triassic-Jurassic boundary. (Cambridge University Press) pp. 35–42.

1986b. Locomotor adaptations in the hindlimb and pelvis of the Thecodontia (Reptilia: Archosauria). Hunteria 1 (2):1–36.

Peabody, F. E. 1952. Petrolacosaurus kansensis, a Pennsylvanian reptile from Kansas. Paleont. Contrib. Univ. Kansas Vertebrata 1:1–41.

Raath, M. A. 1977. The anatomy of the Triassic theropod Syntarsus rhodesiensis (Saurischia: Podokesauridae) and a consideration of its biology. Ph.D. dissertation, Rhodes University.

Reig, O. A. 1963. La presencia de dinosaurios saurisquios en los "Estatos de Ischigualasto" Mesotriasico Superior de las provincias de San Juan y La Rioja (República Argentina). Ameghniana 3:3–20.

Reisz, R. R. 1981. A diapsid reptile from the Pennsylvanian of Kansas. Special Publ. Museum Nat. Hist. Univ. Kansas 7:1–74.

Romer, A. S. 1956. Osteology of the Reptilia. (University of Chicago Press) 583 pp.

1972. The Chanares (Argentina) Triassic reptile fauna. XVI. Thecodont classification. Breviora 395:1–24.

Santa Luca, A. P. 1980. The postcranial skeleton of Heterodontosaurus tucki (Reptilia, Ornithischia) from the Stormberg of South africa. Ann. South African Museum 79:159–211.

Sawin, H. J. 1947. The pseudosuchian reptile Typothorax meadei, new species. Jour. Paleont. 21:201–238.

Thulborn, R. A. 1980. The ankle of archosaurs. Alcheringa 4:261–274.

Walker, A. D. 1961. Stagonolepis and its allies. Phil. Trans. Roy. Soc. London, Series B 244:103–204.

Wellnhofer, P. 1978. Pterosauria. Handbuch der Paläoherpetologie 19:1–82.

Welles, S. P. 1984. Dilophosaurus wetherelli (Dinosauria, Theropoda): osteology and comparisons. Paleontographica (A) 185:85–180.

Wild, R. 1973. Die Triasfauna der Tessiner Kalkalpen. XXIII. Tanystropheus longobardicus (Bassani) (neue ergebnisse). Schweiz. Paläontolische Abhandlungen 95:1–162.

1978. Die flugsaurier (Reptilia: Pterosauria) aud der oberen Trias von cene bei Bergamo, Italien. Bolletino della Societa Paleontólogica Italiani 17:176–256.

Young C. C. 1964. The pseudosuchians in China. Paleontologica Sinica 151:1–205.

27 A Predictive Method for the Identification of Vertebrate Ichnites and its Application to Pterosaur Tracks

DAVID M. UNWIN

Abstract

The traditional method of identifying the creators of fossil prints and tracks consists of deducing their anatomy, stance and gait from the ichnological record and comparing these features with those derived from the osteological record. This method produces reasonable results and is widely used, but suffers from some drawbacks: (i) differing tracks produced by variations in the locomotory patterns of trackmakers are difficult to identify; (ii) there is no satisfactory process for identifying and replacing data lost during or subsequent to the print–forming process; and (iii) misidentifications tend to be compounded rather than revealed.

A solution to these problems, already employed to some extent by paleoichnologists takes the form of a 'predictive' method in which details of anatomy, stance and gait, derived from osteological studies of potential trackmakers, are combined with knowledge of sediment response during the track–forming process to reconstruct prints and tracks. New discoveries can be identified and known material reassessed on the basis of diagnostic features of these reconstructions.

As an illustration of the predictive method, the ichnological record of pterosaurs is compared with reconstructions of their prints and tracks. It is concluded that pterosaurs were not responsible for any of the tracks currently ascribed to them, as they all lack characters diagnostic of a pterosaurian mode of origin.

Introduction

The basic methods employed by vertebrate paleontologists originated in the early 1800's (Sarjeant 1974). Since then, descriptive techniques and interpretation have improved. A good example is provided by the recent derivation (Alexander 1976) and development (Thulborn 1982, 1984; Demathieu 1986) of an expression for determining velocity from an animal's trackway. Knowledge of poten-

tial ichnite producers (osteologic taxa) has also improved. Furthermore the scope of paleoichnological studies continues to increase, as does interest in related subjects. On the other hand, the process of identification of fossil tracks remains largely unchanged since the early 1800's. This contrast prompts the question. How effective is this identification process?

To answer these questions we must begin with the identification process itself. It begins with the discovery or rediscovery of fossil tracks. Details of these and other relevant data are carefully gathered, interpreted and used to reconstruct the anatomy of the printmaker, and, where possible, its stance and gait. These features are then matched against the characteristics of potential printmakers in the osteological record. If an acceptable degree of similarity is found, the prints are usually allocated a name which, though separate from that of the printmaker, is intended to indicate its assumed identity (Sarjeant and Kennedy 1973, Baird 1980), as for example *Pteraichnus saltwashensis* ascribed by Stokes (1957) to a putative pterosaur trackway. This traditional method of identification is best exemplified by the works of Baird (1954), Bock (1952) and, more recently, Olsen and Baird (1986). Only a brief perusal of the literature is needed to appreciate the ubiquity of this method.

Despite its popularity, the traditional method is hampered by a fundamental problem. The data derived from tracks are influenced by a variety of factors (Fig. 27.1) (Padian and Olsen 1984 Fig. 4), each of which may alter the data. Reconstructions must take into account the possible effects of these factors, especially the stance, gait and speed of the printmaker and variations in substrate composition and behavior during print formation (Baird 1957, Peabody 1959, Padian and Olsen 1984). Herein lies the problem. The traditional method lacks a routine accounting for those factors which may have played a role during and after print formation (Sarjeant 1975, Padian and

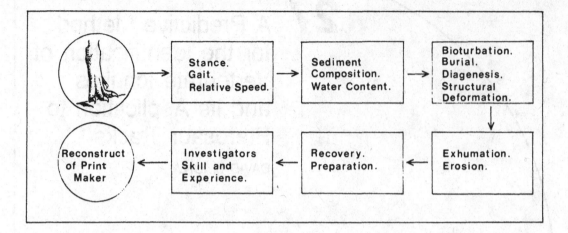

Figure 27.1. Factors operating during and after print formation which may hamper reconstruction and identification of the trackmaker.

Olsen 1984).

This problem manifests itself in a number of ways. Firstly, some tetrapods are capable of a variety of locomotory patterns during which the number and orientation of digits contacting the substrate can vary considerably. Crocodiles for example employ a 'high walk' during rapid locomotion producing a different trackway from that left by their sprawling gait during normal locomotion (Frey 1984). Without prior knowledge of crocodile locomotion, it could be difficult to determine, via the traditional method, that these different tracks were made by the same animal. Secondly, attempts to account for data which are suspected to have been filtered out must rely to a large extent on guesswork, an unsatisfactory procedure at best. Thirdly, errors in identification of ichnites are likely to be compounded rather than revealed since, once identified, correctly or otherwise, tracks are often used in the identification of later discoveries. Without an additional 'independent' basis for comparision, mistakes may be perpetuated and never revealed (Fig. 27.1). Attempts to circumvent this problem (Lull 1953, Baird 1957) have often proved unsatisfactory.

The remainder of this chapter is devoted to suggesting a solution to these problems. This solution, already used to some extent by paleoichnologists, takes the form of a predictive approach wherein diagnostic characters of tracks, reconstructed from an osteological basis, are employed as an alternative data set for identifying fossil tracks.

A Predictive Method of Identification in Vertebrate Paleoichnology

The predictive method of fossil track identification follows a somewhat different path than the traditional method described above. Details of foot structure and limb kinematics, associated with knowledge of sediment behavior during print formation, can be used to reconstruct the tracks of potential trackmakers. These reconstructions can

then serve as an independent data base for the identification of fossil footprints.

If the printmaker is extant, direct experimentation involving the printmaker and a variety of sediments can be carried out. The results of these experiments provide data for comparision with modern and fossil footprints, together with much needed insight into the processes of track formation. The Reverend William Buckland was probably the first to carry out experiments of this type (Sarjeant 1974) and there has been a resurgence of interest in recent years (Schaeffer 1941; Peabody 1959; Padian and Olsen 1984, and this volume; Thulborn this volume; and Thulborn and Wade this volume), but the great potential of this field remains largely untapped.

If the printmaker is extinct, the process is necessarily more complex. To begin with, a careful analysis of the pes and (if quadrupedal) the manus is required, taking into account the number and relative length of the digits and the phalanges they contain, the presence or absence of claws, mobility of joints, and all other relevant details. The next stage consists of fleshing out the skeletal structure. The sizes and distributions of major muscles can be deduced from the location and relative development of their origin and insertion sites, supplemented by careful use of modern analogues (Norman 1985). The position and development of plantar pads can be ascertained using the method proposed by Heilman (1927) and further described by Peabody (1948), Bock (1952) and Baird (1957).

At least four other factors need to be taken into account before a print can be generated from a reconstructed foot. These are: (i) the stance; (ii) the gait; (iii) the relative speed of the printmaker; and (iv) the way in which variation in sediment types and their properties may affect the morphology of the print.

The stance can take many forms ranging from that of a plantigrade sprawler such as a lizard, to that of a fully erect digitigrade such as a bird. Usually the stance can be determined from functional morphological analyses of uncrushed associated skeletons, combined with suitable modern analogues. Many stances are evidenced by a particular suite of characters, represented in the limb girdles, limbs and feet (Charig 1972, Rewcastle 1981, Parrish 1984), though some tetrapods, such as bipedal lizards (Snyder 1962) and crocodiles (Frey 1984) are capable of a variety of gaits, and may display an unusual combination of characters.

Relative proportions of the limbs and degrees of movement possible at each joint can be combined with knowledge of the stance to determine the types of gaits and relative speeds employed by the potential printmaker (e.g., Coombs 1978). The extent (and also the force, calculated from an estimate of the printmaker's weight) with which the pes and (if quadrupedal) the manus came into contact with the ground can be deduced from knowledge of the stance and gait. Using the geometric modifications specified by Baird (1957), prints and thus tracks can be generated for each type of stance and gait employed by the printmaker.

Clearly the morphology of the print will be determined to a degree by the properties of the print–bearing sediment. Unfortunately, the effect of these properties during and after print formation are, as yet, poorly understood (Padian 1986, Unwin 1986, Seilacher 1986), though the situation could be remedied by direct experimentation.

The predictive method should produce a data base containing the reconstructed tracks of each morphologically distinct group of tetrapods. Diagnostic characters of each group's tracks could be built up into a catalog serving as a means of identifying fossil tracks. This would provide the vertebrate paleoichnologist with a systematic method of comparing fossil material with diagnostic characters of each tetrapod group, including those with which he had little or no familiarity, removing the temptation to make intuitive leaps when faced with an unusual track. Furthermore, this method has some potential for isolating and identifying tracks produced by tetrapods as yet unknown in the osteological record.

This method can also be used to identify trackmakers which produced different tracks when moving at different speeds. Moreover, problematic prints can be reassessed against this osteologically derived 'independent' data base, rather than against other fossil prints whose identity may also be in doubt.

The predictive method does have some drawbacks. There is only partial overlap between the ichnological and osteological records (Baird 1980). Although some fossil tracks and traces appear to have no body fossil representatives, most tracks can be attributed to tetrapod groups with osteological remains.

A second problem is posed by the broad degree of similarity present in the morphology, stance and gait of some major tetrapod groups, such as the ornithischian dinosaurs. In these cases it may well be impossible to identify diagnostic features which would enable the paleoichnologist to distinguish, for example, between prints produced by hadrosaurids and iguanodontids.

Even with these drawbacks, the predictive method still offers an important addition to the paleoichnologists' investigative repertoire. The predictive process already forms a component of some studies (e.g., Baird 1957, 1980; Peabody 1959; Padian and Olsen 1984), though it is often brought into play only after decisions concerning the trackmaker's identity have been made. Clearly, much remains to be done if the potential of this approach is to be fully exploited.

Illustration of the Predictive Method
Reconstruction of the Prints and Tracks of Pterosaurs

The remainder of this Chapter is devoted to an illustration of the predictive method outlined above. The prints and tracks of pterosaurs are reconstructed and their diagnostic features used to reassess tracks and traces currently ascribed to pterosaurs. This group was chosen for two reasons: (i) their familiarity to the author and (ii) the need for a review of pterosaur tracks, the identity of some of which, including *Pteraichnus* (Stokes 1957 pers. comm., Padian and Olsen 1984) is disputed.

This is not the first occasion on which pterosaur prints and tracks have been reconstructed, at least in part, on the basis of their skeletal remains. In 1862 Oppel published an illustration of *Rhamphorhynchus* standing on all fours at the end of a short trackway, beneath which are two more reconstructed trackways (Fig. 27.2). Oppel's somewhat inaccurate reconstruction (the pes print has only three digits) was produced to conform with the then prevailing views on pterosaurs' stance and gait; the original idea, however, came from tracks in the Solnhofen Limestone, thought by Oppel to have been made by pterosaurs but later shown to be those of the horseshoe crab *Limulus* (Malz 1964).

More recently Padian and Olsen in their paper on *Pteraichnus* (1984) described, though without illustration, some likely characteristics of pterosaur prints and tracks based on recent analyses of their anatomy and functional morphology (Padian 1983a,b).

Footprint Reconstruction

Pterosaurs undoubtedly produced prints with their pedes; quadrupedal sprawlers could in addition have produced prints or traces with their manus, ventral surface of the body, and, in long–tailed forms, the tail. The latter are dealt with below.

The Hind Limb and Pes

The hind limb in pterosaurs was typically a rather spindly structure (Fig. 27.3). It consisted of a bowed femur, usually with a spherical offset head, an elongate tibia which ranged from one to one and a half times the length of the femur (Padian 1983b) and a reduced, splint–like fibula. All known pterosaurs had a mesotarsal ankle, with two proximal and two or three distal tarsals (Fig. 27.4a,d), though

Figure 27.2. Pterosaur tracks as reconstructed by Oppel (1962). Above, *Rhamphorhynchus* at the end of a short trackway, resulting from a sprawling quadrupedal gait. Below, two hypothetical pterosaur tracks. Redrawn from Wellnhofer (1983).

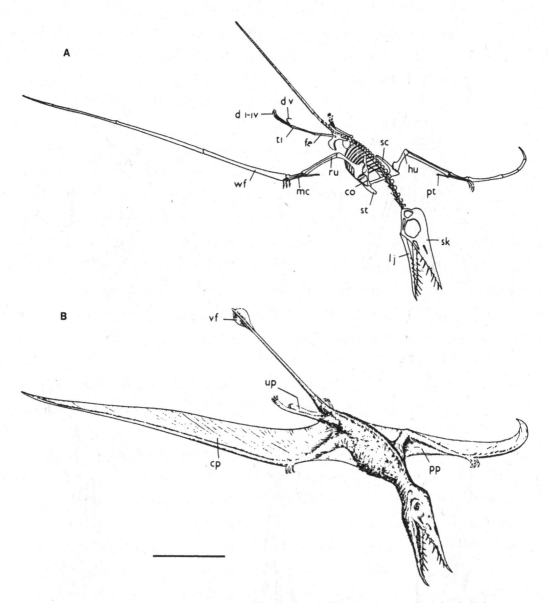

Figure 27.3. *Rhamphorhynchus muensteri*. **a,** skeletal reconstruction. **b,** life reconstruction. Note that here the pes is twisted inward, so that the fifth digit can support the uropatagium. Abbreviations: *co* coracoid, *cp* cheiropatagium, *d* digit, *fe* femur, *lj* lower jaw, *mc* metacarpus, *pp* propatagium, *ru* radius and ulna, *sc* scapula, *sk* skull, *st* sternum, *ti* tibia, *up* uropatagium, *vf* vertical tail flap, *wf* wingfinger (digit four) (from Wellnhofer 1975b). Scale bar 100 mm.

the two proximal tarsals often fused to the distal end of the tibia (Wellnhofer 1978) (Fig. 27.4e). The metatarsus consisted typically of four elongate metatarsals which overlapped proximally and diverged distally, giving the pes a distinctive elongate triangular plantar profile. All pterosaurs bore a 'hooked' fifth metatarsal, which usually articulated with the lateral face of the lateral distal tarsal (Padian 1983a), though in some forms it appears to have had an additional articulation with the calcaneum (Wild 1978) (Fig. 27.4a). In later forms, where the fifth digit was lost altogether, the fifth metatarsal was reduced to a mere nubbin of bone (Fig. 27.4e).

The primitive phalangeal formula of the pes in pterosaurs was 2,3,4,5,2, a configuration present in all rhamphorhynchoids. In the pterodactyloids the fifth digit was greatly reduced, and finally lost in Upper Cretaceous forms such as *Pteranodon* and *Nyctosaurus* (Williston 1903) (Fig. 27.4e). In the most primitive pterosaurs such as *Peteinosaurus* (Wild 1978) (Fig. 27.4a) the fifth toe consisted of two highly elongate phalanges. In some forms such as *Scaphognathus*, the distal phalanx was hooked, in others such as *Dimorphodon*, it remained straight. The significance of this, and the general function of the fifth digit, remains in dispute; some authors incorporate it into the patagium (Wellnhofer

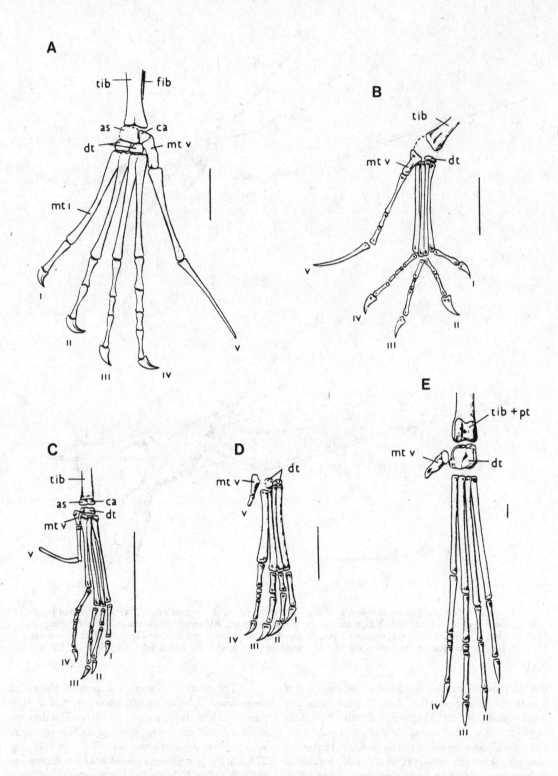

Figure 27.4. The pes in various pterosaurs. **a,** *Pteinosaurus,* left pes anterior view. **b,** *Anurognathus* left pes post. view. **c,** *Rhamphorhynchus* left pes post. view. **d,** *Pterodactylus* left pes post. view. **e,** *Pteranodon* left pes post. view. Abbreviations: *as* astragalus, *ca* calcaneum, *dt* distal tarsals, *fib* fibula, *mt* metatarsal, *pt* proximal tarsals, *tib* tibia. a, c and d redrawn from Wellnhofer (1978), b redrawn from Wild (1978). Scale bar 10 mm.

1975b, Pennycuick 1986) (Fig. 27.3b), others (Padian 1983b) do not. The fifth digit became progressively shorter in more advanced rhamphorhynchoids such as *Campylognathoides*, reduced to a very short structure in early pterodactyloids, and finally lost in later forms (Wellnhofer 1978).

Digits one to four possess, in the context of this study, two important characters. The unguals in most pterosaurs, though not particularly elongate, were somewhat compressed, sharply pointed and gently recurved (Fig. 27.4a–e). They are more like those of climbers (Yalden 1985; Unwin 1987a,b, 1988) than those of typical terrestrial vertebrates, though this characteristic is not particularly marked. A second character relates to the penultimate phalanx which is, in all pterosaurs, markedly more elongate than any of the preceding phalanges (Unwin 1987a, 1988) (Fig. 27.4a–e). This configuration, as far as I am aware, does not occur in any other reptilian pes, fossil or otherwise, and as such provides a useful and important diagnostic character.

The Forelimb and Manus

The pterosaur forelimb was highly modified to form the main wing spar (Fig. 27.3a). The humerus was relatively short, with a large delto–pectoral crest upon which most of the flight musculature was inserted (Wellnhofer 1975b, Padian 1983b). The radius and ulna were elongate cylindrical bones, articulating distally with the carpus, which consisted of between three and five, possibly six, bones (Wild 1978) and bore the pteroid, an element unique to pterosaurs (Fig. 27.5a). The metacarpus consisted of three long slender metacarpals lodged against the medial face of a large, stout fourth metacarpal (Fig. 27.5a). The fourth metacarpal of rhamphorhynchoids remained rather short, forming only about 5% of the total wing length. In the pterodactyloids it became increasingly elongate, forming up to 20% of the total wing length in Upper Cretaceous forms such as *Pteranodon*.

The phalangeal formula of the manus was 2,3,4,4 in all pterosaurs except for *Nyctosaurus*, which lacked the terminal wing phalanx (G. Brown pers. comm.). The first three digits were small, while the fourth was a greatly enlarged highly elongate structure, forming the distal portion of the wing spar. The distal end of the fourth metacarpal bore a large pulley–like joint, around which the first phalanx of the wing finger was protracted and retracted. The wing finger, which was extended perpendicular to the body in flight, could be rotated through an angle of 165° to fold back against the posterior surface of the fourth metacarpal (Wellnhofer 1978). Digit two was slightly longer than digit one and slightly shorter than digit three. All three bore large, highly compressed strongly recurved unguals, with well developed flexor tubercles and in some cases antungual sesamoids (Wild 1978, Padian 1983a) (Fig. 27.5a), suggesting a well developed grasping function. The penultimate phalanx of each digit, like those of the pes, was much more elongate than those of the preceding phalanges. The digits lay in the line of axis of the fourth

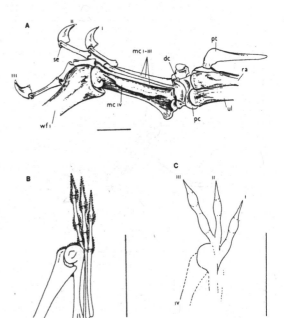

Figure 27.5. Pterosaur manus and reconstructed manus print. **a,** forearm and manus of *Eudimorphodon* (from Wild, 1978). **b,** manus of *Rhamphorhynchus* in contact with the substrate (modified from Wellnhofer 1975a). **c,** reconstructed manus print. Abbreviations: *dc* distal carpal, *mc* metacarpal, *pc* proximal carpals, *pt* pteroid, *ra* radius, *ul* ulna, *wfi* first phalanx of wing finger. Scale bar 10 mm.

metacarpal, and presumably flexed and extended in the dorso–ventral plane (Fig. 27.5b) with restricted movement in the lateral plane.

Reconstructing the Soft Tissues

The second stage of reconstruction consists of fleshing out the pes and manus. The pterosaur pes is unlikely to have been particularly fleshy, because mechanical and aerodynamic selective forces probably concentrated the main mass of muscle in the proximal part of the limb, as they do in most birds. Long thin tendons flexed and extended distal segments. The plantar pads were probably located beneath the joints, rather than beneath the phalanges (Bock 1952, Baird 1957). Phalanges two and three of the fourth digit and phalanx two of the third digit were relatively short, especially in later pterodactyloids (Fig. 27.4d,e). It is reasonable to assume, in these cases, that only a single pad was present between the distal end of phalanx one and proximal end of phalanx three in the third digit, and between the distal end of phalanx one and the proximal end of the fourth phalanx in the fourth digit (Fig. 27.6c).

Some pterosaurs, such as *Rhamphorhynchus* and *Pterodactylus* (Wellnhofer 1970, 1975b), appear to have had webs of skin stretched between the first four digits of the

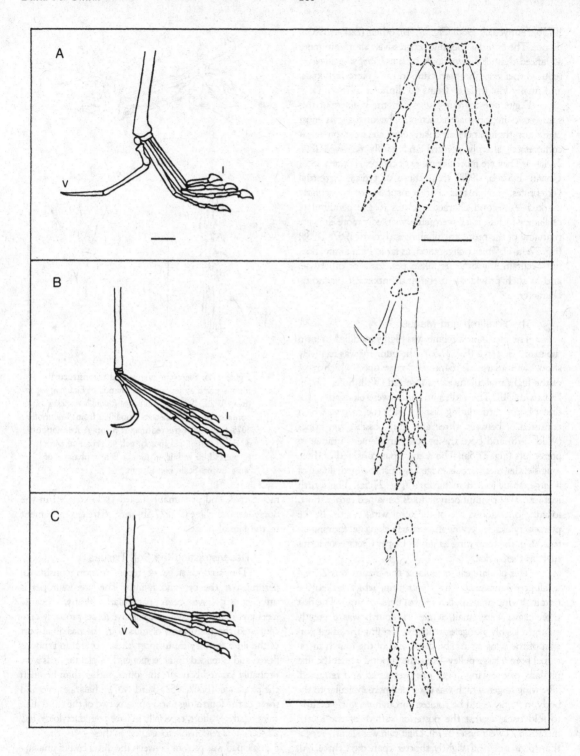

Figure 27.6. Pes and reconstructed pes prints in various pterosaurs. **a,** *Dimorphodon macronyx* in digitigrade pose, modified from Padian (1983a). **b,** *Rhamphorhynchus* in plantigrade pose, modified from Wellnhofer (1975b). **c,** *Pterodactylus* in plantigrade pose, modified from Wellnhofer (1970). Scale bar 10 mm.

pes (Fig. 27.7a–c). Wild (1978) has suggested that the fifth digit was also incorporated into the webbed foot. However, the opposite is indicated by Wellnhofer (1975b, Fig. 43g) in his reconstruction of the pes of *Rhamphorhynchus*.

For similar reasons, it is also unlikely that the pterosaur manus was particularly fleshy. The components of the wing spar — humerus, radius and ulna, fourth metacarpal and wing finger — were encased in the wing membrane so as to present a smooth landing edge profile, from which the first three digits projected (Wellnhofer 1970, 1975b) (Fig. 27.3b).

The Stance and Gait

Before tracks are considered, the possible stances and gaits of pterosaurs should be determined, in order to ascertain the extent to which the plantar surface of the pes and manus came into contact with the substrate. Pterosaurs were capable of a variety of stances (Paul 1987), ranging from sprawling or semi–erect plantigrady (Wellnhofer 1975b, 1978, 1982; Wellnhofer and Vahldiek 1986) (Figs. 27.8c, 27.9a,b) to fully erect digitigrady (Padian 1983a,b) (Figs. 27.8a,b, 27.9c). The predominant form of locomotion used by most if not all pterosaurs consisted of a rather sprawling or semi–erect plantigrade stance and a largely quadrupedal gait (Unwin 1987a,b), possibly reflecting an arboreal ancestry (Wild 1984). Some early pterosaurs may have had a fully erect stance and a good cursorial ability (Padian 1983a,b), though a new species of *Dimorphodon* does not support this (Unwin 1988).

In plantigrade pterosaurs the major point of articulation was at the ankle (Fig. 27.8c). Metatarsals one to four, together with the digits, acted as a functional unit and the whole of the plantar surface was applied to the substrate.

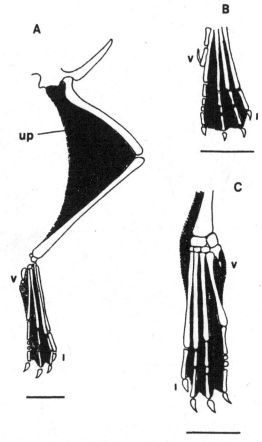

Figure 27.7. Webbed feet of pterosaurs. **a** and **c,** *Pterodactylus,* **b,** *Rhamphorhynchus.* Abbreviations: *up* uropatagium. Redrawn from Wellnhofer (1978). Scale bar 10 mm.

Figure 27.8. Hind limb stance of pterosaurs. **a,b,** Fully erect bipedal digitigrade stance in *Dimorphodon* (from Padian 1983a). **c,** semi–erect plantigrade stance in *Rhamphorhynchus* (modified from Wellnhofer 1975b). Scale bar 10 mm.

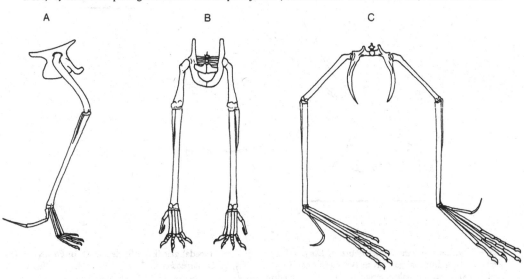

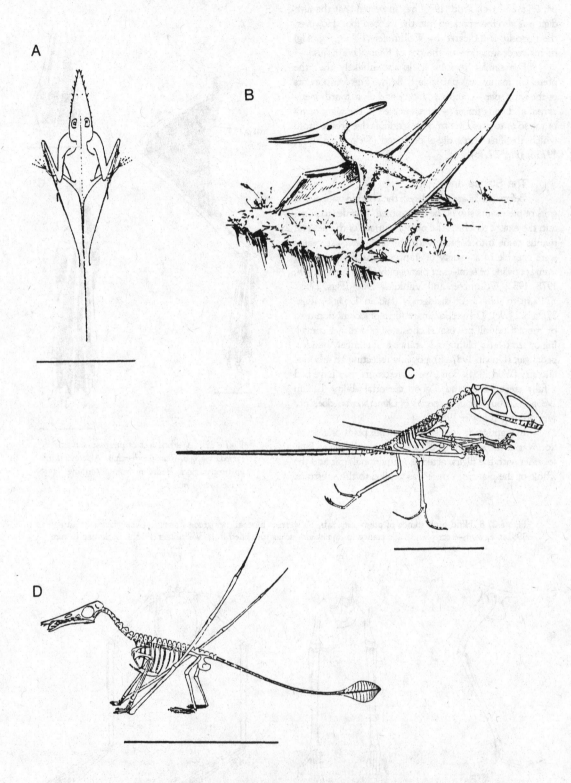

Figure 27.9. Gait in various pterosaurs. **a,** *Dorygnathus* in semi–erect bipedal gait (from Stieler 1922). **b,** *Pteranodon* in semi–erect quadrupedal gait (from Bramwell and Whitfield 1974). **c,** *Dimorphodon* in fully erect bipedal gait (from Padian 1983a). **d,** *Rhamphorhynchus* in fully erect quadrupedal gait (from Seeley 1901). Scale bar 250 mm.

Those pterosaurs with a fifth digit may have used it as a lateral prop as do lepidosaurs (Robinson 1975), in which case it too would have contacted the substrate, though diverging widely from the other digits. In digitigrade pterosaurs, the major point of articulation was between the distal ends of the metatarsals and the proximal ends of digits one to four (Padian 1983a). In this case, only the plantar surface of the digits would have made contact with the substrate (Fig. 27.8a,b). Pterosaurs which bore a fifth digit would, according to Padian (1983b), probably not have brought it into contact with the ground, though what its function was remains a mystery.

Pterosaurs may have used their forelimbs for terrestrial locomotion (Bramwell and Whitfield 1974). However the manus could only be used if the wing was partly or fully extended. In this position most, if not all, of the manus would have come into contact with the ground, as would the distal portion of the metacarpal four, and proximal portion of the wingfinger (Fig. 27.9b).

Reconstructed Footprints

Having determined the skeletal structure of the manus and the pes, applied a 'fleshing–out' process, and taken into account the degree to which they contacted the substrate, footprints can be reconstructed (Figs. 27.5c, 27.6a–e). The pes print of a plantigrade pterosaur would have been tetradactyl with an elongate triangular outline, and flanked laterally in rhamphorhynchoids by the print of the fifth digit (Fig. 26.6b,c). The apex of the triangle is formed by the heel, which, since it would have borne most of the weight during the step, would correspond with the deepest part of the print. The flat bar–like metatarsals would no doubt have left some impression in softer sediments. The plantar pads might present evidence of the elongate penultimate phalanx and, presumably, only three pads would be indicated in the second, third and fourth digits of the pterodactyloid pes print (Fig. 27.6c). The claw marks would have been narrow and fairly deep. Digits one to four would have been gently divergent, while the fifth digit would have ranged from sub–parallel to highly divergent. The markedly elongate, yet clawless, imprint of the fifth digit would form a useful diagnostic characteristic in rhamphorhynchoid pes prints. In those forms which had webbed feet, the webbing would probably have left some indication of its presence.

The reconstructed pes print of a digitigrade pterosaur (Fig. 27.6a) is much squarer in outline than that of a plantigrade pterosaur. In other respects it is rather similar, but would have lacked a heel and a fifth digit print. The metatarsal/phalangeal pads may also have failed to impress if the animal were moving rapidly.

The prints produced by the manus of a pterosaur are likely to have been more varied than those of the pes, since the hind limb was always involved in terrestrial locomotion, whereas use of the forelimb could have varied from brief contacts necessary to maintain balance, through to physically dragging the animal forward over the ground. The manus print would have consisted of three weakly divergent digits rotated outward and ending in large, deep, narrow claw marks (Fig. 27.5b,c). The distal condyle of the fourth metacarpal would have formed a large circular impression just proximal and slightly to the rear of the first three digit prints (Fig. 27.5b,c). The proximal portion of the wingfinger is also likely to have left an impression, though this could occur in a variety of positions, as discussed earlier. It should be noted that considerable "damage" could occur to the manus prints if the pterosaur were to claw its way forward. Drag marks produced by other portions of the wing could also have occurred.

Trace marks could have been produced by the tail and the ventral surface of the body. An elongate tail with a vertical terminal flap was present in most rhamphorhynchoids. In the pterodactyloids and one family of rhamphorhynchoids, the Anurognathidae, the caudal vertebrae were reduced to eleven or fewer in number, forming a structure similar to the pygostyle of birds (Wellnhofer 1978). The long tails of rhamphorhynchoids contained numerous highly elongate bony rods similar to those found in the tail of *Deinonychus* (Ostrom 1969) and which served to stiffen the whole structure. The rather inflexible tail projected posteriorly and was unlikely to have left any traces unless the pterosaur were to have sprawled very close to the ground, in which case a long, narrow and somewhat sinuous trail could have been produced (cf. Oppel 1862) (Fig. 27.2). Some pterosaurs could conceivably have had such a poor terrestrial ability that they flopped onto their bellies after landing (Bramwell and Whitfield 1974), and proceeded to shuffle and drag their way forward. In this case as well as pes and manus prints, a wide median groove could have been produced, possibly, under ideal circumstances, with hair–like impressions along the lateral edges, since it is now known that the body of pterosaurs had a covering of short stiff hairs (Broili 1927, Sharov 1971).

Reconstruction of Tracks

Before pterosaur tracks can be generated using such track reconstructions, the stance, gait and likely velocity of a grounded pterosaur must be considered.

Early authorities on pterosaurs credited them with a variety of terrestrial locomotory abilities, ranging from a dog–like, fully erect quadrupedal gait (Seeley 1901) (Fig. 27.9d) to a bird–like, fully erect bipedal gait (Wagner 1852, Seeley 1901) (Fig. 27.9c), though most opted for some form of bat–like sprawling quadrupedal or bipedal gait (Soemmering 1817) (Fig. 27.9a,b). Current opinion admits two possibilities. Either most pterosaurs were semi–erect quadruped/bipeds, or they were fully erect bipeds; this is linked to the suggestion of an arboreal ancestry for pterosaurs in the former case (Wild 1984), or a terrestrial ancestry in the latter (Padian 1984).

The main advocate of the semi–erect stance is Wellnhofer (1970, 1975b, 1978, 1982; Wellnhofer and Vahldiek 1986). Their reconstruction of pterosaur stance has the following features. The pelvis was open ventrally and the acetabula faced outward and a little upward. The femora were directed obliquely out and downward, the tibiae were oriented vertically, the feet articulated at the ankle amd were plantigrade, and the fifth digit helped to support a uropatagium (Fig. 27.8c).

From this it would appear that pterosaurs proceeded either by an awkward waddle, on the hind limbs alone, or employed the forelimbs to varying degrees as props or to pull the animal forward. Padian (1983b) has demonstrated that, with the forelimbs folded up, the manus could not be brought to bear directly upon the substrate, as illustrated by Seeley (1901, Fig. 56, 63) (Fig. 27.9d). I have been able to confirm this by manipulation of uncrushed wing bones of *Rhamphorhynchus* from the Kimmeridge Clay of Dorset (Wellnhofer 1975a), *Santanadactylus* from the Santana Formation (Aptian) of Brazil (de Buisonje 1980), and *Quetzalcoatlus* from the Javelina Formation (Maastrichtian) of Texas (Lawson 1975, Langston 1981). The manus could only be used to support or pull the pterosaur forward when the wing was partly or fully extended (Fig. 27.9b). Furthermore, the limited degree of fore and aft movement available would only have permitted a very short stride.

The reconstructed trackway of a semi–erect plantigrade pterosaur proceeding on all fours is shown in Figure 27.10a. The tracks are characterized by a plantigrade pes print, in which the digits are rotated inwards, and an angle of 30° or more is formed between the long axis of the pes and the direction of travel. The pes trackway is relatively broad, with a short stride and low pace angulation of about 80°, indicative of a slow sprawling gait. The manus prints lie well to the outside of the pes prints, the broadness of the track depending on the degree to which the wing was folded. The long axis of manus prints produced by a fully extended wing lie almost perpendicular to the direction of travel (Fig. 27.9a); as the wing is folded, this angle decreases to a minimum of about 50°. As noted by Padian and Olsen (1984), the glenoacetabular to foot–length ratio is very low, ranging from about 1.5 to 2.1.

In contrast to Wellnhofer's semi–erect plantigrade pterosaur, it has been proposed by Padian (1983a,b, 1984) that all pterosaurs were fully erect digitigrade bipeds. The main features of this reconstruction are as follows. The pelvis was fused ventrally, the acetabula looked outward and a little downward, and the shank of the femur moved in a parasagittal plane. Much of the step cycle movement took place at the knee, the metatarsus was raised off the ground, and the stance was digitigrade (Figs. 27.8a,b, 27.9c). Both stance and gait were bird–like, the wings being folded over the back and rarely if ever used in terrestrial locomotion.

The tracks produced by this type of stance and gait (Fig. 27.10b), assuming that the pterosaur was moving at a walking speed, are characterized by digitigrade pes prints, with their long axis virtually parallel to the line of travel or even slightly inturned (Padian pers. comm.). The track is very narrow, with a long stride and a large pace angulation of about 150°.

Additional diagnostic features of pterosaur tracks may result from behavior patterns unique to pterosaurs. Most important of these are 'take off' and 'landing' runs, consisting of tracks in which the pace and stride increase quite rapidly, then cease abruptly. Confusion with similar patterns produced by birds should not occur since their pes prints would be very different and easily distinguishable. Small pterosaurs may have been capable of 'spring take offs', trace remains of which may be predicted from those produced by birds. Typically they have relatively deeply impressed pes prints when compared with those leading up to the take–off point, and are often surrounded by marks produced by the wing hitting the substrate during the first wing beat.

The Ichnological Record of Pterosaurs

A number of fossil tracks have been assigned to pterosaurs (Stokes 1957, 1978; Anonymous 1973; Langston 1974; Logue 1977; Stokes and Madsen 1979), each of which was identified largely through the traditional approach discussed earlier. Some, but not all, of these tracks were recently reinterpreted by Padian and Olsen (1984), who suggest that they were made by crocodiles. This identification is disputed (Stokes pers. comm.) and moreover was based on the questionable assumption that all pterosaurs were digitigrade bipeds. A review of these and all other so–called 'pterosaur' tracks is therefore clearly required.

The first documented putative pterosaur trackway, *Pteraichnus saltwashensis*, from the Morrison Formation (Upper Jurassic) of Arizona, was described by Stokes in 1957. The pes prints and trackway (Fig. 27.11a) share some similarities with the reconstructed tetradactyl tracks of a plantigrade pterosaur. The prints have a narrow triangular outline, with a deep heel at the apex. The first four digits are weakly divergent, rotate inward, and are virtually identical in relative lengths to those of the reconstructed pes print. In both cases the stride is short, and the pes track broad. On the other hand, the long axis of the pes in *Pteraichnus* is sub–parallel to the direction of travel, unlike those in the reconstruction trackway. The lack of a fifth digit imprint in *Pteraichnus* is irrelevant, as the track could have been produced by a pterodactyloid pterosaur lacking a fifth digit, or a rhamphorhynchoid pterosaur in which the fifth digit was not employed in terrestrial locomotion.

The manus print of *Pteraichnus* is quite unlike the reconstructed manus print (Figs. 27.5c, 27.10a). It consistently lacks the first digit, though digits 'two' and 'three' do lie medially to the supposed wing finger impression and not laterally as stated by Padian and Olsen (1984). The claw marks are small and shallow, and, although a 'deep central depression' corresponding to the position of the distal condyle of the fourth metacarpal is present (Stokes 1957), its morphology is quite unlike that of the reconstructed fourth metacarpal print. Most significantly, the manus prints frequently lie beneath the pes prints, a position which, even if attainable, would have resulted in the digits of the manus prints pointing forward, and not laterally as in this case. Furthermore, the forelimb stride is much larger than the

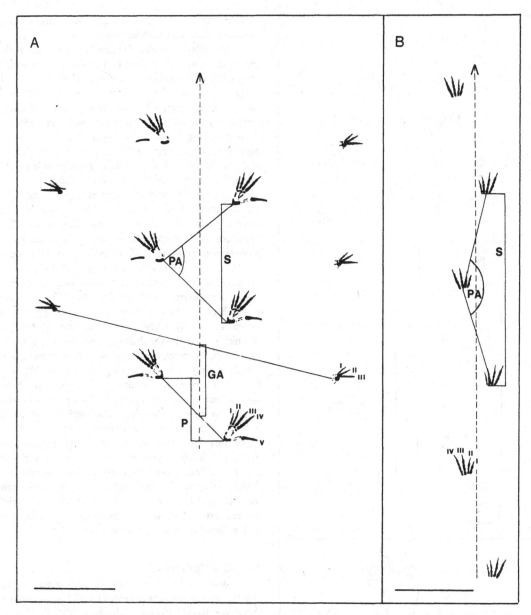

Figure 27.10. Reconstructed tracks of pterosaurs. **a,** track produced by a semi–erect pterosaur with quadrupedal gait, pace and stride relatively short, glenoacetabular to foot–length ratio 2.0, pace angulation 80°. **b,** track produced by a pterosaur with a fully erect bipedal gait, stride relatively long, pace angulation 150°. Abbreviations: *ga* glenoacetabular length, *p* pace, *pa* pace angulation, *s* stride. Scale bar 250 mm.

predicted maximum for this size of animal, and, as Padian and Olsen have noted (1984), the ratio of the glenoacetabular length to foot length in *Pteraichnus* is about 2.6, whilst that of pterosaurs ranges from 1.5 to 2.1. Indeed, since *Pteraichnus* lacks any diagnostic characters in common with the reconstructed trackway, a pterosaurian origin seems unlikely.

In the early 1970's, Lynn Ottinger of Moab, Utah, discovered a second 'pterosaur' trackway in the Navajo Sandstone (Lower Jurassic) of Grand County, Utah (Anonymous 1973). It was mentioned by Stokes in 1978

and briefly described by Stokes and Madsen (1979), but a full description has never been published. The close proximity of the manus and pes prints and the presence of what appear to be five digits in some of the manus prints (Padian pers. comm.) have prompted the suggestion that these tracks belong to the crocodilian ichnogenus *Batrachopus* (Padian 1986). Clearly a pterosaurian mode of origin seems doubtful.

In 1977 Logue published a short note on 'Pterodactyl tracks' from the Sundance Formation (Upper Jurassic) of Alcova, Wyoming, which appear virtually identical to those

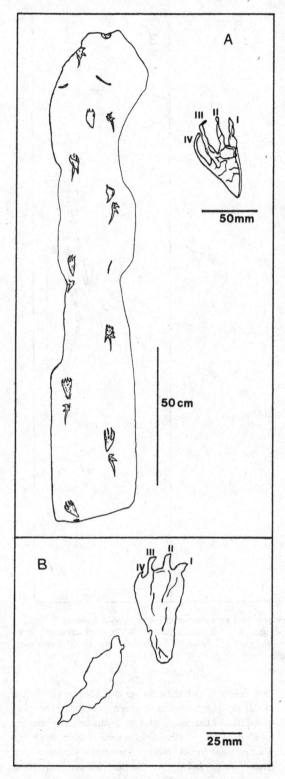

of *Pteraichnus*. Again, a pterosaurian mode of origin seems unlikely (Prince and Lockley this volume).

Langston (1974, plate 4) figured some unusual markings in the Lower Glen Rose Formation of Texas which are possibly attributable to pterosaurs. Parallel striae, occurring in groups of three, could have been produced by a pterosaur manus. But, as Langston notes, the lack of a large depression corresponding to the distal condyle of the fourth metacarpal, together with the apparent lack of any pes print, is rather puzzling. Furthermore the subequal length of the striae does not tally with the disparity in length of reconstructed manus digit prints (Fig. 27.10a).

A single trackway consisting of six prints made by a left manus, discovered in the Dakota Formation of Clayton Lake State Park, northeastern New Mexico, has been tentatively assigned to a pterosaurian trackmaker (Gillette pers. comm., and Gillette and Thomas this volume, Fig. 36.4). The prints generally consist of three digits, associated with other marks, which have been interpreted as part of the fourth metacarpal (Fig. 36.4). The prints are perpendicular to the line of travel, and the length of pace increases markedly in the inferred direction of travel, all of which suggest the possibility of a pterosaur attempting to take off. However, the digits are in most cases of subequal size, and do not show an increase in length, as is typical in the pterosaur manus (Fig. 27.5a–c). Furthermore the claw marks are very shallow, and the proposed metacarpal four imprint, though roughly similar in shape, is much smaller than that of reconstructed tracks. Most importantly, the digits show only a weak convergence proximally, and at least one print may contain four digits of similar size. Both these features are inconsistent with the details of the reconstructed tracks. Hence a pterosaurian mode of origin seems doubtful and a "swimming crocodile" inferred instead (Unwin 1986). See Gillette and Thomas (this volume) for a detailed description.

Conclusions

Based on the predictive approach advocated herein, and the reinterpretations suggested for all known putative "pterosaur tracks" (Padian and Olsen 1984, Prince and Lockley this volume), there is no convincing evidence that true pterosaur tracks are currently known. Those that have been suggested as pterosaurian in affinity have all been reinterpreted, at least once, as crocodilian.

Acknowledgments

I am most grateful to Charles Deeming, Andrew Kitchener, Penelope Milner and Paul Unwin for critically reading this manuscript. I am particularly indebted to Kevin Padian and two anonymous reviewers for their comments and suggestions. Special thanks go to David Gillette for enabling me to attend the Tracks Conference and for generously supplying information on the Clayton Lake Track Site. This work was supported by the Wilkie Calvert Postgraduate Reseasrch Award administered by Reading University. Additional support was given by Mr. H. L. Unwin, to whose memory this paper is dedicated.

Figure 27.11. **a,** *Pteraichnus saltwashensis*. Trackway from Stokes (1957), enlarged view of left pes from Padian and Olsen (1984). **b,** associated manus and pes print of a 'Pterodactyl', redrawn from Logue (1977).

References

Anonymous. 1973. Track of the pterosaur: Probable oldest evidence. *Science News* 104:85.

Alexander, R. McN. 1975. Estimates of speeds of dinosaurs. *Nature* 261:129–130.

Baird, D. 1954. *Cheirotherium lulli* a pseudosuchian reptile from New Jersey. *Bull. Mus. Comp. Zool.* (Harvard University) 111 (4):165–192.

——— 1957. Triassic reptile footprint faunules from Milford New Jersey. *Bull. Mus. Comp. Zool.* (Harvard University) 117 (5):449–520.

——— 1980. A prosauropod dinosaur trackway from the Navajo Sandstone (Lower Jurassic) of Arizona. *In* Jacobs, L. L. (ed.). *Aspects of Vertebrate History* (Musseum of Northern Arizona Press) pp. 219–230.

Bock, W. 1952. Triassic reptilian tracks and trends of locomotive evolution. *Jour. Paleont.* 26:395–433.

Bramwell, C. D., and Whitfield, G. R. 1974. Biomechanics of *Pteranodon*. *Philosophical Trans. Roy. Soc. London* (B), 267:503–581.

Broili, F. 1927. Ein *Rhamphorhynchus* mit Spuren von Haarbedeckung. *Sitzungsberichte Bayer. Akademie Wissenschaften, math. naturwiss. Abteilung:* pp. 29–48.

Buisonje, P. H. de. 1980. *Santanadactylus brasiliensis* nov. gen. nov. sp., a long necked, larger pterosaurier from the Aptian of Brasil. *Proc. Koninkl. Nederland Akad. Wetenschaft* (B) 83 (2):145–172.

Charig, A. J. 1972. The evolution of the archosaur pelvis and hind limb: An explanation in functional terms. *In* Joysey, K. A., and Kemp, T. S. (eds.). *Studies in Vertebrate Evolution* (Edinburgh: Oliver and Boyd) pp. 121–155.

Coombs, W. P. 1978. Theoretical aspects of cursorial adaptations in dinosaurs. *Quart. Rev. Biol.* 53:393–418.

Demathieu, G. 1986. Nouvelles recherches sur la vitesse der vertébrés auteurs de traces fossiles. *Geobios* 19 (3):327–333.

Frey, E. 1984. Aspects of the biomechanics of crocodilian terrestrial locomotion. *In* Reif, W. E., and Westphal, F. (eds.). *Proc. Third Symp. Mesozoic Terrestrial Ecosystems* (Tubingen: Attempto) pp. 93–97.

Heilman, G. 1927. *The Origin of Birds.* (New York: D. Appleton and Co.) 210 pp.

Langston, W. Jr. 1974. Nonmammalian Comanchean tetrapods. *Geoscience and Man* 8:77–102.

——— 1981. Pterosaurs. *Sci. Amer.* February 1981:122–136.

Lawson, D. A. 1975. Pterosaur from the Latest Cretaceous of West Texas. Discovery of the largest flying creature. *Science* 187:947–948.

Logue, T. J. 1977. Preliminary investigation of pterodactyl tracks at Alcova, Wyoming. *Wyoming Geol. Assoc. Earth Sci. Bull.* 10 (2):29–30.

Lull, R. S. 1953. Triassic life of the Connecticut Valley. *Connecticut Geol. Nat. Hist. Surv. Bull.* 81:1–336.

Malz, H. 1964. Kouphichnium walchi die Geschichte eine Fahrte und ihres Tieres. *Natur und Museum* 94 (3):81–97.

Norman, D. 1985. *The Illustrated Encyclopedia of Dinosaurs.* (London: Salamander Books Limited) 208 pp.

Olsen, P. E., and Baird, D. 1986. The ichnogenus *Atreipus* and its significance for Triassic stratigraphy. *In* Padian, K. (ed.). *The Beginning of the Age of Dinosaurs* (Cambridge University Press) pp. 61–87.

Oppel, A. 1862. Uber Fährten im lithographischen Schiefer.

Palaeontol. Mitt. Museum Kgl. Bayer. Staates 1:121–125.

Ostrom, J. H. 1969. Osteology of *Deinonychus antirrhopus*, an unusual theropod from the Lower Cretaceous of Montana. *Bull. Yale Peabody Mus. Nat. Hist.* 30:1–165.

Padian, K. 1983a. Osteology and functional morphology of *Dimorphodon macronyx* (Buckland) (Pterosauria: Rhamphorhynchoidea) based on new material in the Yale Peabody Museum. *Postilla* 189:1–44.

——— 1983b. A functional analysis of flying and walking in pterosaurs. *Paleobiology* 9 (3):218–239.

——— 1984. The origin of pterosaurs. Reif, W. E., and Westphal, F. (eds.). *Proc. Third Symp. Mesozoic Terrestrial Ecosystems* (Tubingen: Attempto) pp. 163–168.

——— 1986. On the track of the dinosaurs. *Palaios* 1986:519–520.

Padian, K., and Olsen, P. E. 1984. The fossil trackway *Pteraichnus*: not pterosaurian, but crocodilian. *Jour. Paleont.* 58 (1):178–184.

——— 1988. Ratite footprints and the stance and gait of Mesozoic theropods. this volume.

Parrish, J. M. 1984. Locomotor grades in the Thecodontia. *In* Reif, W. E., and Westphal, F. (eds.). *Proc. Third Symp. Mesozoic Terrestrial Ecosystems* (Tubingen: Attempto) pp. 169–171.

Paul, G. S. 1987. Pterodactyl habits — real and radio controlled. *Nature* 328:481.

Peabody, F. E. 1948. Reptile and amphibian trackways from the Lower Triassic Moenkopi Formation of Arizona and Utah. *Univ. Calif. Bull. Geol. Sci.* 27:295–468.

——— 1959. Trackways of living and fossil salamanders. *Univ. Calif. Publ. Zool.* 63:1–72.

Pennycuick, C. J. 1986. Mechanical constraints on the evolution of flight. *Mem. California Acad. Science* 8:83–98.

Newcastle, S. C. 1981. Stance and gait in tetrapods: An evolutionary scenario. *Symp. Zool. Soc. London* 48:239–267.

Robinson, P. L. 1975. The functions of the hooked fifth metatarsal in lepidosaurian reptiles. *Colloques International Centenary National du Recherche de Science* 218:461–483.

Sarjeant, W. A. S. 1974. A history and bibliography of the study of fossil vertebrate footprints in the British Isles. *Palaeogeog., Palaeoclimat., Palaeoecol.* 16:265–378.

——— 1975. Fossil tracks and impressions of vertebrates. *In* Frey, R. W. (ed.). *The Study of Trace Fossils* (New York: Springer–Verlag) pp. 283–324.

Sarjeant, W. A. S., and Kennedy, W. J. 1973. Proposal of a code for the nomenclature of trace fossils. *Canadian Jour. Earth Sci.* 10:460–475.

Schaeffer, B. 1941. The morphological and functional evolution of the tarsus in amphibians and reptiles. *Bull. Amer. Mus. Nat. Hist.* 78:395–472.

Seeley, H. G. 1901. *Dragons of the Air. An account of extinct flying reptiles.* (London: D. Appleton and Co.) 239 pp.

Seilacher, A. 1986. Dinosaur tracks as experiments in soil mechanics. (Abs.) *In:* Gillette, D. D. (ed.). *Abst. Prog. First Internat. Symp. on Dinosaur Tracks and Traces* (Albuquerque: New Mexico Museum of Natural History) p. 24.

Sharov, A. G. 1971. New flying reptiles from the Mesozoic of Kazakstan and Kirgisia. *Akademia Nauk SSSR Trudy Paleontolog. Inst.* 130:104–113.

Snyder, R. C. 1962. Adaptations for bipedal locomotion of

lizards. *Amer. Zool.* 2:191–203.

Soemmerring, S. Th. v. 1817. Uber einen *Ornithocephalus brevirostris* der Vorwelt. *Denkschr. Kgl. Bayer. Akademie Wissenschaften* 6:105–112.

Stieler, C. I. 1922. Neuer Rekonstruktionversuch eines liassichen Flugsauriers. *Naturw. Wochensch.* 21 (20) 273–280.

Stokes, W. L. 1957. Pterodactyl tracks from the Morrison Formation. *Jour. Paleont.* 31:952–954.

1978. Animal tracks in the Navajo-Nugget Sandstone. *Contr. Geol. Univ. Wyoming* 16 (2):103–107.

Stokes, W. L., and Madsen, J. H. Jr. 1979. Environmental significance of pterosaur tracks in the Navajo Sandstone (Jurassic) Grand County, Utah. *Brigham Young Univ. Geol. Studies* 26 (2):21–26.

Thulborn, R. A. 1982. Speeds and gaits of dinosaurs. *Palaeogeogr., Palaeoecol., Palaeoclimat.* 38:227–256.

1984. Preferred gaits of bipedal dinosaurs. *Alcheringa* 8:243–252.

1988. The gaits of dinosaurs. This volume.

Thulborn, R. A., and Wade, M. 1988. A footprint is a history of movement. This volume.

Unwin, D. M. 1986. Tracking the dinosaurs. *Geology Today* 2 (6):168–169.

1987a. Pterosaur locomotion. Joggers or waddlers? *Nature* 327:13–14.

1987b. Reconstructing extinct animals: Did pterosaurs walk or waddle? *Anima* 180:39–43 [in Japanese].

1988. A new specimen of *Dimorphodon* and the terrestrial ability of early pterosaurs. *Modern Geology* 13 (1) (in press).

Wagner, A. 1852. New aufgefundene Saurier-Uberreste aus den lithographischen Schiefern und dem oberen Jurakalke. *Abhandlungen Bayer. Akademie*

Wissenschaften 6: 50 pp.

Wellnhofer, P. 1970. Die Pterodactyloidea (Pterosauria) der Oberjura-Plattenkalke Süddeutschlands. *Abhandlungen Bayer. Akademie Wissenschaften*, N.F. 141:1–133.

1975a. Die Rhamphorhynchoidea (Pterosauria) der Oberjura-Plattenkalke Süddeutschlands. I. Allgemeine Skelettmorphologie. *Palaeontographica* A 148:1–33.

1975b. Die Rhamphorhynchoidea (Pterosauria) der Oberjura-Plattenkalke Süddeutschlands. III. Palökologie und Stammesgeschichte. *Palaeontographica* A 149:1–30.

1978. *Handbuch der Paläoherpetologie.* Teil 19, Pterosauria (Stuttgart: Gustav Fischer) 82 pp.

1982. Zur biologie der flugsaurier. *Natur und Museum* 112 (9):278–291.

1983. Solenhofer Plattenkalk: Urvogel und Flugsaurier. *Freundes des Museums beim Solenhofer Aktiens Verein* 59 pp.

Wellnhofer, P., and Vahldiek, B. W. 1986. Ein Flugsaurier-Rest aus dem Posidonienschiefer (Unter-Toarcium) von Schandelah bei Braunschweig. *Palaont. Zeitschrift* 60 (3/4):329–340.

Wild, R. 1978. Die Flugsaurier (Reptilia, Pterosauria) aus der Oberen Trias von Cene bei Bergamo. *Italien Bolletino Societa Paleontologica Italiana* 17 (2):176–256.

1984. Flugsaurier aus der Obertrias von Italien. *Naturwissenschaften* 71:1–11.

Williston, S.W. 1903. On the osteology of *Nyctosaurus (Nyctodactylus)* with notes on American pterosaurs. *Field Columbian Museum* 78 (2):125–163.

Yalden, W. D. 1985. Forelimb function in *Archaeopteryx.* In Hecht, M. K., Ostrom, J. H., Viohl, G., and Wellnhofer, P. (eds.). *The Beginnings of Birds.* (Eichstät: Proc. *Archaeopteryx* conference 1984) 91–97.

VIII Site Reports

"La prolifération des découvertes indique clairement qu'on n'a jusqu'á present
pas prêté suffisamment attention à ces témoins particuliers des faunes reptiliennes."
Jenny et al. (1981 p. 430)

In recent years there has been what Chure (1986 p. 13) has called an "unprecedented surge of research
on all aspects of dinosaur ichnology." This surge of activity has largely been the result of the discovery
of numerous new tracksites worldwide. A large number of these newly discovered sites are located in remote
and thinly populated areas with extensive rock exposures. These regions include parts of South America
(Leonardi this volume), the Great Western Desert of North America (Lockley and Conrad this volume),
and parts of North Africa (Ishigaki this volume). Reports arising from these areas sometimes document
ten or more tracksites in a single paper (e.g., Jenny et al. 1981, quotation above).

When the First International Symposium on Dinosaur Tracks and Traces convened in 1986, one
of the objectives of the forum was a chance to simultaneously report many of the newly discovered sites,
and, rather than scatter reports throughout the literature, document them under one cover. The site reports
which follow represent predominantly new discoveries, many in the last decade, spanning four continents
and the entire "Age of Dinosaurs". Papers in Section VI on Biostratigraphy also deal mainly with new
sites that are of some particular stratigraphic significance.

Progress in site discovery has been so rapid that in the last generation the total number of known
sites has increased from "no more than about thirty locations" (an over–conservative estimate made by
Avnimelech in 1962) to approximately 1000 localities (cf. Introduction this volume). The spate of new
discoveries shows no immediate sign of abating.

References

Avnimelech, M. 1962. Dinosaur tracks in the Lower Cenomanian of Jerusalem. *Nature* 196:264.
Chure, D. 1986. The history of the study of dinosaur tracks in North America (Abst.) *in* Gillette, D. D. (ed.).
    Abst. Prog. First Internat. Symp. Dinosaur Tracks and Traces (Albuquerque: New Mexico Museum of
    Natural History) p. 13.
Jenny, J., Marrec, A. le, and Monbarron, M. 1981. Les empreintes de pas de dinosauriens dans le Jurassique
    Moyen du Haut Atlas Central (Maroc): Nouveaux gisements et precisions stratigraphiques. *Geobios* 14
    (3):427–431.

28 Dinosaur Footprints from the Redonda Member of the Chinle Formation (Upper Triassic), East–central New Mexico

ADRIAN P. HUNT,
SPENCER G. LUCAS AND
KENNETH K. KIETZKE

Abstract

A dinosaur–footprint locality in the uppermost Redonda Member of the Chinle Formation at Mesa Redonda, Quay County, New Mexico has been collected for the past fifty years. The prints are poorly preserved as casts or as molds in siltstone or in limestones that contain a large amount of clastic debris. Most prints are tridactyl with lengths of as much as 11.8 cm and can be assigned to *Grallator* sp. These prints were made by small theropods that were mainly moving east–northeast.

Introduction

Vertebrate footprints are common in many Upper Triassic red bed sequences, but have not been well documented in the Chinle/Dockum depositional complex of the American Southwest (e.g., Lull 1915, Baird 1964, Olsen and Galton 1984, Lockley 1986, Conrad et al. 1987). Ongoing studies of the ichnofauna of the Redonda Member of the Chinle Formation in east–central New Mexico are intended to give greater knowledge of the vertebrate fauna of a unit that has produced relatively few body fossils (Lucas et al. 1985a,b), and to aid in correlation of this unit with the extensive ichnofaunas of the Newark Supergroup of the eastern United States (Olsen 1980). The Redonda Member is the stratigraphically highest unit in the Chinle Formation in east–central New Mexico and is dominantly lacustrine in origin (McGowen et al. 1983, Lucas et al. 1985b). At its type section at Mesa Redonda, it consists of 123.6 m of mudstone, fine–grained sandstone, limestone, siltstone and conglomerate (Lucas et al. 1985b). The presence of an "advanced" phytosaur in the Redonda suggests that the unit is younger than the Petrified Forest Member of the Chinle in northeastern Arizona and may be equivalent in age to the Owl Rock Member (Gregory 1957, 1962, 1972; Colbert and Gregory 1957; Lucas et al. 1985b).

History of Investigation

Vertebrate footprints were first collected from the Redonda Member on Mesa Redonda by Robert Abercrombie in 1934. Four slabs were purchased by the Museum of Paleontology at the University of Michigan (UM), two containing one footprint and the other two containing multiple footprints. One of these slabs is unnumbered, the other three are numbered: UM 16161 ("dinosaur footprint"), UM 16160 ("amphibian footprint") and UM 1644 ("dinosaur footprints"). In 1961, Abercrombie collected a footprint of *Grallator* sp., now lost, and a counterpart for the Royal Ontario Museum (ROM 04569), apparently from the same locality. Gregory (1972 p. 123) noted that Abercrombie's locality was "on the northeastern point of Mesa Redonda" in sec. 27, T9N, R31E. We collected this locality in the SW-1/4 SW-1/4 NW-1/4 sec. 27 (Figs. 28.1, 28.2), and collected three slabs, two with one footprint each (UNM MV-2250 and UNM MV-2251) and one preserving a pace of one and a half prints (UNM MV-2252).

Elsewhere on Mesa Redonda Carl Johnson collected a print (New Mexico Museum of Natural History P1059) identified by M. G. Lockley as *Brachychirotherium*. Vertebrate footprints have also been found at Apache Canyon by parties from Yale University and the University of New Mexico (Gregory 1972).

Stratigraphy

Although Gregory (1972) considered the trackways to come from the lower part of the Redonda, they actually occur near the top of the unit (Fig. 28.3). Three beds have produced footprints. The highest print was found as a cast in limestone (unit 31 of Fig. 28.3), 19.6 m below the contact with the Entrada Sandstone. The main footprint horizons occur at the base of a similar limestone, 22.5 m below the Entrada (unit 25 of Fig. 28.3). These limestones contain abundant fine–grained subangular quartz and a few oolites. One print was found as a mold on the top

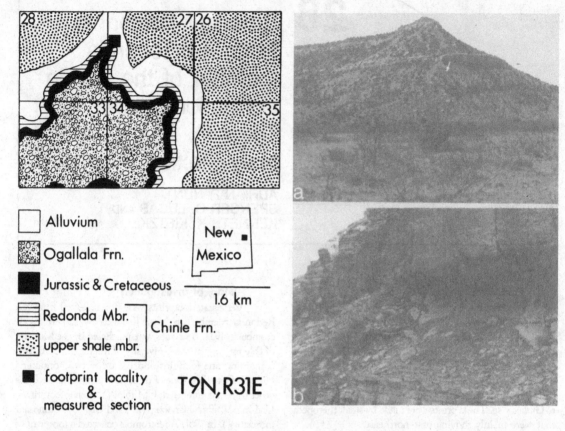

Figure 28.1. Geologic map of the northeastern edge of Mesa Redonda, Quay County, New Mexico showing the location of the dinosaur footprints in the Redonda Member and the location of the measured section in Figure 3; after Dobrovolny et al. (1946).

Figure 28.2. The dinosaur footprint locality at Mesa Redonda. **a**, photograph of point on which dinosaur footprints (arrow) are located. **b**, photograph of footprint-bearing limestone (unit 25 of Figure 28.3) and siltstone (unit 24 of Figure 28.3).

of a siltstone, 23.6 m below the Entrada (unit 24 of Fig. 28.3).

Description of Footprints

The majority of the footprints represent one tridactyl taxon (Fig. 28.4). They are poorly preserved, usually as casts in a limestone or rarely as molds in an underlying siltstone. Preservation is so poor that phalangeal formulas cannot be determined for any prints. Following the terminology of Leonardi (1987), the tridactyl prints are fairly uniform in size, ranging in length from 7.2 to 11.8 cm and in width from 4.8 to 9.1 cm. Digit III is the longest, with the side digits II and IV being subequal in length. Divarication of the side digits varies from 25° to 42°. Claw impressions are discernible in about eighty percent of the tracks. One stride length of 65.0 cm was obtained from a block containing three successive prints, and a pace of 20.3 cm was taken from a sequence of two prints. At least six slabs contain three or more prints with four and six individuals being represented on the larger slabs. The orientation of one trackway and several individual prints were made from in

situ blocks and, although the sample size is small (Fig. 28.5), there appears to be a predominant east–northeast trend.

A similar tridactyl morphology is seen in ROM 04569, and all but one of the approximately seventy footprints noted in the field are of this morphology. In addition, one poorly preserved print has the impressions of four digits. UM 16160 is labelled as amphibian, but is referable to *Brachychirotherium* sp., and probably represents a moderate–sized pseudosuchian.

Discussion

The tridactyl footprints from the Redonda Member are assignable to *Grallator* sp. (Lull 1915 fig. 53–57; Olsen and Galton 1984 fig. 3I, 4A). They preserve only three digits of the pes, with no manus impressions; digit III is the longest and digits II and IV are subequal (Olsen and Galton 1984). Although *Grallator* is conventionally considered the imprint of a small theropod dinosaur (Lull 1904), it may also represent the juvenile of a large theropod (Olsen and Galton 1984). In view of the large number of individuals of subequal size it is likely that the Redonda *Grallator* do indeed

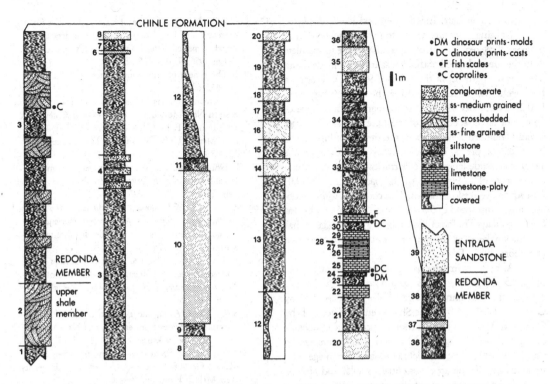

Figure 28.3. Measured stratigraphic section of Upper Triassic (Chinle Formation) and Middle Jurassic (Entrada Sandstone) rocks on northeastern point of Mesa Redonda (see Fig. 28.1).

Figure 28.4. Two footprints of ichnogenus *Grallator* from Mesa Redonda, UNM MV–2250 and 2252.

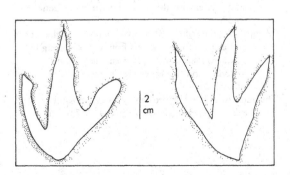

Figure 28.5. Rose diagram of some bearings of footprints at Mesa Redonda.

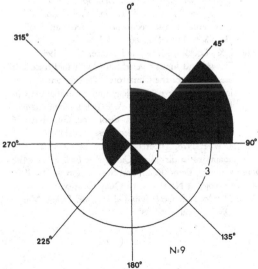

represent a small Late Triassic theropod like *Coelophysis*.

Utilizing measurements from a trackway that preserves three sequential footprints it is possible to speculate as to the size and speed of the dinosaur which made them. The mean length (L) of the prints is 9.1 cm indicating a hip height (H) of 36.2 cm, using the relationship H = L/0.25 (Alexander 1976). This compares with a value of 36 cm for larger specimens of the Late Jurassic *Compsognathus* (Thulborn 1982). However, estimates of hip height, based on footprints, can be seriously in error (Coombs 1978). The stride length for this trackway is 65 cm which gives a speed of 2.24 m/s (8.1 km/hr) based on Kool's (1981) least squares regression analysis. This compares with a probable maximum speed of 5.3 – 7.1 m/s (11 – 26 km/hr) for *Compsognathus* (Thulborn 1982, table IV) and suggests that the trackmaker was moving at a fast walk or slow trot (Thulborn 1982, Table 1).

Although *Grallator* is of limited biostratigraphic utility, *Brachychirotherium* is considered a Late Triassic ichnogenus. The ichnofauna resembles that described by Conrad et al. (1987) from the Sheep Pen Sandstone of the Cimarron Valley. These authors reported a *Grallator*-dominated assemblage with a few *Brachychirotherium* trackways, and assigned the ichnofauna a Norian age. The Redonda ichnofauna appear essentially similar and provide the only known evidence of dinosaurs in this member.

Acknowledgments

We are grateful to J. T. Gregory for information on his collecting in east-central New Mexico, to W. J. Ryan and A. G. Edmund for information about collections at the University of Michigan and Royal Ontario Museum, respectively, and to three anonymous reviewers for helpful comments.

References

Alexander, R. McN. 1976. Estimates of speeds of dinosaurs. *Nature* 261:129–130.

Baird, D. 1964. Dockum (Late Triassic) reptile footprints from New Mexico. *Jour. Paleont.* 38:118–125.

Colbert, E. H., and Gregory, J. T. 1957. Correlation of continental Triassic sediments by vertebrate fossils. *Geol. Soc. Amer. Bull.* 68:1456–1467.

Conrad, K., Lockley, M. G., and Prince, N. K. 1987. Triassic and Jurassic vertebrate-dominated trace fossil assemblages of the Cimarron Valley region: implications for paleoecology and biostratigraphy. *In* Lucas, S. G., and Hunt, A. P. (eds.). *Northeastern New Mexico*. New Mexico Geol. Soc. Guidebook 38. (University of New Mexico Press) 127–138.

Coombs, W. P. 1978. Theoretical aspects of cursorial adaptions in dinosaurs. *Quart. Rev. Biol.* 53:393–418.

Dobrovolny, E., Bates, R. L., and Summerson, C. H. 1946. Geology of Northwestern Quay County, New Mexico. *U.S. Geol. Surv. Oil Gas Inv. Prelim.* Map OM–62, scale 1:62,500.

Gregory, J. T. 1957. Significance of fossil vertebrates for correlation of Late Triassic continental deposits of North America. *Report 20th Session Internat. Geol. Congr. 1956* Section II:7–25.

——— 1962. The genera of phytosaurs. *Amer. Jour. Sci.* 260:652–690.

——— 1972. Vertebrate faunas of the Dockum Group, Triassic, eastern New Mexico and west Texas. *In* Kelley, V. C., and Trauger, F. D. (eds.). *Guidebook of East-central New Mexico* (University of New Mexico Press) pp. 120–123.

Kool, R. 1981. The walking speed of dinosaurs from the Peace River Canyon, British Columbia, Canada. *Can. Jour. Earth Sci.* 18:823–825.

Leonardi, G. (ed.). 1987. *Glossary and Manual of the Tetrapod Footprint Palaeoichnology*. Departamento Nacional de Producao Mineral, Brasilia. (Brasilia: Ministeria das Minas e Energia) 75 pp.

Lockley, M. G. 1986. *A Guide to Dinosaur Tracksites of the Colorado Plateau and American Southwest*. Univ. Colorado Denver Geol. Dept. Mag., Special Issue 1:1–56.

Lucas, S. G., Hunt, A. P., and Bennett, S. C. 1985a. Triassic vertebrates from east-central New Mexico in the Yale Peabody Museum. *In* Lucas, S. G., and Zidek, J. (eds.). *Santa Rosa-Tucumcari Region*, New Mexico Geology Society Guidebook 36. (University of New Mexico Press) pp. 199–203.

Lucas, S. G., Hunt, A. P., and Morales, M. 1985b. Stratigraphic nomenclature and correlation of Triassic rocks of east-central New Mexico: a preliminary report. *In* Lucas, S. G., and Zidek, J. (eds.). *Santa Rosa-Tucumcari Region*. New Mexico Geology Society Guidebook 36. (University of New Mexico Press) pp. 171–184.

Lull, R. S. 1904. Fossil footprints of the Jura–Trias of North America. *Mem. Boston Soc. Nat. Hist.* 5:461–557.

——— 1915. Triassic life of the Connecticut Valley. *State Geol. Nat. Hist. Surv. Connecticut Bull.* 24:1–285.

McGowen, J. H., Granata, G. E., and Seni, S. J. 1983. Depositional setting of the Triassic Dockum Group, Texas Panhandle and eastern New Mexico. *In* Reynolds, M. W., and Dolly, E. D. (eds.). *Mesozoic Paleogeography of the West-central United States*. (Rocky Mountain Section of the Society of Economic Paleontologists and Mineralogists) pp. 13–38.

Olsen, P. E. 1980. A comparison of the vertebrate assemblages from the Newark and Hartford Basins (Early Mesozoic, Newark Supergroup) of Eastern North America. *In* Jacobs, L. L. (ed.). *Aspects of Vertebrate History*. (Museum of Northern Arizona) pp. 35–53.

Olsen, P. E., and Galton, P. M. 1984. A review of the reptile and amphibian assemblages from the Stormberg of Southern Africa, with special emphasis on the footprints and the age of the Stormberg. *Palaeontogr. Africana* 25:87–110.

Thulborn, R. A. 1982. Speeds and gaits of dinosaurs. *Palaeogeogr., Palaeoclimat., Palaeoecol.* 38:227–256.

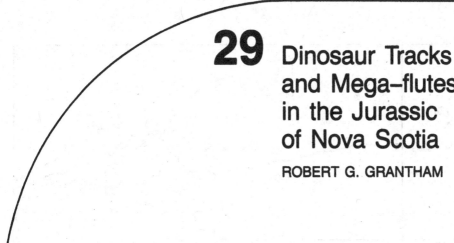

29 Dinosaur Tracks and Mega–flutes in the Jurassic of Nova Scotia

ROBERT G. GRANTHAM

Abstract

A superb specimen of *Otozoum* sp. as well as small footprints and footprint–like, extremely large flute marks have been recently recovered or documented from the Hettangian red quartz wackes and siltstones of the McCoy Brook Formation of Nova Scotia. Located on a wavecut platform adjacent to an extensive tidal flat and being subject to 12 m tides, exploration and collecting require special techniques.

Geological Setting

The Newark Supergroup of eastern North America is exposed in Nova Scotia as the Fundy Group which is composed of the Middle to Late Triassic Wolfville and Blomidon Formations; the Early Jurassic North Mountain Basalt, McKay Head Basalt and the McCoy Brook and Scots Bay Formations. The Middle to Late Triassic sandstones and siltstones are well exposed along the shores of the Minas Basin and in two small outcrops on the north shore of Chedabucto Bay. The Hettangian age basalts, siltstones, red quartz wackes, and minor dolomitic limestones are best exposed in four small outcrops along the Bay of Fundy and along the north shore of Minas Basin. Most of the shoreline of the Bay of Fundy is composed of the North Mountain Basalt (Fig. 29.1).

Footprints

Footprints from the McCoy Brook Formation were first publicized in the early 1970's through the work of Paul Olsen of Columbia University. Trackways occur sporadically and, when they do, must be documented or collected quickly because of rapid erosion by the high ranging Fundy tides. So far all specimens found have been sandstone casts. On the tidal flats, any outcrop where siltstone and sandstone are in contact provides an excellent location to look for fossil tracks. The beds are simply separated and examined. If a track is found, the blocks are left on the shore to be washed by the tides for a few days. In this way the features of the footprint are washed clean and the details become much clearer. Because this is a saltwater shore, the porous sandstones must be washed repeatedly with fresh water in order to remove the salt derived from the salt water which saturates the rock. When the specimen is dry, if salt has crystallized on the surface, an undiluted thick coat of carpenters' water–soluble white glue painted on the surface will remove the salt when the glue is peeled off after drying. We use Weldbond, a concentrated white glue manufactured by Frank T. Ross and Sons Limited of Toronto, Ontario, Canada.

Large Footprints

Recently, some large footprints from this formation have been donated to the Nova Scotia Museum (NSM). They have been identified as *Otozoum* sp. The trackway consists of one manus, one pes and four partial pes (Fig. 29.2). Unfortunately interference of footprints has made the identification to species level very difficult. Well preserved scale impressions are visible on the two right pes (Fig. 29.3). The length of the pes is 430 mm and the length of the poorly preserved manus is 120 mm. The overall length of the slab containing the trackway is 1.44 m.

Small Footprints

Very small footprints have also been found in the McCoy Brook Formation. The preservation on the smaller tracks is superb. One block containing several 2 cm tracks and a tail drag appear to be reptilian, not dinosaurian, as previously thought. Some well preserved small dinosaurian tracks found by Mr. Eldon George and used recently in worldwide news coverage were attributed to a juvenile dinosaur (Anonymous 1986).

Mega–flutes

A report to the museum in the fall of 1985 told of

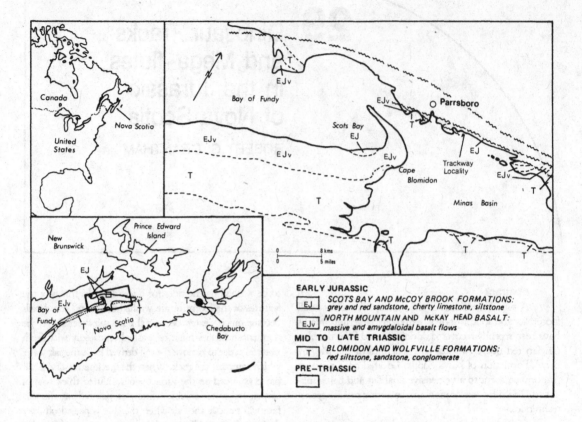

Figure 29.1. Distribution of the Early Jurassic and Late Triassic Fundy Group in Nova Scotia.

Figure 29.2. Sandstone cast trackway of *Otozoum* sp. on the right and the interpreted trackways on the left. Manus track has been partially overprinted by the left pes making it indistinct.

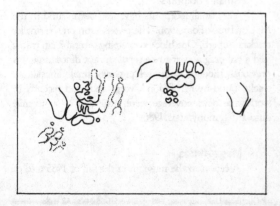

Figure 29.3. Scale impressions on the partial right pes.

Figure 29.4. Possible large footprint when first found by Mr. George.

Figure 29.5. Full feature thought to be a footprint. When fully exposed later, it was identified as a large flute mark.

the existence of a partially exposed set of very large tracks (Fig. 29.4). Photos shown to experts by the finder came back with the report that the features might indeed be large dinosaur footprints. As large footprints had been found before in the same area, there was a possibility that these new features were true footprints. In order to verify the markings as tracks, the museum decided to expose and prepare them for further investigation.

The north shore of Minas Basin is subject to 12 m tides, the highest in the world, and the outcrop containing the "footprints" is below the high tide line and exposed only one hour after high tide. This provided approximately 7 hours per day to excavate the site. The beds dip at a 10 to 15 degree angle into the beach and therefore pneumatic rock drills and a backhoe were required in order to remove the overlying strata. The hole was washed clean using a pump and recycling sea water from a prepared sump. The delicate work of removing the 4 to 5 cm bed immediately above the suspected footprints was done by hand with hammers, chisels and brushes.

When most of the overlying rock was removed, a possible metatarsus impression was clearly visible (Fig. 29.5), giving the feature an overall length of 2 m. As well, a depression interepreted as a possible tail drag between the two "footprints" was uncovered. At this point the author consulted Paul Olsen of the Lamont–Doherty Geological Observatory, Columbia University, and Dale Russell of the Paleobiology Division, National Musuem of Natural Sciences, Canada, and invited them to visit the site to give their evaluation of the feature. The NSM prepared for a major expedition in the event that the "footprint" features were considered worth casting.

The primary infill material was siltstone containing occasional indeterminate plant and pelecypod fossils. After Olsen completed the removal of infill and examined the underlying beds where the undertrack should have been, the consensus was that the markings were not footprints but were very large flute marks of an order of magnitude greater than had ever been reported before. Large flute marks normally do not exceed .5 m in length or 10 cm in amplitude (Pettijohn and Potter 1964). The flutes reported here are relatively isolated, lobate, only slightly inter-tonguing and are 2 m in length and 20 cm in amplitude. The author now terms these mega–flutes. No casts were made.

Conclusion

The resemblance of mega–flutes to footprints is obvious and precautions must be exercised in the future when searching for fossil tracks in the Jurassic of Nova Scotia and elsewhere. Although this exercise did not result in the discovery of large dinosaur footprints, it did provide the NSM with a chance to organize a rapid response casting team. In the fragile sandstones and mudstones on the tidal flats of the Minas Basin, any feature which is exposed today is likely to be eroded away in a month's time. Therefore it is essential that any feature be investigated to its fullest before it is lost to the tides.

Acknowledgments

I should like to thank Mr. Eldon George for drawing our attention to all of the features described above and for donating the superb *Otozoum* specimen to the NSM. I should also like to thank Dale Russell for his keen interest and many hours of great discussion and for coming to visit the site on such short notice. Paul Olsen is also thanked for his quick response and invaluable expertise applied to our investigation.

References

Anonymous, 1986. National Geographic Society News Service, released January 29, 1986.

Pettijohn, F. J., and Potter, P. E. 1964. *Atlas and Glossary of Primary Sedimentary Structures*. (Springer–Verlag New York, Inc.) 370 pp.

30 Dinosaur Trackways in the Lower Jurassic Aztec Sandstone of California

ROBERT E. REYNOLDS

Abstract

A western occurrence of Jurassic footprints has been recognized in the Aztec Sandstone of the Mescal Range, eastern San Bernardino County, California. Tracks, measuring from 2 cm to 3–4 cm in diameter, appear to have been left by quadrupeds. Abundant tridactyl tracks, approximately 10 to 12 cm in diameter, represent three coelurosaurs. One form is similar to *Grallator* sp. and another compares to *Anchisauripus* sp., both from the Newark Basin and ranging through the Late Triassic and Early Jurassic (Olsen and Baird 1982). The third is unidentified. The estimated age of the Aztec Sandstone is Early Jurassic. Dates on volcanic rocks overlying the Aztec Sandstone provide a mimimum age of 155 Ma (Burchfiel and Davis 1971).

Tracks are apparently absent in the underlying red silty "Moenave/Kayenta–equivalent" Sandstone. Within the Aztec Sandstone, tracks occur on multiple horizons within the middle quartz arenite. All beds containing trackways exhibit crossbedding. Tracks have not been found in overlying silty sands and graywackes which interfinger with the lowest volcanic rocks. The structure of the sediments is complex and beds are overturned in places. The tracks are indicators of tops of beds.

Compass bearing differs between sets of trackways. Parallel sets of trackways trend in opposing directions. Individual tracks exhibit sand crescents and concentric slip structures on specific margins that suggest undulating topography. Well–defined trackways end, suggesting change from a moist to a dry dune sand substrate.

Introduction

Previous inventories of California localities producing dinosaur remains have been limited to skeletal material (Deméré 1985). The occurrence of ichnofossils from the Mescal Range of eastern California provides a new locality record for several dinosaur ichnotaxa. These trackways were first mentioned by Evans (1958), and later by Evans (1971) and Reynolds (1983).

The author brought the locality to the attention of the U.S. Bureau of Land Management (BLM) and, because it is situated on federal lands in the California Desert Conservation Area, the agency included the site in the "Mescal Range" Geological Area of Critical Environmental Concern (ACEC) (Bureau of Land Management 1986). The San Bernardino County Museum (SBCM) initiated a detailed program of inventory and replication of the trackways in 1986 in cooperation with the BLM to assist in their management and protection. Ichnofossils were first observed in the Aztec Sandstone of the Delfont Quarry (Evans 1958); subsequent detailed field survey by the San Bernardino County Museum in 1986 and 1987 has allowed them to be traced throughout a half–mile diameter area around the quarry site.

Location

The Mescal Range trackways are located in the Mojave Desert of northeastern San Bernardino County, California. The Mescal Range is 35 miles northeast of Baker, California, and 50 miles south–southwest of Las Vegas, Nevada (Fig. 30.1). The locality (with spatially distinct sets of trackways designated SBCM 1.10.3 through SBCM 1.10.8) is approximately 5 km south of the Bailey Road exit on Interstate Highway 15 at Mountain Pass, within the southeast quarter of section 25 and the northeast quarter of section 36, Township 16 North, Range 13 East, and the southwest quarter of section 30, Township 16 North, Range 14 East, San Bernardino Base and Meridian, as shown on the Mescal Range 15 minute U.S. Geological Survey quadrangle map, edition 1955.

Stratigraphy and Structure
Stratigraphy

The Mescal Range contains an early Mesozoic sequence of carbonate and clastic rocks (Hewett 1956; Evans

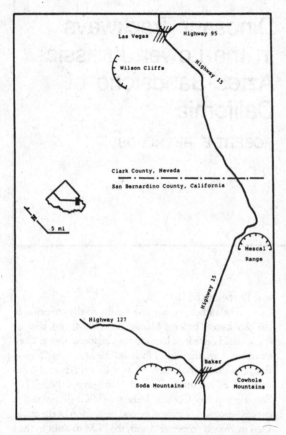

Figure 30.1. Locality map. The Aztec Sandstone is exposed in the Wilson Cliffs of southern Nevada and in the Mescal Range, Cowhole Mountains and Soda Mountains of San Bernardino County, eastern California. Trackways are known only from the Mescal Range.

interfingers with the redder Moenave–Kayenta Sandstone equivalent. In the Mescal Range, a similar situation has been suggested regarding the red sandstones (R. Bereskin pers. comm. 1978), although they were apparently included as the base of the Aztec Sandstone as described by Hewett (1956), Evans (1958, 1971), and Marzolf (1982).

Pending detailed measurement, lithologic description, and further regional comparisons, the author uses the term "Moenave/Kayenta–equivalent" (Wilson and Stewart 1967, Marzolf 1982) to distinguish a sandstone unit of uniform color and texture that sits stratigraphically between the Chinle Formation and the Aztec Sandstone. This unit is uniformly brick red to pale brick red, in contrast to the maroon– to liver–colored sands of the underlying Chinle. Limited exposures suggest that it sits disconformably on the Chinle Formation and basally contains ripped–up clasts of the Chinle. Crossbedded exposures of the "Moenave/ Kayenta–equivalent" appear to have been deposited flat-lying and are less lenticular than the overlying light–colored Aztec Sandstone. The "Moenave/Kayenta–equivalent" ranges from moderately coarse to fine–grained sandstone with laminae from 0.5 cm to 0.5 m in thickness paralleling bedding planes.

The "Moenave/Kayenta–equivalent" may have been deposited in part under fluviatile conditions, in contrast to the partial eolian deposition of the Aztec Sandstone. The contact between the two in the Mescal Range appears to be abrupt and yet conformable (Evans 1958, 1971). The "Moenave/Kayenta–equivalent" in the Mescal Range has yet to produce fossils or ichnofossils.

Mapping of the Aztec Sandstone (Reynolds 1983, Fig. 30.2) suggests that it can be divided into two units. The lowest and most prominent is the resistant sandstone unit, about 100 m thick. It is tan to off–white in outcrops, but a pinkish hue is quite evident in fresh exposures such as the Delfont Quarry. The crossbedded lenses can be easily observed at the quarry and elsewhere along strike. Frosted and pitted quartz grains well–cemented by silica are described by Evans (1958, 1971). The upper and less resistant unit, 200 m thick, consists of alternating white quartz arenites and red to brown silty sands. Beds are approximately 0.7 to 1.8 m in thickness. The white quartz arenites do not exhibit the prominent steep crossbedded lenses that are characteristic of the underlying unit. Toward the top of the section, one of the brown beds contains volcanic detritus. Within one kilometer to the northeast, this volcaniclastic sandstone thickens to a 30 m thick, black volcanic rock which exhibits breccia and flow textures. Overlying this volcanic marker bed is buff sandstone which grades upward into gray and green silty sandstone. The latter beds (SBCM 1.10.11) have produced fossil plants similar to the horsetail *Equisetum*.

No trackways have been located in the upper unit of the Aztec Sandstone. They appear to be restricted to the massive crossbedded sandstone of eolian origin, perhaps because it is well–cemented and resistant to erosion whereas the alternating silty sandstones of the upper unit are easily weathered.

1958, 1971, 1974). The units include, from lowest: the Triassic Moenkopi Formation, characterized by "buff and blue–gray limestone..." (Evans 1971) with limited shale; the Triassic Chinle Formation, "largely red shaley sandstone; red, brown, and green shale; and thin zones of chert and limestone conglomerate..." (Hewett 1956); and the Jurassic (?) Aztec Sandstone, first described by Hewett (1931) as "This great thickness made up of many lenses..., each lens in turn made up of smaller laminae.... The color is uncommonly uniform throughout...." Evans (1971) noted that the sandstone is "brightly colored and shows hues of buff, red, yellow, and gray, as well as off–white". Hewett (1931) attributes this variable coloration near Aztec Tank to leaching of ferric oxide. However, detailed mapping by this author in the Mescal Range indicates that, although leaching may account for some of the color changes in the local Aztec Sandstone, there are facies of distinct color and lithology that reflect discrete environments of deposition.

Recent work in the Wilson Cliffs southwest of Las Vegas (Marzolf 1982, Marzolf and Dunne 1983) shows that the basal portion of the light–colored Aztec Sandstone

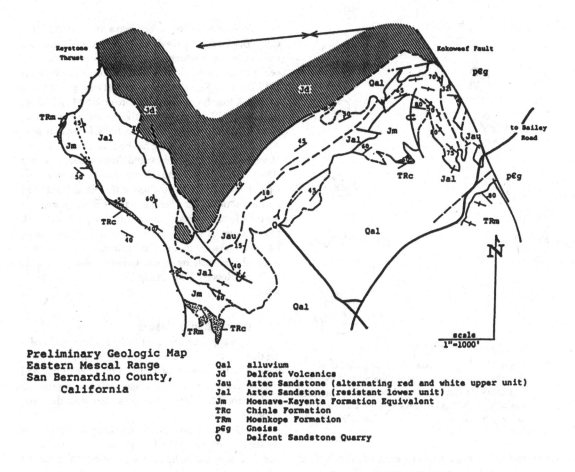

**Preliminary Geologic Map
Eastern Mescal Range
San Bernardino County,
California**

Qal	alluvium
Jd	Delfont Volcanics
Jau	Aztec Sandstone (alternating red and white upper unit)
Jal	Aztec Sandstone (resistant lower unit)
Jm	Moenave–Kayenta Formation Equivalent
TRc	Chinle Formation
TRm	Moenkope Formation
pEg	Gneiss
Q	Delfont Sandstone Quarry

Figure 30.2. Preliminary geologic map, eastern Mescal Range, San Bernardino County, California.

The upper unit of the Aztec Sandstone is overlaid by the Mountain Pass Rhyolite (Evans 1958, 1971; Hewett 1956; = Delfont Volcanics of Burchfiel and Davis 1971). The rhyolite is green, maroon, and red with white orthoclase phenocrysts and xenoliths of sandstone. Evans (1971) estimates that this section is more than 1,200 feet (= 400 m) thick.

Environment of Deposition

The clastic units in the Mescal Range were deposited under a variety of conditions that may, in part, reflect local structural history. The Moenkopi and Chinle formations show a change through time from silty marine carbonates to estuarine? shales and then to arkosic fluviatile deposits with channels. The "Moenave/Kayenta–equivalent" appears to have been deposited as silty fluviatile beds. The lower unit of the Aztec is eolian and characterized by southeastward–dipping crossbeds. The direction of dip correlates with the lower portion of the Aztec Sandstone elsewhere (Marzolf 1982).

In the Mescal Range, the upper unit of the Aztec Sandstone is dissimilar to sections elsewhere. It consists in part of clean fluviatile sands, which may have derived from an eolian source, and silty fluviatile sands which contain a plant fossil suggestive of a stream bank or swamp flora. The Moenkopi–Chinle–"Moenave/Kayenta–equivalent" sequence suggests a retreating ocean margin being overrun by fluviatile sediments.

After establishment of the dune system reflected in the early deposition of the Aztec Sandstone, the area was poorly drained. The basin was suitable for receiving a relatively thick sequence of volcanic debris, the lowest of which interdigitates with the fluviatile sediments of the upper Aztec Sandstone. The rapid change of thickness of the volcanic beds and the presence of sandstone xenoliths in the volcanic rocks suggest a local source for the volcanics.

The trackways appear to be restricted to the resistant lower unit of the Aztec Sandstone in the Mescal Range. The steeply crossbedded lenticular beds of frosted quartz grains (Evans 1971) indicate dune deposits. The imprints themselves provide information about the dune slope and surface (Brady 1960). A crescent–shaped impact mound was formed around each imprint and is oriented downslope. Certain trackways (Fig. 30.3) appear to record a change in slope that is not observable today on the bedding plane surface. The sand crescents often have concen-

Figure 30.3. Impact mounds or sand crescents with concentric scarps form as prints are made. Mounds and scarps are oriented downslope in each case, suggesting that the bipeds were walking on an undulating dune substrate. Scale bar equals 17 cm (8 in.).

tric microscarps, suggesting that moisture allowed the sand mound to hold its shape without slumping. The presence of moist sand or a near-surface groundwater level is shown by isolated tracks or trackways with several paces that end abruptly. This suggests that individuals walked through moist sand where their tracks were preserved and into dry sand which was not competent enough to hold tracks. Portions of the trackways also may have been removed by wind erosion. Prints vary in distinctness, suggesting that the moisture content varied during the time spanned by the passing of different individuals.

Age and Structural Relationships

The structure of the Aztec Sandstone is relatively complex (Fig. 30.2). The clastic sediments and overlying volcanics have been anticlinally folded and overturned along the northwest–trending Kokoweek Fault (Burchfiel and Davis 1971, = Clark Mountain Fault of Evans 1971) at the far eastern margin of the Mescal Range. Fault intersection (South Fault of Evans 1971) has abbreviated the clastic section on the northeast face of the Mescal Range. The western edge of the clastic and volcanic section has been synclinally overturned and apparently underlies the carbonate rocks of the Keystone Thrust (Burchfiel and Davis 1971, = Mescal Thrust of Evans 1971). Tracks and other sedimentary structures are useful in determining the tops of beds. The Keystone Thrust cuts grannite of 138 Ma (Sutter 1968, Burchfiel and Davis 1971). The clastic and

volcanic section has been folded into a west–plunging anticline (Burchfiel and Davis 1971, Evans 1971) and the volcanic section has been duplicated by thrusting that may be a part of the Keystone event. The time constraint of 138 Ma on the Keystone Thrust and its relationship to the volcanics, together with a KAr date of 155 Ma (G. Davis pers. comm. 1978), suggest that the age of the Aztec Sandstone is older, possibly Middle to Early Jurassic in age.

The Aztec Sandstone has been correlated with the Navajo Sandstone by Wilson and Stewart (1967), Marzolf (1982), and Marzolf and Dunne (1983). Peterson and Pipiringos (1979) restrict the age of the Navajo and Nugget sandstones to the Pliensbachian to Toarcian stages of the early Jurassic period. Thus, a date of approximately 180 Ma is suggested for the Aztec Sandstone in the Mescal Range. Trackways in the Navajo Sandstone have been reported previously by several authors (Baker et al. 1936, Faul and Roberts 1951, Brady 1960, Stokes 1978, Baird 1980).

Description of Tracks

Heeding the warning of Peabody (1955) and realizing that the tracks were made in a loose, sandy substrate, the author has been conservative in their description and comparison. Except for the single *Grallator* sp. print, all tracks identified as bipeds consist of more than three sets of prints.

The prints are compared using size, morphology, and the digit divarication (Baird 1957), an approximate angle measured from the middle of the pes along digit III to the lateral digits. This rough measurement may overemphasize foot symmetry and may vary because of the undulating sandy substrate or the animal's locomotor response to the substrate. Many parallel tracks of quadrupeds are present at the Mescal Range site that show close association of the pes to the manus.

Bipedal Ichnotaxa
Grallator sp.

The ichnogenus *Grallator* is represented in the Mescal Range by a single print. The specimen has a length of 15.2 cm, a width of 15.2 cm and has digit divarication of approximately 30°. The size, divarication, and pad morphology distinguish it from the other two bipedal ichnogenera recognized in the Mescal Range.

The Mescal Range specimen (Fig. 30.4A) is slightly larger than but reminiscent of G. *sulcatus* (Baird 1957) in the broad separation of digit IV from digit III compared to the proximity of digits II and III. Also, as in G. *sulcatus*, the termination of digit IV reaches as far anteriorly as the anterior edge of the second phalangeal pad of digit III. Digit II is faint, reaching the middle of the second pad of digit III. There is no evidence of a hallux.

Anchisauripus sp.

Anchisauripus sp. is represented in the Mescal Range by a trackway containing six prints. Two prints are adja-

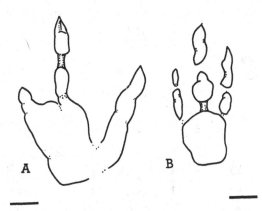

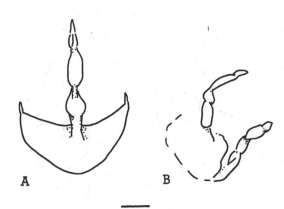

Figure 30.4. **A,** *Grallator* sp. is represented in the Mescal Range by a single large isolated track. **B,** *Anchisauripus* sp. is represented in the Mescal Range by one trackway with six prints. Bars equal 2.45 cm (1 in.).

Figure 30.5. **A,** a large track from the unnamed biped from the Aztec Sandstone of the Mescal Range. **B.** Curved digits in a similar track may be due to foot rotation during withdrawal from sand. Bar equals 2.45 cm (1 in.).

cent and 5 cm apart, suggesting that the animal may have halted and then resumed its pace. The pace averages 34.3 cm and the stride 69.1 cm. When compared to *Grallator* sp. and other ichnogenera from the Mescal Range, the prints are distinctly narrow with a digit divarication of approximately 14° (Fig. 30.4B). The print dimensions average 12.2 cm in length and 7.1 cm in width. There is no hallux apparent. As in *A. milfordensis* (Baird 1957, Olsen and Baird 1982), the Mescal Range specimen has a termination of digit II that reaches as far anteriorly as the crease between the first and second phalangeal pads of digit III. The termination of digit IV reaches anteriorly as far as the middle of the second pad of digit III. Metatarso–phalangeal pad IV is circular and strongly domed. The dimensions of the Mescal Range *Anchisauripus* tracks are longer and slightly wider than *A. milfordensis* and *A. parellelus* (Baird 1957).

Unidentified Biped

The most common bipedal print from the Mescal Range is distinguished from the *Anchisauripus* sp. and *Grallator* sp. described above by the divarication of the lateral digits. The anterior margin of digits II and IV subtend angles of 50°–70° with digit III (Figs. 30.5, 30.6). The actual divarication of the digits may be less, but is certainly greater than 40°. The termination of both digits II and IV reach anteriorly as far as the anterior edge of the second phalangeal pad of digit III. No hallux is present.

Measurements of 27 prints provide an average length of 11.4 cm and width of 10.9 cm. A larger individual (Fig. 30.5A) measures 14 cm by 10 cm. Size differences might be attributed to sexual dimorphism or growth stages. Small individual prints, possibly from a juvenile (Fig. 30.6B), have dimensions of 8.9 cm by 5.1 cm. These smaller prints are 15 cm from those of average–sized tracks of the same type and trend parallel for all of the exposed three paces.

If the marks at the termination of the digits were made by claws, they appear to be mores slender than the broadly acute termination of digits from *Grallator* sp. (40°) and *Anchisauripus* sp. (30°) at the site. The digit imprints are curved in several instances (Fig. 30.5B). The curvature is interpreted to be an artifact produced by foot rotation during withdrawal from the sand.

Several excellent exposures with abundant tracks allowed a pace of 37.9 cm and stride of 78.2 cm to be measured. This is slightly longer than that of the *Anchisauripus* sp. from the site. One trackway with indistinct digits has a pace of 73.7 cm and a stride of 142.2 cm. The difference in stride between similar size ichnnotaxa suggests a difference in gait.

The distinctive digital divarication of the tracks of this ichnogenus and the width of the tracks make the prints appear "partly webbed" (Anderson 1939, Fig. 30.6). It is possible that the animal might have been slightly more digitigrade than the others present, or the apparent breadth of the foot might be a locomotor response to undulating sandy terrain. These alternatives seem unlikely because footprints of many individuals have similar morphology across undulating terrain. The change in slope of the terrain along the trackways is emphasized by the change in direction of the concentric sand crescents around each print.

Tracks of similar size and morphology, described as "... suggestive of a partly–webbed foot", are reported from the lower Cretaceous Lakota Sandstone of South Dakota (Anderson 1939). *Zhengichnus jinningensis* from the lower Jurassic Fengjiahe Formation near Jinning, Kunming, Yunnan is of similar morphology and age but, at more than 20 cm in length, the tracks are almost twice the size of those in the Mescal Range (Zhen et al. 1986).

Quadrupedal Ichnotaxa

Quadruped tracks are abundant in the Aztec Sand-

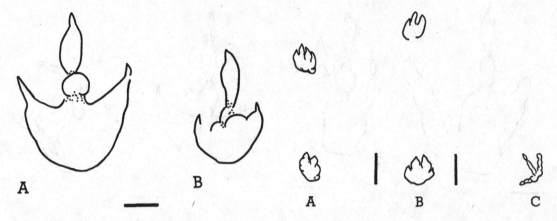

Figure 30.6. **A**, unnamed biped, adult? of moderate size. **B**, unnamed biped, juvenile? print in a series of tracks paralleling the adult (left) and 14.7 cm (6 in.) distant. Bar equals 2.45 cm (1 in.).

Figure 30.7. **A,B**, tracks of quadrupeds with short, broad digits. **C**, quadruped print with gracile digits, Mescal Range, San Bernardino County, California. Bar equals 2.45 cm (1 in.).

stone of the Mescal Range, but research and space limitations preclude extensive discussion herein. The tracks represent forms with short, broad digits (Fig. 30.7A,B), which resemble *Batrachopus* (cf. Padian 1986), and others with gracile digits (Fig. 30.7C).

Direction of Trackways

Although imperfect prints and isolated prints do not allow determination of an individual animal's travel over any extended distance, it was possible to take bearings of the long axis of individual prints, bearings along trackways with multiple prints, and bearings in the apparent direction of travel of bipeds and quadrupeds. Diagrams representing these bearings are shown in Figure 30.8A (bipeds) and Figure 30.8B (quadrupeds). Bearings of individual tracks are weighted the same as bearings of trackways. Graphic representation of the biped tracks/trackways does not suggest obvious direction of travels which might indicate parallelism to, or avoidance of, a natural feature. Diagrams of the quadruped tracks/trackways do suggest a possibility of movement in a southwesterly and northeasterly direction. However, chi-square texts on bearings of both the bipeds and the quadrupeds indicate that tracks/trackways are oriented randomly. Because of the small sample sizes, data were lumped for both groups (quadrupeds into 120° increments and bipeds into 60° increments) to make possible use of the chi-square method. This lumping might have masked significant orientation in the quadrupeds.

Conclusion

Ichnogenera *Grallator* sp., *Anchisauripus* sp., and a third unidentified taxon are reported from the Aztec Sandstone in the Mescal Range of San Bernardino County, California in association with quadrupedal ichnotaxa. Con-

centric sand crescents around prints indicate change in slope and moisture content of the undulating dune sand substrate.

A preferential direction of travel is not obvious at this locality. One exposure has trackways presumed to be of a juvenile biped parallel to those of a larger individual. Tracks occur in the well-indurated quartz arenites of the Aztec Formation but have not been located in interfingering silty facies.

The *Grallator* sp. and *Anchisauripus* sp. from the Mescal Range Aztec Sandstone resemble ichnotaxa from the Newark supergroup of the Connecticut Valley (Lull 1953, Baird 1957). Faunal correlations by Olsen and Galton (1977) suggest that these ichnotaxa range from Late Triassic through Middle Jurassic. The "partly-webbed' unnamed bipedal ichnites from the Mescal Range bear a resemblance to forms in the lower Cretaceous Lakota Sandstone of South Dakota (Anderson, 1939) and lower Jurassic Fenghiahe Formatinon in China (Zhen et al. 1986). A date of 155 Ma (G. A. Davis pers. comm. 1971) on the volcanics that overlie the Aztec Sandstone suggests that the trackways are older, perhaps Early Jurassic in age.

Acknowledgments

This project has been supported with permits, supplies and air photos from the Bureau of Land Management and with personnel and volunteers from the San Bernardino County Museum in Redlands, California. The author thanks Quintin Lake, Thomas Greer, Michael Moore, Barbara Pitzer, Richard Chacon, Scott Packer, Julia Packer, Jed Reynolds, and Kate Reynolds for their services as the field crew in survey and replication; H. Thomas Goodwin and Jennifer Reynolds for their assistance in documentation, preparation, and analysis; and Vicky Hipsley of the San Bernardino County Museum for her help with graphics.

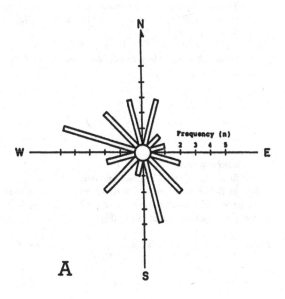

A

Bipeds

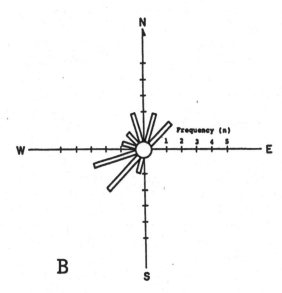

B

Quadrupeds

Figure 30.8. **A,** orientation of biped trackways from the Aztec Sandstone of the Mescal Range, San Bernardino County, California. Bars represent the frequency of tracks/trackways oriented within 30° compass increments (n = 31). **B,** orientation of quadruped trackways from the Aztec Sandstone of the Mescal Range, San Bernardino County, California. Bars represent the frequency of tracks/trackways oriented within 30° compass increments (n = 15).

References

Anderson, S. M. 1939. *Jour. Paleont.* 13 (3):361–364, pl. 38, Fig. 2.

Baird, D., 1957. Triassic reptile footprint faunules from Milford, New Jersey. *Harvard Univ. Museum Comp. Zool. Bull.* 117:447–520, 4 pls.

——— 1980. A prosauropod dinosaur trackway from the Navajo Sandstone (Lower Jurassic) of Arizona. In Jacobs, L. L. (ed.) *Aspects of Vertebrate History.* (Museum of Northern Arizona Press) pp. 219–230.

Baker, A. A., Dane, C. H., and Reeside, J. E. Jr. 1936. Correlation of Jurassic formations of parts of Utah, Arizona, New Mexico and Colorado. *U.S. Geol. Surv. Prof. Paper* 183. 66 pp.

Brady, L. F. 1960. Dinosaur tracks from the Navajo and Wingate sandstones. *Plateau* 32 (4):81–82.

Burchfiel, B. C., and Davis, G. A. 1971. Clark Mountain thrust complex in the cordillera of southeastern California: geologic summary and field trip guide. *In* Elders, W. A. (ed.). *Geological Excursions in Southern California.* Univ. California Riverside Museum Contrib. 1. 182 pp.

Bureau of Land Management. 1986. *Areas of Critical Environmental Concern.* (U.S. Deptartment of Interior) 8 pp.

Deméré, T. 1985. Dinosaurs of California. *Environment Southwest* 509:15–17.

Evans, J. R. 1958. Geology of the Mescal Range, San Bernardino County, California. M.S. thesis. University of Southern California, Los Angeles. 133 pp.

——— 1971. Geology and mineral deposits of the Mescal Range quadrangle, San Bernardino County, California. *California Div. Mines and Geology* Map Sheet 17.

——— 1974. Relationship of mineralization to major structural features in the Mountain Pass area, San Bernardino County, California. *California Geol.* 27 (7):147–157.

Faul, H., and Roberts, W. A. 1951. New fossil footprints from the Navajo (?) Sandstone of Colorado. *Jour. Paleont.* 25 (3):266–274.

Hewett, D. F. 1931. Geology and ore deposits of the Ivanpah quadrangle, Nevada. *U.S. Geol. Surv. Prof. Paper* 162. 172 pp.

——— 1956. Geology and ore deposits of the Ivanpah quadrangle, Nevada and California. *U.S. Geol. Surv. Prof. Paper* 275. 172 pp.

Lull, R. S. 1953. Triassic life of the Connecticut Valley. *Connecticut Geol. Nat. Hist. Surv. Bull.* 81:1–336.

Marzolf, J. E. 1982. Paleogeographic implications of the early Jurassic (?) Navajo and Aztec sandstones. *In* Frost, E. G., and Martin, D. L. (eds.). 1982. *Mesozoic-Cenozoic tectonic evolution of the Colorado River region, California, Arizona, and Nevada* (Anderson-Hamilton volume). (Cordilleran Publishers) 493–501.1.

Marzolf, J. E., and Dunne, G. C. 1983. *Evolution of Early Mesozoic Tectonistratigraphic Environments, Southwestern Colorado Plateau to Southern Inyo Mountains: Field Guide.* (Geological Society of America) 103 pp.

Olsen, P. E., and Baird, D. 1982. Early Jurassic vertebrate assemblage of the Fundy group (Newark supergroup, Nova Scotia, CA). *Geol. Soc. Amer. Northeastern/ Northwestern Sections Abst. Prog.*: p. 70.

Olsen, P. E., and Galton, P. M. 1977. Triassic-Jurassic

tetrapod extinctions: are they real? *Science* 197:983–986.

Peabody, F. E. 1955. Taxonomy and the footprints of tetrapods. *Jour. Paleont.* 29 (5):915–924.

Peterson, F., and Pipiringos, G. N. 1979. Stratigraphic relations of the Navajo Sandstone to middle Jurassic formations, southern Utah and northern Arizona. *U.S. Geol. Surv. Prof. Paper* 1035 B. 43 pp.

Reynolds, R. E. 1983. Jurassic trackways in the Mescal Range, San Bernardino County, California. *In* Marzolf, J. E., and Dunne, G. C. (eds.). *Evolution of Early Mesozoic Tectonistratigraphic Environments, Southwestern Colorado Plateau to Southern Inyo Mountains: Field Guide* (Geology Society of America) 103 pp.

Stokes, W. L. 1978. Animal tracks in the Navajo–Nugget Sandstone. *Contrib. Geol. Univ. Wyoming* 16 (2):103–107.

Sutter, J. F. 1968. Geochronology of major thrusts, southern Great Basin, California. M.S. thesis, Rice University. 32 pp.

Wilson, R. F., and Stewart, J. H. 1967. Correlation of upper Triassic and Triassic (?) formations between southwestern Utah and southern Nevada. *U.S. Geol. Surv. Bull.* 1244–D:1–20.

Zhen S., Li J., and Rao C. 1986. Dinosaur footprints of Jinning, Yunnan. *Mem. Beijing Nat. Hist. Museum* 33. 18 pp.

Editorial Note

Recent publications by Olsen and Baird, and Olsen and Padian in Padian (1986) provide updated information on the ichnospecies *A. milfordensis* (Fig. 30.4) and *Batrachopus*-like (crocodylomorph) quadruped tracks (Fig. 30.7).

Padian, K. (ed.). 1986. *The Beginning of the Age of Dinosaurs* (Cambridge University Press) 378 pp.

31 Dinosaur Footprints of Western Canada

PHILIP J. CURRIE

Abstract

A rapid increase in the rate of discovery of dinosaur tracks in western Canada in recent years has resulted in the documentation of 27 localities, as compared with only four in the previous literature. These tracks range in age from Latest Jurassic to Latest Cretaceous and have yielded a large number of representative specimens, most of which are housed in the Tyrrell Museum of Palaeontology.

The oldest tracks include Jurassic forms like *Anomoepus*, which are quite distinct from the younger Cretaceous ichnofaunas dominated by large ornithopods, theropods and miscellaneous large ornithischians. Sauropod tracks are unknown, apparently for paleoecological reasons.

Western Canadian dinosaur tracksites have also yielded the oldest known bird footprints and the only known Late Cretaceous track with skin impressions. Other rare phenomena include a trackway which makes a 90° turn and reworking of Cretaceous tracks into Pleistocene sediments.

Introduction

Sternberg (1926) gave the first description of dinosaur footprints in western Canada, although McLearn (1923) had previously reported the discovery of dinosaur footprints in the Peace River Canyon. Sternberg (1932) collected in the Peace River Canyon in 1930, and reported on a rich footprint fauna that did not receive any additional professional attention until the construction of the W.A.C. Bennett and Peace Canyon Dams. In response to the discovery and subsequent destruction of footprints during construction of the former dam, the Royal Ontario Museum worked in the Canyon in 1965. Sternberg's original sites in the Canyon and many new localities were worked by the Provincial Museum of Alberta (whose paleontological program formed the core of the Tyrrell Museum of Palaeontology) from 1976 to 1979 as a salvage operation prior to the inundation of the Canyon following completion of the latter dam.

Footprints found in the Mist Mountain Formation of southeastern British Columbia were reported to a number of museums in the 1940's but none of the footprints were ever reported or described in the literature.

In 1960, Langston described a hadrosaur footprint from southern Alberta. Subsequent to his study, a better specimen was found in Drumheller.

Storer (1975) described a specimen from the Dunvegan Formation of northeastern British Columbia that was found by an amateur collector.

Using the knowledge gained from working on the footprints of the Peace River Canyon, staff of the Tyrrell Museum of Palaeontology have substantially increased the number of known footprint sites in western Canada in recent years (Figs. 31.1, 31.2). These new sites are documented here for the first time, but work on some of the major sites will be ongoing for years to come.

Upper Jurassic/Lower Cretaceous Footprint Sites

The oldest dinosaur footprints presently known from western Canada are from Upper Jurassic and Lower Cretaceous (Upper Tithonian, Berriasian) strata of the Mist Mountain Formation. Footprints have been recovered from three localities in the Crowsnest Pass of British Columbia where coal has been actively mined since the last century.

Some of the best footprints were collected from the ceilings of underground coal mines at Michel, B.C. (latitude 49°42'30"N, longitude 114°49'30"W). Samples of specimens from B seam of #3 mine of the Crow's Nest Pass Coal Company were sent to the National Museum of Canada (NMC 8827, 8828) and the Royal British Columbia Museum (RBCM 722) in the early 1940's by H. P. Wilson. Molds were made of other specimens that were retained by the company (now Crowsnest Pass Resources of Calgary) and are cataloged as TMP 79.22.2 (a theropod with a track 23

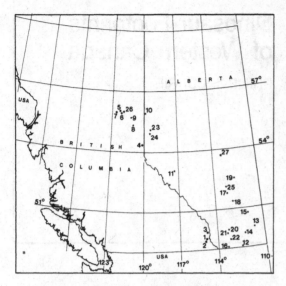

Figure 31.1. Localities of dinosaur footprints in British Columbia and Alberta. **1**, Michel (Mist Mountain). **2**, Flathead Ridge (Mist Mountain). **3**, Eagle Mountain (Mist Mountain). **4**, Narraway River (Minnes). **5**, W.A.C. Bennett Dam (Cadomin). **6**, Peace River Canyon (Gething). **7**, Carbon Creek (Gething). **8**, Murray River (?Dunvegan). **9**, East Pine Bridge (Dunvegan). **10**, Pouce Coupe River (Dunvegan). **11**, Luscar (Cardium). **12**, Red Creek (Milk River). **13**, Redcliff (Foremost Horizon, Judith River Formation). **14**, Grassy Lake (Foremost Horizon, Judith River Formation). **15**, Dinosaur Provincial Park (Oldman Horizon, Judith River Formation). **16**, St. Mary River (Belly River Formation). **17**, Rumsey (Horseshoe Canyon). **18**, Drumheller (Horseshoe Canyon). **19**, Battle River (Horseshoe Canyon). **20**, Barons (St. Mary River Formation). **21**, Oldman River (St. Mary River Formation). **22**, St. Mary River (St. Mary River Formation). **23**, Red Willow River (Wapiti). **24**, Pinto Creek (Wapiti). **25**, Huxley (Scollard Formation). **26**, Peace River Canyon (Pleistocene). **27**, Ft. Saskatchewan (Pleistocene).

cm long and a very narrow, 52° divarication between digits II and IV) and TMP 79.22.3 (two tracks, 15 cm long, of the same type of animal). None of the footprints at Michel are larger than 35 cm in length, and all represent bipedal, tridactylous dinosaurs. Notes made by Sternberg identify NMC 8827 and NMC 8828 as *Irenesauripus* cf. *acutus*. RBCM 772 was made by small theropods (the smallest print is 12 cm long, a second is 14 cm and the largest is 20 cm wide) walking across a carbonaceous mud covered by a mat of fern fronds. It appears that many more footprints were found in the roofing shales about 1/2 kilometer inside the mine. Unfortunately, the mine has now been abandoned and the specimens that were collected have been dispersed.

TMP 79.22.1 is a cast of an original specimen that currently is sitting in storage in a yard of the B.C. Ministry of Mines in Victoria. The specimen was found near Fernie, B.C. on top of Flathead Ridge (latitude 49°20'N, longitude 114°53'W) by Dave Pearson and Mike Welder in 1981. Although the coarse, cross-bedded sandstone block was not found *in situ*, it appears to have come from a level 440 m (1450') above the base of the Mist Mountain Formation. There are four footprints preserved as natural casts on the block, representing at least two individuals that can be identified as *Anomoepus*, a general form of footprint that may be primitive for ornithischians (Olsen and Galton 1984). The footprints are 188–200 mm long with divarications of 76–83°. The toes were long and slender, had distinct phalangeal pads, and were armed with claws. Some additional, poorly-preserved marks may be manus prints.

The third Mist Mountain Formation site is the Fording Coal open pit mine on Eagle Mountain (latitude 50°05'N, longitude 115°00'W) near Elkford, B.C. A single large block (TMP 85.105.1) was discovered by T. J. Wozniak in 1984, approximately 520 m (1700') above the base of the formation. Four footprints of a single individual are preserved as natural casts (Fig. 31.3). The animal was walking across a wet, ripple marked substrate that shows many invertebrate trails and burrows. Using the formula developed by Alexander (1976) to estimate the speed of track-making vertebrates, the dinosaur was accelerating to a speed of 5.6 km/hr. The trackmaker was a tridactylous biped (footprint length is 17 cm), but the feet were relatively wide, and the step is strongly inturned, characteristics that are suggestive of ornithopods rather than theropods. Details of the tracks are not well preserved, but are sufficient to assign them to the ichnogenus *Anomoepus*.

Lower Cretaceous Footprint Sites

A major trackway site in the Lower Cretaceous Minnes Formation (Berriasian/Valangian) has been found along the Narraway River (latitude 54°21'N, longitude 120°03'W). Ironically, the locality is located northwest of Dinosaur Ridge, a topographic ridge named for its sinuous shape. More than two hundred footprints are preserved in at least eight trackways on a single ripple–marked, sandstone bedding plane. Because the site is accessible only by helicopter or a long hike across rough terrain, only an initial survey was completed in 1981 by staff of the Tyrrell Museum of Palaeontology. The majority of footprints were made by small theropods, but the most dramatic trackway was made by a large biped whose feet were more than a half meter in length (Fig. 31.4). This individual came to a stop, turned sharply to the right, paused and then set off in a direction perpendicular to the original route.

The Lower Cretaceous (Barremian–Aptian) Cadomin Formation is a predominately conglomeratic facies that crops out at the base of the Peace River Canyon at the W.A.C. Bennett Dam. The Cadomin grades laterally into the coarse–grained sandstones and coal–bearing beds of the Gething Formation in this region (Stott 1973). During construction of the dam (latitude 56°01'N, longitude 122°12'W), more than fifty footprints were discovered when ledges on

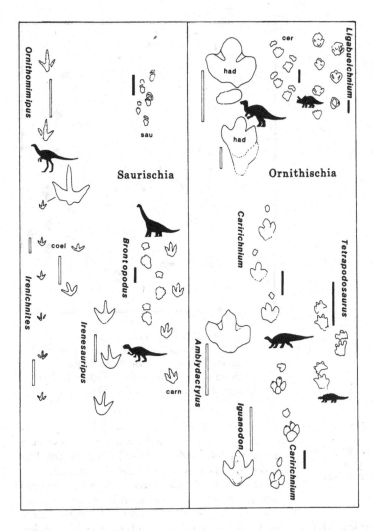

Figure 31.2. Distinctive Cretaceous dinosaur ichnotaxa. *Amblydactylus*, ceratopsians (cer), *Irenichnites,*, *Irenesauripus*, *Ornithomimipus* and *Tetrapodosaurus* are some of the forms known from western Canada. It is reasonable to expect that tracks similar to *Caririchnium* from Brazil and Colorado, and *"Iguanodon"* from England will eventually be identified in western Canada. The absence of *Brontopodus* and other sauropod tracks, and *Ligabueichnium* is probably related to the distribution of these trackmakers during the Cretaceous, and it is not expected that they will ever be found in western Canada. After Lockley in press.

the canyon walls were hydraulically cleaned. The Royal Ontario Museum made molds of two trackways at this site in 1965, and a couple of individual footprints (ROM 4869, 4870). The footprints were made by both theropods (*Irenesauripus*) and ornithopods (*Amblydactylus*). Manus prints associated with the latter showed that ornithopods were quadrupedal at least part of the time. This site is considered by some to be in the base of the Gething Formation.

Dinosaur footprints from the Gething Formation (Aptian–Albian) of the Peace River Canyon (latitude 55°57'30"N, longitude 122°10'W) were first reported in 1923 (McLearn). However the footprints are so well preserved and numerous, it is difficult to imagine that they weren't seen by Sir Alexander MacKenzie when he explored the canyon in 1793. An expedition by C. M. Sternberg in 1930 to the canyon led to the establishment of five new ichnogenera and seven ichnospecies (Sternberg 1932). When construction started on the Peace Canyon Dam, four expeditions by the Provincial Museum of Alberta (the parent institution of the Tyrrell Museum of Palaeontology) collected data and specimens between 1976 and 1979 (Mossman and Sarjeant 1983). Ninety original specimens (including TMP

78.11.17, an *Amblydactylus* track with a double impression, Fig. 31.5) now in the Tyrrell Museum of Palaeontology and the Royal B.C. Museum (Victoria) were collected, and molds were made of more than 200 footprints. All in all, more than 1700 footprints were discovered, most of which were preserved in a hundred trackways. Trackway evidence indicated that large ornithopods were abundant and probably gregarious (Currie 1983). Another ornithopod ichnospecies was described (Currie and Sarjeant 1979), as well as the earliest known bird tracks (Currie 1981). A set of poorly preserved footprints may represent Lower Cretaceous mammal tracks (Sarjeant and Thulborn 1986).

The Gething Formation exposed in the Canyon is approximately 300 m (1000 ft) thick (Stott 1968), and tracks have been recovered throughout the section. Dinosaur footprints were recovered from carbonaceous mudstones, siltstones, sandstones and even coarse grained channel sandstones.

The Peace River Canyon represented one of the best documented major trackway sites in the world. By 1980, the trackway sites had been inundated by the reservoir behind the new Peace Canyon Dam at Hudson's Hope.

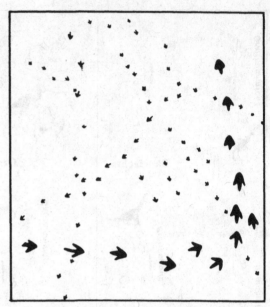

Figure 31.4. Map of a portion of large theropod trackway and associated tracks of small theropods from the Minnes Formation (Lower Cretaceous) exposed on the Narraway River. Each of the large footprints is slightly longer than 50 cm.

Figure 31.3. Trackway of a small ornithopod (TMP 85.105.1) from the Mist Mountain Formation of Eagle Mountain, southeastern British Columbia. Scale = 50 cm.

Footprints will continue to be found in the Gething Formation adjacent to the Canyon, but there are few horizontal bedding planes that are likely to produce lengthy sections of trackways.

In 1981, dinosaur footprints were found on an exposure of the Gething Formation 60 km west of Hudson's Hope on the west side of Carbon Creek (latitude 55°55'59"N, longitude 122°38'40"W). The site is at the northwestern end of an area cleared for a temporary campsite for Utah Coal Exploration, about 900 m south of Little Carbon (Indian) Creek. The largest footprint, which can be identified as *Amblydactylus kortmeyeri*, is more than 600 mm long. Only two of the three digits left an impression, but the "heel" is well defined. Another *Amblydactylus* track illustrated by Busbey (1983) is slightly smaller and appears to be overlain by an *Irenesauripus* track. There are other footprints on the bedding plane, but are not as well preserved and were not photographed.

Upper Cretaceous Footprint Sites

A natural cast of a large left manus print (TMP 81.32.1) of *Tetrapodosaurus* was found by Carl Kortmeyer

on the Murray River approximately twenty-five miles upstream (about 55°N, 121°15'W) from the East Pine Bridge. The manus was 27 cm wide. Unfortunately it was found on an isolated block and it cannot be determined with any certainty whether it originated in the Lower Cretaceous Minnes, Cadomin or Gething Formations or the Late Cretaceous Dunvegan Formation. The location, sandstone coloration and print morphology all suggest that it is probably from the Dunvegan.

In 1975, John Storer reported on the discovery of footprints on the Pine River at East Pine Bridge. Three natural casts of *Columbosauripus ungulatus* were described from a single slab of thinly-bedded siltstone in the private collection of Mr. and Mrs. H. C. Calverly. The original specimen has since been donated to the Tyrrell Museum of Palaeontology (TMP 75.4.1). The locality (55°44'N, 121°12'W) was revisited in August 1977 and numerous footprints were found in blocks at the base of the cliff. A natural cast of a small (20 cm long) pes track of *Tetrapodosaurus* (TMP 77.16.22) was collected.

In 1951, Dr. C. R. Stelck (University of Alberta) collected a block (UALVP 25271) of eight footprints from the Dunvegan Formation along the Pouce Coupe River in Alberta (56°00'N, 119°55'W), about three kilometers downstream from the mouth of the Doe River. Apparently there are many footprints preserved in large blocks of sandstone for a stretch of about two kilometers along the river. The uncataloged specimen on display in the Palaeontology Museum (Geology Department, University of Alberta) has

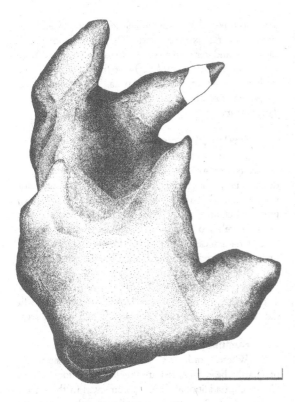

Figure 31.5. *Amblydactylus* track (TMP 78.11.17) from the Gething Formation of the Peace River Canyon of British Columbia, showing a double impression as the animal slid and turned in soft mud. Scale = 20 cm.

small, tridactylous footprints. The most distinct footprint is 3.5 cm long. The divarication of the toes is less than 90°, which suggests that it was made by a dinosaur rather than a bird (Currie 1981). The remaining footprints consist mostly of toe drags made by swimming animals. A curious feature of all prints is that the back of the impression of what appears to be digit II is placed more anteriorly than the backs of digits III and IV. This suggests the possibility that the footprints were made by a functionally didactylous animal, and that only the tip of digit II was touching the ground. Because of the size of the footprints and the peculiar orientation of the toes, it is also possible that the footprints were made by hesperornithiform and/or ichthyornithiform birds. Hesperornithiform birds are best known from Kansas, but ranged as far north as the Arctic Circle (Russell 1967), and *Ichthyornis* has been reported from Lower Turonian beds not far from the footprint site (Fox 1984).

The coal-bearing Upper Cretaceous Cardium Formation (Turonian) between Luscar and Cadomin, Alberta has not been explored for footprints. However, three dinosaur tracks collected by Luke Lindoe from the roadside in this area show that the formation may be rich in trace fossils. The tracks were made by theropods with long slender toes. Divarication between digits II and IV is 65°

in the best preserved footprint, which is 14 cm long, and 55° in a slightly smaller print.

In 1987, two large hadrosaur footprints were found in Red Creek (49°01′N, 112°07′30″W) by Tyrrell Museum staff. Strata in this area are from the Milk River Formation (Lower Campanian). The footprints are more than a half meter long, but are poorly preserved subtraces.

Around 1960, Luke Lindoe discovered dinosaur footprints in the Foremost "Formation" (now the lower part of the mid–Campanian Judith River Formation), on the north shore of the South Saskatchewan River (50°03′N, 110°46′W) near Redcliff, Alberta. The footprints originate from a level about 30 m below the contact between the Foremost and Oldman sections of the Judith River Formation. Seven hadrosaur and two tyrannosaurid tracks were observed in 1979, and the track–bearing layer, a coarse red sandstone, continues into the hillside. The hadrosaur tracks are between 40 and 64 cm in length, the corresponding widths being 44 and 66 cm.

The Foremost is also exposed at a site (49°45′N, 111°40′W) near Grassy Lake, Alberta, and footprints have been found in a coarse sandstone two meters above a mined coal seam (Taber Coal Zone). Only one hadrosaur footprint (TMP 79.9.2) has been recovered to date (33 cm wide), but another half dozen negative and positive footprints were observed in the immediate area, most of which were poorly preserved.

Dinosaur Provincial Park (Judith River &= Oldmané Formation) has produced one of the richest Cretaceous terrestrial faunas known (Currie 1987), based largely on the articulated remains of dinosaur skeletons and disarticulated bones of dinosaurs and other animals. Footprints are rare in the park, although there is ample evidence of massive disturbance of the sediment by "dinoturbation". In 1981, a large hadrosaur footprint (TMP 81.34.1) was discovered in the center of the badlands in the park (50°46′N, 111°28′W). The specimen was excavated (Quarry #155, Danis 1986), and is now on display in the Field Station of the Tyrrell Museum of Palaeontology. A second, incomplete footprint was less than two meters away, and was probably made by the same individual.

The Belly River Formation near the source of the St. Mary River (49°01′N, 113°17′W) has produced one good carnivore track (25 cm long), and three other tracks on a large slab of sandstone. The locality was discovered by Greg Nadon, but specimens have not been collected yet.

Two footprint sites are known in the Horseshoe Canyon Formation near Drumheller. *Ornithomimipus angustus* was described by Sternberg in 1926 on the basis of four footprints discovered southwest (51°48′N, 112°58′W) of Rumsey. Although these were considered to be synonymous with *Irenesauripus* by Haubold (1971), the digits are quite different in that they are broader proximally, and were made by an entirely different sort of theropod. For these reasons, *Ornithomimipus* should remain a distinct ichnogenus.

A large (length = 67 cm, width = 70 cm), well preserved negative of a hadrosaur footprint was collected low in the Horseshoe Canyon Formation close to the mouth of Willow Creek (51°23'N, 112°31'W) by Allan Jensen and his father in 1967, and is on display in the Drumheller and District Fossil Museum. The specimen seems to have been extracted from one of the coal mines in the Red Deer River valley near Drumheller, but its exact provenance is unknown.

An ichnite was found in the Horseshoe Canyon Formation north of Stettler along the Battle River (latitude 52°29'30"N, longitude 112°11'W). This specimen is preserved on a sandstone block that was torn up in a strip mining operation. It is a four–digit impression of a pes of a quadrupedal dinosaur, and resembles *Tetrapodosaurus* (Sternberg 1932) from the Peace River Canyon. The specimen is relatively small (the width is 17.5 cm) and has a prominent metatarsal pad or "heel". It was probably made by a ceratopsian. The specimen has been photographed (TMP PN79.43), but not collected.

The Late Campanian St. Mary River Formation of southwestern Alberta has produced a number of footprint localities. Langston (1960) described a hadrosaur footprint (NMC 9487) near Barons (latitude 49°58'N, longitude 113°05'W), that he cast but was unable to collect. The site was destroyed when the irrigation canal where it was found was upgraded around 1985.

In 1986, Greg Nadon (University of Toronto) reported the discovery of numerous tracks along the St. Mary (49°25'N, 113°01'W) and Oldman Rivers (49°47'N, 113°12'W). More than forty footprints of tyrannosaurids and hadrosaurs have already been found. Most are in blocks of sandstone that have tumbled from the valley walls, but some of these blocks, which represent crevasse splay deposits, are so large that short segments of trackways are revealed. Almost all of the specimens are negatives (natural casts) because most of the footprints (counterparts) were preserved in friable siltstones that break up as soon as they are exposed. A manus (TMP 87.76.7) and a pes (TMP 87.76.6) from a hadrosaur have been excavated. The footprint was covered with skin impressions that are currently under study (Currie, Lockley and Nadon in prep.). This is one of only two Cretaceous dinosaur tracks known to exhibit skin impressions (see Lockley Summary Chapter, this volume).

The Wapiti Formation of northwestern Alberta is a thick unit laid down during Campanian and Maastrichtian times. TMP 86.80.1 is a hadrosaur footprint discovered in 1983 along the Red Willow River (55°03'N, 119°22'W). The specimen is a negative made up of a medium–grained sandstone, but the original footprint appears to have been made in soft mud because the toes are widely spread (width of the footprint is 61 cm compared with a length of 52 cm). Less than a kilometer downstream from this site, a negative theropod track (TMP 86.80.9) was discovered and collected by Darren Tanke (Tyrrell Museum). Finally, a possible ankylosaur track (TMP 87.56.39) was found along Pinto Creek (54°56'N, 119°27'W).

Dinosaur footprints are rare in terminal Cretaceous beds, but the negative (TMP 81.12.3) of a manual impression was collected from the Scollard Formation (Maastrichtian) east of Huxley, Alberta. The manus was pentadactylous and very large (46 cm across), and would have been made by either *Ankylosaurus* or *Triceratops*, which are both recovered from adjacent strata. The specimen was recovered from the ravine (51°55'N, 113°02'W) below a *Tyrannosaurus rex* skeleton.

Reworked Footprints

I am not aware of any previous reports of Pleistocene dinosaurs, but two such sites in western Canada have produced dinosaur footprints. An *Amblydactylus* natural cast was found in 1979 in the Pleistocene gravels of the blocked, preglacial channel of the Peace River. The specimen was unfortunately too large to collect. D. Schowalter also collected a natural cast of a hadrosaur footprint (TMP 80.39.2) from the "Saskatchewan Sands and Gravels" being quarried along the North Saskatchewan River near Fort Saskatchewan (53°38'N, 113°18'W). This specimen (length = 41 cm, width = 45 cm) is probably derived from Maastrichtian beds, which are only shallowly buried in this area.

Summary

Whereas only four dinosaur footprint sites had previously been reported in the literature for western Canada, an additional 23 are documented in this paper. Footprints have been found from all ages of the Cretaceous except Santonian, Coniacian and Barremian (Table 31.1). Although many of these localities have produced relatively poor and/or relatively few footprints, each seems to contribute information that wasn't otherwise available. For example, hadrosaur footprints from Campanian and Maastrichtian sites are from formations that have abundant skeletal remains. This permits a greater degree of confidence in the identification of the footprints, and shows that certain animals represented by both footprints and skeletons actually did live in the region (rather than being carcasses carried in from other environments by the rivers).

Sauropods are known from tracks and trackways of the Lower Cretaceous of Texas (Farlow 1987) and by skeletal material from the Upper Cretaceous of the United States and Asia (Currie in press). Although their distribution goes well beyond this, there would have been no physical restrictions preventing their access to western Canada throughout the Cretaceous. The complete absence of sauropod tracks and bones is a good indication that sauropods did not penetrate into Canada for ecological reasons, as suggested by Lockley (in press) and Lockley and Conrad (this volume).

The Upper Jurassic/Lower Cretaceous Mist Mountain Formation has footprints that are much more primitive in appearance than all other sites in western Canada, and are similar to Jurassic tracks from the United States that have relatively low angles of divarication between the toes and more distinct phalangeal pads. *Anomoepus* tracks are not found in younger beds in western Canada. Differences between Jurassic and Cretaceous footprint types can be added to a growing body of evidence that suggests there

Table 31.1. *Geological ages of footprint sites in western Canada.*

Epoch	Age	Formations	Sites
Late Cretaceous	Maastrichtian	Scollard	Huxley
	Campanian/ Maastrichtian	Wapiti	Pinto Creek, Red Willow River
	Campanian	Horseshoe Canyon	Drumheller, Rumsey, Battle River
		St. Mary River	Barons, St. Mary River, Oldman River
		Judith River (Oldman)	Dinosaur Provincial Park
		Judith River (Foremost)	Redcliff, Grassy Lake
		Belly River	St. Mary River
		Milk River	Red Creek
	Santonian	—	—
	Coniacian	—	—
	Turonian	Cardium	Luscar
	Cenomanian	Dunvegan	Murray River, East Pine Bridge, Pouce Coupe River
Early Cretaceous	Aptian/ Albian	Gething	Peace River Canyon, Carbon Creek
	Barremain/ Aptian	Cadomin	W.A.C. Bennett Dam
	Barremain	—	—
	Berriasian/ Valangian	Minnes	Narraway River
	Tithonian/ Berriasian	Mist Mountain	Michel, Flathead Ridge, Eagle Mountain

was a major faunal turnover at the end of the Jurassic.

Early Cretaceous footprints were made by more primitive dinosaurs than those of the Late Cretaceous. However, the morphological appearance of footprints of the Early and Late Cretaceous are remarkably similar within any major group of animals. The faunal composition of theropods, ornithopods and quadrupedal dinosaurs also seems to remain relatively constant throughout the Cretaceous.

Now that people are becoming more familiar with dinosaur footprints, we expect the number of known sites in western Canada to steadily increase. There are already

unconfirmed reports of localities in the extreme northeastern corner of British Columbia (on the Liard River) and in the foothills west of Calgary. Unfortunately, the proliferation of sites has not been matched by a corresponding increase in qualified people willing to do the work.

Acknowledgments

Many people have been involved in the discovery, documentation and collection of dinosaur footprints in western Canada, and it would be impossible to thank everyone in a paper of this nature. Amongst those that had a role in the preparation of this manuscript, I would like to thank Dr. Martin Lockley for encouraging me to complete the paper, and for providing me with unpublished data on his work. Dr. Donald Stott (Institute of Sedimentary and Petroleum Geology, Calgary) directed me to the site on the Narraway River, the family of the late Carl Kortmeyer (Dawson Creek, B.C.) donated the specimen from the Murray River locality, and Dr. C. R. Stelck (University of Alberta) discovered the tracks on the Pouce Coupe River. Jane Danis (Tyrrell Museum), Dean Wetzel (Archaeological Survey of Alberta), John McMurdo (B.C. Heritage Conservation Branch) and Darren Tanke (Tyrrell Museum) all provided specific information on sites. Dr. Gordon Edmund (Royal Ontario Museum) was kind enough to supply information on their expedition to the Peace River Canyon. Figure 31.2 was prepared by Martin Lockley, Figure 31.3 by Darren Tanke, Figure 31.6 by Elizabeth Garsonnin, and the others by the author.

References

Alexander, R. McN. 1976. Estimates of the speeds of dinosaurs. *Nature* 261:129.

Busbey, J. 1983. Stage II Heritage Resource Impact Assessment, Carbon Creek Coal Development. Unpublished report on file with Utah Mines Ltd., Port Hardy, B.C. 54 pp.

Currie, P. J. 1981. Bird footprints from the Gething Formation (Campanian, Lower Cretaceous) of northeastern British Columbia, Canada. *Jour. Vert. Paleont.* 1:257–264.

——— 1983. Hadrosaur trackways from the Lower Cretaceous of Canada. *Acta Palaeont. Polonica* 28:63–73.

——— 1987. New approaches to studying dinosaurs in Dinosaur Provincial Park. *In* Czerkas, S. J., and Olson, E. C. (eds.). *Dinosaurs Past and Present Volume II* (Los Angeles: Los Angeles County Museum) pp. 100–117.

——— in press. Saurischian dinosaurs of the Late Cretaceous of North America and Asia. *In* Mateer, N., and Chen P. (eds.). *Proc. First Internat. Symp. Cretaceous Nonmarine Correlations.* (Beijing, China: Ocean Press).

Currie, P. J., Lockley, M. G., and Nadon, G. In prep. Dinosaur footprints with skin impressions from the Cretaceous of Alberta and Colorado.

Currie, P. J., and Sarjeant, W. A. S. 1979. Lower Cretaceous footprints from the Peace River Canyon, B.C. Canada. *Palaeogeogr., Palaeoclimat., Palaeoecol.* 28:103–115.

Danis, J. C. 1986. Quarries. *In* Naylor, B. G. (ed.). *Field Trip Guidebook to Dinosaur Provincial Park, Dinosaur Systematics Symposium* (Drumheller: Tyrrell Museum of Palaeontology) pp. 43–52.

Farlow, J. O. 1987. *A Guide to Lower Cretaceous Dinosaur Footprints and Tracksites of Paluxy River Valley, Somer-*

ville County, Texas. (South Central G.S.A., Baylor University). 50 pp.

Fox, R. C. 1984. Ichthyornis (Aves) from the Early Turonian (Late Cretaceous) of Alberta. Canadian Jour. Earth Sci. 21:258–260.

Haubold, H. 1971. Ichnia Amphibiorum et Reptiliorum Fossilium. Handbuch der Paläoherpetologie pt. 18. (Gustav Fischer) 124 pp.

Langston, W. Jr. 1960. A hadrosaur ichnite. National Museum Canada Nat. Hist. Papers 4:1–9.

Lockley, M. G. in press. Cretaceous dinosaur-dominated footprint assemblages: their stratigraphic and paleo-ecological potential. In Mateer, N., and Chen P. (eds.). Proc. First Internat. Symp. Cretaceous Nonmarine Correlations. (Beijing, China: Ocean Press).

McLearn, F. H. 1923. Peace River Canyon Coal area, B.C. Geol. Surv. Canada Summary Reports 1922 (B):1–46.

Mossman, D. J., and Sarjeant, W. A. S. 1983. The foot-prints of extinct animals. Sci. Amer. 248 (1):74–85.

Olsen, P. E., and Galton, P. M. 1984. A review of the reptile and amphibian assemblages from the Stormberg of South Africa, with special emphasis on the footprints and the age of the Stormberg. Palaeont. Africana 25:87–110.

Russell, D. A. 1967. Cretaceous vertebrates from the Anderson River, N. W. T. Canadian Jour. Earth Sci. 4:21–38.

Sarjeant, W. A. S., and Thulborn, R. A. 1986. Probable marsupial footprints from the Cretaceous sediments of British Columbia. Canadian Jour. Earth Sci. 23:1223–1227.

Sternberg, C. M. 1926. Dinosaur tracks from the Edmonton Formation of Alberta. Geol. Surv. Canada Bull. 44:85–87.

——— 1932. Dinosaur tracks from Peace River, British Columbia. National Museum Canada Ann. Rept. 1930:59–85.

Storer, J. 1975. Dinosaur tracks, Columbosauripus ungulatus (Saurischia: Coelurosauria), from the Dunvegan Formation (Cenomanian) of northeastern British Columbia. Canadian Jour. Earth Sci. 12:1805–1807.

Stott, D. F. 1968. Lower Cretaceous Bullhead and Fort St. John Groups, between Smoky and Peace Rivers, Rocky Mountain Foothills, Alberta and British Columbia. Geol. Surv. Canada Bull. 152:1–279.

——— 1973. Lower Cretaceous Bullhead Group between Bullmoose Mountain and Tetsa River, Rocky Mountain Foothills, northeastern British Columbia. Geol. Surv. Canada Bull. 219:1–228.

32 Dinosaur Footprints from the Lower Cretaceous of East Sussex, England

KEN E. WOODHAMS AND
JOHN S. HINES

Abstract

Dinosaur footprints were first reported from the Wealden strata of East Sussex by Tagart (1846). No recent study of Sussex footprints has been published apart from a summary of the work of the two present authors (Delair and Sarjeant 1985). Two of the more recently discovered sites, both coastal exposures, are described here.

At Cooden, near Bexhill, two interesting trackways were exposed on the foreshore late in 1980. They were made up of twenty-two tridactyl imprints of the form generally attributed to the bipedal dinosaur *Iguanodon*. The fine-grained sandstone bearing the footprints occurs in the upper part of the Tunbridge Wells Sand and is of late Valanginian age.

The other footprint site to the east of Hastings consists of several siltstone and sandstone horizons in the Ashdown Beds (Upper Berriasian). Although the cliff section generally yields isolated natural casts rather than trackways, the detail displayed by some is exceptional. Again they are tridactyl, but in addition to the iguanodontids, theropods are also present.

Introduction

Tridactyl footprints have been observed and recorded from the Wealden strata of the Hastings area for the past 140 years, having been first reported by Tagart (1846). They have occurred across the entire width of the Hastings Beds outcrop from near Pevensey Sluice in the west to Cliff End in the east (Figs. 32.1, 32.2), a distance of nearly 25 km (Beckles 1854). To the west of Bexhill, footprints are exposed at intertidal levels, often in sufficient numbers to constitute trackways, whereas, to the east of Hastings, cliff erosion generally reveals only isolated natural casts, a series of three or more consecutive prints being rare.

There have been many important footprint sites between Bexhill and Hastings, at Bexhill (Beckles 1854), Bulverhythe (Ticehurst 1928) and at Galley Hill (Beckles 1854) amongst them, but only one important site to the east of Hastings, Fairlight (Tagart 1846, Beckles 1851). A history and bibliography of British dinosaur footprint discoveries, including those of Sussex, has been compiled by Sarjeant (1974) and Delair and Sarjeant (1985). See also Delair (this volume).

Without doubt, Samuel Beckles occupies a position of prime importance, for his contributions promoted the establishment of the science of fossil ichnology in England (Woodhams in press).

Almost without exception the dinosaur footprints of Sussex have been attributed to the genus *Iguanodon*, but, in view of the poor preservation of many of them, the classification of iguanodontid would be more appropriate. As *Iguanodon* is known to have been abundant in the region during the Early Cretaceous, it has become an almost automatic identification for the footprints of Sussex, seemingly denying the presence of other genera.

Methods of Documentation

From the start of their research the present authors used simple line diagrams to record both natural casts and imprints, drawn to a constant scale allowing immediate comparisons to be made. The drawings are supplemented by comprehensive sets of dimensions (Figs. 32.3–32.5) and are further supported by the extensive use of photography.

Photographing natural casts presents few technical problems, but attempts to record imprints, especially as part of a trackway, can give disappointing results, particularly evident under conditions of low contrast winter lighting. In an attempt to overcome this problem, imprints have been filled with dark soil prior to photographing (Calkin 1968). At Cooden, the present authors sprayed the trackway footprints lightly with black water–soluble paint to clarify both shape and alignment. The effectiveness of this technique can be seen in photographs taken by one of the present authors (JSH) to be seen in Delair and Sar-

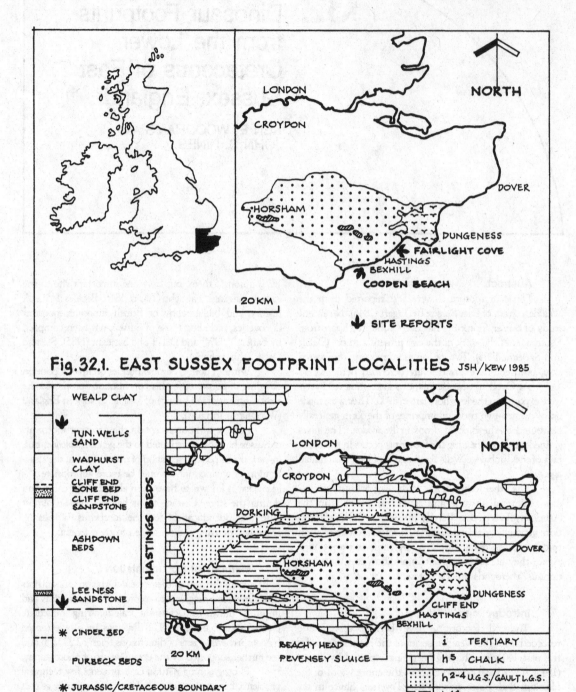

Figure 32.1. East Sussex footprint localities. Figure 32.2. Geology of southeast England.

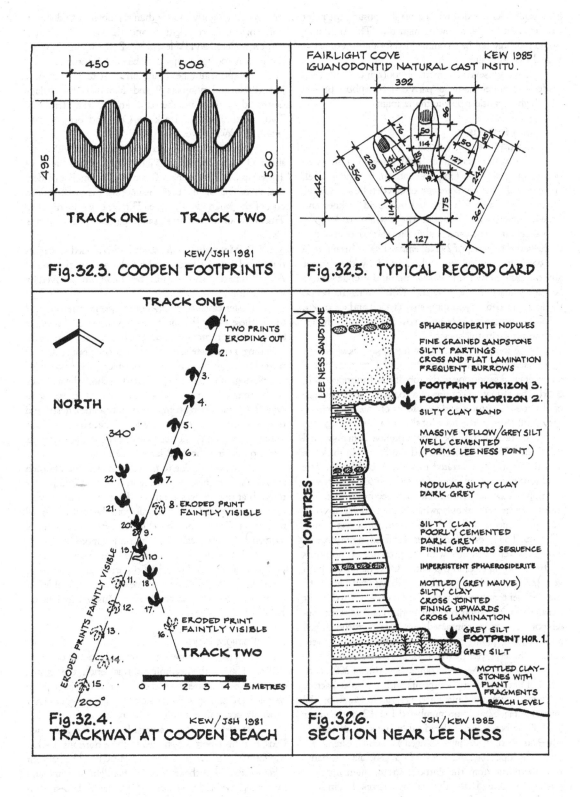

Fig.32.3. COODEN FOOTPRINTS

TRACK ONE   TRACK TWO

450   508

495   560

KEW/JSH 1981

Fig.32.5. TYPICAL RECORD CARD

FAIRLIGHT COVE   KEW 1985
IGUANODONTID NATURAL CAST INSITU.

392

442

Fig.32.4.
TRACKWAY AT COODEN BEACH

TRACK ONE

NORTH

340°

200°

1. TWO PRINTS ERODING OUT
2.
3.
4.
5.
6.
7.
8. ERODED PRINT FAINTLY VISIBLE
22.
21.
20.
19.
18.
17.
16. ERODED PRINT FAINTLY VISIBLE
15.
14.
13.
12.
11.
10.
9.

ERODED PRINTS FAINTLY VISIBLE

TRACK TWO

0 1 2 3 4 5 METRES

KEW/JSH 1981

Fig.32.6.
SECTION NEAR LEE NESS

LEE NESS SANDSTONE

10 METRES

SPHAEROSIDERITE NODULES

FINE GRAINED SANDSTONE
SILTY PARTINGS
CROSS AND FLAT LAMINATION
FREQUENT BURROWS

FOOTPRINT HORIZON 3.
FOOTPRINT HORIZON 2.
SILTY CLAY BAND

MASSIVE YELLOW/GREY SILT
WELL CEMENTED
(FORMS LEE NESS POINT)

NODULAR SILTY CLAY
DARK GREY

SILTY CLAY
POORLY CEMENTED
DARK GREY
FINING UPWARDS SEQUENCE

IMPERSISTENT SPHAEROSIDERITE

MOTTLED (GREY MAUVE)
SILTY CLAY
CROSS JOINTED
FINING UPWARDS
CROSS LAMINATION

GREY SILT
FOOTPRINT HOR.1.
GREY SILT

MOTTLED CLAY-
STONES WITH
PLANT
FRAGMENTS
BEACH LEVEL

JSH/KEW 1985

Figure 32.3. Cooden footprints. Figure 32.4. Typical record card. Figure 32.5. Trackway at Cooden Beach. Figure 32.6. Section near Lee Ness.

jeant (1985). As Cooden was a coastal exposure, the paint was removed by the following high tide. This technique would probably not be recommended for more permanent inland footprint sites.

Casts of several important footprints have been made using either plaster of paris or latex rubber backed with high expansion polyurethane foam.

Site Reports
Cooden Beach, Bexhill
The winter storms of 1980 revealed a series of twenty-two tridactyl footprints at an intertidal level, approximately one kilometer east of Cooden Beach Railway Station. They were first reported by David Tickner and Ernest Withey, who discovered them whilst working on the sea defenses. This prompted the appearance of an article in the Bexhill Observer of 29 November 1980. This multiple trackway site was subsequently studied by the present authors, assisted by Jenny Woodhams and Brenda Graves between January 1st and May 2nd, 1981. During this period the footprint bed was continuously eroded and degraded, eventually becoming once more covered by beach shingle several meters in thickness.

The footprints in two intersecting trackways (Fig. 32.4), referred to as Trackway One and Two, occurred in a fine-grained sandstone at the top of the Tunbridge Wells Sand. This formation consists of a complex cyclic sequence of siltstones, sandstones and mudstones of Late Valanginian age. Until recently the rocks of this area were thought to be part of the Ashdown Beds, but borehole data disproved this (Lake 1975). The footprint bed was 85 mm in thickness and the imprints penetrated to a depth of 75 mm. The sandstone was probably part of a point bar sequence featuring ripple cross-lamination that had been disturbed and intersected by vertical burrowing. Pyrite nodules up to 25 mm in diameter and clay pellet structures were fairly common. The horsetail Equisitites occurred in positions of growth, which correlates with the presence of footprints in that at no time could the water depth here have been very great. The plant growth must have kept pace with the sediment accretion rate. Finely disseminated plant remains occurred throughout the horizon and adjacent contemporaneous strata contained fragments of crocodile skull and lignified plant remains. Dessication cracks were commonly displayed in the adjoining mudstones.

Trackway One, on a compass bearing of 200°, consisted of fifteen tridactyl footprints with a regular pace of 1.295 m. Each imprint measured 495 mm in length by 450 mm wide (Fig. 32.3). Prints numbered 1 and 2, at the upper level of the beach, were already showing signs of erosion when first seen, their shape changed to such an extent that they were displaying similarities of form generally associated with footprints from the Purbeck strata, often inappropriately attributed to the large theropod dinosaur Megalosaurus.

Footprints numbered 3 to 7 were all well-defined imprints, typically iguanodontid in form, with a large bulbous heel and robust toes, the inner appearing to be

stronger and slightly shorter than the outer. The divarication angle between digits II and III was 39°, whereas between digits III and IV it was 36°. Each imprint displayed the pigeon-toed gait typical of bipedal dinosaurs.

Footprint number 8 and prints 11 to 15 were only just visible and numbers 9 and 10 had been partially destroyed by those numbered 19 and 20 from Trackway Two, illustrating that Trackway One had been formed before Trackway Two.

Trackway Two, on a compass bearing of 340°, consisted of seven footprints slightly larger than those of Trackway One, measuring 560 mm by 508 mm, but of identical shape (Fig. 32.3). Six of the imprints were well-defined and of the iguanodontid type. The pace was identical to Trackway One and the same pigeon-toed gait was displayed.

By March 1981, tidal scour had degraded footprints numbered 1 and 2 into circular holes and by the end of April this part of the footprint bed had been removed by wave action.

Plaster casts were taken of footprints numbered 17 to 21, these being the most clearly defined of the whole series, and a thorough photographic record was made, including a complete series of overlapping photographs of the entire length of Trackway Two, which, when assembled as a photographic montage, was used to check the accuracy of the completed measured survey. The survey was considered accurate, being compiled from baselines, offsets and triangulations, and it was only found necessary to make minor corrections to the alignment of two of the footprints when producing the final large-scale diagram.

By May 1981, due to the combined action of beach erosion and shingle replacement, the opportunity for research at this site was at an end. Following the construction of a reinforced concrete wall from the west of Bexhill to Cooden Beach, the possibility of further exposures of footprints at this locality has become remote.

It has subsequently been discovered that a partial exposure of this trackway had occurred earlier and was described by Mr. I. Childs of Bexhill. His unpublished report was deposited at the Bexhill Museum in January 1978.

Fairlight Cove
This report is based upon research carried out from 1979 to 1986. At the beginning of this period, the natural casts were both numerous and of exceptional quality, being revealed by the gradual process of cliff erosion. They generally occur as isolated specimens.

Footprints occur at several layers of the Ashdown Beds, but this report is restricted to three horizons in what was formerly referred to as the Fairlight Clay. The Fairlight Clay is exposed in the core of the Fairlight anticline and is now regarded as being part of the Lower Ashdown Beds (Upper Berriasian).

The lowest footprint horizon (Horizon 1; Fig. 32.6) is a ripple-marked, well-cemented gray siltstone containing carbonaceous plants in growth position. A prominent parting at the footprint horizon indicates a break in deposition, but plant stems are continuous into the overlying siltstone, which again demonstrates that plant growth kept pace with the sediment accretion rate. The regular occurrence of clay laminae within the siltstone indicates that the deposition was pulsatory.

The footprints from Horizons 2 and 3 have always occurred as natural casts, with Horizon 2 being the most prolific. At this level (Horizon 2), the dinosaurs were walking on what is now a silty clay and was probably originally an overbank or lagoonal mudstone. In parts it has been extensively burrowed in the remaining water-filled hollows, prior to final dessication. The fine preservation of detail of the natural footprint casts and the occasional occurrence of heel scrape and slip marks (Fig. 32.8) indicate the plastic and slippery nature of the clay at this stage.

The sediment responsible for the casts is known as the Lee Ness Sandstone and has been described as a lacustrine basin fill (Stewart, 1981). This unit is approximately two meters in thickness and coarsens upwards from a silt to a fine-grained sand. It contains low angle accretion units with bioturbated silty partings at about 20–50 mm intervals. Bands of sphaerosiderite and irregular loading structures are also present.

The upper footprint level (Horizon 3) occurs along a prominent bedding plane 200 mm above the base of the Lee Ness sandstone. The only other distinctive feature so far recorded from this level is the presence of casts of large horizontal burrows.

The main sedimentary features that can be seen in adjacent strata are point bar sequences, channel fill structures with large-scale cross beds, crevasse splays and fossil soil horizons.

With the exception of Figure 32.7e, all of the footprint types illustrated in Figure 32.7 occurred in the form of natural casts.

A wide diversity of footprint configurations have been observed, but the majority are of the types represented by Figures 32.7a,b,c, which can be attributed to the iguanodontid group and reinforce the generally held belief that they were the most numerous of the dinosaurs inhabiting the Weald during the Lower Cretaceous. The presence of several theropod footprint forms (Fig. 32.7e–j) is reported, and they are described for the first time.

The footprint illustrated as Figure 32.7a was observed at the base of the Lee Ness Sandstone (Horizon 2) in 1979 and measured 370 mm between the extremities of digits II and IV, with an overall length of 420 mm. All three toes displayed robust phalangeal pads with hoof-like terminations. Traces of hide pattern in the form of clearly separated hemispherical bumps, or tubercles, 3–4 mm in diameter, were preserved on parts of the bulbous heel. This pattern also accounts for the striations occurring on some of the heel slide marks, a feature observed on several of the deeper footprint casts (Fig. 32.8), and supports earlier descriptions of the integument of dinosaurs (Marsh 1889, Hooley 1917).

The track illustrated as Figure 32.7b occurred at the

uppermost footprint horizon (Horizon 3) and was recorded in 1984. Similar to Figure 32.7a, with bulbous heel and hoof-like terminations to the toes, it had a more robust appearance, and was larger, with length–width dimensions of 495 mm by 368 mm. A plaster of paris copy of this important specimen is in the possession of the authors.

The track shown as Figure 32.7c occurred at the base of the Lee Ness Sandstone (Horizon 2) and was recorded in 1985. Another iguanodontid form similar to Figures 32.7a and 32.7b, but of a lighter build, it has a length of 422 mm and a width of 342 mm. It closely resembles a natural cast in the Bexhill Museum collection.

The many iguanodontid footprints seen at Fairlight have included width measurements ranging from 175 mm to 500 mm.

One footprint (Fig. 32.7d) showed a more angular appearance than the previous three forms, but is still considered to be that of an ornithopod. It occurred in the middle horizon (Horizon 2) and is the only specimen of its kind found to date, measuring 300 mm by 340 mm.

One track (Fig. 32.7e) was one of a series of four making up a trackway in the lowest footprint horizon (Horizon 1). They were similar to the footprints found in the Purbeck rocks of Dorset, usually attributed to *Megalosaurus*. The terminations at the toes indicated the presence of claws, while the sharply pointed heel suggested the possibility of the first toe as a backward pointing spur, typical of many theropods. These positive impressions measured 375 mm long by 400 mm wide, with a pace of 1.0 m along a compass heading of 30°. These are the only imprints recorded from this site.

A natural cast (Fig. 32.7f) of a large theropod footprint measured 445 mm by 420 mm, but, as the specimen was incomplete, the true dimensions would have been slightly greater. The heel lacked the bulbous shape of the iguanodontids and had a flat and wrinkled appearance.

The track illustrated as Figure 32.7g is also attributed to a large theropod; all three toes possessed claws and the presence of a heel claw is also indicated. The footprint cast measured 450 mm by 410 mm.

A rarely discovered 'bird-like' theropod form (Fig. 32.7h) is one of only three specimens of this type seen since 1984. The casts of this type are very shallow, appearing to be multi-segmented. The length–width dimensions were 280 mm by 240 mm.

Another 'bird-like' theropod footprint cast (Fig. 32.7j) is also very shallow. Although the definition is generally poor, the claw on the center toe appears to be very long and must have been strongly curved to produce the gap between the end of the toe and the first sign of the claw. The overall size was 360 mm by 264 mm; it displays similarities to a natural cast in the Museum of the Isle of Wight geology collection.

Figure 32.7k remains unclassified, being a shallow cast somewhat lacking in detail. It could possibly have been cast from a manus impression or from a small five-toed footprint, or even from two small tridactyl footprints overlapping each other.

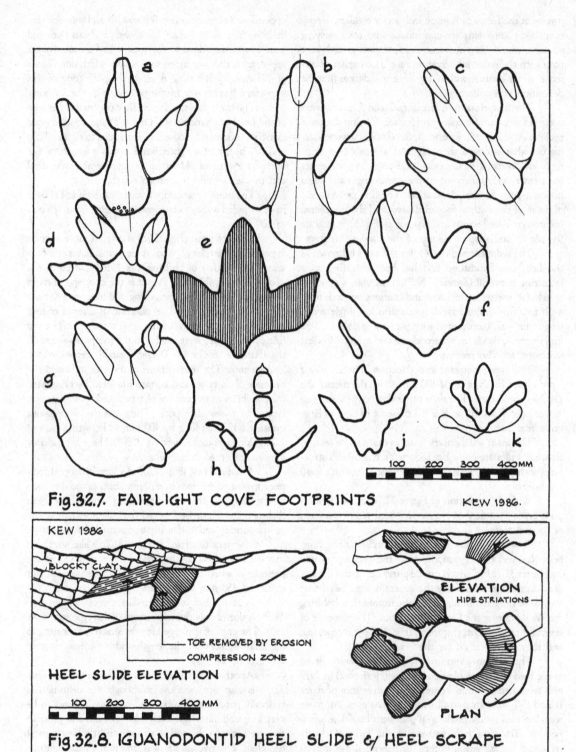

Fig. 32.7. FAIRLIGHT COVE FOOTPRINTS

KEW 1986.

KEW 1986

BLOCKY CLAY

TOE REMOVED BY EROSION

COMPRESSION ZONE

HEEL SLIDE ELEVATION

100   200   300   400 MM

ELEVATION

HIDE STRIATIONS

PLAN

Fig. 32.8. IGUANODONTID HEEL SLIDE & HEEL SCRAPE

Figure 32.7. Fairlight Cove footprints. Figure 32.8. Iguanodontid heel slide and heel scrape.

The tracks illustrated as figures 32.7f–k all occurred at the base of the Lee Ness sandstone (Horizon 2).

Observations

The construction of sea defenses to the west of Bexhill makes it unlikely that any further footprints or trackways will be reported from the Cooden area. To the east of Bexhill, near Galley Hill, there is still the possibility of further discoveries, but the principal footprint site remaining on the Sussex coast is Fairlight Cove.

Many differing footprint configurations have been observed at Fairlight, indicating the presence of a greater diversity of species or groups than are suggested by the scant skeletal record.

Footprint horizons are much more common throughout the Ashdown beds and the Tunbridge Wells Sand than was originally suspected (Allen 1975). Although research needs to be completed before this can be elaborated on, it is unquestionable that they make valuable paleoenvironmental indicators of shallow–water facies undergoing periods of emergence.

The footprints at Fairlight tend to be oriented either in a northerly or a southerly direction, suggesting the presence of a natural barrier, or that the dinosaurs followed regular routes to feeding grounds or water holes. Fossil soil horizons with *Equisitites* are fairly common; these could possibly have been used as grazing areas.

Footprints preserved as casts often retain more detail than imprints. If imprints are to be preserved at all, they generally have to be made in sediments retaining some porosity, allowing a degree of diagenesis to take place, but the coarser grained sediments are not the perfect material for recording fine detail. Footprints that have been made in a fine–grained sediment retain the best detail, but the low porosity and lack of subsequent diagenetic hardening leads to a footprint of low mechanical strength, e.g., the silty clay at the base of the Lee Ness Sandstone (Fig. 32.6). This material formed the edge of an evaporating lagoon when imprinted and was of the right consistency to retain detail. The infill material needs to be slightly coarser to allow some degree of diagenesis and the resulting footprint casts attain a higher mechanical strength than the original imprint, while retaining most of the detail.

This ideal situation occurs occasionally at Fairlight and allows excellent preservation of detail that has emphasized that the simple subdivision of footprint types into either "iguanodontid" or "megalosaurid" is no longer either appropriate or sufficient.

Acknowledgments

The authors have had discussions and exchanged information with many individuals, but are particularly indebted to the following: Justin Delair of Oxford; Paul Ensom, Dorset County Museum; Stephen Hutt, Museum of Isle of Wight Geology; Dr. Martin Lockley, University of Colorado at Denver; Dr. Tony Thulborn, University of Queensland; and the late H. J. Sargeant, Curator Bexhill Museum for information on earlier reports of footprints from the Hastings/Bexhill areas.

We would also thank Jenny Brandon for her assistance in typing the original manuscript.

References

Allen, P. 1975. Wealden of the Weald: a new model. *Proc. Geologists' Assoc.* 86 (4):389–437.

Beckles, S. H. 1851. On the supposed casts of footprints in the Wealden. *Quart. Jour. Geol. Soc. London* VII:117.
1854. On the Ornithoidichnites of the Wealden. *Quart. Jour. Geol. Soc. London* 10:456–464.

Calkin, J. B. 1968. *Ancient Purbeck. An account of the geology of the Isle of Purbeck and its early inhabitants.* (Dorchester: The Friary Press)

Delair, J. B., and Sarjeant, W. A. S. 1985. History and bibliography of the study of fossil vertebrate footprints in the British Isles: Suppl. 1973–1983. *Palaeogeog., Palaeoclimat., Palaeoecol.* 49:123–160.

Hooley, R. W. 1917. On the integument of *Iguanodon bernissartensis*, Boulenger and *Morosaurus becklesii*, Mantell. *Geological Mag.* 148–150.

Lake, R. D. 1975. The stratigraphy of the Cooden Borehole near Bexhill, Sussex. *Report Inst. Geological Sciences* 75/12. 23 pp.

Marsh, O. C. 1889. Comparison of the principle forms of Dinosauria of Europe and America. *Geological Mag.* 204–210.

Sarjeant, W. A. S. 1974. History and bibliogrpahy of the study of fossil vertebrate footprints in the British Isles. *Palaeogeog., Palaeoclimat., Palaeoecol.* 16:265–378.

Stewart, D. J. 1981. A field guide to the Wealden Group of the Hastings area and the Isle of Wight. In Elliot, T. (ed.). *Field Guides to the Modern and Ancient Fluvial Systems in Britain and Spain. Proc. 2nd Internat. Conf. on Fluvial Sediments Dept. Geology Keele Univ.* 3.1–3.32.

Tagart, E. 1846. On markings in the Hastings sands near Hastings, supposed to be the footprints of birds. *Quart. Jour. Geol. Soc. London* 2:267.

Ticehurst, N. F. 1928. Iguanodon footprints at Bulverhythe. *Hastings and East Sussex Naturalist* 4 (2):15–19.

Woodhams, K. E. in press. On the trail of Iguanodon – Samuel Beckles of St. Leonards. *Proc. Croydon Nat. Hist. Scientific Society Limited.*

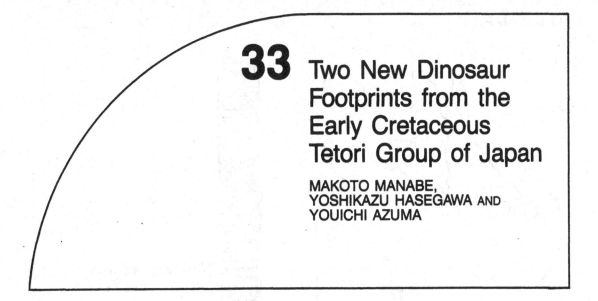

33 Two New Dinosaur Footprints from the Early Cretaceous Tetori Group of Japan

MAKOTO MANABE,
YOSHIKAZU HASEGAWA AND
YOUICHI AZUMA

Abstract

Two isolated footprints and a single dinosaur tooth are reported from the Early Cretaceous Tetori Group, Ishikawa Prefecture, Japan (Fig. 33.1). These are as yet the oldest dinosaur remains from Japan and represent the first dinosaurs known from the northwestern edge of the Japanese Archipelago.

Introduction

Dinosaurs are well known from the Asian continent, yet only recently have such remains been reported from Japan. These Japanese dinosaurs comprise only a few fragmentary bones, teeth, and some apparent trackways. The first report[1] refers to the bones of *Edmontosaurus* from a Cretaceous coal mine on Takashima Island, Nagasaki Prefecture, reported at the meeting of the Palaeontological Society of Japan (Takai 1962), although these remains have not been described and their validity has not been confirmed. A part of a humerus, believed to be that of *Mamenchisaurus*, is known from apparent marine sediments of the Miyako Group (Lower Cretaceous) of Iwate Prefecture (Hasegawa et al. 1982), and a tooth of a medium-sized carnosaur was discovered in the Mifune Group (lower Upper Cretaceous) of Kumamoto Prefecture (Hasegawa and Murata 1984). A sacral vertebra, ascribed to *Gallimimus*, was reported from a black shale representing brackish water conditions in the Sebayashi Formation (Lower Cretaceous) of Gunma Prefecture by Hasegawa et al. (1984). Possible trackways of dinosaurs found near the locality of *Gallimimus* in the Sebayashi Formation were reported by Matsukawa and Obata (1985 and this volume). There is, however, no general consensus as to their proper identification because of the irregularity and indistinctness of the preserved tracks.

In 1981, a group from The Fukuka Prefectural Museum led by Youichi Azuma discovered fossil crocodilian remains in the Cretaceous section of the lower part of the Tetori Group of Katsuyama, Fukui Prefecture. Since that time they have been surveying this area with the hope of discovering the remains of dinosaurs. Subsequently, a tooth of a small-sized carnivorous dinosaur (FPM85050-1) was found in Shiramine, Ishikawa Prefecture, by high school students, and forwarded to Azuma for study. This find led to the discovery at the same locality of two dinosaur footprints of two distinct kinds (FPM 86031-1 and 87003-1) in 1985 (Figs. 33.1, 33.3, 33.5).

Geological Setting

The tracks and the tooth were found in the Kuwajima Formation of the Itoshiro Subgroup. This is an alternating sequence of sandstone and shale units of Berriasian age (Fig. 33.2). These sediments represent a series of environmental changes from brackish to non-marine conditions from the bottom to the top of the subgroup, as evidenced by the occurrence of *Viviparus ogamigoensis* (Gastropoda) and *Unio* sp. (Bivalvia) (Maeda 1961).

Descriptions

FPM85050-1: The tooth[2]

The single isolated tooth (Fig. 33.4) can be identified no more precisely than Carnosauria fam. indet. It is 19.5 mm in maximum length, 10.6 mm in maximum width and 5.6 mm in maximum thickness. It is characterized by a convex labial surface, a flat lingual surface, and an absence of serrations along its mesial edge. The upper third of the distal edge of the tooth exhibits at least 18 small, rounded, pillar shaped denticles. The tooth resembles that referred to Carnosauria from Lufeng, Yunnan, China (NSMP17178-17180).

FPM86031-1: Footprint[3]

This track (Fig. 33.5B) was found on a talus block at the tooth (FPM85050-1) locality. It is an isolated, tridactyl footprint which is preserved in relief as a natural cast of the original print. The maximum width is 134 mm, and the maximum length 144 mm. It is similar to No. 13 of the 23 types of footprints reported from the Cretaceous

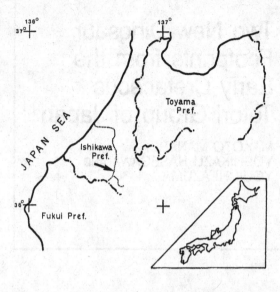

Figure 33.1. Map of the locality, Ishikawa Prefecture, Japan.

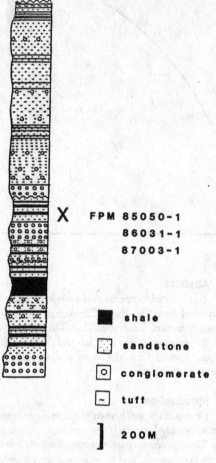

X FPM 85050-1
86031-1
87003-1

■ shale
sandstone
○ conglomerate
~ tuff

] 200M

Figure 33.2. Geologic column of the Tetori Group. After Maeda (1961).

Figure 33.3. Dinosaur remains from Japan.

	Kyusyu	Honsyu	Sakhalin
Upper Cretaceous	Edmontosaurus 7 Megalosaurid 8		Nipponosaurus 1
Lower Cretaceous		Mamenchisaurus 2 Gallimimus 3 Footprints 4 Carnosauria 5 Footprints 6	

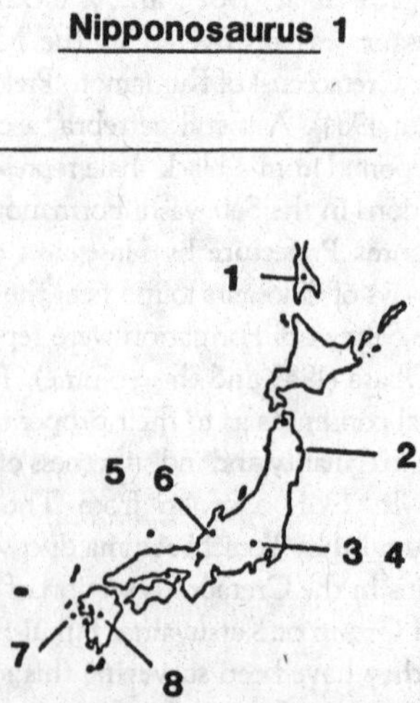

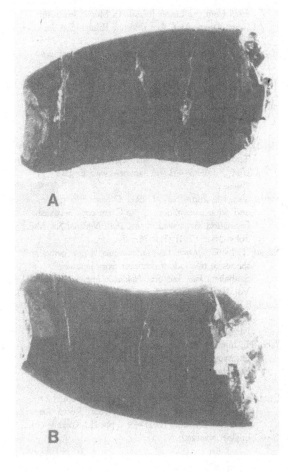

Figure 33.4. Carnosaurian tooth (FPM85050–1). **A,** FPM85050–1, lingual side, length 19.5 mm. **B,** FPM85050–1, labial side, length 19.5 mm.

Figure 33.5. Tridactyl footprints. **A,** FPM86031–1, length 144 mm. **B,** FPM87003–1, length 370+ mm.

Gyeongsang Group of Korea by Yang (1982) and is probably of iguanodontid affinity (cf. Woodhams and Hines, Fig. 32.7 this volume).

FPM87003–1: Footprint[4]

This print was discovered on the lower surface of a sandstone bed located approximately 12 m above the base of the cliff-forming locality. It is approximately the same stratigraphic level as the tooth. The model was made on the bed in September, 1986, and the specimen was cut off from the bed in December, 1986. It is also an isolated, tridactyl relief cast (Fig. 33.5B). The maximum length is 370 mm+, and maximum width is 250 mm. The track is similar in shape but not in size to *Jeholosauripus s–satoi* (Yabe et al. 1940) from the Early Cretaceous of northern China. The angle between its digits is, however, narrower than that of *Jeholosauripus s–satoi*. It is also similar in shape to No. 5 and No. 15 of the 23 types reported from Korea by Yang (1982). There is a deformed fossil leaf on the surface of the track.

Conclusion

The described footprints represent two different species, and there is no pattern or relationship between the two. However, we believe that there is no doubt that these are dinosaur footprints because of their distinct morphology. Unlike the poorly preserved footprint impressions in the Sebayashi Formation, the Kuwajima tracks are preserved as natural casts with clear outlines. The described tracks and teeth from the Itoshiro Subgroup (Berriasian) are as yet the oldest dinosaur remains from Japan. In addition, they are the first reported from the northwestern coast of the Japanese Archipelago. Thus they are relevant to the discussions of the geological and paleontological relationship between China, Korea, and Japan.

There are not many, but some reliable dinosaur fossils of the Cretaceous in Japan (Fig. 33.3). With continued exploration, it is expected that more and better dinosaur tracks and skeletal remains will be found, and that future work will contribute to a greater understanding of poorly known paleogeographical relationships between the Asian Continent and Japan.

Acknowledgments

Several individuals have made possible and facilitated this study. The authors gratefully acknowledge their aid. Our thanks especially go to Miss Aki Matsuda and Mr. Hironobu Matsuda of Sabae, Fukui Prefecture, who found the carnosaur tooth, to Mr. Kenichi Takeyama of Katsuyama High School, Fukui, Fukui Prefecture, and to Mr. Tetsuji Araki of The Fukui Prefectural Institute of Education, who discovered the footprints, to Dr. Philip Currie of Tyrrell Museum of Palaeontology, Alberta, Canada, who gave us useful information and comments on the tooth, and to Dr. Glenn Storrs of The Yale Peabody Museum of Natural History, Connecticut, USA, who reviewed this manuscript. The Fukui Prefectural Museum and Shiramine Village of Ishikawa Prefecture supported this research in every aspect.

Abbreviations

FPM = Fukui Prefectural Museum, Fukui, Japan.
NSMP = National Science Museum, Tokyo, Japan.

Notes

[1]The first report of a Japanese dinosaur sometimes refers to *Nipponosaurus sachalinensis* (Nagao 1936) from Sakhalien which was formerly a Japanese territory.

[2-4]It is traditional in Japan that a significant fossil specimen is given a "Japanese name", a nickname to the general public. FPM85050-1 is named "Kaga-ryu", FPM86031-1 is named "Shiramine-ryu", and FPM87003-1 is named "Kuwajima-ryu".

References

Hasegawa, Y., Hanai, T., and Kase, T. 1982. The vertebrate fossil from the Lower Jurassic in Moshi, Iwaizumi, Iwate, Japan. *Abst. Ann. Meeting Paleont. Soc. Japan* Chiba University. [in Japanese]

Hasegawa, Y., and Murata, M. 1984. First record of carnivorous dinosaur from the Upper Cretaceous of Kyusyu, Japan. *Abst. Ann. Meeting Paleont. Soc. Japan* University of Kyoto.

Hasegawa, Y., Kase, T., and Nakajima, S. 1984. Megavertebrate fossil from the Sanchu Graben. *Abst. 91st Meeting Geol. Soc. Japan.* 219 pp. [in Japanese]

Maeda, S. 1961. On the geological history of the Mesozoic Tetori Group in Japan. *Jour. Coll. Arts Sci., Chiba Univ.* 3 (3):369–442. [in Japanese with English abstract]

Matsukawa, M., and Obata, I. 1985. Dinosaur footprints and other indentation in the Cretaceous Sebayashi Formation, Sebayashi, Japan. *Bull. National Sci. Mus., Tokyo,* Ser. C. 11 (1):9–36.

Nagao, T. 1936. *Nipponosaurus sachalinensis,* a new genus and species of trachodont dinosaur from Japanese Sakhalien. *Jour. Fac. Sci., Hokkaido Imp. Univ.* Ser. IV, Vol. III, No. 2:185–220.

Takai, F. 1962. The hadrosaurian dinosaur of Takashima, Kyusyu, Japan. *Ann. Meeting Paleont. Soc. Japan* Hiroshima University. [in Japanese]

Yabe, H., Inai, Y., and Shikama, T. 1940. Discovery of dinosaurian footprints from the Cretaceous(?) of Yangshan, Chinchou. *Preliminary Note. Proc. Imp. Acad.* Tokyo 16 (10):560–563.

Yang, S.Y. 1982. On the dinosaur's footprints from the Upper Cretaceous Gyeongsang Group, Korea. *Jour. Geol. Soc. Korea* 18 (1):37–48, 2 pls. [in Korean with English abstract]

34 The Briar Site: A New Sauropod Dinosaur Tracksite in Lower Cretaceous Beds of Arkansas, USA

JEFFREY G. PITTMAN AND DAVID D. GILLETTE

Abstract

Thousands of sauropod dinosaur tracks atop a thin limestone bed within the Lower Cretaceous De Queen Formation were discovered in 1983 at Briar Plant Quarry, 20 km north of Nashville, Howard County, Arkansas. Sauropod tracks were observed on two beds within the De Queen Formation; exposure of the upper trackbed is much more extensive than that of the lower bed. The trackways were photographed and some were cast and mapped before the area was destroyed by blasting.

No tracks of juvenile sauropods have been observed at the site. The average stride length of a typical trackway is approximately 2.5 m; rear paces average 1.65 m; and front paces, 1.45 m. Quality and depth of tracks vary from shallow, indistinct U–shaped depressions to deeper tracks exhibiting pad impressions and claw marks. Manus prints are broadly U–shaped, with distinct pad impressions separated by indentations from the sidewall and posterior margin of the track. No claw marks are present in manus prints. Pes tracks are approximately twice as large as manus prints, U–shaped with apex forward, and typically exhibit a laterally directed claw mark of digit I and a posterolaterally directed claw mark of digit II. Claws of digits III and IV were usually oriented in a posterior direction, as suggested by track outlines. In several deeper tracks the claw marks of the digits point in a more anterior direction.

Thousands of tracks densely cover the upper bed in the quarry. Aerial photographs reveal an alignment of hundreds of tracks, recording the movement of a group of at least about 10 sauropods across the lime mud surface in one area. A number of trackways extend in other directions, which, together with the fact that track depth and preservation are variable, indicates the tracks were made over a period of time, not on a single occasion.

Both trackbeds were deposited very nearshore and possess features indicating variable depositional conditions ranging from shallow subtidal to supratidal. They are equivalent to strata of the Glen Rose Formation of Texas
wherein eight other sauropod tracksites are known, in addition to numerous theropod sites and several ornithopod tracksites.

Introduction

The tracks were discovered in November, 1983, on a limestone ledge in Briar Plant Quarry, Howard County, Arkansas (Fig. 34.1), where gypsum is mined from lower levels. In the quarrying operation, overburden above the trackbed is removed by heavy equipment. Because the trackbed is too thick (50 cm) to be removed by this process, it and other beds lying above the gypsum beds are removed by blasting. Thus, a large area of the upper surface of the trackbed is exposed for a variable period of time atop a five meter ledge. The area where the tracks were

Figure 34.1. Outcrop of the De Queen Formation, southwestern Arkansas. Tracks occur at Briar Plant Quarry, Howard County.

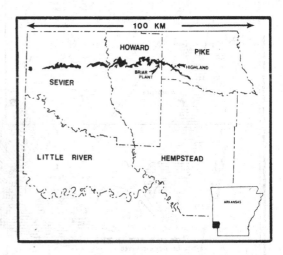

first identified had lain idle for an extended period of time as a parking area for excavation equipment.

The "potholes" had been a nuisance to heavy equipment operators in the quarry for years. To avoid wear and tear on the equipment it was common practice to leave a layer of clay on the surface of the trackbed. Operators recalled hitting the holes on the upper trackbed in all mined out areas of the quarry, which covers approximately 25,000 square meters (Fig. 34.2).

After the discovery, the quarry operator delayed blasting operations where we were working to allow excavation and mapping. However, we were able to uncover only a fraction of the total area exposed at the time. Wet and cold weather hampered our work. Tracks so densely covered the surface of the upper trackbed that trackways of individual dinosaurs were obscured over most of the area. Fortunately, one of the first trackways excavated (herein referred to as #1) could be traced uninterrupted for about 50 m. A map of this trackway was made and a section of it was cast (replicas are on display at the New Mexico Museum of Natural History and on the courthouse lawn in Nashville, Arkansas). Another trackway (#2), oriented at a right angle to #1 and crossing it, was mapped for over 50 m. A number of other trackways were followed for shorter distances.

Sauropod tracks also occur on a bed two meters below the main trackbed. These tracks were observed for a short period of time in a small area. A 15 m section of one trackway (#B1, Figs. 34.12, 34.16) was mapped. The upper bedding surface of this lower stratum is rarely exposed.

Aerial photographs were critical to the study. These photographs reveal the alignment of hundreds of tracks in the area adjacent to trackways #1 and #2, recording the movement of a group of at least 10 sauropods across the area. Aerial photographs also confirm the ocurrence of tracks in other parts of the quarry. Tracks may be seen on these photographs in areas which we visited on foot, but had no time to do extensive work.

This tracksite is the easternmost locality in a suite of Cretaceous tracksites extending across Texas and (now) into Arkansas (Pittman this volume, Fig. 15.1). This site is at the easternmost extent of outcropping Lower Cretaceous strata in the Gulf Coastal Plain; a short distance eastward from the site, Lower Cretaceous strata are covered by Upper Cretaceous and younger sediments. Faunal and floral information from the outcrop and subsurface indicates an Early Albian Age for the track–bearing beds.

Figure 34.2. Map of Briar Plant Quarry. Most of our work was done within the area bracketed by the corner marks. These corner marks denote the boundaries of Figure 34.4.

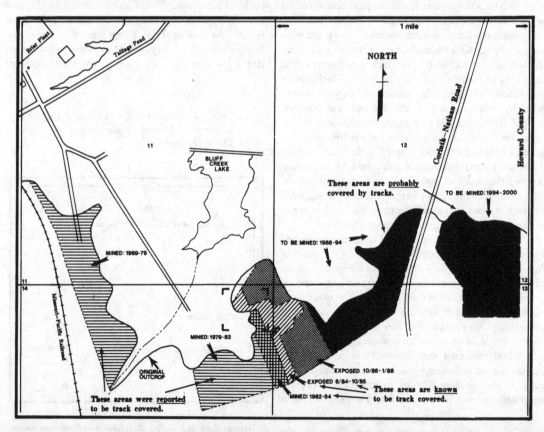

Locality and Geologic Setting

The tracks occur in Weyerhaeuser Company's Briar Plant Quarry, 20 km north of Nashville, Howard County, Arkansas, SE/4, Sec. 14, T&S, R27W, 34°03'57" N latitude, 93°50'43" W longitude. The unit exposed in the quarry is the De Queen Formation, named for a limestone and fine-grained clastic sequence exposed in the townsite of De Queen, Arkansas, near the Oklahoma state line.

About 30 m of strata may be observed in the highest quarry wall. The lower half of the section consists of interbedded claystone, siltstone, limestone, and gypsum beds. The upper half of the section, where the track–bearing beds occur, is composed of interbedded thin clastic, carbonate, and very thin evaporite beds (Fig. 34.3).

The Lower Albian age of the section exposed at the outcrop in southwestern Arkansas is indicated by the occur-

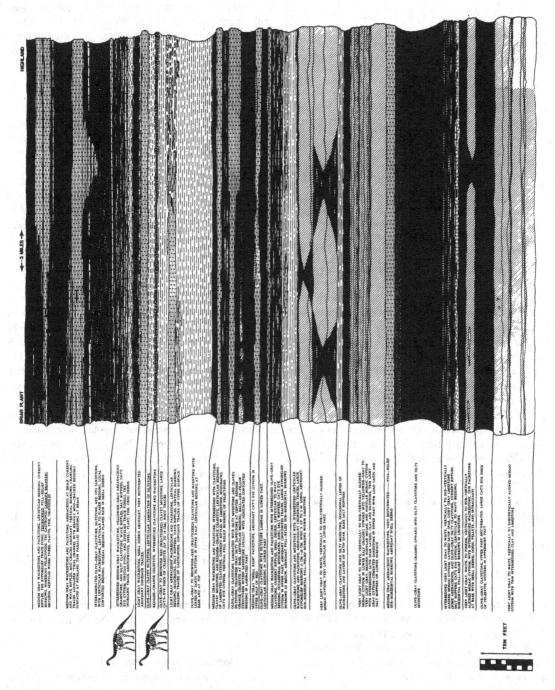

Figure 34.3. Composite measured section from Briar Plant Quarry to Highland Quarry, Pike County. Dinosaur symbols indicate levels of trackbeds. (from Pittman 1985)

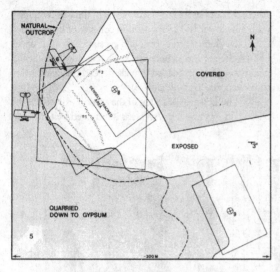

Figure 34.4. Map of the area within the corner brackets in Figure 34.2. The boundaries of aerial photographs (Figs. 34.6–40.9) are indicated. Figures 34.6 and 34.7 were taken from an oblique angle; airplane symbols show position of the photographer. Figures 34.8 and 34.9 were taken from directly above. The arrow in the lower left corner indicates the direction at which the photograph in Figure 34.5 was taken. The black spot is a hole where a pes–manus set was excavated. This hole is visible on aerial photographs shown in Figures 34.6, 34.7, and 34.8.

rence of the ammonite *Douveilliceras* sp. within the lower gypsum sequence (Pittman 1985), by the presence of the foraminifer *Orbitolina texana* in subsurface equivalents of the units (Loucks and Longman 1982), and by the sub-surface occurrence of Albian ammonites above and below the equivalent of widespread gypsum beds in the lower half of the section (Young 1974).

The tracks were made very near the shoreline of the Gulf of Mexico during Early Albian time (see Pittman this volume, Fig. 15.4). This is indicated by the facies relationships, sedimentary structures, and trace fossils in the beds. In addition, the present day foothills of the Ouachita Mountains of Oklahoma and Arkansas are located only a few kilometers to the north of the tracksite. Gently-dipping Lower Cretaceous rocks in this area rest atop steeply–dipping Carboniferous strata which in some areas stood up as promontories and islands in the Early Cretaceous sea (Honess 1923, Miser 1927).

Methods

The tracks were filled with olive gray claystone. Most work to remove this clay was done with broom and shovel, although a portable air compressor was useful for cleaning the surface before casting. A heavy pump truck and a fire truck were used with success on two occasions to wash the clay from the tracks. However, trips through the muddy quarry with a full tank load of water proved to be impracticable.

Once the surface of the bed was clean, the tracks

in a given trackway were mapped with the aid of a grid fashioned with lumber and string. This grid (3 x 5 m) could be dragged from mapped to unmapped areas easily and was kept in line with north–south and east–west base lines. The outline of the tracks was traced onto graph paper. This outline was determined visually in a consistent manner. Thus, the track outline may be regarded as a contour line at the level flush with the anterior rim of each track.

A 15 m section of trackway #1 was cast in silicon rubber and fiberglass. The rubber was applied to the rock surface within each track and on a strip between each track. Fiberglass was applied over the rubber and across the trackway to about 30 cm beyond each side of the trackway. There was some difficulty in removing the rubber and fiberglass from the rock surface, but only minor repair was needed. Two duplicates were made from the cast, one for the New Mexico Museum of Natural History, the other for the city of Nashville, Arkansas. A cast made in only fiberglass with lightweight oil used as a separator proved to be of sufficient quality.

Several generous private citizens were able to remove isolated tracks and several track pairs. Much work had to be done with jackhammers and heavy equipment before the 50 cm bed was cut through. One left set from trackway #1 is on display with the cast in the town of Nashville and a single manus print is on display at the Mid–America Museum, Hot Springs, Arkansas.

Aerial photographs were made of the tracked surface from a lightweight airplane. A number of different cameras and film sizes were used. Black and white 35 mm, 125 ASA, provided good results with a 28 mm lens. The photographs were made of the tracked surface during early morning and late evening, the contrast highlighted by low angle sunlight. Tracks which had been cleaned of clay show up well in these photographs. Over most of the area the clay was not removed from the tracks, but only skimmed off the surface with a road grader. Fortunately there was sufficient contrast between the clay–filled tracks and the bare rock surface between tracks to make the tracks visible on photographs. The best photographs were made of the area a few days after heavy rainfall. When the surface was very wet or very dry, there was little contrast between the tracks and the intertrack surface. Since the area was destroyed by blasting, the aerial photographs are now the only record we have of most tracks and trackways at the site. It is therefore important that we include several of the best photographs in this report. In Figure 34.4 boundaries of the aerial photographs (Figs. 34.6–34.9) are shown on a schematic map of the area which we studied. The exposure of the upper trackbed and several gypsum beds may be seen in a view of this area from the southwest (Fig. 34.5).

Depositional Environments

Depositional environments of beds within the section exposed at the quarry were varied. Gypsum beds of the lower half of the section were precipitated from hyper-

Figure 34.5. Aerial photograph of Briar Plant Quarry taken from the west. The black arrow to the upper left of center indicates the position of the air compressor visible in the center of Figures 34.6 and 34.7.

saline waters of an extensive lagoonal sea that stretched from central Texas, across southern Arkansas, Louisiana and Mississippi, to southern Florida. Individual gypsum beds within this outcropping sequence have been traced into the subsurface (Ferry Lake Anhydrite) where they may be correlated for hundreds of kilometers (Pittman 1985). Much of the lower half of the section, including the gypsum beds, was deposited in a shallow–subaqueous setting. However, some beds were deposited in very shallow water and subaerial exposure occurred periodically.

Water depth and salinity during the deposition of the upper half of the section were highly variable. Several of the limestone beds are probably the product of subtidal deposition. Several of these limestone beds consist entirely of shell debris, perhaps deposited by storm waves. Intertidal conditions are indicated by lenticular lamination of claystone, mudstone, siltstone, sandstone and oscillation ripples. Supratidal and exposed conditions are indicated by mudcracks, vertical burrows, grazing traces, displacive evaporites, and root traces. Depositional environments within a single bed varied considerably, as water depth and salinity fluctuations were rapid and numerous (Pittman 1984).

Throughout most of the section the invertebrate assemblage is largely made up of only two species of pelecypods — *Corbicula arkansaensis* Hill, 1888 and *Anomia texana* Hill 1893. These two species must have been tolerant of the wide range of salinity occurring during deposition of these beds. Only in the uppermost beds of the section does the invertebrate assemblage attain a diversity indicating more normal marine conditions. Present in these upper beds are abundant serpulid worm tubes, numerous species of pelecypods, bryozoa, echinoids, and the abundant gastropod *Cassiope branneri* (Hill), Stanton 1947.

In addition to the tracks of dinosaurs, vertebrates are represented within the section by isolated bones of a small crocodile, turtles, pycnodontid fish, and rare shark teeth.

Depositional Conditions of Trackmaking

Our interpretations of the depositional environment of individual sedimentary beds are often oversimplified, especially of those beds deposited in nearshore areas which are subject to frequent rise and fall of water level and saturation. Notions of supratidal/intertidal/subtidal facies tracts promote "either ... or" interpretations; individual sedimentary beds are usually given such descriptions as shallow subtidal, high intertidal, or supratidal, whereas the sediments of the beds probably accumulated under more varied conditions. In rocks at the Arkansas site, mud cracks and other evidence of exposure may be found on a number of bedding planes within very thin sequences of strata which were probably deposited in a subaqueous setting. Within a given bed, such as the 50 cm thick upper trackbed, texture and components of the rock vary, reflecting the different conditions under which it was deposited. Subaerial exposure was a frequent occurrence during deposition of strata exposed in the quarry walls.

When attempting a reconstruction of the environmental setting at the time a given set of tracks were made,

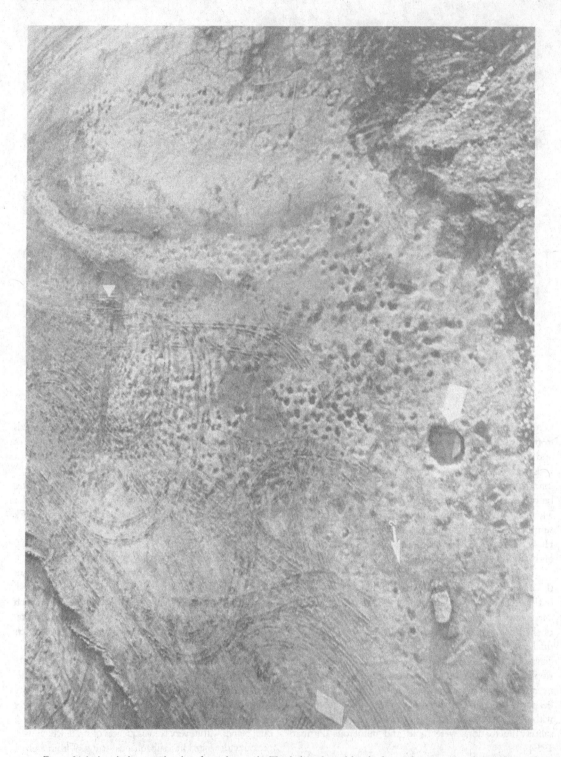

Figure 34.6. Aerial photograph taken from the north. The hole indicated by the large white arrow is also visible on the aerial photographs in Figures 34.7 and 34.8. Trackway #1 occupies the cleaned off area just right of center (the direction of travel is away from the viewer). A "1" marks stride position 1 on this trackway (see text and Fig. 34.16 and 34.17). Several other trackways can be seen near the right edge of the photograph, also leading away from the viewer. The 3 x 5 m grid used in mapping lies at right center atop the beginning of one of these trackways. The triangle just above center points to the air compressor visible on aerial photographs in Figures 34.5 and 34.7. Hundreds of tracks in trackways leading away from the viewer stretch from the foreground and run left of trackway #1. Several tracks at the beginning of trackway #2 are visible just left of the arrow labeled "2".

Figure 34.7. Aerial photograph taken from the northwest (taken the same day as the photograph in Fig. 34.6). White bar in foreground is about 10 m in length. "Fat" arrow marks the hole which is also visible on aerial photographs in Figures 34.6 and 34.8. The 3 x 5 m grid lies just below center. As in Figure 34.6, the triangle points to the air compressor visible in Figures 34.5 and 00.6 and the arrow labeled "2" marks the beginning of trackway #2. The dense area of tracks lies to the upper left of trackway #1. The fracture pattern on the trackbed in the center foreground was caused by weathering (this area lies at the natural outcrop).

one must acknowledge also the following complications: ghost prints, which may reflect a "real" track above or below the particular horizon; modification by current action or bioturbation; and substrate consistency, which is not only controlled by water saturation but also by grain size and components. These factors affect our judgment of timing. In addition, we should not attempt an environmental reconstruction without consideration of aspects of adjacent beds.

Ghost Prints

The clay lying atop the upper trackbed was observed in cross–section within several tracks. The laminations in this claystone followed the gentle contour of the upper surface of the limestone bed within the tracks, so that it is likely that these particular tracks, at least, were made on the uppermost surface of the limestone bed. Many of the poorly preserved footprints on the upper trackbed may be considered overprints — these tracks were apparently made before the well–preserved prints, before deposition of the uppermost sediments of the bed. Several thin clay laminations occur in the upper half of the trackbed. Only the uppermost surface of the limestone bed was exposed by quarrying operations. However, during the excavation of one pes–manus set, the block of rock separated along a

natural fracture, revealing gentle deformation of the laminae within the bed to a depth of about 20 cm (Fig. 34.10). If one of these bedding planes were exposed (which might have occurred if the exposure were natural), broad, almost featureless depressions would be observed, such as the sauropod prints on the lower trackbed (Figs. 34.11, 34.12) and at several sites in Texas.

Substrate Consistency

Track depth varies across the expansive exposures of the upper trackbed. Even in single trackways, a transition of deepening or shallowing could be observed (Fig. 34.13). An individual sauropod walking across the area encountered both firm and soft substrates. But not all variation in track depth is attributed to "spotty" wet and dry areas; the variation in preservation of the very large number of tracks on the upper trackbed indicates that there were probably several cycles of wetting and drying during the tracking of the surface.

Bioturbation

Small full–relief burrows a few millimeters in diameter were observed in the upper trackbed, both within and outside tracks. Some bioturbation of this type may have continued after some tracks were made, destroying

Figure 34.8. Aerial photograph taken from directly above the area of dense tracks and Trackway #2. White bar at the top is about 10 m in length. The hole which is visible on aerial photographs in Figures 34.6 and 34.7 is labeled by the "fat" arrow in the lower left corner. By the time this photograph was taken, trackway #1 had been destroyed. Its position had been along the lower edge of this photograph, just west of the line of drill holes and the densely tracked area. The beginning of trackway #2 is marked by the arrow labeled "2" just to the right of this hole. Through careful observation a trackway (labeled "X") may be seen just to the left of and running parallel to trackway #2. This trackway follows closely the curvature of trackway #2. Perhaps these two trackways were made by sauropods walking in tandem. Other trackways and pes-manus sets may be observed on this photograph. The position of two faint trackways are labeled by the arrows at the upper right and top edge of the photograph.

surface detail and resulting in the featureless U–shaped depressions observed in a few areas. However, if bioturbation was extensive, the detailed structure that we observed should have been destroyed. No deformation of the tracks could be definitely attributed to bioturbation.

Depositional Setting of Trackbeds

The two track horizons at the Arkansas site are two meters apart, separated by a sequence of mudstone, carbonaceous claystone, a continuous thin celestite (strontium sulphate) bed, topped by another claystone bed (Fig. 34.3). In stratigraphic order, the following depositional environments are interpreted as being represented: sand tidal flat (lower trackbed, with dinosaur tracks on upper surface); marsh and exposed supratidal (clay with root casts); intertidal (laminated clay/silt); algal and evaporitic flat (carbonaceous clay and celestite &originally gypsum or anhydriteé); lower intertidal flat and shallow subtidal (laminated clay/silt); subtidal to supratidal (upper trackbed); subtidal to intertidal (clay); tidal to supratidal flat (alternating sands, silts, and clays).

The lower track–bearing bed, which thickens from another quarry to the east, is about 70 cm thick in Briar Plant Quarry and is composed of silty fine–grained sandstone, and arenaceous wackestone and packstone. Sedimentary structures and trace fossils include mudcracks, salt hoppers, wave ripples, lenticular bedding at the base and laterally to the east, soft–sediment deformation in some areas of the quarry, vertical burrows, and grazing traces of gastropods. Root traces occur in the reddish mudstone just above this bed.

The upper track–bearing bed, about 50 cm thick, is an arenaceous, pellet–rich, bioturbated, light gray wackestone, rich in shell debris. Very thin claystone is occasionally seen interlaminated in the bed. Allochems include abundant pellets, silt, and some miliolid foraminifera. Shell debris typically is most abundant in the middle of the bed. Wave ripples and mud cracks may sometimes be observed. Large scale· dessication cracks occur in some areas of the quarry in the top of the bed. As discussed above, tracks on this bed were probably made over a period of time, some perhaps by wading sauropods, others by sauropods walking across the exposed area when the substrate was firm and/or soft.

Figure 34.9. Aerial photograph taken from directly above an area south of the area in the photographs in Figures 34.6, 34.7 and 34.8 (see Fig. 34.4 for location). The bar at the top is about 10 m in length. As in the other aerial photographs, tracks appear as dark "spots" against the lighter colored rock (individual tracks are clay filled). This high density of tracks was observed in all areas of the quarry we visited and has been reported for all mined-out areas of the quarry by equipment operators.

Figure 34.10. Photograph of a block in the process of excavation from the hole common to aerial photographs in Figures 34.6, 34.7, and 34.8. A left pes print occurs in cross section (posterior view). The bed is about 50 cm thick. The upper lamination is obviously curved downward below the track, while the lower lamination (20 cm from the top) is only slightly deformed (see discussion of ghost prints in text). The pes print is deeper in its medial half (left). This is characteristic of Gulf Coastal Plain sauropod prints — compare to the photograph of a Paluxy River pes print in cross section in Farlow (1987, Fig. 16c).

Track Description

Sauropod tracks are known at nine tracksites in the Gulf Coastal Plain. See Pittman (this volume) for a discussion of the stratigraphy and lithology of trackbeds at these sites and other Texas sites, and Farlow et al. (this volume) for a discussion of sauropod prints from several Texas sites. In addition to the Arkansas site, well-preserved sauropod tracks occur in the bed of West Verde Creek, Medina County, and Paluxy River, Somervell County, Texas. Bird (1939, 1941, 1944, 1954, 1985) described these sites, as well as the "swimming sauropod" trackway at the Medina River Site, Bandera County, Texas. Sauropod tracks occur at four other sites in Texas: Blanco River, Miller Creek, Blanco County; South San Gabriel River, Williamson County; and Sidney, Comanche County. The size and form of sauropod tracks at the Arkansas site is consistent with that of tracks at the other sites. Sauropod tracks at all these sites were probably made by the same type of sauropod (tracks at the Briar Site are referred, together with Texas sauropod prints, to the ichnospecies *Brontopodus birdi* by Farlow et al. [this volume]). The varied structure observed may be attributed to different behavior of sauropods walking across varied substrates and to different depositional conditions and preservation.

Pes Prints

A typical pes track is broadly U-shaped with the

Figure 34.11. Photograph of Trackway #B1, atop the lower trackbed. These large, shallow tracks are probably ghost prints. The left manus print in the foreground (direction of travel toward viewer) illustrates that even in poorly preserved trackways an outward rotation of the symmetry of manus prints is observed.

Figure 34.12. Photograph of Trackway #B1 taken from atop a drill rig. The short sectiin of this trackway contains the only tracks observed on the lower trackbed.

apex forward (Figs. 34.14, 34.15, 34.17). Claw impressions of digits I, II, III, and IV variably occur in pes prints. Most common is a claw impression of digit I, usually laterally directed at about 90 degrees to the direction of travel (Fig. 34.14). Frequently, where a claw mark for digit I is preserved, a shallow impression of the claw of digit II occurs on the outside margin of the track, about 10 cm behind the digit I claw mark, parallel to the outside wall of the track. The digit II claw mark is usually oriented about 45° to the direction of travel. Only in the deepest tracks is there a well-preserved impression of the claw of digits III and IV. In one deep track, where claw marks of digits I, II, III, and IV are all preserved, the claw mark of digit I points more anteriorly (Figs. 34.16, 34.17).

The medial half of pes prints is consistently longer (because it is more deeply impressed) than the lateral half of the track. In several tracks in trackways #1 and #2 this form is pronounced (Fig. 34.17f,g,h). This is consistent with the observations by Farlow et al. (this volume) of sauropod tracks in Texas. In several pes prints along trackway #1 a shallow groove lay along the lateral edge of the track, beginning at the position of digit I (Fig. 34.17b). This groove

extends along the lateral edge of several tracks to the posterior margin of the track. In several of these tracks an indentation in the posterior margin occurs at the posterolateral corner of the track (Fig. 34.17i,j).

The arrows along the lateral margin of the pes prints shown in Figure 34.17a–e indicate the direction in which the claws were extended. Depth variation in tracks at the Arkansas site, as well as at West Verde Creek and Paluxy River, Texas, suggests that on firmer substrates the claws were "wrapped" around the lateral margin of the foot, and on softer substrates the claws were often extended forward, probably to provide better traction. The pes print at the Briar Site in which claws are more forwardly directed (Fig. 34.17e) was in a short trackway which we cast (TMM 43041-3). This short trackway was within the mass of tracks lying east of trackway #1 and south of trackway #2 (Fig. 34.6). In this area several trackways are superimposed. The pes print shown in Figure 34.17e was apparently made just as the trackmaker had walked into a muddy area. Perhaps this form may be attributed to an effort for traction, braking, or turning.

In tracks in which the claw marks are well-preserved

Table 34.1. *Trackway #1 data. Measurements in meters and degrees.*

Position	Strides					Paces		Angles	
	Right Rear	Left Rear	Right Front	Left Front	Stride Average	Rear	Front	Rear	Front
1	2.51		2.34		2.43	1.36	1.20	122	130.4
2		2.57		2.51	2.54	1.5	1.38	111.9	113.9
3	2.68		2.65		2.67	1.59	1.62	114.6	116.9
4		2.68		2.71	2.7	1.59	1.49	114.6	119.9
5	2.64		2.58		2.61	1.59	1.64	116.5	119.2
6		2.55		2.54	2.55	1.51	1.34	112.6	116.8
7	2.6		2.69		2.65	1.55	1.63	116.1	119.6
8		2.85		2.71	2.78	1.61	1.48	139.9	132.4
9	2.33		2.57		2.45	1.53	1.48	117.3	114.6
10		2.42		2.47	2.45	1.53	1.58	110.8	117.5
11	2.52		2.5		2.51	1.42	1.31	100.4	120.5
12		2.44	2.54		2.49	1.61	1.57	116.4	134.6
13	2.57		2.52		2.56	1.26	1.18	127.8	128.3
14		2.47		2.54	2.51	1.54	1.61	108.8	111.3
15	2.4		2.5		2.45	1.49	1.46	116.3	126.4
16		2.59		2.5	2.55	1.53	1.34	111.3	125.2
17	2.11		2.55		2.33	1.6	1.48	101.2	118.1
18		2.3		2.11	2.21	1.5	1.49	96.6	98.2
19	2.29		2.38		2.34	1.58	1.3	95.6	101.6
20		2.27		2.67	2.47	1.24	1.74	110.9	125.8
21	2.41		1.67		2.08	1.51	1.24	98.6	110.8
22		2.25		2.08	2.17	1.51	0.76	97.5	124
23	2.33		2.68		2.51	1.48	1.56	106.2	126.6
24		2.27		2.25	2.26	1.54	1.44	100.1	118.8
25	2.61		2.26		2.44	1.42	1.18	104.7	124.7
26		2.37		2.44	2.41	1.51	1.37	107.2	124.9
27	2.29		2.15		2.22	1.43	1.39	107.7	91.9
28		2.65		2.5	2.58	1.8	1.6	111.5	119.5
29	2.36		2.31		2.34	1.39	1.33	112.4	111.2
30		2.32		2.34	2.33	1.37	1.47	122.5	130.9
31	2.15		2.39		2.27	1.27	1.11	108.6	113.7
32		2.28		2.6	2.44	1.62	1.72	102.8	121.5
33	2.37		2.17		2.27	1.29	1.25	99.5	130
34		2.33		2.28	2.31	1.53	1.14	111.1	128.3
35	2.37		2.57			1.29	1.39	109.5	128.3
36		2.29		2.23	2.26	1.6	1.58	95.8	119.3
37						1.48	1.25	95.8	102.8
Minimum:	2.11	2.25	1.67	2.08		1.24	0.76	95.6	91.9
Maximum:	2.68	2.85	2.69	2.72		1.8	1.74	139.9	134.6
Average:	2.43	2.44	2.42	2.45		1.49	1.41	109.9	119.1
Variance:	0.027	0.028	0.058	0.034		0.014	0.038	07.64	91.63
Std. Dev.:	0.164	0.169	0.241	0.185		0.116	0.196	9.36	9.57

Table 34.2. Trackway #2 data. Measurements in meters and degrees.

	Strides					Paces		Angles	
Position	Right Rear	Left Rear	Right Front	Left Front	Stride Average	Rear	Front	Rear	Front
1		2.59		2.17	2.38				
2									
3		2.63		2.22	2.43	1.6	1.51	107	91
4	2.53					1.66	1.61	97	
5		2.51		2.68	2.55	1.71		96.2	
6	2.63					1.66		106.1	
7		2.51		2.52	2.52	1.63	1.38	106	113.2
8	2.65		2.66		2.66	1.51	1.63	109	120.9
9		2.64		2.5	2.57	1.74	1.42	111.3	128.3
10	2.43		2.53		2.48	1.46	1.36	103.2	118.6
11		2.73		2.8	2.77	1.64	1.58	108.2	125.7
12	2.7		2.63		2.67	1.72	1.57	102.1	117.6
13		2.56		2.63	2.6	1.74	1.5	104.2	118.8
14	2.51		2.57		2.54	1.5	1.55	101.6	117.8
15		2.64		2.43	2.54	1.73	1.44	105	118.1
16	2.57		2.59		2.58	1.6	1.39	102.2	121.7
17		2.55		2.68	2.62	1.7	1.57	103.8	125.4
18	2.42		2.5		2.46	1.54	1.45	96	113.4
19		2.55		2.42	2.49	1.71	1.54	97.6	114.7
20	2.41		2.38		2.4	1.69	1.33	89.2	114.8
21		2.12		2.41	2.27	1.76	1.49	87.3	116.9
22	2.11		2.11		2.11	1.29	1.34	94.5	108.8
23		2.48		2.26	2.37	1.57	1.26	100.7	109.6
24	2.45		2.44		2.45	1.65	1.5	91.5	110.1
25		2.68		2.54	2.61	1.78	1.48	100.5	119.6
26	2.61		2.57		2.59	1.7	1.46	95	116.9
27		2.56		2.63	2.6	1.83	1.55	93	114.7
28	2.54		2.58		2.56	1.69	1.57	96.1	120.6
29		2.58		2.53	2.56	1.72	1.4	97	124.5
30	2.48		2.38		2.43	1.72	1.45	96.2	122.5
31		2.41		2.4	2.41	1.6	1.26	97.4	110.4
32	2.45		2.64		2.55	1.61	1.65	100.6	117.8
33		2.53		2.67	2.6	1.58	1.43	103.9	126.4
34	2.49		2.74		2.62	1.63	1.55	105.1	129.3
35		2.52		2.41	2.47	1.51	1.48	102	124.8
36	2.5		2.71		2.61	1.73	1.23	99.6	
37		2.46		2.5	2.48	1.53	1.39	94.1	110.8
38	2.65		2.51		2.58	1.82	1.64	99	119.5
39		2.46		2.5	2.48	1.66	1.25	95.9	122.9
40	2.58		2.63		2.61	1.65	1.59	100.3	121.3
41		2.59				1.71	1.43	96	
42						1.78			
Minimum:	2.11	2.12	2.109	2.17		1.29	1.23	87.3	91
Maximum:	2.7	2.73	2.75	2.8		1.83	1.65	111.3	129.3
Average:	2.51	2.54	2.54	2.49		1.65	1.47	99.8	117.9
Variance:	0.016	0.014	0.022	0.024		0.011	0.013	29.23	50.92
Std. Dev.:	0.126	0.12	0.147	0.155		0.105	0.113	5.407	7.136

Figure 34.13. Photograph of Trackway #1 soon after its discovery. The numbers indicate stride positions as they are labeled on trackway maps in Figures 34.18 and 34.19. Tracks in the foreground are poorly preserved. Note variation in depth and preservation of tracks in different parts of the trackway.

Figure 34.14. Photograph of a typical track set (right) from the Briar Site (from trackway #1). Note the outward rotation of the symmetry of the manus print. Only a claw mark of digit I is present in the pes print (see Fig. 34.17a).

(Figs. 34.16, 34.17e; compare to Fig. 34.17c [West Verde Creek Site] and 34.17d [Paluxy River Site] in Pittman [this volume]), an indentation occurs in the medial side wall of the track at about the position of the metatarsal–astragalar articulation of digit I. This indentation is about one-half the distance from the anterior margin of the track to the posterior margin.

One short trackway at the Briar Site consisted only of deeply impressed pes prints, somewhat triangular in outline (Fig. 34.17o; cast TMM 43401-4). This structure is also reported by Farlow et al. (this volume) at the Paluxy River Site in Texas. Claw marks for digits I, II, and III are present in these Arkansas tracks and those of digits II and III are directed laterally into the sidewall of the track (Fig. 34.17o). This lends support to Langston's (1974) idea that the claws of the pedal digits "under certain circumstances, were rotated so that the morphologically medial surfaces became functionally almost plantar in position."

Manus Prints

Manus prints are slightly less than one-half the size of pes tracks, broadly U-shaped, with distinct pad impressions separated by indentations from the side walls of the track and the posterior margin (Fig. 34.17). No claw marks occur in manus prints. Typically, across the entire width of the front half of the track, is a broadly ovate pad impression presumably representing closely bound digits II, III, and IV. At posteromedial and posterolateral corners of a well-preserved front print, separated from the large front pad impression by indentations of the side wall of the track, are the presumed pad impressions of digits I and V, respectively. Separating these pad impressions of digits I and V, at the corner of the posterior margin of the track is a rather broad shelf sloping from the posterior margin of the track down toward the front, becoming flush with the bottom of the track near its center. In several manus prints in trackway #2, digit V is separated somewhat from the other four (Fig. 34.17p, upper manus print). This structure suggests some flexibility for the digits of the manus. See Pittman (this volume, Fig. 15.12) for a description of a well–

Figure 34.16. Photograph of a pes print (TMM 43401–3) from the cast of the trackway in the densely tracked area between trackways #1 and #2. Impressions of the unguals of digits I, II, III, and IV are well-preserved. The arrow labels the indentation along the medial wall of the print, which was discussed in the text. On the line drawing above, the position of several depresions in the floor of the track are shown. Perhaps the two anterior depressions record the position of metatarsal–phalangeal articulations of digits I and II. A small "cone" of mud apparently was squeezed between the toes of digits II and III.

Figure 34.15. Photographs of a left track set (Set 8/9) from trackway #1 viewed from directly above (lower photograph) and in an oblique anterior view (upper photograph). These photographs illustrate the state of preservation typical in this trackway. The pes print is slightly deeper in its medial half. The axis of symmetry of the manus print is, as usual, rotated slightly outward. The indentation from the rear wall of this manus print is larger than usual (probably partly the result of lateral compression and "pushing" of mud forward by the pes).

preserved manus print from the West Verde Creek Site, Texas.

Trackway Descriptions

Measurements of stride, pace, and pace angle (calculated using law of cosines) were taken from maps of trackways #1 and #2 (Fig. 34.18, Tables 34.1, 34.2). As used here, a stride is the distance from a reference point on successive prints of the same foot and a pace is the distance between a point on successive prints of opposite feet. The point of reference on both manus and pes prints is at the apex of curvature (usually) on the anterior margin of the track. This is also the point where a line bisecting a pes track in a parasagittal plane crosses the anterior margin. On manus prints, the outlines of which are consistently rotated outward relative to the direction of travel, this reference point corresponds to the intersection of a line bisecting the symmetry of the track and the anterior

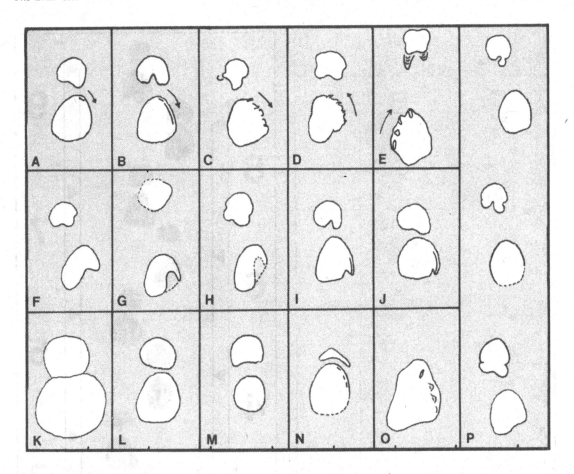

Figure 34.17. Sketches of selected track sets from the Briar Site (A, B, E, F, G, H, I, J, L, M, O, and P) and several track sets from sites in Texas for reference (C, D, K, and N). All sets were drawn with vertical lines parallel to travel direction. All sets except E are from the right side. Tick marks along base line are one meter apart. **A,** set 3/5 from trackway #1; photograph in Fig. 34.14. **B,** set 7/9 from trackway #1. **C,** West Verde Creek Site track set (TMM 40637-1, -2). **D,** Paluxy River Site (after Bird [1939]). **E,** set from cast TMM 43041-3 (see Fig. 34.16). **F,G,H,** sets illustrating shallow posterolateral area of pes prints: **F,** set 27/29 from Trackway #1. **G,** set 34/36 from Trackway #2. **H,** set 24/26 from Trackway #2. **I,J,** sets from Trackway #1 illustrating indentation in posterolateral corner of pes prints: **I,** set 9/11. **J,** set 5/7. **K,** very poorly preserved ghost prints from the Miller Creek Site, Texas. **L,M,** track sets from Trackway #B1. **N,** set from the Blanco River Site, Texas, illustrating deformation of the manus print. **O,** pes print (TMM 43041-4) from the trackway discussed in the text in which no manus prints occur. **P,** sets in sequence from Trackway #2 (sets 16/18, 18/20, and 20/22) illustrating variation in manus print morphology. Note in particular the uppermost manus print.

margin. The term set, as used herein, refers to the sets of pes and manus prints of the same side seen alternating left and right down an individual trackway. On maps of trackways #1 and #2 the numbers refer to stride position, starting at the beginning of the trackway (with reference to the direction of travel). These stride position numbers are also listed in Table 34.1 and Table 34.2. Stride position corresponds to offset pairs of stride measurements (e.g. right rear stride/right front stride = X; left rear stride/left front stride = X + 1; etc.) (Fig. 34.19). At a given stride position, values of rear stride, front stride, rear pace, front pace, rear angle, and front angle, either left or right, are listed (read across Table 34.1 and Table 34.2). These data

were plotted against stride position for trackways #1 and #2 (Fig. 34.21a,b).

Trackway #1

Trackway #1 (Fig. 34.18) is approximately 47 m long. The trackway is straight for most of its length, curving slightly to the right near its end. Over the first three-fourths of its length, the trackway meanders slightly. Left and right track sets alternate down the length of the trackway. The position of rear tracks is more consistent than that of front prints (compare the graph of front and rear pace, Fig. 34.21). Partly because manus prints fall closer to the midline of the trackway, front pace is consistently less than rear pace.

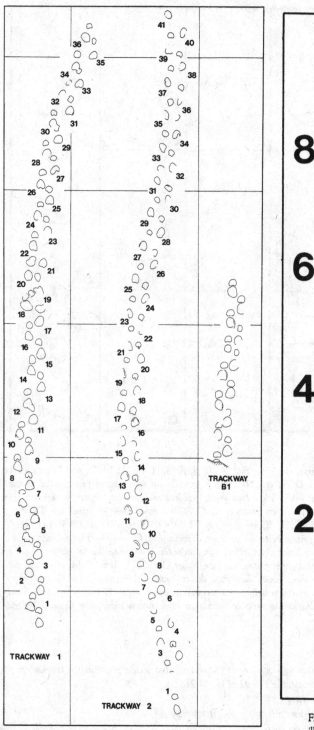

Figure 34.18. Maps of Trackways #1, #2, and #B1. Grid is 10 x 10 m. Numbers refer to stride positions (see Fig. 34.19 and 34.21).

Figure 34.19. Map of the beginning of Trackway #1 illlustrating the convention used here to denote stride position. The values of a rear and a front stride on the same side are associated with a given stride position (e.g., "1"). Triangles mark the lateral projection of reference points used in measuring trackway data — large triangle = rear; small = front. Also, with each stride position, four pace values are associated (two consecutive rear; two consecutive front). This notation is used to facilitate the plotting of trackway data (in Fig. 34.21 stride position lies along the x–axis).

and front strides in the first half of the trackway, whereas rear and front strides are not as highly correlated in the second half of the trackway. If rear and front strides are averaged at each position along the trackway and plotted (upper curve), a simpler pattern emerges.

On the graph of pes versus postion, considerable crossover of the rear and front pace lines occurs. Rear pace values are slightly more consistent down the trackway; the line connecting front pace values shows more variation than that of rear paces. This probably reflects the greater weight-bearing role of hind limbs relative to front; the placement of hind feet in space during locomotion was apparently limited relative to the palcement of front feet. Some of the greater variation of front paces may possibly be related to torque of the forequarters caused by turning of the head and neck during locomotion.

Trackway #2

Trackway #2 (Fig. 34.18) extends in an easterly direction approximately 53 m across the area. This is the only well preserved (or unobscured) trackway that extends across the area in other than a southerly direction (several other eastbound trackways may be observed on the aerial photograph in Fig. 34.8). For most of its length the trackway turns slightly to the right. Again, rear–front track sets alternate left and right down the trackway. Impressions of the manus consistently fall closer to the midline of the trackway than pes prints, but no overstepping by pes prints over opposite manus prints occurs, as it does in Trackway #1.

Graphs of Trackway #2 Data, Figure 34.21b

In general, the graphs of Trackway #2 illustrate a more consistent variation of stride, pace, and pace angle measurements than those of Trackway #1. On the plot of pace angle vs. position, front pace angle is consistently greater than rear angle, and rear pace is greater than front pace. These patterns may be attributed to the position of pes and manus prints relative to the midline of the trackway; pes prints lie farther from the midline than manus prints.

On the plot of rear and front strides vs. position, rear and front strides are highly correlated. On the graph of stride average vs. position, midpoints between rear and front strides at given positions are plotted. Stride average may more clearly document changes in velocity. On this plot generally there is a regular transition from position to position, barring several changes in slope. Across the plot there are successions of regular changes in stride average. This curve apparently records several accelerations and decelerations.

The average rear and front paces of Trackway #1 are 1.49 and 1.41 m, respectively (a difference of about .08 m), and the graph of Trackway #1 paces shows considerable crossover of rear and front pace lines. The average rear and front paces of Trackway #2 are 1.65 and 1.46 m, respectively (a difference of about .2 m), and on the graph of these measurements there is very little crossover of the lines.

Figure 34.20. Photograph of Trackway #1 beginning at stride position 4. Note the outwardly rotated axis of symmetry of the first two manus prints. Grid is 3 m across.

At three points along the trackway right manus prints are partially obscured by impressions of the left pes. At these three points the dinosaur was turning slightly to the right. Near the end of the trackway, in the broad right turn, in addition to the points where the right manus and left pes impressions overlap, right manus prints lie close to the impression of the left pes. As is the case with all Gulf Coastal Plain sauropod trackways, the axis of symmetry of manus impressions is consistently rotated outward (Fig. 34.20).

Just past the midpoint of the trackway (from position 20 to 25) there are several extraneous tracks. We were not able to determine whether these tracks were made before, at the same time as, or after the maker of trackway #1 ambled across the area. Other than the very beginning of the trackway, this is the only place along the trackway where the consistent pattern of alternating track sets is complicated.

Graphs of Trackway #1 Data, Figure 34.21a

On a plot of rear and front strides versus position along the trackway, a good correlation exists between rear

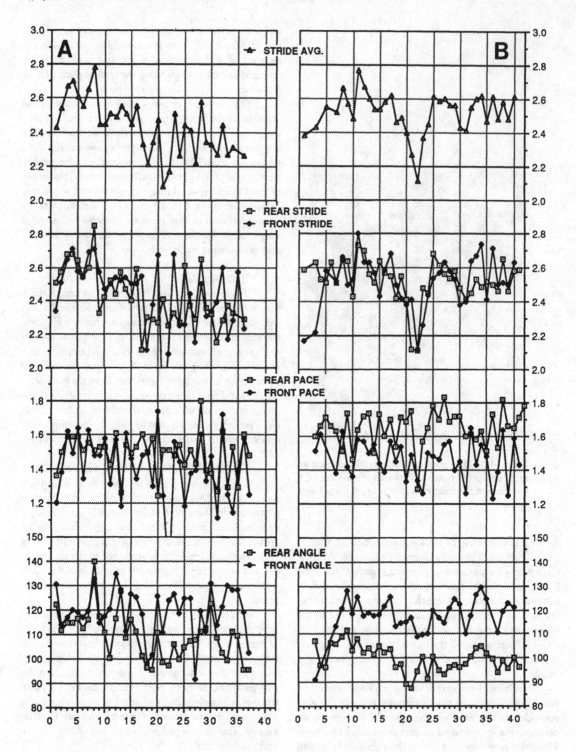

Figure 34.21. Graphs of trackway data. **A,** trackway #1. **B,** trackway #2. Meter and degree scale along the y-axis. Stride position along x-axis. See text for discussion.

Summary

The Briar Site is unique among dinosaur tracksites in the large number of tracks exposed over such an extensive area and for the area of exposure (projected to at least 25,000 m²). The density of theropod and sauropod prints is also great at many sites in Texas where only relatively small areas of bedding surfaces are exposed. Considering the large area of exposure, the lack of carnosaur tracks at the Briar Site is somewhat surprising, because theropod tracks occur at four of the nine sauropod tracksites of the region, and at the Jurassic site in Colorado described by Lockley et al. (1986).

Several long sauropod trails could be mapped across one area. Aerial photographs reveal an alignment of about ten trackways in this same area, perhaps recording the movement of a herd. Many more tracks and trackways occur across the extensive exposure of the upper trackbed at the Briar Site. Sauropod tracks, many poorly preserved, but sufficiently distinct for identification, are known to have densely covered the limestone bed in the area mined between 1982 and the present. Equipment operators reported sauropod tracks in all mined-out areas of the quarry.

Sauropods walked across at least two bedding surfaces during deposition of the upper De Queen Formation. Tracks were made across the upper bed when the substrate was wet (perhaps some of the animals were wading), wet and "dry", and during times when the surface was exposed and completely dry. Tracks were probably made across the nearshore area over a period of time, possibly during several cycles of submergence and exposure.

The trackbeds were deposited near the shoreline. They occur within a sequence of shallow subtidal, intertidal, and supratidal claystone, siltstone, sandstone, limestone, and thin evaporite beds. The depositional environment of a single bed should not be identified by just one prefix (i.e., intertidal); changes in depositional conditions were frequent during the accumulation of the track-bearing sequence.

The trackbeds at the Briar Site are equivalent to strata of the Glen Rose Formation of Texas, probably to upper beds of that unit (see Pittman this volume). The sauropod responsible for the Arkansas tracks also made tracks in similar nearshore environments in the Texas region during Glen Rose time.

Structure of the tracks in Arkansas varies from area to area, from trackway to trackway and within single trackways, recording different behavior of sauropods walking across substrates of differing consistency.

Impressions of the manus indicate that the forefeet were "elephantine"; pad impressions indicate that digits II, III, and IV were tightly bound, while digits I and V were also closely bound to these digits, but digit V could be separated somewhat and was free to move independently, to a limited degree.

Tracks at the Arkansas site add to our knowledge of sauropod pes footprint structure derived from tracksites in Texas. Pes prints are deeper in the medial half of the tracks; most of the weight borne by rear feet fell across this region. Claw marks of digits I and II, and sometimes III and IV occur in pes prints at the Briar Site. As may be inferred from Texas sauropod footprints, structure of pes prints at the Briar Site suggests that claws were "wrapped" along the lateral edge of the foot by sauropods walking across firm or moist substrates, or extended forward for braking, turning, or traction in muddy areas.

Plots of measurements from the two long trackways suggest that these two sauropods ambled across the area in different manners. These plots apparently document several changes in velocity by the trackmakers.

Acknowledgments

We first would like to thank the employees of Weyerhaeuser Company's Briar Plant Gypsum Quarry. Without their cooperation this study would not have been possible. Weyerhaeuser not only helped us in our study, they encouraged visitation by school children from all over southern Arkansas.

Special thanks go to Linda and Bill Moery of Nashville, Arkansas. These two energetic and enthusiastic individuals were instrumental in organizing much of our work in the quarry, in addition to providing muscular support. The many other individuals who helped with the excavation, at times under unpleasant conditions, are too numerous to mention; we sincerely thank them. Several private concerns expended considerable time and effort in the removal of tracks for their respective local schools and, in so doing, helped preserve a record of the tracks. Lynett Gillette and David Thomas did much of the casting work.

Riberglas Company, of Dallas, Texas, provided the fiberglass for the project. Dow Corning Company provided the silicone rubber. The Institute for the Study of Earth and Man, Southern Methodist University, provided logistic support. Our sincere thanks are extended to these institutions.

We thank James Farlow, Wann Langston, Jr., and Timothy Rowe for editing and suggesting improvements to the manuscript.

References

Bird, R. T. 1939. Thunder in his footsteps. *Nat. Hist.* 43:254–261, 302.

— 1941. A dinosaur walks into the museum. *Nat. Hist.* 47:74–81.

— 1944. Did *Brontosaurus* ever walk on land? *Nat. Hist.* 53:61–67.

— 1954. We captured a "live" brontosaur. *National Geographic Mag.* 105:707–722.

— 1985. *Bones for Barnum Brown: Adventures of a Dinosaur Hunter.* Schreiber, V. T. (ed.). (Texas Christian University Press) 225 pp.

Farlow, J. O., Pittman, J. G., and Hawthorne, J. M. Chapter 42, this volume. *Brontopodus birdi*, Lower Cretaceous sauropod footprints from the Gulf Coastal Plain of North America.

Honess, C. W. 1923. Geology of the southern Ouachita mountains of Oklahoma. *Oklahoma Geol. Surv. Bull.* 32:204–264.

Langston, W., Jr. 1974. Nonmammalian Comanchean

tetrapods. *Geoscience and Man* 8:77–102.

Lockley, M. G., Houck, K. J., and Prince, N. K. 1986. North America's largest dinosaur trackway site: implications for Morrison Formation paleoecology. *Geol. Soc. Amer. Bull.* 97:1163–1176.

Loucks, R. G., and Longman, M. W. 1982. Lower Cretaceous Ferry Lake Anhydrite, Fairway Field, East Texas: product of shallow-subtidal deposition. *In Depositional and Diagenetic Species of Evaporites — A Core Workshop. Soc. Econ. Paleont. Miner. Core Workshop* 3:130–173.

Miser, H. D. 1927. Lower Cretaceous (Comanche) rocks of southeastern Oklahoma and southwestern Arkansas.

Amer. Assoc. Petr. Geol. Bull. 11 (5):443–453.

Pittman, J. G. 1984. Geology of the De Queen Formation of Arkansas. *Gulf Coast Assoc. Geol. Soc. Trans.* 34:201–209.

———. 1985. Correlation of beds within the Ferry Lake Anhydrite of the Gulf Coastal Plain. *Gulf Coast Assoc. Geol. Soc. Trans.* 35:251–260.

———. Chapter 15, this volume. Stratigraphy, lithology, depositional environment and track type of dinosaur track-bearing beds of the Gulf Coastal Plain.

Young, K. 1974. Lower Albian and Aptian (Cretaceous) ammonites of Texas. *Geoscience and Man* 8:77–102.

35 Large Dinosaur Footprint Assemblages from the Cretaceous Jindong Formation of Southern Korea

SEONG–KYU LIM,
SEONG–YOUNG YANG AND
MARTIN G. LOCKLEY

Abstract

The Cretaceous Jindong Formation of Southern Korea yields one of the largest dinosaur trackway assemblages hitherto reported. The tracks occur at as many as 160 separate horizons in a lacustrine succession that lacks body fossils. The footprints indicate a fauna dominated by ornithopods and sauropods. Ongoing studies are expected to produce thorough documentation of this unit and reveal the significance of the ubiquitous footprint assemblages.

Introduction

In recent years dinosaur tracks have been reported from a number of localities in southeastern Korea. The most spectacular tracks occur in Jindong Formation exposures in the vicinity of Dukmyeongri, Koseong–gun, Kyungsangnam–do Province (Yang 1982, Lim 1985, Lim and Yang 1987); see Figure 35.1. Tracks also occur in inland exposures of the Jindong Formation and in the underlying Haman Formation (Kim 1987). Both formations are considered to be late Early Cretaceous in age.

The Jindong Formation tracks are of particular interest because they occur at as many as 160 distinct horizons throughout the 300 m of gray shales and siltstones that comprise the formation (Lim and Yang 1987). Although multiple trackbearing horizons are known from many formations (Lockley 1986), few formations reveal the wealth of trackway data yielded by the Jindong succession. To date more than 250 discrete trackways have been recorded and mapped. Preliminary identifications indicate a number of distinct morphotypes attributable to bipeds (predominantly ornithopods) and a few types attributable to quadrupeds (mainly sauropods).

The tracks, which are exposed at a number of adjacent coastal exposures (Figs. 35.1, 35.2) are still being studied intensively. The report given herein is therefore preliminary and brief.

Historical Background

The Jindong Formation comprises a distinctive unit of gray shales and siltstones of probable lacustrine origin. The footprint–rich Dukmyeongri site was first reported by Yang (1982), who noted the presence of multiple track-bearing beds and diverse trackway assemblages. Subsequent work by Lim (1985) and Lim and Yang (1987) has resulted in the recognition of many more trackway horizons and track morphotypes. Recently Kim (1987) erected various ichnotaxa in a personal report submitted to the Korean Science and Engineering Foundation. However, because of the obscure nature of this publication, and the fact that some of the material was discovered by Lim and Yang, and is still under investigation, we can not recognize these names. In view of the wealth of data emerging from the Jindong Formation we avoid any new assignments at this stage.

Preliminary Observations and Lithostratigraphic Notes

Field investigations by one of us (Lim) have resulted in the compilation of a series of measured sections to which maps of trackbearing bedding planes have been keyed (Fig. 35.1). The Jindong Formation dips at about 10 degrees to the southeast comprising a series of well–exposed sea cliffs and an intertidal wave–cut platform of bedding plane exposures (Fig. 35.2). The trackbearing beds are therefore vulnerable to erosion by the sea and by cyclones, which damaged exposures in the summer of 1987.

Although no detailed lithostratigraphic studies have been completed, the succession is known to consist of rather monotonous, thinly bedded gray and dark gray laminated shales and siltstones, with discontinuous nodular limestone beds characterizing the lower part of the formation and thin continuous calcareous beds characterizing the upper part. The lower carbonates are typically subhorizontal branching cylindrical tubes 3–5 cm in diameter. These resist compaction more than the surrounding shales. Locally, the

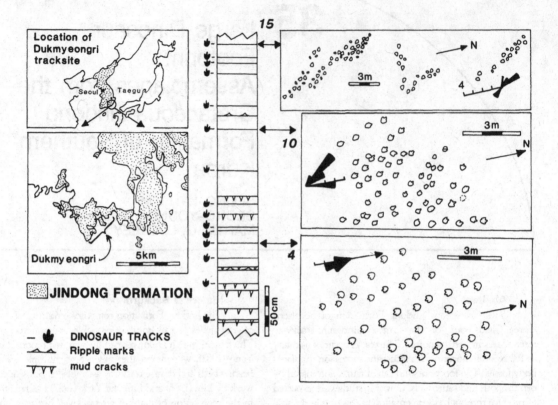

Figure 35.1. Location of Jindong Formation trackbearing exposures at Dukmyeongri (left), with a representative section of the upper part of the formation (Dukmyeongri–Sangjok) showing multiple trackbearing beds (center); modified after upper part of section shown in Yang 1982, Figure 3. Maps of the 4th, 10th and 15th trackbearing beds are shown (right) with the former two exhibiting ornithopod trackways and the latter one exhibiting quadrupedal trackways attributable to sauropods. Note the sub-parallel orientation in all cases, indicating strong NNW–SSE trends.

nodule networks create large polygonal patterns. The superficially monotonous upper part of the sequence contains a large number of thinly laminated carbonate units typically ranging from 5–15 cm in thickness. Weathered surfaces reveal that many of these are stromatolitic in character (Fig. 35.3). They frequently show signs of reworking such as small scale cut and fill structures and flat pebble textures. Small vertical structures usually with irregular fillings probably indicate roots, or possibly *Skolithus*–like burrows.

In the upper part of the section a single meter thick "red bed" unit is noted. It comprises a very thin basal conglomerate grading up through sand to green and red mottled shales and siltstones overlaid by a sandstone unit. Elsewhere in the section, sandstone and conglomerate are rare and usually calcareous. The formation also contains a few thin volcanic ashes.

Mudcracks and symmetrical wave ripples occur throughout the sequence, with a small proportion of asymmetrical and irregular ripples. The formation appears to be devoid of flora and fauna near the coast, although a distinctive nonmarine mollusc fauna is known from near Taegu to the north. Some distinctive traces of invertebrate bioturbation are known, but have not been studied. Thus the vertebrate tracks represent the only significant paleon-

tological evidence associated with this unit in the southern region.

Considered collectively the lithostratigraphy indicates a lacustrine depositional environment characterized by shallow water and frequent (cyclic) shoaling and emergence. The apparent absence of lacustrine biota indicates a hostile physio–chemical paleoenvironment. However, the abundance of dinosaur tracks attributable to herbivores indicates the probable availability of vegetation in the vicinity of the lake basin.

Tracks and Trackmakers

The Dukmyeongri site is unique in terms of trackway statistics. With the possible exception of parts of the Newark Supergroup of eastern North America, it yields more trackways and trackbearing beds than any hitherto known stratigraphic unit. It can be compared with the multiple trackbearing beds of the Sousa Formation of Brazil (Godoy and Leonardi 1985) and various formations on the Colorado Plateau (Lockley 1986). As suggested by Lockley and Conrad (1987 and this volume), such sites are likely to have considerable potential in census paleoecology.

The Jindong tracks and trackways have been studied using a number of methods. Firstly, bedding planes are

mapped to show trackway distributions. Representative individual trackways are recorded by tracing footprint outlines on transparent overlays (cf. Fig 35.4A). Individual footprints have also been cast using plaster of paris. Some of these have been subjected to detailed analysis by placing the inverted cast in a tank and progressively submerging it with water, raising the level in 2 mm increments and projecting the water level onto a transparent lid. The result is an accurate full scale contour map of the track's topography, Figure 35.4. The results are similar to the three dimensional imaging achieved by Ishigaki and Fujisaki (1985 and this volume) in their application of the moiré method to dinosaur footprint analysis. Such methods enhance the reliability of footprint differentiation and identification.

According to Lim and Yang (1987), about 186 bipedal and 65 quadrupedal trackways have been documented to date. The bipeds were dominated by ornithopods, with very few tracks attributable to theropods, and the quadrupeds were dominated by sauropods. Although one small sauropod trackway, the smallest on record, is documented (Fig. 35.4), there are no small tracks currently known. One trackway may be attributable to a non-dinosaurian vertebrate, possibly a large crocodile.

As data are still being accumulated, a detailed synthesis is premature. However, preliminary mapping indicates that parallel trackways exist at many horizons (Fig. 35.1). Moreover trackway orientation patterns at different horizons sometimes, but not always, show coincident modal trends. For example, in the illustrated section, coincident modal trackway trends exist at the 4th, 10th and 15th track-

Figure 35.2. General view of coastal exposures of the Jindong Formation.

Figure 35.3. Detail of stromatolitic bed (coin diameter 2.4 cm).

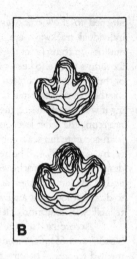

Figure 35.4. **A,** two sections of the trackway of a small sauropod, the smallest on record. Manus pace angulation = 110°. Pes pace angulation = 125°. **B,C,** ornithopod and theropod pes tracks respectively, showing topographic detail revealed by contour method. See text for details.

bearing horizons. Trackways at some of the other horizons show similar trends but are generally more variable or too sparsely distributed to indicate a clear pattern. Coincident trends at successive or closely spaced stratigraphic intervals have been reported by Godoy and Leonardi (1985) and Lockley (1986) as probable indicators of well established shoreline trends persisting for significant intervals of geologic time.

Conclusions

The Jindong Formation trackways at the coastal exposure near Dukmyeongri are among the most spectacular dinosaur ichnites hitherto reported. They warrant detailed analyses of the type currently being undertaken. Preliminary indications suggest that the number of track-bearing beds exceeds that hitherto reported from any other stratigraphic unit of comparable thickness. The ichnites suggest a herbivore dominated fauna (ornithopods and sauropods) frequenting a lacustrine environment in which biotic remains, other than algae, were very sparse.

Future work should reveal more about the depositional environment and allow the relative abundance and orientation of trackways to be compared at successive horizons throughout the succession.

Acknowledgments
Partial support for travel expenses was provided to Lockley by the National Science Foundation (USA), Grant #EAR 8618206.

References

Godoy, L. C., and Leonardi, G. 1985. Direcoes e comportamento dos dinossauros de localidade de Piau, Sousa, Paraiba (Brasil), Formacao Sousa (Cretaceo inferior). *In* Brasil, DNPM. *Coletanea de Trabalhos Paleontologicaos.* Serie "Geologica", Brasilia, 27 (Secao Paleontologica e Estratigrafia, 2), pp. 65–73.

Ishigaki, S., and Fujisaki, T. 1985. Three dimensional representation of the dinosaur footprint by the method of moiré topography. *Jour. Fossil Research* 18:59–64.

Kim H. M. 1987. *Stratigraphy of Lower Cretaceous Jindong Formation.* A report presented to the Korean Science Foundation, pp. 1–88.

Lim S-K. 1985. Cretaceous dinosaur tracks in the Upper Kyungsang Group. (Abs.) *Geol. Soc. Korea* 1:10.

Lim S-K. and Yang S-Y. 1987. On the Cretaceous dinosaur's footprints in Korea. (Abs.) *Pacific Science Assoc. 16th Congress, Seoul, Korea* 1:157.

Lockley, M. G. 1986. The paleobiological and paleoenvironmental importance of dinosaur footprints. *Palaios* 1:37–47.

Lockley, M. G., and Conrad, K. 1987. Mesozoic tetrapod tracksites and their application in paleoecological census studies. *In* Currie, P. J., and Koster, E. H. (eds.). *4th Symposium on Mesozoic Terrestrial Ecosystems* (Drumheller, Alberta).

Yang S-Y. 1982. On the dinosaur footprints from the Upper Cretaceous Gyeongsang Group, Korea. *Jour. Geol. Soc. Korea* 18:46.

36 Problematical Tracks and Traces of Late Albian (Early Cretaceous) Age, Clayton Lake State Park, New Mexico, USA

DAVID D. GILLETTE AND
DAVID A. THOMAS

Abstract

Problematical tracks and traces in the lower track-bearing horizon at Clayton Lake State Park (Lower Cretaceous Mesa Rica Sandstone) include (i) tracks made by at least one small or baby dinosaur, cf. *Irenichnites gracilis*; (ii) a suite of elongate single digit impressions probably made by small avian or coelurosaurian trackmakers; and (iii) a set of six possible pterosaurian left manus impressions that extends over 12 m in a broad arc. Although others have argued that the latter trackway is crocodilian, the evidence more strongly suggests a pterosaurian origin.

Introduction

Numerous dinosaur tracks and traces occur in three horizons of Cretaceous sediments at Clayton Lake State Park, Union County (36°34'35"N, 103°17'32"W), in northeastern New Mexico, USA. Gillette and Thomas (1985) mapped and described the tracks, and regarded the track-bearing horizons as the Dakota Formation (Aptian–Albian age). Subsequent studies of the stratigraphy in the region by Lucas et al. (1986), Lucas et al. (1987), and Lucas et al. (this volume) have assigned the Clayton tracks to the top of the Mesa Rica Sandstone and lower part of the Pajarito Formation, as subdivisions of the Dakota "Group." Associated invertebrate fossils and regional stratigraphy indicate a Late Albian (Early Cretaceous) age for the Clayton Lake tracksite (see Lucas et al. this volume and references therein).

The unusual tracks that are the subject of this report occur on the upper surface of the lower trackbearing horizon (unit no. 1 in the terminology of Gillette and Thomas [1985]), a massive crossbedded sandstone with abundant ornithopod tracks, ripple marks, and mudcracks. The tracks and traces described below occur on the same surface and extend over an area of approximately 50 sq m. The uppermost centimeter of unit 1 is a bioturbated clay that drapes over the massive sandstone and adheres to it

on weathered surfaces. The traces were made on this surface or within a few millimeters of it.

The origins of the traces described below are difficult to determine with confidence. The interpretations and alternatives are presented for the purpose of discussion, especially to expand the background for the discussion on pterosaur tracks developed by Unwin (this volume) in which the putative pterosaur tracks from Clayton are ascribed to a crocodilian origin.

Unwin maintains that all previous reports of pterosaurian tracks, including the Clayton tracks, do not satisfy his criteria and therefore are either not assignable to a known trackmaker or were made by a crocodilian trackmaker. Whether his predictive analysis and conclusions is valid are not the subject of this report; rather, the purpose is to more fully describe the Clayton tracks and present an alternative interpretation for one trackway in context with other problematical features in the same bed. For a full discussion of pterosaurian anatomy, review and analysis of putative pterosaurian tracks, and predictive analysis regarding pterosaurian footprints, see Unwin (this volume) and references therein.

A "Baby" Dinosaur Track

An isolated tridactyl footprint with a broad U-shaped contour is attributable to a small or baby dinosaur (Fig. 36.1). The impressions of the outer digits are 7 cm long, and the impression of the middle digit is 9 cm in length. The outline of the track is asymmetrical. The impressions of the right and central digits are continuous along a single depression; the impression made by the left digit connects to the right of the rear apex of the footprint.

The angles of divarication (as defined in Leonardi [1987]) are II–IV: 53°; II–III: 18°; and III–IV: 37°. According to Currie (1981), the angle of divarication in avian footprints is consistently greater than for dinosaur footprints, which do not exceed 100°. Because the track has a narrow

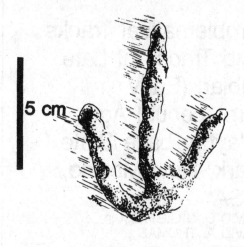

Figure 36.1. Line drawing interpretation of the small or baby dinosaur track, cf *Irenichnites gracilis*. The track is visible in Figure 36.5 in the photograph of the pterosaurian track no. 4; and it is plotted on the map of the pterosaurian trackway in Figure 36.4.

Figure 36.2. ?Avian or small coelurosaur toe impressions.

angle of divarication and is smaller than any hitherto reported from the Dakota Group (cf. Lockley 1987), it seems to have been made by a small or "baby" dinosaur, possibly one of the coelurosaurs known from adult tracks at the site. The single Clayton footprint closely resembles in size and anatomical details the track of the coelurosaur ichnospecies *Irenichnites gracilis* illustrated by Currie (1981, Fig. 1B), but is more robust and differs in proportions. In the absence of additional information concerning variation and trackway characteristics, we assign this track to cf. *Irenichnites gracilis*, rather than describe a new ichnospecies on the basis of limited material.

A cast of this track is deposited in the paleontology collection at the New Mexico Museum of Natural History.

Possible Avian or Baby Coelurosaur Footprints

A set of multiple impressions of isolated toes (Fig. 36.2) is probably attributable to a small slender-toed vertebrate of uncertain affinity. The tracks are well preserved in the mudstone and resemble multiple sets of toe marks and footprints that can be seen along modern shorelines where wading birds have milled about in search of food. The Clayton tracks are not well enough preserved to distinguish between avian and dinosaurian origin. Currie (1981) reported similar traces in coeval sediments in Alberta (Canada), and attributed their origin to an avian source; Mehl (1931) reported bird tracks from the Dakota of Colorado. Because coelurosaurs are known from the site, it is also reasonable to attribute these tracks to small or baby coelurosaurs, whose footprints may resemble those of wading birds. These impressions, numbering in excess of several hundred in this small area, were made by birds or small dinosaurs for which the central digit depressed the substrate sufficiently to produce an impression. Apparently the lateral digits carried less weight and failed to leave an impression, resulting in tracks for only the central digit with each footfall. Such differential digit impression depths sometimes characterize underprints.

Alternatively, these elongate depressions might have been produced by clams whose shell impressions, in living position within a supporting substrate, left elongated bottom marks. This alternative is less attractive than avian or coelurosaurian origins because of lack of bivalves at the site, and because of the generally uniform quality of the impressions.

cf. Pterosaurian Trackway

A set of six problematical tracks (Figs. 36.3–36.8) in a trackway that describes a broad arc covering 12 m occurs in the same vicinity as the multiple elongate depressions (see Fig. 36.2) and the baby coelurosaur track (see Fig. 36.1). The tridactyl tracks are impressions of three digits that made contact with the substrate nearly at right angles to the line of progression (Fig. 36.4). The digit impressions are nearly parallel in each track, rather than divergent as in avian, dinosaurian, and crocodilian footprints. The wide spacing between the footprints, the orientation nearly at right angles to the line of progression, the parallel orientation of the digit impressions, and the anatomy of the individual tracks are best reconciled by assignment of the impressions as left manus of a large pterosaur, possibly in the accelerating phase of a takeoff sequence (as indicated by the increasing stride lengths through the sequence). Although it is not possible to ascertain the identity of the trackmaker with certainty, we favor a pterosaurian origin over the crocodilian origin

Figure 36.3. Photograph of the pterosaurian trackway, showing tracks 2–5. Track 1 is adjacent to the ledge that continues from bottom right of photograph to viewer's left, at projected intersection of the string. Track 6 is too far from track 5 to be visible in this orientation; it is 584 cm from track 5, in line with the string, which marks the arcuate line of progression. Photographs of the individual tracks presented in Figures 36.5–36.8 were taken with the same string and paper markers in place.

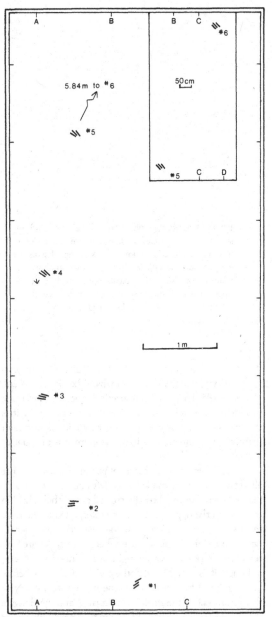

Figure 36.4. Schematic map showing positions of individual tracks in the pterosaurian trackway. The baby dinosaur track is shown in approximate position with respect to track 4. Inset showing position of track 6 is at larger scale. Distance between tick marks and coordinates A, B, C, D is 1 m. The tracks are not drawn to scale.

proposed by Unwin (this volume).

Our orientational terminology follows the interpretation as left manus impressions, progressing in order from the track no. 1 to track no. 6. Alternative interpretations (which might involve identification of the tracks as pes impressions, or attribute this series to a right limb rather than a left) must necessarily be altered and the orientations accounted for if cogent arguments are to be presented.

The most clearly preserved track in the set is no. 4 (Fig. 36.8), which provides the most detail, and may be regarded as typical for the trackway. A cast of this track is deposited in the New Mexico Museum of Natural History. Individual lengths of the three digit impressions, here identified as digits I–III, are 5 cm, 8 cm, and 10 cm, respectively. The depression for each digit is elongate and pointed at the terminus. The impressions for digits II and III are nearly equal in size, and they are nearly parallel in orientation. The impression for digit I is shorter and more robust, and its proximal margin is considerably offset with respect to the proximal margins of digits II and III. This anatomy more closely resembles the predicted anatomy of pterosaurian manus tracks as presented by Unwin (this volume) than the pes or manus anatomy of archosaurian reptiles as summarized by Parrish (this volume). We also note that crocodilians normally exhibit a pentadactyl manus and tetradactyl pes (Padian and Olsen 1984) and are therefore *not* likely to leave tridactyl tracks.

There is no certain indication of contact with the substrate by metapodials associated with these three digits in track no. 4, although a depression in the expected position of metacarpal IV (Fig. 36.8, right arrow) might be attributed to the distal extremity of the metacarpal; and

another depression lateral to digit III (Fig. 36.8, left arrow) might be attributable to the proximal extremity of the first phalanx of digit IV, the wing finger, which in our interpretation was otherwise extended and did not contact the substrate. Other tracks in the set are similar in size and character. Impressions in each track are distinct for digits

Figure 36.5. Photograph of track 1. Line of progression is at right angles to tape measure (cm scale) and upward in this orientation. Hints of digital pads are evident but not sufficiently detailed for analysis. The rounded feature at the proximal end of digit I (the uppermost digit impression) might be a trace of substrate contact by the distal extremity of metacarpal I.

II and III, and generally less distinct for digit I. Digit impressions are longest in track no. 6, in which the digits are 9 cm, 11 cm, and 11 cm in length respectively; because this track cannot be definitely associated with the other five, it is possible that it was made by a larger individual, although slightly longer dimensions in a track are not unusual in a trackway.

The most puzzling feature about this trackway is that all of the tracks were made by the left manus as the trackmaker progressed in a broadly curved arc to the right. No tracks from the opposite limb (here interpreted as the right manus) are preserved with this trackway. Because there is no interruption on the surface where the tracks are preserved, it appears that no right footprints were made, although it is possible that they were destroyed before preservation. Other features in the immediate vicinity of the tracks, such as the small dinosaur track and the elongate digit impressions described above, are preserved with similar detail, indicating that tracks from the opposite limb, if made in the same vicinity (such as would be expected from a walking crocodilian trackmaker) should be preserved. The lack of tracks for the opposite limb may be explained by a great lateral distance between left and right hand falls, on the order of several meters or more, whereby sufficient substrate variation could have caused lack of preservation for the right side. Pterosaur anatomy, with a broad distance between hands being used for pushing off during acceleration and takeoff, can fulfill these dimensional requirements. The possibility of a buoyant or floating trackmaker (cf. McAllister this volume) appears to be militated against by the presence of small walking trackways at this horizon.

According to our interpretation, the pterosaurian trackmaker was pushing off with the left hand, and either

Figure 36.6. Individual photographs of tracks 2–5. The double string follows the line of progression of the trackway, joining individual tracks at their right edge. The siltstone infillings in tracks 3 and 4 have preserved track detail better than in tracks 1 and 5. The baby coelurosaur track (see Fig. 36.2) is barely visible in lower left position of photograph for track 4.

pushing with the right hand as well or not contacting the subsurface with it at all. Lack of contact of the right hand with the substrate, while the left hand pushed in consecutive sequence, could produce a curved path of progression.

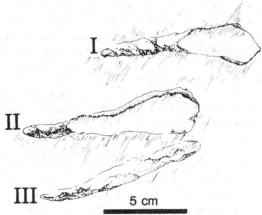

Figure 36.7. Photograph of track 6. This track is so far from track 5 that probably an intervening track is missing. Illumination from upper right. Line of progression at right angles to scale (cm), and upward in this orientation.

The first four stride lengths are 132 cm, 142 cm, 158 cm, and 183 cm respectively. The distance between tracks 5 and 6 is 585 cm; it is probable that an intervening track between these two (nos. 5 and 6 as here labeled) is not preserved. If such a "missing" track is interpolated the sequence becomes quite regular.

This rather regular increase in stride length from track no. 1 to no. 5 and possibly through no. 6 indicates acceleration. The distance between track no. 5 and no. 6 is too great to permit confident assignment of no. 6 as the next track in the progression unless a missing track is inferred. However it may be the next in sequence if it represents the last downward lunge and contact with the substrate prior to takeoff.

Unwin (this volume) argued that the morphology of the Clayton tracks does not fit theoretical constraints that he has developed in his predictive analysis. According to his prediction, during terrestrial locomotion, the wing had to be partly or fully extended (which had to produce a great spread between manus impressions [i.e., long pace and low pace angle]), and most or all of the manus would contact the substrate, including digits I–III, and the distal portion of metacarpal IV and the proximal part of the wing finger, digit IV. The Clayton tracks preserve three digital impressions with overall proportions that are consistent with the proportions of pterodactyloid manus digits I–III and not consistent with the predicted morphology of complete crocodilian or other archosaurian tracks. The circular depressions in several of the Clayton tracks in the position of the distal end of metacarpal IV are insufficiently preserved for confident determination of these features as track impressions rather than unassociated sedimentary features. In the extended flying position, the manus orientation would have digit I anterior, and the successive digits posterior to it. If the manus were used for pushing off, perhaps the rotational motion of the wrist prevented full

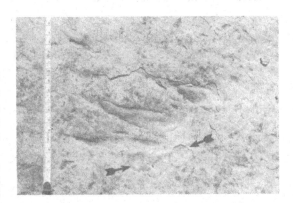

Figure 36.8. Detail photograph and line drawing, slightly enlarged, of track 4. The terminal extremities of digits I and II are depressions that extend from the sand cast positives in the proximal regions of these tracks. The two circular depressions (arrows) at the proximal end of digit III in the photograph, and to the lateral side of the same digit seem to be real features rather than artifacts of weathering; if so, they might represent distal extremity of metacarpal III and proximal extremity of digit IV (the wing finger), respectively. A similar depression at the base of digit III occurs in track 1 (Fig. 36.5), although it is less distinct.

contact with the substrate of digit IV and metacarpal IV. Finally, the orientation of the tracks with respect to the line of progression, at nearly right angles rather than almost parallel as in ordinary terrestrial animals, can be reconciled by pterosaurian manus anatomy, with the hand in extension.

Conclusions

The dinosaur–dominated ichnofauna at Clayton Lake State Park (Gillette and Thomas 1985) also exhibits a number of small and problematic tracks and traces, that should not be overlooked simply because they are inconspicuous or difficult to interpret (cf. Lockley and Gillette this volume). Part of the difficulty associated with interpreting small and problematic tracks arises from observational biases and inadequate descriptive treatments, as well as preservational problems. The controversial pterosaurian trackway provides a good example. When carefully described and analyzed there is little compelling evidence for dismissing a pterosaurian trackmaker or inferring a crocodilian origin (cf. Unwin this volume). The Clayton trackway is significantly different from *Pteraichnus* (Stokes 1957), the putative pterosaur trackway that was reinterpreted by Padian and Olsen (1984) as crocodilian. Ichnologists must be careful not to dismiss all possible pterosaurian tracks as crocodilian simply because the "type specimen" has been reinterpreted. Each track type must be carefully analyzed before either specific or generalized conclusions and inferences are proposed.

Acknowledgments

We are grateful to Martin Lockley for reading the original manuscript and suggesting a number of changes and amendments.

References

Currie, P. J. 1981. Bird footprints from the Gething Formation (Campanian, Lower Cretaceous) of northeastern British Columbia, Canada. *Jour. Vert. Paleont.* 1:257–264.

Gillette, D. D., and Thomas, D. A. 1985. Dinosaur tracks in the Dakota Formation (Aptian–Albian) at Clayton Lake State Park, Union County, New Mexico. *In* Lucas, S. G., and Zidek, J. (eds.). 1985. *Santa Rosa Tucumcari Region. New Mexico Geol. Soc. Guidebook 36th Field Conf.* Univ. New Mexico Press pp. 283–288.

Leonardi, G. (ed.). 1987. *Glossary and Manual of Tetrapod Footprint Palaeoichnology.* Departamento Nacional de Produçal Mineral Setor de Autarquias Norte, Brasília, Brazil. xi + 75 pp., 20 pls.

Lockley, M. G. 1987. Dinosaur footprints from the Dakota Group of eastern Colorado. *Mountain Geol.* 24:107–122.

Lucas, S. G., Hunt, A. P., Kietzke, K. K., and Wolberg, D. L. 1986. Cretaceous stratigraphy and biostratigraphy, Clayton Lake State Park, Union County, New Mexico. *New Mexico Geol.* 8:60–65.

Lucas, S. G., Holbrook, J., Sullivan, R. M., and Hayden, S. N. 1987. Dinosaur footprints from the Cretaceous Pajarito Formation, Harding County, New Mexico. *In* Lucas, S. G., and Hunt, A. P. (eds.). *Northeastern New Mexico. New Mexico Geol. Soc. Guidebook 38th Field Conf.* Univ. New Mexico Press pp. 31–33.

Mehl, M. G. 1931. Additions to the vertebrate record of the Dakota Sandstone *Amer. Jour. Sci.* 21:441–452.

Padian, K., and Olsen, P. E. 1984. The fossil trackway *Pteraichnus* not pterosaurian, but crocodilian. *Jour. Paleont.* 58:178–184.

Stokes, W. L. 1957. Pterodactyl tracks from the Morrison Formation. *Jour. Paleont.* 31:952–954.

37 Dakota Formation Tracks from Kansas: Implications for the Recognition of Tetrapod Subaqueous Traces

JAMES A. McALLISTER

Abstract

Criteria for the recognition of subaqueous tetrapod trails are compiled from the literature and are illustrated by one example of footmarks from the Dakota Formation of Kansas. The Kansas footmarks are most easily recognized by criteria which indicate tracemaker buoyancy. Such footmarks are characterized primarily by posterior overhangs and reflectures of the individual digit impressions; and secondarily by striations and claw marks along their length, and the often incomplete nature of the trails. These swim footmarks would be expected to grade into subaqueous traces formed by more typical terrestrial propulsion and demonstrate less buoyancy as the water becomes shallow, and disappear as the tracemaker becomes fully buoyant in deeper water and the digits no longer reach the substrate.

Introduction

References to fossil footmarks formed under water are scattered throughout the literature. Various criteria for recognition are used by the different authors. Some features are specific to individual trails and their depositional environments. The features I consider diagnostic or common will be summarized, with an emphasis on swim marks created by paddling.

Recently, Lockley (1986) discussed the paleoenvironmental and paleobiological significance of trackways including some from the Purgatoire River in southeastern Colorado. He noted that a set of sauropod tracks increased in depth in areas covered by shallow water compared to subareal tracks. The dryer sediment was firmer than the wet sediment. Additionally the tracks made under water appear to be overprinted as the velocity of the animal changed (Lockley et al. 1986). Lockley (1986) also referred to isolated toe impressions and elongate toe drag traces found in the Chinle Formation of eastern Utah and tentatively attributed them to swimming phytosaurs.

Boyd and Loope (1984) reinterpreted traces from the Triassic Red Beds of Wyoming which were commonly considered to be prod marks of driftwood. They reinterpreted the marks to be produced by swimming tetrapods. The criteria used to identify these traces as swim footmarks include striations on the surface of the mark, similar dimensions, posterior projections (overhangs), regular spacing and orientation, and reflectures (a continuation of the trace originating from the posterior of the track created by retraction of the foot). Webb (1980) and Peabody (1948) discuss similar traces from Triassic beds of the western U.S.A.

Currie (1983) described hadrosaur tracks from the Lower Cretaceous of western Canada. He interpreted some of the tracks as subaqueous by the following features: a decrease in stride length coupled with shallower heel imprints, indicating an increase in water depth; tracks which stop at one point and continue in a line parallel but shifted laterally to the original path, explained as an animal moved by currents while resting; the occurrence of clear isolated tracks, suggestive of a swimming animal which occasionally touched bottom; the exclusive occurrence of large footprints in some areas, possibly implying water too deep for small animals to reach the bottom.

Swim marks of a possible theropod from the Jurassic of Connecticut were described by Coombs (1980). These marks occur in two sizes. The middle digit III was longest and left both a claw and phalangeal pad impression. The lateral digits (II and IV) did not reach the substrate until after the middle digit was impressed to a short depth — they left only scratch marks due to their shorter length. Some scratch marks terminated posteriorly in small mounds accumulated by the plowing action of the digits. Coombs (1980, pp. 1198–1199) describes the action of the tracemakers as follows: "As the animal flexed the pes to propel itself forward, digit III acted as a fixed pivot with digits II and IV sliding slightly to the rear."

Brand (1979) reinterpreted the probable paleoenvironmental setting of the Coconino Sandstone (Permian) tetrapod trails as subaqueous. He compared trails of small living amphibians and reptiles made on dry, damp, wet, and subaqueous sand to the fossil trails. The laboratory trails form continuous traces across all the substrates. The fossil trails characterized by the presence of toe marks, short sole impression (compared to width), and uniform appearance throughout the trail most closely resembled the trails produced by living tetrapods on a subaqueous substrate.

Peabody (1948) considered some traces found in the Moenkopi sediments of the western United States to be swimming footmarks of tetrapods. The footmarks occur as partial impresssions made by the tips of the digits; usually only digits III and IV are impressed. The marks are interpreted to result from paddling limbs which graze the water-covered sediments in extensive shallow pools on a flood plain. Peabody referred to the lack of any clearly defined imprint of a foot, the lack of trail continuity, and the general appearance of the footmarks as being made by the tips of digits, as swim mark characteristics. Other characteristics which Peabody considered to be important for the interpretation include the lack of shrinkage cracks or salt pseudomorphs and the lack of ripple marks, which indicate a quiet and wet environment at these localities. These substrate characteristics are useful as supplementary evidence for the environment of footmark formation.

Bird (1944) described a sauropod trail which consisted almost entirely of forelimb footmarks. This example was interpreted as a floating animal propelling itself with kicks to the substrate with its forelimbs. Coombs (1975) also commented on these footmarks in relation to inferred aquatic habits and adaptations of sauropods (see Ishigaki this volume, and Lockley and Conrad this volume).

Toepelman and Rodeck (1936) mention the possibility that footmarks observed from the Late Paleozoic of Colorado might be subaqueous. These marks were only slightly impressed into what they considered to be very soft sediment and so may have been produced by a buoyant tracemaker.

Willard (1935) also discussed the possibility that some Devonian tetrapod traces from Pennsylvania were subaqueous. He noted isolated tracks and skewed ratios of manus to pes marks, and compared the inferred behavior represented by the fossil traces to the locomotory behavior of the living aquatic axolotl, *Amblystoma tigrinum*.

Experiments with a living salamander, *Triturus* (= *Triton*) *alpestris*, were performed by Schmidtgen (1927) and were used for comparison to a Permian subaqueous trail from Germany. Recent salamander tracks made on soft mud were characterized as clear complete tracks. The tail left a straight drag mark. The trail produced by the same animal on mud covered by a few millimeters of water were different due to the slight buoyancy. For example, the tail drag was more sinuous; the footmarks were mainly long parallel scratches; the hindfoot created eddies during propulsion which left a thin covering of mud on the footmarks; the posterior portion of the forefoot mark (sole impression) was less likely to be preserved compared to the hindfoot

mark due to the shorter length of the foreleg.

Notable references which document the aquatic locomotion of modern reptiles are Zug (1971) and Sukhanov (1974). Although they do not consider trails produced by aquatic locomotion they do describe locomotor changes from the terrestrial to aquatic environment. Zug (1971) noted that the extra buoyancy in bottom-walking turtles allowed an increase of limb speed, a decrease in the support phase of the body, changes in the gait sequence and greater variation in the timing of limb movement. Aquatic locomotion is briefly discussed by Sukhanov (1974) for turtles and various lizards, with a similar theme of major locomotory differences based on environment.

Formation of the Kansas Trails

In 1933 a large set of footmarks was collected from the Dakota Formation (Cretaceous) of Kansas (KUVP 5914). The footmarks have often been interpreted as skid marks of a quadrupedal reptile on a muddy delta flat. The traces were originally impressed into a muddy matrix and later filled in with a fine sandstone. Recent re-examination of these traces (McAllister manuscript) suggests that they were formed by floating animals. The tracemakers are interpreted to be ornithischian dinosaurs based on deduced foot morphology (digit counts and digit lengths) and mode of locomotion (discontinuous propulsion). A brief review of these traces will serve as an example of one set of subaqueous swim marks produced by a paddling tetrapod.

Important differences between locomotion on land and in water can be attributed to buoyancy. In a floating animal the digits can extend farther posteriorly in the propulsive phase without unbalancing (losing the necessary support to maintain posture) the organism. This allows the propulsive force to be on a more horizontal plane and scrape instead of compressing downward into the sediment.

A paddling animal can produce a wide variety of traces with this horizontal motion depending on the degree of foot-sediment contact. For example, it would be expected that, if the digits barely touch the substrate while paddling, the foot would not encounter much resistance from the substrate and would then continue posteriorly along the arc described by the limb. The resultant footmark would be an elongated, striated scratch mark in the substrate, deepest in the middle. If the phalanges extended a little farther into the substrate, the substrate would resist backward movement and the limb would not complete the arc as just described. The Kansas footmarks are interpreted to have been formed in this manner without (Fig. 37.1A) and with (Fig. 37.1B) resistance from the sediment.

Under certain conditions, a mold of the digits can form which points posteriorly to the travel direction. A cast of the mold would have a posterior overhang. The overhang indicates a near-weightless condition in which the digits delivered a propulsive force on a more horizontal plane and extending farther posteriorly than is possible in a terrestrial situation. The trace in Figure 37.2 is interpreted in this manner.

Figure 37.2. Stereopair of left manus mark from the trail block KUVP 5914. The scale bar length is 5 cm.

Figure 37.1. Stereopairs of **A,** right manus mark and **B,** right pes mark from the trail block KUVP 5914. The scale bar length is 5 cm.

Figure 37.3. Stereopairs of **A,** right manus mark (note the short reflectures which form a 45 degree angle with the posterior end of the digit impressions); and **B,** right pes mark from trail block KUVP 5914. Scale bar length equals 5 cm.

Reflectures are another indication of swim traces. These are marks made by the digits during retraction. The aquatic reflectures are different from terrestrial drag marks often seen in tetrapods with unequal and elongate digits. The distinguishing difference is the placement of the extended digits in the footmark from where the retraction phase begins. The tips of the digits are placed at the anterior end of the track in terrestrial locomotion. Drag marks of the digits are formed in front of the track and tend to be slightly lateral to the corresponding impression in the track. In the swim footmarks the terminal ends of the digits are placed posteriorly in the trace at the end of the propulsive phase, so during retraction the drag marks begin farther posterior in the footmark. The buoyant tracemaker can make long impressions with its digits in the substrate during propulsion, and reflectures continue within or alongside these initial impressions during retraction. The track shown in Figure 37.3A is interpreted to have been formed in this manner.

A variation of the reflecture just decribed is the postero–medial extension of digit II observed in some of the Kansas footmarks (Fig. 37.3B). This is interpreted to be a little final push into the sediment by the tip of digit II as the leg ends the propulsive phase. This is similar to using a pole to push a raft. The striations along the imprint of digit II are continuous with these extensions, which confirms the continuity of the imprint with the extensions.

The best criteria for the determination of swim traces are those which indicate buoyancy of the tracemaker. For

example, there would be problems with the interpretation of the Kansas footmarks if they are considered terrestrial because the ratio of manus to pes footmarks is 32 to 57. The ranges of pace angles in the trails and the footmark lengths within a trail are highly variable. Some trails end abruptly and some footmarks are isolated. Although there can be alternative explanations for these problems in a terrestrial setting, e.g., underprints (sensu Lockley and Conrad this volume), an interpretation as aquatic traces also resolves these problems to a significant degree.

Inferences of aquatic adaptations of the tracemaker may be included as corollary evidence for subaqueous trace formation. The Kansas footmarks, especially those created by the pedes, seem to have "web" impressions between the digits (Fig. 37.1B). These are indicated by the indentations in the interdigital regions. There are antero–posterior lineations along these impressions. These impressions may have been made by interdigital webs that scraped across the substrate. Alternatively, the length of the digits may have been impressed to the depth of the metapodials creating striated impressions by the movement of the palmar and plantar surfaces at these joints.

The second hypothesis, or a combination of the two (e.g., a short web), may be more plausible than the presence of an extensive web in the Kansas tracemaker for the following reasons: the digits and hence the distance to the metapodial joint are short; the scratch marks and digit impressions occur anterior to the beginning of the interdigital impressions; and some footmarks lack interdigital impressions. Either case would make subaqueous formation more plausible, either as corollary evidence of aquatic adaptation or as an indication of the path of the metapodial–phalangeal joint in a non-terrestrial manner.

Finally, the sediment and evidence of the paleoenvironment may corroborate the interpretation of subaqueous formation of the traces — for example, other associated subaqueous fauna or flora, or sedimentary features such as ripple marks or scours. These are not infallible indicators as the water level may have changed between the times when the trails were made and associated material was deposited.

Discussion

Variation in the nature of subaqueous traces can be attributed to different water depths. For example the small salamander trail described by Schmidtgen (1927) was produced in water too shallow for the body to be fully buoyant. The salamanders left body and tail drags. Peabody (pers. comm. in McKee 1947, p. 27) refers to the aquatic locomotion of salamanders as follows: "When leaving the water, it swims to shore and then walks out on the land. In water which is too shallow to cover it, the salamander wiggles along with a combination of swimming and walking movements which produce a trackway open to only the broadest interpretation."

Most other subaqueous trails described in the literature were impressed in deeper water, indicating bodies buoyed well up from the substrate. With an increase in water depth there may be such changes in the trail of a generalized tetrapod as: (i) tail and body drags disappear — if the tracemaker typically had them in terrestrial locomotion; (ii) stride length decreases corresponding with difficulty of wading; however an increase in stride length may be expected as the body becomes more buoyant and remains suspended over the substrate or if the paddle interval becomes longer in duration; (iii) the smaller footmarks become fewer as shorter limbs cannot reach the substrate at increasing depth; (iv) prints become poorly impressed as the weight of the tracemaker on the sediment is reduced by buoyancy (Currie 1983); (v) footmarks become more elongate, scratch-like and finally disappear.

In relation to body size there is only a narrow band of water depths in which an animal can create traces. Locomotion in relatively shallow water allows the majority of the body weight to be translated through the feet as in typical terrestrial locomotion (Fig. 37.4A). Recognition may be difficult without sedimentological data or direct comparison to terrestrial trails (Lockley 1986, Lockley et al. 1986, Schmidtgen 1927). As water depth increases the body becomes buoyant (Fig. 37.4B) and characteristics associated with buoyancy and swimming become apparent — especially if the tracemaker is paddling (Boyd and Loope 1984, Coombs 1980, Webb 1980, the Cretaceous footmarks from Kansas). There are no traces when the water depth exceeds the length of the limb from the digit tips to the

Figure 37.4. Discontinuous progression (paddling). **A**, tetrapod in shallow water. Progression is similar to terrestrial locomotion. **B**, deeper water buoys the tetrapod. Extended digits paddle and reach the substrate, resulting in elongate scratches. **C**, water depth exceeds the limbs' reach. No trace from the tetrapod is left on the substrate.

center of buoyancy (Fig. 37.4C) unless the animal is capable of swimming underwater.

No one model for subaqueous trail recognition can be produced due to the wide range of animal adaptations. For example, animals such as the pachyostotic hippos, turtles, and some small tetrapods (Brand 1979) can walk on the sustrate below the water surface. Radical changes in locomotor pattern from terrestrial to aquatic environments can also create exceptions.

Tetrapods also can locomote in a variety of ways (Braun and Reif 1985). However, most tetrapods which would be expected to leave subaqueous traces locomote with either discontinuous propulsion (propulsion with a recovery stroke — paddlers and rowers) or continuous subundulatory propulsion (undulation which has no recovery stroke). The dichotomy of these two broad locomotor patterns would be reflected by differences in the trails produced. Paddlers would tend to progress into the water in a manner similar to the way they walk on land. As the water deepens they would go through a transition to a paddling or rowing pattern. Subundulatory locomotors would tend to locomote in a terrestrial manner up to the water line and then accentuate the undulations of the body with less or no reliance on the limbs for progression. As an example, upon entering the water, crocodylomorphs immediately progress by lateral undulation of the tail and body, tucking their limbs by their sides to reduce drag (Fig. 37.5).

Some factors which are common to aquatic tracemakers and their swim traces are based on buoyancy, despite the great influence of body type, size, water depth, sediment differences, and locomotor pattern. In the paddlers, buoyancy of the tracemaker can be suggested by the presence of posteriorly directed overhangs in casts, and reflectures. The elongated and typically striated nature of the footmarks, and the preferential impressions of the most distal foot parts also can be indicators of subaqueous formation. Buoyancy in the undulators can be indicated by increased sinuosity of the trail, presence of only the distal ends of the digits forming long parallel scratches, and partial impressions of the sole. These latter criteria are less diagnostic than the swim trace criteria as they could also occur in terrestrial trails. They are useful nevertheless, and have been produced in laboratories by living tetrapods for comparison with fossil trails (Schmidtgen 1927, Brand 1979).

Figure 37.5. Tetrapod using continuous subundulatory locomotion.

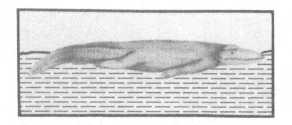

The principal factors which affect the morphology of footprints (tracemaker anatomy, kinematics of locomotion, and substrate condition) have been diagrammatically represented by Padian and Olsen (1984) in a ternary diagram. The apices of the triangle represent the principal factors, and individual trace examples can be placed within the triangle in the approximate position representing the extent to which each factor influenced the trace morphology. In the case of the Kansas swim traces, not only do these three factors exert major influences but the degree of buoyancy of the tracemakers also adds an important dimension to the trace morphology. The addition of another apex, to create a three-cornered pyramid, allows for such special cases where three factors are not sufficient. In the example of the Kansas trails, the degree of buoyancy exerted a dominant influence which also influenced the stride, the frequency of paddling, and the ability of the tracemaker to impress the sediment.

It is important to have well defined criteria for interpreting subaqueous marks. Buoyancy evidence tends to be best, but subaqueous marks can be produced in a continuum from the shoreline to specific depths. Paddlers can produce shallow water trails very similar to terrestrial trails. For example, soft water–logged clay could preserve distorted tracks, slippage marks, have mud backflows and induce irregular gaits. It would be difficult to distinguish these from a shallow subaqueous walking trail if the other diagnostic criteria were not preserved.

Terrestrial trails could also possess many non-diagnostic characteristics of subaqueous trails. For example, they might preserve missing, irregular, or indistinct marks or portions of the trail. These characters could be related to the substrate competency or topography, locomotory shifts, or even postdepositional scouring or weathering. A propensity of the trail to preserve only large footmarks may not always be related to the ability of the tracemaker to reach the subaqueous substrate; alternatively they could be related to the ability of the substrate to resist impressions from smaller individuals, or to produce various underprints (Lockley and Conrad this volume).

Summary

Familiarity with the concept of subaqueous trace formation helps in their recognition. Proper interpretation of footmark formation helps in understanding their distribution, variability, identification of possible tracemaker affinity as well as understanding habitat and behavior. For example, the Kansas footmarks indicate swimming ability of some large reptile tracemaker which could enter relatively deep water and propel itself with paddle motion.

Subaqueous traces of tetrapods vary with the tracemaker and with water depth. The footmarks from Kansas are interpreted as those of a buoyant animal swimming over a subaqueous substrate. The evidence includes the posterior overhang, reflectures of the digits, the sharp increase in depth of the footmarks corresponding to the arc of the limb during propulsion, elongation of the footmarks, and striations parallel to the direction of movement.

I consider the Kansas footmarks to be good examples of buoyant tetrapod swimming trails. However, other types of subaqueous trails are produced differently and so require different recognition criteria. Variation in trails can be expected for different kinds of tracemakers (varying with body size and shape, foot morphology, locomotor pattern and speed), substrates (topography, composition, and consistency), and degree of buoyancy. Despite these variations, the most important criteria for recognition of subaqueous traces are considered to be buoyancy indicators inherent in the footmarks.

Acknowledgments

I thank the following people for their help during this project: Drs. H.-P. Schultze and Frank Cross, University of Kansas, Lawrence, Kansas, for their guidance; Dr. Detlev Thies, University of Hannover, West Germany, for the German literature translation; Deborah McAllister for manuscript assistance; and the anonymous reviewers who played a central role in the improvement of the manuscript.

References

Bird, R. T. 1944. Did *Brontosaurus* ever walk on land? *Nat. Hist.* 53:61–67.

Boyd, D. W., and Loope, D. B. 1984. Probable vertebrate origin for certain sole marks in Triassic Red Beds of Wyoming. *Jour. Paleont.* 58 (2):467–476.

Brand, L. 1979. Field and laboratory studies on the Coconino Sandstone (Permian) vertebrate footprints and their paleoecological implications. *Palaeogeog., Palaeoclimat., Palaeoecol.* 28:25–38.

Braun, J., and Reif, W.-E. 1982. A survey of aquatic locomotion in fishes and tetrapods. *Neues Jahrbuch für Geologie und Palaeontologie* 169:307–332.

Coombs, W. P. 1980. Swimming ability of carnivorous dinosaurs. *Science* 207:1198–1200.

——— 1975. Sauropod habits and habitats. *Palaeogeog., Palaeoclimat., Palaeoecol.* 17:1–33.

Currie, P. J. 1983. Hadrosaur trackways from the Lower Cretaceous of Canada. *Acta Palaeont. Polonica* 28 (1–2):63–73.

Lockley, M. G. 1986. The paleobiological and paleoenvironmental importance of dinosaur footprints. *Palaios* 1:37–47.

Lockley, M. G., Houck, K. J., and Prince, N. K. 1986. North America's largest dinosaur trackway site: Implications for Morrison Formation paleoecology. *Geol. Soc. Amer. Bull.* 97:1163–1176.

McAllister, J. A. Subaqueous vertebrate footmarks from the upper Dakota Formation (Cretaceous) of Kansas, U.S.A. Manuscript submitted to *Occasional Papers, Museum Nat. Hist, Univ. Kansas Lawrence, Kansas.*

McKee, E. D. 1947. Experiments on the development of tracks in fine cross-bedded sand. *Jour. Sedim. Petrol.* 17 (1):23–28.

Padian, K., and Olsen, P. E. 1984. The fossil trackway *Pteraichnus:* not pterosaurian, but crocodilian. *Jour. Paleont.* 58 (1):178–184.

Peabody, F. E. 1948. Reptile and amphibian trackways from the Lower Triassic Moenkopi Formation of Arizona and Utah. *Univ. California Public. Zool.* 27 (8):295–468.

Schmidtgen, O. 1927. Eine neue Fahrtenplatte aus dem Rotliegenden von Nierstein am Rhein. *Palaeobiologica* 1:245–252.

Sukhanov, V. B. 1974. *General System of Symmetrical Locomotion of Terrestrial Vertebrates and Some Features of Movement of Lower Tetrapods.* (New Delhi: Amerind Publishing Company Limited) 274 pp.

Toepelman, W. C., and Rodeck, H. G. 1936. Footprints in Late Paleozoic red beds near Boulder, Colorado. *Jour. Paleont.* 10:660–662.

Webb, S. K. 1980. Early Triassic tetrapod trackways in the Moenkopi Formation of southeastern Utah. *Geol. Soc. Amer. Abst. Prog.* 12:308.

Willard, B. 1935. Chemung tracks and trails from Pennsylvania. *Jour. Paleont.* 9 (1):43–56.

Zug, G. R. 1971. Buoyancy, locomotion, and morphology of the pelvic girdle and hindlimb, and systematics of Cryptodiran turtles. *Misc. Publ. Museum Zool. Univ. Michigan* 142:1–98.

38 Dinosaur Footprints from the Lower Cretaceous of Cameroon, West Africa

LOUIS L. JACOBS,
KATHRYN M. FLANAGAN,
MICHEL BRUNET,
LAWRENCE J. FLYNN, JEAN DEJAX,
AND JOSEPH VICTOR HELL

Abstract

A trackway of four tridactyl bipedal dinosaur footprints from the Koum Basin near the Yola Arm of the Benue Trough are the first reported dinosaur tracks from the Lower Cretaceous of Cameroon. Sediments of the Benue Trough and related structures were deposited concommitant with the opening of the South Atlantic, and therefore have major biogeographic significance. Study of dinosaur–bearing Lower Cretaceous rocks of Cameroon is in its infancy.

This note is to report results of preliminary field work in Lower Cretaceous rocks of Cameroon, West Africa, including the discovery of dinosaur footprints. In Africa, fossils of Mesozoic vertebrates are known from Triassic, Jurassic, and Cretaceous rocks. African Cretaceous terrestrial bones and footprints are known from as close to Cameroon as Niger (Ginsburg et al. 1966, de Broin et al. 1974, de Lapparent 1960, Taquet 1976).

In the past two field seasons, an international project with the acronym PIRCAOC (Projet International de Recherche dans le Cénozoïc/Crétacé d'Afrique de l'Ouest [Cameroun]) has prospected six areas of Cretaceous outcrops in northern and northeastern Cameroon (Fig. 38.1). These are (1) the Hama Koussou Basin, which has a repetition of sandstone, shale, limestone, and some pillow basalts, (2) the Mayo Oulo Basin with predominantly green nearshore shales, (3) the Figuil Basin, the poorly exposed (4) Mbere River and (5) Vina Basins, and (6) the Koum Basin. The fish Lepidotes is abundant in sediments of the Hama Koussou and Figuil basins, including more–or–less complete specimens from the Figuil Basin. The first dinosaur bones discovered in Cameroon came from the Hama Koussou Basin (Brunet et al. 1986). Unfortunately, all are too fragmentary for more precise identification. Other preliminary results are given by Flynn et al. (in press). The

dinosaur footprints reported here, along with osteological remains subsequently discovered, are the first vertebrate fossils reported from the Koum Basin. As such, they con-

Figure 38.1. Location map of Cameroon, West Africa, showing basins with Cretaceous rocks prospected by PIRCAOC field parties in 1985 and 1986: (1) Hama Koussou Basin, (2) Mayo Oulo Basin, (3) Figuil Basin, (4) Mbere River Basin, (5) Vina Basin, and (6) Koum Basin.

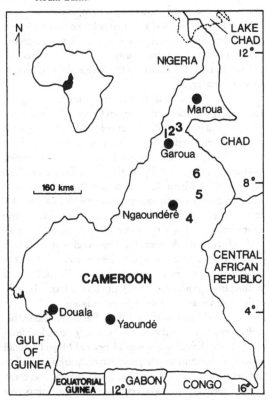

tribute both paleobiological and paleoenvironmental evidence in an area of the world that is extremely poorly known in terms of its terrestrial vertebrate fauna.

A major feature of the geology of West Africa is the Benue Trough, an aborted arm of a rift–rift–rift triple junction (Grant 1971, Mascle 1977 and references therein). Activity on the other two arms led to the opening of the South Atlantic, usually considered to have occurred about 115 million years ago (Aptian). The Benue Trough received up to 5500 meters of Cretaceous mainly marine sediments and rare volcanics. It extends from the Gulf of Guinea and forks into a "Y" shaped structure. One fork of the "Y" heads north into Chad where it is obscured by Quaternary deposits. The other fork of the "Y" is directed more to the east and southeast. The southeastern extension of the Benue Trough is called the Yola Arm, and it stretches across northern Cameroon. The Benue Trough provides a setting for the preservation of Cretaceous fossils. This is particularly significant because of the depositional setting relative to the opening of the South Atlantic.

Most investigations of the Benue Trough have concentrated on the portion that occurs in Nigeria and emphasize marine strata, terrestrial sediments apparently being rare. Few detailed studies relating directly to terrestrial vertebrate fossils have been published. From studies of marine sediments, mainly but not exclusively located outside of Cameroon, it appears that the upper Benue Trough of Nigeria and Cameroon has experienced more than one episode of tectonism in the Cretaceous associated with the opening of the South Atlantic (Wright et al. 1985, Enu 1986). The first of these is apparently the most relevant to the dinosaur–bearing basins of northern Cameroon. Beginning in the Albian, marine and deltaic sedimentation began in the Yola Arm of the Benue Trough (see Enu 1986, Petters 1978).

The Koum Basin (indicated by the 6 on Fig. 1), southeast of the Yola Arm, contains Cretaceous continental deposits (Tillement 1970), and probably formed as a graben associated with rifting during the early formation of the Benue Trough. While we have not observed marine sediments in the Koum Basin, the Doba Trough of Chad lies east of the Koum Basin, and from borehole evidence contains marine Cretaceous sedimentary rocks (Cratchley et al. 1984). The precise chronological, structural, and stratigraphic relationships of the Koum Basin to the main body of the Yola Arm on the one hand and the Doba Trough on the other remain to be determined, but it is clear that the sedimentary regime of the Koum Basin is distinct from that of the Yola Arm and the Doba Trough, at least so far as present evidence indicates.

The sediments of the Koum Basin are reasonably well exposed in a band at least 50 km long extending from the village of Krouc in the west to beyond Mayo Djarendi in the east. Sediments are predominantly arkosic sand to conglomerate and purple, maroon, and red silts and clays. In the course of a paleontological survey near Rhinoceros Camp, south of the village of Koum, a trackway (designated

Locality KB 3) comprising four tridactyl prints of a bipedal dinosaur was found in reddish sandstone. The sandstone is ripple marked in places, demonstrating that the dinosaur printmaker was walking in shallow water. Fresh water branchiopods *(Estheria)* were found nearby. Sedimentary environments in the vicinity of the dinosaur trackway are predominantly fluviatile rather than marine deltaic or coastal. This is consistent with Tillement (1970).

Measurements of the trackway are given in Table 38.1. Figure 38.2 shows the best preserved of the prints. The size of the prints (averaging approximately 39 cm in length) and the tridactyl morphology are consistent with either iguanodont or theropod morphology. The digits are not squat as would be expected in iguanodonts and there appear to be remnants of claw marks on some digits. These

Table 1. *Measurements of the Koum Basin trackway.*

Stride (mean)	= 213 cm
Pace (mean)	= 114 cm
Step angle	= 155° (estimated from photographs)
Trend	= 122°

	Length (maximum)	Width (maximum)
footprint 1	39 cm	25 cm
footprint 2	42 cm	28 cm
footprint 3	37 cm	25 cm
footprint 4	39 cm	27 cm

Figure 38.2. Dinosaur print 3 from the Koum Basin (maximum length = 37 cm, maximum width = 25 cm; scale is a 6 inch and 15.2 cm plastic ruler).

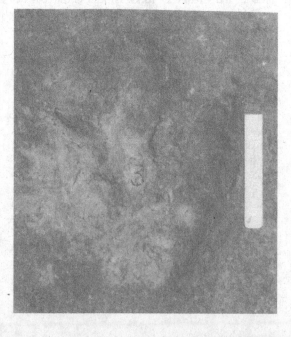

attributes suggest that the printmaker was a theropod. The internal digit of each print is recurved. Both iguanodont and theropod teeth were discovered subsequently at Locality KB 6 near Mayo Djarendi, along the Mayo Rey west of the trackway (KB 3).

Theropod tracks are known from Lower Cretaceous rocks on most continents. The prints from Cameroon provide only general taxonomic information, but they do resemble those that occur in the Lower Cretaceous of Brazil (Leonardi personal communication and 1979), a geographic area that represents the counterpart to Cameroon during the opening of the South Atlantic. An improved footprint record in Cameroon would be useful in terms of better understanding the depositional history of the Koum Basin, the faunal composition of Africa during the Cretaceous, and the relationships of the African fauna to South American.

Thus, the full significance of the Koum Basin tracks must be considered in light of their geologic occurrence. Field work is in its infancy. Well preserved terrestrial vertebrate fossils have only recently begun to be discovered. Sediments in the Koum Basin were deposited in the context of the opening of the South Atlantic. Specifically, the Mayo Rey Basin has terrestrial fossiliferous sediments that provide a window into the Lower Cretaceous of Africa at its western margin.

Acknowledgments

We thank David Pilbeam, Abel Brillanceau, Sevket Sen, and all other members of PIRCAOC. Funding was provided by National Geographic Society Grant 2942-84 and NSF Grant BRS-8419703 to David Pilbeam, NSF Grant BSR-8700539 to LLJ and LJF, by grants from MRE, CNRS, Université de Poitiers to Michel Brunet, and by a grant from IRGM at Yaoundé (MESRES). Air transport was arranged through the cooperation of the international carrier UTA. Assistance was provided by Alisa J. Winkler and John D. Congleton. Special thanks are due the government and people of Cameroon. This paper is a contribution to IGCP-245, Nonmarine Cretaceous Correlations.

References Cited

Broin, F. de, Buffetaut, E., Koeniguer, J. C., Rage, J. C., Russell, D., Taquet, P., Vergnaud-Grazzini, C., and Wenz, S. 1974. La faune de vertébrés continentaux du gisement d'In Beceten (Sénonien du Niger). *Compte-Rendu Acad. Sci. Paris D* 279:469-472.

Brunet, M., Coppens, Y., Pilbeam, D., Djallo, S., Behrensmeyer, K., Brillanceau, A., Downs, W., Duperon, M., Ekodeck, G., Flynn, L., Heintz, E., Hell, J., Jehenne, Y., Martin, L., Mossser, C., Salard-Cheboldaeff, M., Wenz, S., and Wing, S. 1986. Les formations sédimentaires continentales du Crétacé et du Cénozoique camerounais: premiers résultats d'une prospection paléontologique. *Comptes-Rendu Acad. Sci. Paris* Série II 303 (5):425-428.

Cratchley, C. R., Louis, P., and Ajakaiye, D. E. 1984. Geophysical and geological evidence for the Benue-Chad Basin Cretaceous rift valley system and its tectonic implications. *Jour. African Earth Sci.* 2 (2):141-150.

Enu, E. I. 1986. Influence of tectonics and paleoenvironment on Late Cretaceous clay sedimentation in the upper Benue Trough, Nigeria. *Geol. Jour.* 21:93-99.

Flynn, L. J., Brillanceau, A., Brunet, M., Coppens, Y., Dejax, J., Duperon-Laudouneix, M., Ekodeck, G., Flanagan, K.M., Heintz, E., Hell, J., Jacobs, L. L., Pilbeam, D., Sen, S., and Djallo, S. In press. Vertebrate fossils from Cameroon, West Africa. *Jour. Vert. Paleont.*

Ginsburg, L., Lapparent, A. F. de, Loiret, B., and Taquet, P. 1966. Empreints de pas de Vertébrés tétrapodes dans les séries continentales a l'Ouest d'Agadès (République du Niger). *Comptes-Rendus Acad. Sci. Paris* 263 (1966):28-31.

Grant, N. K. 1971. South Atlantic, Benue Trough, and Gulf of Guinea Cretaceous triple junction. *Geol. Soc. Amer. Bull.* 82:2295-2298.

Lapparent, A. F. de. 1960. Les dinosauriens du "Continental intercalaire" du Sahara central. *Mém. Soc. Géol. France* 88A:1-57.

Leonardi, G. 1979. New archosaurian trackways from the Rio do Peixe Basin, Paraiba, Brazil. *Annali dell'Universita di Ferrara* Nuovo Serie 5 (14):239-250.

Mascle, J. 1977. Le Golfe de Guinee (Atlantique Sud): an exemple d'évolution de marges Atlantique en cisaillement. *Mém. Soc. Géol. France* 55 (128):1-104.

Petters, S. W. 1978. Stratigraphic evolution of the Benue Trough and its implications for the Upper Cretaceous paleogeography of West Africa. *Jour. Geol.* 86:311-322.

Taquet, P. 1976. Géologie et paléontologie du gisement de Gadoufaoua (Aptien du Niger). *Cahiers Paléont. Editions du CNRS.* 191 pp.

Tillement, B. 1970. Hydrogéologie du Nord-Cameroun. *Bull. de la Direction des Mines et de la Géologie* 6:1-294.

Wright, J. B., Hastings, D. A., Jones, W. B., and Williams, H. R. 1985. *Geology and Mineral Resources of West Africa.* (London: George Allen and Unwin, Ltd.) 187 pp.

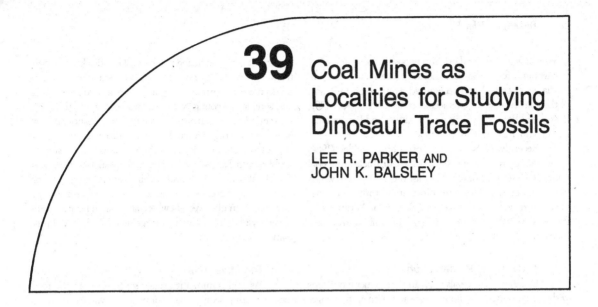

39 Coal Mines as Localities for Studying Dinosaur Trace Fossils

LEE R. PARKER AND
JOHN K. BALSLEY

Abstract

Dinosaur footprint casts have been observed at three major horizons within coal mines in the Upper Cretaceous (Campanian) Blackhawk Formation near Price, Utah. These include the mine floor, the mine roof, and in sediment above the roof seen only in rock–fall areas. Carbonaceous sandstones in the floor were formed as brackish and subsequently freshwater swamps became established on foreshore sediment, close to the shoreline. A small, bipedal animal producing footprints 11 cm long walked in one of these youthful swamps. Sediment exposed in the mine roof, after coal is removed, often has preserved in it thousands of footprint casts of numerous types and sizes of animals which had walked on the peat blanket of much older swamps. In certain areas the soft peat surface had been heavily bioturbated by dinosaur activities with many footprints partly overlapping and obscuring previous tracks. Fossil plants occur with many of these footprints.

Selected coal pillars which support the roof have been removed in some mines, allowing overburden to fall to the floor. These localized rock–fall areas display several levels of fluvial and lacustrine sedimentation some distance above the surface of the coal. It is evident in these units that dinosaurs walked in lake–margin muds, often causing footprint-bioturbation here too. Tail drag features are present in one mudstone. In addition, more than 50 tracks of a large bird which produced footprints 15 cm long occur in a 2 x 2.5 m area of mudstone.

Introduction

This report describes natural casts of dinosaur trace fossils we have observed in coal mines of east–central Utah near the cities of Helper and Price. All mines have been within coals of the Upper Cretaceous (Campanian) Blackhawk Formation. We have been able to examine twelve mines within several coal beds and have seen footprint casts in most of them. However, because of certain mining techniques and various factors of preservation, most mines display few specimens of significance. Only four mines have abundant tracks where thousands have been seen (Balsley and Parker 1983, Parker and Rowley this volume, Parker and Balsley in prep.). They are so common in some areas of mine roof that they overlap and obscure one another. In addition to these in Utah, we have been able to study mines in time-equivalent coals near Rock Springs, Wyoming (Rock Springs Formation), and near Cuba, New Mexico (Menefee Formation), where we have seen similar dinosaur traces.

Previous Work

Peterson (1924) first reported natural casts of dinosaur footprints from Rocky Mountain coal mines. He illustrated a track collected in the roof of the now–abandoned Castlegate Mine (Kenilworth Coal, Blackhawk Formation), north of Helper, Utah, and sketched a bipedal trackway in the now–abandoned Ballard Mine (Neslen Coals, Price River Formation) north of Thompson. He also mentioned a collection from the Panther Mine (Castlegate B Coal, Blackhawk Formation), north of Helper. Subsequently, Lull (Strevell 1932) sketched a bipedal trackway and described 8 species of "Dinosauropodes" from the now–abandoned Standard Mine (Castlegate A and B Coals, Blackhawk Formation) west of Helper. This collection is housed in the Utah State Natural History Museum, University of Utah, in Salt Lake City (Frank L. DeCourten pers. comm.). A popular account of mine footprints has been written by Wilson (1969). Although Lull's study (Strevell 1932) represented an excellent beginning for studying the biology of the Cretaceous swamp fauna, there were no further studies of dinosaur traces in Utah until we began recording them as part of our examination of the geology, paleontology and paleoecology of the Blackhawk Formation (Parker 1976, 1979). During field studies we began using the planar surfaces which coal mine roofs and floors provided as a way to extend outcrop exposures of sedimentary features and fossil plants. Dinosaur trace fossils were immediately observed and recognized as an important part of the paleontology of Cretaceous coals (Parker 1980, 1981; Parker and Balsley 1977, in prep.; Balsley and Parker 1983; Parker and Rowley this volume).

Recently, Lockley et al. (1983) and Lockley (1986) have photographed and mapped dinosaur tracks from coal mines of western Colorado. They list previous studies which have mentioned or illustrated dinosaur footprints in mines of east–central Utah and western Colorado. Additionally, Bass et al. (1955) and Ratkevich (1976) have illustrated single footprints from Colorado mines.

Horizons of Preservation

Dinosaur trace fossils occur in three general horizons within coal mines: the floor, deposited with initial stages of swamp development; rock of the immediate roof which was in direct contact with the coal being mined; and in areas where roof-rock has fallen exposing fluvial and lacustrine units as much as 2 m above the coal surface.

The Mine Floor

Occasionally, large thick slabs of coal and rock from mine floors are warped or buckled upward, sometimes violently, because great amounts of energy are directed into the floor from overburden weight by way of coal pillars. Such features are termed mine "bumps" and are of some significance in mine safety (Osterwald and Dunrud 1966). Three small dinosaur tracks were observed on the lower surface of slabs which had been loosened from the floor of a mine in the Kenilworth Coal, near Kenilworth, Utah, east of Helper. These specimens were part of a sequence of left and right footprints made by a single small three-toed bipedal animal. One specimen (Fig. 39.1) was removed from a carbonaceous sandstone slab 2.3 m long, 2 m wide and 25 cm thick. Broad dicotyledon leaves occur on the surface of this rock close to the footprint (Fig. 39.1). On other slabs, more fossil plants occurred at the same stratigraphic level including fragments of the palm *Phoenocities imperialis*, plus leafy twigs of the conifers *Sequoia cuneata* and *Moriconea cyclotoxon*. About 18 cm below these tracks, leafy twigs of a third conifer, *Araucaria* sp., are abundant, and the only fossil plant collected at that stratigraphic level.

Primary sedimentary features of this mine floor indicate that leafy twigs of *Araucaria* sp. were deposited on the surface of foreshore or beach sands in brackish mangrove-like plant communities, similar to certain extant communities in the Southern Hemisphere where *Araucaria* sp. is the only woody plant preserved (Balsley and Parker 1983). These swamps developed very close to the shoreline. With seaward progradation of the shoreline, these paralic swamps eventually became less influenced by marine conditions. This allowed several plant communities composed of many ferns, conifers, broadleaved trees, and palms to succeed the *Araucaria*-dominated community. They became the

important peat–producing plants of all the Blackhawk coals (Parker 1976, Balsley and Parker 1983). The animal which made these footprints was walking on a mixture of peat and sand in the young freshwater swamp. Sand in this environment is of the same type as that in the foreshore below. It was apparently blown landward into the swamps or deposited from occasional wash–over fans during storms (Balsley and Parker 1983). The swamp surface here was slightly above sea level and within a short distance, 1 or 2 km, of the shoreline. Footprint casts of this size and type have not been observed elsewhere and may have been produced by an animal which was restricted to paralic delta-plane environments.

The Mine Roof

As coal is mined it usually separates from the roof rock, exposing sedimentary features and various types of fossils which occurred on the immediate surface of the Cretaceous swamps. Roof rock is usually of fluvial origin, deposited into the swamps as overbank sediment when local rivers flooded. Footprints in the peat of the swamp surface were filled with this sediment, preserving a record of the most recent activities of the dinosaur fauna. Such features as bipedal walking patterns, feeding behavior around coalified trees, and resting positions are evident (Parker and Balsley in prep.). Thousands of footprints occur at this coal–roof rock contact. In many mines the roof is so covered by footprints overlapping one another that the entire surface for several hundred square meters is completely bioturbated or "dinoturbated". One such roof surface is within the same Kenilworth Coal mine as the small animal track described from the mine floor. Here, the roof has a texture of irregular–shaped lumps extending down from the surface as far as 30 cm (Fig. 39.2). Few distinguishable footprints can be identified, but innumerable toes, heels and partial footprints overlap one another randomly. In a nearby area, footprints are more separated from one another and distinct in shape and size (Fig. 39.3). This roof surface has in it at least eight different track types within an area of about 100 m square. It includes one specimen 96 cm in length, the largest we have seen. In another Blackhawk mine, 14 track types have been collected from the same kind of roof surface (Parker and Rowley this volume). We recognize the probability that footprints of a single animal species would be preserved differently, depending on animal age, behavior (feeding, progressing, resting), and substratum (kind of peat, kind of clastic sediment). Nevertheless, morphological consistency among numerous specimens (often hundreds of specimens) of a single track type leads us to believe that several animals co–existed in certain swamp forests. Dinosaur traces and their relation to other fossils in mine roof surfaces have been described in more detail (Parker and Balsley 1983, in prep.).

Sediment Above the Mine Roof

In older portions of almost all mines, "room and pillar" methods were used to extract coal. In this process,

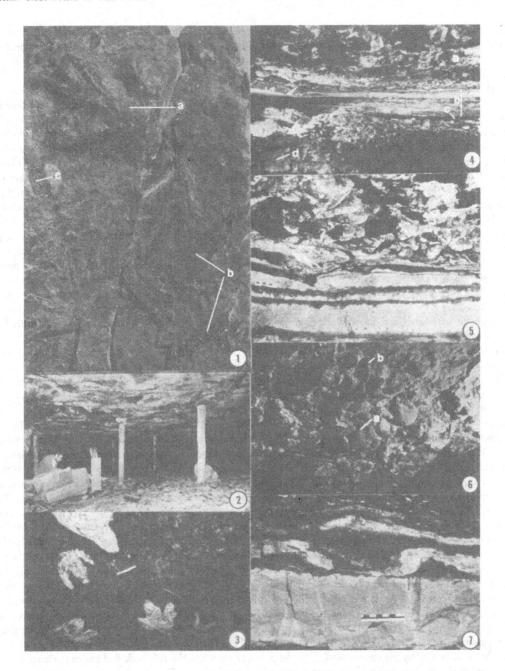

Figure 39.1. Small three-toed footprint (a), dicotyledon leaves (b) and vitranized twig (c) on a slab from mine floor. Footprint is 11 cm long. Figure 39.2. Coal mine, showing numerous overlapping round-bottomed dinosaur footprint casts extending down from the roof. The swamp surface had been heavily bioturbated. Figure 39.3. Footprints in roof surface about 35 m away from the area shown in Figure 39.2. They do not overlap one another and their shape is well-defined. Figure 39.4. View of roof-fall area approximately 14 m wide. Top half of photo (a) shows an oblique view of a horizontal surface. Numerous irregular structures extending downward are overlapping dinosaur footprint casts produced on a muddy surface. Parallel bedding in center (b) is at the edge of the roof fall, showing lateral views of lacustrine bedding. Black layer (c) is the top of the Castlegate A Coal. Rubble which has fallen to the floor includes a large three-toed dinosaur footprint cast (d), 67 cm long. Figure 39.5. Close-up of roof fall area in Figure 39.4, showing footprint bioturbation and lacustrine bedding. Scale is 16 cm long. Figure 39.6. Oblique view of planar surface of Figure 39.4 showing many indistinct overlapping tracks plus two tracks of a three-toed anaimal coming from (a) to (b). Pointers are on the center toe of each cast. They are members of a four-track sequence. Figure 39.6b is 60 cm in length. Figure 39.7. Close-up of roof-fall area near that shown in Figure 39.4. Lateral view of sediment distortion due to dinosaur footprint is evident. Scale is 16 cm long.

pillars of coal about 5 m square were left in place to secure the roof. Final mining activities in certain mines included the removal of alternate pillars, eventually allowing sections of the roof to fall. In this way, sedimentary features are exposed as much as 2 m above the coal surface. The significance of these rock–fall areas is that they show depositional and biological activities which occurred after the termination of the peat–forming swamp. Rocks at this level were deposited at the leading edge of the fluvial coastal plain which prograded seaward over the thick peat blanket (Balsley and Parker 1983).

One such mine is west of Helper, Utah in the Castlegate A Coal. In this locality, the coal is covered by thin, laterally extensive, even–bedded lacustrine units, including some with well–defined laminations. Roof falls in this mine (Figs. 39.4, 39.5, 39.8, 39.12) show sedimentary features in both broad planar and lateral (side) views. Some beds were heavily bioturbated by several dinosaur types (Fig. 39.6). Others have no evident tracks at all. One mudstone is covered by large polyhedral cracks indicating fluctuations in water level and subaerial exposure.

Figures 39.7 and 39.9 are lateral views of lacustrine beds where distortions of certain units can be seen. Both are vertical sections through dinosaur footprints. The animal which produced the structure in Figure 39.7 was walking on a peat surface (now the thin coaly layer), its foot pressing into one soft bed below. However, the dinosaur which made the distorted features in Figure 39.9 was walking on mud.

Occasionally, large, distinctly flat–footed footprint casts occur with a unique straight–sided preservation of the vertical walls. One such specimen is shown in both Figures 39.10 and 39.11. It extends down from the surface 28 cm, and is one of three which were in a bipedal trackway. The animal was walking on a thick, soft mud surface and produced a unique "cookie–cutter" effect by apparently placing its feet straight down and lifting them straight up.

Tail drag structures are rare in mines, although linear depressions which are interpreted as such are seen (Robert L. Rowley pers. comm.). Two depressions, each about 8 cm deep, 2.5 m in length and 0.3 m wide, are present in a lightly bioturbated unit (Fig. 39.12). Both are curved. Their relation to one another suggests that they were produced at the same time by the same animal. Short lateral projections on the tail seem to have been present, making the parallel striations seen on the sides of these depressions. The infrequent occurrence of tail traces in these lakeshore muds suggests that the animals which lived here normally supported their tails above the surface.

Tracks which are different from the morphology of the rest of the footprints we have seen occur in one rock–fall area (Fig. 39.13). They are on the recently exposed surface in Figure 39.8. A map has been prepared (Fig. 39.15), showing at least 50 of these specimens in an area 2 x 2.5 m. They were produced by a three–toed animal with a distinctly asymmetrical foot. One lateral toe is long and slender, almost 2.5 times the length of the median and opposite lateral toes. Total length from the heel to the end of the longest toe is 15 cm. No claw impressions are evident. Both left and right footprints occur. They apparently were produced by the pedes of a bipedal animal, since no footprints of any other shape or size are present. At least seven clusters of tracks occur where a single foot was being picked up, moved slightly and replaced. This shuffling gait, plus their random orientation, appears similar to tracks of extant birds as they feed on mud–dwelling organisms. Interestingly, associated with these tracks are many small 1 x 2 cm depressions and several thin (1.5 cm wide) elongate striations among the tracks, neither of which have been observed elsewhere in the lacustrine beds. Are these depressions beak or "peck" marks made in the mud by the animal as it fed? Lockley (pers. comm. 1988) suggests that, because of the comparatively small size of these odd tracks, plus their widely separated toes, they probably are bird tracks. Currie (1981) studied several features of fossil and modern footprints, and pointed out that the angle of divergence of digits II and IV in small dinosaur tracks does not exceed 100°, while the same angle in bird tracks is greater. The angle in the Blackhawk specimens is 120°, well within that observed in birds.

Several techniques, inherent in the mining process, limit the observable roof area and in some cases destroy the specimens within the roof. The most common is white "rock dust", a powdered limestone, which is applied as a thick layer onto all freshly exposed mine surfaces in order to reduce the quantity of explosive coal dust. This completely covers all roof features or distorts observation of roof topography with a "snow–blindness" effect. Secondly, "top coal", a layer of coal usually from 10 to 50 cm thick, is left in place on the roof to cover certain types of roof shales which are otherwise subject to "air slacking" (the absorption and subsequent expansion of shales to the point that they fall from the roof). Because of these problems, only about 20% of the roof surface in any mine has been visible for study. But the most serious problem to the study of mine–roof paleontology is the recent development and use of long–wall retreat mining. The equipment in this technique removes coal in long 161 m swaths in such a way that the roof is unsupported and collapses within a few hours in the "gob area" behind the support shields. This completely destroys whatever features might have been preserved in the roof rock and makes studying the areas very hazardous.

Because dinosaur footprint casts are abundant in every level of swamp development which we can see underground, why don't they appear in outcrop? We have determined that they are indeed present and common in outcrop but are rarely recognized. For one thing, in lateral view they appear to be load castings which are expected (and present) in these sediments. Additionally, they weather rapidly, destroying characteristic toe and metatarsal features. In fresh exposures, however, and on protected undersurfaces of ledges there are often several intriguing three–lobed structures the same size and shape as those specimens in mine roofs (Fig. 39.14). We believe that many of these are dinosaur footprints, *not* load casts.

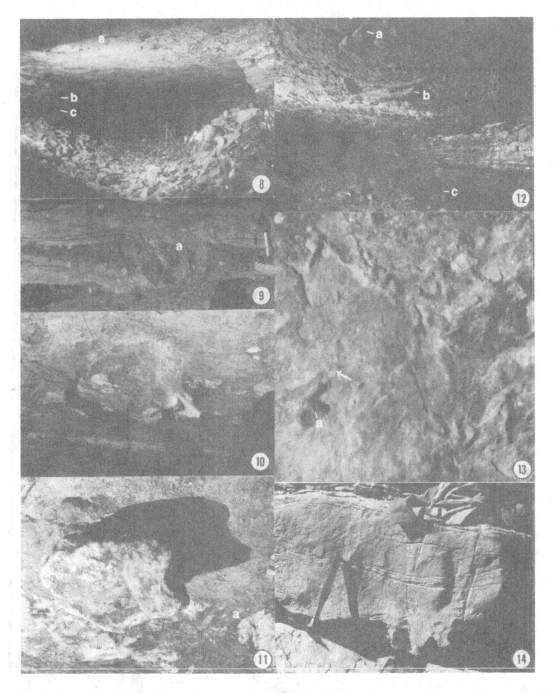

Figure 39.8. View of roof–fall area showing apparent smooth planar surface of lacustrine mudstone (**a**), lateral view of sediment over coal (**b**), top of the Castlegate A Coal (**c**) and rubble on the floor. Tripod in center is 1 m tall. Figure 39.9. Lateral view of block fallen from roof in Figure 39.6. Distortion of bedding due to dinosaur footprint (**a**) is evident. Pen is 1.5 cm long. Figure 39.10 and 39.11. Two views of the same footprint cast in a roof-fall area. Its three–toed outline, flat bottom and straight vertical sides are evident. White area in Figure 39.11(a) is where the next track in sequence has fallen. Arm in Figure 39.10 for scale. Figure 39.12. Roof–fall area showing two apparent tail drag features (**a** and **b**) on a shallowly bioturbated surface. The center one (**b**) is 2.5 m in length. Note vertical edge of roof fall area, the top of a mine support post (**c**) in the position where coal was removed, and rubble. Figure 39.13. Several small dinosaur footprint casts which occur on the exposed surface of the roof fall area in Figure 39.8. Cast in lower left (**a**) shows three–toed, asymmetrical organization. End of longest toe originally extended to arrow tip. Map, Figure 39.15, was made from this surface. Figure 39.14. Cast of three–toed footprint seen in lateral view on undersurface of ledge in outcrop exposure. Rock hammer for scale.

Figure 39.15. Map of numerous randomly oriented three-toed footprint casts. Clusters of two or more footprints seem to have been made by shuffling. Short linear depressions may be "peck marks" of the animal as it fed. Apparent tail drag casts occur in several places. Stippled areas are rock remaining on surface.

It is clear that the fluvial delta and coastal plain peat-forming environments of the Upper Cretaceous of the Rocky Mountain states had a large and varied dinosaur fauna recognized mainly from their tracks. Coal mines in the Blackhawk Formation of east–central Utah provide a unique opportunity to study their diversity, behavior and paleoecology.

Acknowledgments
We are grateful to Aureal T. Cross of Michigan State University for his infectious enthusiasm, help underground and insightful comments.

References
Bass, N. W., Eby, J. B., and Campbell, M. R. 1955. Geology and mineral fuels of parts of Routt and Moffat counties, Colorado. *U.S. Geol. Surv. Bull.* 1027-D:143–250.

Balsley, J. K., and Parker, L. R. 1983. *Cretaceous Wave-dominated Delta, Barrier Island, and Submarine Fan Depositional Systems: Book Cliffs east central Utah.* Amer. Assoc. Petrol. Geol. Field Guide. 279 pp.

Currie, P. J. 1981. Bird footprints from the Gething Formation (Aptian Lower Cretaceous) of northeastern British Columbia, Canada. *Jour. Vert. Paleont.* 1 (3–4):257–264.

Lockley, M. G. 1986. *Dinosaur Tracksites. Univ. Colorado Denver Geol. Dept. Mag. Spec. Issue* No. 1. 56 pp.

Lockley, M. G., Young, B. H., and Carpenter, K. 1983. Hadrosaur locomotion and herding behavior: evidence from footprints in the Mesaverde Formation, Grand Mesa Coal Field, Colorado. *Mountain Geol.* 20:5–13.

Osterwald, F. W., and Dunrud, C. R. 1966. Instrumentation study of coal mine bumps, Sunnyside District, Utah. *Utah Geol. Min. Surv. Bull.* 80:97–110.

Parker, L. R. 1976. Paleoecology of the fluvial coal-forming swamps and associated floodplain environments in the Blackhawk Formation of central Utah. *Brigham Young Univ. Geol. Studies* 22:99–116.

1979. Paleoecology of delta and coastal plain plant communities in the Upper Cretaceous Blackhawk Formation of Utah. *Botanical Soc. Amer. Misc. Ser. Publ.* 157:35.

1980. Paleoecology of palm forests in Upper Cretaceous coal-forming swamps of central Utah. *Botanical Soc. Amer. Misc. Ser.* 158:86–87.

1981. Paleoecology of the Cretaceous coal-forming environments in the Rock Springs Formation of Wyoming. *XIII Internat. Botanical Congr. Sydney Australia.* Abst.:204.

Parker, L. R., and Balsley, J. K. 1977. Paleoecology of the coastal margin coal-forming swamps in the Upper Cretaceous Blackhawk Formation of central Utah. *Geol. Soc. Amer.* Abst.:1125–1126. Seattle.

in prep. Plant fossils and dinosaur footprints in coal mine roof surfaces.

Peterson, W. 1924. Dinosaur tracks in the roofs of coal mines. *Nat. Hist.* 24 (3):388–391.

Ratkevich, R. P. 1976. *Dinosaurs of the Southwest.* (Albuquerque: University of New Mexico Press) 115 pp.

Strevell, C. N. 1932. *Dinosauropodes.* (Salt Lake City: C. N. Strevell and Deseret News Press) 15 pp.

Wilson, W. D. 1969. Footprints in the sands of time. *Gems and Minerals* June 1969:25.

Editorial Note: The bird–like tracks illustrated in Figure 39.15 display morphologies consistent with the asymmetric foot of *Hesperornis* and its relatives.

40 Dinosaur Footprints from a Coal Mine in East–Central Utah

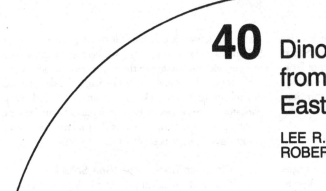

LEE R. PARKER AND
ROBERT L. ROWLEY, JR.

Abstract

Thousands of natural casts of dinosaur footprints occur in the roof surface of a mine in an Upper Cretaceous coal in east–central Utah. Ninety three have been removed for study. They range in length from 28 to 87.5 cm. Most are three–toed forms; only two types are four–toed. About 14 dinosaur taxa were probably responsible for producing the specimens in the collection. These tracks were made by animals which walked in the peat on the surface of a swamp; their footprints were filled by mud, silt or sand during the flooding of a local river. Most footprints are very broad with short thick toes, apparently well–adapted for walking on soft peat. Some footprint casts become loose in the mine roof and can fall, creating a hazardous condition for miners, especially in the case of tracks weighing up to 140 kg. In order to collect quality tracks, they must be chiselled from the roof rock matrix. Footprints from coal mines are well–known in central Utah and are displayed in front of the homes of miners and also in some businesses.

Introduction

For twelve years one of us (Rowley) has been able to examine thousands of natural casts of dinosaur footprints as they occur in the roof surface of the Price River Coal Company mine in Spring Canyon, west of Helper, Utah. Ninety three have been removed for further study. This report briefly describes the conditions involved in the formation and collection of these dinosaur footprint casts and illustrates 14 types which have been collected.

Dinosaur footprints occur in the mine roof surfaces as protrusions which hang down from the roof, sometimes as far as 30 cm. They often occur in trackways made by bipedal animals or are positioned around tree bases; sometimes the tracks are over the top of woody litter or tree roots, indicating that the animals had been walking on that material as it occurred on the swamp forest floor. Frequently, the roof surface is so covered with them that one track oversteps another, similar to tracks of livestock in a corral (Peterson 1924, Parker and Balsley this volume and in prep.). Because of overstepping or incomplete depression of the foot into the peat, few tracks are of the "exhibit" quality which is desirable for removal. The sediment which filled in the original footprints is usually a light–colored fluvial shale or siltstone. The lower surfaces of most tracks are partially or completely covered with a thin layer of hard vitreous coal or a fine–grained carbonaceous siltstone, both of which have a highly polished slickenside–like surface.

The occurrence of natural casts of dinosaur footprints from coal mines is well–known locally. It is common to see them displayed as a front yard ornament at miners' homes or as conversation pieces in reception areas of some businesses in the cities of Helper and Price. Sicne they can be as long as 1 m and weigh hundreds of kilograms, they are an impressive natural curiosity.

The College of Eastern Utah Prehistoric Museum in Price has a good display of several track types. It also contains a part of the W. D. Wilson collection of many types and sizes (Lockley 1986). A collection of 30 tracks, described by Strevell (1932), is housed at the Utah State Natural History Museum, University of Utah, in Salt Lake City (Frank L. De Courten pers. comm. 1986). However, the largest collection of fossil footprint casts from coal mines of the Rocky Mountain area in terms of variety and total specimen numbers is the one illustrated in this report (made by Rowley). It numbers nearly 100 specimens, and includes 14 different footprint morphotypes. Other museums which we know to have one or several specimens include: The American Museum of Natural History, New York; California Polytechnic State University, San Luis Obispo; The Field Institute of Natural History, Chicago, Illinois; Louisiana Polytechnic Institute, Ruston; New Mexico Museum of Natural History, Albuquerque; Peabody Museum of Natural History, New Haven, Connecticut; The Museum of Western Colorado, Grand Junction; San Diego Museum of Natural History, San Diego, California; The Smithsonian Institution, Washington, D.C.; The South Dakota School of Mines and Technology, Rapid City; University of Pennsylvania, Philadelphia (Strevell 1932); University of New Mexico, Albuquerque (Ratkevich 1976) and Utah State University, Logan (Peterson 1924).

Geologic Setting and Paleoecology

The Castlegate D Seam which the Price River Coal Company is mining is one of several coals in the Upper Cretaceous (Campanian) Blackhawk Formation (Doelling 1972). It was developed from a peat-forming swamp on the upper surface of the sheet-like sands of the Aberdeen Member. This member represents one of several major deltas in the formation which prograded into the Cretaceous epicontinental seaway and allowed the development of brackish and freshwater swampy environments when subaerial surfaces occurred (Balsley and Parker 1983).

Animals living on the swamp surface made deep footprints into the peat. Before they became obliterated by the rebounding peat, a local river flooded and deposited overbank sediment into the swamp, filling the footprints and thus preserving them. Later, the peat became coal and has been removed, allowing an examination of the natural casts of these footprints and other fossils which were on the swamp surface (Parker and Balsley this volume and in prep.).

Like many other Blackhawk mine roof surfaces, there are abundant fossil leaves, horizontal logs and trees in growth position which are directly associated with dinosaur footprint casts (Parker and Balsley in prep.). Recently, Rowley has collected several ferns, dicot leaves, petrified tree stumps, pelecypods and gastropods from the roof of the Price River mine, all of which made up a portion of the swamp flora and fauna at the time the footprints were made.

Removal of Tracks from the Roof Surface

As the coal is being mined it normally separates easily from the roof rock, exposing sedimentary features and fossils which might be present. Footprints can be removed where the roof rock is carbonaceous shale or siltstone; sometimes these specimens are loose and are easily pried down.

The best tracks for removal are those which extend down from the roof surface at least 10 cm and have a carbonaceous layer above them. Becuase they are heavy and positioned slightly above head height, it is impossible for one or two persons to safely hold or catch them when they separate from the roof. Therefore, it is necessary to prepare a cushion under them and allow them to fall on it. Mine "brattice", a yellow plastic-impregnated fabric used in mines as a fire retardant and as a drape for directing ventilating air, has been effectively used for this purpose.

The sedimentary matrix around the track is chiselled away with hand tools until a groove or channel is formed around it. Eventually, horizontal chiseling, up behind the track, will loosen it and allow it to drop from the roof. Specimens as large as 140 kg have been obtained in this manner without damage to the tracks. An average weight of the tracks after removal is about 45 kg. Outside the mine, extraneous coal and rock matrix can be cleaned away. Sandstone preserves footprints less frequently, but, because it makes a much harder surrounding matrix, sandstone tracks have been impossible to remove intact.

Dinosaur Tracks and Mine Safety

During the removal process it is important to consider the structural integirty of the surrounding roof to ensure that large blocks of rocks do not fall. In addition, the presence of "black damp" (a deficiency of oxygen, including the build-up of carbon dioxide and carbon monoxide) and methane must continually be checked, since these gases can accumulate very quickly at certain times in a working mine.

Dinosaur footprint casts which extend down from the roof several inches are a nuisance where the coal seam is thin, causing the roof to be low; mine workers continually bump their heads on them. More serious problems have existed with them since mining began in the area in the early part of the century, because they fall and kill or seriously injure mine workers. Therefore, loose footprints are bolted to the roof with a vertical drill designed to drive a 1 to 3 m long steel bolt upward into the roof rock and prevent tracks and blocks of rock from falling (Figs. 40.1–40.3). We are unaware of other lethal trace fossils, nor do we know of other circumstances where dinosaur activity has contributed to the possible death of human beings.

Morphology of the Footprints

The natural footprint casts collected from the Price River Mine seem to have been produced by several animal taxa. A few of the casts we illustrate here (Figs. 40.9–40.23) have similar morphologies, suggesting that they may have been made by the same dinosaur species, depending on age of the animal or activities at the time the footprints were made. However, it should be emphasized that all those 14 morphotypes illustrated are represented in the collection by at least four similar specimens. The only exception is the specimen shown as Figure 40.22, which is unique in the collection. Certain track types are represented by as many as 12 specimens including distinct left and right pes. In addition, it is clear from examinations of primary sedimentary features in the roof surface that all footprints had been pressed into peat; none were made in clastic sediment. Therefore, the morphological consistency of all specimens of each track type, each produced on the same kind of surface, suggests that the dinosaur fauna included at least 14 taxa. Most Price River specimens are three-toed; only two types are four-toed. No five-toed tracks have been observed in this mine, but they are known in other Blackhawk Formation coal mines. These tracks range in length from 28 to 87.5 cm. With a few exceptions, the footprints, which are mostly pedes, have short toes on wide, apparently flat feet (Figs. 40.1, 40.3, 40.7–.20, 40.23). This broad foot structure seems to have been adapted for walking on the soft peat of the swamp surface, similar to a snowshoe. Tracks with narrow toes occur but they are rare, both in the collection and in the actual mine roof (Figs. 40.2, 40.4–.6, 40.21).

Lull (Strevell 1932) gave Latin binomials to eight ichnospecies of the ichnogenus *"Dinosauropodes"* collected from a coal in the now–abandoned Standard mine (Blackhawk Formation, Castlegate A and B Coals, Doelling 1972), although Lockley and Jennings (1987) indicated that these names are not valid. Three of the species collected in the Price River mine are similar in size and shape to those in the Standard mine: *D. bransfordii* (Fig. 40.7); *D. magrawii* (although our specimens are not as large; Fig. 40.9); and *D. osborni* (Fig. 40.23). In addition, at least six species from the Price River mine have been seen in a Kenilworth mine (Blackhawk Formation, Kenilworth Coal, Parker and Balsley in prep.). These have not been described nor given Latin binomials, but include those shown here as Figures 40.1, 40.4, 40.7, 40.10, 40.16, and 40.18.

The fact that certain of these track types occur in three stratigraphically different coals in the Blackhawk Formation indicates that the animals which produced them were part of the Cretaceous swamp fauna for a great length of time. Other types, collected in only one coal bed, may be restricted in time and may prove useful as stratigraphic or paleoecologic indicators.

Diagnostic skeletal material is rarely collected in the Blackhawk Formation. What has been collected includes a carnosaur tooth (Steven F. Robison pers. comm. 1984) and the skull of *Albertosaurus* sp. (James H. Madsen, Jr., pers. comm. 1985; Parker and Balsley in prep.). It is thought that many of the large, flat footprint types were made by unidentified hadrosaurian species (Figs. 40.1, 40.3, 40.7–.9, 40.11–.20) (Strevell 1932, Parker and Balsley in prep.), certain of the narrow–toed forms were probably made by theropods like *Albertosaurus* (Figs. 40.2, 40.4–.6, 40.10, 40.21), and a ceratopsian probably produced the four–toed specimen (Fig. 40.23, cf. Lockley 1986, Lockley and Jennings 1987).

Acknowledgments

References

Balsley, J. K., and Parker, L. R. 1983. Cretaceous Wave-dominated Delta, Barrier Island, and Submarine Fan Depositional Systems: Book Cliffs east central Utah. *Amer. Assoc. Petrol. Geol. Field Guide.* 279 pp.

Doelling, H. H. 1972. Central Utah Coal Fields: Sevier-Sanpete, Wasatch Plateau, Book Cliffs and Emery. *Utah Geol. Min. Surv. Monog.* No. 3. 496 pp.

Lockley, M. G. 1986. *Dinosaur Tracksites. Univ. Colorado Denver Geol. Dept. Mag. Spec. Issue* No. 1. 56 pp.

Lockley, M. G., and Jennings, C. 1987. Dinosaur tracksites of western Colorado and eastern Utah. In Averett, W. R. (ed.). *Paleontology and Geology of the Dinosaur Triangle.* (Grand Junction, Colorado: Museum of

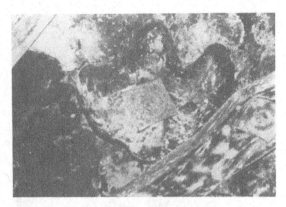

Figure 40.1 (top). Dinosaur footprint in the roof of the Price River Coal Company mine, which has been bolted in place to prevent it from falling. This track, probably produced by a hadrosaur, is 45 cm in length. Note additional steel stripping used to support a hazardous roof area.

Figure 40.2 (middle). Small three-toed dinosaur footprint, probably made by a theropod, is 21 cm in length. Apparent claw casts are present at the ends of the two visible toes. Square washer which bolt passes through is positioned over the median and right toes (as seen in the photo). An additional track (lower left) has fallen, leaving a light-colored scar.

Figure 40.3 (bottom). Bolt with square washer is positioned behind a large three-toed probable hadrosaur footprint. All light-colored areas in the roof rock here are complete or partial dinosaur footprint casts with toes oriented in several directions.

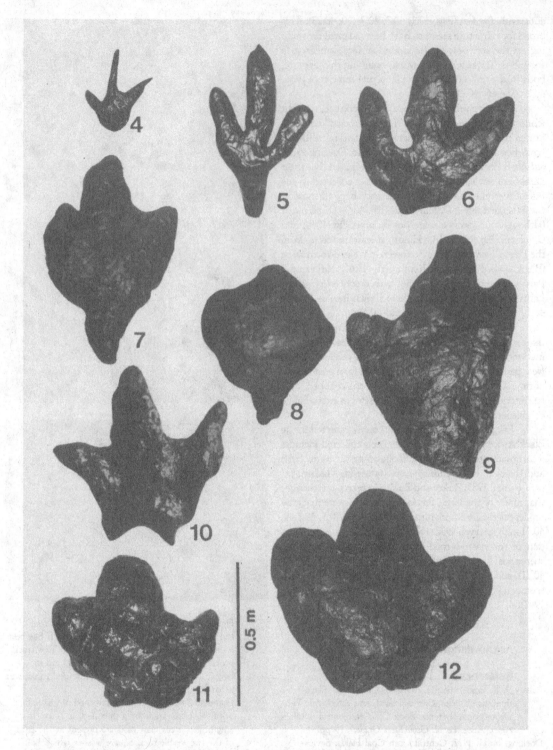

Figure 40.4–12. Natural casts of several dinosaur footprints removed from the Price River Coal Company mine roof. **40.4**, pes with narrow toes, probable theropod, smallest specimen in the collection, 28.5 cm long. **40.5–.6**, pes, narrow-toed, possibly produced by the same theropod taxa, 61 and 59 cm in length respectively. **40.7**, pes, *Dinosauropodes bransfordii* Lull, probable hadrosaur, 73 cm long; note rounded callous pads under toes. **40.8**, pes with no evident toes, 52 cm long; the extended metatarsus like that seen in Figure 40.7 may indicate the same, or similar taxa produced both. **Figure 40.9**, pes, *D. magrawii* Lull, probable hadrosaur, 87.5 cm long. **40.10**, pes with triangular toes, cordate metatarsus, linear callous pads, probable theropod, 61.5 cm long. **40.11–.12**, pedes with thick rounded toes, shallow-cordate metatarsus and rounded callous pads; these may have been produced by the same hadrosaur taxa, 53 and 72 cm long respectively.

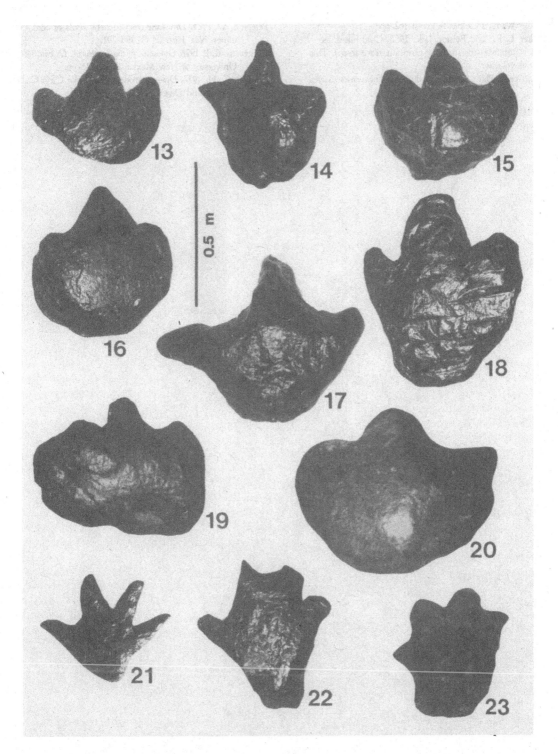

Figure 40.13–.23. Natural casts of dinosaur footprints removed from the Price River Coal Company mine roof. **40.13–.16,** pedes with short, triangular toes and broad round heel; center of foot is always pressed deeper than the toes, no callous pads under toes; these, including Figure 40.20 below, may all have been made by the same or similar hadrosaur taxa, 38, 51.5, 48, and 53 cm long respectively. **40.17,** pes with longer toes, broad rounded heel, made by a probable hadrosaur, 59 cm long. **40.18,** pes, toes of variable size, broad rounded heel, made by a probable hadrosaur, 68 cm long; note parallel striations produced by mining equipment. **40.19,** pes, center toe well-defined, lateral toes obscure, rounded heel, probable hadrosaur, 48 cm long; **40.20,** pes, short toes, rounded heel, probable hadrosaur, 58 cm long. **40.21–.22,** four-toed, 38 and 56 cm long respectively. **40.23,** pes?, *Dinosauropodes osborni* Lull, a possible ceratopsian, 46 cm long.

Western Colorado Press) 162 pp.

Parker, L. R., and Balsley, J. K. 1988. Coal mines as localities for studying dinosaur trace fossils. This volume.

— in prep. Plant fossils and dinosaur footprints in coal mine roof surfaces.

Peterson, W. 1924. Dinosaur tracks in the roofs of coal mines. *Nat. Hist.* 24 (3):388–391.

Ratkevich, R. P. 1976. *Dinosaurs of the Southwest.* (Albuquerque: University of New Mexico Press) 115 pp.

Strevell, C. N. 1932. *Dinosauropodes.* (Salt Lake City: C. N. Strevell and Deseret News Press) 15 pp.

IX Systematic Ichnology

"Perhaps the limited usefulness of ichnites for taxonomic classification may account for their neglect. It seems, however, that their limited usefulness is in part due to their neglect."

Moussa (1968 p. 1433)

Much of the classic early work on dinosaur tracks by Hitchcock was characterized by detailed systematic descriptions. This popular 19th century trend was continued into the early 20th century by ichnologists like Lull and Sternberg, whereas other workers, like Bird, avoided systematic descriptions altogether.

Although a number of leading vertebrate ichnologists of the present generation, including Demathieu (this volume), Haubold (1984), Sarjeant (this volume) and Leonardi (this volume) have elsewhere made substantial ichnological contributions, there has generally been a trend away from comprehensive systematic treatment of dinosaur ichnites. Many studies deal with isolated ichnofaunas and introduce ichnotaxa whose relationships to other footprint taxa are, at best, loosely defined and interpreted. The differences in classification and interpretation of dinosaur tracks proposed by various authors are in no small measure influenced by the absence of clear-cut regulations in the International Code of Zoological Nomenclature (Sarjeant and Kennedy 1973).

Such a state of affairs "is in part due to (the) neglect" of ichnites (see quote above) and should be improved and rectified to a considerable degree as the large volume of new data is "systematically" synthesized.

The two papers in this section illustrate these points well. On the one hand they indicate that ichnologists are now cautious about launching into the field of systematics without firm procedural guidelines, while on the other hand they recognize the need for studies that will fill obvious gaps and amend problematic classifications (e.g., *Brontopodus*, Farlow et al. this section, and illustration above) and amend problematic classifications. If ichnologists follow the type of systematic approach advocated by Sarjeant (cf. Leonardi et al. 1987), "Systematics", "Parataxonomy" or "Ichnotaxonomy" in the strict sense will continue to benefit from improved standardization.

References

Haubold, H. 1984. *Saurierfährten*. (Wittenberg: Ziemsen) 231 pp.

Leonardi, G. (ed.). 1987. *Glossary and Manual of Tetrapod Footprint Paleoichnology. Dept. Nacional de Produção Mineral, Brasil.* 75 pp.

Moussa, M. T. 1968. Fossil tracks from the Green River Formation (Eocene) near Soldier Summit, Utah. *Jour. Paleont.* 42:1433–1438.

Sarjeant, W. A. S., and Kennedy, W. J. 1973. Proposal of a code for the nomenclature of trace fossils. *Canadian Jour. Earth Sci.* 10:460–475.

41 'Ten Paleoichnological Commandments': A Standardized Procedure for the Description of Fossil Vertebrate Footprints

WILLIAM A. S. SARJEANT

Abstract

In the form of ten numbered 'commandments', a standardized procedure for the description and illustration of fossil vertebrate footprints is proposed. Such a standardization would, it is considered, significantly reduce future problems in the definition and comprehension of taxa.

Introduction

When I began first to work on vertebrate footprints in 1966 I sought in vain for any clear published statement of descriptive procedures. More than a decade later (1975), I attempted to summarize the methods of interpretation and Haubold (1984) has provided a similar summary, but there has been no clear agreement on procedures, even yet. The result is much unnecessary confusion and difficulty. The fashion in which fossil footprints are described and named cannot, and should not, be legislated too closely. Nevertheless, I would like to suggest ten considerations — or, maybe, formulate ten commandments! — that might be borne in mind when fossil footprints are described:

I. Quite evidently a trackway — a series of successive footprints of both, or of all four, feet — is the best possible basis for the definition of a footprint ichnospecies. A set of prints — impressions of all four feet, or at least of a manus and a pes — is the next best basis. Single footprints should be given names only when they are markedly different from all described types.

II. No new names for footprints should be proposed until a thorough literature search has been made; this can begin with the perusal of standard compilations (Kuhn 1958, 1963; Haubold 1971, 1984; Sarjeant, 1987) but should also include a check of the references quoted in the latest available papers. If proven necessary, new generic names of footprints should be made distinctive by the utilization of an appropriate suffix such as –ipus, –podus, –podion, –ipes, –pezia, –peda, –dactylus, –pterna. The suffixes –ichnus, –ichnis, –ichnites or –ichnium are also appropriate but, since they are nowadays being applied to invertebrate traces of various kinds, they are perhaps best used when the stem of the name makes it clear that the trackmaker was a vertebrate. The derivation of the name, whether generic or specific, should be stated; paleontologists can no longer be assured to have sufficient knowledge of Latin and Greek to understand the name otherwise.

III. A holotype and, if appropriate, paratypes should be designated for each ichnospecies. Types should be stored in a collection accessible to the public. Ichnofossils are, of course, exempted from the provision of typification under the zoological *Code*: but ichnologists should press for the restoration of this requirement, for footprints at least, and in the meantime strive for a high standard of systematic procedure. However, not just an actual, natural mold or cast of the footprint or trackway, but also an artificial mold or cast, should be considered acceptable as type. In all such instances, the character of the type — whether natural or artificial and, if artificial, of what nature — should be made explicit. If the type is known to be a supertrace or subtrace, this should be stated also.

IV. Photographic illustration is essential. It is desirable that the photographs be taken in oblique illumination under darkroom conditions; but this is, of course, not always possible. However, the direction of illumination should always be indicated, either by arrows on the photograph or in the accompanying caption. The presentation of photographs in stereo–pairs is an approach that may overcome some of the difficulties inherent in comprehension of a two–dimension photograph (see Sarjeant and Thulborn 1986 for an example). Ishigaki and Fujisaki (this volume) have also introduced the moiré topography method for three dimensional representation of dinosaur tracks. The making of bas-relief double exposure prints is a technique that can also be tried.

V. Interpretative drawings of the pes and (where present) manus should be provided, since no single photograph can ever show all the features of a footprint — unless it be of the very simplest type. The drawing should be carefully shaded or contoured, to indicate the elevated areas and declivities. In preparing the drawing, all available impressions of manus and of pes should be taken into consideration, not just the best ones, since other impressions may yield details not obvious from the best ones.

Such drawings are subjective and should be supplemented by photographs, to make clear the author's ideas of footprint morphology. Techniques of computer simulation may provide a degree of objectivity, but it should be borne in mind that the controls on their production can be determined in so many different ways that they may remain quite as subjective as drawings produced by other methods.

VI. A drawing or clear photograph — or, better still, both! — of the pattern of footprints in the trackway should be included.

VII. The dimensions of the individual footprints should be stated with precision and, insofar as possible, those of the trackway also. The positions from which the measurements are taken, or between which interdigital angles are measured, should be made explicit, either in the text or (better still) in explanatory sketches. There have been attempts to establish a standard for the positions in which measurements are made; the proposals of Leonardi et al. (1987, pls. 1-2) should be adopted wherever possible. However, the variations in footprint topography and the subjectivity of deciding the limits of the footprint, where it grades gently into the surrounding sediment, make exact regulation impossible.

VIII. Diagnoses should be sufficiently tight to leave no ambiguity in the mind of the reader and to permit no confusion with types of footprints described earlier. The diagnoses should be limited to identifying characteristics, but should be accompanied by an ampler description. In both diagnosis and description, morphological terms should be used in a consistent fashion; a recently published international glossary (Leonardi et al. 1987) should prove helpful. Comparisons should be made with existing taxa, to clarify similarities and differences.

IX. The type locality and stratigraphical horizon should be stated as accurately as possible. (Where museum specimens are being described or redescribed, this may not be easy!) The lithology should always be described and information given on the paleontological context — the nature of any associated body or trace fossils — and on any associated sedimentary structures, since this information will facilitate environmental interpretation.

X. The name of the collector should be stated, if known and other than the author. It may also be helpful to state the date of collection and to describe the circumstances under which the specimen was collected.

Conclusions

In the past, footprints have been described in arbitrary and random fashion, while illustration has often been painfully inadequate and the location of specimens, in geological terms and in terms of lodgement, has too often been left unclear. If the above "commandments" are followed in the future, the lot of the vertebrate paleoichnologist should be very much happier.

References

Haubold, H. 1971. Ichnia Amphibiorum et Reptiliorum fossilim. In Kuhn, Oskar (ed.). Handbuch der Paläoherpetologie, Part 18. (Stuttgart, Germany and Portland, U.S.A.: Gustav Fischer Verlag) 124 pp.

——— 1984. Saurierfährten. (Wittenberg Lutherstadt, Germany: Die Neue Brehm-Bucherei) 232 pp.

Kuhn, O. 1958. Die Fährten der vorzeitlichen Amphibien und Reptilien. (Bamberg, Germany: Verlagshaus Meisenbach KG) 64 pp.

——— 1963. Ichnia Tetrapodorum. In Westphal, F. (ed.). Fossilium Catalogus I: Animalia. ('s-Gravenhage: Dr. W. Junk) 176 pp.

Leonardi, G., Casamiquela, R. M., Demathieu, G., Haubold, H., and Sarjeant, W. A. S. 1987. Glossary in eight languages (pp. 21-51, pls. 1-20). In Leonardi, G. (ed.). Glossary and Manual of Tetrapod Footprint Palaeoichnology. (Brasilia: Departamento Nacional de Producao Mineral) 75 pp.

Sarjeant, W. A. S. 1975. Fossil tracks and impressions of vertebrates. In Frey, R. W. (ed.). The Study of Trace-fossils. (New York: Springer-Verlag) pp. 283-324.

——— 1987. The study of fossil vertebrate footprints. A short history and selective bibliography (pp. 1-19). In Leonardi, G. (ed.). Glossary and Manual of Tetrapod Footprint Palaeoichnology. (Brasilia: Departamento Nacional de Producao Mineral) 75 pp.

Sarjeant, W. A. S., and Thulborn, R. A. 1986. Probable marsupial footprints from the Cretaceous sediments of British Columbia. Canadian Jour. Earth Sci. 23 (8):1223-1227.

42

Brontopodus birdi, Lower Cretaceous Sauropod Footprints from the U.S. Gulf Coastal Plain

JAMES O. FARLOW,
JEFFREY G. PITTMAN AND
J. MICHAEL HAWTHORNE

Abstract

The name *Brontopodus birdi* is assigned to sauropod footprints from the Glen Rose Limestone (Lower Cretaceous) and equivalents of Texas and Arkansas. The ichnotaxon is characterized by: U-shaped manus prints, probably made by a hand in which digits II–IV were bound together in a pad and set off from digits I and V; no claw marks on manus tracks; large, laterally directed claw marks on digits I–III of the pes; outward rotation of manus and pes tracks relative to the direction of travel; a relatively broad trackway, in which footprints do not impinge on the trail midline, and in which the outer limits of the trackway are defined by the pes footprints, rather than the manus tracks; step angles usually 100–120°; tail marks rare or absent. The trackmaker is interpreted as a brachiosaurid, perhaps *Pleurocoelus.*

Introduction

In August 1934 two residents of Glen Rose, Texas, Charlie and W. B. Moss, discovered fossilized sauropod footprints in the Paluxy River Valley near Lanham Mill (Wilson 1975). Three-toed footprints of bipedal dinosaurs had been found in the area earlier in the century (Nunn 1975) and recognized for what they were (Shuler 1917), although some local residents still believed them to be giant bird tracks. The new footprints were clearly something different, and Charlie Moss regarded them as elephant tracks (L. Moss pers. comm., 9-10-86).

Their dinosaurian nature was established four years later by Roland T. Bird, a fossil collector for Barnum Brown of the American Museum of Natural History. At the end of the 1938 field season Bird was in Gallup, New Mexico to crate a specimen for shipment to New York, and learned that alleged human and tridactyl dinosaur footprints, said to have been collected at Glen Rose, were being sold at local trading posts (Bird 1939, 1985). He concluded that the footprints in question were forgeries, but decided never-

theless to visit Glen Rose on his way back to New York. There he discovered that genuine tridactyl tracks did exist, and someone, either Charlie Moss or James Ryals, told him about the "elephant tracks" (Farlow 1987). Bird saw and identified the brontosaur ichnites, and two years later returned to Texas to find sauropod footprints elsewhere in the state and to collect a portion of spectacular trackway sequence from the Paluxy River for the American Museum and the Texas Memorial Museum (Bird 1985).

Bird published a series of popular articles about his work in Texas (Bird 1939, 1941, 1944, 1954), and discussed it at length in his memoirs (Bird 1985). Toward the end of his life he hoped to write a formal technical description of the sauropod tracks, but failing health prevented his carrying out this project.

In 1980 the senior author began studies of Lower Cretaceous dinosaur footprints of Texas (cf. Farlow 1987), visiting Bird's sites and numerous others, examining the tracks collected by Bird, and interviewing surviving members of the Bird family. Shortly after this the two junior authors initiated studies of their own of the stratigraphy and paleoecology of Lower Cretaceous dinosaur tracksites in Texas and adjoining states (Hawthorne 1983, 1987; Pittman 1984, and this volume; Pittman and Gillette this volume). Because of overlapping interests, we agreed to jointly prepare a formal description of the sauropod footprints. We owe a great debt to the late R. T. Bird, whose family and friends made available to us a wealth of his unpublished charts, notes, and photographs, and we dedicate this paper to his memory.

Stratigraphic Setting

Lower Cretaceous (Comanchean) strata occur in the subsurface around the Gulf Coastal Plain, and are exposed at the surface in southwestern Arkansas, southern Oklahoma, and Texas (Hawthorne 1987, Pittman this volume). Comanchean dinosaur tracksites are most common in the

Lampasas Cut Plain and Edwards Plateau physiographic provinces of Texas. Known trackbearing units include the Glen Rose Limestone and the Twin Mountains Formation of the Trinity Group, the Paluxy Sand and the Fort Terrett Formation of the Fredericksburg Group, and the Segovia Formation of the Washita Group. Tridactyl footprints of bipedal dinosaurs are present at nearly all sites, and are usually the most abundant dinosaur tracks at those sites (Farlow 1987, Pittman this volume).

Sauropod ichnites are found at fewer sites, although they are often abundant at those sites. Most or all of the presently verified sauropod footprints of the Paluxy River sites (Dinosaur Valley State Park, Somervell County) occur in a single layer at the base of the lower member of the Glen Rose Limestone, just above its contact with the Bluff Dale Sand, in what has been interpreted as an intertidal/supratidal flat deposit, perhaps a storm layer (Hawthorne 1983, 1987). The Medina River (Mayan Ranch — Bandera County) "swimming" sauropod trail also occurs in the Lower Glen Rose Limestone, but in what seem to be subtidal deposits, an interpretation consistent with the belief that this trackway was made by a half-floating dinosaur (Bird 1985, Hawthorne 1987, Ishigaki this volume, Pittman this volume). The West Verde Creek (Davenport Ranch — Medina County) site occurs in the upper member of the Glen Rose Limestone, also in a very shallow nearshore setting (Hawthorne 1987, Pittman this volume). Other known or possible sauropod tracksites include Sidney (Twin Mountains Formation, Comanche County), Blanco River (Lower Glen Rose Limestone, Blanco County), the Briar site (De Queen Formation [equivalent to the middle of the Glen Rose Limestone], Howard County, Arkansas), Miller Creek (middle of the Glen Rose Limestone close to the *Corbula* bed, Blanco County), and South San Gabriel River (Upper Glen Rose Limestone, Williamson County); for details, see Pittman (this volume).

R. T. Bird's Tracksite Maps

In 1984 Farlow was given a large map (Fig. 42.1) of the Paluxy River quarry site that Mrs. Alice Erickson, Bird's sister, had found in the attic of the Bird family home in Rye, New York. A card attached to this very detailed chart (even the sandbags of the coffer–dams are depicted!) indicated that it was compiled by Bird in 1942. Bird never specified how he prepared this chart, although in describing the Giant (S5) sauropod trail in his memoirs he says that "I stretched measured string to chart the trail..." (Bird 1985: 175). After the death in 1986 of Bird's lifelong friend, V. T. Schreiber, Schreiber's widow gave Farlow pages from an unpublished manuscript that Bird had written about the Somervell County site; in these notes Bird indicates that time constraints and the frequent flooding kept him from charting the site in as detailed a fashion as he would have liked, but that he carefully measured at least two trails in the sauropod–carnosaur "chase sequence."

Bird also prepared other charts of portions of the Paluxy River quarry. One of these, presumably made about

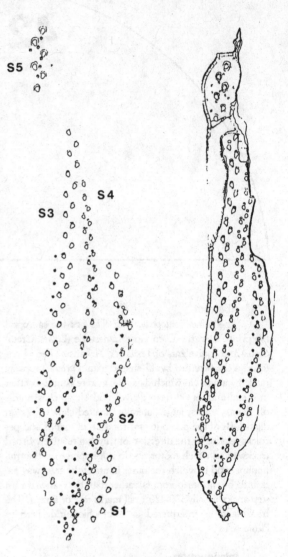

Figure 42.1. The Homestead (left) and Rye (right) master charts of Bird's Paluxy River quarry. The maps are oriented with south (the sauropods' direction of travel) at the top of the page. Trackway labels were assigned by Farlow (1987).

the same time as the Rye chart, depicts the last few footprints in the New York (American Museum of Natural History [AMNH]) chase sequence slab, all of the tracks of the two dinosaurs in the Austin (Texas Memorial Museum) [TMM] slab, and several footprints beyond the limits of the Austin slab; the map ends where both trails disappear beneath the Paluxy River bank. Langston (1974: text–Fig. 8) published a version of this map, which we designate as the Austin chart (Fig. 42.2).

Bird drew two diagrams of the New York slab (published in Farlow 1987, Fig. 31), at least one of which he intended to use in a formal description of the Paluxy sauropod tracks. One of these charts shows the sauropod and carnosaur tracks as they appear in the slab; the other shows the sauropod footprints alone, with the margins of those tracks that were distorted by the theropod's tracks restored to their presumed original shape. Bird planned to name the brontosaur ichnites *Brontopodus*, and we refer to this sauropod trail map as the *Brontopodus* chart (Fig. 42.3).

In 1976 Bird prepared another chart of the chase sequence (included among the manuscript pages Farlow received from Mrs. Schreiber), as well as a second large master chart; Bird described the latter in a letter to Wann Langston, Jr., dated 10 August 1976, saying that the map had been compiled "using various old piecemeal charts I was able to relocate, aided and abetted [by] a large number of 12 x 14 enlargements of photographs made that year, 1940." This chart is at least in part subjective as to the positions of dinosaur footprints, for in describing one of the carnosaur trails Bird wrote in his letter to Langston that "the latter part of this trackway is covered with a heavy pile of spoil in most photographs and remained uncharted. I have filled in from memory the probable location of these tracks." Bird's letter was written from his home in Homestead, Florida, so we call this the Homestead chart (Fig. 42.1).

A prime concern is the accuracy of Bird's maps. The Homestead and Rye charts are consistent in their overall depiction of the site, but there are differences in detail. The Homestead chart shows many more theropod footprints than are in the Rye chart, including the above-mentioned trail as well as long carnosaur trackways associated with the trails of sauropods S3 and S4; only one and two footprints, respectively, of these latter two theropod trails are depicted in the Rye chart.

The numbering system used here to label dinosaur trackways was devised by Farlow [1987]; although in his correspondence and memoirs, Bird referred to sauropod trails by number, it is not always clear which number belongs with which trail. Bird's trail three is obviously the sauropod trackway in the chase sequence, here labeled S2; Bird [1941] referred to this as trail two, however. Bird's trail five is unquestionably the Giant trail, our S5. Our S3 is probably Bird's trail one, the trackway he originally saw in 1938, and our S4 is probably Bird's trail two. For details, see Bird's letters to Brown of 12 March to 1 June 1940 on file in the Department of Vertebrate Paleontology, AMNH, and Bird [1985].

In the Rye chart a manus track of sauropod S4 is drawn superimposed on a footprint of the theropod, while in the Homestead chart the theropod track is atop the sauropod print, a significant difference if one tries to work out the sequence of events that created the tracksite. The Homestead map shows a carnosaur trail closely associated with the Giant trail, made in Bird's opinion by a carnivore pursuing the big brontosaur; in the Rye chart, however, this theropod trackway cuts across the Giant trail, and does not look so much like the trail of one animal following another (cf. Farlow 1987: Fig. 29A). In the Rye chart sauropod trail S4 crosses trail S3, and after the intersection the two trails are close together and nearly parallel. In the Homestead chart, however, there is a greater distance between the two trails after the

Figure 42.2. The Austin (Texas Memorial Museum) slab chart of the Paluxy River carnosaur–sauropod chase sequence, drawn by R. T. Bird. Text–figure 8 of Langston (1974) was based on this original.

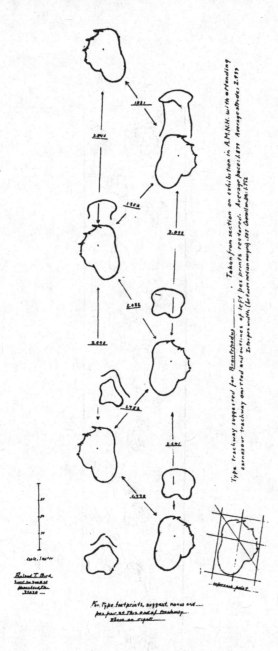

Figure 42.3. Bird's diagram of the New York (American Museum) slab (*Brontopodus* chart). The carnosaur footprints are not depicted (cf. Farlow 1987: Figure 31). The sequence begins with manus S2L (left) and proceeds through manus-pes sets S2M (right), S2N (left), S2O (right), S2P (left), and S2Q (right). The final footprint in the slab is pes S2R (left), the associated manus track of which is the first track in the Austin (Texas Memorial Museum) slab. In the New York slab pes tracks S2P and (probably — see Figure 42.9 K–O) S2R are deformed by footprints C1H and C1J, respectively, of the carnosaur, but in this chart these pes tracks are restored to their presumed original outlines. Footprint labels assigned by Farlow (1987).

crossover, and trail S4 takes a curvilinear path similar to, and nearly parallel to, that of S2. Finally, sauropod trackway S1 in the Rye chart consists of three pes and two manus impressions, but has considerably more footprints in the Homestead chart.

Neither of the master charts attempts to show all of the dinosaur tracks in the vicinity of Bird's quarry, but depicts only those within the confines of the coffer-dams, as is readily apparent if one compares the master charts with John C. Germann's sketch of the tracksite (Bird 1985: 179). In addition, examination of the New York and Austin slabs reveals numerous tridactyl footprints that appear in none of Bird's charts (Figs. 42.4, 42.7).

Locations of peaks and troughs in stride length of S2 and its attending carnosaur are fairly consistent among Bird's various charts and with actual measurements of the Austin and New York slabs, although the Homestead map sometimes differs from the others. Sauropod trackways S1, S3, S4, and S5 are depicted only on the two master charts. Too little of S1 is shown on the Rye chart to compare with the Homestead chart. In contrast to the consistency in stride lengths of the two dinosaurs in the chase sequence among the maps, there is rather poor agreement in peaks and troughs of stride length between the two master charts for sauropod trackways S3, S4, and S5. However, mean stride lengths for the trails as estimated from these two charts are reasonably close.

In the letter to Brown (Thanksgiving Day, 1938) in which he first reported his discovery of the Paluxy River sauropod footprints, Bird gave some measurements of the portion of trail S3 then visible: "The hind foot measures a full yard [91 cm] in length not counting a possible bit of slippage ... the trail itself ... measures about six feet [183 cm] across its lateral margins.... The strides (measurements taken from corresponding rights and lefts) run from 11 feet 5 inches [348 cm] to 12 feet 10 inches [391 cm] — which of course strikes an average over 12 feet [366 cm]." These values are not directly comparable to those given for trackway S3 in Table 3 (the latter being based on the entire trail as exposed in 1940), but are reasonably close. Similarly, measurements of pes track length, pes pace, and pes stride made on the New York and Austin slabs (S2) and the portion of S4 still in situ in the bed of the Paluxy River (Table 42.2) are consistent with those estimated from Bird's maps (except for S4's stride — Table 42.3), although other measurements (step angle, trackway width) show less satisfactory agreement.

In reporting his discovery of the Giant sauropod trail S5 to Brown (9 May 1940), Bird indicated that the pes tracks of this monster, "exaggerated somewhat by the shank of the animal's leg, ... [were] about four and a half feet [137 cm] in length, with an overall, exterior, margin to margin measurement roughly touching four feet ten [147 cm] by forty inches [102 cm]." Using the Rye chart, and allowing for Bird's 'shank' impressions, we estimate a pes track length of 110 cm.

We conclude that the various charts of the Paluxy River quarry site prepared by Bird are, within the limits of drafting error, reasonably accurate depictions of the disposition of the main dinosaur trails at this locality. Where there is disagreement between the two master charts, we follow the more-detailed Rye chart, because the Homestead chart, which was drafted later, is by Bird's own admission at least in part subjective.

Bird also drafted a map of the Davenport Ranch site; Bird (1944) and Langston (1974: text-Fig. 9) both published a version. In 1984 Farlow was given the original map by Mrs. Erickson. For some trails it depicts footprints that were not included in the published version. Measurements of trackway parameters of the Davenport Ranch brontosaurs (Table 42.3) are based on the original version. In a letter to Brown (1 March 1940) Bird gave measurements of "the one and only real promising trail" (for display purposes): pes track length = 26 inches (66 cm), "stride of corresponding rights and lefts" = 9 feet (274 cm), and "width of trackway" about 4 feet (122 cm). This trackway is probably the one labeled trail 8 on Bird's original Davenport Ranch site chart; a right manus-pes set of this trail (cf. Langston 1974: Plate 3, Figs. 2, 3; Farlow 1987: Figs. 5B, 17D, E) was collected for the Texas Memorial Museum (letter to Brown, 23 March 1940). Measurements of trail 8 based on the trackway chart are close to those quoted by Bird (Table 42.3), although pes track length as estimated from the map is a little high. Lockley (1987, p. 84-86) published an interpretation of the Davenport Ranch tracksite in which he studied overlapping trackway relationships and estimated trackway dimensions using the published map. His estimate for the dimensions of trail 8 are 9-13% greater than Bird's unpublished records. However, foot length and mean stride estimates derived from Bird's unpublished map (Table 42.3) average 10-12% greater than Lockley's "overestimates" from the published map.

Footprint and Trackway Measurements

a) Footprint dimensions.

1) Manus. Well-preserved manus tracks have a stacked, double-crescent shape (Figs. 42.6-42.7), with a conspicuously anteriorly-directed indentation in the middle of the posterior margin of the track. Manus track length was measured from the middle of the anterior footprint margin to an imaginary line connecting the posterior edges of the medial and lateral posterior lobes of the footprint (across the indentation). In poorly-preserved manus tracks length could not be measured. Manus track width was measured as the greatest width perpendicular to the above-defined fore-aft footprint axis, and could usually be taken, perhaps not always accurately, even on poorly-preserved tracks.

2) Pes. Length was measured from the middle of the posterior margin of the footprint "heel" to the tip of the claw mark of digit I (or its vicinity in poorly-preserved tracks whose digit marks were effaced). Width was measured roughly perpendicular to this axis, at the point of greatest footprint width, at or near the impression of digit IV.

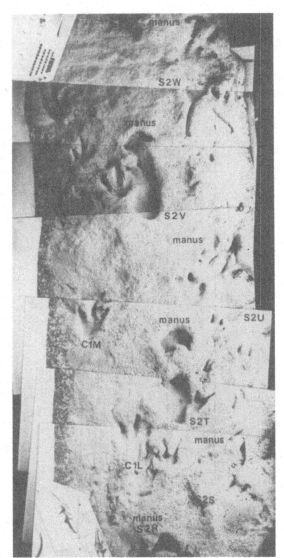

Figure 42.4. Photomosaic of the Austin slab (Texas Memorial Museum 40638-1); original photographs by Daniel Griggs.

b) Angle of pes track long axis with animal's direction of travel (divarication of foot from midline — Leonardi 1987). The compass heading of each footprint was sighted from the "heel" to the tip of the claw mark of digit I. Mean headings of left and right footprints were calculated, and the dinosaur's direction of travel calculated as the mean of the average headings for left and right tracks. The angle of the pes long axis with the direction of travel is the difference between the trackway mean heading and the means of the left or right footprint headings; note that the angle can be calculated in this manner only if the dinosaur's path was linear or nearly so.

c) Manus-pes distance. Ideally this measurement should

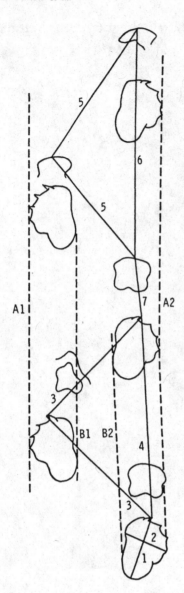

Figure 42.5. Measurements of sauropod footprints and trackways used in this paper; see text for details and additional measurements. **1**, pes footprint length. **2**, pes footprint width. **3**, pes pace. **4**, pes stride. **5**, manus pace. **6**, manus stride. **7**, manus–pes distance. Inner trackway width is the perpendicular distance between line segments B1 and B2. Outer trackway width is the perpendicular distance between line segments A1 and A2.

be made between the centers of the tracks of a manus–pes set, but in many cases the manus footprints were deformed or possibly in some cases even obliterated by overprinting by the pes tracks. Consequently, the manus–pes distance was routinely measured between two reference points (middle of anterior margin of manus, tip of claw mark I of pes), as well as between track centers, where possible; the track center manus–

pes distance is roughly 30–50% larger than the reference point manus–pes distance.

d) Pace and stride. The pace is the distance between successive footprints of the opposite feet, and the stride the distance between successive footprints of the same foot. Paces and strides were routinely measured for both manus and pes tracks using the reference points already described. For comparative purposes we also measured paces and strides from track centers, but consider those measurements less useful. Apart from the problem of finding a track center in a deformed manus print, the center of a sauropod manus or pes footprint is not a clear, unambiguous landmark, and is consequently a rather subjective location, whereas the above-defined reference points are more straightforward. Nevertheless, paces and strides measured by the two methods were quite similar.

e) Step angle (pace angulation — Leonardi 1987) and trackway width. The step angle was calculated from pace and stride measurements by the law of cosines, and was usually similar for reference point- and track center-based determination of pace and stride. Trackway width was calculated from pace, stride, and step angle by the law of sines, and represents the distance, perpendicular to the stride direction, between left and right reference points or track centers of manus or pes footprints. This measurement of trackway width is equivalent to the width of pace of Leonardi (1987). Trackway widths based on manus tracks were similar for reference point- and track center-based measurements, but the track center values of trackway width based on pes tracks were roughly 15% larger on average than their reference point-based counterparts.

f) Inner and outer trackway width. These measurements were determined only for pes tracks. The inner trackway width (interpedes distance — Leonardi 1987) is the distance between two line segments connecting the innermost margins of successive (in a stride) left and right pes tracks. Because the innermost margin of the track is not a fixed anatomical landmark, and can vary with the vagaries of foot emplacement, and slight changes in direction of travel by the animal, the inner trackway width as measured here can diminish to zero even though the inner margins of the footprints do not impinge on the trackway's midline. The outer trackway width (straddle of Halfpenny 1986) is the distance between two line segments connecting the outermost edges of successive (in a stride) left and right pes tracks.

g) Glenoacetabular distance (body length). Estimating the distance from a quadrupedal trackmaker's shoulders to its hips requires making certain interpretations of the animal's gait (Leonardi 1987): (1) If the creature employs the alternate pace gait of primitive tetrapods, the glenoacetabular distance can be calculated as half the stride length plus the manus–pes distance. (2) If, as commonly seen in mammals, the feet in each diagonal limb pair do not move synchronously, the body length is about three–quarters the stride length plus the manus–pes distance; this is also roughly the value of the stride itself (Halfpenny 1986).

Hindfoot tracks made by animals walking in this fashion may impinge on or even obliterate forefoot impressions (Thomas 1986 and pers. comm., Leonardi 1987). (3) If the animal is short–bodied and walks in an amble (pace or rack of Halfpenny 1986, who uses amble in a different sense than Leonardi 1987), the pes impressions may be emplaced ahead of manus footprints; in this case, the body length is equivalent to the distance from a pes footprint to the next manus track in front of it on the same side of the body. (4) In a longer–bodied ambler, the pes may be emplaced just behind the manus impression made during the previous step cycle, and the glenoacetabular distance is calculated as the stride length plus the manus–pes distance. Because sauropod pes tracks frequently overprint manus tracks or otherwise distort them, sauropods probably did not often use the primitive alternate pace gait. None of the Comanchean brontosaur trackway we have seen shows pes tracks in advance of their associated manus tracks (see Kim 1987 for a Korean sauropod trail in which this apparently does occur). Consequently we estimated the body lengths of our sauropod trackmakers by methods 2 and 4 above; where possible, we calculated the gleno-acetabular distance from both reference point- and track–center values; the latter estimate is somewhat larger, and is probably closer to the true value of the body length.

Footprint Morphology (Figs. 42.4, 42.6–42.10)

a) Manus. The smallest manus tracks we have seen were about 30 cm wide, but some of the sauropod manus prints from the Davenport Ranch site were probably smaller. The largest manus prints we have seen (c. 60 cm wide) were made by the Mayan Ranch "swimming" sauropod (cf. Farlow 1987: Fig. 17c; Lockley and Conrad this volume).

Well-preserved manus tracks are slightly wider than broad, and have a "double crescent" shape (Figs. 42.6–42.7). The footprint is deepest at its anterior end and shallowest at the indentation in the center of the track's rear margin. The lateral and medial posterior lobes of the track are separated from the deep anterior crescent of the footprint by marked indentations in the track outline and by shallow areas, and from each other by the shallow region of the track's rear margin. The lateral lobe is usually a little deeper, and sometimes somewhat larger, than the medial lobe. There are no indication of claws on manus tracks.

The hand skeleton of many sauropods is roughly consistent with that which one would infer for the Texas sauropod trackmakers (Marsh 1896; Riggs 1901; Gilmore 1925, 1936, 1946; Janensch 1961; Borsuk-Bialynicka 1977; Fig. 42.8), with a semi–circular or tubular meta-carpus. However, most sauropods have a stout ungual

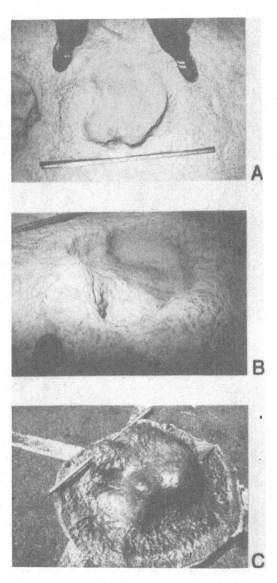

Figure 42.6. Right manus footprint S2M, New York slab (AMNH 3065). Length and width about 50 cm.
A, overhead view. **B,** posterolateral oblique view; note variations in depth. **C,** artificial convex hyporelief (negative copy of the footprint itself) used by the Reproduction Department of the American Museum to make copies of the footprint for sale; note variations in depth.

on digit I of the manus, in marked contrast with the Texas tracks. Beaumont and Demathieu (1980) suggested that sauropods may have walked on the knuckles of their hands, their fingers bent backward, but this seems far-fetched. Lockley et al. (1986) describe sauropod manus tracks from the Morrison Formation that do show claw marks, undermining the interpretation of Beaumont and Demathieu. Thulborn (this volume) pro-

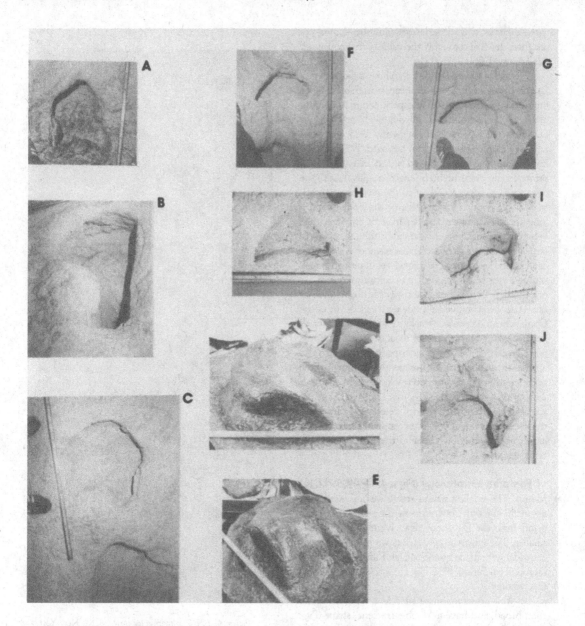

Figure 42.7. Various manus footprints of type trackway S2, New York (AMNH 3065) and Austin (TMM 40638-1) slabs. **A,** track S2L (left), New York slab. Length 44 cm, width 35–40 cm. **B,** medial oblique view of Track S2N (left), New York slab. Measurements not taken due to irregular shape of track as preserved. **C–E,** track S2O (right), New York slab. Length 53 cm, width 51 cm. **C,** overhead view of the footprint. **D** and **E,** oblique views of artificial convex hyporelief, American Museum. **F,** track S2P (left), New York slab. Width about 35 cm. **G,** track S2Q (right), New York slab. Width about 43 cm. The manus is superimposed on a small tridactyl footprint (estimated length about 40 cm). **H,** track S2R (left), Austin slab. Width about 39 cm. An unpublished photograph of this footprint in situ shows a large bulge of rock projecting forward into the rear margin of the track; this bulge is not so conspicuous in the footprint as reassembled in the Austin slab. **I,** track S2T (left), Austin slab. Length about 40 cm, width about 46 cm. **J,** track S2V (left). Width about 36 cm.

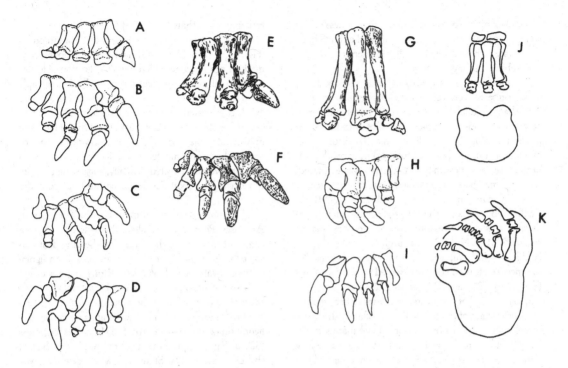

Figure 42.8. Hand and foot skeletons of various sauropod dinosaurs; not to scale. **A,** right manus of *Apatosaurus excelsus* (Carnegie Museum 563); redrawn from Gilmore (1936: Fig. 35). Width about 73 cm. **B,** right pes of *Apatosaurus excelsus* (Carnegie Museum 89); redrawn from Hatcher (1901: Fig. 22). Width about 80 cm. **C,** reconstruction of right pes of *Diplodocus hayi* (Texas Memorial Museum 41425–1; original specimen is Carnegie Museum 662) by W. Langston, Jr. Width about 65 cm. **D,** left pes of *Barosaurus africanus* (No. 28); redrawn from Janensch (1961: Plate 23, Fig. 6). Width about 50 cm. The pes of *Barosaurus* may have had three unguals, rather than two as shown here (J. S. McIntosh pers. comm.). **E,** right manus of *Tornieria robusta* (No. 5); redrawn from Janensch (1961: 194, Fig. 2a). Width about 45 cm. **F,** right pes of *Tornieria robusta*; redrawn from Janensch (1961: Plate 23, Fig. 8). Width about 60 cm. **G,** right manus of *Brachiosaurus brancai* (No. S II); redrawn from Janensch (1961: 194, Fig. 1a). Width about 62 cm. **H,** left pes of *Camarasaurus lentus* (U.S. National Museum 13786); traced from a photograph by W. Langston, Jr. Width about 38 cm (M. Brett–Surman pers. comm.). **I,** left pes of *Pleurocoelus* sp. (Field Museum PR 977); redrawn from Gallup (1975: Fig. 13). Width 45–50 cm. **J,** diagram of a Texas sauropod manus footprint and the hand skeleton reconstructed from the track by R. T. Bird. **K,** R. T. Bird's interpretation of the skeleton of the left rear foot of the Texas sauropod trackmaker, superposed on the outline of a pes track. Widths of hand and foot skeletons given above are in most cases dimensions across images of the hand or foot projected into two dimensions, and thus are not directly comparable to footprint widths.

posed that sauropods may have carried the pollex claw above the ground by dorso–medial hyperextension. Although not inconceivable for the Texas trackmakers, the great depth of some of the Paluxy River manus footprints (20–30 cm) suggests that some indication of "cocked" manus claws would have been preserved, and for these trackmakers an alternative interpretation appears more plausible. *Brachiosaurus* has a very small manus claw (Fig. 42.8G), and the Texas manus footprints can be readily interpreted as having been made by a brachiosaur-like hand in which the claw was either completely imbedded in the foot pad or lost altogether.

In his notes R. T. Bird provided his own interpretation of the hand skeleton of the Texas sauropod trackmaker (Fig. 42.8J). We follow his reconstruction,

with a slight modification. The configuration of well-preserved manus tracks suggests that digits II–IV were bound together in a pad to form the deep anterior crescent, and that digits I and V were slightly offset distally from the three central digits to form the medial and lateral posterior lobes, respectively, of the manus footprints (cf. Paul 1987: Fig. 2G). In contrast to the condition inferred by Gilmore (1936: 22) for *Apatosaurus*, in which "the principal weight of the body was supported by the manus on the inner side of the foot," the depth variations seen in the Texas tracks suggest that the dinosaur's weight was more evenly distributed across the hand (cf. *Alamosaurus* fidé — Gilmore 1946: 39), but with slightly more weight carried by the conjoined digits II–IV, and by digit V, than by digit I. The indentation at the back of the footprint, and the shallowness of this region, suggest that the manus in life had a horseshoe-like shape, with relatively little development of a fibrous pad behind the metacarpals.

Most manus footprints of Texas sauropods are poorly preserved, crescent-shaped depressions. In some cases the poor quality of manus track preservation may be due to collapse of a track after foot withdrawal, or modern erosion, but it is likely that the deformation of many manus footprints occurred during production of the associated pes tracks. As the pes was emplaced, sediment may have been pushed forward, encroaching upon the manus track from the rear. In the AMNH-TMM slab trackway, the reference point manus–pes distance in sets with reasonably well–preserved manus tracks (S2M, S2N, S2O, S2T, and S2V) ranges 45–81 cm, $\bar{x}$ = 66.8 cm; the reference point manus–pes distance in sets with poorly preserved manus footprints (S2P, S2Q, S2S, and S2W) ranges 32–63 cm, $\bar{x}$ = 49.4 cm, suggesting that the greater the distance between the anterior margins of the two footprints of a set, the better the chance of good preservation of the manus track. The "Wet" sauropod trail (Ozark Bronto trail A) of Dinosaur Valley State Park extends from some distance without clear signs of manus prints (Fields 1980, Farlow 1987); perhaps in this trackway the pes tracks completely overprint and obliterate their associated manus footprints.

The manus tracks of the famous Mayan Ranch "swimming" sauropod trail are of poor quality, but not due to any interference from the pes. These footprints are large, ovoid depressions with faint indications of the indentation at the rear track margin (Farlow 1987; Fig. 17C); the tracks are very shallow (at most a couple of centimeters deep), as compared with maximum depths of 20–30 cm for some of the Paluxy River manus tracks. The poor quality and shallowness of the Mayan Ranch tracks may be due to the dinosaur's having been semi-floating in deep water. The large size of these tracks suggests a fairly large sauropod. It is also possible, on the other hand, given the indistinct nature of these foot-prints, that sauropod hands of more modest size were thrust against the substrate, forcing sediment outward to form tracks of exaggerated dimensions. However, the great width of this trackway suggests that the trackmaker was indeed a big sauropod (Table 42.2). Conceivably the Mayan Ranch manus footprints are ghost tracks, which would account for their faint outlines, and also be consistent with their preservation in a subaqueous setting (M. G. Lockley pers. comm.).

b) Pes. Some of the Davenport Ranch trackmakers were very small sauropods, with pes footprint lengths of as little as 40–50 cm. In contrast, the Paluxy River S5 (Giant) trail was made by an enormous creature with pes track lengths in excess of 100 cm (see above). The AMNH-TMM slab sauropod was among the larger Texas brontosaurs whose trails we have seen, but still

rather smaller than S5.

Pes footprints are considerably longer than broad (Figs. 42.4, 42.9). Digits I–III have well–developed claw marks. Digit IV, where there is any indication of it, has a small nail mark or depression made by a very small ungual or foot callosity. Digit V, where it can be seen at all, is indicated by a small lateral lobe of the track. The three large inner claw marks are directed laterally; in vertical projection against the footprint's anterior margin they sometimes have a semi–circular shape, with the concave side directed medially, suggesting a rotary motion of the foot during its withdrawal from the sediment (Fig. 42.9F).

Sauropod pes footprints from the Briar site in Arkansas (Pittman and Gillette this volume) suggest a degree of mobility in the three claw–bearing digits. In some hindfoot tracks the outer toe marks show a signifi-cant backward as well as outward turning, associated with an indentation in the posterolateral track margin, between the "heel" and the position of digit V, that gives these footprints a slight U–shape. A similar con-figuration occurs in one of the TMM slab pes footprints (S2S in Fig. 42.4) and in one of the hindfoot tracks of the Dinosaur Valley State Park Main Site trackway (Farlow 1987: Fig. 15B).

In pes footprint S2R (the "Tommy Pendley" track) of the American Museum slab (Fig. 42.9 K–O), three large claw or toe marks occur in a somewhat trifur-cating fashion along the front edge of the track. R. T. Bird believed these marks to have been made by a car-nosaur thought to have been following sauropod S2, and not by the brontosaur itself (Farlow 1987: Fig. 31); as previously noted, the carnosaur stepped into other footprints of sauropod S2 as well. Although there is no other suggestion that a carnosaur print is superimposed on pes track S2R, we provisionally accept Bird's inter-pretation; the position and arrangement of the toe marks in this footprint together constitute a configura-tion quite different from that seen in other Comanchean sauropod pes tracks. If, however, the three toe marks were in fact made by the sauropod, this would con-stitute even stronger evidence for digital mobility in the trackmaker's hindfoot.

The Dinosaur Valley State Park "Wet" (Ozark A) trail has pes footprints with a remarkably triangular shape, with rather sharper "corners" (inner toe region, outer footprint region, and "heel"), and straighter edges between these corners, than in other trackways (Farlow 1987: Fig. 15G). One of the Briar site trackways similarly has slightly triangular hindfoot tracks (Pittman and Gillette this volume). Pes footprints completely obliterate manus footprints in both of these trails; whether this is connected with the unusual shape of the pes tracks is unknown.

Pes tracks are deepest along their inner margin, in the "heel," and in the region of the large inner claw mark (Figs. 42.9, 42.10). Footprints are shallowest at the

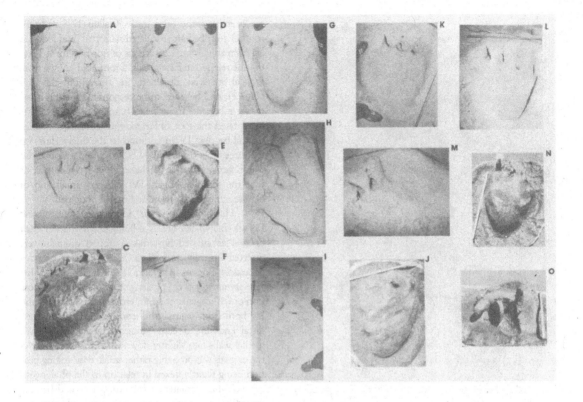

Figure 42.9. Various pes footprints of trail S2, New York slab (AMNH 3065). **A–C,** footprint S2M (right). Length 87 cm. **A,** overhead view. **B,** posteromedial oblique view. **C,** posteromedial oblique view of artificial convex hyporelief, Reproduction Department, American Museum. Note well-developed claw marks on digits I–III, lobed projections in digit positions IV and V. **D–F,** Footprint S2N (left). Length 87 cm. **D,** overhead view. **E,** overhead view of artificial convex hyporelief, Reproduction Department, American Museum. **F,** posteromedial oblique view. Note semicircular, concave–inward shape of claw marks in vertical projection against anterior margin of the footprint. **G,** footprint S2O (right). Length 88 cm. **H,** posterior oblique view of footprint S2P (left; length 87 cm) and associated manus. **I** and **J,** footprint S2Q (right). Length 94 cm. **I,** overhead view. **J,** overhead view of artificial convex hyporelief, Reproduction Department, American Museum. **K–O,** footprint S2R (left). Length about 88 cm. The three claw marks were believed by Bird to be a superimposed track of a passing carnosaur, and not part of the sauropod footprint itself. As suspected by Farlow (1987), unpublished notes written by Bird indicate that this is the "Tommy Pendley" track. **K,** overhead view. **L,** posterior oblique view. **M,** lateral oblique view. **N** and **O,** posterior oblique and anterior views of artificial convex hyporelief, Reproduction Department, American Museum.

lateral edge, in the region of digits IV and V. This shallow area extends medially as a slight "rise" along the inner margin of the footprint, separating the deep anteriormost moiety associated with the claw mark of digit I from the deep region of the "heel." In some tracks these regional depth variations are subdued, and in others (e.g. the AMNH-TMM slab footprints) quite pronounced, presumably due to variations in substrate firmness.

Contributing to the great depth of the antero-medial corner of the track in some footprints (e.g. S2M) is an elliptical depression on the floor of the track near the base of claw mark I, perhaps an impression of the metatarsophalangeal joint of the inner digit (Fig. 42.9A). Shallow grooves sometimes extend back from the anterior edge of the track, behind claw marks II and III, perhaps indicating the positions of digits II and III.

A variety of sauropods have an asymmetric, entaxonic foot skeleton that could produce pes tracks similar to those seen in Texas and Arkansas (Marsh 1896; Hatcher 1901; Gilmore 1925, 1932, 1936; Janensch 1961; Coombs 1975; Gallup 1975; Borsuk-Bialynicka 1977; Fig. 42.12). Metatarsal I is the most massive of the foot bones; sauropods presumably carried most of their weight on the inner side of the foot (cf. Hatcher 1901: 51–52; Gilmore 1936: 241; Gallup 1975: 92), an interpretation consistent with the depth variations seen in the Texas and Arkansas tracks.

Most sauropod hind feet apparently bore three stout claws that decreased in size from digits I–III, also

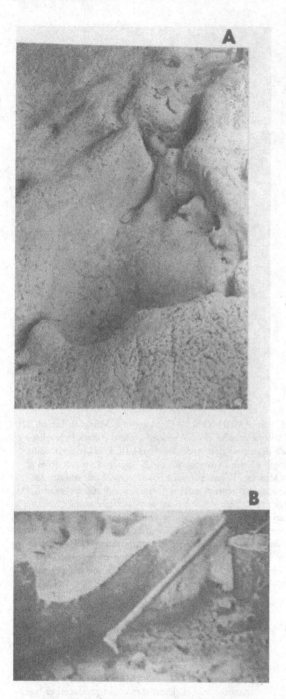

Figure 42.10. Depth variations in pes footprints, trail s2. **A,** posterior oblique view of right pes track S2U, Austin slab. Note the great depth of the "heel" and the anteromedial corner of the footprint, and the shallowness of the outer toe region. **B,** image from a film made by R. T. Bird of his 1940 Paluxy River quarry operation. Above the wood handle of the pick is a nearly transverse section across a sauropod right pes footprint. The dinosaur was moving away from the viewer. Note the great depth and vertical edge of the medial (left) side of the track, and the shallowness and gentler slope of the lateral (right) edge.

consistent with the Texas tracks. Gallup (1975 and pers. comm.) described an articulated sauropod hind foot and leg from the Lower Cretaceous of north–central Texas that he referred to *Pleurocoelus*, a form also known from skeletal material elsewhere in Texas (Pittman this volume), and believed to be responsible for the brontosaur footprints from Texas (Langston 1974). Gallup interpreted the foot of *Pleurocoelus* as bearing a small ungual on digit IV in addition to the larger claws on the three inner digits (Fig. 42.8I). If so, and if *Pleurocoelus* was in fact the Texas trackmaker, the pes tracks indicate that the putative ungual IV was very small and/or almost completely embedded in the foot pad.

Hatcher (1901: 50, 52), Gilmore (1936: 238), and Borsuk–Bialynicka (1977: 49) suggested that in their normal articulated positions sauropod unguals were directed laterally as well as anteriorly; for *Opisthocoelicaudia*, Borsuk–Bialynicka (1977: 49) concluded that "the asymmetry of the phalanges causes the oblique position of the unguals, their antero–posterior plane slanting both from sagittal and from vertical planes; their distal ends are directed antero–laterally, while their medial walls face slightly downwards. The mobility of the unguals was probably rather small, their resting position being slightly flexed in relation to the phalanges." A somewhat similar foot construction would be consistent with the nature of the Texas sauropod pes tracks, although the rearward turning of the toemarks seen particularly in Arkansas sauropod pes tracks suggests greater flexibility than in Borsuk–Bialynicka's interpretation of *Opisthocoelicaudia*.

The "heel" of the track was probably impressed by a thick pad of fibrous tissue that supported the rear of the foot (cf. Norman 1985: 84). The large size and great depth of impression of the "heel" indicate that it bore a sizeable fraction of the dinosaur's weight.

Footprint S2W of the TMM slab has two large grooves that project backward from the heel in a peculiar, diverging fashion (Fig. 42.4). Conceivably these were made by two of the dinosaur's pedal claws as the foot was brought forward, but the great posterior divergence of the ends of the grooves makes it difficult to visualize how this could have happened. In any case, the odd "wishbone" rear margin of this track makes it an important landmark in identifying trackway sequences in Bird's charts and photographs.

Possibly related features were reported by Fields (1980: 11), associated with some right pes tracks of the "Wet" (Fields' Ozark A) trail, described as "short grooves in the rock, 50–60 cm long.... The grooves are at a slight angle to the line of the trail." None of these marks shows the wishbone shape seen in TMM slab pes track S2W. Fields interpreted these furrows as tail drag marks; G. Kuban (pers. comm.) interpreted them as drag marks of pes claw I. Neither of these interpretations is impossible, but given the widespread occurrence of linear scour marks in the Paluxy's bed (see below) we cannot confidently endorse either interpretation of these features as trace fossils.

Trackway Pattern (Tables 42.1–42.3)

Combining our field observations with measurements estimated from R. T. Bird's Paluxy River and Davenport Ranch charts, we have trackway data for 31 individual sauropods from Texas and Arkansas. As already noted, the brontosaur trackmakers of the Gulf Coastal Plain spanned a considerable size range. Some of the Davenport Ranch dinosaurs had trackway widths of as little as 50–60 cm, while the maker of trail S5 had a trackway width of about 165 cm. The AMNH–TMM slab trackmaker had a trackway width of about 100 cm and a glenoacetabular length of about 3–4 m; this dinosaur's glenoacetabular length was probably roughly comparable to that of *Apatosaurus*, to judge from mounted skeletons at the Yale Peabody Museum, American Museum, Carnegie Museum, and Field Museum. Because of the great size range of the Comanchean brontosaurs, some parameters are not directly comparable among trackways; for comparative purposes these are standardized by dividing them by an index of animal size, the length of pes footprints.

Values of trackway parameters calculated separately for manus and pes footprints show reasonably good agreement. In the following statistical analyses all trackway parameters are based on pes tracks (n = 25 trackways). Because of the small sample size, and because some trackway measurements are estimates made from Bird's site maps, all interpretations based on the statistical analyses are only tentative.

Unsurprisingly, stride length is strongly correlated with pes footprint length (r = 0.85, p < 0.0001); one would expect larger animals to take longer steps than smaller animals, all other things being equal. Relative stride length (stride length/pes track length) decreases with increasing footprint length (r = –0.58, p = 0.0022); the stride length ranges from about two to five times the footprint length.

Trackway width is strongly correlated with footprint size (r = 0.90, p < 0.0001 [n = 23]) but there is no good indication that relative trackway width (standardized against pes track length) changes with increasing size (r = 0.31, p = 0.15). Relative trackway width ranges about 1–1.5. Although reference point trackway widths are similar for manus and pes track determinations, much of a manus track is usually closer to the trackway midline than much of a pes track (cf. Table 42.1 for track center–based manus and pes estimates of trackway width of the AMNH–TMM slab trail).

Most values of the step angle fall in the range 100°–120°. The step angle decreases with increasing dinosaur size (r = –0.62, p = 0.0009), and increases with increasing relative stride length (r = 0.59, p = 0.0018). Because relative stride length and body size may themselves be negatively correlated (see above), it is possible that the relationships of these two variables with step angle are two different ways of describing the same overall relationship.

In contrast to the trackways of bipedal dinosaurs, in which footprints usually toe inward in a pigeon–toed fashion (cf. Farlow 1987), the sauropod pes tracks angle outward (positive rotation) by about 10–15° with respect to the dinosaur's direction of travel. In many (most?) cases the manus also appears to be turned outwards. Sauropod manus footprints from the Purgatoire Valley (Morrison Fm) of Colorado also tend to toe outward (M. G. Lockley pers. comm.).

Bird's charts of the three long Paluxy River trails (S2, S3, and S4) suggest that these sauropods walked in a slightly sinuous fashion (particularly S3). The trackways of the three dinosaurs show marked changes in pace and stride length over their courses. Particularly in S3 and S4 there is a tendency for the trackway to show a succession of long–short–long–short paces; sometimes it is the pace ending in a right foot, and sometimes that ending in a left, that is the longer, in both trails. These long–short cycles of pace length are not obviously associated with changes in trackway direction (e.g. the above–mentioned sinuosity), although in one trackway interval of trail S3 the longer paces end in the foot on the outside of a slight bend in the trail. The "East" (Ozark C) and "Wet" (Ozark A) (cf. Fields 1980: 46–47) trails in Dinosaur Valley State Park also show the long–short pace pattern, as do sauropod trackways from the De Queen Formation (Lower Cretaceous) of Arkansas (Pittman and Gillette this volume).

Occasionally associated with sauropod footprints in the Paluxy River are curvilinear grooves that local residents interpret as tail drag marks (Farlow 1987: Fig. 15D). While we cannot rule this interpretation out, we have our doubts. The carbonate rock in which the footprints are impressed is soft, and readily erodes into a variety of irregular and/or elongate shapes (including some of the controversial "man tracks"). At other places in central Texas (e.g., McKinney Falls State Park near Austin and Hondo Creek in Bandera County), river scouring has created furrows in the river bed that are similar to the alleged Paluxy River tail drags (Farlow 1987), and so the latter may not be trace fossils at all. The sauropod trails at Dinosaur Valley State Park unfortunately trend in a direction parallel to river flow, making it difficult to distinguish erosion scours from tail marks, even if the latter do exist.

Bird (1944) identified a groove associated with some poorly preserved putative sauropod tracks at the Davenport Ranch as a tail drag mark. Farlow examined this site in 1984, and located what appeared to be the same groove. It is not in the same bedding plane as the main footprint assemblage, but is in a lower ledge (cf. Bird 1954: 718). According to Bird, the groove ran "squarely between a set of prints" (Bird 1985: 163). This was not obvious to Farlow, although the site may have deteriorated since 1940. However, the groove is oriented roughly parallel to the course of the adjacent West Verde Creek, in which similar grooves, also oriented parallel to the current direction, occur. Thus we suspect, contra Bird, that the "tail drag" is an erosion scour. In any case, if there really are tail marks associated

Table 42.1. *Detailed measurements of the American Museum – Texas Memorial Museum Paluxy River sauropod trackway (trail S2).*

Trackway Interval	Pace Length (cm): Manus Tracks: Reference Point	Track Center	Pes Tracks: Reference Point	Track Center	Trackway Interval	Stride Length (cm): Manus Tracks: Reference Point	Track Center	Pes Tracks: Reference Point	Track Center
S2L – S2M	175.3	—	—	—	S2L – S2N	254.0	—	—	—
S2M – S2N	157.5	—	186.7	177.8[b]	S2M – S2O	271.8	—	273.1	264.1[b]
S2N – S2O	168.9	—	167.6	175.2[b]	S2N – S2P	290.8	—	303.5	304.8[b]
S2O – S2P	176.5	—	194.3	208.2[b]	S2O – S2Q	322.6	—	331.5	309.8[b]
S2P – S2Q	207.0	—	190.5	185.4[b]	S2P – S2R	—	—	307.3[a]	294.6[b]
S2Q – S2R	—	—	163.8[a]	198.1[b]	S2Q – S2S	—	—	—	—
S2R – S2S	167	173	—	—	S2R – S2T	298.5	294.6	—	—
S2S – S2T	178	178	178 (175)[c]	183.5	S2S – S2U	284.5	280.7	307 (303.5)[c]	315.0
S2T – S2U	150	150	175 (176)[c]	186.0	S2T – S2V	288.3	289.6	314 (308.6)[c]	318.8
S2U – S2V	200	197	195 (186)[c]	200.0	S2U – S2W	328.9	—	336 (335.3)[c]	335.3
S2V – S2W	192	—	199 (193)[c]	199.0					
Mean Values:	177.2	174.5	183.3	190.4		292.4	288.3	310.3	306.1

Trackway Interval	Step Angle: Manus Tracks: Reference Point	Track Center	Pes Tracks: Reference Point	Track Center	Trackway Width (cm): Manus Tracks: Reference Point	Track Center	Pes Tracks: Reference Point	Track Center
S2L – S2N	99°	—	—	—	107	—	—	—
S2M – S2O	113°	—	101°	97°[b]	90	—	113	117[b]
S2N – S2P	115°	—	114°	105°[b]	93	—	98	116[b]
S2O – S2Q	114°	—	119°	104°[b]	103	—	98	121[b]
S2P – S2R	—	—	120°[a]	100°[b]	—	—	88[a]	123[b]
S2Q – S2S	—	—	—	—	—	—	—	—
S2R – S2T	120°	114°	—	—	86	95	—	—
S2S – S2U	120°	117°	116° (120°)[c]	117°	81	84	98 (88)[c]	97
S2T – S2V	110°	112°	121° (117)°[c]	111°	98	94	87 (95)[c]	109
S2U – S2W	114°	—	117° (124°)[c]	114°	107	—	103 (88)[c]	108
Mean Values	113°	114°	115°	107°	96	91	98	113

[a]Values of stride and pace measured using track heel as reference point.

[b]Values of stride and pace taken from R. T. Bird's *Brontopodus* chart.

[c]Replicate values

Mean pes footprint length = 86.5 cm; pes footprint width = 60–65 cm

Mean angle of pes footprint long axis with animal's direction of travel: 13° outward (American Museum slab),
    9° outward (Texas Memorial Museum slab)

Inner width of trackway: 38.0 cm (American Museum slab), 23.5–32.5 cm (Texas Memorial Museum slab)

Outer width of trackway: 182.5 cm (American Museum slab), 165–170 cm (Texas Memorial Museum slab)

Length of well-preserved manus tracks: c. 50 cm

Distance from manus track to associated pes track of set:

    a) Reference points of tracks: 32–81 cm; mean = 58.1 cm

    b) Track centers: 61–93.5 cm; mean = 76.1 (Texas Memorial Museum slab)

    c) Anterior margin of manus to heel of pes; 117–147 cm; mean = 132.1 cm (Texas Memorial Museum slab)

with any Comanchean sauropod trails, they are not common (cf. Bird 1985: 163).

Comparisons with Other Sauropod Ichnites

There is a striking lack of unanimity among ichnologists as to the proper philosophy of naming and classifying trace fossils (reviewed by Sarjeant and Kennedy 1973, Sarjeant 1975). Some workers believe that ichnotaxa should, insofar as possible, be named and classified according to the nature of their makers, either incorporated into the standard systematic framework erected for body taxa or placed in a parallel parataxonomy. Advocates of this view see trace fossils as manifestations of organism anatomy, even if indirect and viewed through the murky filter of preservation as sedimentary structures. Other ichnologists emphasize that trace fossils *are* sedimentary features, and not anatomical structures; adherents of this view contend that one could validly apply formal names to different types of traces made by the same animal during different activities (e.g., *Rusophycus* and *Cruziana* of trilobites).

We see some validity in each of these viewpoints, and in the absence of a universally accepted code of ichnotaxonomy, adopting one or the other philosophy is a matter of personal preference. That being the case, it behooves us to make our own biases explicit. Our views are perhaps colored by the fact that we work on dinosaur footprints. Such fossils were generally made by animals in transit from one place to another, and so represent only one of the behavioral categories of trace fossils commonly recognized by ichnologists. Furthermore, many dinosaur tracks reflect the foot structures of their makers with remarkable fidelity, permitting comparisions of the foot skeletons of dinosaur skeletal taxa with interpretations of pedal structure based on footprints. The situation is more complicated for other kinds of trace fossils, such as burrows, feeding traces, or nests; such trace fossils may preserve little or no detailed information about the morphology of their makers.

We see no need for any comprehensive classification of all the behavioral categories of ichnofossils, and we feel that footprints in particular are of systematic (as opposed to behavioral or paleoecological) significance primarily as indirect "stand-ins" for body fossils. Thus we contend that formal names should be given only to those dinosaur footprints that are preserved well enough to provide significant information about the pedal structures of their makers. We will apply this philosophy in our discussion of sauropod trace fossils. Although we will argue that some previous workers have improperly named brontosaur ichnites, our remarks do not necessarily mean that we consider those ichnologists' studies to have been flawed; those workers may merely have had different outlooks on ichnotaxonomy from ours.

Brontosaur tracks have now been reported from sites around the world (Ginsburg et al. 1966; Kaever and Lapparent 1974; Antunes 1976; Friese and Klassen 1979; Hendricks 1981; Jenny et al. 1981; Campbell 1983; Leonardi 1984, 1985, 1986; Godoy and Leonardi 1985; Casanovas-Cladellas and Santafe-Llopis 1986; Ishigaki 1986, and this volume; Ishigaki and Haubold 1986; Kim 1987; Lockley 1986; Lim et al. this volume). Unfortunately, few of these ichnites have been described in any detail. Some have received formal names in the absence of diagnostic or even discernible morphological features, and the sauropod nature of some is open to question. Many footprints ascribed to sauropods are little more than rounded or ovoid depressions, identified as brontosaur tracks mainly by their size and arrangement in a trackway pattern (e.g., *Elephantopoides barkhausensis* Kaever and Lapparent 1974; *Neosauropus lagosteirensis* Antunes 1976); we question the validity and usefulness of naming such ichnites — the procedure is reminiscent of assigning names to isolated dinosaur teeth.

Dutuit and Ouazzou (1980) gave the name *Breviparopus taghbaloutensis* to a sauropod trackway from the Middle Jurassic of Morocco; Ishigaki (1986 and this volume) provides additional information about the trail. The trackmaker, with pes tracks some 115 cm long, may have been comparable in size to the maker of trail S5 of the Paluxy River. Manus tracks of the Moroccan sauropod are crescent-shaped depressions without any indication of toes, as often seen in Comanchean sauropod manus footprints. The very short manus–pes distance in *Breviparopus* may mean that pes tracks to some extent obliterate the posterior parts of manus tracks, but forefoot impressions in Moroccan trails thought to have been made by half-floating sauropods are very similar to those in the type trackway (Ishigaki this volume), suggesting that distortions created during pes emplacement may not be serious. Some of the pes impressions have an overall shape reminiscent of that seen in U.S. Gulf Coast sauropod pes footprints. The claws of at least some of the *Breviparopus* pes tracks seem to be directed forward, with little lateral inclination; other hindfoot impressions in the trail do seem to be directed outward in a manner similar to that seen in most Comanchean brontosaur pes tracks (Ishigaki this volume). The "heel" region of pes footprints is relatively shallow (none of the footprints is very deep in absolute terms), unlike Texas tracks. In contrast to most U.S. Gulf Coast sauropod trails, the Moroccan trail has manus tracks whose centers are frequently farther from the trackway mid-line than those of pes tracks (the single manus track preserved in the Paluxy River "Middle" [Ozark B] trail does have a rather *Breviparopus*-like relationship to its associated pes track, however). In consequence, the outer limits of the trackway are often defined by the manus footprints, rather than the pes tracks, and the pace of manus tracks is considerably greater than that of pes tracks. The angle formed by the long axis of the pes tracks with the dinosaur's direction of travel is larger in the Moroccan than in the Comanchean trackways (the Dutuit and Ouazzou definition of this angle differs slightly from ours, but as best we can tell from their trackway diagram, the angle would be larger than in Comanchean trails even if Dutuit and Ouazzou's method of determining the angle had been the same as ours). The trackway midline intersects the posteromedial edges of pes

Table 42.2. *Summary measurements of sauropod trackways from the Lower Cretaceous of Texas. Unless otherwise stated, measurements of manus–pes distance, pace and stride, and calculations of step angle and trackway width, were made using track reference points rather than track centers. Numbers in parentheses indicate the range of values.*

Site	Trackway Label	Trackway Bearing[a]	x̄ Manus Track Width (cm)	x̄ Pes Track Length (cm)	x̄ Angle of Pes Track Long Axis with Direction of Travel	Manus–Pes Distance (cm)
Paluxy River, Somervell County	AMNH–TMM Slabs (S2)	South	43.7	86.5	9°–13° outward	58.1 (32–81)
	Main Site Trail	187°	42.5[b]	78.8[c]	17° outward	48[b, d]
	"East" (Ozark C)	193°	28.5	76.5	8° outward	47.0 (25–70)
	"Wet" (Ozark A)	199°	—	90.0	17° outward	—
	"Middle" (Ozark B)	204°	36.0[b]	62.0	14° outward	32.0[b]
	Bird Site[i]	198°[j]	47.5[b]	78.3[j]	12° outward[j]	65.0[b]
South San Gabriel River, Williamson County	Trail 1 (Pittman 3)	179°	48	c.74[b]	—	79.0 (75–82)
	Trail 2 (Pittman 2)	180°	52	c.72[b]	—	61.3 (55–66)
	Pittman 1	South	c.47	.c.66	—	77.0 (76–78)
	Pittman 4	South	c.46	c.70	—	64.0 (60–68)
	Pittman 5	South	c.42	c.63	—	78.7 (69–84)
	Pittman 6	South	c.44	c.66	—	55[b]
	Pittman 7	South	c.44	c.63	—	47.5 (45–50)
Mayan Ranch, Bandera County	"Swimming" Sauropod	East-Northeast[k]	62.6[l]	—	—	—

Site	Trackway Label	Pace Length (cm): Manus Tracks	Pace Length (cm): Pes Tracks	Stride Length (cm): Manus Tracks	Stride Length (cm): Pes Tracks
Paluxy River, Somervell County	AMNH–TMM slabs (S2)	177.2 (150–207)	183.3 (163.8–199)	292.4 (254.0–328.9)	310.3 (273.1–336)
	Main Site Trail	—	179[b, e]	—	—[f]
	"East"	156.7 (137.2–175.3)	163.8 (154.9–172.7)	255.7 (221.0–273.1)	262.6 (248.9–284.5)
	"Wet"	—	161.7 (149.9–170.2)	—	267.5 (243.8–294.6)
	"Middle"	—	131.6 (127.0–142.2)	—	225.4 (221.0–231.1)
	Bird Site[i]	—	199.0[j] (180.3–233.7)	—	388.6[b]
South San Gabriel River, Williamson County	Trail 1 (Pittman 3)	136.3 (124.5–142.2)	132.9 (129.5–137.2)	222.3 (218.4–226.1)	225.2 (215.9–231.1)
	Trail 2 (Pittman 2)	133.8 (127.0–142.2)	136.5 (124.5–142.2)	229.9 (226.1–233.7)	226.9 (223.5–228.6)
	Pittman 1	161.0 (154–168)	154[b]	275[b]	—
	Pittman 4	153.5 (147–160)	148.3 (145–150)	252[b]	255 (255–255)
	Pittman 5	123.7 (115–130)	134.7	215.3 (210–223)	220.5 (217–224)
	Pittman 6	135[b]	140.0 (130–150)	—	230[b]
	Pittman 7	126[b]	133[b]	—	—
Miller Creek,[q] Blanco County		167.9 (137.2–188.0)	185.1 (160.0–200.7)	281.3 (226.1–307.3)	323.3 (289.6–411.5)
West Blanco River, Blanco County		169.7 (164–175)	204.0 (197–210)	283.0 (272–292)	280.0 (270–292)
Mayan Ranch, Bandera County		178.7 (170–185)	—	274.4 (260–288)	—

Site	Trackway Label	Step Angle: Manus Tracks	Pes Tracks	Trackway Width (cm): Manus Tracks	Pes Tracks	Inner Trackway Width	Outer Trackway Width
Paluxy River, Somervell County	AMNH–TMM slabs (S2)	113° (99–120°)	115° (101–121°)	96 (81–107)	98 (87–113)	38.0, 23.5–32.5	182.5, 165–170
	Main Site Trail	—	—[g]	—	—[h]	36	163
	"East"	103° (97–110°)	108° (100–117°)	96 (95–97)	95 (86–105)	—	—
	"Wet"	—	111° (99–123°)	—	91 (80–104)	—	—
	"Middle"	—	119° (114–131°)	—	66 (53–73)	—	—
	Bird Site[i]	—	137°[b]	—	74[b]	—	—
South San Gabriel River, Williamson County	Trail 1 (Pittman 3)	108° (105–110°)	116° (111–120°)	81 (76–86)	70 (67–74)	c.9	117
	Trail 2 (Pittman 2)	115° (114–117°)	110° (107–114°)	72 (72–73)	79 (73–85)	15	135
	Pittman 1	117°[b]	—	84[b]	—	c.19	c.140
	Pittman 4	110°[b]	120° (120–120°)	88[b]	74 (74–74)	10	140
	Pittman 5	123° (122–125°)	110° (110–111°)	58 (55–62)	77 (75–79)	12	115
	Pittman 6	—	110°[b]	—	80[b]	12	148
	Pittman 7	—	—	—	—	c.12	c.117
Miller Creek,[q] Blanco County		115° (93–129°)	114° (106–121°)	89 (73–105)	99 (92–110)	—	203–206
West Blanco River, Blanco County		110° (109–111°)	88° (83–93°)	97 (97–97)	145 (138–152)	—	203
Mayan Ranch, Bandera County		99° (94–103°)	—	118 (111–129)	—	—	—

Site	Trackway Label	Glenoacetabular Length[m] (cm): Reference Point	Track Center	Number of Footprints Seen: Manus	Pes
Paluxy River, Somervell County	AMNH–TMM Slabs (S2)	284–359	299–373	6L, 6R	5L, 6R
	Main Site Trail	—	306–384[n]	1L, 1(2?)R	3L, 4R
	"East"	241–306	—	3L, 2R	4L, 3R
	"Wet"	—	—	None	5L, 5R
	"Middle"	c.201–257	—	1R	3L, 3R
	Bird Site[i]	c.356–454[o]	—	1L	3L, 2R[j]
South San Gabriel River, Williamson County	Trail 1 (Pittman 3)	247–303	—	2L, 2R	2L, 2R
	Trail 2 (Pittman 2)	233–290	—	2L, 2R	2L, 3R
	Pittman 1	c.283–352[p]	—	2L, 1R	1L, 1R
	Pittman 4	254–317	—	1L, 2R	2L, 2R
	Pittman 5	242–297	—	2L, 3R	2L, 2R
	Pittman 6	c.227–285[b]	—	1L, 1R	2L, 1R
	Pittman 7	—	—	1L, 1R	1L, 1R
Miller Creek,[q] Blanco County		—	—	5L, 5R	5L, 4R
West Blanco River, Blanco County		—	—	3L, 2R	3L, 2R
Mayan Ranch, Bandera County		—	—	3L, 4R	None

[a]0° – Magnetic North.

[b]Single measurement.

[c]Width of pes tracks = 65 cm.

[d]72 cm as measured from track centers (single measurement).

[e]188 (172–200) cm as measured from track centers.

[f]312 (300–335) cm as measured from track centers.

[g]111° (101–118°) as measured from track centers.

[h]107 (91–125) cm as measured from track centers.

[i]This is probably the trackway labeled S4 on the Rye and Homestead charts, and called trail two in Bird's field correspondence with Barnum Brown.

[j]Measurements are for the main part of the trail, located adjacent to and below the Park Ledge site, as well as a possible upstream (downtrail) continuation near the Paluxy River ford in Dinosaur Valley State Park.

[k]The trackway follows a curvilinear path, changing from northeast to more nearly east.

[l]Manus track lengths = 48–60 cm.

[m]Glenoacetabular length calculated by methods 2 (first value) and 4 (second value), as described in the text.

[n]The calculated values are probably too high; compare values of pes track length and glenoacetabular length for this trackway and trail S2. This dinosaur was taking rather long strides for its size, and there was only one measurement of the manus–pes distance.

[o]The calculated values are probably too high. There was only one measurement each for stride and manus–pes distance.

[p]Probably too high; note the relatively high manus–pes distance, and that there was but a single stride measurement.

[q]Measurements approximate.

tracks, suggesting a dinosaur of narrower gauge than the Gulf Coast brontosaurs. As in sauropod trails from the U.S. Gulf Coastal Plain, the *Breviparopus* trackway shows slight oscillations along its course and lacks tail marks.

Judging from photographs published by Dutuit and Ouazzou (1980), the *Breviparopus* footprints are not particularly well preserved, making it difficult to infer the pedal morphology of their maker, but we provisionally accept the ichnotaxon as valid. If the manus impressions have not been altered very much during emplacement of pes footprints, then the maker of *Breviparopus* may not have had medial and lateral indentations in the sidewall of the manus footprint as conspicuous as those in Comanchean forefoot impressions, suggesting that the Moroccan trackmaker's metacarpals may have formed a single tubular structure, with digits I and V not separated from the middle three. Most of the differences between *Breviparopus* and sauropod ichnites from Texas and Arkansas occur in the trackway pattern; if the geometry of the *Breviparopus* trackway was typical of trails made by its trackmaker, then the dinosaur walked in a fashion different from Comanchean sauropods. For now, then, we do not feel that the Gulf Coast brontosaur ichnites can be referred to *Breviparopus*.

Jenny and Jossen (1982) and Ishigaki (1986 and this volume) reported other sauropod ichnites from the Lower and Middle Jurassic of Morocco. Some of these are reminiscent of *Breviparopus*, but one trackway is rather different; it is relatively wide, such that the footprints do not impinge on the trackway midline. Like *Breviparopus*, manus tracks of this trackway have no claw marks. In these features the trail resembles those from the Gulf Coastal Plain. However, pes footprints in this trackway are very different from those of Texas and Arkansas; there are four large claw marks that are directed radially outward along the anterior margin of the track, and there is a conspicuous concavity along the medial edge of the footprint outline.

Ginsburg et al. (1966) described unnamed sauropod tracks from the Jurassic–Cretaceous of Niger. Some of these are somewhat elephant–like footprints with four digits; their width (60 cm) considerably exceeds their length (45 cm). If these are sauropod pes tracks, their gross shape is quite different than in Comanchean pes (or even manus) footprints.

At another locality, Ginsburg et al. (1966) found a long sauropod trail with well–preserved manus and pes footprints. The trackmaker was smaller than most known Comanchean trackmakers, leaving pes tracks some 36 cm long, as measured by Ginsburg et al. (1966); measured in our way (Fig. 42.5), these tracks would be about 55 cm long. Pes tracks have four very large (seemingly larger relative to track size than in Texas and Arkansas sauropod footprints) claw marks, and possibly a very small, posteriorly–directed, fifth claw mark. Manus footprints appear to be closer to the trackway midline than pes tracks, unlike *Breviparopus* and more like the Comanchean sauropod trails. Manus tracks apparently bear a large inner claw, and one manus footprint is said to have three distinct, medially–directed claw marks and two outer digital pads — very different from U.S. Gulf Coast manus tracks.

Kim (1986, 1987) and Lim et al. (this volume) report sauropod trackways from the Lower Cretaceous Jindong Formation of southern Korea. Kim (1986, 1987) applied the name *Hamanosauripus ungulatus* to the trackway of a small sauropod (pes tracks only 33 cm long). Manus tracks are egg–shaped, with no traces of individual fingermarks, but showing two clear metacarpal impressions, and are not obviously deformed by associated pes tracks. Pes tracks are elliptical, with three toe marks situated along the flat anterior rim of the footprint. The claw mark on the inner toe (identified by Kim as digit II, but more likely digit I) is large and sharp. Unlike claw marks in Comanchean sauropod pes footprints, those of *Hamanosauripus* do not seem to be laterally directed. The trackway is relatively broad, with both manus and pes footprints well away from the trail midline, and both manus and pes impressions appear to angle outward with repect to the dinosaur's direction of travel, rather like the U.S. Gulf Coast brontosaur trails. On the left side of the trackway, the centers of manus and pes tracks seem to be about the same distance from the trackway midline, but on the right side the manus footprints are further away than the pes tracks, at least in the portion of the trail diagrammed by Kim (1982). The step angle is rather less (manus 70°, pes 95°) than in Comanchean sauropod trackways. The pes footprints seem to have been emplaced in front of their associated manus tracks, an unusual situation in sauropod trails. There is no tail drag mark.

Kim (1986, 1987) named the trail of a much larger sauropod (pes tracks about 90 cm long) *Koreanosauripus cheongi*. Neither manus nor pes footprints show distinct digit marks. The manus is somewhat crescent–shaped, with backward–directed medial and lateral corners. Manus tracks angle outward, and in at least some sets of manus tracks are further from the trail midline than pes footprints. Pes tracks are nearly round, and situated close behind (possibly deforming?) their associated manus impressions. The trackway is relatively narrower than *Hamanosauripus*, with pes tracks almost intersecting the trail midline. The step angle (102°) is closer to that seen in sauropod trackways from the Gulf Coast region than that of *Hamanosauripus*. There is no trace of tail marks.

Lim et al. (this volume) question the appropriateness of the manner in which Kim published his sauropod ichnite names. Regardless of the merits of their concerns, we have other reservations. First of all, the sauropod footprints described by Kim do not seem to be very well preserved, judging from his photographs (those of *Koreanosauripus* are low–angle views that do little to convey footprint shape), and do not provide much information about the pedal skeletons of their makers. Like Dutuit and Ouazzou (1980), Kim does not discuss the degree to which his named trackways are likely to have been typical ichnites of their makers; it would be interesting to know, for instance, whether the pes overstep pattern seen in *Hamanosauripus* is a consistent feature of its maker's trail. We are not certain that Kim's ichnites are well enough preserved to merit formal names. Nonetheless neither *Hamanosauripus* nor *Koreanosauripus* appears enough like the Comanchean brontosaur ichnites to warrant assigning the U.S. Gulf Coast footprints to either of those ichnotaxa.

Hendricks (1981) assigned the name *Rotundichnus münchehagensis* to brontosaur ichnites from the Lower Cretaceous of northwestern Germany. The trackmaker left footprints roughly comparable in size to those of the AMNH–THM slab dinosaur. As in the Texas and Arkansas trackways, the *Rotundichnus* trail is wide–gauge, with manus and pes tracks some distance from the trackway midline. Also as in the American trails, the German trail's pes footprints are deepest on their inner sides, but, unlike the Texas and Arkansas trackways, the pes tracks are shallow in the "heel" region. The position of manus and pes footprints with respect to the trackway midline seems similar to that of Comanchean sauropod trails.

We could quite happily apply the name *Rotundichnus* to U.S. Gulf Coast brontosaur footprints were it not that the German tracks show no sign of toes on manus or pes tracks, probably due to poor preservation of these relatively shallow (10–15 cm deep) ichnites — perhaps they are ghost tracks. We are uncomfortable with Hendricks' essentially naming a trackway pattern, the constituent footprints of which are devoid of interpretable morphology. Even though we suspect that the Comanchean and German sauropod footprints could be assigned to the same ichnogenus, we do not feel that *Rotundichnus* (in our opinion a *nomen vanum*) is an appropriate name for it.

Spectacular sauropod ichnites are now known from the Morrison Formation (Late Jurassic) of Colorado (Lockley 1986, Lockley et al. 1986). Two distinct types of brontosaur ichnites may occur in the Purgatoire Valley: a larger, broad–footed form with distinct claw marks, and a smaller, narrow–footed form lacking clear digit impressions. Tail drag marks are absent in both trackway types. In both forms, manus and pes tracks are very close to, or intersect, the trackway midline, unlike Comanchean trails. Some manus footprints apparently have digit impressions.

Garcia-Ramos and Valenzuela (1977: Fig. 8) illustrated a large unidentified footprint from the Upper Jurassic of Asturias, Spain. The track is round, except for an indentation in one edge that gives the footprint a horseshoe–like shape; the track is 33–35 cm across, and is quite likely a sauropod manus impression, but one without the medial and lateral sidewall indentations characteristic of Comanchean manus tracks.

Weems (1987) gave the name *Agrestipus hottoni* to small (15 cm long) footprints from the Late Triassic of Virginia. The footprints are entaxonic, as in well–preserved sauropod pes tracks, with three and perhaps four blunt toe marks. The trackway is wide, with a relatively short pace and stride and low step angle. Weems suggested that *Agrestipus* was made by a bipedal animal, or a quadruped whose pes tracks completely overprinted manus impressions — perhaps a sauropod precursor. Weems' interpretation is not unreasonable, but the much larger size and well–developed pedal claw marks of the Comanchean sauropod pes tracks make them quite different from *Agrestipus*.

Some of the ichnogenera from the Late Triassic-Early Jurassic Stormberg Group of southern Africa have been questionably identifed as trace fossils of very early sauropods (*Sauropodopus, Deuterosauropodopus, Pentasauropus*; Ellenberger 1972; Haubold 1984). *Pentasauropus* tracks are more likely to have been made by dicynodonts, and at least some *Deuterosauropodopus* are probably referable to the thecodont ichnogenus *Brachychirotherium* (Olsen and Galton 1984). We doubt that any of the Stormberg tracks were made by sauropods; all of the *Sauropodopus* and *Deuterosauropodopus* ichnites probably represent pseudosuchians or

Table 42.3. *Trackway measurements as estimated from R. T. Bird's site maps. Measurements are for pes tracks only. Numbers in parentheses indicate the range of values.*

Site	Trackway	Chart	x̄ Pes Track Length (cm)	Pace (cm)	Stride (cm)
Paluxy River	S2 (AMNH–TMM slab trackway; Bird's trail 3)	Rye	c.91	203 (173–245)[a] 200 (173–223)	304 (229–356)[a] 306 (244–356)
		Homestead	—	189 (161–206)[a] 191 (170–206)	303 (259–343)[a] 311 (274–336)
		Austin slab		192[b] (167–214)	282[b] (231–320)
	S3 (Bird's trail 1)	Rye	c.98	204 (170–282)	336 (256–424)
		Homestead	—	216 (197–242)	335 (305–362)
	S4 (Bird site trackway; Bird's trail 2)	Rye	c.84	179 (160–192)	269 (234–300)
		Homestead	—	178 (142–202)	279 (247–331)
	S1	Rye	c.92	210[c]	—
		Homestead	—	206 (190–228)	354 (343–372)
	S5 (Giant trail)	Rye	c.110	261 (254–271)	405 (391–420)
		Homestead	—	239 (211–266)	371 (353–394)
Davenport Ranch[e]	Bird's trail 8	Davenport	83[d]	166 (139–186)	263 (245–272)
	A (Lockley 3)	Davenport	51	119 (101–136)	202 (187–226)
	B (Lockley 2)	Davenport	49	117 (100–138)	213 (203–220)
	C (Lockley 1)	Davenport	c.60	125 (119–133)	215 (206–222)
	D (Lockley 4)	Davenport	61	136 (119–158)	227 (215–236)
	E (Lockley 6)	Davenport	88	169 (155–181)	258 (247–274)
	Bird's trail 7 (Lockley 5)	Davenport	43	100 (83–119)	173 (150–200)
	Bird's trail 1 (Lockley 7)	Davenport	69	150 (150–150)	249 (244–258)
	H (Lockley 9)	Davenport	63	130 (107–155)	205 (193–228)
	I (Lockley 23)	Davenport	70	160 (158–162)	257[c]

crocodiloid archosaurs (but see Demathieu and Weidmann 1982). In any case, none of them is very similar to the Texas sauropod tracks.

Hatcher (1903) published a photograph of a presumed dinosaur footprint found in Morrison beds near Canyon City, Colorado. Although Hatcher offered no interpretation of the trackmaker, the footprint is sometimes attributed to a sauropod (Haubold 1971), perhaps because Hatcher's monograph dealt with the osteology of *Haplocanthosaurus*. The footprint does not look very sauropod-like, and may be an ornithopod manus print (D. Baird pers. comm.) — or a crocodile track (M. G. Lockley pers. comm.).

Systematic Paleontology

Class Reptilia
  Order Saurischia
    Suborder Sauropoda
      Family: We think it likely that the Texas trackmakers belonged to the camarasaurid–brachiosaurid branch of the Sauropoda — perhaps closer to the brachiosaurids, given our interpretation of the manus. We are not confident enough in this conclusion, however, to assign the Texas sauropod ichnites to the Family

Brachiosauridae, or to name an ichnofamily for the Texas tracks in such a way as to make an explicit identification with the brachiosaurids. Furthermore, given the paucity of adequate descriptions of sauropod ichnites, we believe it premature to name ichnofamilies for brontosaur trace fossils.
Ichnogenus: *Brontopodus nov.*

    Type species: *Brontopodus birdi nov. gen. nov. sp.*

Holotype: American Museum of Natural History 3065 *and* Texas Memorial Museum 40638–1, the trackway (S2) collected by R. T. Bird in 1940. Using our numbering system, the type trackway includes left manus track S2L through left pes tracks S2R (New York slab), and left manus track S2R through right manus–pes set S2W (Austin slab).

Referred specimens: Texas Memorial Musemm 40637–1 and 40637–2, a right manus and pes track, respectively, of a set (part of Bird's trail 8) from the Davenport Ranch site; New Mexico Museum of Natural History (portion of trackway one, Briar site [Pittman and Gillette this volume]). Although not formally included here, our concept of the ichnotaxon is also based on uncollected

Site	Trackway	Chart	Step Angle	Trackway Width (cm)	Inner Trackway Width (cm)	Outer Trackway Width (cm)
Paluxy River	S2 (AMNH–TMM slab trackway; Bird's trail 3)	Rye	97° (78–110°)[a]	133 (118–168)[a]	c.16	c.192
			100° (85–110°)	128 (118–133)		
		Homestead	107° (89–121°)[a]	110 (94–131)[a]	—	—
			110° (104–114°)	109 (98–122)		
		Austin slab	96° (78–107°)	127 (112–145)	—	—
	S3 (Bird's trail 1)	Rye	111° (92–130°)	113 (94–144)	c.20–47	c.180
		Homestead	102° (92–109°)	135 (121–153)	—	—
	S4 (Bird site trackway; Bird's trail 2)	Rye	97° (88–107°)	119 (105–129)	c.16	c.168
		Homestead	106° (96–130°)	105 (75–128)		
	S1	Rye	—	—	—	—
		Homestead	117° (111–120°)	108 (102–116)	—	—
	S5 (giant trail)	Rye	102° (99–106°)	165 (159–172)	c.60–66	c.228–234
		Homestead	101° (95–110°)	151 (129–165)	—	—
Davenport Ranch	Bird's trail 8	Davenport	103° (97–111°)	104 (92–120)	c.18–42	c.144–168
	A	Davenport	122° (115–127°)	57 (48–72)	c.12–36	c.84–96
	B	Davenport	123° (121–126°)	57 (52–61)	c.12–18	—
	C	Davenport	120° (113–122°)	62 (58–68)	—	c.102–114
	D	Davenport	111° (103–118°)	77 (65–91)	c.24–42	c.120–138
	E	Davenport	101° (98–104°)	105 (101–109)	c.18–30	c.150–168
	Bird's trail 7	Davenport	119° (104–131°)	50 (38–64)	c.0–48	c.54–102
	Bird's trail 1	Davenport	109°[c]	88[c]	c.0–30	—
	H	Davenport	102° (99–105°)	81 (74–92)	c.24–36	c.114–132
	I	Davenport	107°[c]	96[c]	c.36	c.132

[a]Upper row of values is for the entire trackway; lower row is for the AMNH–TMM slabs portion of the trackway (cf. Table 1 for the actual values).

[b]Values calculated from Bird's chart using the empirical (11.1x) scale factor; for a 12x factor, $\bar{x}$ pace = 207 cm, $\bar{x}$ stride = 305 cm.

[c]Single measurement.

[d]The actual measurement of a single footprint from this trackway = c.70 cm.

[e]Lockley (1987: Table 1) independently estimated trackway parameters from published versions of Bird's Davenport Ranch site map; his labels for particular trails are given after ours.

sauropod tracks *in situ* in the bed of the Paluxy River (Farlow 1987) and elsewhere).

Locality: Holotype — Bed of the Paluxy River, in what is now Dinosaur Valley State Park, Glen Rose, Somervell County, Texas, USA; Referred specimens — West Verde Creek (Davenport Ranch, near Bandera in extreme northern Medina County, Texas, USA), Weyerhaeueser Company Briar Plant Quarry (near Nashville, Howard County, Arkansas, USA).

Horizon: Trinity Group, Comanche Series (Aptian–Albian) Lower Cretaceous (see Pittman this volume for details).

General diagnosis and discussion: At present it is uncertain which features of the Comanchean sauropod ichnites are diagnostic at the genus level, and which at the species level; this cannot be ascertained until more complete descriptions of other sauropod trace fossils become available.

Sauropod ichnites of small to large size, known pes footprint length ranging 50 to over 100 cm. Manus foot-print length and width about the same in well-preserved tracks; manus tracks clawless, somewhat U–shaped, with digit impressions I and V slightly separated from the impression of conjoined digits II–IV. Pes tracks longer than broad, with large, laterally directed claw marks at digits I–III (diminishing in size from I to III), a small claw, nail, or callosity mark at digit IV, and a small callosity or pad mark at digit V; digit marks IV and V only seen in well–preserved footprints. Manus tracks often (usually?) rotated outward with respect to direction of travel. Manus track medial to a line through pes track long axis, such that manus track centers are somewhat closer to the trackway midline than pes track centers. Trackway broad, with left and right manus and pes footprints often well away from the trackway midline; trackway width roughly 1–1.5 times pes track length. Outer limits of trackway defined by pes tracks. Manus–pes distance 0.5–1.2 times pes footprint length. Stride length roughly 2–5 times pes track length. Step angle generally 100–120°. Glenoacetabular length c. 3–4 times pes track length. Tail drag marks rare or absent.

Most of the Davenport Ranch sauropods were much smaller than the Paluxy River trackmakers, and even the largest Davenport Ranch dinosaurs were considerably smaller than the biggest Somervell County animals, judging from Bird's site maps and letters to Brown from the field (Farlow 1987). Was the Davenport Ranch trackway assemblage made by a herd (cf. Lockley 1987) of juveniles and subadults, or juveniles tended by adults of the smaller sex? Conceivably the Davenport Ranch sauropods were a species different from those of the Paluxy River, but we are reluctant to make that interpretation without evidence besides footprint size to support it. In any case, we see no morphological grounds for putting the Davenport Ranch tracks in an ichnospecies different from the larger Comanchean sauropod footprints.

Differential diagnosis: Differs (at least usually) from *Breviparopus taghbaloutensis* in having pes tracks farther from the trail midline than manus tracks, manus and pes paces of similar magnitude, probably a smaller angle between the pes long axis and the direction of travel, a broader trackway gauge, with pes tracks generally not impinging on the trail midline, pes claw marks directed strongly laterally (and sometimes even backward) as much or more than forward, and a deep pes track "heel." The manus construction may have been somewhat different in the maker of *Breviparopus* than in the *Brontopodus* trackmaker, but this is not certain. It is possible that these differences reflect differences in gait and substrate at the times the Comanchean and Moroccan brontosaurs left their footprints, rather than real anatomical differences, but the various U.S. Gulf Coast trials are consistent enough in these features to make us believe them to be valid taxonomic distinctions.

Brontopodus birdi differs from *Hamanosauripus ungulatus* and *Koreanosauripus cheongi* (assuming the validity of the Korean names) in gross manus and pes footprint shape and trackway pattern. *B. birdi* differs from *Agrestipus hottonoi* in having much larger footprints and well-developed claw marks on the pes. We do not believe any other named ichnotaxa of sauropods or their relatives to be based on adequate material, given our philosophy of trace fossil nomenclature.

Etymology: The generic name *Brontopodus* (Greek: "thunder foot") was coined by the late Roland T. Bird, but was never published or even defined by Bird except in his diagram of the American Museum slab (Fig. 42.3; previously published as Farlow [1987: Fig. 31]). Bird should receive all credit for having created the name. The specific name *birdi* is to honor him as the collector of the holotype, and for his meticulous drafting of the Paluxy River and Davenport Ranch site maps.

Acknowledgments

We thank Alice Erickson, Hazel Bird, Peggy Bird, Tom Bird, Wann Langston, Jr., and the late V. T. Schreiber and his wife for making available to us unpublished materials of R. T. Bird in their possession. We also thank J. S. McIntosh, G. Leonardi, G. Demathieu, D. Baird, L. Moss, B. P. Baker, L. C. Mansfield, M. Gallup, F. W. Johnson, C. Holton, M. G. Lockley, G. Kuban, W. Fields, R. Bonem, O. T. Hayward, N. Wilson and M. Brett-Surman for discussions and/or assistance. Most diagrams were prepared or reproduced by C. Bishop and E. Denman. The senior author was supported by grants from the American Philosophical Society and the Office of Sponsored Research, Indiana University–Purdue University at Fort Wayne.

References

Antunes, M. T. 1976. Dinossáurios Eocretácicos de Lagosteiros. *Universidad Nova de Lisboa, Ciencias da Terra* 33 pp.

Beaumont, G. de, and Demathieu, G. 1980. Remarques sur les extrémités antérieures des Sauropodes (Reptiles, Saurischiens). *Société de Physique et d'Histoire Naturelle de Genève, Compte Rendu des Séances* 15:191–198.

Bird, R. T. 1939. Thunder in his footsteps. *Nat. Hist.* 43:254–261, 302.

1941. A dinosaur walks into the museum. *Nat. Hist.* 47:74–81.

1944. Did *Brontosaurus* ever walk on land? *Nat. Hist.* 53:61–67.

1954. We captured a "live" brontosaur. *National Geographic* 105:707–722.

1985. *Bones for Barnum Brown: Adventures of a Dinosaur Hunter.* Schreiber, V. T. (ed.). (Texas Christian University Press) 225 pp.

Borsuk-Bialynicka, M. 1977. A new camarasaurid *Opisthocoelicaudia skarzynskii* Gen. N., Sp. N. from the Upper Cretaceous of Mongolia. *Palaeontologica Polonica* 37:5–64.

Campbell, K. E. 1983. Trackways: clues to passing dinosaurs. *Terra* 21 (3):12–13.

Casanovas–Cladellas, M., and Santafe–Llopis, J.–V. 1986. Dinosaur footprints in Spain with special reference to Lower Cretaceous from "Sierra de los Cameros" (La Riojá, Spain). *In* Gillette, D. D. (ed.). *First International Symposium on Dinosaur Tracks and Traces, Abstracts with Program.* (New Mexico Museum of Natural History) p. 12.

Coombs, W. P., Jr. 1975. Sauropod habits and habitats. *Palaeogeog., Palaeoclimat., Palaeoecol.* 17:1–33.

Demathieu, G. 1970. *Les Empreintes de Pas de Vertébrés du Trias de la Bordure Nord–Est du Massif Central.* (Cahiers de Paléontologie, Editions du Centre National de la Recherche Scientifique, France) 211 pp.

Demathieu, G., and Weidmann, M. 1982. Les empreintes de pas de reptiles dans le Trias du Vieux Emosson (Finhaut, Valais, Suisse). *Ecologae Geological Helvetiae* 75:721–725.

Dutuit, J.–M., and Ouazzou, A. 1980. Découverte d'une piste de Dinosaure sauropode sur le site d'empreintes de Demnat (Haut–Atlas marocain). *Mem. Soc. Geolog. de France* N.S. 139:95–102.

Ellenberger, P. 1972. Contribution à la Classification des Pistes de Vertébrés du Trias: les Types du Stormberg d'Afrique du Sud (I). *Palaeovertebrata Mem. Extraordinaire* Montpellier, France 117 pp.

Farlow, J. O. 1987. *A Guide to Lower Cretaceous Dinosaur Footprints and Tracksites of the Paluxy River Valley, Somervell County, Texas.* Field trip guidebook, 21st Annual Meeting, South–Central Section, Geol. Soc. Amer., Waco, Texas. 50 pp.

Fields, W. 1980. *Paluxy River Explorations (1977–1979)* revised edition. (privately published) 48 pp.

Friese, H., and Klassen, H. 1979. Die Dinosaurierfährten von Barkhausen im Wiehengebirge. *Veröffentlichungen des Landkreises Osnabrück,* Heft 1.36 pp.

Gallup, M. R. 1975. Lower Cretaceous dinosaurs and associated vertebrates from north-central Texas in the Field Museum of Natural History. Master's thesis, University of Texas, 159 pp.

García-Ramos, J. C., and Valenzuela, M. 1977. Huellas de pisada de vertebrados (Dinosaurios y otros) en el Jurásico Superior de Asturias. *Estudios Geologicos (Instituto de Investigaciones Geologicas "Lucas Mallada")* 33:207–214.

Gilmore, C. W. 1925. A nearly complete articulated skeleton of *Camarasaurus,* a saurischian dinosaur from the Dinosaur National Monument, Utah. *Mem. Carnegie Museum* 10:347–384.

———. 1932. On a newly mounted skeleton of *Diplodocus* in the United States National Museum. *Proc. United States National Museum* 81:1–21.

———. 1936. Osteology of *Apatosaurus,* with special reference to specimens in the Carnegie Museum. *Mem. Carnegie Museum* 11:174–271.

———. 1946. Reptilian fauna of the North Horn Formation of central Utah. *U.S. Geol. Surv. Prof. Paper* 210–C:28-53.

Ginsburg, L., Lapparent, A. F. de., Loiret, B., and Taquet, P. 1966. Empreintes de pas de Vertébrés tetrapodes dans les séries continentales à l'Ouest d'Agades (République du Niger). *Comptes Rendus Acad. de Sciences Paris* 263:28–31.

Godoy, L. C., and Leonardi, G. 1985. Dirações e comportamento dos dinossauros de localidade de Piau, Sousa, Paraíbo (Brasil), Formação Sousa (Cretáceo inferior). In *Brasil,* DNPM. *Coletânea de Trabalhos Paleontológicos, Série "Geologia,"* Brasília, 27 (Seção Paleontologia e Estratigrafia 2):65–73.

Halfpenny, J. 1986. *A Field Guide to Mammal Tracking in Western America.* (Boulder, Colorado: Johnson Books) 164 pp.

Hatcher, J. B. 1901. *Diplodocus* (Marsh): its osteology, taxonomy, and probable habits, with a restoration of the skeleton. *Mem. Carnegie Museum* 1:1–64.

———. 1903. Osteology of *Haplocanthosaurus* with description of a new species, and remarks on the probable habits of the Sauropoda and the age and origin of the *Atlantosaurus* beds. *Mem. Carnegie Museum* 2:1–72.

Haubold, H. 1971. *Ichnia Amphibiorum et Reptiliorum Fossilium.* Handbuch für Paläoherpetologie 18. 124 pp.

———. 1984. *Saurierfährten.* (East Germany: A. Ziemsen Verlag) 131 pp.

Hawthorne, J. M. 1983. Stratigraphy and depositional environment of the dinosaur track-bearing Glen Rose Limestone in the Paluxy Basin, Texas. Bachelor of Science thesis, Baylor University. 126 pp.

———. 1987. The stratigraphy and depositional environments of Lower Cretaceous dinosaur track–bearing strata in the Edwards Plateau and Lampasas Cut Plain physiographic provinces of Texas. Master's thesis, Baylor University.

Hendricks, A. 1981. Die Saurierfährte von Münchehagen bei Rehburg-Loccum (NW–Deutschland). *Abhandlungen aus dem Landesmuseum für Naturkunde zu Münster in Westfalen.* 43:1–22.

Ishigaki, S. 1986. *Dinosaur Footprints in Morocco* (Tokyo: Tsukiji Publishing Company) 264 pp. [in Japanese]

———. 1988. Footprints of swimming sauropods from Morocco. this volume.

———. in press. Les Empreintes de Dinosaures du Jurassique Inférieur du Haut Atlas Central. To be published in *Mines Geologie et Energie* (Moroccan Ministry of Energy and Mines).

Ishigaki, S., and Haubold, H. 1986. Lower Jurassic dinosaur footprints from the Central High Atlas, Morocco. In Gillette, D. D. (ed.). *First International Symposium on Dinosaur Tracks and Traces* (New Mexico Museum of Natural History) p. 16.

Janensch, W. 1961. Die Gliedmaszen und Gliedmaszengürtel der Sauropoden der Tendaguru-Schichten. *Palaeontographica* Suppl. 7:177–235.

Jenny, J., Le Marrec, A., and Monbaron, M. 1981. Les empreintes de pas de Dinosauriens dans le Jurassique moyen du Haut Atlas central (Maroc): nouveaux gisements et precisions stratigraphiques. *Geobios* 14:427–431.

Jenny, J., and Jossen, J.-A. 1982. Découverte d'empreintes de pas de Dinosauriens dans le Jurassique inférieur (Pliensbachien) du Haut–Atlas central (Maroc). *Comptes Rendes Acad. de Sciences Paris* 294:223–226.

Kaever, M., and Lapparent, A. F. de. 1974. Les traces de pas de Dinosaures du Jurassique de Barkhausen (Basse Saxe, Allemagne). *Bull. Soc. Géolog. de France* 16:516–525.

Kim, H. M. 1986. New Early Cretaceous dinosaur tracks from Republic of Korea. In Gillette, D. D. (ed.). *First International Symposium on Dinosaur Tracks and Traces* (New Mexico Museum of Natural History) p. 17.

———. 1987. *Stratigraphy of Lower Cretaceous Jindong Formation, Korea.* [in Korean]

Langston, W. Jr. 1974. Nonmammalian Comanchean tetrapods. *Geoscience and Man* 8:77–102.

Leonardi, G. 1984. Rastros de um mundo perdido. *Cienciahoje* 2:48–60.

———. 1985. Vale dos Dinossauros: uma janela na noite dos temos. *Revista Brasileira de Tecnologia* 16:23–28,

———. 1986. An inventory and statistic study of 40 dinosaurian ichnofaunas of South America and their paleobiological meaning. In Gillette, D. D. (ed.). *First International Symposium on Dinosaur Tracks and Traces* (New Mexico Museum of Natural History) p. 18.

——— (ed.) 1987. *Glossary and Manual of Tetrapod Footprint Palaeoichnology.* República Federativa do Brasil, Ministério das Minas e Energia, Departamento Nacional de Produção Mineral, Brazil. 75 pp.

Lim S.-K., Yang S.-Y., and Lockley, M. G. 1988. Large dinosaur footprint assemblages from the Cretaceous Jindong Formation of southern Korea. This volume.

Lockley, M. G. 1986. A guide to dinosaur tracksites of the Colorado Plateau and American Southwest. *Univ. Colorado Denver Geol. Dept. Mag.,* Spec. Issue no. 1, 56 pp.

1987. Dinosaur trackways. *In* Czerkas, S. J., and Olson, E. C. (eds.). *Dinosaurs Past and Present* Vol. I. (Natural History Museum of Los Angeles County/University of Washington Press) pp. 80–95.

Lockley, M. G., Houck, K. J., and Prince, N. K. 1986. North America's largest dinosaur trackway site: implications for Morrison Formation paleoecology. *Geol. Soc. Amer. Bull.* 97:1163–1176.

Marsh, O. C. 1896. The dinosaurs of North America. *Ann. Rept. U.S. Geol. Surv.* 16th, 1894–1895, part 1:133–244.

Norman, D. 1985. *The Illustrated Encyclopedia of Dinosaurs.* (Crescent Books) 208 pp.

Nunn, W. C. 1975. *Somervell: Story of a Texas County.* (Texas Christian University Press) 258 pp.

Olsen, P. E., and Galton, P. M. 1984. A review of the reptile and amphibian assemblages from the Stormberg of southern Africa, with special emphasis on the footprints and the age of the Stormberg. *Palaeontologia Africana* 25:87–110.

Paul, G. S. 1987. The science and art of restoring the life appearance of dinosaurs and their relatives: a rigorous how-to guide. *In* Czerkas, S. J., and Olson, E. C. (eds.). *Dinosaurs Past and Present* Vol. I. (Natural History Museum of Los Angeles County/University of Washington Press) pp. 4–49.

Pittman, J. G. 1984. Geology of the De Queen Formation of Arkanss. *Gulf Coast Assoc. Geol. Soc. Trans.* 34:201–209.

1988. Stratigraphy, lithology, and depositional environments of dinosaur track-bearing beds of the Gulf Coastal Plain. This volume

Pittman, J. G., and Gillette, D. D. 1988. The Briar site: a new sauropod dinosaur trackway in Lower Cretaceous beds of Arkansas. This volume

Riggs, E. S. 1901. The fore leg and pectoral girdle of *Morosaurus. Field Columbian Museum, Geology Series,* Publication 63, 1:275–281.

Sarjeant, W. A. S. 1975. Fossil tracks and impressions of vertebrates. *In* Frey, R. W. (ed.). *The Study of Trace Fossils* (Springer–Verlag) pp. 283–324.

Sarjeant, W. A. S., and Kennedy, W. J. 1973. Proposal of a code for the nomenclature of trace-fossils. *Canadian Jour. Earth Sci.* 10:460–475.

Shuler, E. W. 1917. Dinosaur tracks in the Glen Rose Limestone near Glen Rose, Texas. *Amer. Jour. Sci.* 44:294–298.

Thomas, D. A. 1986. Gaits and tracks. *In* Gillette, D. D. (ed.). *First International Symposium on Dinosaur Tracks and Traces* (New Mexico Museum of Natural History) p. 25.

Thulborn, R. A. 1988. The gaits of dinosaurs. This volume.

Weems, R. E. 1987. A Late Triassic footprint fauna from the Culpepper Basin Northern Virginia (U.S.A.). *Trans. Amer. Philos. Soc.* 77 (1):1–79.

Wilson, N. 1975. *Lanham Mill Community.* (Privately published) 41 pp.

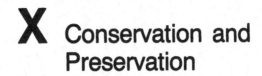

X Conservation and Preservation

"The ... footprints are of prime scientific value. They ... open a gateway to the spirit of eternity. ... That spirit, emanating from the imprints on the rock, may enlarge the outlook of our citizens, and of our youth not least. It is a public duty to guarantee the preservation of the site and make it accessible to all who wish to learn and enrich their minds."

Avnimelech 1966 p. 17,
on the subject of the Beth Zayit tracksite
in Israel.

Tracks are exposed at the surface by natural processes of erosion, or in some cases by deliberate excavation. Once exposed they are highly vulnerable to further erosion and may be destroyed almost instantly by flash flooding and vandalism. Even where weathering and erosion are less intense, deterioration is inevitable and few sites can be expected to survive more than a decade or two without significant degeneration.

For these reasons scientists involved in the discovery and documentation of tracksites have frequently been involved, to varying degrees, in attempts to secure their preservation. This section outlines various methods which facilitate the documentation, conservation and preservation of tracksites. These include photographic documentation techniques (e.g., Ishigaki and Fujisaki), molding and casting techniques, and overall site conservation and management (Agnew et al.).

The Lark Quarry site in Australia is a fine example of an excavated site which has been thoroughly and responsibly documented, protected, and managed. The scientists involved with conservation at this site raise thought-provoking issues regarding scientific and community responsibility. Other examples of well managed tracksites include Rocky Hill State Park in Connecticut (Ostrom 1968).

Although this section does not deal directly with preservation in the primary geological sense (cf. Lockley and Conrad p. 121), Kuban (Chapter 49) demonstrates, for the first time, that some sites require long term monitoring in order to observe color changes and the additional information such changes reveal.

References

Avnimelech, M. A. 1966. Dinosaur tracks in the Judean Hills. *Israel Acad. Sci. Humanities Proc. Ser. Sci.* 1:1–19.
Ostrom, J. H. 1968. The Rocky Hill Dinosaurs. *Connecticut Geol and Nat. Hist. Surv. Guidebook* no. 2:1–12.

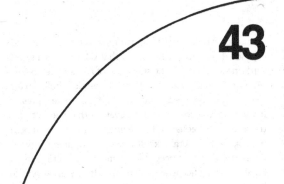

43 Strategies and Techniques for the Preservation of Fossil Tracksites: An Australian Example

NEVILLE AGNEW,
HEATHER GRIFFIN, MARY WADE,
TERENCE TEBBLE AND
WARREN OXNAM

Abstract

Have paleontologists been sufficiently concerned with the long–term preservation of fossil tracksites and other types of field sites which, for one reason or another, need to be saved? No, because for economic and professional reasons they are unwilling to sacrifice their own research time to the demands of field conservation. The existing void between paleontological field work and field conservation to save such field sites from erosion and vandalism needs to be addressed. The instigation for preservation must come from scientists who work on sites.

Conservation measures at the Lark Quarry dinosaur trackways site in central Queensland, Australia, are presented in a chronological case history, from the excavation in 1977 to the present time, to demonstrate the need for preservation. The conservation program divided itself into three phases — diagnosis of the causes of deterioration, testing and laboratory work, and the implementation of preservation measures. Measures included monitoring the condition of the surface (photographically, by installation of dust and crack monitors); construction of protective roof, fence and walkway; infilling of cracks with silicone elastomer and consolidation of dusty areas around the site with polyvinyl acetate emulsion; and interpretive signs for visitors.

Procedural guidelines and recommendations are presented for the prevention of damage at sites and for their long–term preservation. Techniques used at Lark Quarry should find application at similar sites elsewhere.

Introduction

One of the goals of the First International Symposium on Dinosaur Tracks and Traces was "to generate direct and forthright communication among scientists and interested members of the lay public" (Gillette 1986). Taking as the operative word "forthright", one may ask — as scientists involved, in this instance not with the interpretation or analysis of fossil tracks, but with their preservation in the field — how well the international community of paleontologists has performed in saving such sites from destruction by weathering, pilfering and vandalism. It is probably true to say that scientists who specialize in the field of analysis of tracks and in the study of fossil remains are interested primarily in the information their discoveries yield, and only secondarily in the preservation of the material evidence. That may be, indeed, a sweeping statement, capable of being refuted in many particular instances, and, yet, in the literature on fossil tracksites, one may ask also how many papers have been concerned with preservation in situ, as distinct from methods of preservation applied to material collected and brought to the laboratory.

The answer to these questions, it is suggested, is that the performance of the scientific community has generally not been good at all. Yet to imply that paleontologists are not concerned with the need for field preservation would be untrue, since clearly they are; and frequently the first question asked after discovery is "How do we preserve the tracks?" (Gillette 1986). But it seems the reasons for skirting the issue of preservation in the field are largely practical and economic ones. Frequently the site may be too large to be collected in toto, or, for local reasons it may not be politic to cart it away — consequently a representative sample may be taken for study and institutional display and the rest recorded and perhaps cast, after which the site is, in effect, abandoned. Field preservation, with the requirements of at least fencing the site and possibly roofing it as well may just be too expensive for busy research scientists to go seeking funds. Who can blame them when it is such a time–consuming activity to protect a site? There is a real gap here that needs to be addressed and the impetus to fill it, to provide staff and money for preservation work, must come from those who undertake the field work, those who discover, excavate and study tracksites. Documentation of a site is the first step towards its preservation but should not stop at this point.

The catalog of loss or deterioration of field sites is

extensive — to cite but one area, that of the Colorado Plateau and American Southwest (Lockley 1986): loss of 10,000 m² of the Purgatoire River site to erosion since 1939; destruction by vandals of sauropod tracks from the Cimarron Valley; 'enhancing' of tracks by engraved outlines, and so on.

Given that the usual functions of field investigation of fossil tracksites have been carried out: excavation, sampling or part collecting, documenting, molding, perhaps re–covering, it may be asked whether it should not be accepted that inevitably the site will be lost through the forces of weathering or the activities of people and animals. The answer to this should be a categorical 'no'. Fossil sites in general (as well as geological and archaeological ones) represent an important heritage resource which should be kept for future study by techniques more advanced than those presently available — methods which could yet yield a rich harvest of knowledge. These sites have educational value for young and old alike; they have interpretive value in terms of the present day, and past landscape and climate. The dramatic impact and immediacy of first seeing a fossil site in the field far outweighs that of museum display. It is for these reasons that field sites need to be preserved.

An analogy may be drawn between the present methods and attitudes of field paleontology and that of archaeology up to the early 20th century, where the approach taken was one of recovery of artifacts, sometimes accomplished by destruction of sites. It is only in recent times that the importance of the archaeological site per se and the need for its protection and conservation, both during and after excavation, has begun to receive more considered attention (Stanley Price 1984, 1986). Paleontological sites should be treated likewise.

The purpose of the foregoing polemic has not been to adopt a 'holier than thou' posture in view of the considerable expenditure of effort and money that has been undertaken at several sites, including the Lark Quarry trackways of central Queensland. Rather it is to pose a blunt question for thoughtful consideration: Are paleontologists, in the main, abrogating their responsibility to preserve field sites? It is not being suggested that all sites be preserved, but only those of unique or special scientific character or value. Assessment of this value, and the decision whether or not it would be desirable to save the site, belongs with the appropriate specialist.

It is hoped that our experience and documentary presented here of the many inimical influences that attend a field site, particularly one that is in a remote location and not manned, will offer insight to what may need to be taken on if a site is to be preserved. Especially important, we feel, are (i) procedural mechanisms that need to be established for preservation soon after discovery of a site, or early in its excavation; and (ii) adoption of a conservation plan which will identify causes of deterioration and propose countervailing measures.

Lark Quarry Site

Lark Quarry dinosaur trackways were excavated 10

years ago. An area of 210 m² running into the hillside was exposed. It was realized early on that the site was a significant one as it records the stampede of at least 150 coelurosaurs and ornithopods by a large carnosaur (Thulborn and Wade 1984). Between 4000 and 5000 tracks are present. It was noted at the time of excavation that the nature of the mid–Cretaceous mudstone and sandstone, both being high in clay minerals, kaolin, and iron minerals, would make the surface susceptible to rapid weathering in the harsh climate of central Queensland. Its preservation thus became a matter of urgency, and measures were taken to re–cover the surface while funds were being sought for its long–term preservation and discussions were being held with other organizations regarding its future.

Access to the site did, and still does, pose logistical problems. Lark Quarry is in a remote area two full days' drive by motor vehicle from Brisbane over roads that become impassable in wet weather. It is this factor, probably more than any other one, that has led, indirectly, to deterioration and consequently to the expensive measures that have had to be adopted to attempt to reverse the damage. Had the site been more accessible, frequent monitoring of its condition, surveillance, and supervision of construction work would have been possible.

Within three years of the excavation by the Queensland Museum (QM), two additional statutory authorities had become involed with the site. Queensland National Parks and Wildlife Service (QNPWS) became responsible for establishing an environmental park of 374 hectares, open to visitors, in the area surrounding the site; for the legal protection of the trackways and the surrounding landscape, vegetation and fauna; for funding and planning development; for the design and supply of information brochures; and for physical structures at the site. The other body, the Winton Shire Council (WSC), based in the small town of Winton 115 km away, undertook the regular maintenance and cleaning at the site, occasional surveillance, and the construction of an access road. The QM retained the roles of scientific advisors, concerned with study, conservation of the site, and the provision of scientific information to the QPNWS for its information brochures to the public. Details of the development of a conservation plan for Lark Quarry and the evaluation of deteriorative causes have been presented elsewhere (Agnew and Oxnam 1983).

Chronology

A chronology of the site is an instructive way of identifying problems that arose and how, and under what circumstances, damage occurred. If, as may be likely, there is in the future an increase in the discovery, world–wide, of fossil tracks, then the lessons learnt from our experience at Lark Quarry could be of value in the development of a 'whole–site' preservation methodology. Since paleontologically significant sites yet to be discovered — ones, that is, that require elaborate protection and preservation techniques — may well be found in equally remote areas; then these lessons may be all the more important.

In September 1976 the QM undertook preliminary

excavation of the trackways, followed by cleaning the surface and molding it in latex. The site was then covered with soil and rock fragments from the excavation as a protective measure to prevent environmental damage and souvenir collecting. The area received a higher than average rainfall that summer (between 450 and 500 mm) and, by the following season, most of the soil covering had been washed off by storm water flooding down the hillside and over the site. In May 1977 the main excavation of the site and molding of the surface was done. Afterwards it was left uncovered until mid–1978, while tenure and management of the site were being resolved. Considerable deterioration and physical damage occurred during this period, as shown by curved cracks not related to rock jointing (Fig. 43.1). QNPWS then agreed to declare the site an environmental park, with QM and WSC as joint trustees. The site was then covered (by QNPWS) with spread straw under plastic sheeting as a protective measure until funds could be obtained for the design and construction of a roof. The effect on the surface of the microclimate created under the plastic is not known but no further deterioration was visible when it was inspected (M. Wade pers. comm.).

By 1979 funding for the roof had been secured and construction began. Unfortunately the work was not supervised by either QM or QNPWS staff and damage occured where the 12 support columns were set in concrete around the periphery (Fig. 43.2). The straw and plastic covering was not removed from the surface during construction; it caught fire during welding and was apparently allowed to burn unchecked. Discoloration and exfoliation of the thin limonite layer on the surface resulted (Fig. 43.3). The roof is open on the sides and the design does not prevent wind-blown rain and dust from getting onto the surface (Fig. 43.4).

In 1982 a steel walkway was erected by QNPWS to stop visitors from walking on the surface, and a gutter was made on the uphill site to prevent rainwater and mud flooding onto the surface. Again, this work was not adequately supervised (from a conservation viewpoint) and further damage to the site and destruction of some important tracks near the edges occurred. The final physical protection measure — a fence — became necessary when kangaroos used the roofed site as a shelter. Their droppings, urine and scratching on the surface were harmful and several animals died and left intractable stains on the surface. This has been dubbed the 'kangaroo factor' — solving one problem results in another that may not be foreseen.

These were the measures of physical protection afforded the site after excavation. Unfortunately they were implemented too long after excavation so that by then serious damage had occurred.

Figure 43.2. Roof support column set in concrete on the surface of the trackways site.

Figure 43.3. Loss of limonite layer (lighter areas) from the surface.

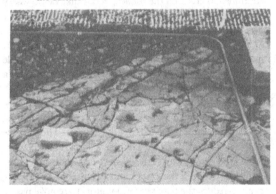

Figure 43.1. Cracking of the surface.

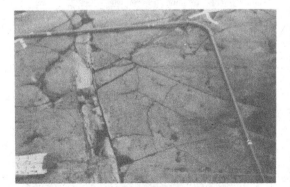

Figure 43.4. Roof and raised walkway.

In mid–1980 when one of us (NA) joined the QM as conservator, he was requested to undertake preservation work on the site. By this time — five years after the start of the excavation — the condition of the surface was causing concern to the QM curator (MW) who had done the excavation. During 1981 and in subsequent years a range of conservation measures was instigated (Agnew and Oxnam 1983, Agnew 1984). These measures were, firstly, the development of a conservation plan, which covered every identifiable cause of deterioration, and a proposed countermeasure. The plan encompassed regular photographic monitoring of the site condition, laboratory analysis and testing of the rock, and testing of consolidants and adhesives. The first measure to be implemented in the field was coating of the surface with a solvent–based silicone resin to prevent absorption of water during periods of rain and to attempt consolidation of the exfoliating surface limonite layer.

Figure 43.5. Carnosaur track before repair work.

Silicone resin was chosen because its breakdown products are silica and substances such as carbon dioxide, water, and small organic molecules. Unlike any clear varnish — which is sometimes suggested by laymen as a conservation measure — these degradation products do no damage, whereas various varnishes can be seriously harmful (Winkler 1975, Amoroso and Fassina 1983, Torraca 1982).

The product, Silicone 18, was applied to the cleaned site by brush as a single flooding coat. On drying it is quite undetectable to the eye. However, the limonite layer has continued to crack off the surface, and it may be necessary to accept loss of the layer in view of the evidence that it was deposited after the tracks were made.

Figure 43.6. Breaking away of edges of the trackways surface.

Figure 43.7. Installing concrete buttress — note white polyethylene foam bond–breaker.

The major preservation undertaking, begun in mid–1983, was infilling of the cracked surface with flexible polyethylene foam rod which was then covered with a pigmented silicone elastomer–sand mixture (Agnew 1984). This technique was borrowed from the building industry. The foam rod functions as a bond–breaker and allows the silicone sealant to move freely with the joint. The purpose was several–fold: to create a cohesive surface and prevent further breaking away at the edges of cracks; to stop water and dust entering the cracks and causing further expansion; to reduce the temptation to visitors to remove pieces; and to effect cosmetic repair. The silicone infilling has particular benefits. It is easily repairable in the event of damage, reversible if desired (it can be cut with a knife and stripped out) and is environmentally stable, with an expected life in excess of 25 years and probably much longer under a roof. In 1984–85 some of the individual dinosaur tracks, particularly those of the carnosaur, which were badly cracked (Fig. 43.5) were restored and reconstructed in part using silicone resin.

To stabilize and halt breaking away of the edges of the site on the downhill side (Fig. 43.6), a concealed concrete buttress was installed. Thin (1.5 mm) flexible polyethylene foam sheet was used between the concrete and the mudstone as a bond–breaker to avoid stress on the edges (Fig. 43.7). Concrete adheres well to the mudstone,

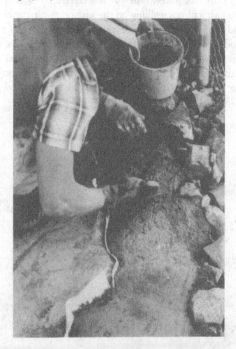

and, being much stronger and with a different coefficient of thermal expansion, it would accelerate breaking away of the edges of the site without the bond–breaker. The buttress is wedge–shape in section (250–300 mm in width) and was cast just below the level of the trackways so that it could be covered with backfill. An acrylic–modified concrete, reinforced with galvanized mesh, was used.

Inconspicuous crack monitors, comprising sets of three small center–punched stainless steel pins in triangular configuration, were epoxy–glued flush with the surface at various points on the site. Distances between pins (about 500 mm) were measurable, with calipers, to an accuracy of less than 0.5 mm. In the three years since stabilization of the surface and buttressing of the edges of the site, no movement greater than 1 mm has occurred and no new cracks have opened.

One of the most difficult problems at Lark Quarry has been that of dust. It blows onto the surface, settles in the tracks and obscures them. This degrades the visual quality of the site, which may tempt the visitor to get off the walkway for a closer look. Vacuum cleaning the site about twice a year has been necessary to prevent accumulation of dust, but, as this damaged the fragile limonite layer, other measures were tried, beginning in 1984. Simple dust monitoring tubes, hung at increasing heights on a pole, were installed for a 12-month period. These showed, within a short time, that most dust drifted from the uphill side and was carried onto the surface at a height of a few inches. To remedy this, a commercially available polyvinyl acetate emulsion (Crustex — used as a stabilizer for mine tailings, roadworks and so on) was applied, both as a surface coating and by deep impregnation, to the uphill area outside the site. This product and other chemical stabilizers of soils have been evaluated (Morrison and Simmons 1977). The treatment has proved quite effective and has about halved the rate of deposition of dust. The final measure being applied is the erection of woven horticultural shade cloth (70% density) on the perimeter fence. While the effectiveness of this inexpensive measure is still to be evaluated, it should further limit dust on the surface by breaking the force of the wind at the fence line. An added advantage will be the elimination of wind–driven rain sweeping under the roof and wetting the surface.

Discussion

We are now at the conclusion of a protracted effort to stabilize and preserve Lark Quarry. The final phase — that of the creation of some 13 interpretive signs along the raised walkway (Fig. 43.8, 43.9) — is about to be done. Interpreting an unmanned site for the public by means of illustrated signs is difficult and many hours were spent in a joint committee of QM and QNPWS staff arguing the merits of different approaches. With text and artwork complete, the production of the signs has begun. Assessment of the interpretive effectiveness of the signs will be by means of the visitors' books at the site. Visitors' books, if regarded

as being expendable, can in our opinion also go some way towards mitigating the potential for vandalism at a site since they may provide an outlet for the graffito exponent.

It can be seen that the main causes of deterioration and damage to the site have been due to human activity (unsupervised construction work, walking on the surface, and to a lesser extent souvenir collecting and vandalism) and environmental factors (exposure to the weather, fire, dust, water and kangaroos).

With hindsight the lessons to be learnt are clear. The recommendations given below, had they been implemented early on, would have prevented much damage.

1. There is a need to appoint a field conservator to examine every aspect of possible damage or deterioration. This person would, in effect, serve as guardian of the site and would recommend procedures, and confer with the architect or designer of any protective shelter or walkway to ensure that the design complied with the preservation needs of the site. An important duty would be to be present on–site during actual construction work. Adherence to conservation specifications would have to be written into the builder's contract.

2. When disparate statutory authorities (in this case, QM, QNPWS, and WSC) are involved with a site, it would seem essential to set up a standing committee of members from each organization. The committee would vet all proposals, plans and so on relating to the site and would control its preservation, development and management. The committee's main purpose would be to ensure that arbitrary decisions or action were not taken by one organization acting alone.

3. Ideally, excavation should not proceed beyond a certain point without assured funds being available, within an acceptable period, to enable protection measures to be adopted. After money has been found for this work, excavation and protection should proceed nearly concurrently.

These proposals may seem to be matters of common sense, but our experience shows the necessity to establish, early on, a logical operational framework with defined roles and functions for the participants in the preservation work. If this is not done and if, instead, the roles of the statutory bodies (with staff member turnover) are allowed to simply 'evolve', the situation is fraught with the likelihood of arbitrary and unilateral decision. In the long run, the site will suffer.

Analysis, Testing and Materials

Laboratory testing was completed before any preservation work, other than erection of physical structures, was undertaken. The purpose of the investigation was to characterize the sediments, to determine their behavior towards water, and to determine the most suitable coating, adhesive, consolidant and other materials to use on the surface.

Dinosaur stampede

Nearly 200 small running dinosaurs left their tracks on this site
as they fled from a carnosaur.

This is the world's only recorded dinosaur stampede.

About 5000 small dinosaur tracks are seen here. They are almost
all the known tracks of running dinosaurs in the world.
On the far side of the site
11 large carnosaur prints
can be seen.

Help preserve Lark Quarry. Keep to the walkway. Don't touch or walk on the surface.

You can buy replicas of the tracks at
Quantilda Historical Museum, Winton,
Winton Tourist Centres and
Queensland Museum, Brisbane.

QUEENSLAND NATIONAL PARKS AND WILDLIFE SERVICE   QUEENSLAND MUSEUM   WINTON SHIRE COUNCIL

Figure 43.8. Artwork and text for one of the signs to be erected. This is the first sign to be seen by the visitor to the site and it carries a message that we hope will be heeded.

Figure 43.9. Artwork and text for one of the interpretive signs.

When the carnosaur came

The carnosaur made at least two changes of direction to head off
attempts to escape led by the larger ornithopods. Some dinosaurs took
to the water - but it was so shallow only the smallest could swim; larger
ornithopods found it too shallow to swim, too muddy to run in. As the
carnosaur lunged at its victim the other dinosaurs started to stampede.

Sediments

The lacustrine sediments are soft, medium–grained arkosic sandstones, reddish–buff in color and interlayered with seams of vertically jointed indurated pink sandstone with conchoidal fracture. Footprints are impressed in a 8–10 cm thick seam of mudstone, which itself is covered with a thin (0.3–1.0 mm), hard and brittle layer of limonite (Thulborn and Wade 1984). The limonite is a secondary deposit formed after the trackways horizon was buried in seasonal or periodic sheet–flooding.

Infrared spectra indicated that the mudstone and sandstone were of the same composition (Fig. 43.10). X–ray diffraction and elemental analysis confirmed this. The main difference is in grain size. Reddish color is due to the iron minerals goethite and limonite. Table 43.1 shows the results.

Physical characteristics were determined by recommended methods (Brown 1981). Parameters measured were concerned mainly with behavior towards water. Table 43.2 shows the results which confirm field observations on the porosity of the sediments and their rapid absorption of water. Clay–bound water at 3600–3700 cm^{-1} (Fig. 43.10) is progressively lost on heating over a micro–bunsen. This water is tightly bound as there is no exchange water with deuterium oxide on prolonged immersion and repeated evaporation to dryness at room temperature.

The high kaolinite content of the sediments and the evidence from thin sections of cracking and micro–fissuring, as a result of wetting and drying cycles, indicate a weathering mechanism dependent, at least in part, upon such alternating cycles. As a long–term preservation measure, it appeared necessary to prevent or limit the absorption of water by the trackways surface.

Comparision was made between the rates of absorption of samples of mudstone and sandstone which were uncoated or coated with Silicone 18 resin 6% solution (w/v) in white spirit (industrial naphtha). Samples were sectioned to 6x6x6 cm size, cleaned and dried at 105°C. Coating was carried out by dipping for 5 seconds followed by draining, air–drying and curing for 24 hours at 105°C. Graduated cylindrical tubes of 15 ml capacity and 1.23 cm^2 cross–section were glued with epoxy to the surface of the samples and the rates of absorption of de–ionized water were measured at 40+5% RH and 25°C over periods which varied, with sample, from 1 to 19 days. The results (Table 43.3) show rapid absorption by the sandstone, slower absorption by the mudstone and a very slow rate of transmission through the limonite layer.

The method of coating samples described above simulated the eventual method of application on the trackways. Results of the void index determinations indicate little penetration of the silicone, especially into the mudstone. That is to say only surface–water repellency was achieved with sufficient absorption across the coating to allow the rock to "breathe".

Experiments on the leaching of solubles from powdered sediments (0.6% over 6 weeks), and on the mass loss from bulk samples due to leaching and

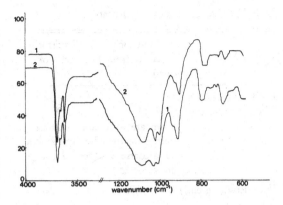

Figure 43.10. Infrared spectra of powdered mudstone (1) and sandstone (2) indicating essentially the same composition.

Table 43.1. *Chemical and mineral composition of sediments*

	Chemical composition (%)	
	Mudstone	Sandstone
SiO_2	58.5	65.8
Al_2O_3	21.6	17.0
Fe_2O_3	8.2	6.8
K_2O	0.3	0.3
CaO, Na_2O, MgO, MnO	≤ 0.1	≤ 0.1
SO_4^{2-}	0.14	0.24
Cl^- (ppm)	350	900
H_2O*	10.0	7.8
	98.84	98.04

*Loss on ignition to 1000°C. This includes absorbed water, clay lattice–water, water of crystallization and organic matter liberated as an oily smoke.

	Mineral composition (%)*		
	Mudstone	Sandstone	Limonite
Kaolinite $(2H_2O.Al_2O_3.2SiO_2)$	33	29	—
Quartz (SiO_2)	48	53	trace
Goethite $(Fe_2O_3.H_2O)$ and Limonite $(2Fe_2O_3.3H_2O)$†	10	8	95
Gypsum $(CaSO_4.2H_2O)$	trace	trace	—

*Approximate percentage by mass as estimated from chemical analysis.
†Limonite is amorphous and is undetected by X–ray diffraction techniques. It has variable absorbed and capillary water. No distinction will be drawn here between limonite and goethite.

Table 43.2. *Characterization of sediments*

	Mudstone	Sandstone
	(Averages from 11 samples of each)	
Hardness (Schmidt rebound hammer)	13	9
	(Compare with values obtained from a weathered concrete floor 50, a coarse sandstone 25, a limestone 25.)	
Bulk density (g cm⁻³)	1.72	1.56
Grain density (g cm⁻³)	2.42	2.44
Water content (%) (heated at 105°C to constant mass)	11.4	14.0
Porosity (%) (ratio of void volume to bulk volume as percentage)	28.6	36.1
Void ratio (ratio of void volume to grain volume)	0.4	0.57
Void index (%) (mass of water in sample after 1 hour immersion as percentage of initial dry mass)	11	17

Table 43.3. *Water absorption rates (ml day⁻¹) (average of 4 determinations) and Void Index determinations on Silicone 18 coated and uncoated sediments.*

	Mudstone		Sandstone		Limonite
	# *	‖ *	# *	‖ *	Layer
Uncoated	2.2	1.4	12.0	10.0	0.5
Period (days)	3	3	1	1	19
Coated	1.9	1.1	3.9	3.3	0.3
Period (days)	3	3	1	1	19
Uncoated void index (%)	11		17		Not determined
Coated†	11.4		14		Not determined

In all cases the initial rates were slightly higher than those towards the end of the run. No attempt was made to prevent evaporation of absorbed water from the samples. Relative humidity was maintained at 40+5%.

*Absorption tube glued perpendicular (#) and parallel (‖) to the bedding plane.

†Samples were lightly brushed during immersion to dislodge adhering bubbles.

crumbling on soaking in water (0.6–1.3% over a week), together with the foregoing results, indicated the necessity to coat the trackways surface to limit absorption of rainwater via cracks.

Peeling and flaking of the limonite layer may be due to absorption and drying cycles of the underlying mudstone. However, this is conjecture since the flaming effect is most severe where the discoloration due to the fire which burnt on the trackways is most obvious. It was hoped that the low viscosity of the Silicone 18 solution, combined with capillary action, would draw the silicone between the limonite and mudstone, especially in places where it had started to flake, and in so doing act as an adhesive. However, since coating the surface in 1981 there is no evidence that the limonite has been successfully bound as it continues to lift from the surface and crack off.

Weathering

Some kaolinite pseudomorphs after feldspar were observed in thin section and indicate that alteration occurred in situ, possibly by acid hydrolysis (Liss and Raiswell 1982):

$$2KAlSi_3O_{8(s)} + 9H_2O_{(l)} + 2H^+_{(aq)} \rightarrow$$
$$Al_2Si_2O_5(OH)_{4(s)} + 2K^+_{(aq)} + 4H_4SiO_{4(aq)}$$

Limonite and goethite may have formed by the oxidation of iron silicate or pyrite:

$$4FeS_{2(s)} + 15O_{2(g)} + 14H_2O_{(l)} \rightarrow$$
$$4Fe(OH)_{3(s)} + 8SO^{2-}_{4(aq)} \, 16H^+_{(aq)}$$

Both the sandstone and mudstone contain angular to rounded grains of quartz in the kaolinite matrix. Sorting is poor. Rocks such as these, consisting predominantly of clay minerals and quartz, are chemically stable under surface conditions. Physical weathering is effected by alternate water absorption and dehydration by the clay minerals, with swelling and shrinking causing fragmentation and powdering of the rock.

Materials Testing

The use of silicon–derived materials on stone and buildings has been discussed (Roth 1982). Silicone 18 is formulated as a gloss–free masonry water repellent. It is an oligomeric alkylalkoxy siloxane, from Wacker GmbH, West Germany designated 090L. According to information from the supplier, the oligomer is catalyzed to ensure completeness of reaction and it incorporates also some medium molecular weight silicone resin to afford immediate water repellency after application. By comparison with methylsilicone resins, the oligomer is stated to have improved depth penetration, higher alkali resistance, and to be suitable as a universal impregnant for natural stone.

On the basis of comparative tests, as discussed below, Dow Corning neutral cure silicone RTV 1080 was chosen for spot–gluing fractured pieces of trackways mudstone in place. This work preceded infilling of cracks with foam rod and silicone elastomer. Since it was desirable to keep the number of different materials and chemicals used in the preservation of the site to a minimum, the satisfactory performance of this material as an adhesive and for infilling of fissures was fortunate.

To evaluate adhesives against each other, a simple testing procedure was devised. Mudstone samples were sectioned (6x6x3 cm thick) perpendicular to the bedding plane. The samples were washed in neutral detergent for one hour at 50°C, followed by rinsing and drying overnight at 50°C. Samples were glued by coating both sawn faces with adhesive, squeezing the two pieces together firmly and allowing the adhesive to set for 24 hours. Half the samples were pre–coated with Silicone 18 resin for 20 minutes followed by draining and drying for 24 hours at 50°C.

Samples were cycled for 8 hours in boiling water and 16 hours in an oven at 50°C until the bond failed (Table 43.4).

After failure, the silicone layer, from both coated and uncoated rock, could be peeled off the sample surface as an elastic film with good strength and resilience. There was no discoloration. By comparison the epoxies failed with softening, yellowing, and the development of flexibility.

Choice of material for infilling the cracks presented some difficulties. Clearly a flexible sealant was required to accommodate any movement in the trackway sediments. Important features sought in the elastomeric material were good physical and chemical stability, recovery from extension and compression, and low modulus. A low modulus material is useful for two reasons — other than its ability to accommodate extremes of movement. Firstly, it puts little stress on the rock, thus eliminating any cracking or breaking of edges in very hot weather when the sealant may be in compression. Secondly, it puts little stress on the bond itself.

Commercially available polysulphides and silicone sealants were the obvious choices. The former were rejected on the basis of poorer weather resistance than silicones and the possibility of long term incompatibility and poor adhesion to the silicone resin coating already on the trackways surface. The final choice was a Dow Corning

grade of clear, neutral cure silicone elastomer designated Silastic RTV 1080. (For obvious reasons silicones that generate acetic acid on curing were not considered.) This grade is stated by the manufacturer to have excellent physical and chemical stability. The salient characteristics of RTV 1080 are given in Table 43.5. Silicone elastomers and resins may be susceptible to alkaline conditions. A recent study has examined aging by sodium sulphate and sulphur dioxide (Mavrov 1983). These substances are not present at Lark Quarry.

In order to match as closely as possible the texture of the infilled silicone with that of the trackway surface, the RTV 1080 was mixed with washed white quartz sand (120 mesh) free of chloride and organic matter, in the proportion 300 ml silicone to 270 ml sand, and colored with Bayer iron oxide pigments (90–95% Fe_2O_3, 3–5% SiO_2 and Al_2O_3). The mix adhered well to mudstone and cured to a hardness (Shore A durometer value of 60) more suitable to that of the rock itself than the unfilled cured silicone (Shore A Value 23).

Table 43.5. *Properties of Dow Corning Silastic 1080 RTV elastomer.*

Silastic 1080 RTV (Room Temperature Vulcanizing) neutral–cure adhesive/sealant.

Physical form:	As supplied ..... Non-slumping paste
	As cured .... Tough, rubber solid
Cure:	At ambient temperatures on exposure to water vapor in the air. Oxime system.
Features:	Non corrosive cure mechanism; good tear strength; resistant to ozone and ultra–violet, moisture weathering and temperature extremes (> 200°C). Life expectancy > 30 years.
Color:	Translucent
Hardness:	Durometer, Shore A scale 23
Tensile strength:	kPa 2000
Elongation:	% 500
Brittle point:	°C −70
Shrinkage:	% max.1
Peel strength:	kg cm⁻¹ 4.5
Joint movement capability:	Extension % 100
Joint movement	Compression % 50
	Extension

Table 43.4. *Adhesives testing on mudstone.*

	Number of cycles (days)		
	RTV 1080	Araldite (5 min. cure)	Araldite (24 hr. cure)
Uncoated (4 samples)	7	2	2–3
Coated (4 samples)	4	1–2	1–2

Figure 43.11. Infilling the cracks with silicone elastomer (carnosaur tracks are on the right of the picture).

Figure 43.12. Infilling — after completion. Contrast between the silicone and rock is less marked than it appears to be in the photograph. Over time it has become even less obvious as dust adheres to the silicone.

Work Procedure

Infilling and gluing broken fragments was done in 1983. It took one person 5 months.

A 2 m square grid, subdivided with wire into 1 m squares, was constructed from rigid plastic tubing. This was laid on the trackway and the area so covered was given a number (marked also on the site plan) and photographed before work started. Infilling was completed one square at a time (Fig. 43.11). In this way an orderly progression of work was achieved, with photographic recording before and after.

Preparation of the squares involved gentle vacuum cleaning, brushing and blowing to remove dust and debris from the surface and fissures, while taking care not to lose small fragments that could be glued into place. Fragments that had broken away cleanly were cleaned with acetone and glued with RTV 1080. Spot-gluing was used to prevent extrusion of adhesive on the surface. When this had been completed for each square, backer rod of appropriate diameter was inserted into the crack. RTV 1080–sand–pigment was blended by extruding the contents (300 ml) of a cartridge onto a board, adding sand and pigment, and mixing rapidly. The mix was then loaded into empty cartriges. By working quickly no appreciable curing of the silcone occurred during blending.

The mix was extruded with a caulking gun as a bead along the crack and carefully tooled level with the surfce. If necessary, pigment was dusted over the infilling to match the surface color. After 15 minutes, when the silicone had lost its tackiness, one of a selection of small silicone molding–rubber pads, previously prepared, was pressed onto the silicone to give a surface texture and pattern like that of the rock. Figure 43.12 shows one of the infilled squares.

Acknowledgments

Many people have worked on the conservation of Lark Quarry. D. P. Le (QM) spent four months in 1983 on the site doing the silicone infilling. Ed Power (QNPWS) and his staff based

in Rockhampton as well as other officers of the QNPWS have at various times been responsible for overseeing the environmental park; Robert Allen (QM) did the artwork for the signs and made valuable suggestions. Rosslyn and Arthur Wallis of Cork, some 20 km from the site, have been unfailingly hospitable whenever field work was carried out. The funding for the work came from the budgets of QM and QNPWS and from a National Estate Grant of $AUS 32,000.

References

Agnew, N. H. 1984. The use of silicones in the preservation of a field site — the Lark Quarry dinosaur trackways. In Brommelle, N. S., Pye, E. M., Smith, P., and Thomson, G. (eds.), Adhesives and Consolidants, (The International Institute for Conservation of Historic Works) Preprints of the Contributions to the Paris Congress, 2–8 September, 1984. pp. 87–91.

Agnew, N. H., and Oxnam, W. 1983. Conservation of the Lark Quarry dinosaur trackways. Curator 26 (3):219–233.

Amoroso, G. C., and Fassini, V. 1983. Stone Decay and Conservation. Materials Science Monographs, II. (Elsevier).

Brown, E. T. (ed.). 1981. Rock Characterization Testing and Monitoring, International Society for Rock Mechanics Suggested Methods. (Oxford: Pergamon)

Gillette, D. D. 1986. Foreword. Abst. Prog. First Internat. Symp. Dinosaur Tracks and Traces. Albuquerque: New Mexico Museum of Natural History)

Liss, P. S., and Raiswell, R. W. 1982. Environmental chemistry: The earth–air–water factory. Endeavor (New Series) 6 (2):66.

Lockley, M. G. 1986. A Guide to Dinosaur Tracksites of the Colorado Plateau and American Southwest. Univ. Colorado Denver Geol. Dept. Mag. Special Issue no. 1, 56 pp.

Mavrov, G. 1983. Aging of silicone resins. Studies in Conservation 28:171–178.

Morrison, W. R., and Simmons, L. R. 1977. Chemical and Vegetative Stabilization of Soils. Bureau of Reclamation Technical Report No. REC–ERC–76–13.

Roth, M. 1982. Siliconates, silicone resins, silanes, siloxanes.

Baugewerbe (Organ des Zentralverbandes des Deutschen Baugewerbes Verlagsgesellschaft Rudolf Muller Gmbth and Co., Koln) 2:3–7.

Stanley Price, N. P. 1984. *Conservation on Archaeological Excavations.* ICCROM, 13 Via di San Michele, 00153 Rome, Italy.

———— (ed.). 1986. *Preventive Measures During Excavation and Site Protraction.* ICCROM conference, Ghent, 6–8 November 1985. ICCROM, 13 Via di San Michele, 00153 Rome, Italy.

Thulborn, R. A., and Wade, M. 1984. Dinosaur trackways in the Winton Formation (Mid–Cretaceous) of Queensland. *Mem. Queensland Museum* 21 (2):413–517.

Torraca, G. 1982. *Porous Building Materials — Materials Science for Architectural Conservation,* 2nd ed. ICCROM, 13 Via di San Michele, 00153 Rome, Italy.

Winkler, E. M. 1975. *Stone: Properties, durability and man's environment,* 2nd ed. (Springer–Verlag).

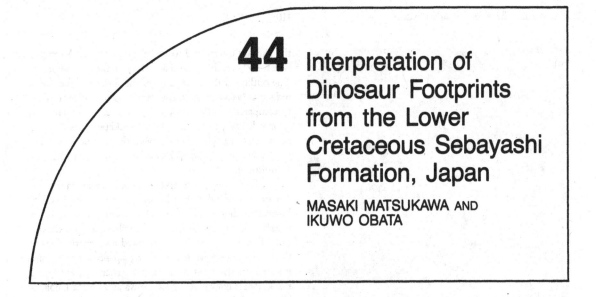

44 Interpretation of Dinosaur Footprints from the Lower Cretaceous Sebayashi Formation, Japan

MASAKI MATSUKAWA AND
IKUWO OBATA

Abstract

Lower Cretaceous tracks from Seboyashi form the basis around which the Nakasato Dinosaur Museum was built. Although the geologic interpretation of the tracks remains difficult, the site has been very effectively displayed for public interpretation. In this and the following paper respectively we discuss i) ongoing efforts to interpreet the trackbearing bed and ii) the investments made in replicating the surface for public display. The indistinct outlines of footprints from the Lower Aptian Sebayashi Formation are discussed. The reasons for the indistinctness are: (i) the footprints may be "ghost prints" beneath the mud layer bearing the original prints, and (ii) the deposits contained much water when dinosaurs walked on them. The sedimentological conditions were clearly different from those under which distinct footprints such as those from South Korea were formed.

Introduction

The Sebayashi locality is an excellent example of a small site which has been very effectively interpreted for public viewing and education. On-site signs have been erected, replicas made, and the Nakasato Dinosaur Museum built nearby. While the public benefits from educational displays at Sebayashi and Nakasato, work continues on the geological interpretation of the trackbearing layer.

Initially, some imprints were interpreted as dinosaur footprints, based on their morphology, size, resemblance to footprints, the regular space between two prints of the same kind, the width of the tracks, and reference to the sedimentary environment of the deposits containing the footprints (Matsukawa and Obata 1985a,b). This paper presents a description of the stratigraphy and the character of the surface of the cliff and discusses the problems involved in interpreting indistinct tracks.

Observations

The cliff on which the dinosaur footprints are pre-

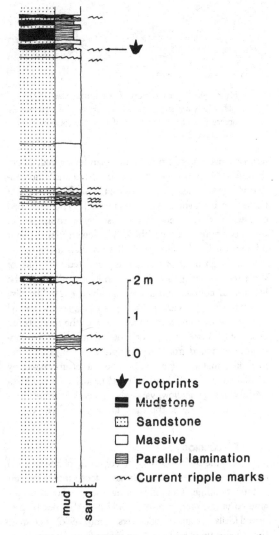

Footprints
Mudstone
Sandstone
Massive
Parallel lamination
Current ripple marks

2 m

1

0

mud sand

Figure 44.1. Columnar section with the indication of the horizon of the cliff which preserves footprints.

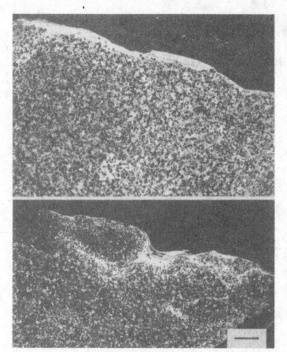

Figure 44.2. Cross-section of the surface part of the cliff. Muds are shown in white part and are exclusively observed under the bottom of tiny pits (below). Scale bars show 2 mm.

served is made up of a fine–grained sandstone bed, 21 cm thick (Fig. 44.1): average grain size is 2.3 phi and mud content is 8.55%. The bed composing the surface of the cliff is overlain by a shale bed 16.5 cm thick. The shale bed consists of thin layers; the lowest layer, composed of numerous films, is 1.5 cm thick. The lowest film is about 0.5 mm thick. The film partly lies on the surface of the cliff, although most of the film is exfoliated. The film of shale rests conformably on the sandstone, although the lithological contact appears abrupt (Fig. 44.2). The film covers the footprints and the tiny pits which are interpreted to be made by some arthropod. A few of the tiny pits are observable in one of the footprints. Where the tube continues downward from the tiny pit, it is plugged by the mud film material (Fig. 44.2). The tracks impressed by animal Ba, animal Bb and others (Matsukawa and Obata 1985a) comprise shallow, variously shaped footprints forming a narrow zone at the right lower part of the cliff.

Discussion

On the surface of the cliff, consisting of fine–grained sandstone, many of the dinosaur footprints have an indistinct outline. If a mud layer had not covered the fine–grained sands, such imprints would probably not be preserved (Seilacher and Lockley pers. comm. 1986). The shale bed is made up of several layers; the surface of one of these layers (possibly 1.5 cm above the sands) is considered to have been the surface of the sediments when dinosaurs walked on it. Accordingly, the imprints on the surface of

the cliff are interpreted to be "ghost prints" on substrata which lay beneath the mud layer bearing the original prints. The outline of the footprints on the surface of the cliff is indistinct because the "ghost print" has an unclear outline in comparison with the original print (e.g., Sarjeant and Leonardi 1987, Langston 1986). In addition, because the sands retaining "ghost prints" beneath the mud layer were uncohesive at that time, the outline of the original prints is indistinct.

Taphonomical interpretation of imprints on the surface of the cliff is shown at left in Figure 44.3. The deposits having the dinosaur footprints have been presumed to be a part of a mouth bar in a delta, reflecting flood deposition. The flood probably represented a southerly flowing fluvial system, showing a southerly opened deltaic plain, inferred on the basis of the asymmetrical current ripple marks (Matsukawa 1983; Matsukawa and Obata 1985a,b). The mouth bar is interpreted as having dried up, at least during ebb tide, and/or spring tide (Matsukawa and Obata 1985a,b). It is known that linguoid type ripple marks can be produced experimentally under conditions of 20 to 60 mm depth (Arai et al. 1958).

There are two kinds of lebensspuren on the surface of the cliff: one type was formed by the vertical burrowing that produced tiny pits; the other lebensspuren is a horizontal trail on the sands. The mud film was punctured by the tiny pits which were subsequently filled with mud. The tiny pits are observable in one of the footprints. The lebensspuren occurring as tiny pits are considered to have been made after the footprints were formed because the footprints were impressed on the mud layer, and would have destroyed pre-existing lebensspuren. Accordingly, the following five phases are envisaged: (i) formation of ripple marks, with deposition of sand bed and thin mud film; (ii) emplacement of lebensspuren occurring as trails; (iii) deposition of overlying mud layer; (iv) imprinting of the footprints and (v) formation of the lebensspuren occurring as tiny pits. On the basis of stratigraphic evidence, the footprint bed corresponds to the maximum regression phase of the Barremian stage (Matsukawa 1983). Furthermore, the climate at that time has been interpreted as a dry period, according to the terrestrial flora in the shale bed beneath the footprint bed (Kimura and Matsukawa 1979, Kimura 1984).

The taphonomical meaning of the indistinct outlines of footprints is different from that of the distinct outlines seen in the footprints from southern Korea (right of Fig. 44.3), because of variations in lithofacies and water content of the deposits. The footprints of Korea (Yang 1982) are probably original prints, impressed into firm silt, marl and/or limestone layers. The difference of taphonomical conditions of the imprints between the two areas is probably based on the sedimentary environment and inferred paleogeography of those areas: the sediments at Sebayashi, Japan, were part of a mouth bar of a delta, while those at southern Korea were on the margin of a lake. Mouth bars in deltas, as compared with the deposits at the margin of a lake, are easily influenced by strong currents; for

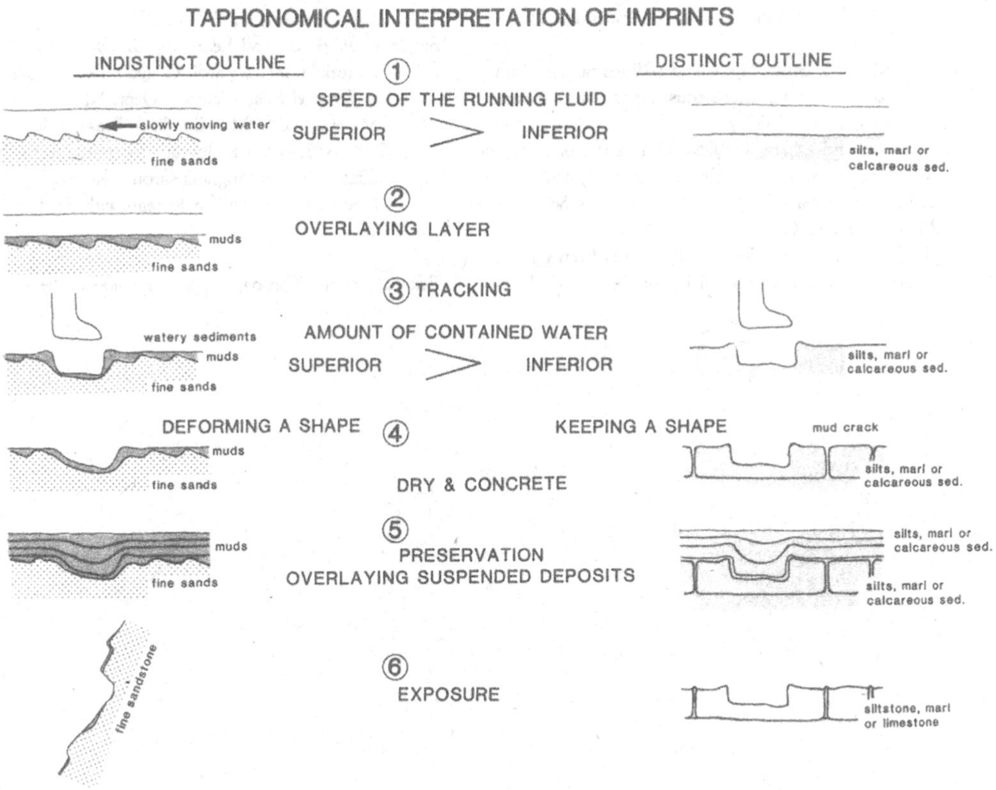

TAPHONOMICAL INTERPRETATION OF IMPRINTS

Figure 44.3. Taphonomical interpretation of imprints. Indistinct outline of imprints based on footprints from Sebayashi, Japan, and distinct outlines are from southern Korea.

instance, by fluctuation of tidal level.

Conclusions

The stratigraphic section of the cliff that contains the dinosaur footprints consists of fine-grained sandstone covered by a shale bed made up of numerous layers. The lowest layer is 1.5 cm thick and is composed of numerous films. The lowest of these films is 0.5 mm thick. A mud layer is considered to have covered watery, fine sands when dinosaurs walked on it. The indistinct outlines of footprints at Sebayashi suggests that they are "ghost prints" and even the original prints are considered to have been less than clear.

The taphonomical processes of imprint preservation were quite different from those in which the distinct footprints from southern Korea were formed. This points to the importance of differences in grain size, mud content and amount of contained water in various deposits. The difference of the taphonomical condition of the imprints between the two areas is related to sedimentary processes, sedimentary environment, and paleogeography.

Acknowledgments

We thank Professor Martin G. Lockley (University of Colorado at Denver) for his fruitful discussion of some problems, critical reading of the manuscript, and on-site observations. We are also indebted to Professor Son-Young Yang and Mr. Seong-Kyu Lim (Kyungpook National University) for their useful discussion and kind help during our field trip to observe the footprints in Korea, and to Professor Yasuhiko Makino (Ibaraki University) for his discussion on sedimentological problems in Sebayashi. The study was financially supported in part by the Research Fund of the Tokyo Geographical Society (Tokyo Chigaku Kyokai).

References Cited

Arai, F., Takei, K., Hosoya, H., Hayashi, S., and Takahashi, K. 1958. Description and some considerations concerning the ripple marks discovered in the Sanchu Graben. *Earth Science (Chikyu Kagaku)* 40:1–12. [in Japanese with English abstract]

Kimura, T. 1984. Mesozoic floras of east and southeast Asia, with a short note on the Cenozoic floras of southeast Asia and China. *In* Kobayashi, T., Toriyama, R., and Hashimoto, W. (eds.). *Geology and Palaeontology of Southeast Asia.* (Tokyo University Press) 25:325–350.

Kimura, T., and Matsukawa, M. 1979. Mesozoic plants from the Kwanto mountainland, Gunma Prefecture, in the Outer Zone of Japan. *Bull. National Sci. Mus. Tokyo C* 5 (3):89–112.

Langston, W. Jr. 1986. Stacked dinosaur tracks from the Lower Cretaceous of Texas – A caution for ichnology. *In* Gillette, D. D. (ed.). *Abstracts with Program of First*

International Symposium on Dinosaur Tracks and Traces
p. 18.

Matsukawa, M. 1983. Stratigraphy and sedimentary environment of the Sanchu Cretaceous, Japan. *Mem. Ehime
University* D 9 (4):1–50.

Matsukawa, M., and Obata, I. 1985a. Dinosaur footprints and
other indentation in the Cretaceous Sebayashi
Formation, Sebayashi, Japan. *Bull. National Sci.
Museum, Tokyo,* C 11 (1):9–36.

1985b. Discovery of the dinosaur footprints from the
Lower Cretaceous Sebayashi Formation, Japan. *Proc.*

Japan Academy 61 B 3:109–112.

Sarjeant, W. A. S., and Leonardi, G. 1987. Substrate and
footprints. *In* Leonardi, G. (ed.). *Glossary and Manual
of Tetrapod Palaeoichnology.* Dept. Nac. Produção
Mineral – DNPM – Brasilia – Brasil p. 53, pl. 6.

Yang S. Y. 1982. On the dinosaur footprints from the Upper
Cretaceous Gyeongsang Group, Korea. *Jour. Geol. Soc.
Korea* 18 (1):37–46. [in Korean with English abstract]

Editorial footnote: This paper is a companion article to Chapter 45.

45 Replicas of Dinosaur Tracks, Using Silicone Rubber and Fiberglass-reinforced Plastics

IKUWO OBATA, HIROMI MARUO,
HIROYUKI TERAKADO,
TSUKUMO MURAKAMI,
TOSHINORI TANAKA AND
MASAKI MATSUKAWA

Abstract

We have successfully made a large mold of silicone rubber (6 m x 9 m) on a natural cliff surface, which dips at 70° and exposes dinosaur trackways. This was done under severe winter conditions in 1985. As the casting method in the laboratory we employed "hand layup" method for the working of fiberglass-reinforced plastics (FRP). We explain the methods, advantages and disadvantages of our procedure.

Methods

Some replication techniques are useful for the study and preservation of dinosaur tracks. Reproduction contributes much to the preservation of paleontological evidence, because it reduces the expense and labor, in comparison with quarrying out slabs on which the footprints are found. Ideally, replicas reproduce with complete fidelity. As the molding method with silicone rubber, and the "hand layup" method of fiberglass-reinforced plastics for casting

Figure 45.1. Working processes of molding and casting in our methods.

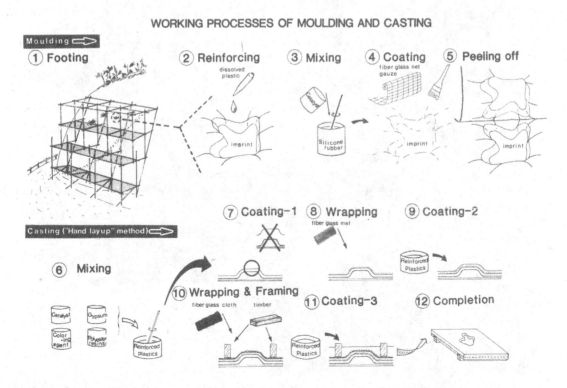

Figure 45.2. Molding scene in the field (1–4) and one of the casts (5) in the laboratory. **1**, coating the cliff surface with mixed silicone rubber. **2**, foothold. **3**, mixed silicone rubber from a bucket into a small bowl (the above left man) for coating the cliff surface. **4**, reinforcing the weathered cliff surface with percolation of dissolved plastics in acetone (the right man).

are considered superior, we employed these methods for making a replica. This paper introduces and evaluates the molding and casting methods with special reference to the Sebayashi specimen.

The partly weathered imprints from Sebayashi, Japan, have been exposed on a steep cliff surface for thirty-three years (Matsukawa and Obata 1985a,b, and this volume). An ingenious contrivance was needed for molding (1, 2 in Fig. 45.1): a foothold for work, percolation of dissolved plastics in acetone into the weathered cliff surface for reinforcing, and covering of partly destroyed imprint surfaces by tin foil (0.02 mm) for protection. The molding was carried out in the following order (3, 4, 5 in Fig. 45.1): (i) mixing silicone rubber with catalyst; RTV silicone rubber 180 kg and catalyst, CE611, 1.8 kg were used; (ii) coating the cliff surface with mixed silicone rubber; first with thin coatings, using a brush and then covering rubber with fiberglass net or gauze for reinforcing; (iii) measuring the curvature of the cliff surface before peeling it off, to reproduce an accurate replica of the natural cliff surface. We made many wooden rulers which are sculptured in order to represent the natural curvature of the cliff surface, so, when we made a replica on the surface, the unevenness of surface was reproduced easily. As we could not peel off one large mold, 6 m in breadth and 9 m in length, without dividing it into smaller parts, we separated five parts of about 1.8 m breadth each, and rolled them up for transport. Our molding work was successfully completed using much catalyst (20 drops for 500 cc silicone). We worked under hard winter conditions, with temperatures as low as minus 10° C although the optimum condition for the work is about 20° C.

In the laboratory, we made the replica in two parts (2 m x 6 m; 2 m x 4 m) by the "hand layup" method using fiberglass reinforced plastics. The working processes are listed as follows: (i) establish the mold with curvature of the natural cliff surface by using wooden rulers made in the field; (ii) prepare the reinforced plastics by mixing polyester resins (140 kg), gypsum or powder of carbonate of lime (60 kg), and coloring agent; (iii) coat the mold with a thin layer of reinforced plastics using a brush; (iv) wrap fiberglass mat for reinforcement; (v) coat by using reinforced plastics again; (vi) wrap fiberglass cloth; (vii) frame with timber; (viii) coat whole part including timber frame by using reinforced plastics again; (ix) remove the mold; (x) apply finishing touches to the surface of coating and repair the surface with talc; and (xi) color the surface. This way casts were produced with complete fidelity. Figure 45.2 shows a molding scene, and one of the casts.

We have now preserved five molds (1.2 m x 1.8 m respectively) and two replicas (2 m x 6 m); 2 m x 4.5 m) in the National Science Museum, Tokyo. We summarize the advantages of our methods as follows: (i) complete fidelity and no shrinking, (ii) no limit of size of fossil, and (iii) keeping a prototype. However, the comparatively high price, one and a half million yen (about U.S. $10,000) and technical skill are probably difficult problems for individual workers without institutional sponsorship.

Acknowledgments

We thank Professor Martin G. Lockley (University of Colorado at Denver) for his critical reading of the manuscript.

References

Matsukawa, M., and Obata, I. 1985a. Dinosaur footprints and other indentation in the Cretaceous Sebayashi Formation, Sebayashi, Japan. *Bull. National Sci. Museum, Tokyo*, C 11 (1):9–36.

1985b. Discovery of the dinosaur footprints from the Lower Cretaceous Sebayashi Formation, Japan. *Proc. Japan Academy* 61 B 3:109–112.

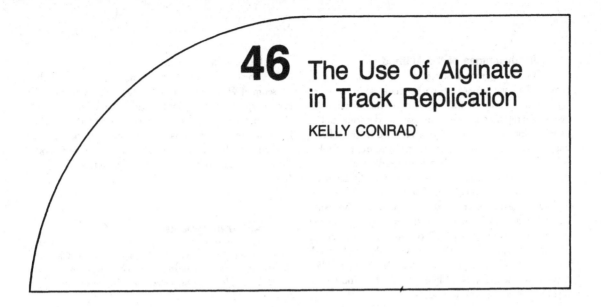

46 The Use of Alginate in Track Replication

KELLY CONRAD

Abstract

Alginate can be used in conjunction with plaster of Paris to make footprint replicas in the field or lab. The benefits of this compound are that it sets up quickly, and produces high fidelity casts from which excellent replicas can be made.

Introduction

The CU Denver Dinosaur Trackers Research Group has used dental alginate as a quick and convenient method for making high quality replicas in the field. The alginate itself is not durable, setting with the consistency of firm 'jello', so it can be used only in conjunction with plaster of Paris for the first step of a two-step process.

In principle the method is similar to the production of a plaster replica from a latex mold. However the method is quicker and can be effective in situations where latex moldmaking is not possible.

Method

After identifying the footprint to be cast with alginate, cut a suitable cardboard rim to act as a confining wall for the fluid alginate mixture when poured. A rim or confining wall of clay can also be used either in isolation or to hold a cardboard or other rim in place. It is important that the alginate does not leak out of the confined area before it has set up. Add alginate to cold water in a suitable container, stirring or mixing gently to avoid creating lumps. Mixture will look like pancake mix. If too viscous it may set up too quickly, but if too fluid it will set up slowly and lack strength. Before pouring into track make sure the track is wet. A film of water is the perfect separator. This is particularly important in hot weather when rock surfaces reach high temperatures, and it may be desirable to shade the track in order to reduce rapid evaporation and maximize the cooling effect of the water. Once these precautions have been taken and the alginate has been poured to a thickness of at least 2 cm above the highest point on the track-bearing surface being cast, it will set up in about 5-10 minutes, providing the mixture is not too thin. When the alginate has the consistency of firm 'jello' it can be turned over and prepared for the second stage in the replication process. The best procedure is to remove the cardboard and/or clay confining wall, then *with great care* lift the alginate cast, making sure to support its weight by carefully sliding a supportive hand underneath. Alginate can tear very easily if a rough object is pushed against it or if one side of the cast is lifted while the remainder is unsupported.

Once lifted, turned over and placed on a flat surface, the same cardboard retaining wall can be placed around it to act as a retainer for fluid plaster. Until the plaster has been mixed and made ready to pour, the alginate should be splashed with water to keep it moist, and kept from undue exposure to strong sunlight. Again water should be used as a separator prior to pouring the plaster.

Advantages of the Method

The alginate method is relatively quick and it produces high quality replicas. It can be used in situations where latex can not be employed, for example where the track-bearing surface is moist, or where insufficient time is available to produce a latex cast. As alginate is used by dentists, forensic scientists and others to produce high fidelity molds and replicas, one can expect high quality replication of such details as claw marks, skin impressions and other fine details of the texture of the surface of tracks. It is useful for replicating deep tracks, particularly those with overhanging topography, which are impossible to replicate with a one-step plaster to rock casting.

Our research group also found that alginate casts made excellent padding for stacking and transporting plaster replicas before they had dried and hardened fully.

Disadvantages of the Method

Because alginate is soft it has to be handled with care, and plaster replicas must be made soon after the alginate has set up. If this is not done the alginate will deteriorate rapidly through shrinkage and dessication. The deterioration process can only be retarded through keeping the alginate moist and at low temperaure, preferably below about 5°C. The exothermic reaction associated with the setting up of plaster of Paris also contributes to deterioration of alginate by speeding up the process of dessication and shrinkage. However, despite these drawbacks we found that, using firm cool alginate, we often successfully obtained two high quality plaster replicas from an alginate cast before any noticeable change in the alginate could be observed. However the second replica was often slightly smaller. For this reason the first replica should be labelled and be the one used in any scientific study.

Examples

The CU Denver footprint collection housed at the Museum of Western Colorado contains a large number of plaster replicas (not casts) made with the method outlined above. Because of the necessity of using a cardboard retaining wall, the replicas made with this method are all aesthetically pleasing in appearance as they have smooth circular or oval outlines and no ragged edges.

Acknowledgments

Materials used in track replication were paid for in part by grants from C.U. Denver, the Colorado Mountain Club and the National Science Foundation. The project was supervised by Dr. M. G. Lockley, who also helped with the preparation of this text.

47 Field and Laboratory Moldmaking and Casting of Dinosaur Tracks

PEGGY J. MACEO AND
DAVID H. RISKIND

Abstract

A method for on-site casting of dinosaur tracks using silicon rubber caulk as a molding compound is presented, and includes moldmaking, release agents and materials. Techniques using paper mache as a casting material are discussed. Molding and casting supplies and sources are provided.

Silicon Rubber Mold

Silicon caulk/sealant as a molding compound has proved successful for field use in the casting of dinosaur tracks. It requires no weighing or mixing and is applied easily. Silicon caulks are extremely durable and their thick, non-slumping consistency makes for easy handling and control in field situations.

To apply: clean track of mud, debris, and loose rock. Apply a generous coat of modeling wax or Johnson's paste wax with a bristle brush. Any deep crevices or surface irregularities that cannot be prepared with mold release must be filled with clay or modeling wax. Coat clay and wax fillings with paste wax. Proper preparation of the track with release agent is imperative. Allow to dry 20 minutes. If the rock matrix is porous, apply a second coat (if the temperature is high, above 30°C, a second wax coat is recommended to keep the rock from absorbing the wax). Let dry 20 minutes, then apply a medium coat of petroleum jelly cut 5% with kerosene. Uncut petroleum jelly will also do. Coat the entire surface, wax and clay included. Using a caulk gun, slowly apply a large bead of caulk, pushing down sufficiently to flatten the bead against the surface of the rock, approximately 0.5 to 1 cm thick (Fig. 47.1A). Apply a second bead, pushing tightly against the first. Continue until three or four beads are down, then spread them evenly to approximately 0.5 cm wide with a palette knife or tongue depressor. Continue until the entire surface is covered.

For added strength, gauze can be pushed into the

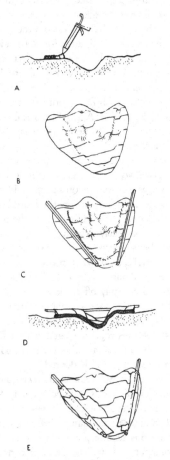

Figure 47.1. Diagrammatic casting of single track using silicon rubber mold. **A,** caulk is applied in a flattened bead on carefully prepared track surface. **B,** cured silicon caulk is covered with overlapping plaster bandages. **C,D,E,** rigid wood supports are added to jacket and covered with additional plaster bandages.

silicon using a tongue depressor. This must be done a section at a time as the caulk is put down. Waiting until the entire track is coated will not work because parts of the caulk surface will have already formed a "skin." Let cure 24 hours.

Silicon caulk cures by pulling moisture from the air while giving off acetic acid (vinegar smell). The cure can be accelerated by the addition of water. Fill a disposable unwaxed cup with approximately six ounces of caulk. Add one teaspoon water. Stir until water disappears and apply silicon immediately. This mixture will cure within one or two hours, regardless of the thickness. However, we strongly recommend experimenting with this method before attempting it in the field. Too much water will keep the caulk from sticking to the track. It will only stick to the tool being used because it will have already cured too much for use.

Any thin strings of silicon left from smoothing can be trimmed off after the silicon has cured. The smoother the surface, the easier the release from the plaster jacket. Once the silicon has cured, a rigid jacket made of plaster bandages can be applied. Use extra fast setting bandages in five- to six-inch widths. Cut into 12" to 14" lengths. Submerge two lengths at a time in a water container and squeeze out excess. Push firmly onto silicon, overlapping lengths by at least 30%. Apply in vertical section, then horizontal (Fig. 47.1B). The plaster jacket should be at least six layers thick to cover a 60 x 60 cm area or under. If the track is longer than 60 cm, supports of wood should be added. Place wood supports on at least four layers of plaster bandages, not directly on silicon mold. Plaster over wood supports using the same method. Add at least four layers of bandages. These will help keep the plaster jacket rigid, which is extremely important. If left extended from the edges, the supports make convenient handles for removal (Fig. 47.1C,D,E). When plaster feels warm and dry (approximately 30 minutes), remove. The silicon mold can then be peeled carefully off track.

Silicon caulk is not a suitable material where the surface to be cast is crumbly and unstable, or where the surface cannot be properly coated with release agents because silicon caulk is an adhesive. It is also unsuitable where exact surface detail is desired.

When exact surface detail is required, Room Temperature Vulcanizer (RTV) silicon casting compounds should be used. These materials are self-leveling and flow readily into and over even minute surface features. Because the RTV silicons require precise mixing of catalyst and casting compound, as well as 24 hour curing, they are unsuitable where rapid fieldwork is indicated. However, if fine details are required and time is not critical, RTV's are excellent casting compounds.

Paper Mache Cast

Paper mache, available from taxidermy suppliers, is an excellent material for making casts. It is very hard, lightweight, nontoxic, easy to use and inexpensive. The texture of paper mache resembles stone, even unpainted.

To apply: Wipe entire mold surface with a very thin coat of petroleum jelly and wipe away excess. Cut mache one-third with plaster (art plaster is preferable). Mix dry, then add water for a very soupy mix (always add water to mache, not mache to water). Mix thoroughly, push into mold surface. This first coat cannot be very thick because of the soupy mix. It will slump. A coat just enough to cover the surface will suffice. Let harden (approximately 20–30 minutes). Apply a second coat the consistency of whipped cream. One-half to three-quarters of an inch depth is sufficient. A large or deep track may need burlap or hardware cloth reinforcement. Cut burlap into 10" x 10" squares and push into second coat. If very large tracks are being cast, this will require many small batches. Do not mix more than can be applied in 20 minutes. Mache can be colored before mixing with universal colorants or dry tempera mixes or can be painted after drying with acrylic or oil paint.

The single track casting method we have detailed yields good quality, durable molds. The method is adaptable for either field or laboratory. If the surface is carefully prepared and release agents are properly applied, this method poses no threat to the integrity of fossil tracks.

Acknowledgments

We thank Bernie Rittenhouse for her patience and editorial skills and the Information Services Section at Texas Parks and Wildlife Headquarters for preparing the final manuscript.

Table 47.1. *Materials List*

- General Electric and Dow Corning silicon caulk/ sealant, available at local hardware stores (cheaper brands are not suitable). Dow Corning 999 can be purchased in cases at a great savings. We recommend clear, not white. A 305 ml tube will cover an area approximately 20 x 20 cm to a thickness of 0.5 cm.

- paper mache (Robert Ruozzi Taxidermist and Supplies, R.D. #1, Irwin, PA 15642 (1-412-446-5943)

---

48 Three Dimensional Representation of *Eubrontes* by the Method of Moiré Topography

SHINOBU ISHIGAKI AND
TOSHIHIDE FUJISAKI

Abstract

The method of moiré topography is suitable for three dimensional representation of dinosaur footprints. Using this method, we can project the contour–like lines directly on the surface of footprints. From moiré pictures, it is possible to get numerical and topographical relief data and computer generated data perspective views.

Introduction

Dinosaur footprints have usually been represented by hand sketches based on observations by naked eyes, and camera pictures. However both these methods have the following shortcomings:

Sketches: (i) The boundaries between "the footprint" and "the normal surface of a bedding plane" are not always clear. Generally, the margin of the footprint is determined at the abrupt turning point of the surface inclination. But, since the margin often changes gradually, determination of the observer is more or less subjective. (ii) It is difficult to show accurately the depth and complicated surface topography of footprints. And (iii) among the footprint literature, figures are often only outline sketches with inadequate detail, especially with respect to the inside relief of the footprint, making comparisons difficult.

Photographs: (i) Camera pictures taken under inappropriate light conditions are of little value. (ii) Photographs taken without shadows show little or no relief. (iii) Photographs taken under dark conditions with artificial light from oblique angles are not always sufficient because the contrast may exceed the dynamic range of film. Moreover, the images are variable depending on the lighting angles and on the types of specimens (positive or negative).

To avoid these problems it is desirable to show precise sketches illustrating surface morphology and two or three photographs taken under the different oblique light angles. Sketches and photographs are defective in representation of large and deep footprints because numerical data that indicate depth and inside relief cannot be recorded by these methods.

Stereo photography is excellent for showing more realistic morphology, and it is possible to get numerical and topographical data from stereo photographs. However stereo photography accomplished through the use of a special apparatus like an autograph is very expensive and time consuming.

Furthermore, there are serious physical and financial limitations to collecting specimens or making replicas in the field. The technique of moiré photography is a new method of three dimensional representation of specimens which cannot be removed from tracksites, as documented in the following description of *Eubrontes*.

Material and Method of Representation

The specimen we used is the concave shaped plaster mold of *Eubrontes* taken from the tracksite at Rocky Hill, Connecticut, U.S.A. It is a white, round shaped mold, 6cm in thickness. The hand–drawn sketch (Fig. 48.1) shows the general features of the footprint, the track of the left pes with three massive digits (II, III, IV). Footprint length is 38 cm. Length of digits II, III, IV are respectively 20, 24, 28 cm. Divarication of digits II and III, 36°; III and IV, 25°. Digits are broad with claw impressions. Digit III has phalangeal pads, but they are not well defined (Figs. 48.2a,b). Camera pictures (Figs. 48.2a,b) are taken in the different oblique light angles indicated by black arrows in each picture. Figures 48.3a and 48.3b are stereo photographs.

Moiré topography is a three dimensional representation method using interference fringes of light. Its optical theory has been explained by Takasaki (1970, 1971, 1972, 1973a, 1973b) and Meadow et al. (1970). Its geometro-optical theory and equipment were described by Terada (1973) and Ikeda (1976). This method has been applied in the fields of medicine, textile design and anthropology.

422

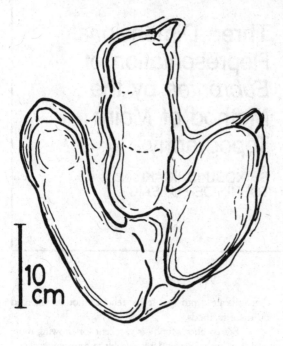

Figure 48.1. Hand-drawn sketch of concave plaster mold of *Eubrontes*.

Using this method the contour–like fringes are projected directly on the surface of the object materials. Moiré pictures provide three dimensional information on a single piece of photographic paper. Therefore, it is easy to record and to preserve the data.

Figure 48.4 shows how to take the moiré picture of a footprint. The grating we used is made of Teflon fishing line stretched every 2mm between the frames. The fishing line is 1mm in diameter and coated by black matting paint (Velvet Coating 101C. 3M Company). Light passing through the grating makes parallel shadow lines and they transform as in Figure 48.5 when they are projected on the surface of the footprint. If we look at these transformed lines through the grating, contour–like fringe lines can be seen (Fig. 48.6). Figure 48.7a was taken with an ordinary 35mm camera following this method. Figure 48.7b was taken with moving grating in the laboratory. The interval of contour–like line is 4mm in Figure 48.7a, 6mm in Figure 48.7b. This interval depends on (i) the pitch of grating, and (ii) the distance of dimensions a, b and c in Figure 48.4.

Fujisaki et al. (1982) described a technique of computerized data analysis and processing of moiré pictures in which a vertically oriented view of the footprint is produced. Figure 48.8 shows the automatically simplified fringes of Figure 48.7 as center lines of each black and white fringe. Figure 48.9 shows the three dimensional plot of a footprint measured by the automatic computer analysis system. It shows the positive form but it is reconstructed directly from the moiré picture of the concave plaster mold.

Figure 48.2. Normal camera pictures of the concave plaster mold. Black arrow shows the lighting direction.

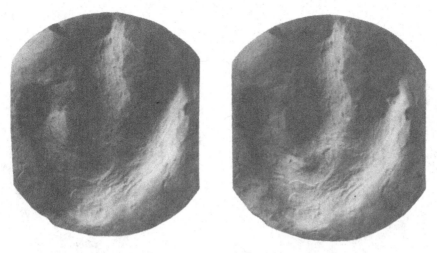

Figure 48.3a and b. Stereo photographs of the concave plaster mold.

Figure 48.4. How to take the moiré pictures.

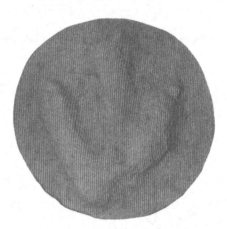

Figure 48.5. Transformed shadow of grating on the surface of the footprint.

Figure 48.6. Looking at the plaster mold through the grating. In this case, the values of a, b and c of Fig. 48.8 are 8 cm, 250 cm, 140 cm respectively.

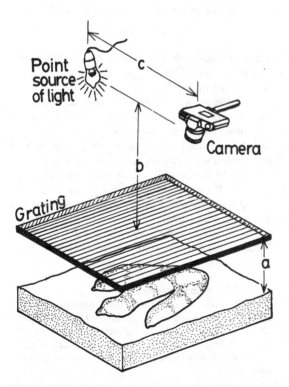

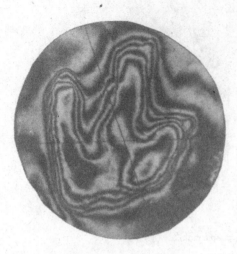

Figure 48.7a and 7b. Moiré pictures of *Eubrontes*.

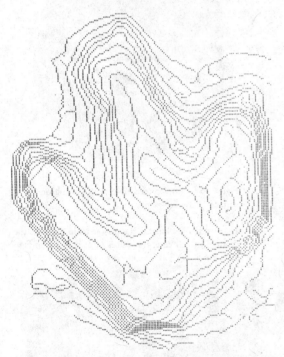

Figure 48.9. Three dimensional plot of the footprint measured by the automatic computer analysis system.

Three dimensional plot of a footprint (Fig. 48.4) shows the perspective view. Using this method, a positive view can be made from a moiré picture of a negative specimen.

Moiré topography has the following advantages compared to other three dimensional representation methods like autograph or autoscanning.

(i) Ease and economy: The instruments are grating, lamp and camera, with ordinary black and white film.

(ii) Portable: It is possible to take moiré pictures during fieldwork. If the sky is clear and the distance between a grating and a footprint is rather small, we can substitute the sunlight for a lamp.

(iii) When taking a moiré picture, the position of contour-like fringe lines can be observed directly, permitting accuracy in the photographic representation.

Figure 48.8. Simplified moiré fringes of Figure 48.7. (Taking the center of black and white fringes.)

Discussion

The moiré picture can improve some of the defects of usual hand–drawn sketches and camera methods. From the contour–like fringes, it is possible to determine the depth of footprint and profiles of inside relief at any position. If the fringe interval is small enough, the small surface structures such as phalangeal pads can be represented (Fig. 48.7).

Acknowledgments

We wish to express sincere thanks to Mr. Andrew Main of Dinosaur State Park (Connecticut) and Mr. Masahiro Tanimoto (Osaka, Japan) who provided the plaster mold of *Eubrontes*. We also thank Professors Sachiyo Doi and Yayoi Fukui of Kyoto Women's University for their kind assistance with taking moiré pictures.

References

Fujisaki, T., Terada, H., and Ikeda, T. 1982. Pattern recognition and automatic analysis on moiré photographs of human bones. *Jour. Anthrop. Soc. Japan* 90:161–168. [in Japanese]

Ikeda, T. 1976. Moiré apparatus by use of parallel light projection. *Kitasato Med.* 6:192–199. [in Japanese]

Meadow, D. M., Johnson, W. O., and Allen, J. B. 1970. Generation of surface contours by moiré pattern. *Applied Optics* 9:942–947.

Takasaki, H., 1970. Moiré topography. *Applied Optics* 9:1457–1472.

——— 1971. Moiré topography. *Shashin Sokuryo* 10:1–12. [in Japanese]

——— 1972. Moiré topography (2). *Gazo Gijutsu* 3:34–46. [in Japanese]

——— 1973a. Moiré topography. *Keisoku to Seigyo* 12:390–399 [in Japanese]

——— 1973b. Moiré topography. *Applied Optics* 12:845–850.

Terada, H. 1973. The moiré contourgraph: An apparatus for recording contour lines. *Kitasato Med.* 3:210–220. [in Japanese]

49 Color Distinctions and Other Curious Features of Dinosaur Tracks Near Glen Rose, Texas

GLEN J. KUBAN

Abstract

Numerous dinosaur tracks on the Taylor Site, near Glen Rose, Texas, are well-defined by their contrasting color and texture from the surrounding substrate, even though in many cases these tracks exhibit little or no topographic relief. The colorations range from blue-gray to rust-brown, in contrast to the ivory to tan color of the surrounding substrate. Many of the color distinctions occur on elongate tracks that were once claimed by some creationists to be human; in many cases the colorations clearly define the shape of dinosaurian digits as well as a metatarsal segment, even where the associated track depressions are indistinct. Other colorations occur on digitigrade tracks with little depression or even slight positive (raised) relief.

The blue-gray color evidently represents a secondary sediment infilling of the original track depressions; the rust color evidently represents oxidation of iron on the surface of the infilling material. This is supported by studies of core samples from the color-distinct tracks and by the observation that the blue-gray tracks occur primarily on the lower parts of the site, and the largely rust-colored tracks on the higher areas (which are exposed to the air more frequently, thus encouraging oxidation). Color-distinct tracks with little topographic relief also have been observed at sites in the western U.S.

Introduction

This paper describes various dinosaur tracks bearing distinct color and texture features, and discusses their possible origin and significance. Numerous color-distinct tracks occur at the Taylor Site, located about 3 miles west of Glen Rose, Texas, in the Paluxy River bed, near the base of the Glen Rose Formation (Lower Cretaceous) (Perkins and Langston 1979). Another paper (Chapter 7 *Elongate Dinosaur Tracks*) in this volume includes maps of the Taylor Site trackways, and summarizes past work on the site.

Initial Observations of Color Distinct Tracks, 1984

While clearing sediment and water from sections of the Taylor Site in August of 1984, I observed several tracks bearing curious rust-brown to blue-gray colorations and a smooth texture, which contrasted with the ivory to tan color and coarser texture of the surrounding substrate. In many cases the color and texture features more precisely defined the details of foot shape than did the track depressions alone. Similar features were independently noticed by coworker Ron Hastings on other parts of the site later the same month. In September of 1984, Hastings, several students from Waxahachie High School, and I cleared sediment from a large area of the site and thoroughly washed the track surface, revealing color and texture distinctions on almost all of the known tracks on the site. Further, many previously undocumented tracks bearing little topographic relief were discovered by virtue of these color distinctions.

Most of the tracks observed in 1984 exhibited combinations of blue-gray and rust-brown coloration, with the blue-gray color predominating, and the rust color largely confined to the higher parts of the site, and the higher parts of individual tracks. On the Taylor (IIS) and Ryals (RY) Trails, both containing elongate tracks formerly claimed by some creationists to be giant human tracks, the colorations followed the posterior (metatarsal) portion of the tracks, but also extended into the shallower anterior region, where they clearly indicated the shape of dinosaurian digits, even on tracks whose digit impressions were indistinct. Track IIS+4 (Fig. 49.1A–C) showed vivid blue-gray and rust coloration even though the entire track exhibited little relief.

In 1984 a few tracks on the Taylor Site were largely or entirely rust-colored. These included a series of newly recognized, small, digitigrade, sharp-toed, tridactyl tracks named the R Trail, which showed slight positive relief (Fig. 49.2A). Also showing distinct rust-brown coloration in 1984 were track RY+5 and a track overlapping it, the latter

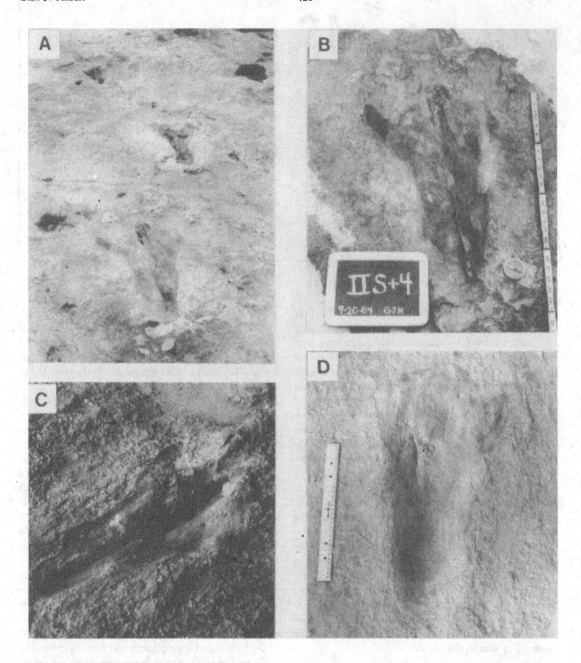

Figure 49.1. Taylor Trail, 1984. **A,** western portion of the Taylor Trail (track IIS+4 at bottom). **B,** track IIS+4, with surface moistened to bring out color contrast. In 1984 this track showed a combination of blue–gray and rust–brown color. Note the clearly defined left digit and metatarsal segment, with only slight topographic relief. **C,** track IIS+4, photographed at a low, oblique angle to highlight the texture contrast and shallow relief features. **D,** track IIS+3. Note the smooth texture of the color–distinct (slightly darker) areas. This is perhaps the most "man like" track in the trail (having an unusually rounded anterior end), but indications of dinosaurian digits are indicated at the upper right and upper left of the track. Compare 1985 photographs (Fig. 49.4).

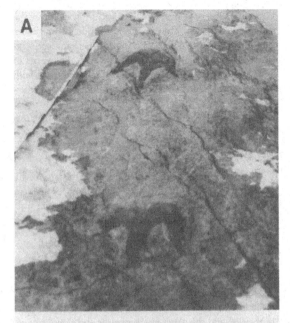

Figure 49.2. Tracks largely rust-colored in 1984, and exhibiting vivid contrast with the surrounding substrate. **A,** Trail R (track R3 at bottom). **B,** two overlapping elongate tracks: RY+5 and IIDW,-11. RY+5 progresses from lower left to upper right (showing two clear digits with little relief). IIDW,-11 progresses from lower left to upper left, and has indistinct digits. The nature of the overlap is more obvious when the rest of each trail is seen (see chapter 7, Figure 7.5, this volume).

showing clear dinosaurian digits delimited primarily by coloration (Fig. 49.2B),

The rust-brown portions of the tracks showed stronger contrast with the surrounding substrate than the blue-gray colorations, but both were easily visible when the track surface was well cleaned, and especially when the surface was clean and damp. The color-distinct material, where bluish, was noticed to be somewhat softer than the surrounding substrate, and had a slippery, claylike feel. However, where rust-brown, the color-distinct material was at least as hard as the surrounding material, which might account for the raised relief tracks (the limestone might be eroding around the rust-colored material).

Initially, I speculated that the color and texture distinctions may have been due to a pressure or compaction phenomenon; envisioned was a dinosaur walking on a firm substrate, making shallow depressions, but compacting the sediment under each step. The compacted areas might show a smoother texture, and perhaps weather differently than the surrounding rock, leading in time to the color contrast. However, further study of the tracks and core samples (discussed below) has largely discounted this idea, and provided much evidence that the color distinctions are primarily due to a secondary infilling of the original depressions with a fine-grained sediment. The blue-gray material evidently represents the main infilling sediment, and the rust color, where present, evidently represents oxidation of iron on the surface of the infilling material.

The infilling concept is consistent with the observation that some of the color-distinct material appears to have been removed from some tracks by erosion or other means. For example, track IIS+1 (Fig. 49.3A) has a deep hole at the anterior end, contrasting to the relative shallowness of most other tracks in the trail. All of track IIS,-7 (Fig. 49.3B) is very deep, even though no track was observed at this spot when the site was first excavated. Creationist workers during the late 1970's related that they chiseled out "hard clay" from this spot, and considered the resulting cavity a "probe hole" rather than a track (Fields 1980). However, the anterior end has a tridactyl shape, and evidently what the early workers really did (unknowingly) was remove the infilling material from an elongate dinosaur track.

The accessible tracks in the Turnage Trail (another alleged "man" trail), which are in a lower part of the site, showed all blue-gray coloration, smooth texture, and shallow tridactyl indentations. Turnage Track IIN3 showed an oblong area of greater depression on the left side (evidently the part interpreted as a "man track"), which appears to be due either to incomplete infilling or later sloughing-out of some of the infilling material (Fig. 49.3C).

Among the trackways newly documented in 1984 by virtue of the color distinctions (besides the previously mentioned R Trail) was the A Trail, a long series of digitigrade, tridactyl tracks showing wide, well-rounded posteriors and short digits; this is one of the few trackways near Glen Rose that can be attributed to an ornithopod

Figure 49.3. Color-distinct tracks showing evidence of partial or complete removal of infilling material. **A,** Taylor Trail track IIS,+1, bearing a deep hole at the anterior end. **B,** at bottom is track IIS,-7, called a "probe hole" by early creationist workers who removed "hard clay" (the infilling?) from this spot. Progressing from left to upper right is the IID Trail, the only deep digitigrade trail on the Taylor Site. **C,** Turnage Trail track IIN3, a semi-elongate track exhibiting all blue-gray coloration, and occurring in a low area of the site (usually under water). The oblong depression at the left side of the track prompted some creationists to consider it man-like; however, on close inspection one can see texture and shallow relief features indicating a tridactyl, dinosaurian digit pattern.

dinosaur. Observed near Trail A, and progressing in the same direction, was Trail G, a short sequence of smaller but similarly shaped color-distinct tracks, perhaps representing a juvenile walking with an adult. Also discovered in 1984 was Trail C, comprising four small tridactyl tracks with blue-gray coloration and little depression; and an extension of the Giant Run (GR) Trail (another alleged "man" trail), which showed color distinctions clearly indicating dinosaurian digits. Other short trails and some isolated prints also were found which showed distinct coloration but little relief.

Unlike most tracks on the site, all of the large digitigrade tracks comprising the IID Trail (Fig. 49.3B) are very deep (most over 12 cm deep) and appear to contain little if any infilling material. Possible reaons for this are discussed further below.

1985 Developments

The Taylor Site remained under water throughout 1985 and 1986. However, in late summer of 1985 the water level dropped to only a few centimeters at the higher parts of the site, and the water was very clear and slow-moving. This allowed an even greater area of the site to be cleaned than in 1984, resulting in the observation of additional tracks and track features. Many of the tracks that were partially blue-gray in 1984 had become more rust-colored in 1985, and contrasted with the outside substrate more vividly. In the Taylor Trail, track IIS+4 (Fig. 49.4A) had become almost all rust-colored, and several other tracks in the trail had become more rust-colored along the rims and in the digit region (the higher parts of the tracks), with the blue-gray color largely confined to the deeper metatarsal sections of the tracks and the tracks in deeper water — some of which were entirely blue-gray (Fig. 49.4D). Much of the "rusting" at the higher areas may have occurred after the clean, exposed surface was abandoned in the fall of 1984. With increasing rains late in 1984, the higher parts of the site (Fig. 49.5) would have been frequently moistened, but the last areas to be completely submerged as the water level gradually rose. During the period of intermittent exposure and wetness, oxidation of iron may have been accelerated.

Among newly observed tracks in 1985 was the extension of the IIDW Trail, which was formerly thought (based on its vague depressions) to be a short sequence of eroded digitigrade tracks, but which now shows color distinctions revealing it to be a long trail of large metatarsal tracks (Fig. 49.6). The coloration patterns on some IIDW tracks are so distinct that in some cases sharp claw marks at the ends of the digits are clearly visible (Fig. 49.6D). IIDW tracks largely delimited by the colorations remain recognizable even where they cross deep erosional scours at the middle of the river (Fig. 49.6C), suggesting that the original track depressions (before infilling) were fairly deep (which is supported by other evidences, discussed later). Also found in 1985 were additional tracks in the Ryals Trail, including RY,-1, a digitigrade track with raised relief (Fig. 49.7).

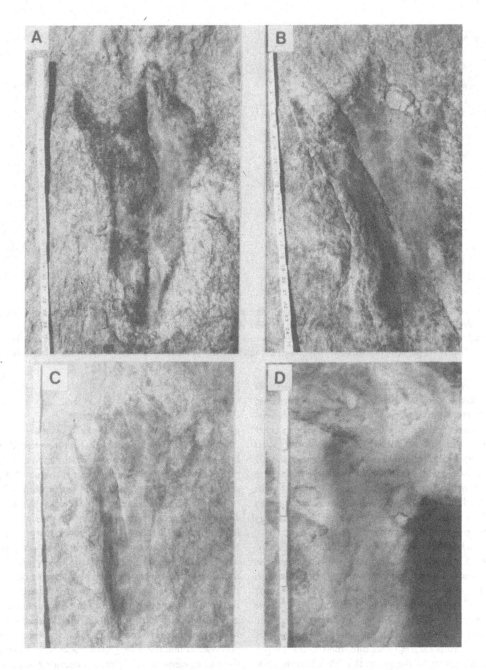

Figure 49.4. Taylor Trail tracks, under shallow water, 1985. **A**, track IIS+4 (a left), showing more rust-brown color, and thus more vivid contrast than in 1984 (compare to Fig. 49.1). This track is approximately 60 cm long. Most other tracks in the trail also exhibiting full metatarsal impressions and indications of digits (by coloration, depression, or both) are likewise between 55 and 65 cm long. (The tape measure is in inches.) Note that the outside digit (digit IV, on the left) is more prominent than the inside digit — a commonly seen feature on elongate tracks in Glen Rose, including **C** and **D** below. **B**, track IIS+2 (a left), showing rust-brown coloration at the highest parts of the track (the sides and digits) and blue-gray coloration in the deeper middle region. **C**, track IIS,+3 (a right), showing rust-brown coloration at the anterior, and bluish coloration in the deeper metatarsal region. The rounded front of the track is evidently due to the way the mud was pushed by the "ball" of the foot, and/or to a wide pad at the base of the middle digit (digit III). Other tracks in the trail, such as IIS+4 and IIS+1, also suggest such a feature. The narrow left digit mark is probably due to partial mud-collapse prior to infilling. **D**, track IIS,-4 (a right), under several centimeters of water, showing all blue-gray coloration.

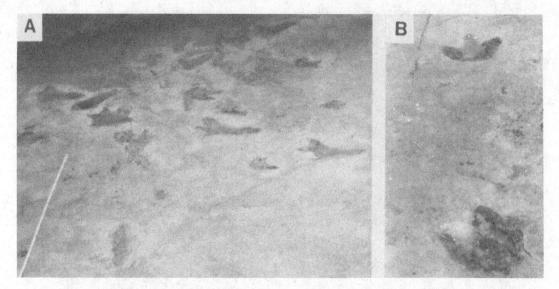

Figure 49.5. The central, highest portion of the Taylor Site, under very shallow water, 1985. Most tracks in this area of the site showed vivid rust-brown coloration in 1985. **A,** the Taylor (IIS) Trail (proceeding from upper right to middle left) intersects the Giant Run Trail (from lower left to top), the IIDW Trail (right to upper left), and Trail A (middle left to upper right). **B,** portion of Trail A, one of the few probable ornithopod trails in the Glen Rose area.

In some cases small fissures or cracks in the substrate were noticed to coincide with the coloration boundaries, evidently representing places where the infilling material has separated slightly from the outside material. In other places apparent "infilled cracks" (delimited by coloration rather than indentation) were observed; these evidently represent ancient cracks in the original substrate which were infilled along with the track depressions (see upper right of Fig. 49.6D). Some wider, ill-defined areas of coloration outside the track borders were also observed, and may represent residual infilling material.

Since the color distinctions added further, dramatic support to previously documented evidences that the elongate tracks on the Taylor Site were metatarsal dinosaur tracks, I invited creationist John Morris (1980) and other "man track" promoters to come to Glen Rose to re-examine the evidence. In response, Morris, accompanied by Paul Taylor and others involved in the making of *Footprints in Stone*, a film featuring the Taylor Site "man tracks" (Taylor 1973), met with me at the site in late 1985. Shortly afterward, Morris and Taylor published statements acknowledging that the tracks could no longer be regarded as unquestionably human (Morris 1986, Taylor 1986), although Morris suggested that several "mysterious" points remained. Most of the alleged involved minor points of omission or misstatement by Morris, which I subsequently addressed (Kuban 1986). Morris also hinted that the colorations might be fraudulent; however, this is strongly refuted by the evidences presented herein (especially the core samples, discussed below). Further, indications of the color distinctions and shallow, tridactyl digit impressions can be seen on several of the Taylor Site "man tracks", even in some early creationist photographs and film footage (for example,

Fields 1980, pp. 33, 34).

1986 Field Observations

Some of the tracks that were predominantly rust-colored in 1985 were noticed to have reverted to more blue-gray coloration when observed in the late summer of 1986 (although most were still rustier than in 1984, and some that were rust-colored in 1984 remained rust-colored in 1985 and 1986). Evidently the rust color is in some cases a transient phenomenon, developing on the surface of the blue-gray infilling material during low water periods, and becoming partially scoured away during high water periods. This does not necessarily conflict with the hypothesis that the highly rusted areas are more resistant to erosion than the blue-gray material or the "outside" material; it may be that a sufficient degree of rust must form atop the blue-gray material for it to substantially resist erosion, or to remain permanently rusted. This is consistent with the observation that the more rusted tracks typically occur at the higher elevations (subject to alternating wet and dry periods), and the all blue-gray tracks are in the lower, continually submerged areas.

Core Sample Analysis

Core samples of several Taylor Site tracks were taken by creationists in late 1985. Morris (1986) indicated that some cores showed evidence of infilling and that others were "inconclusive," but he has not published a detailed report.

In September, 1986 the writer and Ron Hastings took small core samples from tracks IIS+4; IIS,-4; and GR,-1 (one core each); and RY+4 and IIDW4 (two cores

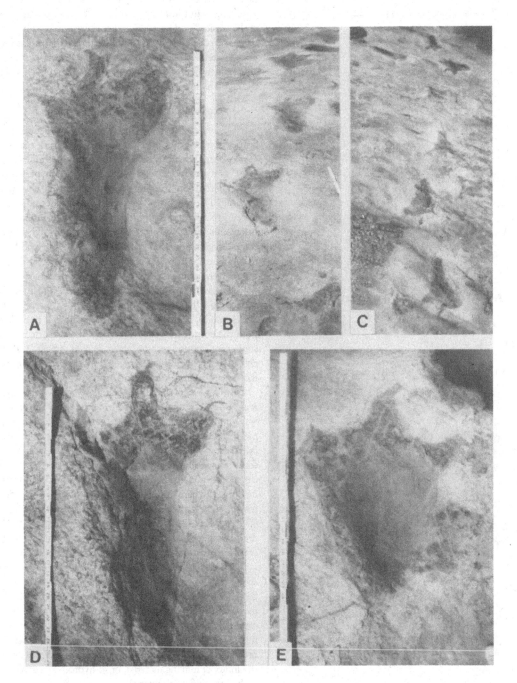

Figure 49.6. The IIDW Trail, a long series of large, color–distinct metatarsal tracks, under shallow water, 1985. **A,** track IIDW3, showing dramatic rust–brown coloration, but little relief. The depressed area at the middle–right is a feature shared by other tracks in the trail, and may relate to the way the river (flowing from west to east, or left to right in these photographs) acted upon the infilling material. This track is 66 cm long. **B,** the western portion of the IIDW Trail, intersecting the Taylor (IIS) Trail at upper right. Track IIDW1 at lower right. **C,** the eastern portion of the IIDW trail, intersecting the Ryals (RY) Trail at top. Note the continuation of the colorations through deep undulations in the substrate (lower part of the picture). **D,** track IIDW2, showing sharp claw marks. **E,** track IIDW4, the only track in the IIDW Trail not showing a full metatarsus. Two core samples were taken from IIDW4 (see Fig. 49.9).

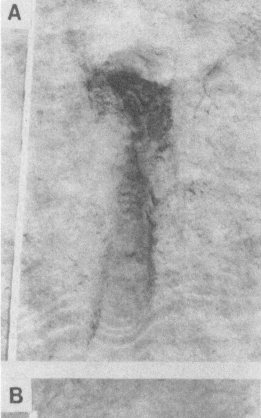

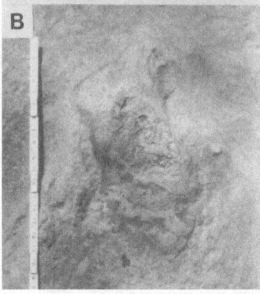

Figure 49.7. Ryals Trail. **A,** track RY+4, an unusually long, narrow track (almost 1 m long), possibly representing a combination metatarsal track and slide. Following this track is a large hole reported to be the spot where a "man track" was removed by local resident Jim Ryals in the 1930's. Core samples were taken from the edge of the right and middle digits. **B,** track RY,-1, the only digitigrade track in the Ryals Trail, showing rust-brown coloration and raised relief.

each). Each core was taken at one of the digits, either straddling or just inside the coloration border, so that some of the "inside" and "outside" material was included in each core. The cores were drilled with a diamond-tipped bit and flexible extension shaft on a portable drill. A glass-bottomed aquarium was used to obtain a clear view of each track through shallow water while each core was drilled.

Gross Appearance of the Cores

Each core sample showed a definite subsurface border between the "inside" and "outside" track material that corresponded precisely with the surface borders (Fig. 49.8). The cores confirmed that the rust color, where present, occurred on the surface of the deeper, blue-gray material. The rust color was typically concentrated in the upper 1 to 3 mm of the inside material, and faded rapidly below this (merging into the blue-gray area). The rust-colored region appeared identical to the underlying bluish material except in color, and shared the same border from the coarser, lighter-colored outside material. In most cases the blue-gray inside material extended well into the subsurface (past the bottom of some cores, which were 2.5 to 3.7 cm deep), and retained a smooth, distinct boundary from the outside material throughout the subsurface. The subsurface boundary typically angled inward as it descended, as would be expected if it represented the surface of an original track depression.

Only where the surface boundary between the inside and outside materials was somewhat indistinct or mottled was the subsurface border also somewhat indistinct or mottled, but even in these cases each dark area at the surface corresponded to a dark area under the surface. Thus, the gross appearance of the cores strongly supports the infilling model, and contradicts a compaction model or staining hoax (where deep and distinct subsurface borders matching the surface borders would not be expected). Figure 49.9 shows where four of the cores were taken, and interprets them in terms of infilled tracks. Micro-analysis of the cores further supports the infilling model (discussed below).

Some cores also contained relatively large, dark specks (later revealed to be microfossils), which in some cores were fairly abundant and concentrated near the inside/outside border, but in other cores were rare or more randomly oriented. Those concentrated near the border might reflect an accumulation of microorganisms on the original track surface (especially in moist depressions) before the influx of infilling material, or to hydrodynamic sorting within the infilling material.

Microscopic and X-Ray Analysis of Core Samples

When viewed under a light microscope, the inside and outside regions of the cores both appeared to represent carbonate muds. However, the bluish inside material was very fine-grained, compact, and homogeneous (except for the microfossils within it); whereas the lighter outside material appeared more porous, coarse and heterogeneous.

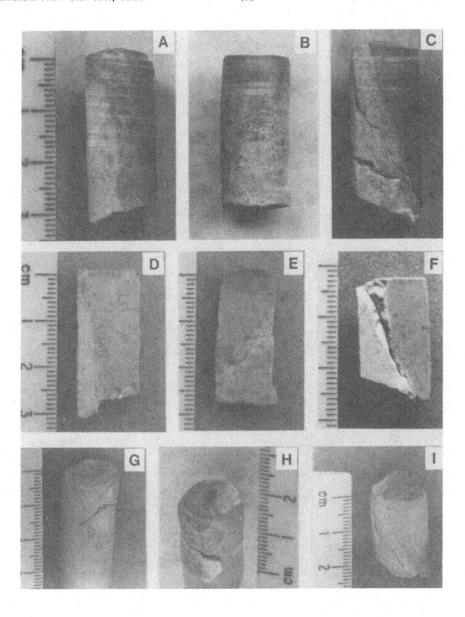

Figure 49.8. Core samples. **A,** core RY+4M, taken just inside the edge of digit III on track RY+4. Note the distinct contrast between the blue–gray inside material and the lighter outside material. The surface of the "inside" region is rust-brown. The larger, darker specks are microfossils. This core and those shown in B and C were moistened to bring out the color contrast. **B,** core IIS+4, taken just inside the coloration border on the outside digit of IIS+4. The outside (lighter) portion of this core was less homogeneous than others (showing a slightly darker, crystalline area near the bottom of the core), but the entire outside area showed a definite boundary from the bluish, fine-grained inside material. The surface of the inside region is rust-colored. **C,** core IIDW4T, taken near the end of digit III, straddling the coloration boundary. This core cracked while being drilled, and on one side (the opposite of that shown) the crack followed the color border. **D, E, F,** same cores as shown in the top row, cross sectioned transversely in preparation for thin section work. **G,** core IIS,-4 taken from the right digit of track IIS,-4 (an all–blue track). Approximately 3/4 of core surface was inside the color boundary. This core cracked in two places during drilling, but shows a definite boundary between the bluish inside and lighter outside region (both at and below the surface). **H,** core IIDW4M, taken near the base of the digit III on track IIDW4, straddling the coloration boundary. Although the subsurface boundary is somewhat uneven, each dark area under the surface corresponds to a dark area at the surface. **I,** core IIDW+4T, showing the correspondence of the surface and subsurface boundaries.

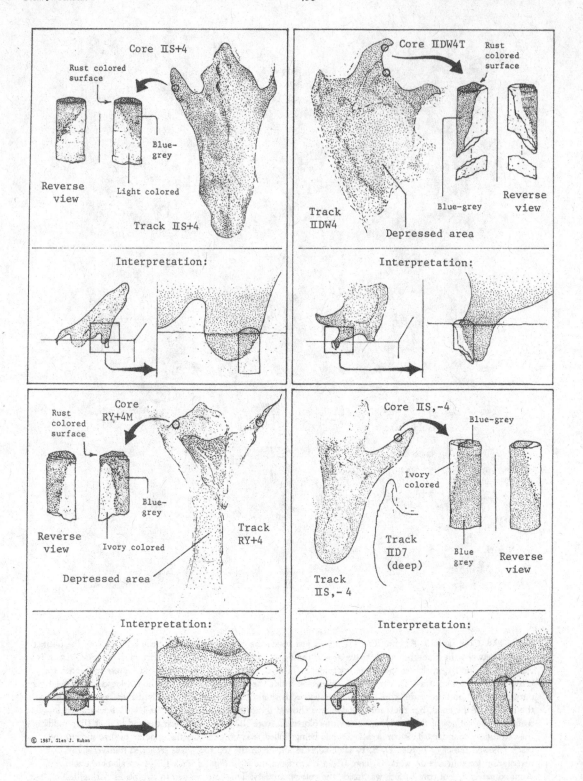

Figure 49.9. The above illustrations show the locations of four core samples taken by Ron Hastings and me, and interprets them in terms of the infilling model. The lower part of each drawing shows how each core relates to a hypothetical cross section of the track at the point where the core was taken.

Where present, the rust color atop the bluish material appeared (as it did to the unaided eye) merely a color phase (iron–oxidized region?) of the inside sediment. The micro–fossils seen in some cores were observed to consist largely of forams and/or ostracods.

With application of weak HCl, the outside material fizzed violently, whereas the inside material fizzed only weakly. This suggested that the outside sediment was highly calcitic, and the inside material more dolomitic, which was confirmed by thin section and X–ray analysis (discussed below). The microfossils were etched very deeply by the acid, suggesting that they were largely calcite, which also was confirmed by X–ray analysis.

The Department of Earth and Space Sciences, Indiana/Purdue University at Fort Wayne, conducted X–ray analysis of core RY+4M, and examined thin sections of cores RY+4, IIS+4T, and IIDW4 (the latter two thin sections were stained with alizarin red S and potassium ferricyanide). Both the thin section and X–ray analysis indicated that the inside material was largely dolomite, with a small amount of calcite, whereas the reverse was the case with the outside material. Small quantities of quartz were found in both the inside and outside areas. Some iron was also indicated in both regions (based on thin section analysis), but was not obviously greater in one region than another — although the already rusted surface of the inside material might have obscured any effect of the potassium ferricyanide there.

Augmenting the above work was an energy–dispersive scanning electron microscope analysis performed on core RY+4M by Scott Pluim of Tulsa, Oklahoma, which indicated the proportion of heavy elements in each area. The inside area was found to contain only a slightly greater percentage of magnesium than the outside material (14.7 inside vs. 10.9 outside), and almost identical percentages of calcium (47.9 vs 48.8). This does not necessarily conflict with the above X–ray diffraction results indicating a great difference in dolomite content, due to variations in the way the elements might combine. The X–ray analysis indicated little difference in iron content between the inside and outside areas, although the reading was taken well below the rusted surface. Thus, the propensity for the inside material to rust more than the outside material may relate to factors other than an inherent difference in iron content. The amount of rust on the inside material seems to be closely related to the degree of elevation (and thus the frequency of exposure to air), but the same does not seem to hold for the outside material (which is generally lighter in color even at the higher regimes, and, where rusted, shows little relationship to degree of elevation). It may be that the iron in the inside material is in a form that more freely oxidizes, or that the inside material is in a form that more freely oxidizes, or that the inside material more readily absorbs iron from the river water than the outside material.

The core analyses performed to date thus indicate that the inside matrix is essentially a finely crystalline dolomite, and the outside material is a more porous, less homogeneous limestone. These basic compositional differences add further support to the infilling model. However, more research is needed to determine why the surface of the inside material "rusts" more readily than the outside material; why it is softer than the limestone where not rusted; and even why it is bluish–gray (that it is more dolomitic does not in itself answer these questions). Small chips taken at the margin of track IIDW4 in 1985 were studied with an electron micro–probe analyzer at the University of Texas and found to contain some kaolin, iron pyrite, and celestite (Wann Langston, Jr., pers. comm.) but this study did not distinguish between the inside and outside material.

Complete cross sections of one or more tracks may further clarify the nature and history of the colorations, but this was not done initially in order to minimize damage to the tracks.

An Infilling Model and Possible Variations

Figure 49.10 is offered as a plausible scenario to account for the color–distinct tracks and core sample data, as well as the unusually deep, apparently non–infilled IID tracks on the same site. Variations of this model are possible (discussed below), but share a common "infilling" or "infilling/erosion" eposide.

The scenario in Figure 49.10 begins with several dinosaurs making fairly deep tracks in soft carbonate mud (A). This may have been followed by a short period of partial drying (B), during which time microorganisms may have accumulated in the moist track depressions. The brief partial drying would minimize mixing of the original substrate with later sediments, accounting for the distinct boundary between the inside and outside materials, and also explain the apparent mud cracks in some areas of the substrate. However, the original substrate probably was not exposed for long — probably a matter of days, weeks, or months at most; otherwise the tracks would have been destroyed by weathering or, if the substrate remained dry, it may have become too firm for the latter IID tracks to be made (discussed further below). Shortly after scene B (still in Cretaceous times) a fine–grained sediment (the "infilling") was deposited over the site (C). If this secondary sediment influx was slight, it may have filled only the depressions on the site. However, it alternatively may have covered the entire site (as is shown in scene C), but was later partially removed by erosion (D), leaving only the depressions filled. In either case, the infilling material must have been largely confined to the depressions alone (by slight deposition and/or erosion) in ancient rather than modern times, since, when parts of the site were first excavated from under undisturbed strata in the late 1960's, the topography of the track depressions was similar to that seen today, and indications of coloration/infilling were already present.

After the infilling (or infilling/erosion) event, the robust IID dinosaur may have walked on the site (D), accounting for the deep, non–infilled IID tracks (other possible explanations for the IID tracks are discussed below).

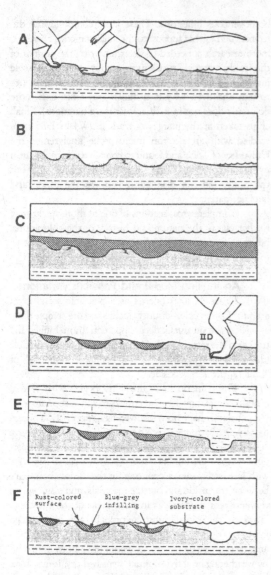

Figure 49.10. A proposed sequence of events leading to the color- and texture-distinct tracks at the Taylor Site. Scenes A through E depict Cretaceous events; F is the current condition. A, bipedal dinosaurs make tracks in moist carbonate mud. B, possible period of partial drying. C, influx of a secondary layer of fine-grained sediment. D, if the deposition in scene C buried the entire site, a period of erosion probably took place, reducing the secondary layer to the depressions alone, leaving "infilled" tracks. A heavy carnosaur walks (IID) on the site while the substrate is still pliable, leaving deep, non-infilled tracks. E, successive depositions of carbonate and carbonate/clastic sediments bury the entire site. F, the modern Paluxy River and human excavators remove the overlying strata, re-exposing the infilled depressions. Exposure and weathering cause slight changes in track relief, and accentuate the color and texture contrast between the original substrate and the infilling material. See text for further details of this model and possible variations.

Also after — or perhaps during — the infilling event, changes in water depth, water chemistry, microbial activity, and/or other factors may have fostered the dolomitization of the infilling material. A sandy clay then buried the entire site, followed by successive layers of other carbonate and clastic/carbonate sediments during the remainder of the Cretaceous (scene E). In modern times (F), the Paluxy River and human excavators reexposed the track horizon, revealing the largely infilled tracks. Recent river scouring then removed some of the infilling from some tracks or parts of tracks, while exposure and weathering caused the infilling surface of others to become rusted and hardened, creating some tracks with slight positive relief.

Variations beyond those discussed above or alternate models are also possible, but I consider them less likely. One possibility is that before the dinosaurs walked on the site the infilling to-be was already present in the form of a shallow layer of watery, fine-grained mud overlying a firmer base of lime mud; so that, as the dinosaurs walked, the overlying material immediately flowed into the depressions. However, if this were the case, one would expect the color borders to be less distinct than those observed (due to partial mixing of the moist sediments). Another variation postulates a two-layered original substrate where both layers were firm but pliable. When the dinosaurs walked, both layers were depressed; the higher areas were later leveled off by erosion, leaving infilled tracks. However, it seems unlikely that the sediments could be soft enough to allow the upper layer to be pushed deeply into the lower layer and yet not distort the boundaries between the two materials (which would contradict the smooth, distinct borders seen in most cores). A more remote possibility is that a very soft layer occurred *under* a firm top layer; so that, as a dinosaur stepped, its feet punched through the top layer to the soft underlying sediment, which welled up from below, filling the depressions.

A number of alternate explanations also may be proposed to account for the apparent lack of infilling in the deep IID tracks. As depicted earlier, the IID dinosaur may have walked after the infilling/erosion episode (while the secondary material covered the entire site), but, if the secondary material at that time was shallow and firm, but pliable, the IID dinosaur may have pushed it down as its foot impressed, leaving the deep IID tracks. The bottoms of the IID tracks are bluish, but the coloration is less distinct than on other tracks, and may represent an underlying bluish clay rather than infilling. (Cores of the IID tracks have not been taken yet.) Last, the present depth of the IID tracks might be due simply to their great initial depth, so that the infilling material filled them less completely than the other tracks.

Other Causes of Coloration on Tracks

The infilling model seems to best account for the origin of the color-distinct Taylor Site tracks and others of similar appearance (discussed further below), but should not be confused with other phenomena that can produce track colorations, most of which are less distinct and less significant.

Colorations unrelated to infilled material include black stains caused by humus and decaying vegetation which often accumulate in track depressions; green films of algae which may cover track surfaces in damp or submerged areas; and simple iron stains (on a single substrate), generally due to differences in frequency or degree of wetness. In most cases these superficial types of coloration are readily recognized as such, since they (i) are often poorly defined; (ii) usually can be scrubbed off; (iii) do not reveal details of foot shape beyond the track depressions; (iv) except for algal coverings, generally show no texture contrast with the surrounding substrate; (v) do not include "raised" tracks; and (vi) do not exhibit deep and distinct subsurface borders corresponding to the surface borders (as revealed by core samples or cross sections).

In some cases such colorations may occur along with (or on top of) infilling-related colorations (algae and humus stains often occur along the deeper colorations at the Taylor Site), but careful inspection and core samples, if necessary, usually will allow one to distinguish among them. The development of rust color on the surface of some infilled tracks is evidently influenced by some of the same factors that produce rust color on infilled tracks (such as relative elevation and dampness), but on infilled tracks (at least those at the Taylor Site), the role of infilling material in selectively concentrating and delineating the rust color is very apparent. Further, such infilling-related colorations are visible even where the rust color is not present (as seen in the blue–gray tracks).

Other Color Distinct and/or Raised Tracks Near Glen Rose

A few tracks with color distinctions and/or raised relief occur within about a kilometer on either side of the Taylor Site. Although these have not been studied in detail (they are usually underwater), at least some appear to be related to an infilling material (Figs. 49.11A,B). Some tracks in other areas of the Paluxy have rust-brown colorations that appear to be due to differential iron oxidation alone (on a single substrate); most show no textural contrast with the surrounding substrate or other evidence of an infilling material.

Color Distinct Tracks in Other Areas

At least one trackway with distinct colorations similar to those on the Taylor Site has been observed outside Glen Rose. This is a trail of four blunt-toed ornithopod tracks with blue–gray and rust colorations and slight raised relief, which occurs in the Dakota Group (Cretaceous), along the Alameda Parkway, near Denver, Colorado (Fig. 49.11C). Some sauropod tracks in the Purgatoire River, Colorado (Morrison Formation, Jurassic) reportedly show purplish coloration and little relief. Other color distinct tracks have been observed on sites in Colorado and Utah, although most are less distinct than those at the Alameda

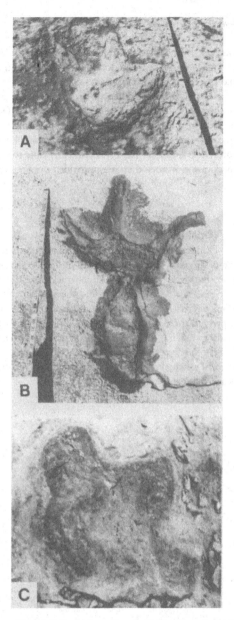

Figure 49.11. **A,** one of several tridactyl tracks with raised relief, located just west of the Taylor Site. This track did not show distinct colorations, but the track was not well-scrubbed. **B,** metatarsal track showing apparent infilling, from a trail located about 300 m NE of the Taylor Site. Water was applied to accentuate the track features. Color contrast was not observed, but may have been obscured by an algal crust that covered much of the track surface here (the infilling here also may ahve been similar to the original substrate, reducing the color contrast). **C,** one of a trail of four probable ornithopod tracks from the Alameda Parkway, near Denver, Colorado, showing blue–gray to rust-brown coloration and slight positive relief. This trail was situated above the horizon of most other tracks in the area, and on a fairly steep slope. Photograph by James Farlow.

Parkway site (Lockley 1987 and pers. comm.)

It is possible that many additional color–distinct tracks exist, even on known tracksites, but perhaps have been largely overlooked. On dry sites, any color contrasts would be reduced, and on wet sites colorations might be obscured by algal films or other superficial coverings. Track researchers are encouraged to clean track surfaces well (even in areas lacking depressions), so as not to miss any color distinctions that may be present, any new tracks or track features the colorations may reveal, and any clues to track formation and history they may provide.

Conclusions

Distinct color and texture features may define the shape of dinosaur tracks even where little or no depression exists. Many such color–distinct tracks occur at the Taylor Site, near Glen Rose, Texas; others occur on sites in the western U.S. Evidently those on the Taylor Site, and perhaps those in other areas, are due primarily to a secondary infilling of the original track depressions, although other factors, including frequency of exposure and differential weathering, evidently have influenced variations in track relief and coloration.

Acknowledgments

I would like to thank Scott Argast, Scott Pluim, Lynton Lande, and Joe Hannibal for laboratory work and consultation; Ron J. Hastings for extensive field assistance; and James O. Farlow for permission to use his photograph from the Alameda tracksite.

References

Fields, W. 1980. *Paluxy River Exploration.* (Joplin, Missouri: self-published), and personal correspondence, 1980–1982.

Kuban, G. J. 1986. Review of ICR Impact article 151. *Origins Research* 9 (1):11–13.

Lockley, M. G. 1987. Dinosaur footprints from the Dakota Group of eastern Colorado. *Mountain Geol.* 24:107–122.

Morris, J. D. 1980. *Tracking Those Incredible Dinosaurs and the People Who Knew Them.* (Creation Life Publishers) 240 pp.

——— 1986. The Paluxy River mystery. *Impact* 15 (1).

Perkins, B. F., and Langston, W., Jr. (eds.). 1979. *Lower Cretaceous Shallow Marine Environments.* Field Trip Guidebook, American Association of Stratigraphic Palynologists, 12th Annual Meeting, Dallas. Revised 1983. pp. 9–12, 49–52.

Taylor, P. S. 1986. Footprints in stone: the current situation. *Origins Research* 9 (1):15.

Taylor, S. 1973. *Footprints in Stone.* (film) (Films for Christ Association).

50 Summary and Prospectus

MARTIN G. LOCKLEY

The objective of this volume has been to present and summarize a large amount of mainly new data on dinosaur tracks and traces from around the world. From this exercise it appears that dinosaur ichnology has much to offer the soft rock sciences of vertebrate paleontology, biostratrigraphy, paleoecology, sedimentology and taphonomy. We do not presume to recognize all the issues which have been adequately resolved at this stage or predict where the science will go in future. However, we do offer the following suggestions on what has been learned and on what areas are in need of further serious attention.

There can be little doubt that the recent revival of interest in dinosaur tracksites, eggsites and nestsites has demonstrated that such finds are much more common than previously supposed. In addition to dispelling the myth that such sites are rare, the unprecedented spate of ichnological discovery has also exploded the myth that dinosaur tracks and traces only occur or become "useful" where body fossils are rare or absent. It is true that they provide a different type of information, but in this age of interdisciplinary studies, it is often information that complements, enhances, refines or recasts the pre–existing paleontological data and its interpretation. The purpose of this chapter, however, is to indicate new directions for further study. Many of the areas in most need of further attention have only been recognized in the last few years as a result of the research of contributors to this volume.

A Systematic Approach to Tracks and Trackmakers

Since one of the main objectives of vertebrate ichnology is to improve our understanding of tracks and trackmakers, it is desirable to adhere to standardized methods of track description and be rigorous in the criteria used to infer trackmaker identity. Fortunately such trends are now becoming established (Sarjeant this volume, Leonardi et al. 1987). The increase in ichnological research

on dinosaurs has not resulted in a proliferation of new names for recent discoveries. This caution appears to reflect a mature understanding of systematic problems by dinosaur ichnologists. Until the factors that affect variation in footprint morphology are adequately understood, ichnologists should avoid establishing new names, except in cases like *Brontopodus* (Farlow et al. this volume) where distinct and well–preserved ichnites are clearly in need of formal description. A number of recent systematic studies (e.g. Padian and Olsen in Padian 1986) indicate the benefits of careful revision of ichnotaxa, not only to elucidate trackmaker affinity but also to clarify biostratigraphy.

It is also becoming clear from examples like *Brontopodus* that footprint morphology is often very different from foot skeleton morphology, especially in the case of larger animals which bore more flesh and padding. The common practice of superimposing foot skeletons on tracks (cf. Parrish this volume) should be only a first step. Skeletons should be studied with a view to reconstructing foot musculature before making comparisons with tracks. Similarly, tracks can be used to help reconstruct tissue anatomy on feet and verify reconstructions based on skeletal studies.

Because footprints vary as a result of substrate, behavior and morphological variation among the trackmakers (Thulborn this volume), it is important to base systematic determinations, comparisons and interpretations on tracks that best reflect the trackmakers' foot morphology, such as those where clear digit outlines exist or where claw, pad, and skin impressions are apparent. If such traces are absent or indistinct it is likely that original foot morphology is obscured by suboptimal track preservation conditions in the original substrate. Tracks may also appear indistinct if one is observing underprints, footprint infillings, tracks that were overprinted before substrate lithification or those that have been recently eroded or weathered following exhumation.

Understanding of the context and preservation of footprints is undoubtedly one of the most fundamental areas in need of further study. In our opinion it is an area of research where progress is badly needed if ichnologists, and paleontologists and geologists in general, are to avoid many of the serious interpretative mistakes and conflicts that have been evident in the past, and not confuse preservational and taxonomic issues. Assuming that ichnologists make progress in discriminating between well preserved tracks and those of inferior quality, we can expect to dispense with a number of invalid ichnotaxa and establish a more streamlined ichnotaxonomy based only on distinctive, well preserved morphotypes.

Context and Preservation

Generally we still need to overcome some important conceptual problems pertaining to how we preceive footprints in the fossil record. The first can be called "two-dimensional track perception." Because we see contemporary tracks on modern substrate surfaces, we are conditioned to think of them as characteristic of planar or subplanar horizons. Consequently, we tend to view trackbearing bedding planes as the exhumed substrate on which the tracks were made, when in fact they are just as likely to represent underprint layers or overlayers. Tracks are three-dimensional features. This fact may be of limited significance to observers of modern tracks, who generally do not concern themselves with the impact of footprints on underlayers or subsequent infilling processes. However, the three-dimensional nature of tracks is of considerable geological significance.

The relative frequency of true tracks, undertracks, and infilling layers has yet to be adequately documented for known trackbearing bedding planes. For instance, as mentioned, we normally wrongly assume that fossil tracks are true tracks. If this is the case, then it is hard to explain the extreme rarity of tracks with skin impressions; we know of only a few good examples among the tens of thousands of documented dinosaur tracks currently known (Fig. 50.1). Similarly, we might expect a higher frequency of well-preserved tracks with crisp digit, pad and claw outlines. Since true tracks are not universally evident, it follows that a large proportion of those preserved in the fossil record are underprints, infillings or otherwise imperfect expressions of the trackmaker's original footprint.

In order to approach some of these problems and emphasize the three-dimensional nature of tracks we provide two examples of how tracks and undertracks may appear in the geological record. The first is based on a theropod trackway from the Entrada-Morrison Formation stratigraphical transition in eastern Utah (Fig. 50.2). It shows the impact of tracks, which are 5 cm deep, on a 3 cm sand bed and an underlying ripple-marked bed. Clearly, the tracks on the ripple-marked bed are undertracks, though this might be hard to establish unequivocally if the overlying track-bearing layer, where the animal walked, was not exposed. In such cases it helps to understand the deposi-

tional history of trackbearing beds. True tracks are relatively rare on ripple marked surfaces and other beds deposited under conditions of high energy.

The second example is partly schematic but based on observations at a large nubmer of tracksites. It shows the differential depth of tracks and resultant impact on underlayers which results from the activity of different animals over a short span of time when substrate conditions are changing, in this case drying out progressively (Fig. 50.3). The first formed tracks are relatively deep, penetrating at least two underlayers, whereas those of the second generation are only of intermediate depth and affect only one underlayer. The last formed tracks impress only the surface layer. The result is three trackbearing beds in stratigraphical succession. These may appear to show the sequence of trackmaking events extended through the period of time represented by the deposition of several beds, when in reality the tracks were all made during a single depositional hiatus.

The above example is simplistic and assumes steady change in substrate conditions. Many other situations may exist as for example where the substrate remains saturated for a long time, as in coal swamp environments, or where a fluctuating water table periodically changes substrate consistency. Where conditions conducive to track formation prevailed for long periods, extensive, variably "dinoturbated" or vertebrate bioturbated beds may be created (Lockley and Conrad this volume). Such variation may complicate the picture of footprint formation and preservation, but it does not necessarily make the understanding of track formation an impossible task. All tracksites exhibit geological features, which help constrain the sequence of trackmaking events. These include the facies architecture, bed thickness, lithology, physical sedimentary structures and in some cases other fossil content.

Some simple methodological guidelines are still needed before many track studies can be considered complete. In addition to two-dimensional length and width measurements, three-dimensional track depth and bed thickness data should always be presented. If track depth approaches, equals or exceeds bed thickness, undertracks will almost invariably occur.

It has been suggested (Lockley and Conrad this volume, Demathieu in Leonardi 1987) that, within the same trackway, manus and pes tracks may show differential depths. Most dinosaurs had extreme heteropody (differential fore and hind foot size) as shown in the example of sauropod tracks (Fig. 50.4). Here the area of five manus-pes sets has been calculated to show a 20/80% ratio, which is similar to the weightbearing ratio independently calculated by Alexander (1985) using a different approach. Such agreement on sauropod weight distribution using different methods suggests great potential for using tracks to verify inferences based on skeletal studies, and vice versa. If the ratio of weight on the front and back feet was any different from the foot area ratio, then one foot would sink in deeper than the other. In the examples given, an even

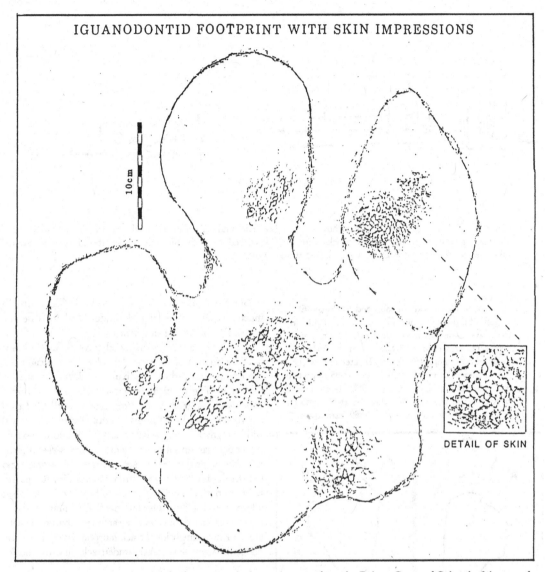

IGUANODONTID FOOTPRINT WITH SKIN IMPRESSIONS

10cm

DETAIL OF SKIN

Figure 50.1. A rare example of a footprint with skin impressions from the Dakota Group of Colorado (Museum of Western Colorado Specimen No. 201.1, after Lockley 1988).

Figure 50.2. A theropod trackway from the Entrada– Morrison Formation transition beds in eastern Utah, showing the relationship of true tracks (overlying bed) to undertracks in an underlying ripple-marked bed.

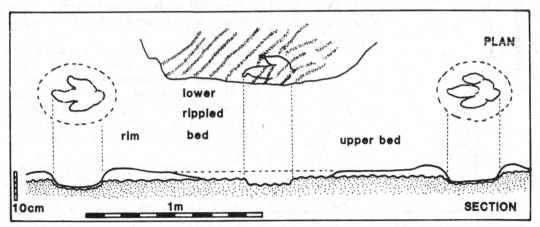

PLAN

lower
rippled
bed

rim   upper bed

10cm   1m   SECTION

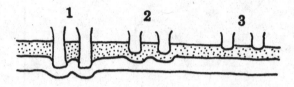

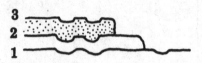

Figure 50.3. Schematic representation of how tracks may impact underlayers to give a false or complicated picture of trackmaking episodes. In this case, all tracks were made when bed 3 was the substrate surface, though some appear to have been made when beds 2 and 1 represented the substrate.

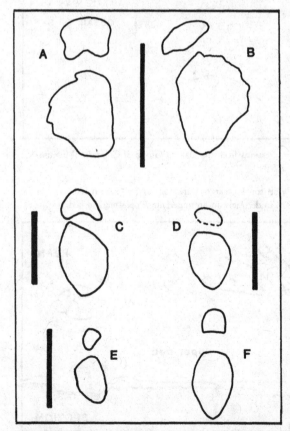

Figure 50.4. Heteropody in sauropod tracks (A–E) suggests the potential for differential track depths in trackways made by the same animals. The average manus/pes area ratio is 20/80%, as in the high-heeled shoe example (F). A and B after Farlow (1987 Fig. 6) show m:p ratios of 26:74 and 15:85 respectively; C after Dutuit and Ouazzou (1980) shows m:p ratio of 22:78; D and E after Lockley et al. (1988 Fig. 7) show m:p ratios of 16:84 and 20:80 respectively.

weight distribution would cause the front feet to sink in much deeper on yielding substrates, and increase the probability of leaving undertracks.

There are of course many other conceptual problems faced by students of vertebrate ichnology, and, because few precedents exist, they are hard to define in specific categories. The problem of two-dimensional vs. three-dimensional awareness and its implications for the geological record is far-reaching for any facet of research concerned with trackmaker size, weight and identification, and any aspect of organism substrate interaction that seeks to determine the original track-marked substrate (or horizon) and the depth of underlying substrate impacted by footprints. Both Langston (1986) and Seilacher (1986), in their addresses to the First international Symposium on Tracks and Traces, called for careful research into this problematic area of three-dimensional track morphology and differentiation between true tracks, undertracks and overtracks.

Regional and Global Scale Track Distributions

In addition to the challenges which lie ahead in this area of research there are also a number of larger scale problems which can be addressed using tracks. These fall into the category of regional or global scale stratigraphic phenomena.

Traditionally tracks have been regarded as localized phenomena, confined to relatively small areas of bedding plane. As more information emerges, it is clear that tracks may be laterally persistent, at particular horizons, for hundreds or even thousands of kilometers (Pittman this volume, Figs. 15.2, 15.5). They may also recur with high frequency in vertical stratigraphic sequences (Lim et al. this volume).

A clear distinction must be made between biostratigraphic correlations of isolated tracks between widely separated localities (e.g. the intercontinental correlations of Olsen and Galton 1984) and the type of stratigraphic correlation where a trackbearing bed or beds can be followed more or less continuously from outcrop to outcrop at a particular stratigraphic level. The former example of traditional biostratigraphical correlation using a zone or index ichnotaxon requires only a few, albeit distinctive tracks, at horizons which are broadly age equivalent but not necessarily at exactly the same stratigraphic level, or even in formations that have any geological or genetic relationship. The latter example (e.g. Fig. 50.5) represents a marker bed or stratigraphic unit of local, regional or even provincial extent.

The example shown in Figure 50.5 is based on detailed correlations of the stratigraphically well studied Dakota Group along the Colorado Front Range. Stratigraphic sections examined every few kilometers along strike reveal a thin package of beds (thickness 1–2 meters), in which several trackbearing beds (typically two to five), recur regularly at almost every exposed locality between Roxborough and Boulder (a distance of approximately 65 km). This particular package of beds, locally known as the Van Biber shale member, represents the Delta Plain facies of the Dakota. At the time of deposition (Latest Albian – Early Cenomanian), immediately before the widespread transgression, extensive coastal plain deposits existed at or near sea level. The importance of such environments in preserving abundant tracks is supported by several lines of evidence, including the widespread occurrence of tracks in similar delta coastal plain deposits such as the Wealden (Woodhams and Hines this volume) and the coalbearing Mesa Verde Group (Parker and Balsley this volume). Pittman (this volume, Fig. 15.2) has also provided an example of a regionally extensive trackbearing "zone" in the well known carbonate platform seqence of the Gulf Coastal Plain.

An important and hitherto unstudied factor in track formation relates to depositional cyclicity. Wherever the water table (base level) coincides with the depositional surface, as in the coastal plain–delta plain environments, cited above, the potential exists for the formation and preservation of abundant tracks. This fact has considerable significance for the study of stratigraphy and ancient depositional environments. In theory, tracks may be abundant and widespread in any situation where the water table has passed up through the depositional surface (a transgressive event) or down (a regressive event). The principle which applies on a regional scale explains the presence of tracks at or near major lithosome boundaries (Fig. 50.6), as well as helping to explain the presence of tracks at bedding plane boundaries, which reflect minor depositional cycles such as lake–level fluctuation and individual flood or storm events.

On the broad scale it is now becoming possible to tie regional trackbearing beds to sea level curves (Lockley et al. 1988). These facts help ichnologists make valuable applied contributions to sedimentological and stratigraphic research. They also allow further sites to be found through simple prediction. Increasingly, ichnologists may use tracks

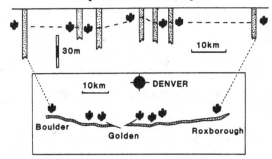

Figure 50.5. Stratigraphic correlation of trackbearing beds, Dakota Group, Colorado.

for correlation and to help define the lateral extent of lake basins and other terrestrial depocenter environments. Similarly, by studying the vertical distribution and frequency of trackbearing beds they can add much to a broader understanding of the history of fluctuating water tables in such depocenter environments. The situation is somewhat analogous to the study of paleo–water tables through examination of the physical characteristics of paleosol sequences.

Eggs, Egglayers, Nests and Nesting Grounds

In just the same way as trackbearing beds are proving to be ubiquitous, recent egg and nest site discoveries are demonstrating that, far from being rare, such sites are common. Discoveries of embryos (Horner and Weishampel 1988) and hatchlings in close association with eggs and nests (Mohabey 1987, Fig. 50.7 herein) have done much to unequivocally establish the affinity of certain egglaying species. As our knowledge of the egg–egglayer relationship improves, it should be possible to determine the extent to which eggshell morphology might be useful in evolutionary studies. In principle, the potential for a significant contribution is possible, just as it is with evolutionary inferences derived from footprint evidence.

Many of the remaining unanswered questions, regarding egg and nest site distribution, may prove to relate to broader interdisciplinary problems, and have much to do with general paleoenvironmental and preservational issues of the type that are coming to be important in the understanding of tracks. Some of the important questions that need to be addressed are, for example, why are the majority of nest sites of Late Cretaceous age? Does the explanation relate to paleobiology and evolution, to paleoenvironmental and preservational considerations, or to a combination of factors? Similarly, how frequently do multihorizoned or stacked nest sites occur and what contributes to their preservation? How laterally extensive are nest sites? Sahni (this volume) and Mohabey (written comm. 1988)

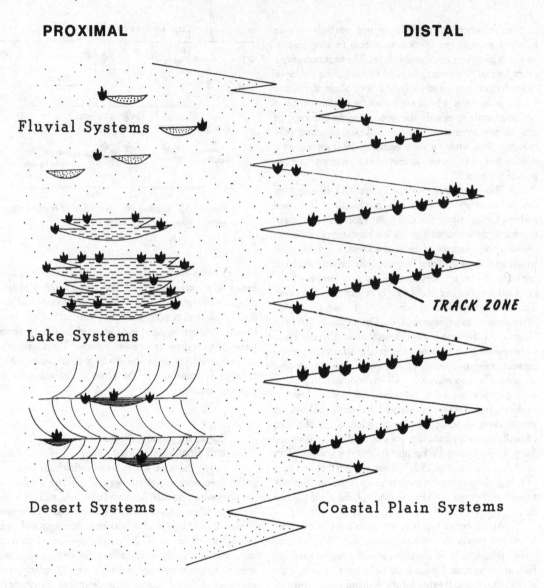

PROXIMAL   **DISTAL**

Fluvial Systems

Lake Systems

TRACK ZONE

Desert Systems   Coastal Plain Systems

Figure 50.6. Model showing distribution of trackbearing beds in different depositional environments. Note relationship to paleo water table. Tracks often occur sporadically in fluvial and desesrt systems, and may be vertically stacked in lake basin deposits. Laterally extensive trackbearing zones sometimes occur in coastal plain deposits.

report a site that extends for 20 to 30 km laterally. Could such a site have some value in local stratigraphic correlation or be coeval with other extensive sites at more distant locations?

Conclusions

Dinosaur ichnology is a rapidly emerging field. We are still caught in a phase of bewildering discovery, with more and larger sites being reported daily. Traditional taxonomic studies are no longer in vogue, although some critical work still needs to be completed on distinctive ichnotaxa,

like *Iguanodon* (Woodhams and Hines this volume) and other tracks that display good preservation.

Because paleobiology and paleoecology are in vogue, ichnologists are increasingly inclined to propose that tracks give insight into locomotion, behavior, paleoecology etc. However current evidence suggests that much is still equivocal. Tracks do not always prove particular gaits (Thulborn this volume) or provide unequivocal evidence of swimming (Ishigaki this volume, Lockley and Conrad this volume). Similarly paleoecological census data may be biased by local dinosaur activity patterns.

In the area of paleoenvironmental interpretation of trackbearing beds, even greater ambiguity needs to be

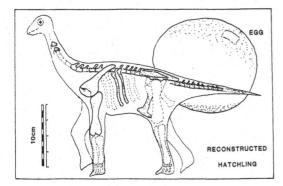

Figure 50.7. A sauropod hatchling and egg, based on Mohabey (1987) and reconstructed with the author's permission.

addressed. Unfounded paleobiological interpretations may arise if the context and preservation of tracks is not adequately understood. Vertical and lateral track distributions may provide radical insights into depositional systems and cycles. Such information should be of considerable interest to sedimentologists and stratigraphers, otherwise unconcerned with the paleobiological implications. In the future, vertebrate ichnology can offer as much to paleoenvironmental analysis as it does to paleobiology.

Rapid recent advances in tracksite and eggsite documentation and analysis have brought dinosaur ichnology into the mainstream of vertebrate paleontology and soft rock geology in general. Many of the more exciting discoveries, particularly those of large and regionally extensive sites have been made in the last two or three years. Such large and numerous sites obviously offer considerable potential for future study by ichnologists and others in the field of earth science. The immediate future of dinosaur ichnology looks bright and the wealth of available material promises to keep researchers active for many years.

Acknowledgments

We thank Dr. D. M. Mohabey for examining the reconstruction of the baby sauropod he described and suggesting the improvements, incorporated in Figure 50.7.

References

Alexander, R. McN. 1985. Mechanics of posture and gait of some large dinosaurs. *Zool. Jour. Linn. Soc.* 83:1–25.

Horner, J. R., and Weishampel, D. B. 1988. A comparative embryological study of two ornithischian dinosaurs. *Nature* 332:256–257.

Langston, W. 1986. Stacked dinosaur tracks from the Lower Cretaceous of Texas — A caution for ichnologists. *In* Gillette, D. D. (ed.). *Abst. Prog. First Internat. Symp. on Dinosaur Tracks and Traces* (Albuquerque: New Mexico Museum of Natural History) p. 18.

Leonardi, G. (ed.). *Glossary and Manual of Tetrapod Footprint Paleoichnology.* (Dept. Nac. Prod. Mineral, Brazil) 75 pp.

Lockley, M. G. 1988. Dinosaurs near Denver. *Geol. Soc. Amer. Field Guide for Centennial Meeting, Denver, Colorado. Colorado School Mines Publ.* (in press).

Lockley, M. G., Conrad, K., and Jones, M. 1988. Regional scale vertebrate bioturbation: new tools for sedimentologists and stratigraphers. *Geol Soc. Amer. Abst. Prog.* 20 (7) in press.

Mohabey, D. M. 1987. Juvenile sauropod dinosaur from Upper Cretaceous Lameta Formation of Panchmahals District, Gujarat, India. *Jour. Geol. Soc. India* 30:210–216.

Olsen, P. E., and Galton, P. M. 1984. Review of the reptile and amphibian assemblages from the Stormberg of Southern Africa, with special emphasis on the footprints and the age of the Stormberg. *Paleont. Africana* 25:87–110.

Padian, K. 1986. *The Beginning of the Age of Dinosaurs.* (Cambridge University Press) 378 pp.

Seilacher, A. 1986. Dinosaur tracks as experiments in soil mechanics. *In* Gillette, D. D. (ed.). *Abst. Prog. First Internat. Symp. Dinosaur Tracks and Traces* (Albuquerque: New Mexico Museum of Natural History) p. 24.

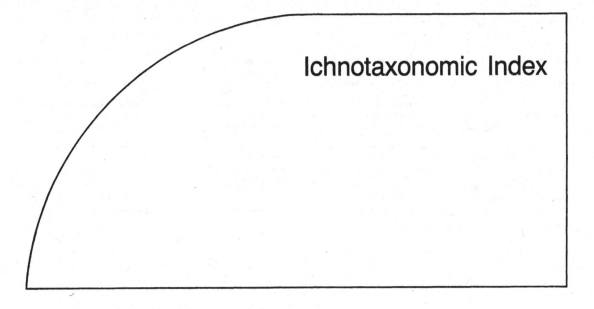

Ichnotaxonomic Index

This index refers only to taxa based on vertebrate tracks, a few invertebrate trace fossils, and egg remains. Osteological taxa are given in a separate index, immediately following.

The letter "f" following a page number refers to a figure on that page; "t" refers to a table.

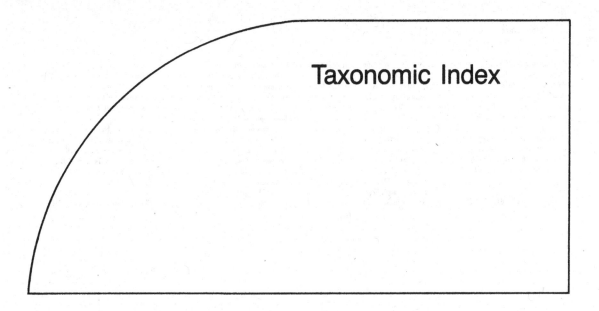

Taxonomic Index

This index refers only to osteological taxa (body fossils) and mineralized egg remains. Ichnotaxa are given in a separate index. The letter "f" following a page number refers to a figure on that page; "t" refers to a table.

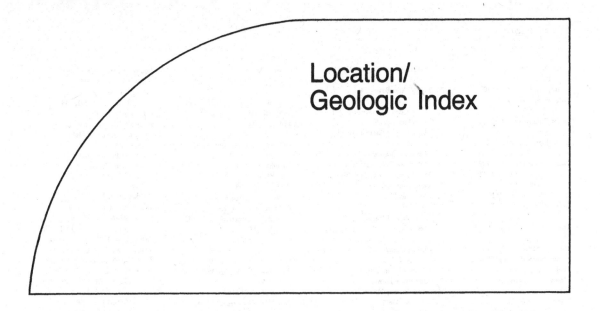

Location/
Geologic Index

This index lists formations, geologic structures, sites, areas, and larger regions. It is not exhaustive, but includes most of the easily recognized names which will be identified by the reader.

Printed in the United States
By Bookmasters